ENGINEERING PRINCIPLES OF GROUND MODIFICATION

ENGINEERING PRINCIPLES OF GROUND MODIFICATION

Manfred R. Hausmann

University of Technology, Sydney

McGraw Hill Education (India) Private Limited

CHENNAI

McGraw Hill Education Offices

Chennai New York St Louis San Francisco Auckland Bogotá Caracas Kuala Lumpur Lisbon London Madrid Mexico City Milan Montreal San Juan Santiago Singapore Sydney Tokyo Toronto

McGraw Hill Education (India) Private Limited

ENGINEERING PRINCIPLES OF GROUND MODIFICATION

Chapters opening are credits appear on page xxiii and on this page by reference.

McGraw Hill Education (India) Edition 2013

Third reprint 2018
RCLARRXERQBCX

Reprinted in India by arrangement with McGraw-Hill Global Education Holdings, LLC, New York

Sales territories: India, Pakistan, Nepal, Bangladesh, Sri Lanka and Bhutan

Library of Congress Cataloging-in-Publication Data

Hausmann, Manfred R.
Engineering principles of ground modification / Manfred R. Hausmann.
p. cm.
Bibliography: p.
Includes index.
ISBN 0-07-027279-4
1. Soil stabilization. 2. Soil mechanics. I. Title.
TA710.H34 1990
624.1'5136—dc19 89-2321

ISBN-13: 978-93-5134-187-1
ISBN-10: 93-5134-187-9

03 04 05 06 07 RAJ 22 21 20 19 18

Published by McGraw Hill Education (India) Private Limited, 444/1, Sri Ekambara Naicker Industrial Estate, Alapakkam, Porur, Chennai 600 116, Tamil Nadu, India, and Printed and bound in India at Rajkamal Electric Press, Kundli 131 028, Haryana

Visit us at: www.mheducation.co.in; (Toll free in India): **1800 103 5875**
Write to us at: info.india@mheducation.com

CIN: U22200TN1970PTC111531

ABOUT THE AUTHOR

Manfred R. Hausmann graduated from the Swiss Federal Institute of Technology, Zurich (1961), received an M.Sc. in Civil Engineering from the University of Alberta, Edmonton (1964), and was awarded a Ph.D. from the University of South Wales (1978).

After gaining experience as a professional engineer with research organizations, consulting firms, and government departments he joined the University of Technology, Sydney, where he developed the geotechnical subject areas for the degree course in civil engineering. Today, engineering principles of ground modification is offered as an elective in the undergraduate program, and attracts many students.

The author has strong ties with the United States having held visiting appointments at the University of California, Los Angeles (1976–77) and Drexel University, Philadelphia (1986).

Professor Hausmann is a member of the Institution of Engineers, Australia, as well as the American Society of Civil Engineering. He is actively involved in other national and international societies and committees concerned with soil mechanics, foundation engineering, and geosynthetics.

CONTENTS

FOREWORD

Ground modification techniques have become a major part of civil engineering practice over the last 30 years, and their use is growing rapidly as worldwide development poses an increasing demand for land reclamation and the utilization of soft or unstable soils. Soil, one of our four most abundant natural resources, deserves to be treated as wisely for engineering purposes as for agricultural uses and this book by Manfred Hausmann, sets out to show how.

Dr. Hausmann has established a considerable reputation in the field of soil stabilization, and readers of this book will soon understand why. He treats his subject in a clear and practical manner; his coverage of the field and in-depth explanations in one volume are unexcelled. Even more importantly, we are given state-of-the-art material, which is no mean feat considering the veritable avalanche of literature on the various specialized aspects of soil stabilization which has appeared in journals and conference proceedings over the past few years.

Engineering Principles of Ground Modification is a well-organized and readable book; and the exercises associated with each chapter will be welcomed by higher-level students and teachers alike. It is best suited to post-graduate and consultant use; but it would also be appropriate for an undergraduate elective course and should certainly be part of the prescribed reading for any civil engineering honors degree or 5-year undergraduate degree course. As one who has worked in this field, in all its aspects, for nearly the last 30 years when these techniques were first explored and brought into use, I am happy to see such a collected treatment now available to the civil engineering profession.

Owen G. Ingles
Swan Point, Tasmania

PREFACE

The idea of writing this book grew out of a long standing interest in methods of soil improvement for engineering purposes, specifically those involving reinforcement by strips and meshes. Added incentive was the recognition of the need to inform civil engineering students of the traditional as well as the many new and exciting geotechnical construction processes involved. The book finally took shape from several generations of lecture notes prepared for a course in ground modification developed by the author at the University of Technology, Sydney.

Most commonly, introductory courses in soil mechanics/soil engineering or geotechnical engineering which form part of all civil engineering curricula are followed by advanced courses emphasizing numerical methods in geomechanics, either as part of an optional undergraduate or a specialized postgraduate program. Such an approach allows ever more complex problems to be solved but often contributes little to reinforce basic knowledge of soil properties and behavior. Mathematical skills therefore have to be complemented by more practical subjects, such as the in-depth treatment of experimental methods in the laboratory or the field.

A course in the principles of ground modification combines and extends the objectives of advanced study as defined above by:

- reaffirming the relevance of basic laws of physics and chemistry to soil mechanics,
- reviewing and enhancing elementary knowledge of soil properties with respect to phase relationships, strength, compressibility, water flow, and more,
- introducing new analytical techniques, and
- demonstrating how theoretical knowledge and observation of engineering performance assist in the rational application of ground modification procedures.

The reader can gain competence in properly devising alternative solutions to difficult foundation and earth construction problems and in evaluating their effectiveness before, during, and after construction. A study of the many different approaches to ground modification broadens the mind of any engineer and inspires creativity and innovation in geotechnical construction and related fields.

The author expects that a first course in ground modification would concentrate on the topics of compaction, dewatering, surface stabilization with admixtures, and tensile reinforcement. Grouting, reinforcement by anchorage and confinement, preloading with vertical drains, the use of geosynthetics, soil heating and freezing, and/or electroosmosis could be added as selected topics or may represent a follow-up course. Much depends on how effectively these techniques were already covered in previous subjects.

At present, a coherent, systematic treatment of ground modification techniques as part of a university course program is still quite rare. Often these topics are spread over a variety of subjects, hidden away in advanced geotechnical engineering, ground water hydrology, construction, retaining structures, dams, pavement design, underground excavations, geosynthetics, and more. After many discussions with academics and practitioners, the author is convinced that the subject area of ground modification has sufficiently matured to be covered in a separate course of its own. This approach better equips an engineer to make an informed decision on which technique to use in a particular situation. It is hoped that this text will assist in such a task.

Inspiration was derived from many sources, publications, companies, and individuals—academics, designers, and construction engineers. Wherever possible detailed references are made to the origin of ideas, concepts, and numeric or graphic information. Please accept an apology for any omission and the inevitable occasional error.

Two people provided much more than just assistance and encouragement: R. M. Koerner (Drexel University), who not only helped in developing the framework for this book, but also provided an example of how to maintain productivity and fitness; and O. G. Ingles (formerly of the University of New South Wales), whose knowledge and experience had a great influence on this text.

Helpful criticism and comments were received from many friends, colleagues, and students. In particular I would like to mention G. J. Ring, University of Technology, Sydney; R. Hryciw, University of Michigan; I. W. Johnston, Monash University; M. S. Boyd, Reinforced Earth Pty. Ltd.; B. R. Fishburn, Department of Main Roads, N.S.W.; K. Forrester and M. D. O'Brien, The Electricity Trust of South Australia; J. P. Welsh, GKN Hayward Baker Inc.; and T. Wilmot, Stabilised Pavements of Australia Pty. Ltd. My thanks also go to the reviewers of the final manuscript; Turgot Demirel, Iowa State University; Hsai-Yang Fang, Lehigh University; Terje Preber, South Dakota School of Mines; William Wolfe, Ohio State University; and Tien Wu, Ohio State University.

Much appreciated during the preparation of the manuscript was the help received from the librarians of the University of Technology, Sydney.

The author has had the privilege of learning from many eminent teachers, either as an undergraduate or graduate student, as a practicing engineer, or during periods of study leave as an academic. This happened in many different parts of the world: At the Swiss Federal Institute of Technology (G. Schnitter, R. Haefeli, J. Huder,

F. Balduzzi, and Ch. Schaerer); in Canada, at the University of Alberta in Edmonton (R. M. Hardy and S. Thomson) and in consulting (C. F. Ripley and E. J. Klohn); at the University of California, Los Angeles (K. L. Lee); and in Australia, at the University of New South Wales (I. K. Lee).

The format of the book was designed to suit students of civil engineering at senior undergraduate or graduate level. However, the practicing engineer will also gain from reading the text and completing the problems at the end of each chapter. Most of these problems are intended to test comprehension and emphasize specific points made in the text. Other problems represent real decision-making situations and serve to acquire practical skills. Some of the discussion topics could well challenge the reader's knowledge of geotechnics and may require researching other texts and original source material.

The author hopes to receive many comments (including criticism!) from readers. Very useful would also be additional information, problems, and illustrations from teachers, researchers, and practitioners alike. Such contributions would add great value to any future editions of this book.

The author also welcomes notices of any audio-visual material which could be made available to educational institutions offering a course in ground modification. It is known, for example, that the U.S. National Committee on Tunneling Technology has produced a set of slides on Rock Bolting Practice. Another excellent set of slides and additional material complementing this text is available from GKN Hayward Baker, Odenton, Maryland; this firm has pioneered many of the techniques described in this book, and in the United States the words "Ground Modification" represent its service mark.

Finally I would like to acknowledge the enthusiastic support received from the editorial staff of McGraw-Hill, especially B. J. Clark, and the project supervisor of The Total Book, Kate Scheinman.

Manfred R. Hausmann

R. Gaskin and Dr. Schaefer; in Canada, at the University of Alberta in Edmonton (R. M. Hardy and S. Thomson) and in consulting (C. F. Ripley and E. J. Klohn); at the University of California, Los Angeles (K. L. Lee); and in Australia, at the University of New South Wales (I. K. Lee).

The format of the book was designed to suit students of civil engineering at senior undergraduate or graduate level. However, the practicing engineer will also gain from reading the text and completing the problems at the end of each chapter. Most of these problems are intended to test comprehension and emphasize specific points made in the text. Other problems represent real decision-making situations and serve to acquire practical skills. Some of the discussion topics could well challenge the reader's knowledge of construction and may require researching other texts and original source material.

The author hopes to receive many comments (including criticism) from readers. Very useful would also be additional information, problems, and illustrations from practice, researchers, and practitioners alike. Such contributions would add greatly to any future editions of this book.

The author also welcomes notices of any audio-visual material which could be made available to educational institutions offering a course in ground modification. It is known, for example, that the U.S. National Committee on Tunneling Technology has produced a set of slides on Rock Bolting Practice. Another excellent set of slides and additional material complementing this text is available from GKN Hayward Baker, Odenton, Maryland; this firm has pioneered many of the techniques described in this book, and in the United States the words "Ground Modification" represent a service mark.

Finally I would like to acknowledge the enthusiastic support received from the editorial staff of McGraw-Hill, especially B. J. Clark, and the project supervisor of the total book, Kate Scheinman.

Manfred R. Hausmann

CHAPTER OPENING ART CREDITS

Chapter

1 Graphical arrangement of names of recent conferences on topics related to ground modification.

2 Smooth steel roller. (*Picture based on photograph published by Jaques Construction Equipment Div., Cnr. Griffiths and Palmer Streets, Richmond, Vic. 3121, Australia.*)

3 Padfoot roller. [*Adapted from an illustration by Pösch and Ikes (1975).*]

4 Vibro-compaction. (*Based on a photograph by GKN Hayward Baker, 1875 Mayfield Rd., Odenton, MD 21113.*)

5 Cone-penetration-test rig. (*Based on an illustration in a brochure by Hogentogler & Company, Inc., 4 Meem Ave., Gaithersburg, MD 20877.*)

6 Self-propelled padfoot roller. (*Scanned from an illustration in* The Earthmover and Civil Contractor Journal, *June 1985.*)

7 Skid-mounted mine dewatering pump. (*Courtesy Hanson Sykes Pumps Pty. Ltd.*)

8 Schematics of an automatic self-priming centrifugal pump used for dewatering. (*Based on a brochure by Hanson Sykes Pumps Pty. Ltd.*)

9 Well point dewatering of pipe trench. (*Courtesy Hanson Sykes Pumps Pty. Ltd.*)

10 Drains lined with TYPAR Geotextile (a Du Pont product).

11 Alidrain Stitcher ADS. (*Scanned from an illustration by Burcan Industries Ltd., Canada.*)

12 Cathode and well installed for electroosmotic dewatering. [*Scanned from an illustration by Fetzer (1967).*]

13 Single-rotor stabilizer for road construction. (*Scanned from a brochure by Stabilex, an operation of Bitupave Ltd., 49 Bridge Street, Rydalmere, NSW 2116, Australia.*)

14 Pattern of grouting below Göscheneralp Dam. (*Based on a drawing by Electro-Watt Engineering Services Ltd., Zurich, Switzerland.*)

15 Section of frozen earth wall. (*Based on a photograph by the Geofreeze Corp., Washington, D.C.*)

16 Interlocking structural elements envisaged by Henri Vidal, the inventor of Reinforced Earth. (*Based on an illustration in the Australian patent No. 286332, Appl. No. 42239/64.*)

17 (*Illustration adapted from a brochure distributed by Reinforced Earth Pty. Ltd.*)

18 Placing a Reinforced Earth face panel. (*Scanned from a photograph by Reinforced Earth Pty. Ltd.*)

19 Laying Polyfelt geotextile. (*Scanned from a brochure provided by Polyfelt Australia.*)

20 Rockbolt photoelastic pattern. (*Based on photographic records.*)

21 Gabion weir. [*From Audova (1978).*]

PART I

INTRODUCTION

CHAPTER 1

INTRODUCTION TO ENGINEERING GROUND MODIFICATION

In Situ's Blacksburg

Soil Reinforcement Paris

Compaction Technology London

Geotextiles Paris Las Vegas Vienna

Sixth Ash Utilization Symposium Reno

New Horizons in Construction Materials

International Conference on Compaction

In Situ Soil and Rock Reinforcement Paris

4th ANZ Geomechanics Conference Perth

Ground Improvement Techniques Bangkok

10th International Conference SMFE Stockholm

Grouting in Geotechnical Engineering New Orleans

Soil Reinforcing and Stabilizing Techniques Sydney

Theory and Practice of Earth Reinforcement Kyoto

Durability and Ageing of Geosynthetics Philadelphia

Australian Road Research Board Conference Adelaide

Symposium on Environmental Technology Bethlehem

European Symposium on Penetration Testing Stockholm

Ground Freezing Bochum Trondheim Hanover Sapporo

European Conf on Soil Mechanics and Foundations Helsinki

1.1 THE GROUND MODIFICATION OPTION IN DEALING WITH DIFFICULT SOILS

1.1.1 The Need for Engineered Ground Improvement

Where a project encounters difficult foundation conditions, possible alternative solutions are

- *Avoid the particular site*. Relocate a planned highway or development site.
- *Design the planned structure accordingly*. Some of the many possible approaches are to use a raft foundation supported by piles, design a very stiff structure which is not damaged by settlement, or choose a very flexible construction which accommodates differential movement or allows for compensation. The solution will depend on the geotechnical performance criteria stipulated, which generally relate to stability, deformation, and/or seepage.
- *Remove and replace unsuitable soils*. Removing organic topsoil, which is soft, compressible, and volumetrically unstable, is a standard precaution in road or foundation construction.
- *Attempt to modify the existing ground*.

Similar options must be considered in the case where there is a lack of good-quality granular materials needed for the construction of dams, embankments, roads, or foundations.

As more and more land becomes subject to urban or industrial development, good construction sites and borrow areas are difficult to find and the soil improvement alternative more frequently becomes the best option, technically and economically.

1.1.2 Classification of Ground Modification Techniques

In this book, four groups of ground improvement techniques are distinguished:

Mechanical modification. Soil density is increased by the application of short-term external mechanical forces, including compaction of surface layers by static, vibratory, or impact rollers and plate vibrators; and deep compaction by heavy tamping at the surface or vibration at depth.

Hydraulic modification. Free-pore water is forced out of the soil via drains or wells. In coarse-grained soils this is achieved by lowering the groundwater level through pumping from boreholes or trenches; in fine-grained soils the long-term application of external loads (preloading) or electrical forces (electrokinetic stabilization) is required. Traditional techniques have benefited from the development of geosynthetics, as in the case of vertical drains.

The establishment of impermeable barriers, such as diaphragm walls, sheet piles, and geomembranes, and the use of compressed-air techniques (caissons, tunneling) for seepage control are considered outside the scope of ground modification and are therefore not discussed here in detail.

Physical and chemical modification. Stabilization by physically mixing additives with surface layers or columns of soil at depth is discussed under the heading of modification by *admixtures* in Chap. 13. Additives include natural soils, industrial by-products or waste materials, and cementitious and other chemicals which react with each other and/or the ground. When additives are injected via boreholes under pressure into voids within the ground or between it and a structure, the process is called *grouting*. Soil stabilization by heating the ground and by freezing the ground are both considered thermal methods of modification. *Heating* evaporates water and causes permanent changes in the mineral structure of soils; *freezing* solidifies part or all of the water and bonds individual particles together.

Modification by inclusions and confinement. Reinforcement by fibers, strips, bars, meshes, and fabrics imparts tensile strength to a constructed soil mass. In situ reinforcement is achieved by nails and anchors. Stable earth-retaining structures can also be formed by confining soil with concrete, steel, or fabric elements (including crib and bin walls and sandbags).

Conventional pile foundations are not included under this heading, although they are sometimes described as ''compressive reinforcement''; it is felt that the principal purpose of pile foundations is to transfer the load to a stronger stratum or simply to a greater depth for improved structural support, bypassing some or all of the soft foundation soils.

Any classification of ground modification techniques is to a degree arbitrary, particularly where one or more of the possible physical, chemical, hydraulic, or mechanical processes are combined. In its broadest sense, any interference in the natural state of soil and rock could be termed ground modification, but this book concentrates on the soil and rock mechanics principles underlying the techniques of improving the engineering performance of foundations and earth structures through modifying and complementing their basic particulate constituents.

1.1.3 Suitability, Feasibility, and Desirability

The choice of a method of ground improvement for a particular object will depend on many factors including:

Type and degree of improvement required
Type of soil, geological structure, seepage conditions
Cost (the size of the project may be decisive)
Availability of equipment and materials and the quality of work required
Construction time available
Possible damage to adjacent structures or pollution of groundwater resources
Durability of the materials involved (as related to the expected life of the structure for the given environmental and stress conditions)

Toxicity or corrosivity of any chemical additives (government regulations may restrict the choice of additives)

Reversibility or irreversibility of the process

Reusability of components

Reliability of methods of analysis and design

Feasibility of construction control and performance measurements

Only one method of improvement is clearly applicable to all types of soil, provided it is moist: ground freezing. Other methods of ground modification are most suitable for particular groups of soils, usually either cohesive or cohesionless soils. Other conditions may also be important, such as whether the ground is

Saturated or unsaturated

Normally consolidated or overconsolidated

Of a special nature: highly organic, a sanitary landfill, or an industrial-waste deposit

The feasibility of a particular method is strongly related to the type of problem in hand: a foundation, an embankment on soft ground, an unstable slope, an excavation, an earth-retaining structure, or a leaking dam or reservoir.

The desirability of a particular method of ground improvement is largely perceived in terms of environmental impact and energy consumption. The use of toxic grouts and experimental techniques such as densification by electrical shock or modification of soils by radiation are low on the list of desirability.

Where relevant, aspects of suitability, feasibility, and desirability will be discussed with each of the methods of ground modification introduced in subsequent chapters.

1.2 RECENT FORUMS

Engineering ground modification has been a prominent topic at many recent conferences, but by far the most important conferences have been the following:

1981 Tenth International Conference on Soil Mechanics and Foundation Engineering (ICSMFE), Stockholm.

1982 Symposium and short course held at the Asian Institute of Technology (AIT), Bangkok. Proceedings were entitled "Recent Developments in Ground Improvement Techniques," were edited by Balasubramaniam et al., and were published by A. A. Balkema in 1985.

1983 Eighth European Conference on Soil Mechanics and Foundation Engineering (ECSMFE), Helsinki. This conference was dedicated to the topic "improvement of ground." Proceedings were edited by Rathmayer and Saari and were published by A. A. Balkema in 1983.

1984 Fourth Australia–New Zealand Geomechanics Conference, Perth. Proceedings were published by the Institution of Engineers, Australia, 1984.

A total of 284 papers on ground modification were presented at the above four conferences alone. The Eleventh International Conference on Soil Mechanics and Foundation Engineering held in San Francisco in 1986 added more information on soil improvements, with specialty sessions on soil reinforcement, geotextiles, and other soil-improvement topics.

These major international conferences, which covered a multitude of different geotechnical processes, were complemented by specialty conferences on soil reinforcement, geotextiles and geomembranes, compaction, grouting, ground freezing, the use of admixtures in road construction, ash utilization, engineering geology, and more, as referred to in this text. Also significant were forums discussing in situ measurement of soil properties, such as IN SITU'86, held in Blacksburg, Virginia, in 1986, and sponsored by the American Society of Civil Engineers (ASCE).

Very recent contributions to the development of ground modification can be found in the proceedings of the ASCE 1987 session on "Soil Improvement—A 10-year update" [Welsh (1987)], and of the International Geotechnical Symposium on "Theory and Practice of Earth Reinforcement," held in Fukuoka Kyushu [Yamanouchi et al. (1988)].

Examining the articles published, it appears that in recent years major developments have occurred in the areas of precompression and dewatering, deep compaction, and soil reinforcement. Considerable attention has also been given to stabilization using admixtures, on the surface or at depth, while interest in high-energy consuming methods such as soil heating seems to be declining. Grouting is probably the first soil-improvement process which has been automated with computers, but because of a slowdown in dam construction in developed countries, its importance has decreased. Ground freezing is maintaining an important role in shaft sinking and tunneling, while the interest in electrokinetic stabilization is sporadic.

1.3 TRADITIONAL OBJECTIVES AND EMERGING TRENDS

Traditionally the aims of improving soils as foundation or construction materials have been one or more of the following:

1. Increase strength, reduce erodibility
2. Reduce distortion under stress (increase stress-strain modulus)
3. Reduce compressibility (volume decrease due to a reduction in air voids or water content under load)
4. Control shrinking and swelling (improve volume stability)
5. Control permeability, reduce water pressures, redirect seepage
6. Prevent detrimental physical or chemical changes due to environmental conditions (freezing/thawing, wetting/drying)
7. Reduce susceptibility to liquefaction
8. Reduce natural variability of borrow materials or foundation soils

These benefits are directly related to the performance of soils in foundations and soil structures. Today, however, engineering soil improvement is no longer seen in only a narrow technical sense, where it assists in solving well-defined problems related to stability, deformation, and seepage. Newly recognized problems (such as those associated with waste disposal), recognition of the need for energy conservation, and innovative techniques and materials have opened up new perspectives for ground modification. Examples of these kind of developments are the emergence of the discipline of environmental geotechnics, the rediscovery of soil as a material for building dwellings, and the availability of a rapidly growing variety of geosynthetics.

1.3.1 Environmental Geotechnics

The task of environmental geotechnics is defined as identifying and understanding the environmental significance of geotechnical processes and predicting their consequences. These processes include making cuts and fills, creating reservoirs, excavating at the surface or underground, dewatering, and redirecting seepage.

Examples of desirable ground modification activities within the framework of environmental geotechnics are

Constructive use and, if necessary, modification of waste materials

Stabilization of newly exposed soil surfaces so as to reduce dust in the air and suspended solids in the runoff

Prevention of subsidence due to mining

Preservation of quality and flow patterns of groundwater

Environmental geotechnics was firmly established by the illuminating "Environmental Geotechnics—State-of-the-Art Report" presented at the Eleventh ICSMFE by Sembenelli and Ueshita (1981). Five years later, the first major specialty symposium on environmental geotechnology took place at Lehigh University (April, 1986).

1.3.2 Containment and Constructive Use of Waste Materials

Ground modification techniques are also increasingly being applied in the rehabilitation of hazardous-waste disposal areas. There, emphasis may be on the fixation of dangerous pollutants, but at the same time a geotechnically engineered design should impart minimal strength and volume stability to the waste mass. In this regard a case can be made for codisposal (or coutilization) of industrial wastes, such as mixtures of fly ash and SO_x scrubber sludge.

Especially difficult and intractable is the treatment of wastes transported and disposed of in slurry form. To solve that problem, many studies are currently being undertaken; some of these are quite novel. They include cement stabilization, the formation of cement crystals (cement bacilli), physicochemical methods, even the use of biochemistry (see Mitchell's "Soil Improvement—State-of-the-Art Report" presented at the Eleventh ICSMFE).

Of increasing concern to environmentally conscious engineers is the constructive use of high-energy waste materials such as slag and fly ash. Part of this concern arises from the estimated increase in ash production in the years to come. In the United States, which is said to have about 40% of the world's coal reserves, current ash production is on the order of 70 to 80 million tons per annum. It has been predicted that by early next century coal burning will produce 200 million tons of ash per year! Current costs of fly ash disposal are staggering, considering that fly ash could be used as structural fill on its own, or, in combination with lime or cement, as a stabilizing agent for road bases.

1.3.3 Low-Cost Building With Soil

Principles of soil stabilization can be applied to the making of low-cost earth walls and building blocks. In recent years, the use of adobe and mud brick construction techniques has seen a revival in interest. Selection of the appropriate type of clayey soil, compaction at the optimum moisture content, and inclusion of straw or saplings were the key to soil construction methods used in the last century in Australia and elsewhere. Soil modification with cement or lime and mechanical pressing are a newer development.

Rather than low initial costs, earth-covered housing offers low life-cycle costs. It is claimed that energy required for heating and cooling can be reduced by as much as 80% if the thermal insulation properties of soil are taken advantage of. Research and development in this area has so far mainly concentrated on physical, financial, and aesthetic aspects and on the use of soil as a passive constituent rather than as a construction material, as the proceedings from several symposia dedicated to this topic indicate [e.g., see Baggs (1983)]. There appears to be a considerable scope for the use of engineered soil improvement in the building of earth-contact structures.

1.3.4 Metallic Reinforcement and Geosynthetics

Initiated by the invention of Reinforced Earth, highly innovative methods of ground modification by inclusions and confinement have emerged in the last two decades allowing considerable economies in building retaining walls and bridge abutments and solving difficult problems in slope stabilization and excavations for buildings.

Synthetic fabrics are now used for construction in virtually every part of the world. The major functions of geotextiles are filtration, drainage, separation, and reinforcement. Geotextiles, geogrids, and related products have found numerous applications in the construction of roads, railways, embankments, retaining walls, drainage systems, erosion protection schemes, and many other areas. An enormous range of products is available, allowing for many novel solutions to engineering problems.

Besides having to consider external or internal stability and deformation of structures incorporating reinforcement or synthetic filters and drainage layers, the designer is confronted with the need to assess the durability of the system components. More is known about the corrosion of metals buried in soil than about the degradation of synthetics under the same conditions; but the prediction of their performance over the normally expected life span of civil engineering structures still requires a substantial extrapolation of existing data for both. To give careful attention to the possibility of long-term loss of strength and deformation under load of ground inclusions represents a new challenge to the civil engineer, who is more familiar with traditional construction elements made of concrete, steel, and timber.

1.4 CURRENT AND FUTURE DEVELOPMENTS

The basic principles of engineering ground modification are not new; indeed, some of the techniques used today may be more than 2000 years old. However, significant advances have been made in recent times. They cover many aspects.

1. Development of new machinery, particularly for deep compaction
2. Availability of new construction materials, such as geosynthetics
3. Emergence of better guidelines for determining the suitability of specific techniques for certain types of soils and site conditions
4. Better understanding of the geotechnical processes involved and appreciation of the significance of the construction sequence
5. Refinement of methods of analysis and computer modeling techniques
6. Advances in the techniques of performance evaluation of modified soils, such as pressuremeter tests and seismic shear-wave velocity measurements

The results of a Delphi survey reported by the ASCE Committee on Placement and Improvement of Soils (1978) included among the 10 most innovative suggestions for future developments the following, presented in order of perceived feasibility:

Embankments of soil encased and stacked in tough, durable membranes to eliminate the need for drying and compaction

Membranes for osmotic dewatering in boreholes and trenches

In situ fusion stabilization of soils to form impermeable barriers and reinforcing elements

Probes driven and exploded to release self-dispersing grout

Stabilization with bacteriological agents

Parallel to newly evolving techniques, the search will go on for the magic low-cost catalyst which makes sandstone out of sand and claystone out of clay.

Any review of recent applications of the technology of ground modification will underline its economic and technical significance in construction related to mining, bulk handling facilities, and waste disposal, besides its acknowledged importance in foundation, road, and dam engineering. In addition, from a concern for the environment, new tasks have arisen for the geotechnical engineer dealing with ground modification.

PART II

MECHANICAL MODIFICATION

CHAPTER 2

INTRODUCTION TO MECHANICAL MODIFICATION

2.1 TERMINOLOGY AND AIMS OF MECHANICAL MODIFICATION

2.1.1 Terminology

Mechanical ground modification refers to soil densification by external forces. In most practical applications mechanical modification is synonymous with compaction. In its classic sense, *compaction* means densification of an unsaturated soil by a reduction in the volume of voids filled with air, while the volume of solids and the water content remain essentially the same. Compaction implies that soil particles are packed closer together by the application of sudden heavy loads or dynamic forces; crushing of some of the soil grains or rock particles may assist this densification.

The above definition of compaction holds true for most surface compaction methods but must be extended for deep compaction techniques. The latter may also involve saturated soils, water jetting, and partial replenishment of the in situ soil but are nevertheless included here under mechanical modification.

The geotechnical engineer makes a clear distinction between the processes of compaction and those of stabilization and consolidation.

For the nonspecialist, *stabilization* may refer generally to an increase in strength or a reduction in the deformation of a soil mass. When used in road engineering and soil mechanics, it usually implies soil improvement by physicochemical reactions caused by additives or induced by environmental changes.

Consolidation is a process where the volume of a soil mass is reduced by the expulsion of water. As modeled by the classic consolidation theory, it involves stress transfer from the water to the solid phase. It is usually achieved by the long-term application of static loads or electric forces to saturated soils. Consolidation may also be induced or accelerated by temporary liquefaction caused by impact forces or vibration—thus the term "dynamic consolidation" in relation to heavy tamping, a deep compaction technique (see Sec. 2.2.3.2).

2.1.2 Compaction Purpose and Strategies

The major aims of compacting soil are to

1. Increase shear strength
2. Reduce compressibility
3. Reduce permeability
4. Reduce liquefaction potential
5. Control swelling and shrinking
6. Prolong durability

It could be added that properly managed engineered compaction may reduce the variability of engineering soil properties in a natural deposit or a human-constructed fill.

Compaction is the most commonly used method of ground modification. Significant early advances in the knowledge of compaction principles were made in the first quarter of this century by engineers building roads for increasingly bigger traffic numbers and wheel loads. Later, in the 1930s, the construction of large earth dams, whose failure would have catastrophic consequences, forced the adoption of a scientific approach to compaction. Today, compaction is of equal importance in highway, airfield, and marine construction; in the preparation of foundation soils; and in the backfill behind abutments, walls, and in trenches.

Improvement of engineering properties by densification is possible for natural soils as well as for soils stabilized with chemicals such as lime and cement. The principles and techniques of compaction also have relevance for other soil-related construction materials, such as asphalt surfacing and concrete.

In soil engineering, strategies developed for optimizing the densification process may include some or all of the following steps.

1. In the case of human-constructed fills, specify placement conditions (water content, density, depth of layers, etc.).
2. Select appropriate equipment (roller, vibro-compactor, tamping) and method of operation (number of passes, patterns of tamping, etc.).
3. Set up adequate control procedures (type and number of tests, statistical evaluation, etc.).

To achieve efficient compaction requires knowledge of the available equipment, the principles of compaction, the properties of compacted soils, and control procedures.

2.2 METHODS OF COMPACTION

2.2.1 Laboratory Procedures

The aim of laboratory compaction tests is to simulate field compaction procedures. Their results should aid in the optimization and control of placement conditions. The most common type of test is the standard compaction test, in which a steel rammer is dropped on loose soil placed in a mold. It employs dynamic compaction as against kneading or static compaction.

2.2.1.1 DYNAMIC COMPACTION. The standard compaction (or Proctor) test is described in the Australian standard AS 1289.E1.1-1977; equivalent procedures are described in other national standards [American Society for Testing and Materials (ASTM), American Society for State Highway and Transportation Officials (AASHTO), British Standard (BS), Australian Standard (AS), etc.]. The test is carried out on that portion of the soil which passes the 19-millimeter (mm) sieve. Five or more samples are prepared at different moisture contents and allowed to cure. Every sample is compacted in three layers, each receiving 25 blows with a 2.7-kilogram (kg) rammer dropping 300 mm. The diameter of the flat rammer head is approximately half the diameter of the mold.

In the modified compaction (or Proctor) test (AS 1289.E2.1) the compactive effort is increased so that the densities achieved are closer to those obtained by the very heavy compaction equipment as employed in airport and road engineering. The

TABLE 2.1
Compaction apparatus and procedures

Detail	Standard compaction	Modified compaction
Mold volume, cm^3	1000	1000
diameter, mm	105	105
height, mm	115.5	115.5
Rammer diam., mm	50	50
drop, mm	300	450
mass, kg	2.7	4.9
Number of layers	3	5
Blows per layer	25	25
Energy input, kJ/m^3*	596	2703

*Kilojoules per cubic meter.

rammer that is used weighs more and is dropped from a greater height than in standard compaction, and five, rather than three, layers are subjected to 25 blows. Details of both tests are given in Table 2.1.

In order to obtain consistent results, care has to be taken that sample preparation and test procedures meet the appropriate standard. An automatic compaction apparatus may be used provided that essential dimensions are adhered to. Small variations in cylinder size, number of blows, etc., will not cause significant error provided that the total energy expended per cubic meter of compacted soil remains the same. In some cases it may be desirable to use an extra large mold so that particles larger than 19 mm can be included in the laboratory test. Results from such tests should not normally be expected to correspond exactly to those obtained in the standard mold.

Compaction test results are plotted in terms of dry density versus moisture content on a diagram as shown in Fig. 2.1. The dry density γ_{dry} is calculated from the total (or wet) density γ_{tot} as follows:

$$\gamma_{dry} = \frac{\gamma_{tot}}{1 + w} \tag{2.1}$$

where w represents the moisture content (or water content). The density γ may be expressed in mass units [metric tons per cubic meter (t/m^3) or grams per milliliter (g/mL)] or force units [kilonewtons per cubic meter (kN/m^3)]. Some engineers prefer the symbol ρ instead γ if mass units are used.

It is highly recommended that a compaction curve be accompanied by the zero-air-void (ZAV) curve. This facilitates the drawing of the compaction curve and assists in identifying erroneous results. The ZAV curve represents dry densities corresponding to a saturation $S = 100\%$ at given water contents and for a particular specific gravity G_s of the soil solids. With γ_w designating the unit weight of water, points of the ZAV curve can be calculated from

$$\gamma_{dry} = \frac{G_s\gamma_w}{1 + wG_s/S} = \frac{G_s\gamma_w}{1 + e} \tag{2.2}$$

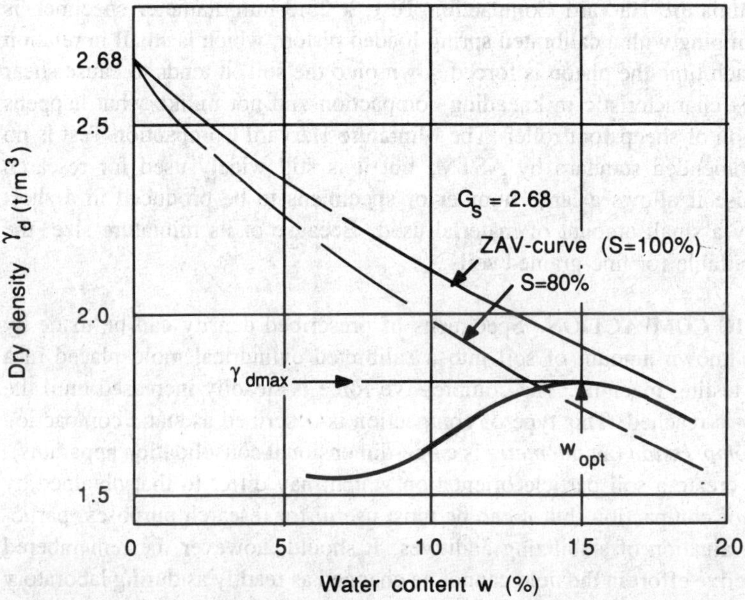

FIGURE 2.1
Dry density versus moisture content.

where *e* is the void ratio. A line similar to the ZAV curve but representing, e.g., 80% saturation could be drawn; this information would help in estimating the degree of saturation of the compacted soil specimen.

Most soils, in particular cohesive soils, show a distinct peak in dry density, probably in the range of 80 to 90% saturation. This peak is called the *maximum dry density* (MDD) γ_{dmax}, and the corresponding water content is the *optimum moisture content* (OMC) w_{opt}. This means that given a certain compactive effort, maximum dry density (and, correspondingly, high shear strength and low compressibility) is only achieved if the soil is at its optimum moisture content.

For certain types of soil, the concept of a single optimum moisture content may not be applicable, nor relevant. For example, highly permeable granular soils may densify best if completely dry or completely saturated. With organic soils, such as peat, the dry density may simply decrease with increasing moisture content and not even be significantly affected by the compactive effort.

2.2.1.2 KNEADING COMPACTION. The California Division of Highways employs kneading compaction for the preparation of specimens used in the stabilometer test. The samples are placed in molds, about 102 mm in diameter and 127 mm high and are compacted by being kneaded 100 times at 2413 kilopascals (kPa). It is said that the soil structure created by this type of compaction closely resembles that obtained with compaction equipment typically used for fine-grained soils in the field, such as sheepsfoot and tamping rollers.

In the Miniature Harvard Compaction Test, a 25.3-mm-diameter specimen is produced by tamping with a calibrated spring-loaded piston, which is small in relation to the mold. Each time the piston is forced down onto the soil, it tends to cause shear failure which is characteristic in kneading compaction and not unlike what happens in the application of sheepsfoot roller. The Miniature Harvard Compaction Test is no longer a recommended standard by ASTM, but it is still widely used for research purposes because it allows a large number of specimens to be produced in a short time, with only a small amount of material used. Because of its miniature size, the mold is only suitable for fine-grained soils.

2.2.1.3 STATIC COMPACTION. Specimens of prescribed density can be made by compressing a known amount of soil into a calibrated cylindrical mold placed in a universal-type testing machine. The compressive force is steadily increased until the desired density is reached. This type of compaction is described as static compaction or *odometric compression* (an *odometer* is a one-dimensional consolidation apparatus). It is known to create a soil particle orientation which may differ to that obtained by other methods of compaction, but it can be most useful for research purposes, particularly in the evaluation of stabilizing additives. It should, however, be remembered that the compactive effort in the field cannot be changed as readily as during laboratory static compaction.

2.2.1.4 LABORATORY COMPACTION USING STRESS PATH SIMULATION. If stresses in a soil mass under the influence of specific compaction machinery are known, an attempt can be made to reproduce these stresses in the laboratory in order to predict the densities which are going to be obtained in the field for particular placement conditions. There is no generally valid stress-strain law available for soils, but under certain conditions, linear elastic theory provides reasonable values for the stresses generated, and these can serve as a guide to laboratory simulation.

As an example, linear elastic theory, or more specifically the Boussinesq theory, is said to yield good results for the compaction stresses generated beneath a rubber-tired roller, particularly in a firm cohesive soil. Elastic solutions are also available for the effect of a rigid wheel. Even vibratory compaction stresses have been analyzed using linear stress-strain theory.

Fry (1980) illustrated the in situ stress path below a roller and how it is approximated in conventional laboratory tests (Fig. 2.2). He concluded that the odometric stress path does not give a good representation of what happens below a roller. One result quoted by him shows that in order to achieve 100% γ_{dmax} (standard Proctor test) for a highly plastic clay at a water content 4% below optimum, a single application of a compressive stress of around $\sigma_1 = 1000$ kPa was required. In the same proceedings, Biarez (1980) gave evidence that an all-around triaxial stress of only say 400 kPa may achieve the same density.

Fry (1980) feels that a repeated triaxial test ($\sigma_3 =$ constant) is suitable to simulate roller compaction in clay. For cohesionless soils, repeated compression-extension triaxial tests are thought to be more representative of field compaction. These tests induce a rotation of the principal axes. If a repeated compression-extension

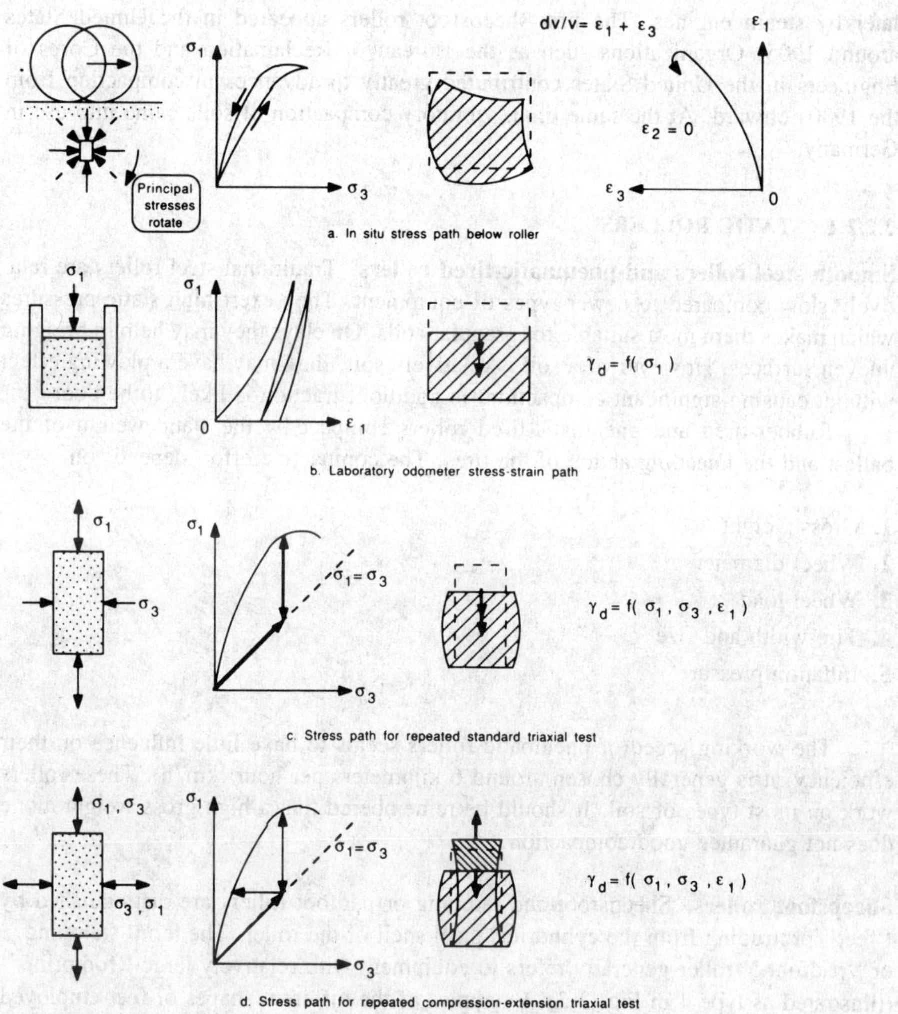

FIGURE 2.2
Stress paths below roller and in laboratory tests. [*Adapted from Fry (1980).*]

test is performed so as to induce repeated shear in a clay soil, this process comes close to the concept of kneading compaction.

2.2.2 Shallow Surface Compaction

Compaction by surface equipment is achieved by static pressure and/or dynamic pressure caused by impact or vibration.

There is a long history in the development of surface compaction machinery resulting in a wide variety of equipment, differing in size, shape, and mode of operation. Records from the last century show rollers pulled by horses and oxen and later by steam engines. The first sheepsfoot rollers appeared in the United States around 1905. Organizations such as the Bureau of Reclamation and the Corps of Engineers in the United States contributed greatly to advances in compaction from the 1930s onward. At the same time, vibratory compaction of soils came into use in Germany.

2.2.2.1 STATIC ROLLERS

Smooth steel rollers and pneumatic-tired rollers. Traditional steel rollers are relatively slow compared to newer types of equipment. They exert high static pressures which makes them most suitable for granular soils. On clays they may help in bridging uneven surfaces. However, if a soil is relatively soft, they may have a plowing effect without causing significant compaction; in addition, traction is likely to be poor.

Rubber-tired and pneumatic-tired rollers compact by the static weight of the ballast and the kneading action of the tires. The compactive effort depends on

1. Gross weight
2. Wheel diameter
3. Wheel load
4. Tire width and size
5. Inflation pressure

The working speed of pneumatic rollers seems to have little influence on their efficiency; it is generally chosen around 6 kilometers per hour (km/h). These rollers work on most types of soil. It should be remembered that a high gross weight alone does not guarantee good compaction.

Sheepsfoot rollers. Sheepsfoot and tamping or padfoot rollers are distinguished by "feet" protruding from the cylindrical steel shell of the roller. The term "tamping," or "padfoot," roller generally refers to equipment with relatively large "footprints" (illustrated as type 4 in Fig. 2.3). Examples of the different shapes of feet employed are shown in Fig. 2.4. Generally, the wetter and softer the soil, the larger the contact area (footprint) required for optimum compaction. Sheepsfoot rollers have proved more suitable for cohesive soils than other rollers. They exert high pressures on the soil, first compacting lower layers and then gradually working to the surface as the soil underneath gains in strength. When the soil yields no further, the sheepsfoot roller is said to "walk out" of the lift ("Compaction Data Handbook"). Blending of the material is assisted by the sheepsfoot action. Moisture control is made easier because of the pockmarked surface during compaction. Steel rollers may be used to level off areas worked by sheepsfoot or rubber-tired rollers.

Grid rollers. Grid rollers have drums covered or consisting of a heavy steel grid. This creates high contact pressures while preventing excessive shear deformation responsible for the plastic wave ahead of the roll. Grid rollers are suitable for compacting

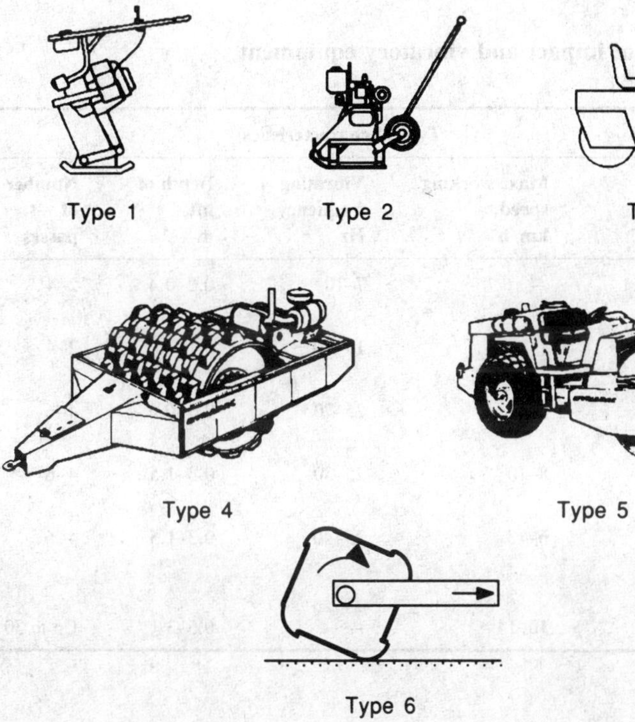

FIGURE 2.3
Vibratory and impact compactors for shallow compaction (refers to Table 2.2).

weathered rock, such as sandstone, by breaking and rearranging gravel and cobble-size particles. Clayey soils, however, may clog the grid and make it ineffective. A relatively high operating speed assists in the breakdown of material, while a lower speed enhances the densification effect.

2.2.2.2 IMPACT AND VIBRATORY EQUIPMENT. Table 2.2 summarizes typical characteristics of vibratory equipment commonly used for surface compaction today. The equipment discussed is illustrated in Fig. 2.3.

Tampers, rammers, and plate compactors. Vibrating *tampers* or *rammers* and vibrating *plate compactors* are used in confined areas such as on backfill in trenches, around pipes, and behind retaining walls and bridge abutments. Rammers may have a stroke length of 30 to 70 mm and therefore mainly work on the impact principle. Plate compactors have a smaller amplitude of vibration and, for the same weight, would be less efficient at depth for most soils. By changing the position of the rotating weights, the compacting force can be adjusted and the plate compactor can be made to move forward or backward.

TABLE 2.2
Typical characteristics of impact and vibratory equipment for shallow compaction*

Type no. and name	Typical characteristics: Mass, t	Max. working speed, km/h	Vibrating frequency, Hz	Depth of lift, m	Number of passes
1. Vibrating rammer	0.3–0.1	—	7–10	0.2–0.4	2–4
2. Light vibrating plate	0.06–0.8	1	10–80	0.15–0.5	2–4
3. Light vibrating roller	0.6–2	2–4	25–70	0.3–0.5	4–6
4. Heavy towed vibrating roller	6–15	8–10	25–30	0.3–1.5	4–6
5. Heavy self-propelled vibrating roller	6–15	6–13	25–40	0.3–1.5	4–6
6. Impact roller	7	10–14	—	0.5–3	Up to 30

*See Fig. 2.3 for illustrations.

Vibrating rollers. Lightweight vibratory rollers have little impact effect; the vibrational amplitude is on the order of 1 or 2 mm. In order to achieve the same depth effect as rammers and vibrating plates, they have to be considerably heavier.

The heavy vibrating drums of towed or self-propelled vibratory rollers are isolated from the frame by rubber shock absorbers. The mass given in Table 2.2 includes the frame and drum. Vibrations are caused by rotating weights. As stated by Forssblad (1977), the compactive effort of vibrating rollers is primarily dependent on

1. Static weight
2. Frequency and amplitude
3. Roller speed
4. Ratio between frame mass and drum mass
5. Drum diameter

The centrifugal force is a function of the moment of the eccentric weight (*mr*) and the frequency *n*:

$$\text{Centrifugal force} = mr4\pi^2n^2 \qquad (2.3)$$

where m is the weight and r is the eccentricity. The actual instantaneous force exerted on the ground will also depend on the properties of the soil and its support.

Impact rollers. Impact rollers consist of a noncircular mass which is towed along the ground. As its center rises and falls, its mass exerts a high impact force causing compaction of the soil. Clifford (1980) described a new type of impact roller developed

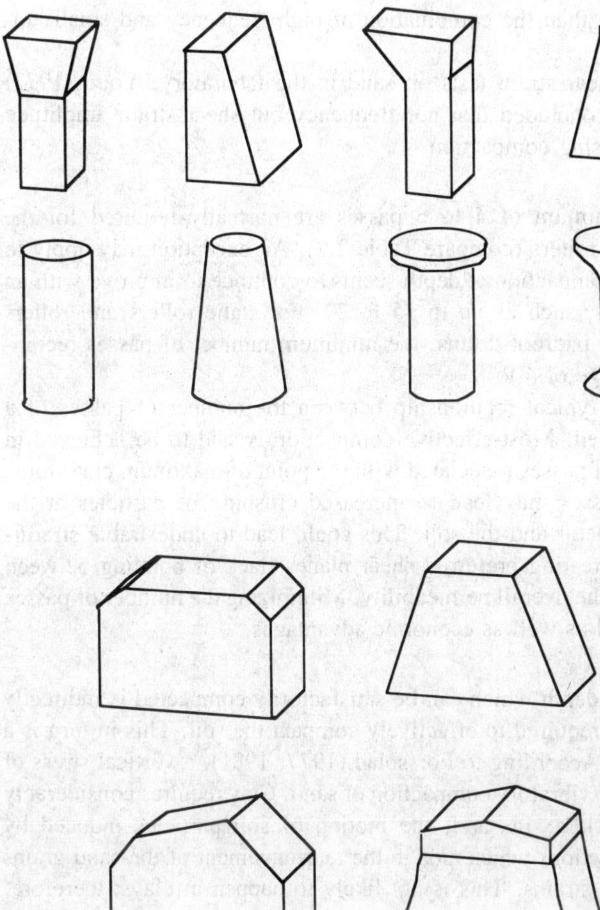

FIGURE 2.4
Sheepsfoot and padfoot shapes. [*After Poesch and Ikes (1975).*]

in South Africa. It consists of a 1.5-m-thick "square" roller with rounded edges. It was found suitable for natural ground and fill. Because the impact roller leaves an uneven surface, it is recommended for subgrades and earth fills rather than for surfacing works.

2.2.2.3 OPERATIONAL ASPECTS OF SHALLOW COMPACTION

Operating frequency. The frequency of vibration of heavy vibratory rollers is usually between 25 and 30 cycles per second (Hz); however, the compactive effort does not appear to vary significantly in the range of 25 to 50 Hz. According to Forssblad (1977), with respect to vibrating rollers, a combination of a large amplitude and a frequency just over the resonance frequency (say 25 Hz) normally results in a better compaction and depth effect than the combination of high frequency and small amplitude.

Carrying out cyclic shear strain tests on sand in the laboratory, Youd (1972) found a similar result. He concluded that not frequency but shear strain amplitude was the dominant factor causing compaction.

Number of passes. A minimum of 4 to 6 passes are normally required for the economical use of vibratory rollers (compare Table 2.2). An exception may apply to saturated sands, where the compaction at depth seems to continue to improve with an increasing number of passes, such as up to 15 to 20. For static rollers and rollers equipped with sheepsfoot or padfoot drums, the minimum number of passes recommended is usually in the range of 4 to 8.

Figure 2.5 shows the typical relationship between the number of passes of a roller and the density obtained. Most-effective compaction is said to be achieved in the range up to the number of passes associated with the point of maximum curvature.

A high number of passes may lead to increased crushing of particles at the interface between the compactor and the soil. This could lead to undesirable stratification of the fill, e.g., by creating preferred shear planes (lack of bonding between adjacent layers) or affecting the overall permeability. Minimizing the number of passes may therefore have technical as well as economic advantages.

Depth of layers. The layer depth which can be satisfactorily compacted is indirectly proportional to the pressure required to effectively compact the soil. This in turn is a function of the type of soil. According to Forssblad (1977, 1981), a vertical stress of 50 to 100 kPa is sufficient for vibratory compaction of sand. Clay requires considerably more pressure: 400 to 700 kPa. In sand, the motion of soil particles induced by vibration reduces internal friction, which aids in the rearrangement of the sand grains under the influence of shear strains. This is not likely to happen in clays; therefore, higher compressive and shear stresses are needed for densification. Figure 2.6 illustrates the depth effect of different types of compactors; superimposed is an indication of the stress range required for effective compaction of sands and clays.

Compaction at freezing temperatures. Because they are so strongly bonded, frozen soils are difficult to compact effectively. If winter compaction is unavoidable, Forssblad (1981) recommends the following strategies:

1. Use dry coarse materials for construction, such as crushed rock or coarse gravel.
2. If fill can be obtained in the borrow area in an unfrozen state, place and compact it without delay, exposing the least possible surface area to freezing. This is helped by using relatively thick lifts and careful planning of construction stages. It should be realized that at −10°C gravel may freeze to a depth of 50 mm within two hours.
3. Recompact and regrade the surface during the following summer, after the entire fill has thawed. Where it was impossible to avoid frozen fill being placed, large settlement may be evident.

Adding $CaCl_2$, which lowers the freezing point of the pore water, may assist compaction, provided there are no unwanted side effects with respect to the environment or the engineering properties of the soil (see Sect. 13.2.4).

Special considerations for specific materials. Some special aspects related to the compaction of specific soil or pavement materials are worth mentioning.

The effect of the degree of saturation manifests itself differently in free-draining sands and gravels than in clays. Partially saturated cohesionless soils have apparent cohesion because the surface tension forces in the pore water cause suction which increases frictional resistance against compaction. It is therefore better to compact these soils either when they are completely dry or fully saturated.

A problem peculiar to cohesionless soils is that compaction close to the surface is difficult because of a lack of confinement. This is demonstrated in Fig. 2.7 which shows the typical variation of density with depth before and after compaction.

Clays and silty clays are very sensitive to the placement water content, as is

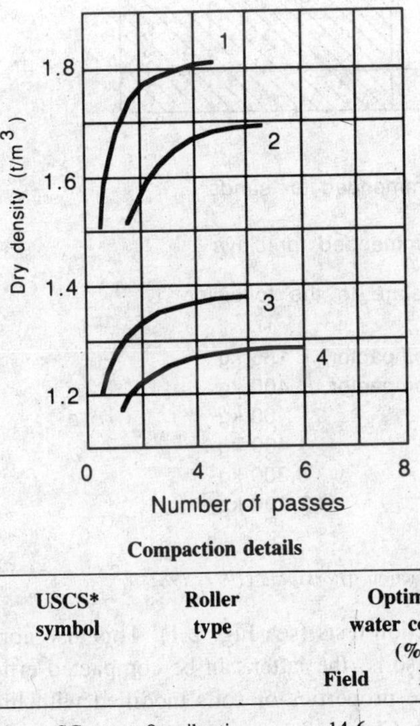

Compaction details

Curve no.	USCS* symbol	Roller type	Optimum water content (%) Field	Optimum water content (%) Lab. std.
1	SC	8t, vibrating	14.6	16.2
2	SC	1.5t, static	16.5	16.2
3	CH	10t, sheepsfoot	27	24.3
4	CH	1.2t, static	31	24.3

*For an explanation of symbols used in the Unified Soil Classification System (USCS) see Apendix A.

FIGURE 2.5

Typical relationship between the number of passes of a roller and the density obtained. [*Adapted from Kyulele (1983).*]

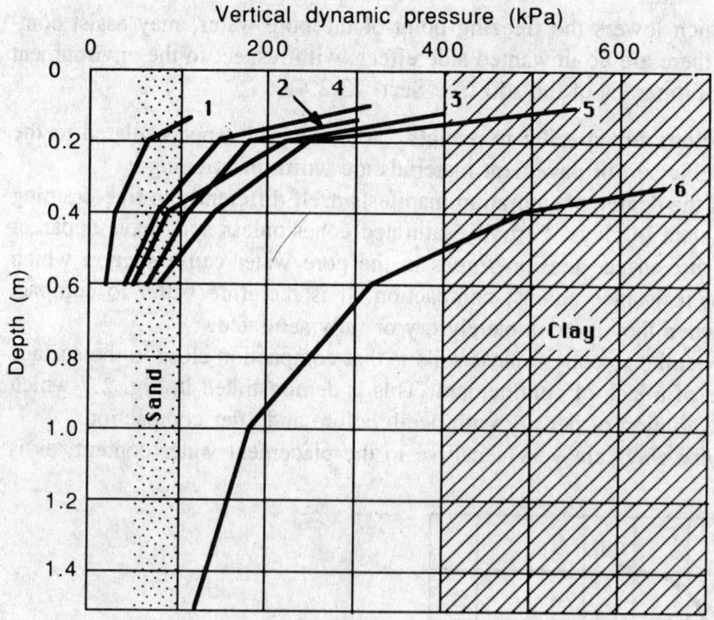

Notes:

Pressure range recommended for sands

Pressure range recommended for clays

Pressure measurements relate to the following equipment:

1	Vibrating plate compactor	135 kg
2	Vibrating plate compactor	400 kg
3	Vibrating tamper	60 kg
4	Vibrating roller	1,400 kg
5	Vibrating roller	3,300 kg
6	Vibrating roller	13,000 kg

FIGURE 2.6
Dynamic pressures at various depths during compaction. [*Forssblad (1977, 1981).*]

well demonstrated in laboratory compaction tests (see Fig. 2.1). This also applies to lime-stabilized soils. Just like cohesive soils, the latter can be compacted efficiently using padfoot or sheepsfoot rollers. The properties of soils modified with lime and other admixtures are further discussed in Chap. 13.

Basic principles of compaction have also been established for manufactured materials such as asphalt, macadams and ''rolled'' concrete, and special applications such as treating railway ballast. For further information in these areas, the reader is referred to the specialist literature on pavement materials and railway engineering.

Applicability and production rate. A guide to the applicability of different types of compaction equipment is given in Table 2.3. It identifies the most- and least-suitable

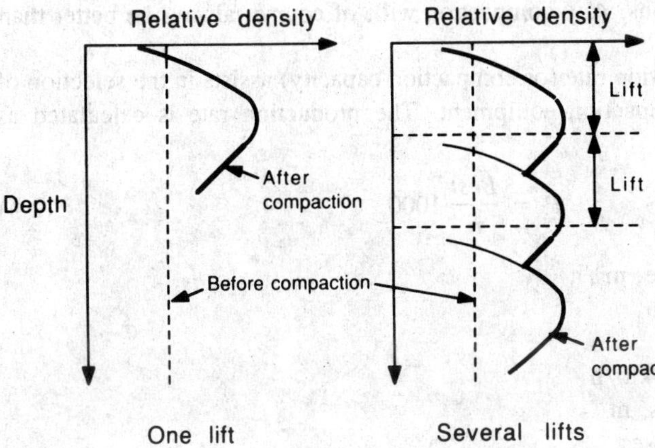

FIGURE 2.7
Density in sand before and after compaction (one or more lifts).

TABLE 2.3
Applicability of compaction equipment

Equipment	Most-suitable soils	Typical applications	Least-suitable soils
Smooth wheel rollers, static or vibrating	Well-graded sand-gravel mixtures, crushed rock, asphalt	Running surface, base courses, subgrades for roads and runways	Uniform sands
Rubber-tired rollers	Coarse-grained soils with some fines	Road and airfield subgrade and base course proof-rolling	Coarse uniform cohesionless soils, and rock
Grid rollers	Weathered rock, well-graded coarse soils	Subgrade, subbase	Clays, silty clays, uniformly graded materials
Sheepsfoot rollers:			
Static	Fine-grained soils with more than 20% fines	Dams, embankments, subgrades for airfields, highways	Clean coarse-grained soils, soils with cobbles, stones
Vibrating	As above, but also sand-gravel mixtures	Subgrade layers	
Vibrating plate (light)	Coarse-grained soils, 4 to 8% fines	Small patches	Cohesive soils
Tampers, rammers	All types	Difficult-access areas	
Impact rollers	Wide range of moist and saturated soils	Subgrade earthworks (except surface)	Dry, cohesionless soils

soils and typical applications. Any compaction will, of course, always be better than no compaction.

Knowing the production rate (or compaction capacity) assists in the selection of the most economical compaction equipment. The production rate is calculated as follows:

$$P = \frac{Best}{n} 1000 \tag{2.4}$$

where P = production rate, m^3/h
B = drum width, m
e = efficiency
s = rolling speed, km/h
t = layer thickness, m
n = number of passes

The efficiency factor, equal to 0.75 to 0.85, allows for overlap between adjacent passes and the time required to change direction, stop, and start.

2.2.3 Deep Compaction Techniques

Densification of deep soil deposits is achieved by the following techniques:

Precompression. A site is preloaded by means of a surcharge or by lowering the groundwater level, causing the ground to consolidate. After restoring original stress levels, future structures built on this site will settle less than those on the untreated ground.

This technique is usually reserved for cohesive soils. Consolidation of these soils is a long-term process, unless the existing longest drainage paths are shortened by the installation of sand columns, paper wicks, or geocomposite drains. Because the success of precompression is essentially dependent on the hydraulic parameters of the soil, precompression is considered a method of hydraulic modification of the soil in this text (see Chap. 11).

Explosion. Explosives are detonated on the surface or, more likely, in an array of boreholes, causing a loose soil structure to collapse which leads to a denser arrangement of the particles. The final density may not be achieved immediately, as the dissipation of excess pore pressures generated may take some time.

Heavy tamping. A large mass is dropped onto the ground surface, causing compaction and possibly long-term consolidation, thus the term "dynamic consolidation."

Vibration. Densification is achieved by a vibrating probe or pile, possibly aided by water jets or pressurized air and the addition of granular material, possibly with added cementing agents.

Compaction grouting. "Zero-slump" mortar is injected into the ground under high pressure, displacing and compacting the surrounding soil. This technique is discussed in Chap. 14.

Vibration is most suitable for free-draining cohesionless soils. Impact loading by explosion and heavy tamping is also suitable for less-pervious silty sands; it may even find application for clayey silts and sands. Precompression may be the only technique feasible for clayey soil and is likely to be less economical than other methods for permeable soils.

The following sections give a brief introduction to compaction by explosion, heavy tamping, and vibro-compaction. Reference is made to some of the field evaluation methods, such as penetration testing, which are discussed in greater detail in either Chap. 4 or 5.

2.2.3.1 EXPLOSION. Explosives can be employed to modify sands, loose rock, and special soils such as loess, which is characterized by relatively high porosity and a distinct soil skeleton.

Explosion of charges on the ground surface or in deep boreholes causes shear stresses in the soil which break down the soil structure, resulting in a reorientation of soil particles and subsequent volumetric compression. In saturated soils temporary high pore pressures are set up, causing liquefaction. These excess semidynamic pore pressures (in excess of hydrostatic pressures) are essential for effective densification ensuing from subsequent consolidation. Installation of vertical drains may assist the explosion-induced consolidation process.

Care must be taken that structures adjacent to the blasting site are not affected and that no large-scale slip or similar shear failure is induced.

Excess pore pressure and settlement due to explosion are related to the ratio

$$N_h = \frac{W^{1/3}}{R} \tag{2.5}$$

where N_h = Hopkinson's number
W = weight of explosives, equivalent kilograms of TNT
R = radial distance from point of explosion, m.

If N_h is less than the range of 0.09 to 0.15, little or no liquefaction is said to occur [Barendsen and Kok (1983)]. This relationship can be used to estimate a "safe" distance from the explosion.

According to Barendsen and Kok (1983), the ratio of excess pore pressure Δu over the effective overburden pressure σ', as well as the ratio of surface settlement Δh to the height h of the soil layer affected by the explosion, are both related to N_h. Experience with sandy soils in the Netherlands suggested the following relationships obtained from a statistical analysis of field results:

$$\frac{\Delta u}{\sigma'} = 1.65 + 0.65 \ln N_h \tag{2.6}$$

and

$$\frac{\Delta h}{h} = 2.73 + 0.9 \ln N_h \tag{2.7}$$

where N_h is calculated using units of kilograms for W and meters for R.

Barendsen and Kok state that for optimum densification, a ratio

$$\frac{\Delta u}{\sigma'} > 0.8 \qquad (2.8)$$

is required. For lower ratios, only partial liquefaction may occur, resulting in lower compaction efficiency.

In a typical application in the Netherlands, a charge equivalent to 10 kg of TNT was used in each of the 15-m-deep holes, spaced 16 m apart, resulting in the use of about 100 g of TNT per 1 m^3 of soil.

Dembicki and Kisielowa (1983) describe the compaction of a deposit of sand and silt with thick layers of organic material in Gdansk harbor using explosives. The use of 0.125 kg of explosives per cubic meter of soil decreased the volume of sand by 6% and that of the organic mud by 4% over a total depth of 18 m. The density index (relative density) of the sand increased from 0.35 to over 0.8. The deposit was observed to consolidate for 2 months after the blasting work was completed.

Ivanov (1980) gave details of 12 USSR projects involving consolidation of saturated soils by explosives. A considerable variety of conditions were encountered, indicated by the amount of explosives used, which ranged from 8 to 220 g/m^3. A typical application may have involved a 5-m-thick soil layer, densified by 7 kg of TNT located 3 m down each borehole (spaced 7 m apart), resulting in 0.3 m surface settlement. These figures are given only as a guide to the order of magnitude of the determinants involved. It is obvious that judicious placement and timing of the charges coupled with performance measurements (piezometer readings, penetration tests, etc.) can lead to significant economies on a large project.

2.2.3.2 HEAVY TAMPING AND DYNAMIC CONSOLIDATION. Heavy tamping and dynamic consolidation, also called "dynamic compaction," refer to the compaction method where a heavy weight is dropped onto the ground surface from a great height. The term "dynamic consolidation" was introduced by Menard and usually refers to very heavy equipment with characteristics such as

Tamper mass	Up to 170 t
Fall	Up to 22 m
Compaction effect	To 40 m depth
Spacing	To 14 m

The use of a smaller mass falling from a lower height, say 12 t dropping 12 m, is normally just called heavy tamping, rather than dynamic consolidation; typically, the drops would be spaced 2 to 3 m apart, causing compaction to about 6 m. However, it would seem appropriate to let "heavy tamping" describe the construction "technique" and reserve the term "dynamic consolidation" for explaining the geotechnical "process" which may accompany heavy tamping. Dynamic consolidation could then be defined as the process of densification of a saturated or nearly saturated soil caused by sudden loading, involving shear deformation, temporarily high pore pressures (possibly liquefaction), and subsequent consolidation.

Heavy tamping may also assist in establishing better drainage in a soil layer, thus speeding up the process of consolidation due to the soils own weight or added surcharge. Field observations after heavy tamping show that pore pressures in excess of hydrostatic pressures may exist for hours or days in sand and silty soils and for longer times in clays.

Heavy tamping has also proved effective for the rehabilitation of waste disposal areas by densifying highly variable, loosely dumped material, possibly containing large voids. Rubbish tips can thus be made into storage areas, playing fields, etc., with less problems due to long-term settlements.

A simple rule of thumb suggests that the depth D, in meters, to which heavy tamping is effective can be estimated conservatively by

$$D = 0.5\sqrt{WH} \tag{2.9}$$

where W is the mass of the falling weight in metric tons and H is the height of fall in meters. According to Mayne et al. (1984) the degree of soil improvement peaks at a "critical depth" which is roughly one half of the maximum depth of influence D. A typical set of test results before and after heavy tamping is presented in Fig. 5.17, as part of an introduction to the control of deep compaction with penetrometers and pressuremeters.

Dynamic consolidation can be combined with a static surcharge and assisted by vertical drains. It can also be used to form sand or gravel pillars in soft soil by punching sections of granular fill placed on the surface into the ground.

An example of a time-settlement record of static (fill) and dynamic loading is shown in Fig. 2.8. The vertical steps in the curve are due to repeated passes of a super-heavy tamper.

As for other deep compaction techniques, the ground improvement achieved is most commonly checked by static or dynamic penetrometers, pressuremeters, dila-

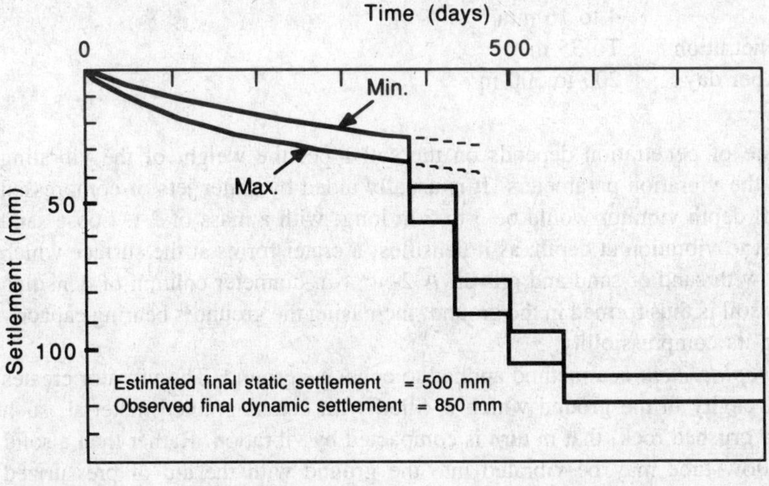

FIGURE 2.8
Time settlement record for precompression combined with dynamic consolidation. [*Gambin (1983)*.]

tometers, or other methods as discussed in Chap. 5. Although the plate load test (Sec. 5.1.5) is more likely to be called on for the control of surface compaction, it is sometimes undertaken for evaluating the effect of heavy tamping, particularly where the job objective is an improvement of the bearing capacity of footings. Since the stress induced by a loaded plate is only significant to a depth equal to about $1\frac{1}{2}$ times its width, the test must be carried out in a trench or large-diameter borehole in order to yield a result which is representative for soil conditions below a large footing.

Significant vibrations are generated during heavy tamping, and this may represent a serious limitation in the applicability of this method of ground modification. Dobson and Slocombe (1982) recorded peak particle velocities on several sites, and their findings, which suggest clearances of at least 30 m, are presented in Fig. 2.9. Additional hazards reported were flying fragments or lumps of soil, so, for safety reasons, a clearance of at least 60 m is preferred.

2.2.3.3 VIBRO-COMPACTION AND VIBRO-REPLACEMENT. Depth vibrators have been used in construction since the 1930s. Some consist of vibrators attached to the top of steel sections which are lowered into the ground. Other types transmit the vibrations from the bottom end of extension tubes lowered into the ground. In standard vibration systems, the dominant direction of vibration is vertical. *Vibro-flotation* is a term coined for systems where the vibrating unit is inserted to the desired depth and vibrated horizontally.

Typical machine characteristics may be as follows [adapted from Jebe and Bartels (1983)]:

Motor output	35 to 120 kilowatts (kW)
Speed	1800 to 3000 revolutions per minute (r/min) (30 to 50 Hz)
Centrifugal force	160 to 220 kN
Amplitude	4 to 16 mm
Depth of penetration	To 35 m
Total depth per day	200 to 500 m

The rate of penetration depends on the soil type, the weight of the vibrating system, and the vibration parameters. It is usually aided by water jets or compressed air. A typical depth vibrator would be 3 to 5 m long, with a mass of 2 t. Loose sand responds best to vibration at depth; as it densifies, a crater forms at the surface which is backfilled with sand or sand and gravel. A 2- to 4-m-diameter column of densified cohesionless soil is thus formed in the ground, increasing the ground's bearing capacity and reducing its compressibility.

Vibro-replacement is a method applied to cohesive ground. The vibrator creates a cylindrical cavity in the ground which is filled with coarse-grained material, such as gravel and crushed rock, that in turn is compacted by vibration. Rather than a solid probe, a hollow tube may be vibrated into the ground with the aid of pressurized water or air. As it is withdrawn, crushed stone or coarse gravel is fed through the tube and compacted. The end result is a column of dense sand and/or gravel with a

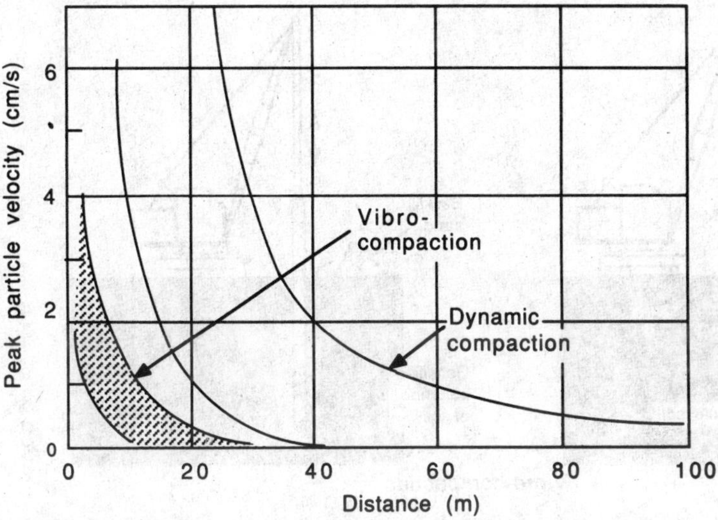

FIGURE 2.9
Comparison of general vibration levels induced by vibro-compaction and heavy tamping (or dynamic compaction). [*Dobson and Slocombe (1982).*]

load-carrying capacity on the order of 100 to 400 kN. This "bottom-feed" method is increasingly being preferred to other vibro-replacement techniques.

Vibro-compaction and vibro-replacement are illustrated in Fig. 2.10. Vibro-compaction is most successful in loose sandy soils, typically with an original SPT value of 5 to 10 near the surface and is not applicable to clays. Depending on the spacing, relative densities of up to 85% can be achieved (Fig. 2.11). In contrast, vibro-replacement is most effective in cohesive soils with an undrained shear strength in the range of 20 to 60 kPa.

The firm of GKN Hayward Baker (1986) summarized the relative effectiveness of vibro-compaction and vibro-replacement as follows:

Type of soil	Vibro-compaction	Vibro-replacement
Sands	Excellent	Not applicable
Silty sands*	Good	Excellent
Silts	Poor	Good
Clays	Not applicable	Good
Mine spoils	Good	Excellent
Dumped fill	Depends on nature of fill	Good
Garbage	Not applicable	Not applicable

*Say less than 20% fines (comment added by the author).

Vibro-compaction is considerably less hazardous from a personnel and structural safety point of view. The magnitude of vibrations felt on or adjacent to the site are significantly less than are experienced with heavy tamping, as illustrated in Fig. 2.9.

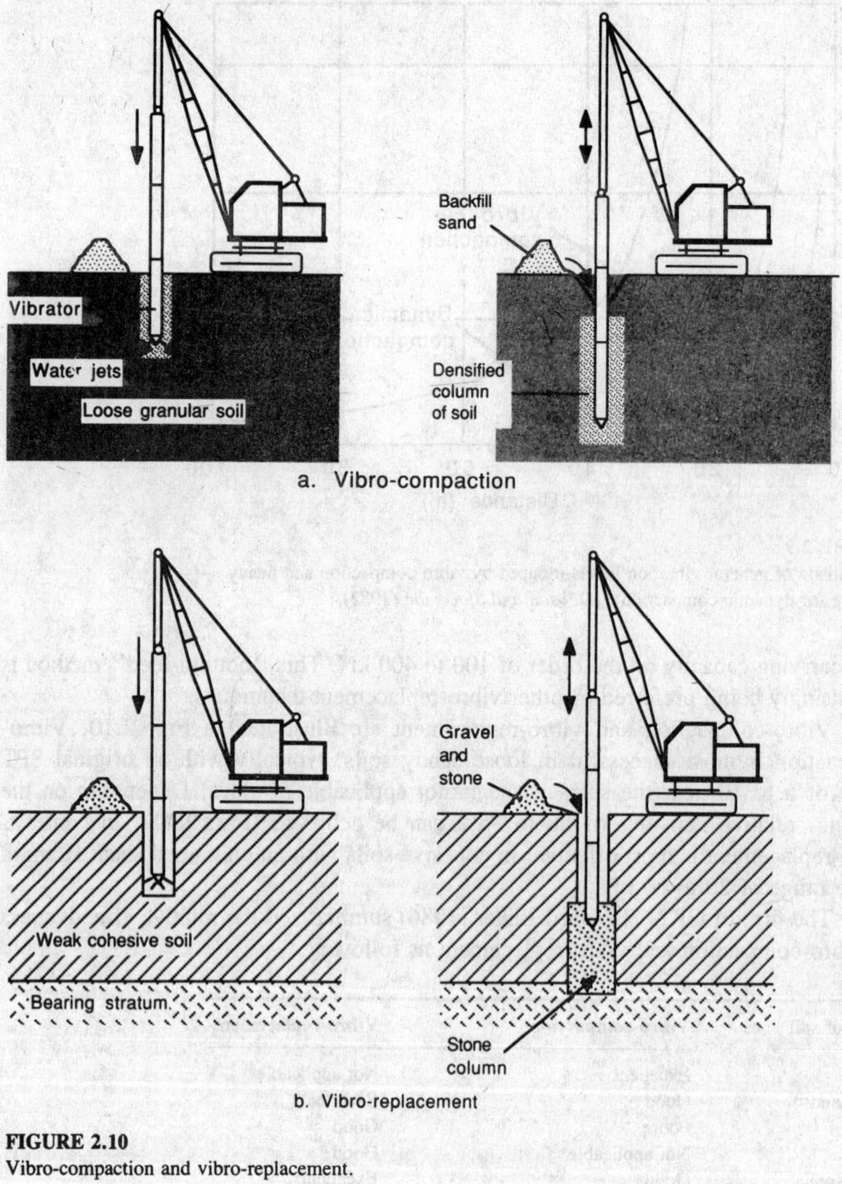

FIGURE 2.10
Vibro-compaction and vibro-replacement.

Vibro-replacement with the addition of stabilizing agents is more appropriately classified as compaction grouting. Such construction methods include *mortared stone columns*, formed by injecting mortar into a stone column, and *concrete vibro-columns*, created by pumping concrete into ground cavities created by a vibrator.

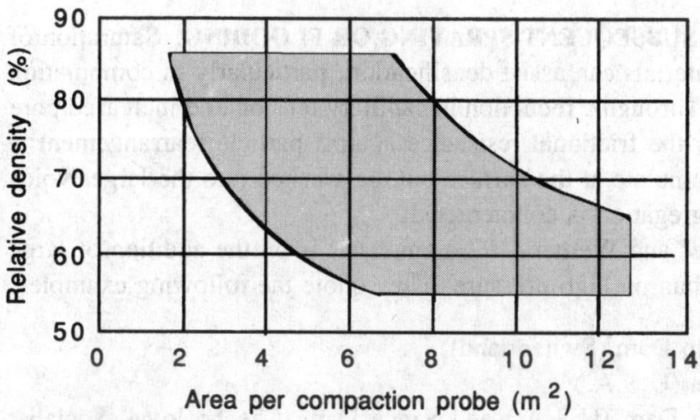

FIGURE 2.11
Envelope for spacing of vibro-centers in clean granular soils. *(Courtesy GKN Keller.)*

2.2.4 Hydromechanical Compaction

Water can assist deposition as well as compaction of cohesionless silts, sand-gravel mixtures, and rock fill.

2.2.4.1 HYDRAULIC FILL. The hydraulic fill method is applicable to predominantly sandy soils. The fill material is pumped through pipelines onto the construction site and is discharged with immediate drainage or fed into a pool at the core of the embankment being formed. Without additional compaction, relative densities of only 50 to 60% are obtained.

The hydraulic fill method is no longer used for major dams because of the danger of liquefaction failures, but it finds application in building minor embankments, flood levees, and fly-ash and mine-waste disposal structures. Turnbull and Mansur (1973) investigated a number of hydraulic fill projects completed in the period from about 1940 to 1970 and concluded that a well-controlled hydraulic fill should satisfy the following requirements:

1. The true water table must be below the surface of the fill so that there is a downward drainage at all times.
2. Material must not be bulldozed into low places without subsequent sluicing; otherwise even lower densities will result.
3. A uniform flow must be maintained over the fill surface by use of bulldozers and shear boards to direct the flow of water.
4. Deposition into pools of water must be prevented in order to eliminate the accumulation of fines and low densities.

Relative densities above 50 or 60% can only be reached if additional surface or deep compaction methods are employed.

2.2.4.2 DRY FILL WITH SUBSEQUENT SPRAYING OR FLOODING. Saturation of free-draining coarse fill materials can assist densification, particularly in combination with vibratory equipment. Through a reduction in capillary tension and increased pore pressures during vibration, the frictional resistance against particle rearrangement is reduced. Fines cannot accumulate at the surface but are washed into the larger voids of the fill material, and segregation is counteracted.

According to Striegler and Werner (1973) emphasis is on the addition of large quantities of water rather than on high pressure. They quote the following examples:

150 $1/m^3$	Göscheneralp Dam (Switzerland)
250 $1/m^3$	Quoich Dam (U.S.A.)
500 $1/m^3$	Lewis-Smith Dam (U.S.A.) and Sance Dam (Czechoslovak Socialist Republic)

2.2.4.3 COMPACTION OF ROCK FILL WITH WATER JETS. Construction of rock fill dams and embankments with conventional surface compaction is rarely done in lifts exceeding 2.5 m. Using water jets for compaction allows placement of up to 60 m (!) of rock in one operation [Striegler and Werner (1973)], with corresponding economic advantages.

The types of pumps used deliver a 50- to 70-mm water jet at up to about 80-kPa pressure and at a rate of up to 150 liters per second (L/s). Several water jets may be in action at the same time, aimed at the material being dumped, and normally oriented in the down-the-slope direction. The amount of water used is on the order of 2 to 4 m^3 per 1 m^3 of rock fill.

Good results have been achieved with the kind of material whose fines do not fill the larger voids completely. In that situation it has been observed that sharp edges of boulders and cobble-size particles are broken off by impact forces during placement, resulting in a denser fill. This process may be aided by the water addition, partly because it may reduce the strength of the rock itself.

One of the disadvantages of high fills compacted with water jets is that segregation can occur, causing a predominance of large rock pieces at the bottom of the fill and a relatively high percentage of smaller particles in the upper layers. Compaction in small lifts by surface rollers, if at all feasible, is likely to result in higher overall densities than hydromechanical compaction of high fills, particularly where segregation occurs.

PROBLEMS

Prefixes indicate problem type: C = calculations, M = multiple choice, D = discussion.

Calculations

C2.1. The following measurements were taken in a laboratory water content determination:

Mass of wet soil + container = 120 g
Mass of dry soil + container = 100 g
Mass of container = 20 g

What is the water content of this soil?

C2.2. The total (or wet) unit weight of a compacted well-graded gravel is 22 kN/m^3. The water content is 10%. Calculate
(*a*) The dry unit weight.
(*b*) The saturation, using $\gamma_w = 9.81$ kN/m^3 and $G_s = 2.65$.
(*c*) The error involved if the saturation is computed using $\gamma_w = 10$ kN/m^3.

C2.3. Calculate points on a ZAV curve for a specific gravity of soil solids of 2.7 and water contents of 10, 15, 20, 25, 30, and 35%. In addition, calculate the dry density for the same conditions but with a saturation of 80%.

C2.4. Standard compaction of a highly plastic tropical black clay (liquid limit = 55%, plasticity index = 30) at various water contents produced the following dry densities:

w (%) =	16.0	18.5	22.0	23.5	25.0	27.5	31.0
γ_{dry} (kN/m^3) =	1.47	1.50	1.54	1.54	1.50	1.45	1.41

The specific gravity of the soil solids is 2.71.
(*a*) Plot the ZAV curve for the appropriate water content and density range. (Choose the scales so that the ZAV curve is inclined at approximately 45°.)
(*b*) Plot the results and draw the compaction curve.
(*c*) Determine the optimum moisture content, maximum dry density, and corresponding saturation.

C2.5. Calculate the production rate (m^3/h) for a roller with the following characteristics:

Drum width = 2.14 m
Efficiency = 80%
Speed = 8 km/h
Layer thickness = 0.6 m
Number of passes = 6

C2.6. A 5-m-deep deposit of sand and silt containing organic layers is to be compacted using explosives placed in boreholes located 8 m apart on a square grid. It is intended to create pore pressures which are at least equal to 80% of the effective overburden pressure. Estimate how many kilograms of TNT have to be distributed in each borehole in order to achieve this degree of liquefaction? How much is that per cubic meter of compacted soil? Refer to the experiences by Barendsen and Kok in the Netherlands.

C2.7. Estimate the "safe" distance from an underground explosion of 10 kg of TNT, beyond which little or no soil liquefaction is likely to occur.

C2.8. To what depth is heavy tamping effective for
(*a*) A 10-t mass dropping 10 m?
(*b*) A 170-t mass dropping 22 m?

C2.9. If a structure adjacent to a site is in danger of being damaged by a peak particle velocity in excess of 2 cm/s in the ground, what is the minimum safe distance for heavy tamping?

C2.10. What is a typical range of plan area (m^2) densified by one application of a vibrocompactor, if a relative density of 80% is to be achieved in clean granular soil?

C2.11. Speculate on the usefulness of elephants as soil compactors, given the following information:

Soil type: Sandy clay

Standard Proctor test results: $\gamma_{dmax} = 1.87$ t/m^3, $w_{opt} = 14.5\%$

Elephant: weight = 2000 kg, foot print = 175 cm^2 (three legs on the ground while walking)

Consider the contact area and pressure, number of passes needed, and achievable density. Also remember that elephants are intelligent animals when it comes to repeatedly traversing soft ground. [For answers refer to the delightful article by Meehan (1967)].

Multiple Choice

M2.12. In the standard compaction test, the volume of the mold is
(*a*) 100 cm^3.
(*b*) 1 L.
(*c*) 500 mL.
(*d*) $\frac{1}{30}$ L.

M2.13. The ratio of the energy input in the modified compaction test over the energy input in the standard compaction test is
(*a*) 1:1.1.
(*b*) 1.1:1.
(*c*) 1.5:1.
(*d*) 4.5:1.

M2.14. The best laboratory simulation of the effect of a rubber-tired roller on clay is obtained in
(*a*) A standard Proctor test.
(*b*) A modified Proctor test.
(*c*) Static compaction (odometric compression).
(*d*) A repeated triaxial test.

M2.15. A typical operating frequency of a vibrating roller is
(*a*) 10 Hz.
(*b*) 30 Hz.
(*c*) 30 MHz.
(*d*) 240 volts (V).

M2.16. Economical use of vibratory rollers usually involves
(*a*) 1 pass.
(*b*) 1 to 3 passes.
(*c*) 4 to 6 passes.
(*d*) 8 to 12 passes.

M2.17. A static sheepsfoot roller is most effective in compacting
(*a*) Gravel.
(*b*) Sand.
(*c*) Clay.
(*d*) Weathered rock.

M2.18. An impact roller is least suitable for compacting
(*a*) Unsaturated clay.
(*b*) Wet sand.
(*c*) Dry sand.
(*d*) Rubble fill.

M2.19. Depth vibrators typically operate at speeds of
(*a*) 20 Hz.
(*b*) 20 r/min.
(*c*) 200 r/min.
(*d*) 2000 r/min.

M2.20. Sandy and silty soils which benefit most from deep vibration have a standard penetration resistance *N* in the range of
(*a*) 5 to 10.
(*b*) 10 to 20.
(*c*) 20 to 50.
(*d*) > 50.

M2.21. Stone columns are said to be possible (and beneficial) in thin layers of cohesive soil with an undrained cohesion of
(*a*) < 10 kPa.
(*b*) 10 to 20 kPa.
(*c*) 20 to 60 kPa.
(*d*) > 200 kPa.

M2.22. Maximum relative densities achieved in a hydraulic fill of cohesionless material are in the range of
(*a*) 30 to 40%.
(*b*) 50 to 60%.
(*c*) 60 to 80%.
(*d*) 80 to 90%.

M2.23. Compaction of rock fill with water jets typically requires the following quantity of water, expressed in liters per cubic meter of fill:
(*a*) 20 to 40
(*b*) 200 to 400
(*c*) 2000 to 4000
(*d*) 4000

Discussion

D2.24. What other design or construction alternatives would be considered besides soil improvement, such as compaction, in order to overcome a difficult foundation problem?

D2.25. Name four methods of engineering soil improvement other than compaction.

D2.26. Explain the difference between compaction and consolidation in soil mechanics.

D2.27. List five major aims of compaction (with respect to modifying soil properties).

D2.28. To achieve the goals set for compaction, the engineer has to make decisions responding to three basic problems (1. Specify . . . , 2. Select . . . , 3. Set up . . .).

D2.29. Give examples of static, dynamic, and kneading compaction in the laboratory and the field.

D2.30. List five factors which affect the compactive effort of
(*a*) A pneumatic-tired roller.
(*b*) A vibrating roller.

D2.31. Suggest equipment suitable for the compaction of weathered sandstone up to cobble size.

D2.32. What special problems may be associated with the compaction of
(*a*) Free-draining sands and gravels?
(*b*) Soils stabilized with lime or cement?

D2.33. Name five basic construction techniques used for densifying deep soil deposits (Pre . . . , Ex . . . , He . . . , Vi . . . , Co . . .).

CHAPTER

3

PRINCIPLES OF SOIL DENSIFICATION

3.1 MOISTURE CONTENT

The moisture content at which a soil is compacted significantly affects the densification achieved. This is true for coarse-grained cohesionless soils as well as for fine-grained cohesive soils, although the water may play a different role in the compaction process.

Field compaction of free-draining coarse materials shows best results for either dry or completely saturated conditions. If partially saturated, sandy soils may exhibit apparent cohesion due to capillary tension in the pore water, creating attraction forces between the particles which in turn cause frictional resistance against their rearrangement. Surface tension forces are also the reason why some loose dry soils may show densification upon wetting without the application of external forces. The "collapse" of soils upon wetting is further discussed in Secs. 4.1.2.1 and 4.2.3. For some types of soil complete saturation combined with high impact compaction may be most efficient. Laboratory compaction may not show all these effects because it is usually not possible to maintain full saturation during the test.

When a fine-grained soil is densified with a constant compactive effort but varying water contents, a typical moisture–dry-density relationship as shown in Fig. 2.1 becomes apparent. Section 3.5 will review the evolution of compaction theories which attempt to explain the shape of the compaction curve using the concepts of effective stress, pore water and pore air pressures and permeabilities and shear stress and strain.

As a consequence of the different effect water has on the compaction of coarse- as opposed to fine-grained soils, field control of the placement water content is generally only carried out for the latter, while the effectiveness of the compaction of coarse-grained soils is evaluated using in-situ determinations of strength and stress-strain modulus.

3.1.1 Empirical Relationships

3.1.1.1 FINE-GRAINED SOILS. Generally, the optimum water content increases and the maximum dry density decreases with increasing plasticity of the soil, as defined by the Atterberg limits. The "Design Manual" [U.S. Navy (1962)], gives the following rules of thumb in relation to the parameters determined in standard laboratory compaction [standard (std.) not modified proctor compaction]:

$$w_{opt}\,(\text{std.}) = \begin{cases} w_p - 5 & \text{at } w_{opt} = 10\% \quad (3.1) \\ w_p - 2 & \text{at } w_{opt} = 30\% \quad (3.2) \end{cases}$$

where w_p is the plastic limit. Alternatively w_{opt} and γ_{dmax} for standard compaction can be estimated from the liquid limit w_l and the plasticity index I_p, defined as the difference between the liquid and plastic limit:

$$w_{opt}\,(\text{std.}) = 6.77 + 0.43w_l - 0.21I_p \quad \% \qquad (3.3)$$

$$\gamma_{dmaz}\,(\text{std.}) = 20.48 - 0.13w_l + 0.05I_p \quad \text{kN/m}^3 \qquad (3.4)$$

Various authorities have prepared charts based on relationships of the kind given above, but the accuracy of the charts should not be overestimated, unless they have been prepared for soils of the same geological origin. For example, Hammond (1980) studied three groups of soils occurring in Ghana and performed a linear regression analysis of the relationships between w_{opt} and either w_p, w_1, I_p, w_s (shrinkage limit), or (% fines). Some of the expressions derived are:

For lateritic soils (predominantly clayey and sandy gravels):

$$w_{opt} = 0.42w_p + 5 \quad (3.5)$$

$$= 0.6 \times (\%\ \text{fines}) + 5.9 \quad (3.6)$$

For micaceous soils (clayey silty sands, with Atterberg limits of the fines plotted below the A line):

$$w_{opt} = 0.45w_p + 3.58 \quad (3.7)$$

$$= 0.5w_L - 6 \quad (3.8)$$

For black cotton soils (silty clays):

$$w_{opt} = 0.96w_p - 7.7 \quad (3.9)$$

$$= 0.15 \times (\%\ \text{fines}) + 6.4 \quad (3.10)$$

Naturally, multiple linear regression may yield better results for specific soil groups, particularly if clay mineralogy and the nature of the cation exchange complex were also taken into account. The purpose of such an exercise would, however, no longer be a quick and economical means of estimating w_{opt}, but more an effort to understand fundamental soil behavior.

The degree of saturation at w_{opt} is usually on the order of 80 to 85%, while the maximum saturation reached when compacting above optimum moisture content is on the order of 90 to 95%, with the compaction curve running parallel to the ZAV curve.

3.1.1.2 COARSE-GRAINED SOILS (<10% fines smaller than 80 μm). Attempts to relate maximum dry density of coarse-grained soils to their grain size distribution date back a long time [e.g., Davidson and Gardiner (1949)]. More recently, a graph relating γ_{dmax} as well as the minimum and maximum void ratio (e_{min} and e_{max}, respectively) to the quotient D_{60}/D_{10} has been prepared by Roper, as quoted by Biarez (1980), and is reproduced in Fig. 3.1. The D_{60} is the diameter corresponding to 60% passing on a grain size distribution curve. According to Biarez, a slight percentage of fines lowers the maximum density approximately as given by

$$\gamma_{dmax\ (std.)} = 2.15 - 0.005 \times (\%\ \text{fines smaller than } 80\ \mu\text{m}) \quad (3.11)$$

For soils containing a significant amount of both coarse- and fine-grained soils (intermediate soils), Biarez (1980) postulated that w_{opt} could be estimated based on

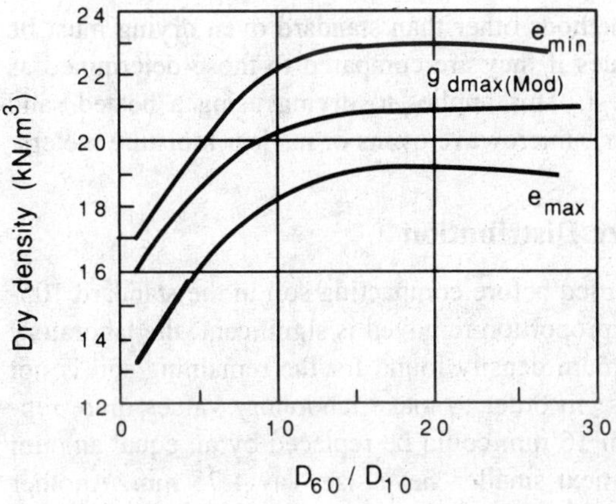

FIGURE 3.1
Effect of grain size distribution on the maximum dry density (modified compaction) and densities corresponding to minimum and maximum void ratios. [*After Ropers, as quoted by Biarez (1980).*]

w_p and the percentage of the fraction <0.4 mm, which happens to be that portion of a mixed soil on which Atterberg limits are determined (Fig. 3.2).

3.1.2 Moisture Content Determination

According to the standard method of testing, the moisture content of a soil is determined by drying the soil in an oven at a temperature of 105 to 110°C. At this temperature all free water in the soil evaporates, but chemically bonded water, and water which forms part of the crystal structure of minerals, remains. Additional significant weight losses may occur if soils, particularly clays, are heated to 400 to 500°C or more. Even drying at 105°C may make the interpretation of the result of the water content determination difficult if the soil contains organic matter, bituminous stabilizing agents, or gypsum.

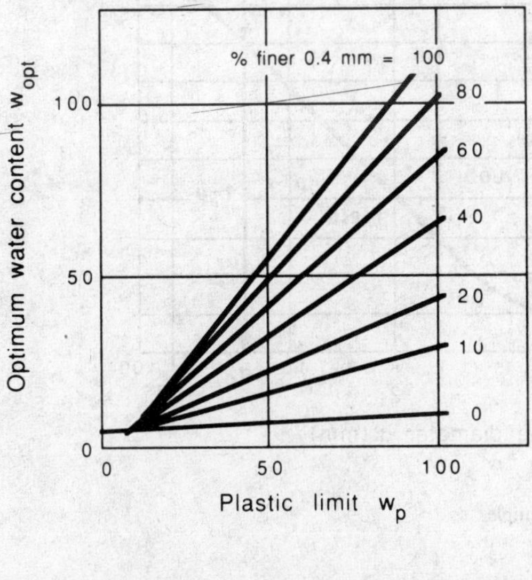

FIGURE 3.2
Generalized relationship between optimum moisture content (standard compaction) and plastic limit of intermediate soils with a known percentage of particles smaller than 0.4 mm. [*As presented by Biarez (1980).*]

Water contents measured by methods other than standard oven drying must be converted to equivalent standard values if they are compared to those determined as part of the laboratory compaction test. This applies to drying using a heated sand bath, carbide (Speedy Moisture Meter), microwave ovens or nuclear moisture meters.

3.1.3 Correction for Grain Size Distribution

Material coarser than 19 mm is discarded before compacting soil in the standard 105-mm-diameter laboratory mold. If the proportion removed is significant, the laboratory optimum moisture content and maximum density found for the remaining soil is not directly comparable with field values. In order to make laboratory values more representative, the proportion larger than 16 mm could be replaced by an equal amount of material between 16 mm and the next smaller sieve size, say 4.75 mm. Another approach is to calculate equivalent field values by considering the moisture content and specific gravity of the discarded coarse material.

Figure 3.3 shows a sketch of the grain size distribution curves of the soil as found in the field and of the sample used in the laboratory compaction test. The material larger than diameter d_0, representing a fraction p of the original soil sample, is removed before testing. Corresponding phase diagrams are shown in schematic form in Fig. 3.4.

If we assume that the coarse material (large gravel, cobbles) contains no moisture and the soil solids have a unit weight of γ_s, the field values corresponding to the laboratory water content and dry density can be calculated. Following the notation

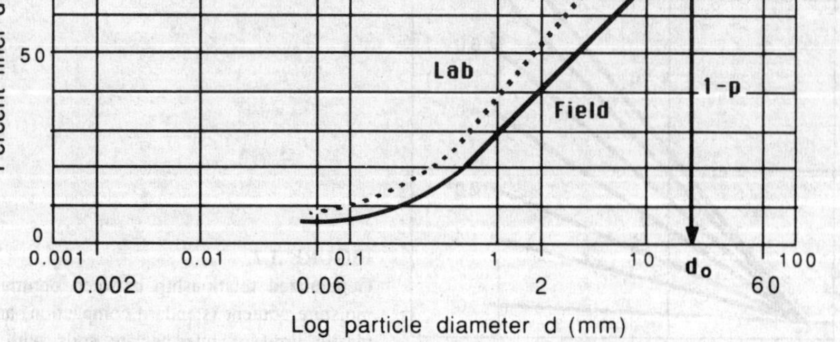

FIGURE 3.3
Grain size distribution of field and laboratory sample.

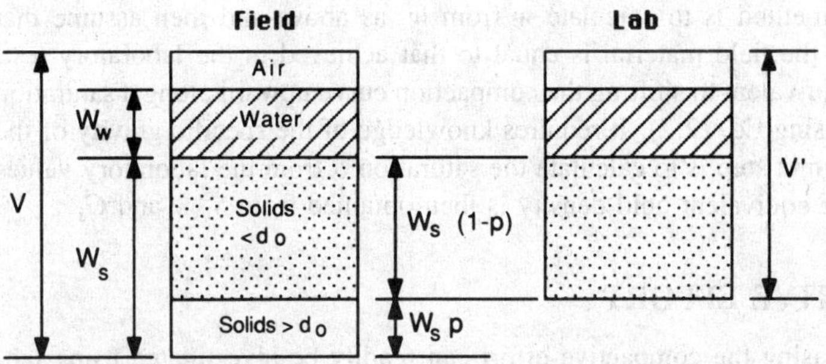

FIGURE 3.4
Schematic representation of the solid, liquid, and air phases of the field and laboratory sample.

used in Fig. 3.4, the laboratory dry density γ_d' and water content w' can be expressed as

$$\gamma_d' = \frac{W_s(1-p)}{V'}$$

$$w' = \frac{W_w}{W_s(1-p)}$$

Equivalent field values are calculated as follows:

$$\gamma_d = \frac{W_s}{V}$$

where $V = \dfrac{W_s p}{\gamma_s} + V'$ and $V' = \dfrac{W_s(1-p)}{\gamma_d'}$

and finally

$$\gamma_d = \frac{1}{p/\gamma_s + (1-p)/\gamma_d'} \tag{3.12}$$

and $$w = \frac{W_w}{W_s} = w'(1-p) \tag{3.13}$$

Assuming zero moisture in the coarse fraction may lead to overestimating the field density, which may not be desirable. Other methods of relating laboratory compaction results to field conditions are therefore also in use.

In one of these alternative approaches, the water content w_c and dry density γ_{dc} of the coarse fraction is estimated and the field values computed as weighted averages of those of the coarse material and the laboratory determined w' and γ_d':

$$w = w_c p + w'(1-p) \tag{3.14}$$

$$\gamma_d = \gamma_{dc} p + \gamma_d'(1-p) \tag{3.15}$$

Typically the γ_{dc} chosen is 90% of the density of the soil solids.

Yet another method is to calculate w from w_c as above and then assume that the saturation S of the field material is equal to that achieved in the laboratory test. This treatment is equivalent to shifting the compaction curve upward along a saturation line as calculated using Eq. (2.2). It requires knowledge of the specific gravity of the soil solids G_s. The first step is to calculate the saturation S from the laboratory values γ'_d, w' and G_s. The equivalent field density is then obtained from S, w and G_s.

3.2 COMPACTIVE EFFORT

The effect of increasing the compactive effort can readily be investigated using laboratory tests. Figure 3.5*a* shows the results of modified, standard, and 50% standard compaction tests on a highly plastic clay (for test procedures see Sec. 2.2.1.1). With increasing compactive effort, the maximum dry density increases and the optimum moisture content decreases. Compaction requirements are usually expressed as a percentage of γ_{dmax} and an allowable deviation from w_{opt} as determined in the standard or modified compaction test. For the compaction of granular layers forming part of the pavement structure of major highways or airfields, engineers usually refer to modified compaction. The compaction of small dams for water storage, subgrade for urban roads or backfill is usually specified in terms of the standard test, as originally proposed by Proctor (1933).

If the maximum density is plotted versus the log of the compactive effort, a more or less straight line is obtained (Fig. 3.5*b*). The increase in maximum dry density with compactive effort is less pronounced for well-graded coarse-grained soils (which exhibit highest densities) than for fine-grained soils. It should be noted, however, that if the water content is appreciably above the respective optimum value, increasing the compactive effort beyond standard Proctor will no longer assist in obtaining a significantly higher density.

The effect of increasing the compactive effort varies with the type of soil. For granular soils, the results of standard and modified compaction show less difference than for cohesive soils. According to Yoder (1959), the maximum dry density obtained in standard compaction is in the range of *85* to about *97%* of that due to modified compaction. The optimum moisture content is *2* to *5%* lower with modified rather than standard compactive effort, the larger differences exhibited by the more plastic soils. Figure 3.6 shows the typical range of the relationship between standard and modified maximum dry density.

The experience with field compaction is similar. Increasing the weight of a roller or increasing the number of passes generally shows either higher density or greater depth effect, or both (Figs. 2.4 and 2.5). There are, however, limits to the compaction which can be achieved. These depend largely on the confinement of the soil mass being compacted and the shear strength of the soil. It is, for example, practically impossible to achieve 100% modified density for some types of clay using mechanical compaction.

Particularly for cohesive soils, it is not only the compactive effort but also the mode of compaction (dynamic, static, kneading) which affect the performance of a compacted soil mass, as will be shown in Chap. 4. Other equipment-related deter-

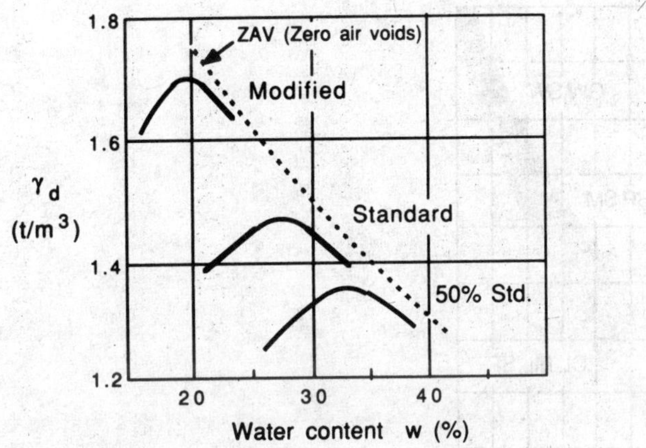

a. Sample result of 50 percent standard, standard, and modified compaction of a highly plastic clay

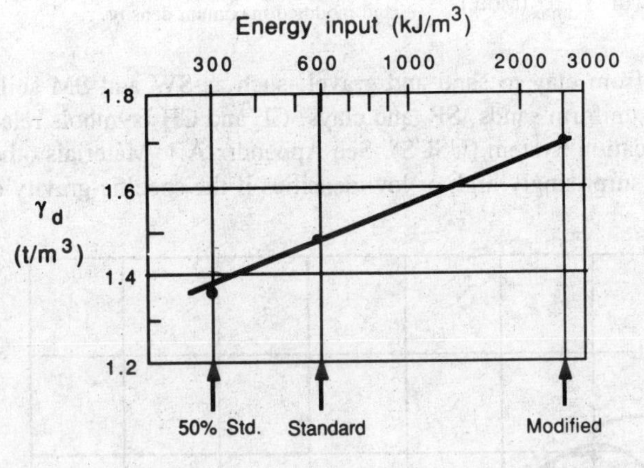

b. Maximum density versus compactive effort on semilog plot

FIGURE 3.5
Results of 50% standard, standard, and modified laboratory compaction of a highly plastic clay.

minants, such as operating procedures may have an equally significant influence in the field.

3.3 SOIL TYPE AND PREPARATION

The preceding discussion of compaction procedures and of the significance of the moisture content has already made it clear that a soil's response to compaction depends on its type and the method used. Figure 3.7 shows typical results for a range of soil types subjected to standard compaction. Well-graded soils, containing particle sizes

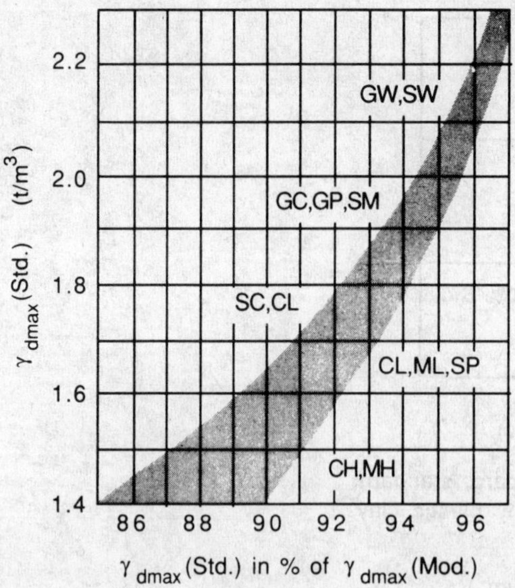

FIGURE 3.6
Approximate relationship between standard and modified maximum density.

covering the whole range from clay to sand and gravel, such as SW and SM soils, reach higher densities than uniform sands, SP, and clays, CL and CH [symbols relate to the Unified Soil Classification System (USCS). See Appendix A.]. Materials other than natural soils can give surprisingly high or low densities if the specific gravity of

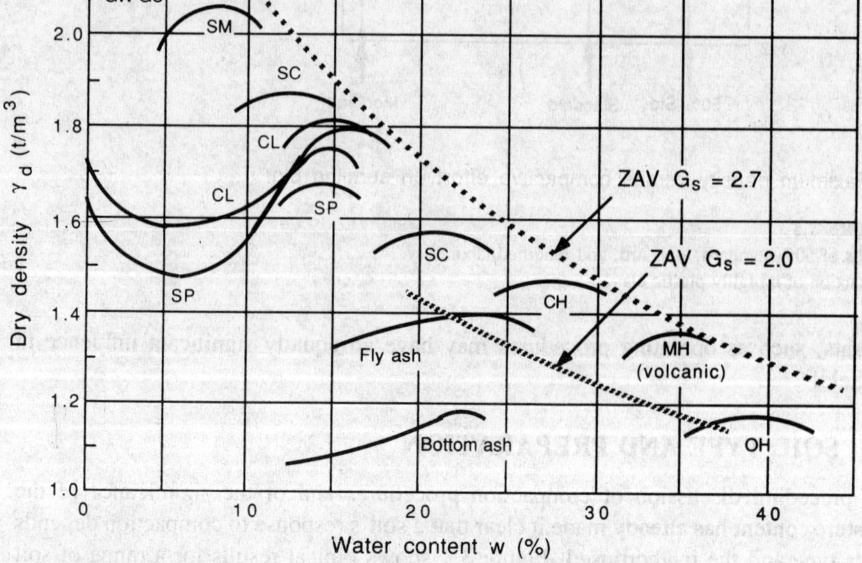

FIGURE 3.7
Arbitrarily selected examples of laboratory test results.

all or some of the constituents is outside the normally expected range of 2.6 to 2.8. As an example, Fig. 3.7 also contains compaction curve for flyash and bottom ash, derived from the combustion of coal, which has a specific gravity in the order of 2. In order to be able to judge whether the test results are possible or not, it is helpful to draw the corresponding zero-air-void curve.

Analyzing large numbers of compaction test results for a particular geographic area or geological formation, it may be found that the compaction curves have characteristic shapes which can be represented by a family of typical curves. Knowledge of these typical curves would make it possible to determine the maximum dry density and optimum water content from a single laboratory compaction test giving a wet (total) density at a particular water content. Using this procedure, the typical curve belonging to this wet density is identified and the corresponding maximum dry density and optimum moisture content is then read from a table [see e.g. Joslin (1959), as quoted by Krebs and Walker (1971)]. This method can be extended further by relating the water content to the penetration resistance of a Proctor Penetration Needle. Thus the time consuming water content determination can be eliminated. Considering the natural variability of soils, this approach is, however, rarely feasible.

Chemical additives can be used to alter the compaction characteristics of a soil. Dispersive agents generally increase the maximum density and decrease the optimum moisture content. Flocculating agents may increase the shear resistance of the soil and make compaction more difficult; similarly, stabilizing additives such as lime and cement generally reduce the maximum dry density achieved and increase the optimum moisture content. This effect can be taken advantage of in the situation where the field water content is above the recommended level: Lime is added and compaction can occur at optimal conditions; the soil may not be at the originally specified density, but because of the cementing action of the lime, the strength criterion is likely to be satisfied.

With careful laboratory performance of compaction tests, a high uniformity of moisture content and a homogeneous soil sample can be achieved. This may not be that easy to control in the field. Proper preparation of the soil for a major embankment, for example, begins in the borrow area. An appropriate excavation and transport strategy can greatly assist in obtaining maximum uniformity in material properties. On the construction site, adequate supervision should ensure consistent layer thickness and compaction procedures. These steps will ensure that the "soil type" is as expected and will make control measures meaningful.

Adverse climatic condition may make it impossible to prepare the soil at the desired moisture content. In these situations, the planning of earthworks has to take account of specific environmental conditions, such as wet and dry seasons.

3.4 CONFINEMENT

The first layer of an embankment placed on soft foundation soil cannot be satisfactorily compacted. Similarly, pavement materials spread on a yielding subgrade will not reach densities required for a base course. This illustrates the importance of confinement for effective compaction, a factor which has also been highlighted by Ingles and Metcalf (1972). Compressive and shear stresses in the soil must be confined in order to produce densification rather than distortion. Practical experience suggests that when constructing an embankment or pavement on soft ground, the density and strength of successive layers must be built up gradually unless the subsoil is so bad that it has to be replaced or stabilized.

Where coarse material, like crushed rock for railway ballast, has to be placed on fine-grained soil, traditional construction requires a transitional layer of intermediate grain size distribution. This layer improves the bearing capacity of the subsoil and prevents the punching of the rock particles into the underlying clay or silty clay during compaction or because of traffic loading. The use of geotextiles as a separator between coarse and fine materials, instead of the more traditional layer of sand, is becoming more common. The geotextile has additional beneficial effects: It provides reinforcement because of its tensile strength and may act as a filter and drainage layer. As a separator, the geotextile prevents mixing of fill and subsoil. As a reinforcement layer, the geotextile provides confinement during compaction; it also increases the bearing capacity of the soil structure being built and reduces its deformation due to external loads. The geotextile may also assist compaction of the fill and consolidation of the subsoil by its drainage and filter capacity.

Both woven and nonwoven geotextiles are used in conjunction with compacted earth structures. Both work well as a separator. Where emphasis is on high strength and a high stress-strain modulus, woven fabrics may be most suitable. Where a fabric must be able to withstand high local deformations without tearing, such as when placed on irregular ground with sharp-edged rocks or mangrove roots, and in situations where filter action and in-plane drainage is required, it is likely that a nonwoven fabric is chosen for the job. At present, more nonwoven than woven fabrics are used for geotechnical applications.

The role of geotextiles as reinforcement is covered in Chap. 19; geogrids are discussed in Chap. 18. The hydraulic functions of geotextiles are topics of Chap. 10.

3.5 EVOLUTION OF COMPACTION THEORIES

Theories of soil compaction seek to explain the typical moisture-density relationships as represented by the compaction curve obtained in laboratory tests (Fig. 2.1) or field compaction. Many interpretations of the basic phenomena have been put forward since Proctor (1933) did his pioneering studies. They began with the lubrication concept and proceeded to examining pore water and air pressures, and finally, the soil microstructure. Each of the theories has its merits, although it may have to be placed in the context of the state of development of soil mechanics at the time and the soil types and methods of compaction used in obtaining experimental data.

3.5.1 Water Films and "Lubrication"

Proctor (1933) stated that the effectiveness of any method of compacting a soil is limited by the friction between the particles. His theory was that in a very dry soil, a thin water film, held in place by surface tension, surrounds each particle. This capillary moisture develops high frictional resistance between the particles, making compaction difficult. The addition of water reduces the capillary forces, decreases friction, and causes slight expansion. Adding more water is thought to have a lubricating effect, aiding particle rearrangement. This effect continues until the moisture becomes just sufficient to fill almost all the voids when the compaction process is complete: the soil now has the greatest density and lowest void ratio that the particular method of compaction can obtain. If more water has to be accommodated in the soil particle structure, a reduction in dry density occurs, the soil softens and may no longer sustain the compactive forces.

Proctor's theory relates to cohesive soils and was put forward at a time when the concept of effective stress and the significance of pore pressures were not widely known. Today, the concept of lubrication as used by Proctor is no longer considered appropriate.

For Hogentogler (1936), lubrication also played a role in compaction, although it was seen as just one of four processes used to explain the moisture-density curve. He assumed that the moisture films around the soil particles have different characteristics in the innermost and outermost layer, the cohesiveness of the water decreasing as the distance from the soil solid increases. The character of these films is affected by the size and shape of the particles as well as the chemistry of the soil and the pore water.

Hogentogler saw four stages of wetting, represented by four straight lines on a plot of moisture content versus dry density (Fig. 3.8), if the moisture content is measured in terms of weight of water over total volume (soil and water). In the hydration stage, water is absorbed by the soil particles and attached to their surfaces as thin cohesive films. In the lubrication stage, part of the contained water is acting as a lubricant, facilitating the rearrangement and closer association of particles without excluding all the air. Water in excess of the optimum water content causes the soil mass to swell with the amount of air remaining approximately constant. With the addition of more water practically all the air is displaced during compaction and the soil becomes saturated (saturation stage).

Another early approach to explain soil compaction is that by Buchanan (1942). His theory is similar to that of Proctor. In a very dry soil, thin water films are thought to bind particles together into a series of arches or groups, like flocculated sediment or bulked sand. Accompanying the addition of initial moisture, the arching effect is

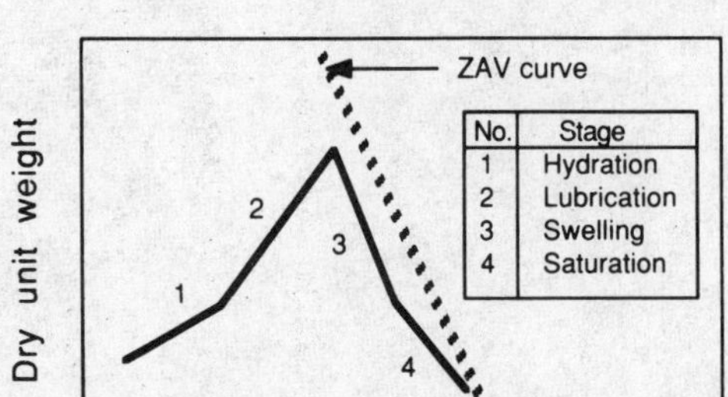

FIGURE 3.8
Compaction and stages of wetting according to Hogentogler (1936).

pronounced, as indicated by the abrupt decrease in density as shown in Fig. 3.9. Note that the compaction curve considered starts at zero moisture. As more water is added, the films of water on the particles become thicker, the capillary action increases and arching is diminished. At the optimum moisture content, the supply of water is just sufficient to neutralize the surface tension and the full effect of the compacting force is utilized in the rearrangement of the soil grains to form a dense mass.

3.5.2 Pore Water and Air Pressures

Hilf (1956), as summarized by Winterkorn and Fang (1975), gave the first modern type of compaction theory, employing the concept of pore water pressures and pore air pressures. He suggested that the compaction-soil moisture relationship be presented in terms of void ratio versus water void ratio (volume of water to volume of solids), as shown in Fig. 3.10. A curve similar to the conventional compaction curve results, with the optimum moisture content corresponding to a minimum void ratio. The ZAV curve is a straight line and so are the saturation lines, all originating at zero void ratio and zero moisture content. Points representing soil samples with equal air void ratios (volume of air to volume of solids) plot on lines parallel to the ZAV or 100% saturation line. According to Hilf, dry soils are difficult to compact because of high friction due to capillary pressure. Air, however, is expelled quickly because of the larger air voids. With increasing water content the tension in the pore water decreases, reducing friction and allowing better densification until a maximum density is reached. Less-effective compaction beyond the optimum moisture content is attributed to the trapping of air and the building up of pore air pressures.

Later work by Olson (1963) confirmed that the air permeability of a soil is dramatically reduced at or very close to the optimum moisture content. At this point, high pore air pressures and pore water pressures minimize effective stress allowing adjustments of the relative position of the soil particles to maximum density. At water contents below optimum, Olson attributes resistance to repeated compactive forces to the high negative residual pore pressures, the relatively low shear-induced pore pressures, and the high residual lateral total stress. On the wet side of optimum, Olson explains the reduced densification effect by pointing out that the rammer or foot penetration during compaction is larger than in drier soil, which may cause temporary

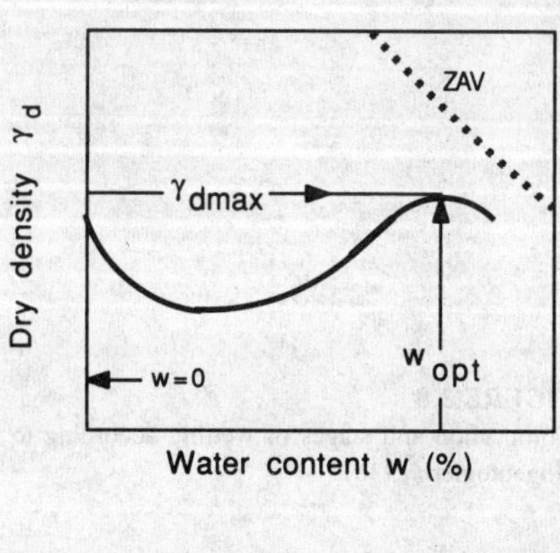

FIGURE 3.9
Compaction curve extended to zero water content.

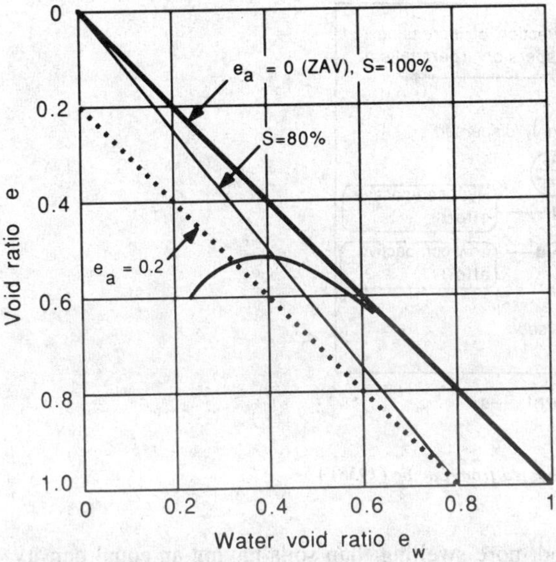

FIGURE 3.10
Presentation of compaction test results as proposed by Hilf (1956).

negative pore pressures known to be associated with large strains in overconsolidated soils; in addition the soil resists compaction by increased bearing capacity due to the depth effect.

The decrease of air permeability beyond the optimum water content is so pronounced that Leflaive and Schaeffner (1980) have proposed to use the measurement of air permeability to evaluate the compactibility of soils.

3.5.3 Microstructure

After examining the findings from optical examinations of the arrangement of clay particles in compacted soils, Lambe (1958) proposed a physicochemical interpretation of their properties. At low water contents, attractive forces between clay particles predominate, creating a *flocculated* structure with more or less random orientation of the platelike particles. In this state many positively charged edges are in contact with negatively charged faces, causing low density. The addition of water increases repulsion between particles leading them to assume more parallel orientation near the optimum moisture content. If compacted wet of optimum parallel orientation is further increased leading to what is described as *dispersed* structure. Lambe's view of the structure of compacted soils is illustrated in Fig. 3.11.

Physicochemical and soil structure considerations may not, on their own, explain the compaction curve, but they certainly help in understanding strength and volume change characteristics of compacted soils and in evaluating different types of densification. Soils with a flocculated structure exhibit higher strength at low strains, greater

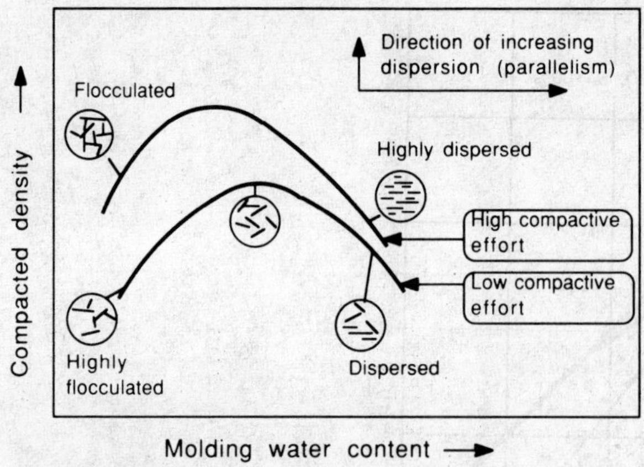

FIGURE 3.11
Effect of compaction on soil structure. [*Adapted from Lambe (1958).*]

permeability, lower shrinkage, and more swelling than soils having an equal density and moisture content but a dispersed structure. The large strains that are characteristic of kneading compaction appear to cause more pronounced dispersion than impact type compaction. Dispersion counteracts the increase in strength gained with higher densities, which may explain why some soils lose strength when *overcompacted* (see Section 4.1.1.3).

3.5.4 Macrostructure: Compaction-Induced Planar Shear Zones

According to Sherard et al. (1984), there have been a number of instances where concentrated leaks have developed through earth dams, typically soon after the first reservoir filling. In these apparently well designed and constructed embankments, the leaks are usually attributed to either cracks caused by differential settlement, seepage along the boundary between the earth fill and rock abutments or foundations, or some form of hydraulic fracturing. The obvious remedy to the problem of concentrated leaks is the incorporation of an effective filter zone downstream of the impervious core.

Discussing Sherard et al.'s paper, McDonald (1988) drew attention to the formation of shear planes parallel to the surface of the compacted soil, described as *pancaking*. The phenomenon observed by McDonald was attributed mainly to the compactive effect by the earth carriers, since it was concentrated at the locations of temporary haul roads.

One of the few studies of shear surfaces induced by compaction was carried out by Whyte and Vakalis (1987). These authors focused on slip planes which have been observed to develop at the interface of compacted layers and which have been variously described as "slickensided," "smooth," or "polished" surfaces. Whyte and

Vakalis postulated two mechanisms associated with these shear planes: bearing failure under heavy construction machinery and aggregated soil movement at the junction between two compacted layers.

The formation of shear planes is also known to road engineers in Australia. In the Sydney Basin area, many urban roads are built up with crushed sandstone, with a grading smaller than 75 to 100 mm. Excessive compacting and leveling of this material can induce a so-called *false pavement*, a laminar separation of the top 25 mm; this prevents proper adherence of the surface seal, resulting in potholes.

In dams, plant-induced shear planes seem to be associated with overcompaction of clays wet of optimum, inferring softening due to a changed microstructure (see Sec. 4.1.1.3). The problem may, however, be more widespread and possibly also occurs when rollers or heavy vehicles break down granular soils very close to the surface.

Shearing parallel to the surface during compaction is not likely to be obvious during construction; only careful excavation or block sampling may provide evidence before unexpected leakages or slip failures occur. If shear zones are indeed discovered, it is recommended that smooth ground be scarified before another lift is placed and that highly trafficked haul roads be ripped up and recompacted.

The significance of the anisotropic and inhomogeneous macrostructure of compacted fills is one of the least-known aspects of ground modification by mechanical means and deserves much more study.

3.5.5 Future Developments

The moisture-density relationship as revealed in the standard Proctor laboratory test and its dependence on compactive effort forms the core of the principles of compaction. The density achieved in dynamic laboratory compaction cannot, however, simply be translated into a mechanical property as the designer of earth structures may wish. Soil type, preparation, and mode of compaction have a demonstrable effect on engineering properties such as strength, volume change, and permeability.

Examination of the microstructure offers at least the possibility for explaining certain aspects of the behavior of compacted soils. Because of the difficulty in measuring soil structure, quantitative analysis based on microstructure and interparticle forces remains elusive.

Microstructure and soil moisture aspects are of particular importance for cohesive soils. This is the main reason why in the following chapters the properties of cohesive and cohesionless soils are treated separately. Additional reasons are that different emphasis may be placed on the objectives of compaction for the two classes of soils, and methods of testing and evaluation may differ.

A case could be made for dealing with intermediate soils (sands and gravels with a significant amount of plastic fines) separately. Indeed, special soils, such as dry uniform sands, loess, and organic soils, may deserve special attention. It is conceivable, that compaction theories of the future will focus more and more on particular soils or groups of soils and specific equipment. Further refinement of compaction machinery and operation will go parallel to this trend. The already talked about remotely controlled "intelligent" robot, which can sense the degree of compaction achieved, may not be too far away.

PROBLEMS

Prefixes indicate problem type: C = calculations, M = multiple choice, D = discussion.

Calculations

C3.1. Using empirical relationships, estimate the optimum water content and maximum dry density for a clay with a liquid limit of 40% and a plasticity index of 20% using Eq. 3.1 to 3.4.

C3.2. Prepare a graph which represents the relationships between w_{opt} and w_p for lateritic, micaceous, and black cotton soils as given in Eq. 3.5 to 3.10. Also plot w_{opt} vs. fines for lateritic and black cotton soils.

C3.3. A silty clay, with 100% of its particles finer than 0.4 mm and 90% of its particles able to pass through a #200 sieve, has a liquid limit of 36% and a plastic limit of 17%. A standard compaction test gave a maximum dry density of 1.805 t/m^3 at an optimum moisture content of 16.4%. Which one of the empirical rules quoted in this text would have given the best estimate of the compaction test results, and what error would have been involved?

C3.4. A lateritic sandy gravel has a D_{60} = 8 mm and a D_{10} = 0.6 mm. Estimate e_{min}, e_{max}, and the maximum dry density according to Fig. 3.1.

C3.5. A dam construction material contains 38% of particles larger than 10 mm. A laboratory compaction test carried out on the soil fraction smaller than 10 mm yielded the following results:

$$w_{opt}\ (< 10\ \text{mm}) = 10\%$$

$$\rho_{dmax}\ (< 10\ \text{mm}) = 1.92\ \text{t/m}^3$$

The specific gravity G_s of the soil solids is 2.7. Estimate the optimum water content w_{opt} and maximum dry density ρ_{dmax} in the field making the following alternative assumptions:

(*a*) The fraction > 10 mm has a water content of 2% and a dry density of 2.6 t/m^3. (The large particles are envisaged to "float" in the finer material).

(*b*) The saturation S of the fraction < 10 mm is the same as the actual field material (including the large particles), and the water content of the fraction larger than 10 mm is 2%, as assumed in (*a*).

C3.6. Kyulule (1983) gave the following laboratory results for a tropical black clay:

Test	γ_{dmax}	w_{opt}
Modified standard	1.56	24.6
Standard	1.40	29.0
50% standard	1.34	31.6

(*a*) Plot the laboratory results on a graph of γ_{dmax} versus log compactive effort.

(*b*) Express, in percent of standard maximum density, the compaction achieved for

Eight passes of a human-towed 1.2-t concrete roller, given $\gamma_{dmax} = 1.27$ and $w_{opt} = 31\%$.
Six passes of a 10-t sheepsfoot roller, given $\gamma_{dmax} = 1.37$ and $w_{opt} = 27\%$.
Human feet compaction, given $\gamma_{dmax} = 1.18$ and $w_{opt} \approx 37\%$. (Remember the elephants of Prob. C2.11, who achieved 87% standard Proctor density).
Hand rammer compaction, given $\gamma_{dmax} = 1.19$ and $w_{opt} \approx 33\%$.

(*c*) Estimate the compactive effort of human feet in terms of standard compaction.

C3.7. Laboratory standard compaction produced a maximum dry density of 1.5 t/m³. Estimate the result which would have been obtained using modified standard compaction.

C3.8. Standard compaction on volcanic ash (USCS classification MH, $G_s = 2.69$) produced the following results:

γ_d (t/m³) =	1.27	1.31	1.35	1.32	1.29	1.24
w (%) =	25.8	27.3	30.0	32.0	34.0	36.1

(*a*) Plot these results on a conventional γ_d versus w diagram
(*b*) Plot the results again on a void ratio e versus water void ratio e_w diagram. Show lines of 80, 90, and 100% saturation as well as lines defining 0, 10 and 20% air voids.

Multiple Choice

M3.9. The optimum water content of a clay is likely to be
(*a*) Equal to the plastic limit.
(*b*) 2 to 5% below the plastic limit.
(*c*) 2 to 5% above the plastic limit.
(*d*) 2 to 5% below the liquid limit.

M3.10. The modified compaction test is different from the standard test because of the use of a different
(*a*) Mold size.
(*b*) Hammer shape.
(*c*) Type of compaction.
(*d*) Compactive effort.

M3.11. The addition of lime to a soil generally
(*a*) Increases γ_{dmax} and strength.
(*b*) Decreases γ_{dmax} and w_{opt}.
(*c*) Decreases γ_{dmax} and increases w_{opt}.
(*d*) Increases γ_{dmax} and decreases w_{opt}.

M3.12. Comparing two clay specimens compacted according to standard Proctor procedure to the same dry density, but at water contents above and below optimum, we find:
(*a*) The wetter soil is likely to have a flocculated rather than a dispersed soil structure.
(*b*) The wetter specimen will have higher unsoaked strength.
(*c*) The wetter specimen will have a higher stress-strain modulus.
(*d*) The wetter specimen will have lower swelling potential.

M3.13. Compaction theories have been put forward by Proctor, Hogentogler, Buchanan, Hilf, Lambe, and others. Which one of the following descriptions of conditions of optimum water content can be attributed to Hilf?

(*a*) Adding water has a lubricating effect, until the moisture becomes just sufficient to fill almost all the voids when the compaction process is complete.
(*b*) Optimum moisture is attained in between stages of moisture lubrication and soil swell.
(*c*) At optimum moisture content, the supply of water is just sufficient to neutralize the surface tension in the pore water.
(*d*) As the water content increases to optimum, tension in the pore water decreases, reducing friction; beyond optimum, air is entrapped and pore air pressure increases, making compaction less effective.

Discussion

D3.14. The four most important factors affecting compaction are sometimes classified as the "principles of compaction." What are they?

D3.15 What is the value of empirical relationships between Atterberg limits and compaction characteristics?

D3.16. In what way can soil preparation assist the homogeneity of an earth dam?

D3.17. How can climatic conditions affect compaction in road construction?

D3.18. Discuss the concept of lubrication used in some compaction theories.

D3.19. Compaction of a clay below and above optimum moisture content may produce the same dry density, but the soil structure created could be different. In this respect, explain the significance of Lambe's interpretation of the properties of compacted clays.

CHAPTER 4

PROPERTIES OF COMPACTED SOIL

The key to the understanding of the behavior of compacted cohesive soil lies in the analysis of the results of carefully controlled laboratory tests, such as cylindrical compression tests, California bearing ratio (CBR) determinations, and the measurement of swelling. These laboratory test procedures are designed to simulate field conditions, in particular with respect to the compaction method, stress path, and possible submergence and saturation. However, this simulation may be far from perfect, and specific laboratory findings should be confirmed by correlation with field performance.

Because of the problem of measuring density and recovering undisturbed samples of cohesionless soils, direct measurement of strength and deformation properties or indices in the field (such as by penetration tests, pressuremeter tests, and plate bearing tests) is common. In many cases, in order to correlate field and laboratory measurements, recourse is taken to empirical relationships, established by large-scale model tests or based on prototype performance.

Following the classical approach in soil mechanics, the most important characteristics of cohesive and cohesionless soils will be discussed separately, but Table 4.1, which is based on information presented in "Design Manual 7.2." [U.S. Navy (1982)], gives typical ranges of w_{opt} and γ_{dmax} and selected engineering properties for all types of soil, grouped according to the Unified Soil Classification System (USCS).

4.1 PROPERTIES OF COMPACTED COHESIVE SOIL

4.1.1 Strength of Cohesive Soil

4.1.1.1 TYPES OF LABORATORY TESTS. The most common strength tests are illustrated in Fig. 4.1.

The *unconfined compressive strength,* usually designated as UCS or q_u, is the parameter obtained in the simplest and cheapest cylindrical compression test. It may also be expressed in terms of undrained shear strength s_u or undrained cohesion c_u, if it is assumed that ϕ_u, the undrained angle of internal friction, is zero. This assumption is, however, usually not justified for unsaturated compacted soil, and the Mohr-Coulomb envelope is not necessarily a straight line. The specimen height/diameter ratio recommended for cylindrical compression tests is 2:1. Shorter specimens may lead to an overestimation of the undrained strength. Note that the soil in a standard Proctor compaction test is formed into a cylinder with a height/diameter ratio of 1.1:1.

In the *standard triaxial* test, the soil specimen is first subjected to an all-around pressure before being brought to failure by an increase in the axial load (or deviator stress). Special triaxial tests may be appropriate where the soil is to be subjected to a particular stress path and where lateral deformation is to be measured and/or controlled. The classification of compacted soil based on triaxial strength is discussed in Sec. 4.1.1.6.

The *Hveem stabilometer* test is a form of triaxial test where a short cylindrical sample, prepared by kneading compaction, is subjected to an axial load. The resultant

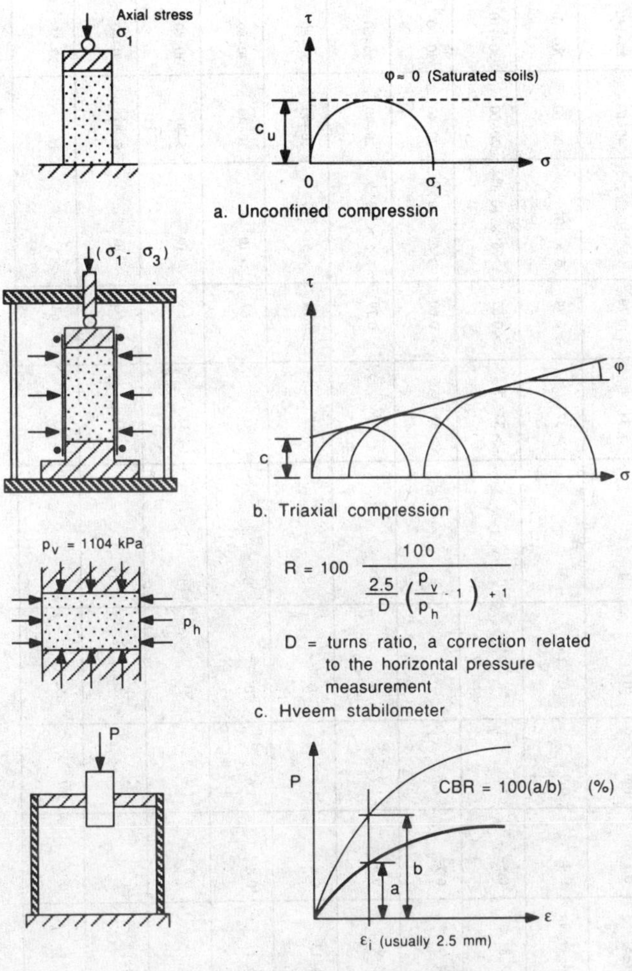

FIGURE 4.1
Types of strength tests used for compacted cohesive soils.

horizontal pressure is measured, and the strength is expressed in terms of a resistance value R.

In the *CBR test* the soil is compacted in the same way as in the standard or modified compaction test, the only difference being the size of the mold (AS 1289.F1.1). After compaction, a piston is forced into the confined soil at a standard rate. The penetration resistance is recorded and plotted against deflection. The curve is "corrected" if because of improper seating of the piston, or other causes, the deflection curve is concave downward or irregular at low pressures. The resistance to penetration at a specified deflection is divided by the resistance offered by a standard

TABLE 4.1
Typical properties of compacted soils

				Typical value of compression		Typical strength characteristics						
Group symbol	Soil type	Range of max. dry unit weight, t/m^3	Range of optimum moisture, %	At about 140 kPa, % orig. height	At about 350 kPa, % orig. height	Cohesion (as compacted), kPa	Cohesion (saturated), kPa	ϕ' (effective stress envelope), degrees	$\tan \phi'$	Typical coeff. of permeability, m/s	Range of CBR values	Range of subgrade modulus k, × 1000 kN/m^3
GW	Well-graded clean gravels, gravel-sand mix	2.0–2.2	11–8	0.3	0.6	0	0	> 38	> 0.79	10^{-5}	40–80	80–140
GP	Poorly graded clean gravels, gravel-sand mix	1.8–2.0	14–11	0.4	0.9	0	0	> 37	> 0.74	5×10^{-5}	30–60	70–110
GM	Silty gravels, poorly graded gravel-sand silt	1.9–2.2	12–8	0.5	1.1	—	—	> 34	> 0.67	$> 5 \times 10^{-10}$	20–60	30–110
GC	Clayey gravels, poorly graded gravel-sand clay	1.8–2.1	14–9	0.7	1.6	—	—	> 31	> 0.60	$> 5 \times 10^{-11}$	20–40	30–80
SW	Well-graded clean sands, gravelly sands	1.8–2.1	16–9	0.6	1.2	0	0	38	0.79	$> 5 \times 10^{-7}$	20–40	55–80
SP	Poorly graded clean sands, sands, sand-gravel mix	1.6–1.9	21–12	0.8	1.4	0	0	37	0.74	$> 5 \times 10^{-7}$	10–40	55–80
SM	Silty sands, poorly graded sand-silt mix	1.8–2.0	16–11	0.8	1.6	50	20	34	0.67	$> 10^{-8}$	10–40	30–80
SM-SC	Sand-silt clay mix with slightly plastic fines	1.8–2.1	15–11	0.8	1.4	50	14	33	0.66	$> 10^{-9}$	5–30	30–80
SC	Clayey sands, poorly graded sand-clay mix	1.7–2.0	19–11	1.1	2.2	75	11	31	0.60	$> 10^{-10}$	5–20	30–80
ML	Inorganic silts and clayey silts	1.5–1.9	24–12	0.9	1.7	65	9	32	0.62	$> 5 \times 10^{-9}$ or less	15	30–55

TABLE 4.1 *(Cont.)*

ML-CL	Mixture of inorganic silt and clay	1.6–1.9	22–12	1.0	2.2	65	22	32	0.62	$> 10^{-10}$	—	
CL	Inorganic clays of low to medium plasticity	1.5–1.9	24–12	1.3	2.5	85	13	28	0.54	$> 5 \times 10^{-11}$ or less	15	15–55
OL	Organic silts and silt-clays, low plasticity	1.3–1.6	33–21	—	—	—	—	—	—	—	5 or less	15–30
MH	Inorganic clayey silts, elastic silts	1.1–1.5	40–24	2.0	3.8	70	20	25	0.47	$> 10^{-10}$	10 or less	15–30
CH	Inorganic clays of high plasticity	1.2–1.7	36–19	2.6	3.9	105	11	19	0.35	$> 5 \times 10^{-11}$	15 or less	15–40
OH	Organic clays and silty clays	1.0–1.6	45–21	—	—	—	—	—	—	—	5 or less	5–30

Source: Adapted from "Design Manual 7.2," U.S. Navy, 1982.

Notes:

1. All properties are for condition of standard Proctor maximum density, except values of k_s and CBR which are for modified Proctor maximum density.
2. Typical strength characteristics are for effective strength envelopes and are obtained from U.S.S.R. data.
3. Compression values are for vertical loading with complete lateral confinement.
4. ">" indicates that the typical property is greater than the value shown.
5. "—" indicates that insufficient data is available for an estimate.

crushed rock at the same penetration: This ratio, expressed in percent, is called the California bearing ratio. This test is normally conducted on a laboratory compacted sample, although it can also be carried out on undisturbed field samples or in situ (AS 1289.F1.2 and F1.3). The soil may be tested as compacted or after 4 days of soaking in water, simulating submergence and saturation which may occur in the field. Measurement of swell during soaking is a useful side result.

The CBR and stabilometer resistance *R* were conceived for pavement design. The unconfined and triaxial compression tests have universal applications.

4.1.1.2 CBR AND INITIAL MOISTURE CONTENT. The CBR strength of a compacted soil is a function of the initial moisture content (as molded) as well as the density. Figure 4.2 presents the CBR of soil compacted with a constant compactive effort as a function of the initial water content. Unsoaked specimens exhibit highest strength if compacted dry of optimum; only after soaking does compaction for maximum dry density show an advantage. As expected, specimens at a higher initial saturation show less swell during soaking.

4.1.1.3 STRENGTH AND DENSITY. Within reasonable limits, a soil with a given initial moisture content can be molded to higher and higher densities if the compactive effort is increased. As shown in Fig. 4.3, the (unsoaked) strength of a cohesive soil compacted dry of the relevant optimum is likely to increase with density (curve *A*). Where the initial moisture content approaches or exceeds the optimum moisture content for the particular compactive effort, the strength may fall off (curve *B*). A similar finding relates to soaked strength (Fig. 4.4), except that the specimens initially compacted on the dry side may, over a certain density range, show less strength than an originally wetter sample.

Loss in strength with increasing density is a phenomenon sometimes referred to as *overcompaction* or *overstress;* it has been verified by construction engineers in the field who became aware that better strength can sometimes be obtained for highly saturated clays by compacting them with lighter equipment to a lower density than by compacting them with heavy equipment to a higher density. The decrease in stability with increasing dry density has been attributed to the high pore pressures induced by compaction. Lambe (1958), however, may explain the phenomenon by saying that it is the formation of a dispersed soil structure which reduces the strength of a clay in these conditions (see Sec. 3.5.3).

For highly plastic clays, the highest CBRs after soaking may be found within a specific range of initial water contents and dry densities, as shown in Fig. 4.5 [Seed (1959) using data from Bell (1956)]. For less-plastic soils, maximum soaked CBRs may be obtained simply by compacting the soil to the highest practicable density at a water content close to the optimum for the particular compactive effort. Seed nevertheless states that "conditions producing highest strength vary widely with different types of soils and can only be determined by an adequate soil testing program." The relevance of laboratory testing alone will, however, depend on how accurately field compaction can be simulated in the laboratory, with respect to type and effort of compaction.

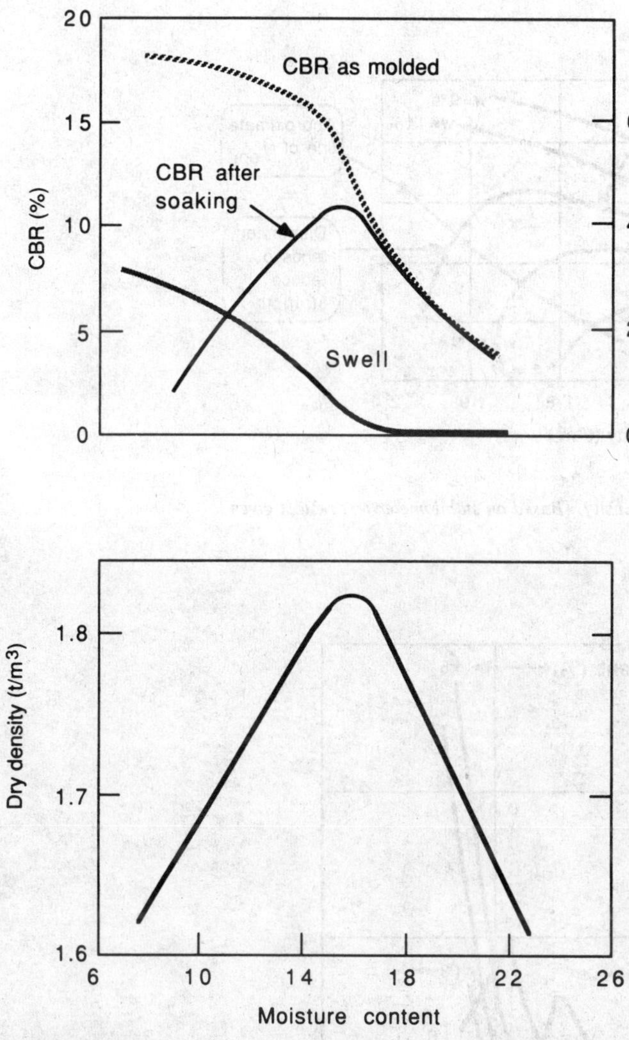

FIGURE 4.2
Density and CBR as a function of initial water content for a typical silty clay (CL). [*After Yoder (1959). Copyright John Wiley & Sons.*]

4.1.1.4 STRESS-STRAIN BEHAVIOR. Lee and Haley (1968) prepared spècimens at equal densities, in between standard and modified maximum dry density, but with moisture contents either dry or wet of optimum. The stress-strain curves in unconfined compression are given in Fig. 4.6. The "dry" specimen showed a high peak strength and a small strain at failure, as is typical for brittle materials. The "wet" specimens showed lower strength and higher deformations at failure, particularly when kneading rather than static compaction was used, the latter probably creating a more pronounced dispersed, rather than flocculated, structure.

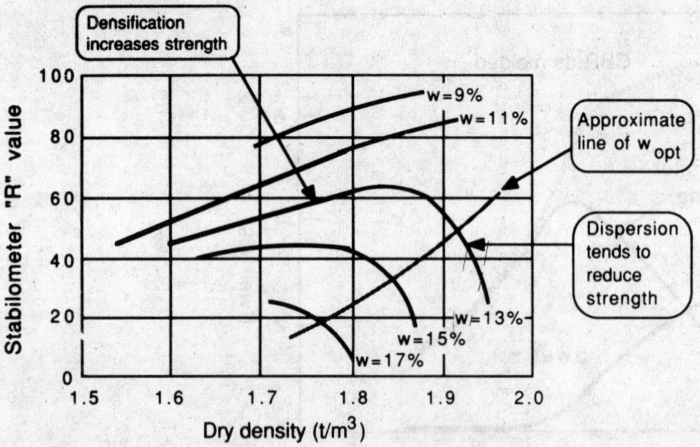

FIGURE 4.3
(Unsoaked) strength versus dry density. [*Based on stabilometer test results given by Seed (1959).*]

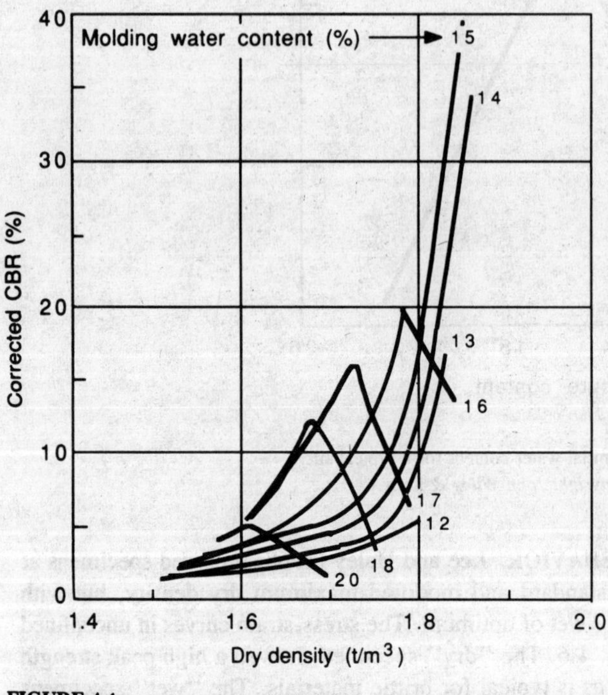

FIGURE 4.4
Soaked strength versus dry density. [*Based on soaked CBRs by Turnbull and Foster (1958).*]

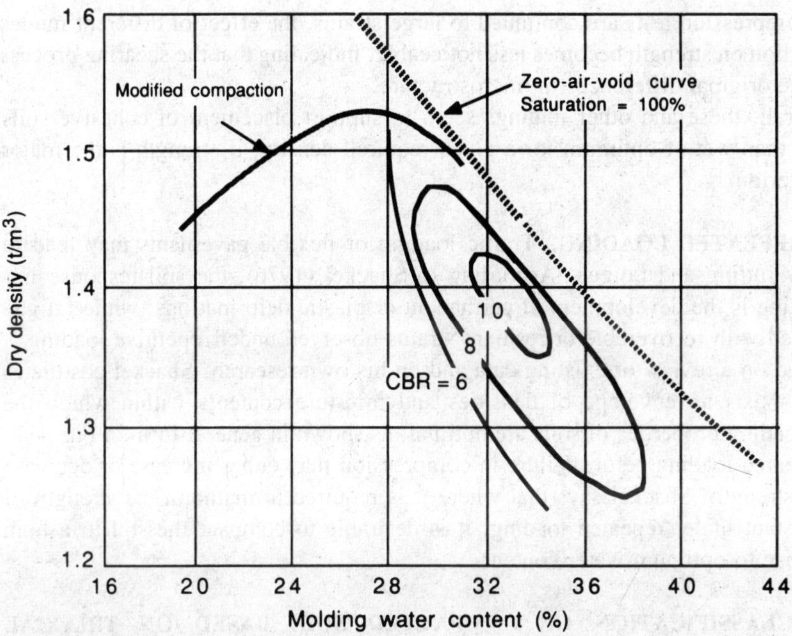

FIGURE 4.5
Soaked CBRs as a function of placement conditions of a highly plastic clay. [*Seed (1959) using data from Bell (1956)*.]

Similar strength differences were also observed in unconsolidated undrained (UU) tests, although the "dry" specimen no longer showed brittle behavior. When the samples were consolidated (with saturation) prior to shear in an undrained condition (CU tests), only small differences in strength and stress-strain behavior remained.

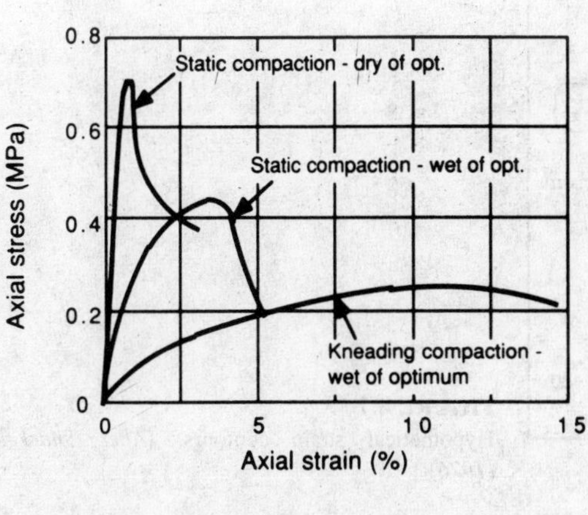

FIGURE 4.6
Stress-strain curves in unconfined compression. [*Lee and Haley (1968)*.]

If compression tests are continued to large strains, the effect of different modes of compaction on strength becomes less noticeable, indicating that the shearing process reduces the original differences in microstructure.

Overall, these and other findings seem to support placement of cohesive soils dry rather than wet of optimum for a given required density, if strength is the major design criterion.

4.1.1.5 REPEATED LOADING. Traffic loading of flexible pavements may lead to failures by rutting and fatigue. According to Shackel (1976), the soil response initiating rutting is the development of permanent or plastic deformations, while fatigue is associated with recoverable or resilient strains observed under repetitive loading.

Based on a review of existing data and on his own research, Shackel postulates that there exists an envelope of densities and moisture contents within which the repeated loading properties of soils are optimal, as shown in general form in Fig. 4.7.

Repeated loading before failure in compression may either increase or decrease the soil's strength. Shackel says that where it is required to maintain the strength of a soil constant under repeated loading, it is desirable to compact the soil to a high density close to optimum water content.

4.1.1.6 CLASSIFICATION OF COMPACTED SOIL BASED ON TRIAXIAL STRENGTH. Compared with the CBR test and the Hveem stabilometer test, triaxial testing offers better control of in-service stress simulation and pretest conditioning, such as drying back or wetting to a particular presumed equilibrium moisture content. It is suitable for all kinds of soil but has the disadvantage of being relatively time-consuming and expensive.

A widely accepted method of ranking soil according to shear strength based on the triaxial test was developed by the Texas Highway Department (1964). This method was adapted and refined for local conditions by the Main Roads Department of Western

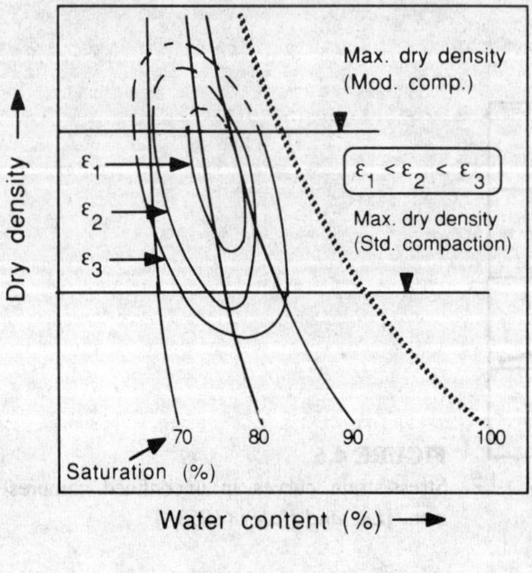

FIGURE 4.7
Hypothetical strain contours. [*After Shackel (1976).*]

Australia, where it is now known as the West Australian Confined Compression Test, WACCT (Main Roads Dept. of Western Australia, 1962). Because it is a good example of how a strength classification system for compacted soils can be made to suit local requirements, it is described in some detail below.

According to the WACCT procedure, five samples are prepared using a standard or modified compactive effort in a 177.8-mm-high mold with an internal diameter of 101.6 mm. The first four specimens are tested in compression at 0, 20, 35, and 70 kPa lateral pressure, respectively. The lateral pressure is applied using an air-inflated rubber sleeve mounted inside a metal cylinder and placed around the test specimen. A stress-strain curve is recorded, and correction is made for the increase in cross-sectional area during compression, yielding a stress-strain modulus and peak deviator stress. The fifth sample is tested in indirect tension (Brazilian method) by placing it on its side in the compression machine. The tensile strength T is calculated according to

$$T = \frac{2P}{\pi dh} \quad \text{usually in kPa} \tag{4.1}$$

where d = diameter of specimen
h = height of specimen
P = maximum applied load

All five strength results are used to plot the best-fit Mohr envelope (usually curved) on a normal stress versus shear stress diagram. This envelope is then transferred onto the classification chart (Fig. 4.8). From the relative position of the failure envelope to the class boundaries the material is classified to the nearest 0.1 class number. Classification is determined by the weakest point of the envelope, that is, where the class number is highest.

In Western Australia, a class number of less than 2 indicates a soil which is suitable as a base course material for most roads. Further considerations are appropriate for marginal materials with a class number between 2 and 3. Like CBR values, the class number, together with wheel load and traffic information, is also used for estimating the required aggregate thickness in pavement design.

Triaxial testing for a variety of compaction conditions allow contours of equal class numbers to be drawn on a dry-density–moisture-content graph, as shown in Fig. 4.9. Steep and closely spaced contours indicate a soil which is highly sensitive to moisture content. It has been proposed that the following ratio be used as an index of moisture sensitivity (MSI) of the strength of the compacted soil [Ladner and Hamory (1974)]:

$$\text{MSI} = \frac{\begin{array}{c}\text{slope of the class number contours}\\ (\text{kg/m}^3 \text{ per } 1\% \text{ change in water content})\end{array}}{\begin{array}{c}\text{separation of contours (percent water content}\\ \text{per unit change in class number)}\end{array}} \tag{4.2}$$

MSI value of less than 3.2 is considered low; a value greater than 9.6 is high.

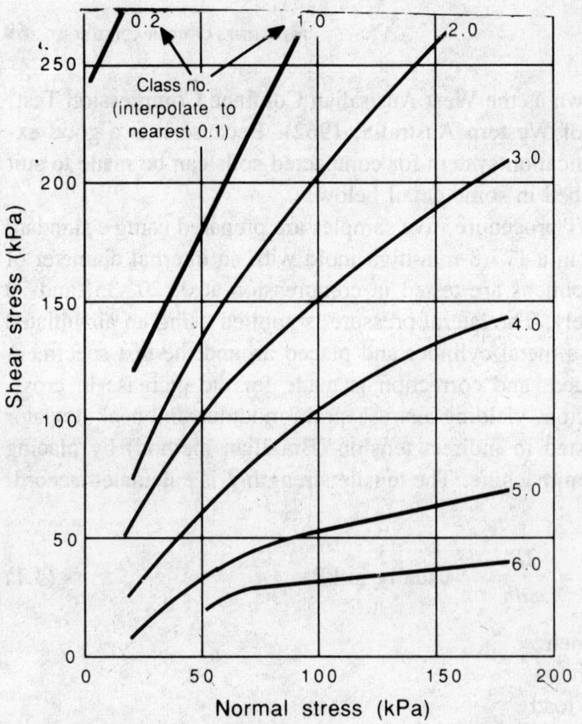

FIGURE 4.8
Western Australian Confined Compression Test classification chart (only major contours are shown). *(Adapted from Main Roads Department of Western Australia, 1962.)*

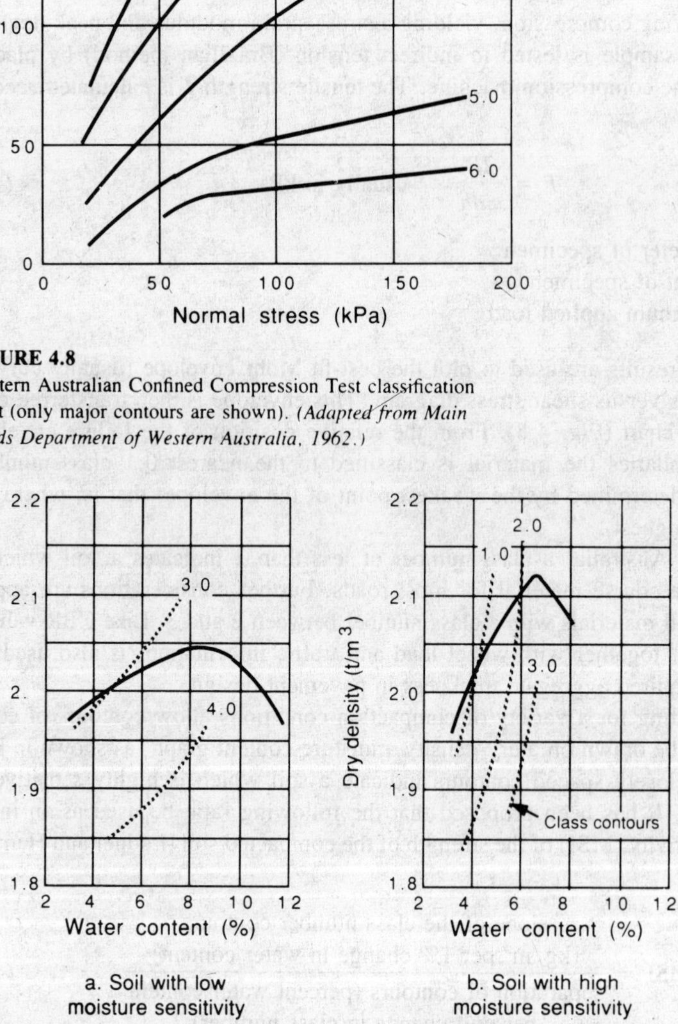

FIGURE 4.9
Examples of class number contours. [*Ladner and Hamory (1974).*]

4.1.2 Volume Change

An increase in external loading or a change in saturation can cause an increase in effective stress, which may force a compacted cohesive soil to compress. Alternatively, clayey soils under low stresses may swell when inundated. Knowledge of the magnitude of volume change is important in most geotechnical problems, including embankment construction and pavement design.

4.1.2.1 COMPRESSIBILITY. Huder (1964), Hilf (1975), and others have reported results obtained on compacted cohesive soils subjected to confined compression in the ordinary consolidation apparatus. These experiments confirmed that a compacted soil's compressibility depends not only on its density but also on its initial moisture content. Significant variations exist from one soil to another, mainly attributed to the grain size distribution and mineralogical characteristics of the fines. An appreciation of the soil structure, e.g., whether it is flocculated or dispersed, assists in the understanding of the compressive behavior of cohesive soils.

Comparing the degree of compression of cohesive soils compacted with equal effort, it is generally found that:

A soil compacted dry of optimum shows less settlement than a soil compacted wet of optimum, if it is left unsaturated.

If soils are saturated after being subjected to a significant compressive load, additional settlement may occur. This ''collapse'' settlement is most significant for soils compacted dry to optimum.

Total settlement due to saturation and compression is least in the vicinity of the optimum water content, where maximum dry density was achieved during compaction.

These typical characteristics are illustrated by Fig. 4.10 which shows settlement, expressed as a percentage of the original height, as a function of the initial water content under loadings of approximately 100, 200, 400, and 800 kPa.

Ingles and Williams (1980) studied saturation settlements of an Australian silty sand and a silty clay compacted dry of optimum. They found these settlements to be relatively small (on the order of 1 to 3%) but pointed out that even a settlement of 1% would be sufficient to open up a seepage flow path in an earth dam which could lead to eventual failure by piping, particularly in dispersive soils.

Collapse settlement of noncohesive soils can be considerably larger. This is discussed in Sec. 4.2.3.

4.1.2.2 SWELLING. Compacted clays subjected to low pressures, say less than 100 kPa, may show significant expansion when inundated. Soils containing a large percentage of clay with predominantly expansive lattice-type minerals, such as montmorillonite, have been identified as being most prone to swelling. How much volume expansion occurs or how much swelling pressure is developed also depends on the soil structure, which in turn is affected by compaction conditions.

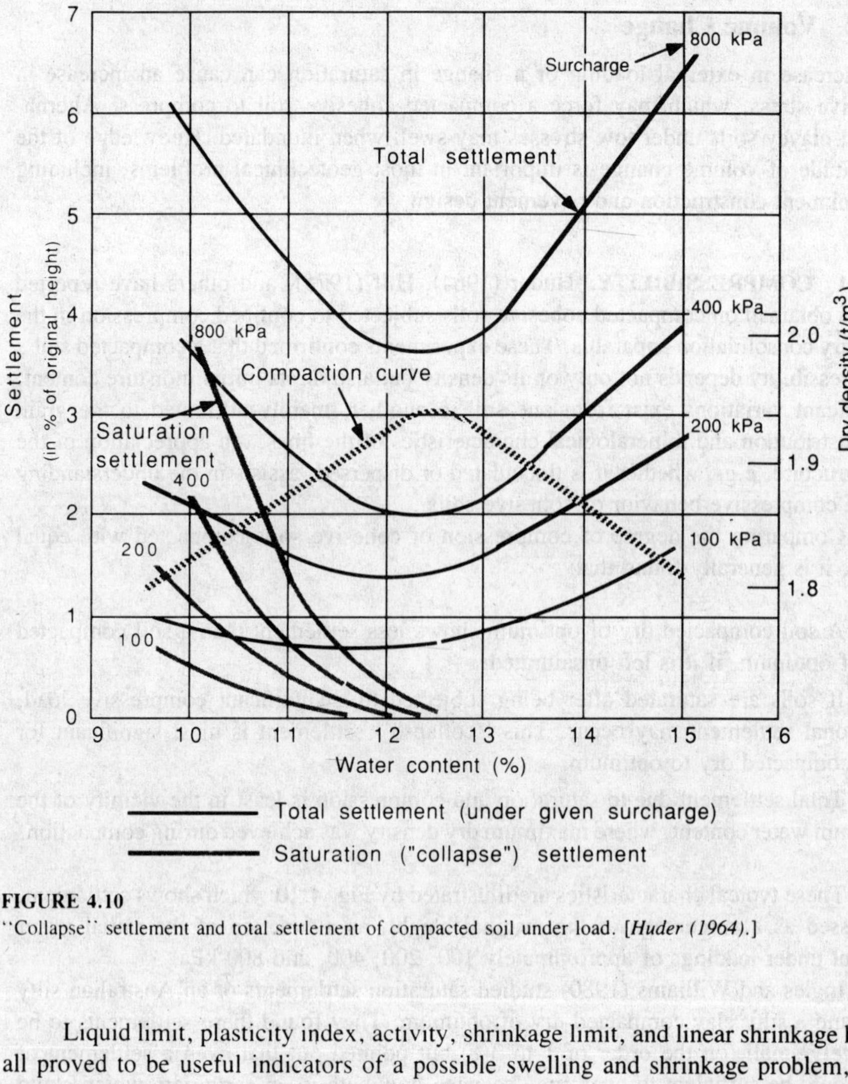

FIGURE 4.10
"Collapse" settlement and total settlement of compacted soil under load. [*Huder (1964).*]

Liquid limit, plasticity index, activity, shrinkage limit, and linear shrinkage have all proved to be useful indicators of a possible swelling and shrinkage problem, but direct measurement of compression or expansion under the expected stress conditions in the field is the most reliable test.

Swelling characteristics are usually studied using a one-dimensional consolidation-type apparatus, where the soil specimen is confined laterally. The vertical expansion of the compacted soil when it is submerged in water is reported as percent swell. *Free swell* is the percent swell with practically zero surcharge. The pressure required to inhibit any volume change upon inundation is termed the *swelling pressure*. The percent swell of a soil compacted as in the standard compaction test and subjected to a 6.9-kPa surcharge before being immersed in water is called the *swelling potential*. A split-type version of the standard Proctor mold with a 25.4-mm-high built-in ring has proved useful in the preparation of specimens for the determination of the swelling potential.

A soil with a high swelling potential usually also exhibits a high swelling pressure if no expansion is allowed, but there does not seem to be a generally valid relationship between the two.

Holtz and Gibbs (1956) investigated the effect of compaction on the swelling potential of a clay used for canal embankments. Their results (Fig. 4.11) show the swelling potential of this soil for a wide range of initial dry density/moisture content combinations. After swelling, dry density/moisture content data points of all specimens were found to plot along the 0% swell curve, for the standard surcharge used. It can be concluded from this work, that, in order to minimize swelling, an expansive-type soil should be placed with a high degree of saturation at a relatively low density. A highly saturated soil has, however, not only low shear strength but it is also difficult to work with in the field. The design engineer will, therefore, most likely have to compromise between high strength and possible volume change.

To deny access of water to a potentially swelling soil mass is an alternative way of preventing expansion. In road engineering, this has been successfully achieved by using semipermeable or impermeable membranes. One of these techniques is called MESL, which stands for *m*embrane *e*ncapsulated *s*oil *l*ayer.

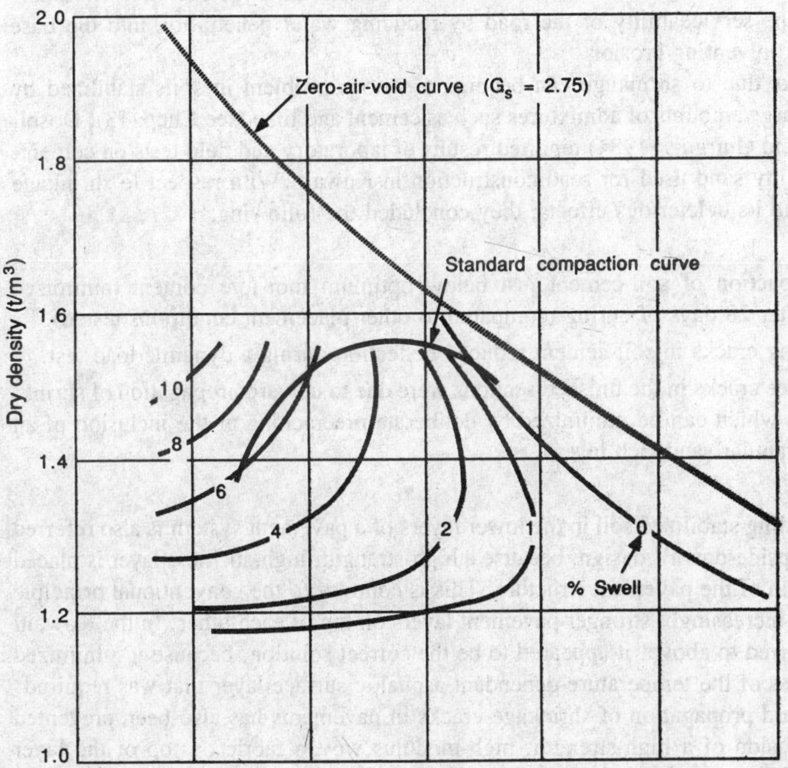

FIGURE 4.11
Percent swell related to placement conditions. [*Holtz and Gibbs (1956).*]

4.1.2.3 SHRINKAGE. Shrinkage is the reverse process to swelling. Soils which are prone to swelling when submerged are usually also subject to shrinkage upon drying. Generally, the higher the density and the lower the moisture content, the less shrinkage will occur in a dry environment.

Although soil shrinking may lead to differential settlement along the perimeter of a building, swelling adjacent and below structural foundations is potentially more damaging than shrinking, because of the enormous forces which can result from swelling pressures. However, in road engineering, shrinking of base course material may cause cracking which is reflected in the asphaltic surface course, initiating rapid deterioration.

Damage due to shrinkage is avoided by appropriate construction procedures and by controlling environmental influences affecting soil moisture. The following two case studies illustrate this approach.

Ladner and Hamory (1974) describe a project in northwestern Australia where shrinkage cracking of a sand-clay base course was minimized by compacting the base course at the optimum moisture content (in order to achieve a high density) and subsequently drying it back to what was considered a long-term equilibrium moisture content before sealing it with asphalt. Bitumen stabilization of the road shoulders improved the serviceability of the road by reducing water penetration into the base course and preventing erosion.

Cracks due to shrinkage can become a serious problem in soils stabilized by relatively large amounts of admixtures such as cement and lime (see Chap. 13). Doshi, Mesdary, and Guirguis (1984) reported results of laboratory and field tests on cement-stabilized silty sand used for road construction in Kuwait. With respect to shrinkage cracking and its deleterious effects, they concluded the following:

Compaction of soil-cement just below optimum moisture content minimized cracking after 28 days of curing (compared to other placement conditions tested).

Sealing cracks in soil-cement reduced deflection during a dynamic load test.

Surface cracks in the finished sections were due to upward propagation of shrinkage cracks, which can be minimized by deliberate precracking or the inclusion of an unbound granular sandwich layer.

Including stabilized soil in the lower layers of a pavement system is also referred to as an "upside-down" design, because a high-strength, high-stiffness layer is placed at the bottom of the pavement structure. This is contrary to the conventional principle of building increasingly stronger pavement layers on top of each other. In the Kuwaiti project referred to above, it appeared to be the correct solution, because it minimized the thickness of the temperature-dependent asphaltic surface layer that was required.

Upward propagation of shrinkage cracks in pavements has also been prevented by the inclusion of a high-strength, high-modulus woven fabric on top of the layer which is subject to cracking. This is however a fairly recent development, and design guidelines are still being evaluated.

4.1.3 Permeability

Lambe (1958), as quoted by Seed (1959), reported that permeability values for a silty clay can be decreased by a factor of 100, if it is compacted wet of optimum rather than say 2% dry of optimum. This finding should be remembered when designing relatively impervious zones in a dam or clay liners for canals, ponds, or waste disposal areas.

As part of a review of the engineering properties of compacted clay, Reséndiz (1980) identified the void ratio, degree of saturation, and microstructure as the most important internal state variables affecting permeability. Obviously, permeability increases with void ratio and saturation, if other nonrelated parameters are equal. With respect to microstructure, Reséndiz states, "At a given void ratio, permeability is much lower for high degrees of preferential orientation of particles." Compaction at high water contents and kneading mode of compaction are known to increase particle parallelism (see Fig. 3.11). Reséndiz also noticed that permeability increases with time elapsed after compaction, another phenomenon attributed to changes in microstructure.

4.1.4 Frost Susceptibility

Freezing of soil may lead to surface heave and increased lateral pressures, and subsequent thawing may significantly reduce soil strength. Frost heave is due to the formation of ice lenses in the soil. The transition of water from the liquid to the solid phase alone brings about a volume increase; what makes it worse is when water rises to the frozen zone through capillary action, increasing the size of the ice lenses. Top-down thawing in the spring leads to weak spots where ice lenses are melting, causing pavement failures or slope instability.

Unlike sands, gravels, and quarried rock, which are the least frost-susceptible soils (because of their large pore size and high permeability), silty soils are highly frost damage prone. With clays, much depends on the duration and severity of the frost. Low permeability may prevent the formation of ice lenses due to reduced capillary flow of water.

In pavements and railway embankments, frost damage is prevented by the inclusion of a frost-insulating layer below the surface course. This layer consists of weathering-resistant coarse granular material. It prevents capillary rise of water, it acts as a filter and drainage layer, and at the same time it forms a high-strength, high-modulus bearing stratum in the pavement structure; it must therefore be compacted to the highest standards. European standards may require over 100% laboratory maximum dry density and a reloading modulus E_2 in excess of 100 MPa in the plate load test (see Secs. 5.1.5 and 6.2.3). Drainage layers made of geosynthetics also have a potential to act as capillary breaks.

Other techniques for the prevention or reduction of frost damage include the provision of good surface drainage, impermeable membranes, and stabilization with admixtures (to increase the effective grain size).

Freezing temperatures during construction could also adversely affect compaction. Adding salts to soils may prevent ice formation but may have deleterious effects on soil strength and volume stability.

4.2 PROPERTIES OF COMPACTED COHESIONLESS SOIL

Typical ranges of densities of cohesionless soils are given in Table 4.2. Highest densities can be achieved in well-graded gravels; in contrast, poorly graded (uniform) sands, and sands with cohesionless fines, may exhibit densities which are very low compared with other noncohesive soils. The table also indicates the likely range of maximum dry densities obtained in the Proctor test using standard compactive effort.

Engineering properties of cohesionless soils show better correlation with relative rather than absolute densities. In addition, many cohesionless soils are fairly insensitive to the water content during compaction—they are often best compacted either dry or fully saturated. For these reasons, and because assessing undisturbed density in the field is difficult, design criteria and control tests are mostly based on in situ measurements of strength and deformation parameters or on related index properties, rather than on actual density and moisture values. Common surface tests include static or dynamic plate load tests, needle penetration tests, and nuclear density tests. The best-known tests at depth are dynamic standard penetration tests (SPT), static cone penetration tests (CPT), pressuremeter tests, and seismic measurements. The dilatometer test (DMT) or nuclear measurements at depth are less frequently used. Chapter 5 gives more details on compaction control tests.

4.2.1 Compactibility and Relative Density (Density Index)

Terzaghi (1925) proposed to describe the *compactibility* of a soil by the *factor*

$$F = \frac{e_{max} - e_{min}}{e_{min}} \tag{4.3}$$

TABLE 4.2
Typical ranges of densities in cohesionless soils

Soil classification	Range of densities, t/m³		
	Very loose state	Laboratory std. compaction	Very dense state
GW	1.8–1.9	2.0–2.2	2.2–2.3
GW-GM, GM, GW-GP, GP-GM	1.7–1.9	1.8–2.1	2.1–2.3
GP	1.8	1.8–2.0	2.2
SW	1.5–1.7	1.8–2.1	2.1
SW-SM, SP-SM, SM	1.3–1.6	1.8–2.0	1.9–2.1
SP	1.4–1.6	1.6–1.9	1.8–2.0

where e_{max} is the void ratio of the soil in its loosest state and e_{min} is the void ratio of the soil in its densest state.

The factor F generally varies from 0.5 to about 2. Poorly graded gravels or sands, which are the most difficult soils to compact, have low compactibility values: $F = 0.5$ to 1. Well-graded soils containing some fines respond well to compaction; their compactibility factors will probably be on the order of 1.5 to 1.8.

A more useful parameter for cohesionless soils has proved to be the *relative density* D_r; it is also described as the *density index* and is designated I_D (AS 1289.E5.1-1977). It is defined as

$$I_D = \frac{e_{max} - e}{e_{max} - e_{min}} \tag{4.4}$$

where e is the actual void ratio of the soil being evaluated. The density index can also be expressed in terms of dry density rather than void ratio:

$$I_D = \frac{\gamma_{dmax}\,(\gamma_d - \gamma_{dmin})}{\gamma_d(\gamma_{dmax} - \gamma_{dmin})} \tag{4.5}$$

The determination of minimum and maximum dry densities is described in AS 1289.E5.1-1977. The loosest possible packing is said to be the one obtained by pouring the dry granular material carefully, with minimum free fall, from a funnel (or scoop) into a measuring cylinder of a diameter and height appropriate for the grain size of the soil being tested. Maximum density is attained by vibrating the fully saturated soil in the mold, first without and then with a prescribed surcharge. Testing procedures adopted by different authorities may vary somewhat, particularly with respect to the determination of the maximum density.

Terzaghi (1925) distinguished the following ranges of relative density for sands:

Loose	$0 < I_D < \frac{1}{3}$
Medium dense	$\frac{1}{3} < I_D < \frac{2}{3}$
Dense	$\frac{2}{3} < I_D < 1$

The most generally accepted ranges and descriptive terms used today for classifying the compaction of sand are those given in Table 4.3, which also gives associated values of penetration resistance, density, and friction angle [after Mitchell and Katti, as quoted by Douglas (1983)]. The relationship between relative density, shear strength, penetration resistance, and other soil properties will be discussed further in subsequent sections.

4.2.2 Shear Strength and Density

The angle of internal friction is not a constant for a cohesionless soil but depends on the density. For example, the friction angle of a sand may increase from 30 to 40° if the relative density is varied from 0 to 100%. Besides density, grain size and shape significantly affect shear strength and stress-strain behavior. Overall, friction angles

TABLE 4.3
Sand properties related to the density index

Sand properties	Density index,* % 0–15 (very loose)	15–35 (loose)	35–65 (medium dense)	65–85 (dense)	85–100 (very dense)
N value, blows/300 mm	< 4	4–10	10–30	30–50	> 50
CPT resistane,† MPa	< 5	5–10	10–15	15–20	> 20
Dry unit weight, kN/m³	< 14	14–16	16–18	18–20	> 20
Friction angle, degrees	< 30	30–32	32–35	35–38	> 38

Source: Quoted by Douglas (1983).

*Freshly deposited normally consolidated sand.

†At an effective vertical overburden pressure of 100 kPa.

reported for a variety of cohesionless soils range from about 29 to 46°. Reviewing findings for rock fill only, Hilf (1975) concluded that its friction angle does not vary much more than 8°; within this range well-graded material containing rounded particles performed best.

Data concerning the relationship between friction angles and relative density of sands and gravels has been presented by Hilf (1975); Schmertmann (1978) proposed definite relationships between the triaxial peak angle of internal friction and the relative density for a range of cohesionless soils. Schmertmann's (1978) diagram is reproduced in Bowles (1988).

Laboratory tests indicate that, at low confining pressures, dense sands dilate and exhibit a brittle-type stress-strain curve (Fig. 4.12). The peak value of strength, recorded as the *peak* friction angle, can in these conditions be considerably higher than the *residual* value, corresponding to large strains. Residual shear strength is sometimes also referred to, in order of preference, as *post-peak, ultimate,* or *critical* shear strength. In contrast, loose sands sheared at the same low confining pressure may show a decrease in volume, a low stress-strain modulus, and no pronounced peak value. Regardless of the initial density, the final density of the sheared soil is likely to be similar, depending mainly on the confining pressure. Casagrande (1936) defined the void ratio of a soil which does not undergo any volume change during shear to be the *critical void ratio*. Later studies confirmed that the critical void ratio is a function of the confining pressure: The critical void ratio decreases with increasing confining pressure. Taking a different point of view, it can also be said that a soil of a particular void ratio has a *critical confining pressure,* at which there is no volume change at failure.

At high confining pressures, sands and other cohesionless soils are likely to show more plastic, rather than brittle, behavior, characterized by high failure strains.

From these observations it can be concluded that shear strength of cohesionless soils is due to actual friction between particles and resistance due to dilatancy. The first is more or less constant; the second is strongly influenced by initial density and confining pressure. In addition, strength can be affected by density changes due to

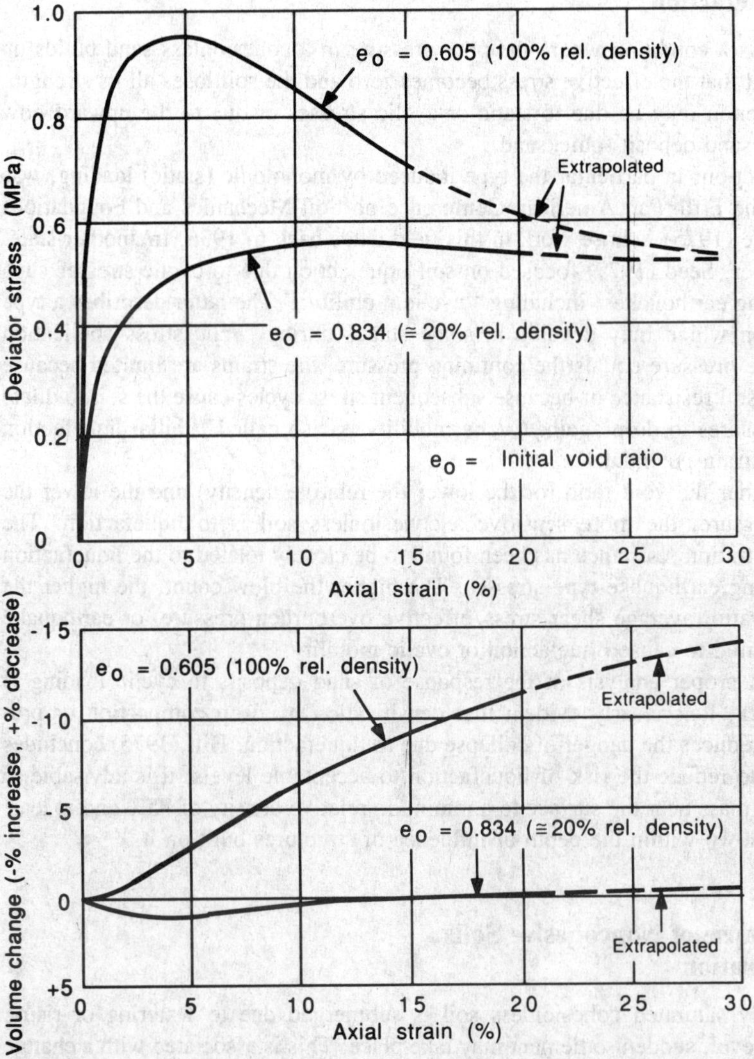

FIGURE 4.12
Typical stress-strain volume change characteristics for a medium fine sand. [*After Taylor (1948) as presented by Lambe and Whitman (1969). Copyright John Wiley & Sons.*]

particle rearrangement following the crushing of sand grains at high confining pressures.

Volume change during shear is of particular significance to saturated noncohesive soils. If there is a tendency for a volume decrease during shear, such as is produced in a loose soil mass during an earthquake, temporarily high pore pressures may lead to liquefaction.

4.2.3 Liquefaction

Liquefaction is a condition where the pore pressure in a cohesionless sand builds up to such a level that the effective stress becomes zero and the soil loses all its strength. This phenomenon may be due to static or cyclic stresses or due to the upward flow of water in a sand deposit (quicksand).

Liquefaction, in particular the type induced by monotonic (static) loading, was reviewed at the Fifth Pan American Conference on Soil Mechanics and Foundations by Casagrande (1975), whose work in this field dates back to 1936. In another state-of-the-art paper, Seed (1979) focused on soil liquefaction due to cyclic stresses such as occur during earthquakes, including "cyclic mobility." The latter describes a type of liquefaction which may develop at some stages during cyclic stress application when the pore pressure equals the confining pressure; the strains are limited because of remaining soil resistance or because subsequent stress cycles cause the soil to dilate and pore pressures to drop again. Cyclic mobility is also called "initial liquefaction with limited strain potential."

The higher the void ratio (or the lower the relative density) and the lower the confining pressure, the more sensitive a cohesionless soil is to liquefaction. The standard penetration resistance has been found to be closely related to the liquefaction potential during earthquake-type stresses. The higher the blow count, the higher the cyclic stress ratio (average shear stress/effective overburden pressure) or earthquake intensity required to cause liquefaction or cyclic mobility.

While a proper analysis of the response of sand deposits to cyclic loading is rather complex, it is clearly evident that densification by deep compaction or pre-compression reduces the danger of collapse due to liquefaction. Hilf (1975) concludes that in order to reduce the risk of liquefaction to acceptable levels, it is advisable to densify a soil mass near the surface to a minimum relative density of 85% and at least 70% further down within the depth of influence of structures built on it.

4.2.4 Collapse of Noncohesive Soils due to Saturation

When partially saturated cohesionless soil is submerged due to watering or rising groundwater level, sudden settlement may take place. This is associated with a change in soil structure and is known to occur in fine uniform soils, loess, crushed chalk, and soils containing soluble salts. The basic mechanisms appear to be a loss of apparent cohesion (which is due to surface tension forces in the pore water) and, less commonly, the destruction of the cementing action of salt solids.

Soil collapse can be demonstrated in an odometer test. If water is allowed access to the specimen at a particular applied load, a sudden increase in settlement is noticed (Fig. 4.13). This settlement may only be 1 to 2% in dense sands, but could reach 5 to 10% in loose soils.

Rizkallah and Hellweg (1980) traced the study of soil collapse back to 1933 in Germany. They identified the following factors influencing the amount of collapse settlement:

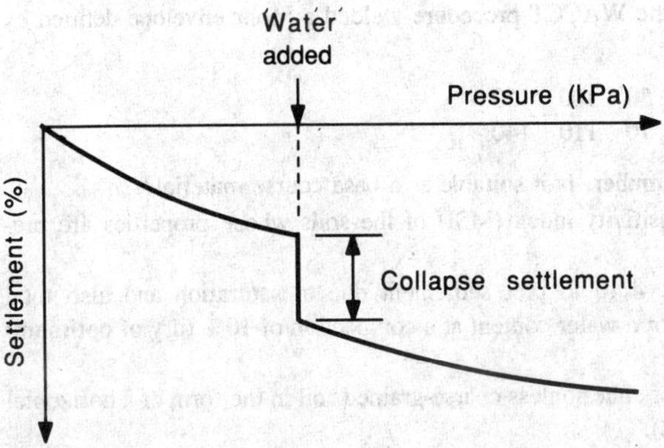

FIGURE 4.13
Demonstration of collapse settlement due to saturation in a one-dimensional consolidation test.

a. Kind of testing apparatus and testing method
b. Grain size distribution and grain shape
c. Density (compactness) of the soil
d. Amount of effective vertical stress
e. Water content before and after flooding
f. First or repeated flooding after drying

Knowledge of the collapse mechanism can be taken advantage of in the compaction of fine uniform sands which are difficult to treat by mechanical means alone.

Settlement upon flooding also occurs in cohesive soils compacted dry of optimum (Sec. 4.1.2.1).

PROBLEMS

Prefixes indicate problem type: C = calculations, M = multiple choice, D = discussion.

Calculations

C4.1. A fine-grained soil has a liquid limit of 12% and a plastic limit of 6%. Determine its group symbol according to the Unified Soil Classification System and indicate typical strength properties according to Table 4.1 after compaction.

C4.2. Grain size distribution and Atterberg limits identified a sample as an SC soil.

(*a*) How much of the soil is coarse (> 0.06 mm), how much is fine, and what is the plasticity of the fines?

(*b*) Give the range of typical properties of this type of soil if it is compacted (Table 4.1), and rank its suitability as a fill material for a water-retaining dam (Table 6.1 on pp. 118–119).

C4.3. Triaxial tests according to the WACCT procedure yielded a Mohr envelope defined by the following points:

Normal stress (kPa)	=	0	50	100	150
Shear stress (kPa)	=	15	70	110	140

Determine the soil's class number. Is it suitable as a base course material?

C4.4. Determine the moisture sensitivity index (MSI) of the soils whose properties are presented in Fig. 4.9.

C4.5. Use the data given in Fig. 4.10 to plot settlement due to saturation and also total settlement versus pressure for a water content at a compaction of 10% (dry of optimum) and 15% (wet of optimum).

C4.6. Plot the ranges of densities of cohesionless coarse-grained soil in the form of a horizontal bar chart (refer to Table 4.2).

(*a*) Which type of soil shows the smallest range of possible densities?
(*b*) Which soils compact to the highest absolute densities?

C4.7. A well-graded sand has a maximum dry density of 2.05 t/m^3 and a minimum dry density of 1.65 t/m^3, determined according to standard test procedures. Assuming $G_s = 2.65$, calculate:

(*a*) Corresponding minimum and maximum void ratios.
(*b*) Compactibility factor according to Terzaghi.
(*c*) The relative density of this sand if its absolute dry density is 1.95 t/m^3.
(*d*) Describe the relative density as calculated in part (*c*) according to Table 4.3 and give a likely corresponding standard penetration resistance (*N* value).

Multiple Choice

M4.8. Which one of the following statements is wrong?

(*a*) Well-graded gravels compact to higher densities than poorly graded gravels.
(*b*) The optimum moisture content of CH soils is higher than that of CL soils.
(*c*) When compacted to equal standards, an inorganic clayey silt is likely to have a lower CBR than a well-graded sand.
(*d*) The subgrade modulus of poorly graded sand is less than that of a low-plastic clay if both are compacted at optimum conditions because the sand has no cohesion.

M4.9. Which one of the following statements is wrong?

(*a*) If left unsoaked, a soil compacted dry of optimum will show less settlement than one compacted wet of optimum.
(*b*) Total settlement due to saturation and compression is smallest if the soil has been compacted at or near optimum water content.
(*c*) If compacted dry of standard optimum, swelling can be reduced by producing the strongest, densest soil possible.
(*d*) Fine-grained soils are more susceptible to frost damage than coarse-grained soils.

M4.10. Which one of the following statements is wrong with respect to cohesionless soils?

(*a*) A soil with a void ratio above the critical void ratio will densify if subjected to shear, according to Casagrande's concept.
(*b*) Liquefaction means an increase in the water content up to the liquid limit.
(*c*) Liquefaction is associated with zero effective stress.
(*d*) Cyclic mobility is also called initial liquefaction with limited strain potential.

Discussion

D4.11. What are the most common laboratory strength tests carried out on compacted soils?

D4.12. Is the CBR a strength or deformation value? Does it have relevance with respect to the long-term performance of road subgrades under repeated loading? (Further reading may be required to discuss this question fully.)

D4.13. Shrinkage of a subgrade can cause pavement damage. What can be done to reduce deleterious effects of shrinking subgrade soils?

D4.14. Define the critical void ratio, liquefaction, and cyclic mobility.

CHAPTER 5

COMPACTION CONTROL TESTS

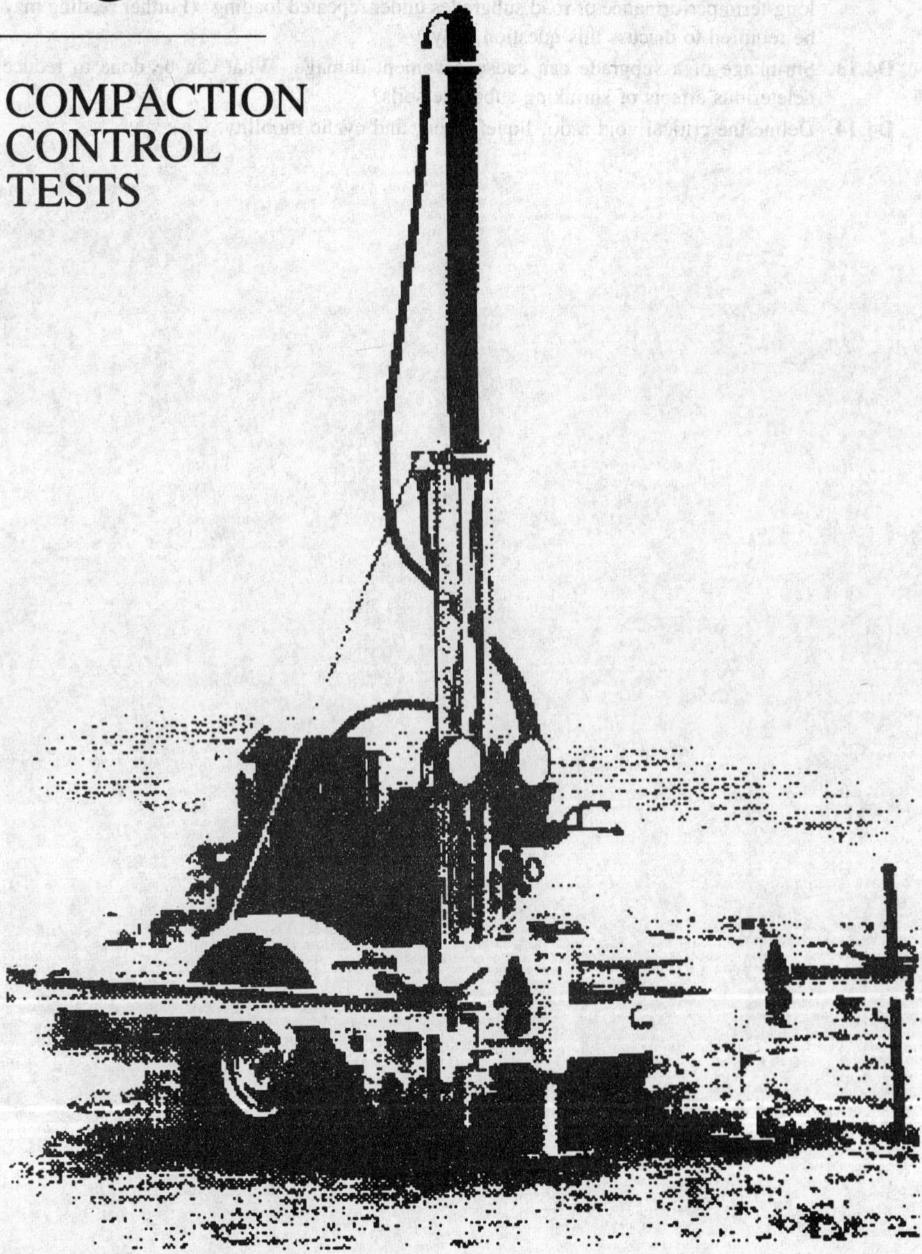

Control tests are designed to cheaply and effectively check whether the objectives of modifying the ground have been achieved. Ideally, actual performance parameters corresponding to the design criteria should be measured, but this is rarely feasible and is usually too expensive. The recourse taken, therefore, is to measure the most basic soil properties, such as water content and density, or determine index properties, such as penetration resistance. In order to subsequently predict actual strength or stress-strain behavior, empirical relationships are then relied upon.

There is a great variety of tests possible for evaluating the effect of mechanical ground modification. In this chapter, they are arbitrarily divided into two groups: those which are carried out at or very close to the surface, and those which can be performed at depth. This division is, however, not necessarily very distinct. For example, the cone penetration test, which could easily reach a depth of 20 m in loose soil, may also serve to evaluate the effectiveness of a heavy roller, as it could well give the compaction effectiveness from 0.5 to 3 m in sandy soils. Similarly, nuclear density determinations could be performed in a borehole, rather than at or very near to the ground surface which is more common in compaction control.

5.1 SHALLOW SURFACE COMPACTION CONTROL TESTS

5.1.1 Direct Density and Water Content Determination

The traditional way of checking whether the field density and water content meet specifications is to test the compacted soil in situ. A small sample of soil is dug out of the compacted layer and weighed. Then it is oven-dried and weighed again. The in situ water content can then be determined. The more difficult problem is to measure the volume of the hole from which the soil was excavated.

One method of determining the volume of the sampled soil is to fill the hole with a selected uniform sand, whose poured dry unit weight is determined by calibration tests. The sand used to fill the hole is weighed and its volume calculated; these values, together with those of the weight of the sample of the compacted soil and its moisture content are used to determine the field dry density. This procedure is described in AS 1289.E3.1 (''Sand Replacement Method Using a Sand-Cone Pouring Apparatus'') and AS 1289.E3.2 (''Sand Replacement Method Using a Sand-Pouring Can'').

Another way of measuring the volume of the hole from which the sample of compacted soil was removed is to seal it with a rubber or plastic membrane and measure the amount of water it can hold, as described in AS 1289.E3.5 (''Water Replacement Method'') and AS 1289.E3.4 (''Balloon Densometer Method''). In the latter method, water from a calibrated container is forced into a rubber balloon until it completely fills the hole. The difference in the volume readings before and after excavating the hole is equal to the hole's volume.

In fine-grained cohesive soil, the in situ density may be assessed by driving a cylindrical steel core cutter of a known volume into the soil, carefully digging it out again, measuring its weight, and determining its moisture content (AS 1289.E3.3).

5.1.2 Hilf Rapid Method

A rapid evaluation of field compaction is significantly delayed, usually for at least 24 h, if the field water content is determined by conventional laboratory testing. Hilf (1959) proposed a method of rapidly assessing compaction without accurate knowledge of the optimum water content. It has been adopted as an Australian Standard (AS 1289.E7.1) although the growing acceptance of nuclear meters and alternative control tests has diminished its importance. The Hilf method, however, still maintains the advantage that the field density is compared with laboratory compaction results on exactly the same soil. This would be particularly important where considerable soil variability exists.

The Hilf procedure is as follows:

1. The field total (or wet) density γ_f is determined using, e.g., the sand replacement or balloon method. The field water content w_f is unknown at this stage.
2. Loose soil excavated at the test site is placed in a watertight container and taken to the laboratory, where three samples are prepared. A first sample is left at the field water content; then 2% water (in terms of total weight) and 4% water are added to a second and third sample, respectively. These three samples are compacted using standard or, if required, modified standard procedure. The resultant total densities γ_L are recorded.
3. So-called converted densities are calculated according to

$$\gamma_c = \frac{\gamma_L}{1 + Z} \tag{5.1}$$

where Z is the percent water added in step 2. The results are plotted as shown in Fig. 5.1 and allow a maximum converted total density γ_{cm} and corresponding added

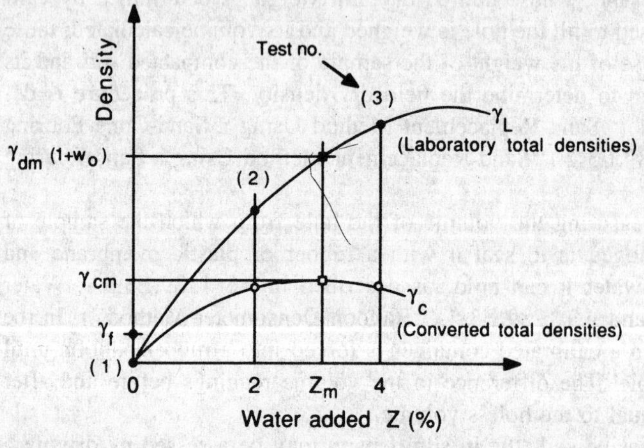

FIGURE 5.1
Hilf rapid method: Laboratory and converted total densities plotted versus water percentage added.

water percentage Z_m to be found, graphically or by fitting a parabola to the test results. Note that if the soil has been compacted wet of optimum, the field sample may have to be dried (Z negative) if a reliable estimate of the maximum converted density has to be made. It can be shown that γ_{cm} is related to the laboratory maximum dry density γ_{dm} according to

$$\gamma_{\mathrm{cm}} = \gamma_{\mathrm{dm}}(1 + w_f) \tag{5.2}$$

4. The following two measures of the effectiveness of field compaction can now be calculated:

Hilf density ratio
$$R_{\mathrm{HD}} = \frac{\gamma_f}{\gamma_{\mathrm{cm}}} = \frac{\gamma_{\mathrm{df}}}{\gamma_{\mathrm{dm}}} \tag{5.3}$$

Roller efficiency
$$E_R = \frac{\gamma_f}{\gamma_c(\text{at } w_f)} = \frac{\gamma_f}{\gamma_c(1)} = \frac{\gamma_f}{\gamma_L(1)} \tag{5.4}$$

5. The laboratory maximum total density γ_{Lm} is computed from known values using the expression

$$\gamma_{\mathrm{Lm}} = \gamma_{\mathrm{cm}}(1 + Z_m) \tag{5.5}$$

but the maximum dry density γ_{dm} and optimum moisture content w_o are still unknown. Hilf suggests that w_o be estimated from an empirical relationship between the maximum total density γ_{Lm} and the optimum moisture content w_o, as can be found for a particular compactive effort. A polynomial fit to data presented by Hilf (1975) yields the relationship

$$w_o = 7.2257 - 7.3062\gamma_{\mathrm{Lm}} + 1.876\gamma_{\mathrm{Lm}}^2 + 0.18975\,\gamma_{\mathrm{Lm}}^3 - 0.091171\gamma_{\mathrm{Lm}}^4 \tag{5.6}$$

It would be desirable to develop a similar relationship for a specific project or group of soils in order to reduce the error inherent in such an estimate.

The difference between the optimum water content and the field water content can now be obtained from

$$w_o - w_f = \frac{Z_m(1 + w_o)}{1 + Z_m} \tag{5.7}$$

Alternatively (AS 1289.E7.1), we determine this difference from

$$w_o - w_f = Z_m + \text{correction value} \tag{5.8}$$

where the correction value can be estimated from Fig. 5.2, based on the added moisture Z_m and the maximum converted density γ_{cm}.

6. The day after the above tests and calculations, which provided a quick evaluation of the compaction achieved in the field, are made, the field water content w_f is

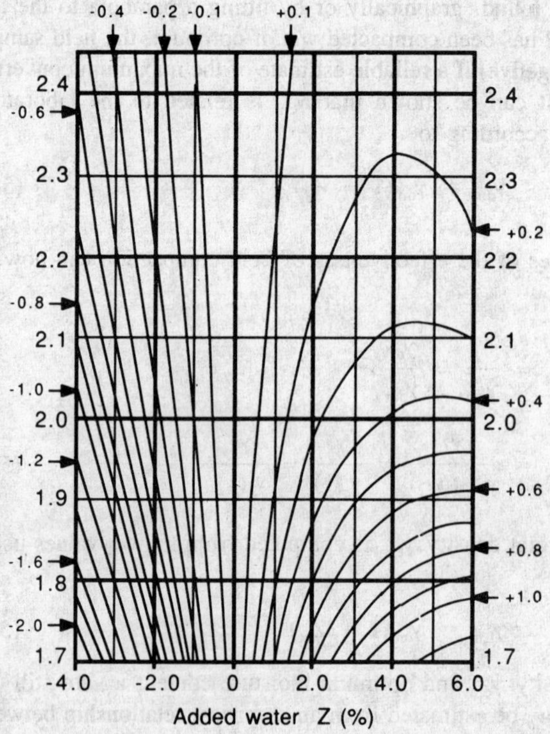

FIGURE 5.2
Correction value based on maximum converted density and added water content. *(AS 1289.E7.1.)*

available and field dry density $\gamma_{\rm df}$ and laboratory maximum dry density $\gamma_{\rm dm}$ as well as the optimum moisture content w_o can be determined:

$$\gamma_{\rm df} = \frac{\gamma_f}{1 + w_f} \tag{5.9}$$

$$\gamma_{\rm dm} = \frac{\gamma_{\rm cm}}{1 + w_f} \tag{5.10}$$

$$w_o = w_f + (1 + w_f)Z_m \tag{5.11}$$

7. The estimated difference between the field water content and the optimum water content can now be compared with the actual difference. As a check, the laboratory dry density achieved at field water content can be calculated and the Hilf density ratio can be verified.

5.1.3 Nuclear Meters

Originally developed for density and hydrocarbon measurements in oil exploration, nuclear meters have been gaining increasing acceptance in compaction control. Speed and consistency of results have made nuclear density and moisture meters an economic alternative to traditional test methods.

One possible problem involved with nuclear meters is the excessive concern, if not fear, of the uninformed public and some transport authorities when sighting the radioactivity warning labels on the equipment.

The use of nuclear meters is covered by government regulations relating to radiation safety. Only trained and licensed personnel are allowed to use them. Care has to be taken during storage and transport. Operators must wear film badges which measure their exposure to radiation, just like x-ray technicians. Over 30 years of experience with nuclear density and moisture meters has proven them to be safe if they are handled properly by trained staff.

For *density measurements* a radioactive source such as cesium 137 which emits gamma rays is placed on the surface of or at a shallow depth into the compacted soil. Geiger-Müller tubes are used to detect how many photons making up the gamma rays are reflected to the surface (backscatter mode) or are transmitted from the depth source to the surface (direct transmission mode) rather than being absorbed by the soil (Fig. 5.3). Absorption of gamma rays is largely a function of the total density of the soil or rock. It also depends on the constituents of the soil being tested, and this is why

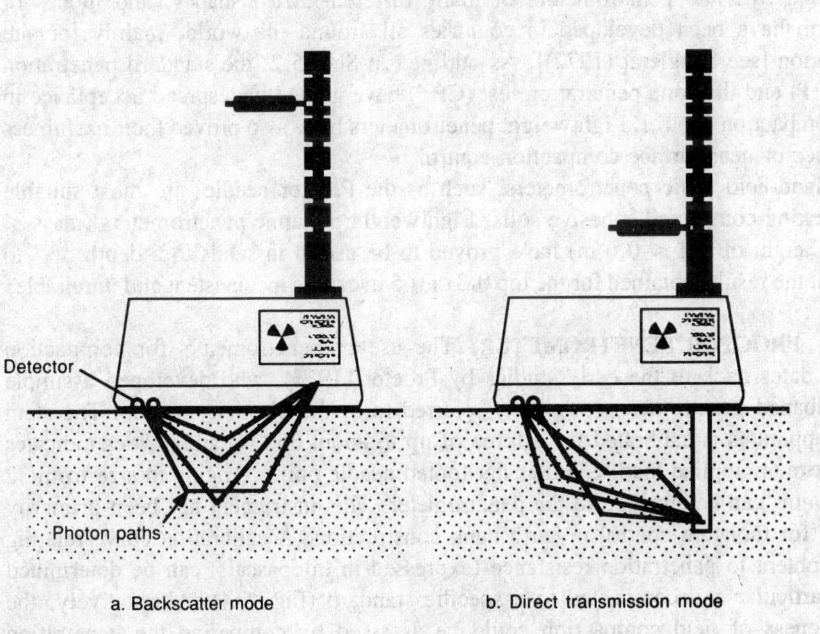

FIGURE 5.3
Nuclear soil density measurement. *(Based on an illustration by Campbell Pacific Nuclear.)*

the meter should be calibrated using typical construction materials. Gauge readings can then be automatically converted into total density values using microprocessors.

For *moisture content measurements* a source of high-velocity neutrons, such as americium 241:beryllium, is employed in the backscatter mode. When fast neutrons collide with hydrogen nuclei, they are slowed down. The quantity of slow neutrons can be measured by boron trifluoride or helium-3 detection tubes. The moisture content is related to the total hydrogen content of the soil and gives a good evaluation of the water content per unit volume (*note:* Not per mass of soil solids) provided there is no organic matter, oil, bitumen, or other hydrogen-containing material or certain other chemical elements present. It is advisable to verify nuclear moisture meter readings with water contents determined by standard laboratory methods for each of the typical materials encountered on a site.

Studies by the Main Roads Department of Western Australia [Hamory (1982)] have indicated that density calibration of nuclear meters is best achieved using standard blocks of uniform rock. This organization uses six different blocks ranging in density from 1.3 (calcareous eolianite) to 2.66 (granite). An accuracy of at least 0.01 t/m^3 is expected over this density range. For calibration of moisture measurements, standard blocks are made from dry-sand–polystyrene blends with the hydrogen of the polystyrene simulating that of the water.

5.1.4 Shallow Penetration Tests

Static and dynamic penetrometers of many different sizes, shapes, and modes of operation have been developed in countries all around the world, mainly for site exploration [see Sanglerat (1972)]. As outlined in Sec. 5.2, the standard penetration test (SPT) and the cone penetration test (CPT) have gained widespread acceptance in deep compaction control. Lightweight penetrometers have also proved their usefulness in surface or near-surface compaction control.

Hand-held static penetrometers, such as the Proctor needle, are most suitable for checking compacted cohesive soils. Lightweight dynamic penetrometers (mass $<$ 10 kg, height of fall $<$ 0.6 m) have proved to be useful in sands to a depth of 4 to 5 m, but the results obtained for the top 0.3 or 0.5 m can be inconsistent and unreliable.

5.1.4.1 PROCTOR PENETROMETER. The use of penetrometers for compaction control dates back to the early studies by Proctor (1933), who developed a simple spring-loaded penetrometer, originally referred to as the plasticity needle. The stem of the apparatus is calibrated for a force of up to about 650 N, and a rider indicates the maximum applied load. Circular flat-footed needle points ranging in size from 32 to 645 mm^2 can be attached to the penetrometer. This instrument has been primarily devised for moisture control in earth bank construction. A calibration curve relating water content to penetration resistance (expressed in kilopascals) can be determined for a particular soil compacted to a specific standard (Fig. 5.4). Alternatively, the effectiveness of field compaction could be assessed by comparing the penetration resistance of a sample compacted according to standard laboratory procedures with the field penetration resistance at the same water content.

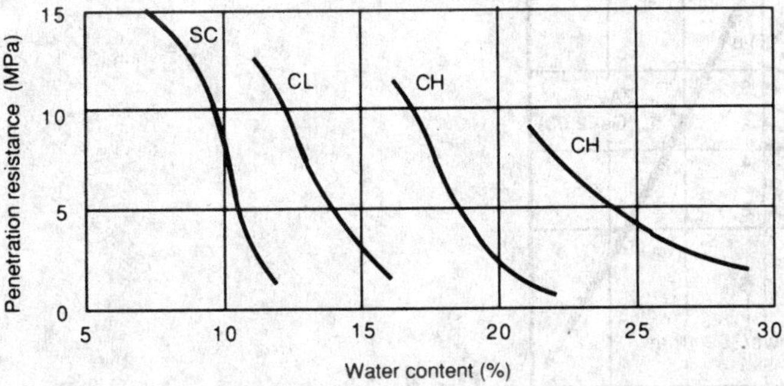

FIGURE 5.4
Examples of Proctor penetrometer calibration curves.

The Proctor needle is only suitable for fine-grained or sandy soils. The presence of gravel-size particles makes the test unreliable. The Proctor penetrometer is standardized by ASTM D1558. The value of penetration resistance expressed in kilopascals and divided by 350 gives an approximate value for the CBR of a cohesive soil, provided the CBR value is less than 10.

5.1.4.2 PERTH SAND PENETROMETER. The Perth sand penetrometer is similar to the Scala penetrometer originally used by the Road Construction Authority in Victoria [Scala (1956)] but differs from this and many other similar penetrometers in the shape of the tip: it has a blunt end, rather than a conical tip. Its dimensions (Fig. 5.5) have been accepted as an Australian Standard (AS 1289.F3.3-1984). It has proved very suitable for the medium-sized sands common around Perth, in Western Australia, where it is extensively used to assess the density of sand fill in newly developing

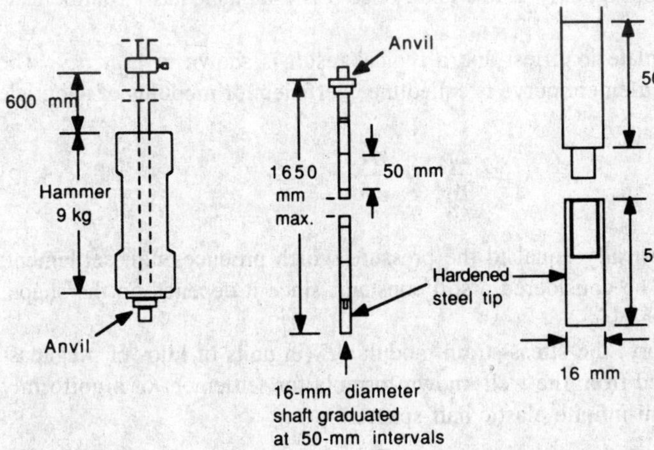

FIGURE 5.5
Dimensions of the Perth sand penetrometer. *(AS 1289.F3.3-1984.)*

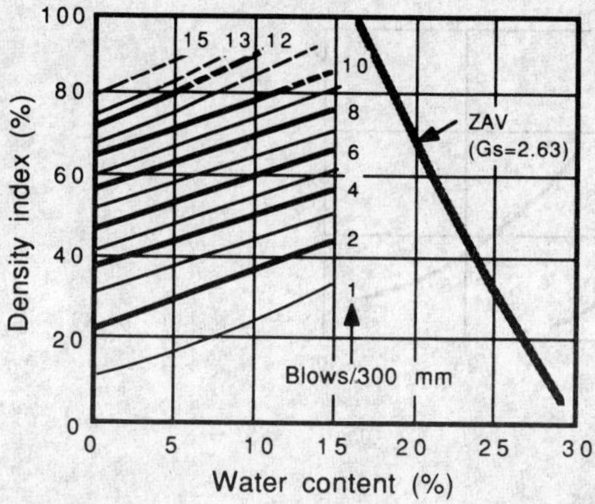

FIGURE 5.6
Calibration chart for Perth sand. [*Glick and Clegg (1965).*]

areas. Glick and Clegg (1965) established a calibration chart relating the blow count to the density index (relative density) of Perth sand at a particular water content (Fig. 5.6).

5.1.5 Plate Bearing Tests

Plate bearing tests allow a *direct evaluation* of the stiffness and strength of a compacted soil layer, which is most valuable in road construction. These tests may be performed on or just below the surface of the subgrade, base course, or surface course. Other advantages of the plate load test are

The results are available immediately.
No laboratory compaction is required.
It is suitable for variable materials.
It can be carried out quickly at low cost, once it is established as a routine test.

The setup for a plate load test and a typical result is shown in Fig. 5.7. The slope of the pressure-settlement curve is called the coefficient or modulus of subgrade reaction:

$$k_s = \frac{\Delta p}{\Delta s} \quad \frac{kN}{\text{m}^3} \tag{5.12}$$

Its value is numerically equal to the pressure which produces unit settlement; the slope k_s should not be considered a soil constant, since it depends on the shape, size, and stiffness of the plate.

Using elastic theory, the stress-strain modulus E (in units of kilo- or megapascals) can be back-figured from the well-known formula for settlement of a uniformly loaded footing on a semi-infinite elastic half-space:

$$s = \frac{IpD(1 - \upsilon^2)\alpha_c}{E} \tag{5.13}$$

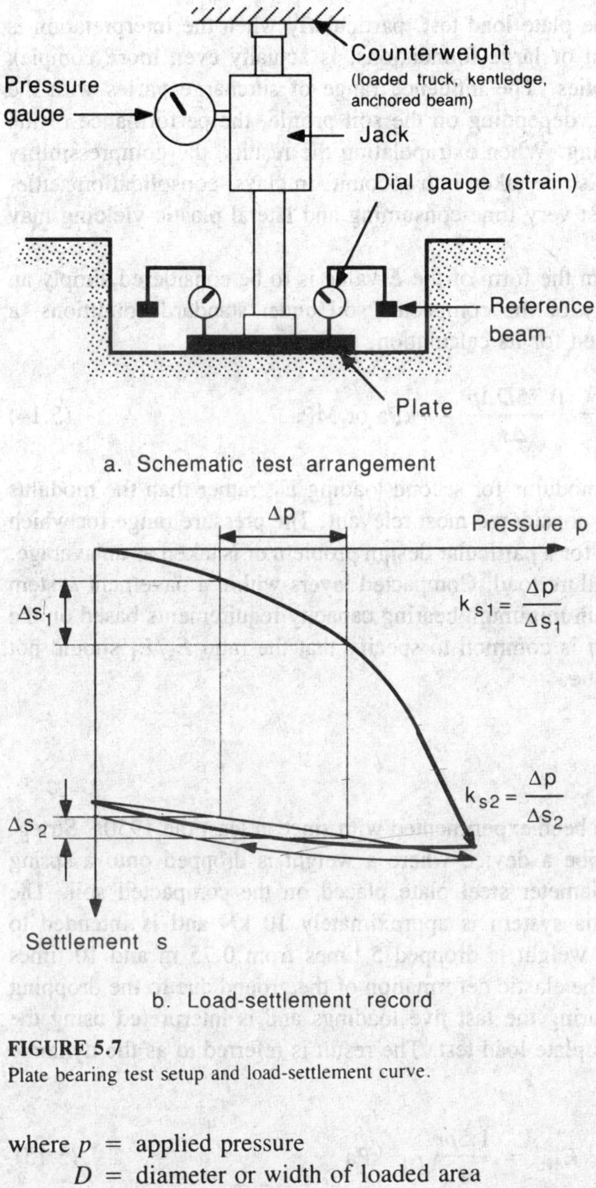

FIGURE 5.7
Plate bearing test setup and load-settlement curve.

where p = applied pressure
D = diameter or width of loaded area
υ = Poisson's ratio of soil
I = influence value
α_c = correction factor

The value of I depends on the shape of the footing and its rigidity: For a circular rigid plate $I = 0.79$ ($\pi/4$). The coefficient α_c is a correction factor for the depth of the embedment of the footing [see, e.g., Wilun and Starzewski (1972)]; for $D = 0$, $\alpha_c = 1$.

The interpretation of the plate load test, particularly when the interpretation is used to predict the settlement of large foundations, is actually even more complex than the above equation implies. The influence range of surcharge varies with the width of the loaded area and, depending on the soil profile, the performance of the small plate could be misleading. When extrapolating the results, the compressibility of the particular soil should also be taken into account. In clays, consolidation settlement is likely to make the test very time-consuming and lateral plastic yielding may distort the result.

If the result of the test in the form of the E value is to be considered simply an indicator of the deformability of the compacted soil under standard conditions, a simplified formula may be used for its calculation, such as

$$E = \frac{0.75D\Delta p}{\Delta s} \quad \text{kPa or MPa} \tag{5.14}$$

For compaction control, the modulus for second loading E_2, rather than the modulus for first loading E_1, is usually considered most relevant. The pressure range for which E is determined is either fixed for a particular design problem or is taken as an average, say from 30 to 70% of the failure load. Compacted layers within a pavement system may also have to satisfy certain minimum bearing capacity requirements based on the plate load test. In addition, it is common to specify that the ratio E_2/E_1 should not exceed a given maximum value.

5.1.6 Impact Tests

Dynamic plate load tests have been experimented with since at least the 1950s. Striegler and Werner (1973) describe a device where a weight is dropped onto a spring connected with a 350-mm-diameter steel plate placed on the compacted soil. The dynamic force exerted by this system is approximately 10 kN and is intended to simulate traffic loading. The weight is dropped 5 times from 0.75 m and 10 times each from 1 m and 1.25 m. The elastic deformation of the ground due to the dropping of the weight is measured during the last five loadings and is interpreted using the same formula as for the static plate load test. The result is referred to as the dynamic soil modulus E_{dyn}:

$$E_{dyn} = \frac{1.5pr}{s} \quad \text{kPa} \tag{5.15}$$

where r = radius of plate
s = elastic settlement
p = applied dynamic pressure

The value p_{dyn} is calculated from the theoretical dynamic force P:

$$P_{dyn} \approx \sqrt{2ghmc} \tag{5.16}$$

where g = acceleration of gravity
h = drop height
m = drop mass
c = spring constant

Although there is no certain relationship between the static modulus determined in the standard plate bearing test and the dynamic modulus E_{dyn}, it can be said that E_{static} is always smaller than E_{dyn}. It has been proposed that E_{dyn} be used as a design value; but E_{dyn} is more likely to serve as a control parameter, particularly if it can be related to the results of more standard tests, such as in situ density tests, on a particular site.

The advantage of the dynamic test is that it is quick and economical and thus allows a larger number of control tests to be carried out than by using more traditional procedures. The dynamic test is, however, considered to be more suitable for cohesionless soils than for cohesive soils. Because of the generation of pore pressures during short-term loading of clays, the dynamic test may indicate higher than actual densities and stiffness for this type of soil.

In Australia and the United States, the *Clegg impact soil tester* is still gaining in popularity. It consists of a 4.5-kg hammer dropping 450 mm. The hammer is similar in size to the piston used in the CBR test. An accelerometer attached to the hammer generates a signal upon impact which is displayed on a digital meter. The factory calibration can be checked before the tester is used. The device is used in the field as well as in the laboratory, e.g., on soil compacted in a CBR mold. According to the manufacturer, percent compaction may be estimated by determining the "impact value" needed to achieve the desired density level for the given material and field moisture content. It is even suggested that this test be used as an alternative to the CBR test.

5.2 IN SITU EVALUATION OF DEEP COMPACTION

The effectiveness of heavy tamping or depth vibration is readily apparent to an observer of the construction process. Settlement records, measurements of impact craters, requirements of imported fill to achieve a certain grade, and energy consumed by the equipment are just some of the indicators which assist a project engineer in the control of mechanical ground modification. Nevertheless, it is desirable to perform tests which yield basic soil performance parameters, either directly or through empirical correlations.

The most common tests which can be carried out at depth are penetration tests (standard and cone penetration tests), pressuremeter tests, and dilatometer tests. Seismic-wave velocity measurements before and after ground improvement can also give an indication of the effectiveness of the procedure and the extent and uniformity of the treated zone.

The following sections give a description of these in situ tests and present the basic correlations needed for translating index-like results into performance predictions, such as settlement calculations. The primary application of all the tests is in general site investigations, but this discussion places emphasis on aspects related to the evaluation of deep compacted layers.

5.2.1 Deep Penetration Tests

Penetration tests can generally be categorized into either dynamic or static tests, depending on whether the standard probe is driven or slowly pushed (quasi-static) into the ground. Many different shapes and sizes are used around the world, even on the moon (Luna 13, in 1966), but two types of test have achieved the greatest degree of standardization and are in most widespread use: the dynamic *standard penetration test (SPT),* where the resistance is expressed in blows per 300 mm (originally blows per foot); and the static *Dutch cone test,* where pressure measurements are recorded, usually in units of MPa. The SPT has been accepted as an Australian Standard (AS 1289.F3.1-1977) because of this method's simplicity and the usefulness of the results. Dimensions of the typical split-tube sampler used in the SPT are shown in Fig. 5.8. Standardization of the basic static cone test has been attempted by ASTM (D.18.02.07-1971) and in Europe, where it is now generally referred to as the *cone penetration test (CPT),* although the pioneering work of the Dutch engineers is well-recognized. Figure 5.9 gives the details of the electrical friction-cone penetrometer tip used by the Road Construction Authority in Victoria.

The SPT is used extensively in the English-speaking world, particularly for sandy soils. In recent times, more effort has gone into improving and extending the CPT rather than the SPT. The cone penetration test not only has the advantage of giving a continuous record, but also allows additional information to be measured, such as the frictional resistance of a sleeve around the rod above the cone and the pore pressure in the soil.

Although penetration testing has been around for some time, the correlation of penetration resistance to relative density, strength, and compressibility of soils is still controversial. Much can be learned from the book by Sanglerat (2d ed., 1979) and the proceedings of specialty conferences such as the First and Second European Sym-

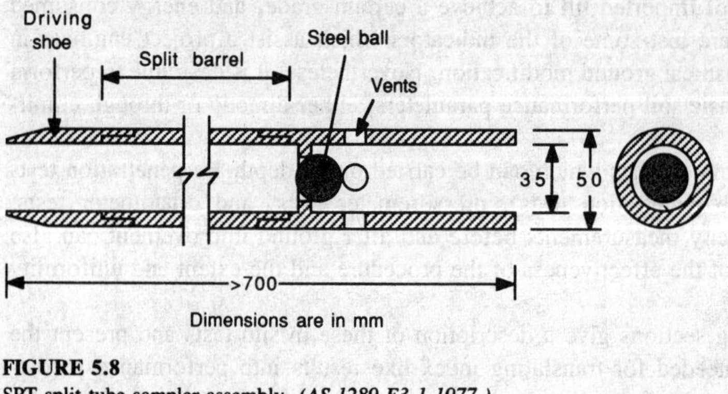

FIGURE 5.8
SPT split-tube sampler assembly. *(AS 1289.F3.1-1977.)*

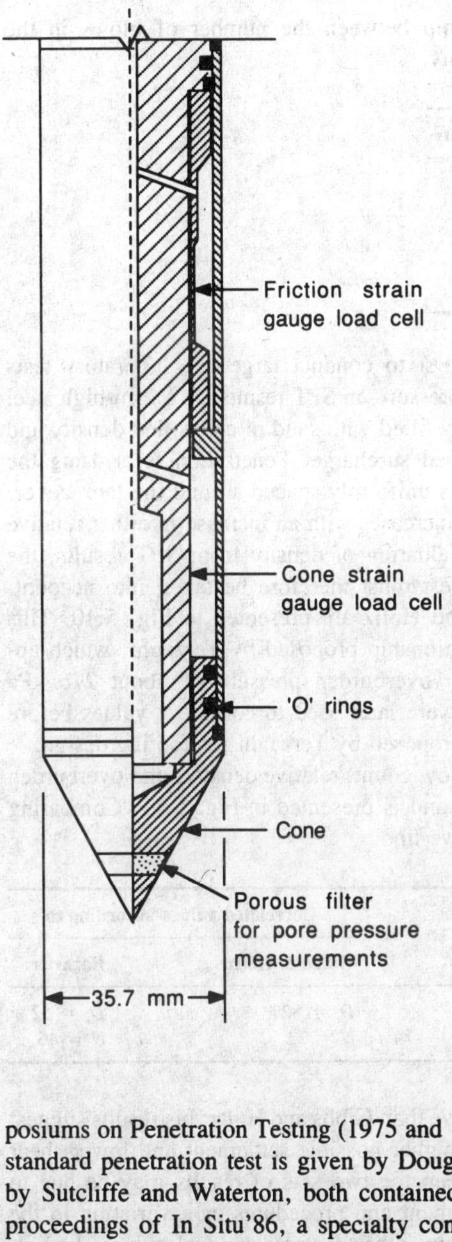

FIGURE 5.9
Typical electrical friction cone penetrometer tip.

posiums on Penetration Testing (1975 and 1982, respectively). A recent review of the standard penetration test is given by Douglas, and of quasi-static penetration testing by Sutcliffe and Waterton, both contained in the text edited by Ervin (1983). The proceedings of In Situ'86, a specialty conference in the use of in situ tests in geotechnical engineering, provides further valuable information on tests relevant for the evaluation of the results of ground modification [Clemence (1986)].

5.2.1.1 STANDARD PENETRATION RESISTANCE, RELATIVE DENSITY, AND OVERBURDEN PRESSURE. Recognizing that the relative density of sand strata has a decisive influence on the angle of internal friction of the sand, Terzaghi and Peck

(1948) suggested the following relationship between the number of blows in the standard penetration test and relative density:

N, blows per 300 mm	Relative density
0–4	Very loose
4–10	Loose
10–30	Medium
30–50	Dense
Over 50	Very dense

Gibbs and Holtz (1957) were the first to conduct large-size laboratory tests investigating the effect of the overburden pressure on SPT results. A 1.2-m-high steel tank with a diameter of 0.9 m was carefully filled with sand of controlled density and water content and subjected to a prescribed surcharge. Penetration tests using the standard split spoon were made in six holes uniformly spaced around the tank cover. They found that the penetration resistance increases with an increase in either relative density or overburden pressure. In the evaluation of density from SPT results, the overburden pressure at the depth of the test must therefore be taken into account. Some of the results obtained by Gibbs and Holtz are presented in Fig. 5.10. This graph also shows, approximately, the relationship proposed by Terzaghi, which appears to closely follow the curve for an overburden pressure of about 276 kPa (40 lb/in^2). These findings, by the way, were later used to correct N values before using some of the empirical design rules proposed by Terzaghi for footing design.

A new set of relationships between blow count, relative density, and overburden pressure was proposed by Bazaraa (1967) and is presented in Fig. 5.11. Comparing Gibbs and Holtz and Bazaraa guidelines, we find:

Given conditions	Correlated values according to	
	Gibbs & Holtz	Bazaraa
N = 40; effective vertical pressure = 192 kPa	D_r = 82%	D_r = 62%
D_r = 70%; effective vertical pressure = 140 kPa	N = 23	N = 46

Bazaraa is seen to be more conservative than Gibbs and Holtz; his results suggest lower relative densities and consequently higher possible settlement and lower shear strength. Some of the discrepancies between the two sets of results may be due to slightly different penetration testing equipment and procedures or a variation in the evaluation of the relative density, but many other factors could also have had an influence on the results.

It is also known that the in situ horizontal pressure affects the penetration resistance [Lacroix and Horn (1973) and others]. According to these findings, Gibbs and Holtz's values would be most appropriate for normally consolidated soils, and Bazaraa's relationships would be more accurate for overconsolidated sand, with higher

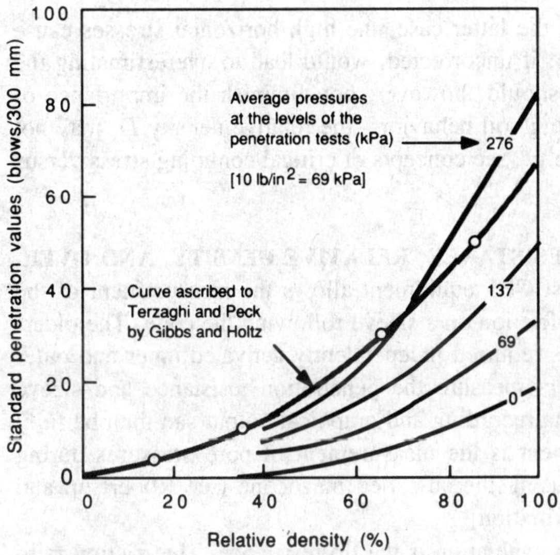

FIGURE 5.10
Relationship between standard penetration resistance, overburden pressure, and relative density for dry sand or saturated coarse sand. [*After Gibbs and Holtz (1957).*]

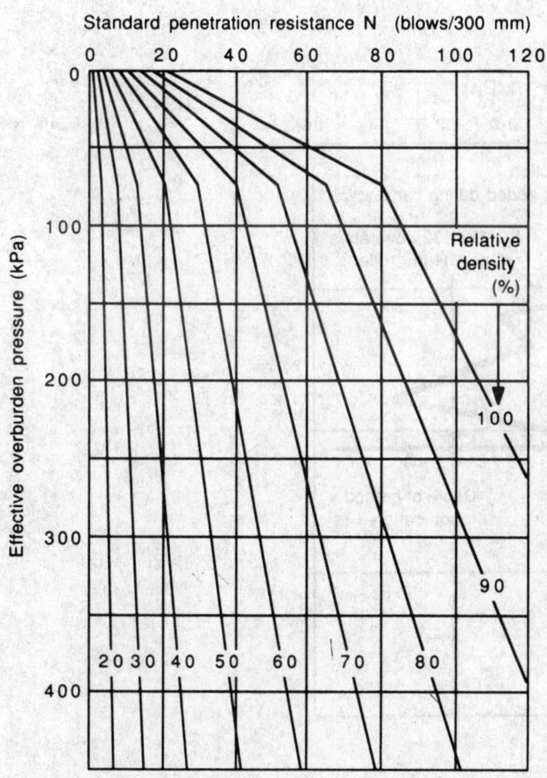

FIGURE 5.11
Relationship between standard penetration resistance, relative density, and overburden pressure. [*After Bazaraa (1967).*]

than normal horizontal stresses. In the latter case, the high horizontal stresses cause higher penetration resistance which, if uncorrected, would lead to overestimating the relative density. These comments should, however, not diminish the importance of confining stress per se in determining soil behavior—the relative density D_r may not be the only parameter to consider (e.g., see concepts of critical confining stress versus critical void ratio in Sec. 4.2.2).

5.2.1.2 CONE PENETRATION RESISTANCE, RELATIVE DENSITY, AND OVERBURDEN PRESSURE. Today, most CPT equipment allows the measurement of the cone penetration resistance and the friction on a sleeve following the cone. The older-style mechanically operated devices required independently activated inner and outer rods. Newer types of penetrometers measure the penetration resistance and sleeve friction using electric load cells; data recording and graphical output can then be fully computerized. The latest development is the measurement of pore pressures during the quasi-static cone penetration, using the so-called piezocone [see Robertson and Campanella (1986) for further information].

Most relevant for compaction evaluation is the tip resistance. The friction ratio

$$F_R = \frac{\text{sleeve friction}}{\text{cone resistance}} \tag{5.17}$$

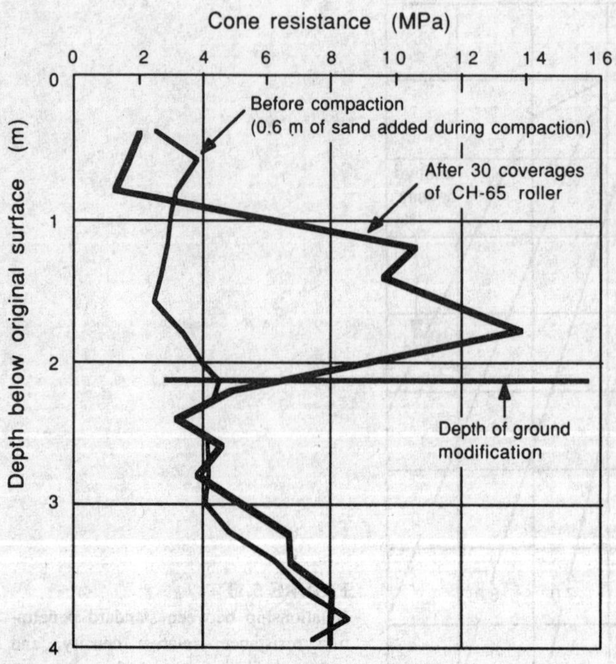

FIGURE 5.12
Static cone penetrometer definition of effect of roller compaction in fine sand below the water table. [*Hogentogler & Company (1985).*]

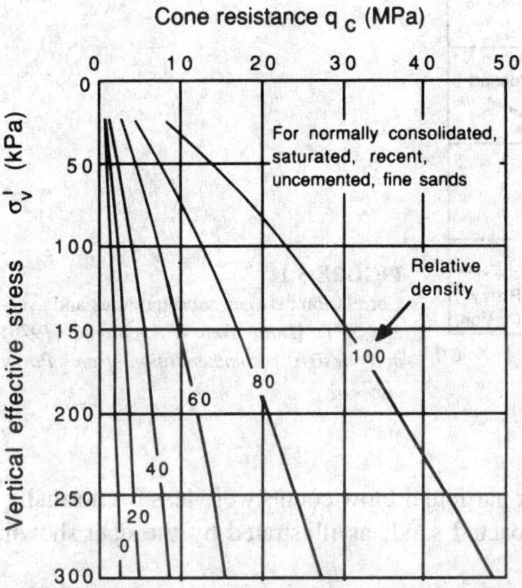

FIGURE 5.13
Cone resistance and relative density. [*Schmertmann (1978).*]

may or may not increase after mechanical modification; the magnitude of any change could significantly depend on the soil characteristics and type of penetrometer tip used [Solymar and Reed (1986)].

The friction ratio assists, together with the cone resistance, mainly in the identification of the soil profile during the investigation stage, and so does the pore pressure ratio, which is defined as

$$P_R = \frac{\text{pore pressure during penetration}}{\text{cone resistance}} \tag{5.18}$$

Sand exhibits lower P_R values (typically around 0.1) than clays (about 1). Because of dilatancy during shear, dense sands may even exhibit negative pore pressures during penetration.

The CPT gives a continuous record with depth and is thus well suited for evaluating the effect of deep compaction by vibration, tamping, or explosion. Closely spaced measurements make it also suitable for evaluating the depth effect of heavy rollers (Fig. 5.12).

Figure 5.13, produced by Schmertmann (1978), makes it possible to estimate the relative density from the cone resistance and the effective overburden stress for normally consolidated sands. Overconsolidation and cementation increases the cone resistance; for these conditions the use of Fig. 5.13 may therefore lead to an overestimation of the relative density.

5.2.2 Compressibility Estimates from Penetration Tests

5.2.2.1 SOIL MODULUS AND SPT RESULTS. D'Appolonia (1970) correlated Young's moduli, back-figured from the observed settlement of footings, to the average

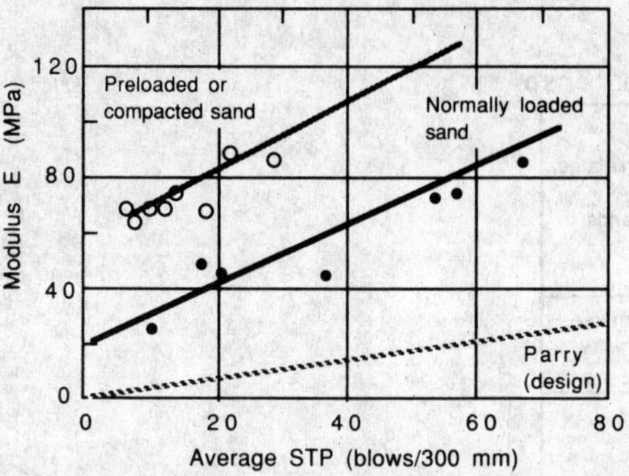

FIGURE 5.14
Correlation between sand modulus and average SPT. [*Data from D'Appolonia (1970) and design recommendation from Parry (1977).*]

SPT values. The E values determined for particular blow counts were less for normally loaded sand than for preloaded or compacted sand, as illustrated by the data shown in Fig. 5.14.

Parry (1977) suggested that settlements for footing design purposes be estimated according to

$$s = \frac{0.3qB}{N} \tag{5.19}$$

where s = settlement, mm
B = width of footing, m
q = applied load, kPa
N = blows per 300 mm

which corresponds approximately to estimating Young's modulus from

$$E = 2.8N \quad \text{MPa} \tag{5.20}$$

This relationship is also shown in Fig. 5.14.

5.2.2.2 STRESS-STRAIN PARAMETERS FROM CONE PENETRATION RESISTANCE. Schmertmann (1978) suggested that the compressibility of normally consolidated sands found below footings is directly related to the cone resistance q_c as follows:

$$E_s = \begin{cases} 2.5q_c & \text{for } L/B = 1 \text{ to } 2 \quad (5.21) \\ 3.5q_c & \text{for } L/B > 10 \quad (5.22) \end{cases}$$

where E_s = constrained modulus, MPa
L = length of footing
B = width of footing

Uncertainties still exist with respect to the compressibility of overconsolidated soils. Sand deposits compacted using deep compaction methods are likely to behave like overconsolidated soils. This makes the use of indicators such as relative density and compressibility, deduced from the same graphs for conditions before and after compaction, somewhat questionable. Nevertheless this is currently an accepted procedure.

Errors inherent in the elastic analysis of settlements make it understandable that correlations between E values and penetration resistance are different for footings and other foundation types. Douglas (1982) reports that settlements of auger-grouted piles during load tests suggest a relationship like

$$E = 5q_c \quad \text{MPa} \tag{5.23}$$

The cone resistance value can also be used to compute settlements according to an equation derived from Terzaghi's consolidation theory:

$$\frac{\Delta h}{h} = \beta \frac{\sigma'_{v0}}{q_c} \log \frac{\sigma'_{v0} + \Delta\sigma_v}{\sigma'_{v0}} \tag{5.24}$$

where Δh = compression of soil layer of thickness h
σ'_{v0} = effective overburden pressure
$\Delta\sigma_v$ = increase in vertical stress
β = empirical factor

The compression index C_c can thus be expressed as

$$C_c = \frac{\beta\sigma'_{v0}}{q_c}(1 + e_0) \tag{5.25}$$

where e_0 is the initial void ratio.

Analyzing results of settlement measurements of full-scale structures, Thorne and Burman (1969) concluded that for estuarine deposits of sands β has an average value of about 1 ($\pm$10%). (For clays, the average β was 0.7.)

5.2.3 Stress-Strain Modulus from Pressuremeter Tests (PMT)

The pressuremeter test is a load test carried out inside a borehole. As pressure is applied to the walls of the borehole, its deformation is monitored. Analysis of the data yields information on the stress-strain modulus, failure strength, and the in situ stress state of the soil or rock mass.

The concept of borehole pressure tests dates back to about 1930, when Kogler [see Baguelin et al. (1978)] reported the use of a cylindrical bladder inflated by gas pressure. The modern type of pressuremeter was developed by Menard in the 1950s; in 1955 the first pressuremeter patent was registered in France. Using Menard's device, the expansion of the borehole is deduced from volume readings of the pressurized measuring cell. Later versions of the pressuremeter developed elsewhere used an electrical displacement transducer to record radial expansion; an example is the Coffey pressuremeter described by Ervin et al. (1980). Also, in order to minimize soil disturbance, self-boring pressuremeters were developed.

5.2.3.1 MENARD PRESSUREMETER. A schematic diagram of the equipment used is given in Fig. 5.15. It has two components: the probe and the volumeter. The probe is a radially expandable cylindrical probe made up of three independent cells. The central cell is filled with water from the volumeter under controlled pressure. The two guard cells are inflated with gas automatically maintained at a slightly lower pressure. This arrangement ensures radial expansion and uniform pressure at the level of the central measuring cell.

Before the results are plotted as shown in Fig. 5.16, a number of corrections are applied. Pressure readings are corrected for

Water head between gauge and probe (hydrostatic pressure)
Pressure required to inflate probe (membrane resistance)

Volume readings are corrected for

Expansion of tubing and measuring equipment
Compression of the probe membrane
Compression of the measuring fluid

A typical plot of corrected pressure and volume readings is shown in Fig. 5.16
where p = pressure acting on soil (corrected for hydrostatic pressure and membrane resistance)
V = volume of water in measuring system (corrected for expansion and compression of equipment)
V_c = volume of measuring cell
AB = elastic range of deformation (straight-line portion of test)
V_0 = initial size of cavity
p_0 = pressure at start of straight-line portion of curve, often assumed to be equal to at-rest pressure
V_f = volume reading at end of elastic range
p_f = creep pressure
V_m = volume reading at midpoint of elastic range
C = point of curve denoting failure to soil
p_1 = limit pressure (at which actual or extrapolated curve is horizontal); corresponds to failure in soil

Menard finds p_1 by stipulating that the volume change $(V_1 - V_0)$ between points A and C is equal to the initial size of the cavity, unless creep occurs earlier. This is equivalent to defining failure in terms of limiting deformation.

The stress-strain modulus E is computed from the expression describing the radial expansion of a cylindrical cavity in an elastic medium, which reads:

$$\frac{\Delta r}{r} = \frac{1 + v}{E} \Delta p \tag{5.26}$$

where v is Poisson's ratio.

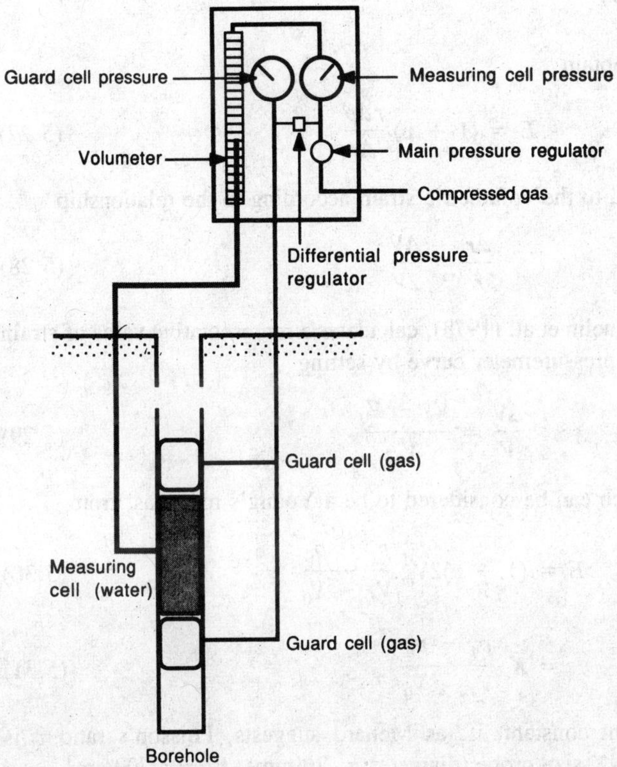

FIGURE 5.15
Schematic diagram of Menard pressuremeter.

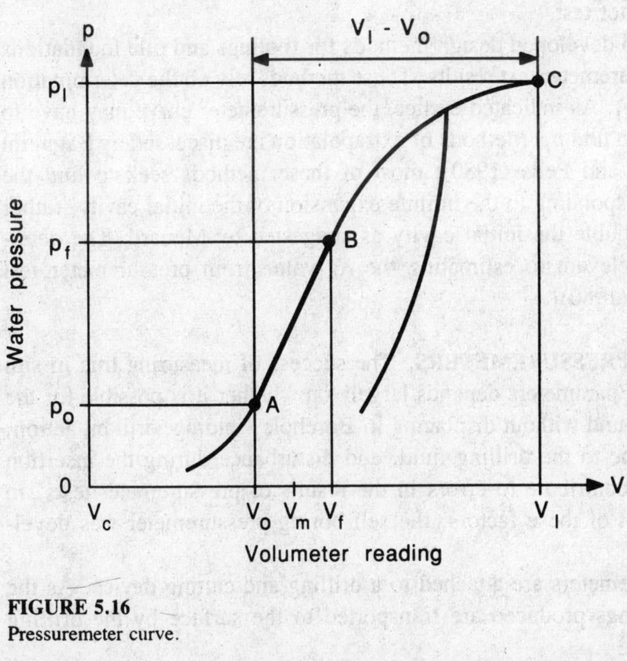

FIGURE 5.16
Pressuremeter curve.

Solving for E, we obtain

$$E = (1 + v)\frac{r\Delta p}{\Delta r} \tag{5.27}$$

The radial strain is related to the volumetric strain according to the relationship

$$\frac{\Delta r}{r} = \frac{\Delta V}{2V} \tag{5.28}$$

Menard, according to Baguelin et al. (1978), calculates a representative value of strain for the *AB* portion of the pressuremeter curve by setting

$$\frac{\Delta V}{V} = \frac{V_f - V_0}{V_m} \tag{5.29}$$

We then compute E, which can be considered to be a Young's modulus, from

$$E = (1 + v)2V_m \frac{p_f - p_0}{V_f - V_0} \tag{5.30}$$

$$= K_e \frac{p_f - p_0}{V_f - V_0} \tag{5.31}$$

where K_e is an equipment constant, if, as Menard suggests, Poisson's ratio v is assumed to be $\frac{1}{3}$. For an NX-size probe (diameter = 76 mm), K_e = 2700 cm^3.

The modulus may be determined for first loading, rebound, or repeated loading. According to Ervin (1983) it is common to include an unload-reload cycle when performing a pressuremeter test.

Menard (1975) also developed design methods for footings and pile foundations which are based on pressuremeter test results. These methods rely on the determination of the limiting pressure p_1. As indicated earlier, the pressuremeter curve may have to be extrapolated in order to find p_1. Methods of extrapolation are discussed by Baguelin et al. (1978) and Davis and Pells (1980); most of these methods seek to find the theoretical pressure corresponding to the infinite expansion of the initial cavity, rather than expansion to just double the initial cavity as suggested by Menard. The determination of p_1 is also relevant to estimating the K_0 value from pressuremeter test reports [Davis and Pells (1980)].

5.2.3.2 SELF-BORING PRESSUREMETERS. The success of measuring true in situ pressure and deformation parameters depends largely on whether it is possible for the probe to penetrate the ground without displacing it. Borehole yielding, drill bit action, pore pressure changes due to the drilling fluid, and disturbance during the insertion of the pressuremeter all contribute to errors in the results of pressuremeter tests. In order to reduce the effect of these factors, the self-boring pressuremeter was developed.

Self-boring pressuremeters are attached to a drilling and cutting device. As the probe is advanced, cuttings produced are transported to the surface by the drilling

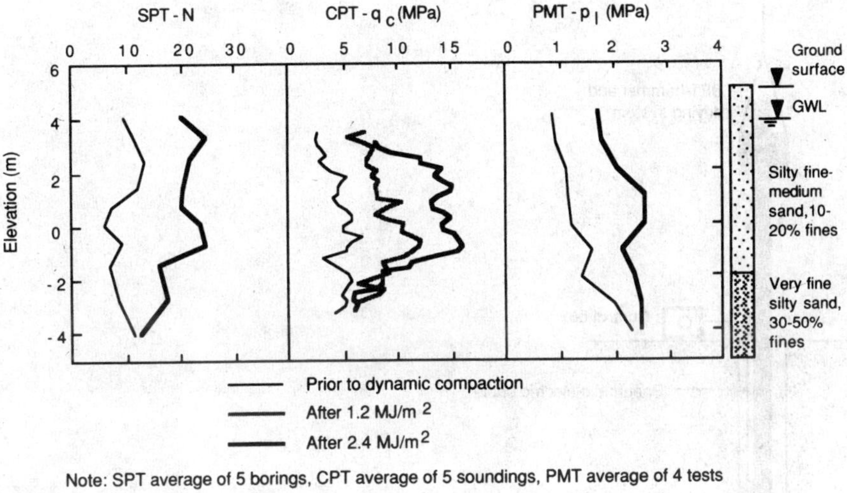

FIGURE 5.17
Degree of ground improvement achieved by dynamic compaction at Massey coal terminal, Newport News, Virginia. [*Mayne et al. (1984)*.]

fluid through hollow drill rods at the center of the apparatus. Such a device is described by Baguelin et al. (1978). Erwin (1983) gives details of the Camkometer (Cambridge K_0 meter) which works along similar principles.

Measurements are usually presented in the form of radial strain versus pressure, because most pressuremeters now record lateral deformation with linear displacement transducers. This eliminates complications associated with volume measurements. The pressuremeter curve looks similar to Fig. 5.16 except that under ideal conditions the initial section of the curve up to point A would be missing, since no deformation of the soil should occur until the in situ horizontal stress is reached.

Figure 5.17 shows the results of pressuremeter tests as well as SPTs and CPTs before and after dynamic compaction on a site in Virginia, United States, as presented by Mayne et al. (1984). In this example, the ground modification is given in terms of the limit pressure; plotting the stress-strain modulus would have revealed a similar pattern.

5.2.4 Flat Dilatometer Tests (DMT)

The flat dilatometer was developed in Italy by Marchetti (1980). It consists of a 14-mm-thick flat stainless-steel blade with a thin flat circular expandable steel membrane on one side, typically with a diameter $D = 60$ mm. The dilatometer, which looks similar to the blade of an oar but with a cutting edge at the end, is pushed into the ground at a rate of 20 to 40 mm/s by a cone penetrometer rig or is driven in by SPT equipment (Fig. 5.18). There are two stages to the test:

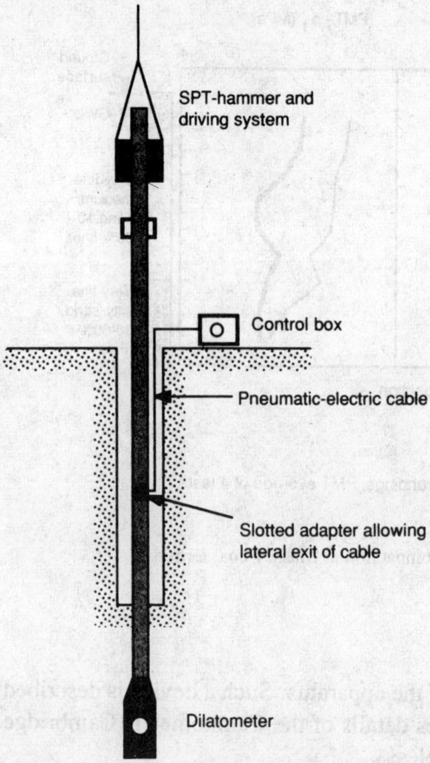

FIGURE 5.18
Dilatometer advanced with standard penetration test.

Penetration. The blade is pushed into the soil at the prescribed rate. This represents a complex loading test on the soil, which could be modeled as the expansion of a flat cavity, such as might be caused by two vertical rigid strip footings.

Expansion. The circular membrane is inflated using pressurized gas. The following pressures are determined: p_0, the pressure required to just begin to move the membrane towards the soil (corrected for membrane stiffness), and p_1, the pressure (corrected) which moves the center of the membrane by $s_0 = 1.00$ mm toward the soil.[1]

With E denoting Young's modulus of the soil and v denoting Poisson's ratio, the dilatometer modulus E_D is, according to Marchetti (1980), calculated from

$$E_D = \frac{E}{1 - v} = \frac{2D}{s_0} \Delta p \tag{5.32}$$

For $D = 60$ mm and $s_0 = 1$ mm,

$$E_D = 38.2 \Delta p \tag{5.33}$$

[1]Some test procedures use $s_0 = 1.1$ mm; then Eq. (5.33) changes to $E_D = 34.7\Delta p$.

Two additional indices are calculated from the dilatometer test results:

Material index
$$I_D = \frac{\Delta p}{p_0 - u_0} \tag{5.34}$$

Horizontal stress index
$$K_D = \frac{p_0 - u_0}{\sigma'_v} \tag{5.35}$$

where u_0 is the original pore water pressure and σ'_v vertical effective stress (prior to testing). Experience shows that the material index is related to the grain size distribution of a soil and that the horizontal stress index varies with K_0, the coefficient of lateral stress at rest, and the overconsolidation ratio of a soil.

Conceptually, the dilatometer test is seen to be a penetration test, rather than a pressuremeter-type test; but it causes less disturbance of the in situ soil than either the SPT or CPT, and therefore stress-strain and strength parameters derived from the readings should be more reliable.

The dilatometer test has been successfully used to evaluate the effect of vibroflotation on the characteristics of a hydraulically placed sand fill in Saudi Arabia [Marchetti (1980)]. More case histories were presented at the In Situ'86 conference [Clemence (1986)].

5.2.5 Shear-Wave Velocity Tests

Measurement of seismic-wave propagation in soils is routinely carried out for dynamic response analyses such as predicting stresses induced by earthquakes. Relatively new is the application of this and other seismic techniques for evaluating the effectiveness of deep compaction.

One method is to determine the shear-wave velocity by *cross-hole tests*, before and after densification. The shear modulus G is related to the shear-wave velocity C_s and the soil density γ by

$$G_{\max} = C_s^2 \gamma \tag{5.36}$$

Assuming a Poisson's ratio of 0.3, the following relationships between the constrained modulus M and Young's modulus E hold:

$$M = 3.5G = 1.34E \tag{5.37}$$

The correlation of static and dynamic parameters for a nonideal material such as soil, which is usually inhomogenous, nonelastic, and anisotropic, is, however, fraught with danger, and the above equation should be treated with caution.

Typical results of shear-wave velocity measurements for a fine to medium sand are given in Fig. 5.19 [after Massarsch and Broms (1983)].

Ohya (1986) also reviews other techniques to measure in situ P (pressure) and S (shear) waves which may potentially be used to evaluate the effectiveness of deep compaction. These include:

Down-hole method. An impulse generated at the ground surface is picked up by geophones clamped in the borehole at various depths.

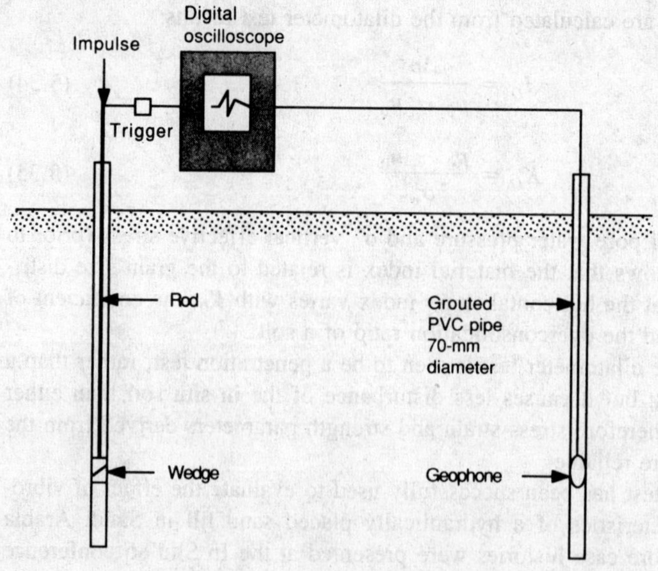

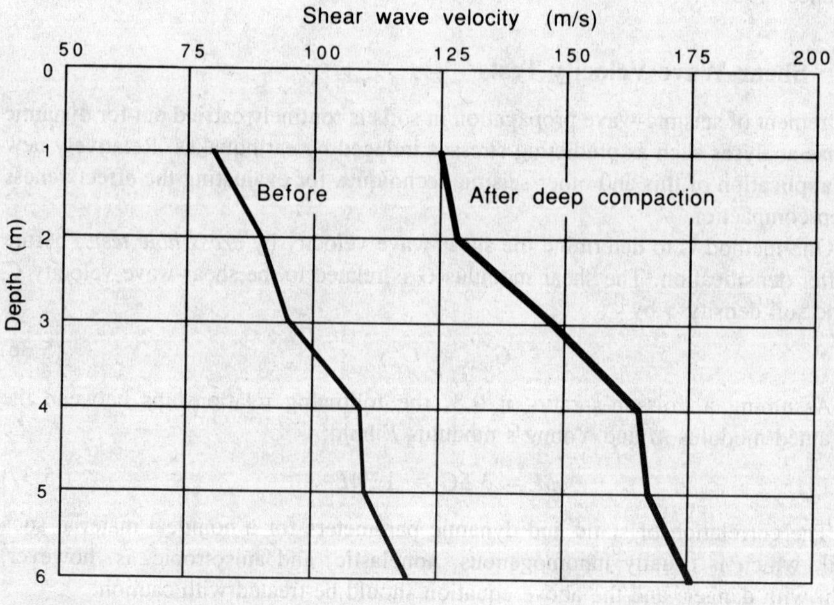

FIGURE 5.19
Shear-wave velocity observed before and after deep compaction. [*Massarch and Broms (1983).*]

Seismic cone method. Essentially a down-hole method where the geophones are built into a cone penetrometer.

Suspension PS logging. Geophones and the vibration source are incorporated in the same probe. This technique requires a fluid-filled borehole and is generally used at depths greater than are influenced by even deep compaction.

Simple seismic refraction techniques would not normally be useful in assessing compacted layers because interpreting the results would be difficult where loose (low velocity) layers may be located between a compacted surface layer and bedrock.

PROBLEMS

Prefixes indicate problem type: C = calculations, M = multiple choice, D = discussion.

Calculations

C5.1. Standard penetration tests were carried out in a saturated sand deposit. At a depth of 10 m, a blow count of $N = 40$ was recorded. Estimate the relative density according to

(*a*) Gibbs and Holts.
(*b*) Bazaraa.

C5.2. In a deposit of fine sand, the cone resistance measured with standard equipment was 20 MPa. Estimate the relative and absolute density if the test was carried out at a depth producing a vertical effective stress of 100 kPa.

C5.3. Estimate the settlement of a 2-m-wide footing subjected to 200 kPa of pressure if the standard penetration resistance of the subsoil is 10 blows per 300 mm. Use the Parry formula.

C5.4. A 20-m-diameter tank exerts a pressure of 100 kPa on a 5-m-thick layer of sand. The groundwater level is near the surface. The average cone resistance recorded was 15 MPa. Estimate the settlement as suggested by Thorne and Burman.

C5.5. A field density test produced the following data.

$$\text{Mass of soil removed} = 1600 \text{ g}$$

$$\text{Volume of soil removed} = 740 \text{ cm}^3$$

The water content was determined in the laboratory as $w = 15.6\%$.

(*a*) Calculate the field dry density and water content.
(*b*) The maximum dry density determined in a standard compaction test was 1.82 t/m^3. Calculate the relative compaction, also called the Hilf density ratio (or simply dry density ratio).
(*c*) The soil collected during the field density test was compacted in the laboratory at field moisture content, using standard procedure. The dry density achieved was 1.75 t/m^3. Determine the roller efficiency as defined by Hilf.

C5.6. Using the Hilf rapid method for compaction control, the following results were recorded:

Field total density = 1.811 t/m^3
Laboratory compacted soil:

Water added (%)	= 0	2.0	4.0
Total densities (t/m^3)	= 1.949	2.040	2.065

(*a*) Determine the Hilf density ratio and roller efficiency.
(*b*) Estimate the deviation from the optimum water content.

C5.7. Using the Perth sand penetrometer, a resistance of 5 blows per 300 mm was recorded. If the moisture content of the sand is 10%, what is the estimated density index (or relative density)?

C5.8. Using a 300-mm-diameter plate, additional settlement of 4 mm was recorded for a pressure increase of 100 kPa. Calculate the coefficient of subgrade reaction (kN/m^3) and estimate the corresponding Young's modulus (MPa).

Multiple Choice

M5.9. Cone penetration resistance
(*a*) Is measured in blows per 300 mm.
(*b*) Is measured in terms of newtons.
(*c*) Is measured in terms of the friction ratio.
(*d*) Is measured in terms of megapascals.

M5.10. An SPT blow count $N = 20$ is typical for
(*a*) A very loose soil.
(*b*) A sand with a void ratio of 1.2.
(*c*) A sand with a friction angle of 32 to 35°.
(*d*) A sand at an effective overburden pressure of 200 kPa.

Discussion

D5.11. How is the result of a pressuremeter test presented, and what design parameters can be deduced from it?

D5.12. What seismic method is most useful in evaluating deep compaction? Describe the procedure.

D5.13. With respect to the Hilf rapid method of compaction control, discuss:
(*a*) Its advantages over other control methods.
(*b*) The actual testing involved and equipment required.
(*c*) Whether this method is being superseded by new technology.

D5.14. With respect to nuclear meters, discuss
(*a*) Calibration.
(*b*) Safety precautions.
(*c*) The difference between backscatter and direct transmission mode.

D5.15. Discuss advantages and disadvantages of the plate load test for compaction control.

CHAPTER 6

SPECIFICATION OF COMPACTION REQUIREMENTS

In this chapter, emphasis is put on compaction specification for fill material placed in layers and compacted by surface equipment. This area of mechanical modification is not only the most researched and developed, but is economically also the most important.

An indication of possible ways of specifying deep compaction was given in Sec. 5.2, which covered the in situ evaluation of deep compaction. This activity is usually very much site-specific and generally valid guidelines would be difficult to formulate.

However, even for surface compaction there are many different specifications used around the world, often reflecting prevalent soil and climatic conditions and equipment available. Before discussing individual approaches in detail, it is appropriate to recall the purposes of compacting soil, which is one or more of the following:

To increase shear strength and thus increase bearing capacity and lateral stability

To reduce compressibility, which may cause settlement and lateral movement

To reduce permeability in order to reduce water losses from reservoirs and canals

To control volume change (swelling and shrinking)

To reduce liquefaction potential with respect to short-term dynamic loads

To improve other engineering properties

The most rational way to specify compaction would be to refer to actual performance criteria as set by the designer. However, this may be rather difficult and is most likely very expensive. Strength and deformation tests, whether done in the laboratory or in the field, are costly and time-consuming. Performance criteria are therefore usually translated into density and moisture requirements, either by experiment or experience. Since even direct density and water content determinations can be cumbersome to carry out, they are often replaced by mechanical or nuclear index tests, which are calibrated to yield results in the conventional way.

Where compaction requirements are expressed in terms of the results to be achieved, it is most common to refer to densities and water content (Sec. 6.1). Particularly in European countries, reference is frequently made to plate bearing tests. This method and other alternatives are discussed in Sec. 6.2, which also mentions specifications defining the method of execution, rather than the result of mechanical modification.

An experienced construction engineer would add to these introductory remarks that a good specification involving control testing should be complemented by good overall supervision of the construction procedures. This will increase the confidence in the test results.

6.1 SPECIFICATIONS IN TERMS OF DENSITY AND WATER CONTENT

6.1.1 Alternative Accept-Reject Criteria

Specification of compaction requirements in terms of maximum dry density and optimum moisture content (as obtained in a normal laboratory compaction test) is the most common way of ensuring that earthworks perform adequately. Figure 6.1 illustrates three ways of defining acceptance criteria for compaction.

Method A simply states that the soil has to be compacted above a certain minimum dry density, e.g., 95% modified maximum dry density (meaning 95% of the

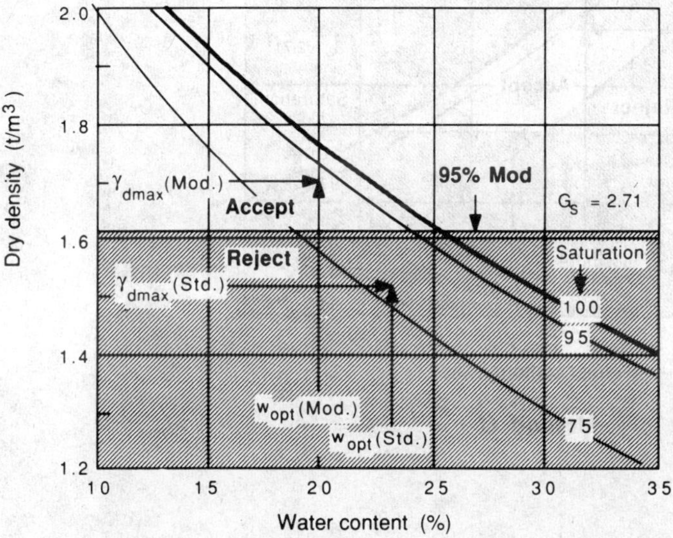

Method A

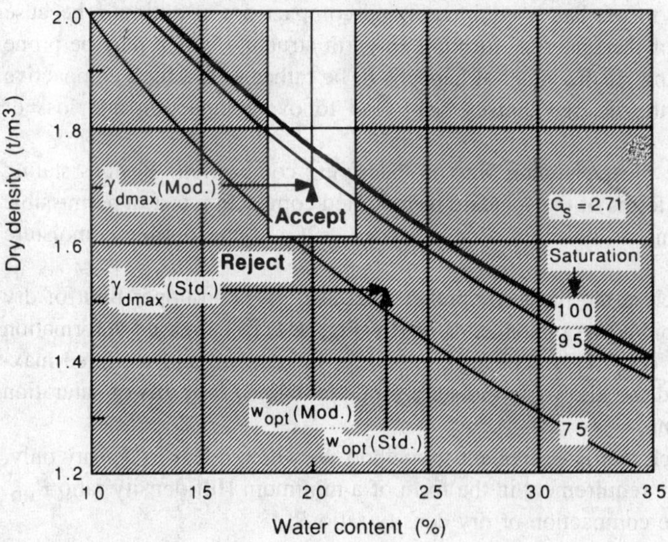

Method B

FIGURE 6.1
Alternative specifications of density and water content.

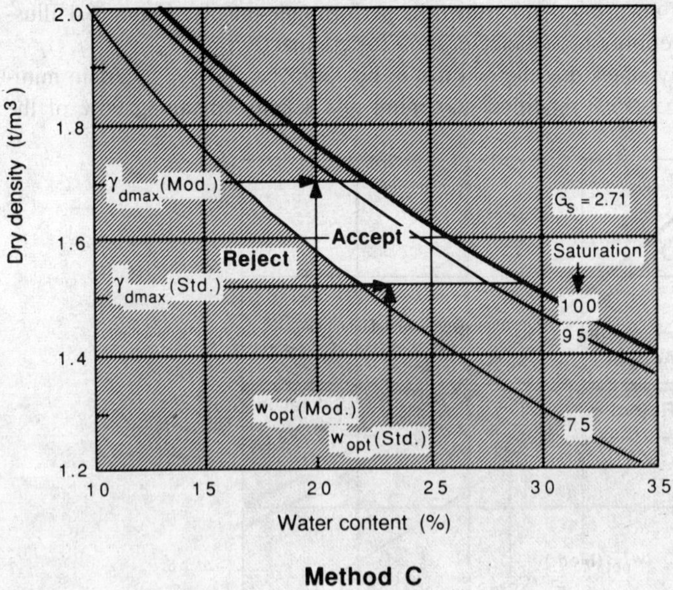

FIGURE 6.1 (continued)

maximum dry density obtained in the modified Proctor compaction test). In such cases, a contractor may tend to use heavy equipment and compact dry of optimum because of better workability of the material, forming an earth structure which may be prone to cracking and swelling. If the material happens to be rather wet, a high compactive effort used to reach the desired density may lead to overcompaction and loss of strength.

Method B is the most common way of specifying compaction. Besides stating a minimum density in terms of the standard or modified compaction test, a permissible range of water content is indicated, e.g., -2 to $+2\%$ of the optimum moisture content.

Method C, which is the most comprehensive, also gives an upper limit of dry density in order to avoid a loss of soaked strength (Fig. 4.4) or increased deformation after repeated loading (Fig. 4.7). Density limits refer to standard and modified maximum dry density, and the allowable moisture range is defined in terms of saturation at placement conditions.

Another approach that is equivalent to method A, which refers to density only, defines the compaction requirement in the form of a minimum Hilf density ratio R_{HD}, also called the relative compaction or dry density ratio R:

$$R_{HD} = R = \frac{\gamma_d(\text{field})}{\gamma_{d\max}} \tag{6.1}$$

The maximum dry density is determined according to the standard or modified laboratory compaction test. Specifying relative compaction could also be combined with limits in allowable moisture content.

It should be noted that where soil is compacted in lifts, the density achieved will vary within each layer. Within a 40-cm-thick layer the density may vary as much as 5 to 10% [Leflaive, (1980); also see Leflaive and Schaeffner (1980)]. Specifications should account for this variability in addition to random variability.

6.1.2 Suitability of Soils as Fill and Conventional Compaction Requirements

The performance of a human-built earth structure depends as much on the suitability of the soil as fill material as on the competency of compaction. Table 6.1, reproduced from the "Design Manual 7.2." (U.S. Navy, 1982), indicates the relative desirability of different types of soil as compacted fill. Major uses considered are fills for dams, canals, foundations, and roadways.

Where high strength and low compressibility are required, but seepage and erodibility are not significant, coarse granular fills are most suitable. As impermeable liners for canals or as core material for dams, clayey gravels and poorly graded gravel-sand-clay mixtures are ranked the highest. Silty soils and dispersive clays, even if compacted well, are vulnerable to erosion by surface runoff or internal seepage. Soils containing organic matter are unsuitable for engineering fills, because of their high compressibility under loads and large volume changes due to environmental influences. Soil classification according to the Unified Soil Classification System can obviously serve as a good initial guideline for assessing the suitability of soil as fill material, but a more extensive testing program is appropriate for large projects.

Table 6.2 gives a range of typical compaction requirements in terms of density and moisture content. This table is not very precise and only provides general guidance. On a particular job, many project-specific conditions may have to be considered, such as the uniformity of the soils, swelling and shrinking characteristics, impermeability requirements, and susceptibility to cracking and piping. In addition, for any compaction program, the importance of confinement should be remembered: Fills on soft ground require a gradual buildup of density with height; it may even be necessary to create a working platform of stabilized soils or to use geotextiles or geogrids.

Compaction requirements for roadworks are probably the most difficult to generalize. Soil type, wheel loads, and expected traffic would be major factors considered, as well as the overall design of the pavement structure, possibly consisting of subgrade, subbase, base course, surface course, and seal.

Earth embankments for highways and structures tend to be compacted to higher densities than water storage dams. In reservoir dams, differential settlement, and the resulting cracking, may lead to seepage and piping and could have catastrophic consequences. For this reason, dam designers are likely to specify densities in the range of maximum dry density corresponding to standard rather than modified laboratory compaction. In addition, the allowable moisture range may be shifted toward wet of optimum, rather than being specified around optimum or on the dry side (the latter is what would be preferred by road engineers!). For dam engineers, higher pore pressures

TABLE 6.1
Relative desirability of soils as compacted fill

Group symbol	Soil type	Relative desirability for various uses (no. 1 is considered the best; no. 14 is least desirable)									
		Rolled earth fill dams			Canal sections		Foundations		Roadways		
		Homog. Embankment	Core	Shell	Erosion resistance	Comp. earth lining	Seepage important	Seepage not important	Frost heave not poss. (fills)	Frost heave possible (fills)	Surfacing
GW	Well-graded gravels, gravel-sand mixtures, little or no fines	—	—	1	1	—	—	1	1	1	3
GP	Poorly graded gravels, gravel-sand mixtures, little or no fines	—	—	2	2	—	—	3	3	3	—
GM	Silty gravels, poorly graded gravel-sand-silt mixtures	2	4	—	4	4	1	4	4	9	5
GC	Clayey gravels, poorly graded gravel-sand-clay mixtures	1	1	—	3	1	2	6	5	5	1
SW	Well-graded sands, gravelly sands, little or no fines	—	—	3 if gravelly	6	—	—	2	2	2	4
SP	Poorly graded sands, gravelly sands, little or no fines	—	—	4 if gravelly	7 if gravelly	—	—	5	6	4	—

TABLE 6.1 *(Cont.)*

SM	Silty sands, poorly graded sand-silt mixtures	4	5	—	8 if gravelly	5 erosion critical	3	7	6	10	6
SC	Clayey sands, poorly graded sand-clay mixtures	3	2	—	5	2	4	8	7	6	2
ML	Inorganic silts and very fine sands, rock flour, silty or clayey fine sands with slight plasticity	6	6	—	—	6 erosion critical	6	9	10	11	—
CL	Inorganic clays of low to medium plasticity, gravelly clays, sandy clays, silty clays, lean clays	5	3	—	9	3	5	10	9	7	7
OL	Organic silts and organic silt-clays of low plasticity	8	8	—	—	7 erosion critical	7	11	11	12	—
MH	Inorganic silts, micaceous or diatamaceous fine sandy or silty soils, elastic silts	9	9	—	—	—	8	12	12	13	—
CH	Inorganic clays of high plasticity, fat clays	7	7	—	10	8 volume change critical	9	13	13	8	—
OH	Organic clays of medium-high plasticity	10	10	—	—	—	10	14	14	14	—

Source: "Design Manual 7.2," U.S. Navy, 1982.

"—" = not appropriate for this type of use.

TABLE 6.2
Typical compaction requirements

Compacted fill for:	Percent modified max. dry density	Moisture range about optimum
Roads		
Depth of 0 to 0.5 m	90 to 105*	−2 to +2
Depth of >0.5 m	90 to 95*	−2 to +2
Small earth dam	90 to 95	−1 to +3
Large dam	95	−1 to +2
Railway embankment	95	−2 to +2
Foundation for structure	95	−2 to +2
Backfill behind walls, in trenches	90	−2 to +2
Canal linings (clay)	90	−2 to +2

*Depending on soil type, traffic loading, and function of soil layer.

Note: The relationship between standard maximum and modified maximum dry density is discussed in Sec. 3.2. As a rough guide:

Sand: γ_{dmax} (std.) = 0.95 γ_{dmax} (mod.)

Clay: γ_{dmax} (std.) = 0.90 γ_{dmax} (mod.)

during construction (which can be monitored by instrumentation), may be preferable to the possibility of cracking within the dam body. The latter is more likely to occur in soils compacted dry rather than wet of optimum and is most difficult to detect before filling of the reservoir.

Special consideration may apply to backfill behind walls, where generally coarse-grained free-draining fill is specified in order to keep hydraulic pressure to a minimum. Where structural walls cannot yield, compaction of the backfill may increase earth pressures from those corresponding to the active state to those characteristic for at-rest conditions or higher, with passive earth pressures representing a theoretical maximum. (Also see discussion and reference given in Sec. 17.2.4.) Proper compaction of fill adjacent to structures is most important where the fill forms part of access ramps or abutments, because of the possibility of excessive differential settlement; but it is not advisable to allow heavy compaction equipment near low walls.

Compaction of backfill in trenches around pipes is governed by the design objectives of reducing the stresses and/or deformations in the conduit to a minimum and preventing differential settlement of the restored ground surface with time. Careful preparation of the pipe bedding and ensuring lateral support by tamping fill along the pipe are considered good practices. Special considerations may apply where so-called negative projection or imperfect trench conduits are installed, where loose fill is placed immediately above the pipe in order to reduce the vertical stresses caused by the overburden. For an explanation of these terms see Spangler and Handy (1973).

6.1.3 Variability of Test Results

The usefulness of specifying placement density and water content depends on their correlation with performance criteria and on how closely laboratory tests simulate field compaction. It also depends on the accuracy of test procedures and the natural variability of the soil properties.

The accuracy of the test results describes how close the reported value is from the true value. Despite the precision with which a result may be presented, it may not be accurate because of random or systematic errors associated with equipment and the operator.

For example, direct field density evaluation relies on exact measurement of the volume of the hole from which a test sample is excavated. It is likely that if the balloon method is employed, the volume of a hole with a rough surface will be slightly underestimated, leading to average densities which may be higher than actual values. On the other hand, using the sand replacement method, vibration from construction equipment or carelessness of the operator may tend to increase the density of the measuring sand, leading to lower than true densities of the compacted fill.

Moisture content determinations could show a higher variation for small samples than for large samples taken from the same soil, largely depending on the soil's grain size distribution and homogeneity.

Because of the many variables involved, compaction test results tend to be normally distributed. The parameters used to describe the variability of a finite set of n test results $x_1, x_2, \ldots, x_n$ usually are the following:

Mean
$$\bar{x} = \frac{1}{n}\sum_{i=1}^{n} x_i \tag{6.2}$$

Variance
$$V = \frac{1}{n-1}\sum_{i=1}^{n} (x_i - \bar{x})^2 \tag{6.3}$$

Standard deviation
$$s = \sqrt{V} \tag{6.4}$$

Coefficient of variation
$$CV = 100\,\frac{s}{\bar{x}}\ \% \tag{6.5}$$

Typically, a homogeneous well-compacted layer should show coefficients of variation CV not exceeding

$$CV = \begin{cases} 5\% \text{ for dry density} \\ 30\% \text{ for moisture content} \end{cases}$$

for a sample of at least 10 tests. The above coefficients are maximum recommended values, which could serve as a basis for initial sampling programs. Lower values have been obtained in practice; for example, using nuclear meters for density and water content control in field compaction trials on a sand-clay base course, Ladner and Hamory (1974) found CVs such as 1.5% for dry density and 8.5% for moisture content.

Hilf (1975) quotes dam construction experience, by the U.S. Bureau of Reclamation, which indicates that moisture control in terms of the deviation of the field water content from the laboratory optimum water content ($w_0 - w_f$) can be based on a standard deviation of less than about 1.5%; similarly, the standard deviation of the dry density ratio R (field dry density/laboratory dry density) is expected to be less than 3%.

6.1.4 Statistical Approach to Specification and Control

Specifications and control tests are intended to ensure adequate performance of the foundation or embankment of compacted soil according to the chosen design criteria. In order to assist in reaching these objectives, control tests have to be

Relevant. Density and water content have to be related to stability, volume change, etc.

Cost-effective. Testing expenses must be reasonable in relation to construction costs and consequences of failure.

Representative. Sample size should be related to the known or estimated variation of the soil properties being evaluated.

Only the last characteristic of control tests can reasonably be easily assessed in probabilistic terms. Relevance, cost, and resulting utility are more difficult to quantify. Nevertheless, experience and engineering judgment suggest approximate minimum numbers of field density and moisture content tests as implied by the following table [adapted from ''Design Manual 7.2'' (1982), Striegler and Werner, (1973), and other sources]:

Earth structure	Volume of fill per test, m^3
Embankments	500–2000
Impermeable liners	200–1000
Subgrade	500–1500
Base course	500–1000
Backfill	100– 200

Because of the wide range of test frequencies reported in the literature, most texts refrain from giving typical values such as those above.

The ''Design Manual 7.2'' (U.S. Navy, 1982) also indicates that it would be reasonable to obtain a supplementary compaction curve for every 10 to 20 field tests, depending on the variability of materials.

Other practical rules suggest that extra tests are appropriate if visual inspection reveals a change in material, moisture control, or effectiveness of compaction. Varying weather conditions or changes in machinery, operators, or supervision could also have an influence on the quality of compaction; for this reason, project managers may require one test every shift or every 24 h of construction time.

Acceptance-rejection criteria in earthworks are increasingly formulated in statistical terms, rather than in absolute minimum allowable values. This approach requires that the earthworks be divided into parts or lots which are subject to a single compliance decision based on random testing. Specifications normally require that the entire lot be produced by the same works process throughout and that all work on the lot be completed at the same time. The lot should appear to be of constant quality without changes in matérials or other attributes and where possible should be representative of the earthworks as a whole. Any lot whose compaction quality is rejected, must be completely reprocessed before being resubmitted for inspection. The Main Roads Department of Western Australia uses such an approach for density specification. Details are given in Table 6.3. This table reflects assumed variances and desirable confidence limits.

For a specific project, compaction specification may be simplified, such as in the following example from a dam project [Hilf (1975)]:

Moisture content $w_0 - 3.5 \leq w_f \leq w_0 + 1$

$w_0 - 3.0 \geq w_f$ for not more than 20% of the tests

$w_0 + 0.5 \leq w_f$ for not more than 20% of the tests

$w_0 - 1.0 \leq w_f \leq w_0 - 0.5$

TABLE 6.3
Density requirements

Works component	Number of tests per lot	Multiplier k	Characteristic dry density ratio R_c, %	Quality level
Embankment foundation and levees	6	0.5	88.0 or greater 87.9 or less	Accept Reject
Embankment construction	6	0.5	91.0 or greater 90.9 or less	Accept Reject
Subgrade preparation	6	0.5	92.0 or greater 91.9 or less	Accept Reject
Subbase incl. bridge abutment embankment to within 600 mm of top surface	9	0.6	94.0 or greater 93.9 or less	Accept Reject
Base course and top 600 mm of bridge abutment embankment	9	0.6	97.0 or greater 96.9 or less	Accept Reject
Cement-modified base course	9	0.6	95.0 or greater 94.9 or less	Accept Reject

Source: Main Roads Department, Western Australia.

Note: Densities are determined with nuclear meters.

Definitions:

R = mean dry density ratio (%)
= field dry density/maximum dry density (modified compaction)

k = multiplier as detailed in table

s = standard deviation of the results of dry density ratio tests on the lot being assessed

R_c = characteristic dry density ratio
$= R - ks$

Density control	$R \geq 96\%$	
	$R \geq 97\%$	for not less than 80% of the tests
	$R_{avg} = 100\%$	

For road construction, Swiss standards (SNV 640 585a) specify:

Depth of 0 to 600 mm	Field dry density >100% laboratory maximum dry density (std. compaction).
Depth of >600 mm	Field dry density >97% laboratory maximum dry density (std. compaction).
Tolerance	One in five test results may be below this requirement, but by no more than 5%.

Statistics and probability theory can assist in developing rational specification and control procedures; in particular, they aid in

Determining the mean, standard deviation, and coefficient of variation of a set of results (normal distribution)

Determining confidence limits of the mean of a sample (*t* distribution)

Giving confidence estimates of the variance (X^2 distribution)

Determining sample size for specified confidence limits of the mean, for a given coefficient of variation

Evaluating the significance of differences in variances between two samples (*F* distribution)

Providing methods of improving successive appraisals of a state of nature based on new data available (Bayes' rule)

Establishing the probability distribution of a function of several randomly varying parameters (probabilistic design)

6.2 SPECIFICATION IN TERMS OTHER THAN DIRECT DENSITY TESTS

Although specification of placement density and moisture content is most common, it is possible to specify the end product in terms of index tests or standardized performance tests as described in Chap. 5. In addition it may be possible to achieve adequate compaction simply by specifying compaction equipment and operation (method specification). A combination of control techniques may also be chosen. For example, it could be decided that plate load tests should be used to obtain stress-strain moduli, and nuclear density and moisture measurements should be used for assessing the uniformity of the compaction achieved (thus guaranteeing the homogeneity of the earth structure created).

6.2.1 Proof Rolling and Settlement Observations

Visual inspection during test rolling (or proof rolling) with a heavy smooth wheel roller is often used to quickly identify weak spots due to soft subgrade or inadequate compaction. Care has to be taken that excessive plastic deformation of relatively wet soils doesn't lead to overcompaction during proof rolling and a resultant loss in strength.

Precision settlement observations have been used by Brandl (1980) to determine whether the number of passes of a compaction machine over coarse-grained (rock) fill have been adequate or not. He formulated his acceptance criterion as follows:

$$\Delta s_n \leq a \sum_{1}^{n-1} \Delta s_i \tag{6.6}$$

where Δs_n = settlement increase of layer during last (nth) pass of compaction machine
a = 0.05 to 0.1 depending on type of rock
n = number of passes
Δs_i = medium settlement increase of layer during ith compaction pass

6.2.2 Compactor-Mounted Control Devices

The Dynapac and Geodynamik corporations of Sweden have developed an *electronic compaction meter* suitable for vibratory rollers. It consists of an accelerometer which registers the vibration of the drum. Mean values for periods of 5 or 30 s are recorded. This has proved useful for instant and continuous assessment of the effectiveness of compaction. It only gives relative information about the compacted soil, but on specific projects it may be possible to correlate the dynamic data with standard compaction control tests [Forssblad (1980) and Thurner and Sandstrom (1980)]. Another accessory available from this company is a frequency meter, which works according to the vibrating fork principle. It is placed on the ground next to the roller when measuring the frequency of the roller that is being used.

6.2.3 Plate Load and Other Strength Tests

In Europe it is not uncommon to have compaction specified in terms of the reloading modulus E_2 determined by the plate bearing test. Typical minimum requirements would be:

	E_2, MPa	
Layer	**Cohesionless soil**	**Cohesive soil**
Surface course	>120	—
Base course	> 60	>30
Subgrade	> 45	>20

One plate load test per 5000 to 10 000 m^3 appears to be a reasonable approach, if fairly uniform material is being treated.

On some cohesive soils the plate load test results may not be adequate (or indeed not suitable) as compaction criteria, because they depend on the water content, the weather, the soil characteristics, and other factors. The dynamic plate load test and similar procedures, such as the Clegg impact test, lend themselves for quick assessment of large areas and the identification of zones of poor compaction. Most engineers, however, only consider them suitable for testing cohesionless soils whose water content does not vary appreciably.

Besides plate load test parameters, field CBR values are also specified in many countries as strength-related performance indicators. They have immediate application in many pavement design procedures.

6.2.4 Compaction Trials

For projects where large quantities (say >20 000 m^3) of relatively homogeneous borrow materials have to be handled, compaction trials may be used to determine the most effective compaction methods. In particular, these trials would indicate the following:

The most cost-efficient compaction equipment
The type and depth effect of the densification
The minimum number of passes needed to obtain the required density

The information gained could be used to prepare the compaction specification by stipulating the type of roller, number of passes, and layer thickness rather than by demanding that a certain density or stress-strain modulus be achieved. This would represent a ''method'' specification (or ''equipment'' specification) as against an ''end-product'' specification and could lead to lower contract costs. Because of the high cost of compaction trials it would be desirable to set up a database for all types of soils and machinery and continually update it as new results become available and the compaction equipment is improved further.

6.2.5 The Future

In the future, computers aboard compaction machinery may not only control its movements, making it into a robot, but attached sensors may automatically adjust speed, vibration frequency and amplitude, and other characteristics in order to optimize the energy use of the equipment. Theoretically it may one day be possible to predict the effect of a particular compactor on a certain soil. This in turn may lead to compactors being designed for a specific project.

PROBLEMS

Prefixes indicate problem type: C = calculations, M = multiple choice, D = discussion.

Calculations

C6.1. The following are the results of laboratory compaction tests on a lateritic soil:

ρ_{dmax} (t/m^3) =	1.72	1.71	1.72	1.68	1.65	1.77
w_{opt} (%) =	15.6	16.8	15.0	16.5	15.5	15.8

Calculate the mean, standard deviation, and coefficient of variation for both the maximum density and optimum water content.

C6.2. The additional settlement of coarse-grained fill measured after subsequent passes of a roller were as follows:

Pass number	1	2	3	4	5	6
Settlement (mm)	65	15	8	3	2	1.5

How many passes are needed to satisfy Brandl's criterion for adequate compaction? Use $a = 0.1$ in Eq. (6.6).

Multiple Choice

M6.3. Of the following soils, characterized by their USCS group symbol, which one is most suitable for a homogeneous earth dam?
(*a*) SM
(*b*) GC
(*c*) GW
(*d*) CL

M6.4. Of the following soils, characterized by their USCS group symbol, which one is most suitable as fill for a road embankment?
(*a*) SM
(*b*) GC
(*c*) GW
(*d*) CL

M6.5. Of the following soils, characterized by their USCS group symbol, which one is most suitable as fill for an unpaved road surface?
(*a*) SM
(*b*) GC
(*c*) GW
(*d*) ML

M6.6. Compaction tests tend to have which type of distribution?
(*a*) Normal
(*b*) Log-normal
(*c*) Gamma
(*d*) Beta

M6.7. The expression $\sqrt{\frac{1}{n-1}\sum_{i=1}^{n}(x_i - \bar{x})^2}$ represents the

(*a*) Mean.
(*b*) Variance.
(*c*) Standard deviation.
(*d*) Coefficient of variation.

M6.8. In the construction of a homogeneous earth dam, the coefficient of variation for dry densities should not exceed

(*a*) 2.
(*b*) 5.
(*c*) 10.
(*d*) 30.

M6.9. For an embankment, a reasonable test program is likely to be in the range of one field density test per

(*a*) 500 to 2000 m^3.
(*b*) 200 to 1000 m^3.
(*c*) 500 to 1000 m^3.
(*d*) 100 to 200 m^3.

M6.10. In the determination of the confidence limits of the mean of a set of measured densities, calculations refer to the

(*a*) Normal distribution.
(*b*) t distribution.
(*c*) X^2 distribution.
(*d*) F distribution.

Discussion

D6.11. Discuss advantages and disadvantages of the plate load test for compaction control.

D6.12. Describe alternative compaction specifications based on water content and density values.

D6.13. Outline special aspects of compaction of backfill in pipe trenches.

D6.14. Discuss compaction wet and dry of optimum for the case of

(*a*) An earth dam.
(*b*) A road subgrade.
(*c*) A foundation subsoil.

D6.15. Discuss method specification versus end-product specification in compaction.

PART III

HYDRAULIC MODIFICATION

CHAPTER 7

INTRODUCTION TO HYDRAULIC MODIFICATION

7.1 OBJECTIVES AND TECHNIQUES

In its most general sense, dewatering means modifying ground by lowering the water table, redirecting seepage, or simply reducing its water content. In coarse-grained soils, dewatering can be achieved by gravity drainage into sumps, ditches, and wells. In fine-grained soils, gravity drainage is slow or ineffective; for this type of ground, the process of dewatering becomes synonymous with forced consolidation (induced by preloading or electroosmosis).

Dewatering of soil (or fissured rock) in civil engineering or mining projects is carried out for one or more of the following reasons:

To provide a dry working area, such as in excavations for building foundations, dams, and tunnels

To stabilize constructed or natural slopes

To reduce lateral pressures on foundations or retaining structures

To reduce the compressibility of granular soils

To increase the bearing capacity of foundations

To improve the workability or hauling characteristics of borrow materials

To prevent liquefaction due to an upward gradient

To reduce the liquefaction potential during earthquakes

To prevent soil particle movement by groundwater (leading to piping)

To prevent surface erosion

To prevent or reduce damage due to frost heave

Every engineer dealing with soil is acutely aware of the significance of the water phase in the soil, yet each engineer's perspective may differ significantly, depending on his or her specialist background. A groundwater hydrologist may emphasize the characteristics of water flow in terms of quantities and directions. The geotechnical engineer is more aware of what benefits reducing pore water pressures has for soil strength (and the consequent increase in bearing capacity and slope stability). A road engineer's interests may extend to evaluating equilibrium moisture conditions in a subgrade. In pavement design an engineer may also be concerned with the transmission of water in vapor rather than just liquid form. However, here in our discussion of hydraulic modification of ground, we will only deal with the control and management of free water in soils and rocks.

The installation of drainage systems and wells could be considered an age-old building problem. Nevertheless, drilling and pumping technology has been improving steadily, and today construction procedures are highly efficient and may include the use of geotextiles and geomembranes.

Dewatering techniques based on traditional gravity drainage and pumping from sumps and wells are briefly introduced in this chapter, which also includes a review of fundamental soil-water relationships. The hydraulics of slots and wells is covered in Chap. 8; its principles underlie the design of dewatering systems for excavations and slopes (Chap. 9). Because they are relatively new but rapidly growing in importance, geosynthetics and their applications for filtration, drainage, and seepage control are treated separately in Chap. 10. The concept of preloading as a means of dewatering fine-grained soils is covered in Chap. 11; this topic is now closely associated with geosynthetics, because today vertical drains are almost exclusively made of synthetic materials. Chapter 12 recalls electroosmosis as a means of consolidating (= dewatering) fine-grained soils. This process has been known for many years but has seen relatively few practical applications. Nevertheless its potential for modifying difficult ground and improving the performance of foundation elements within it is well recognized.

Besides the traditional dewatering by gravity and the more involved preloading or electroosmosis, other geotechnical processes can be resorted to for eliminating or controlling groundwater. These include compressed air techniques in caisson construction and tunneling, and various cutoff systems (diaphragm walls, sheet piling, geomembranes). Although these methods may be important in combination with dewatering, they are considered outside the scope of ground modification in this text.

It should be pointed out that physical and chemical modification (Chap. 13–15) also affects ground water and permeability. However, because aspects other than dewatering are the dominant features of these techniques, they are treated as a separate group (Part IV).

The suitability of soils for the more traditional groundwater control techniques depends largely on their grain size distribution as illustrated in Fig. 7.1.

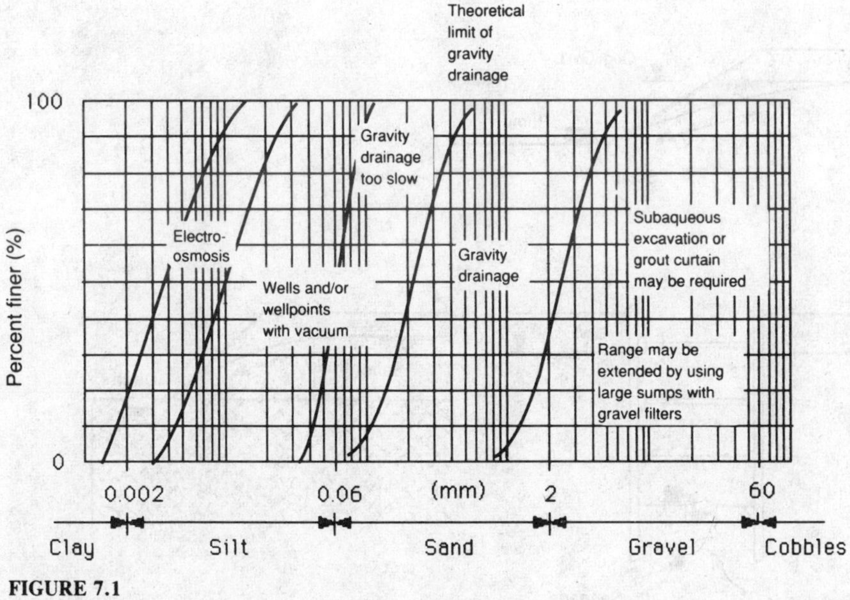

FIGURE 7.1
Dewatering methods applicable to various soils. [*After Mansur and Kaufmann as found in Leonards (1962).*]

7.2 TRADITIONAL DEWATERING METHODS

7.2.1 Open Sumps and Ditches

Collecting seepage water in open sumps and ditches and removing it by gravity flow or pumping is the most common and cheapest dewatering method. This technique works well in relatively shallow excavations in dense, well-graded coarse soils, in rock, and in the case of permeable soils overlying impermeable strata. It may also be considered in situations where floaters or other obstructions do not allow the sinking of wells. When used for sheeted and braced excavations, there is the danger of slumping, wall collapse or bottom instability due to an upward seepage gradient. The latter is referred to as a "quick" condition (Fig. 7.2).

Dewatering of clayey slopes may possibly be accomplished by a combination of a toe drain and gravel-filled lateral slots (Fig. 7.3). Stabilization is then effected not only by gradual dewatering but also by the supporting (reinforcing) effect of the buried gravel walls. A similar concept applies to the use of "sand piles" or "gravel columns" in foundations.

7.2.2 Gravity Flow Wells

If the water level in a borehole is lowered by steady pumping, groundwater will flow into the well under gravity until the phreatic surface, the level at which the water

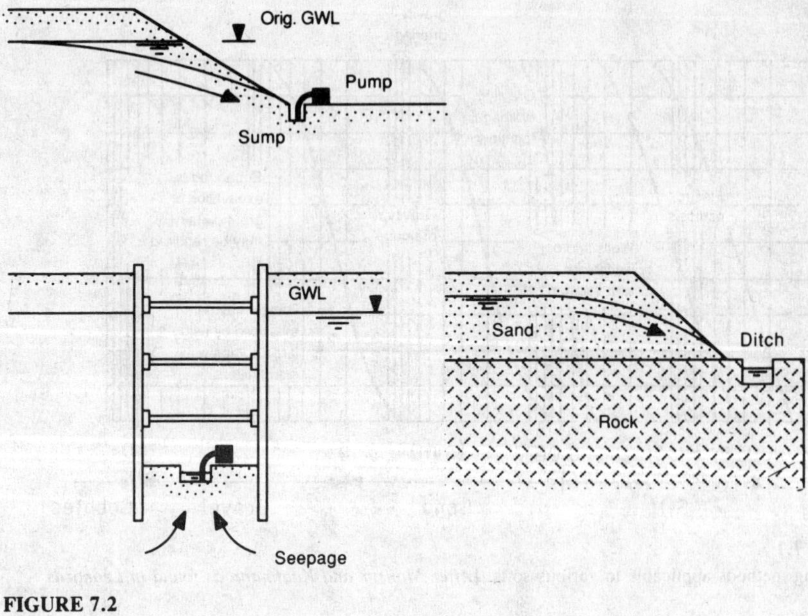

FIGURE 7.2
Dewatering by open sumps and ditches.

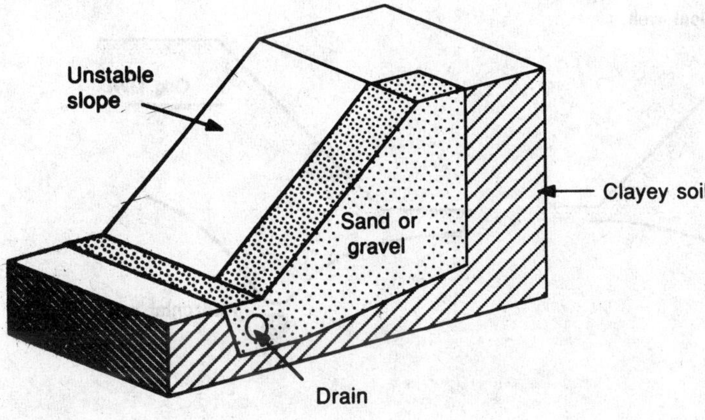

FIGURE 7.3
Lateral drainage slots.

pressure is atmospheric, has dropped to a new equilibrium position. This process can be used to lower the water level for construction purposes as illustrated in Fig. 7.4. The usual aim of dewatering is to lower the phreatic line to a level at least 0.5 m below the base of an excavation in gravel and coarse sand; in fine sands, the water level should be lowered further, preferably more than 0.7 m below the base of the excavation.

Standard bored wells involve the insertion of an inner casing (with a screen section at the bottom) into the cased borehole. After dropping appropriately graded filter material into the space between the inner and outer casing, the latter is withdrawn. Water is pumped to the surface through a riser pipe lowered into the inner casing. The casing and well screens may extend to a depth of 10 to 20 m; usual diameters are 150 to 200 mm. One-stage installations generally allow the lowering of the water level by a maximum of about 3.5 to 4 m near the center of a building excavation. Although more complicated, multistage installations are frequently used (Fig. 7.5).

The use of eductor wells, working on the Venturi principle, allows considerably deeper wells than in the standard arrangements. As demonstrated by many deep large-diameter wells in operation for municipal and industrial water supply, submersible pumps have virtually no limit as far as depth and diameter of practical wells are concerned.

A system of closely spaced single-pipe wells of small diameter with a common header pipe and a central pump installation is commonly referred to as a *well-point system*. The well points (or spears) are driven or jetted into the ground. Several patented systems are available (Fig. 7.6). Generally, after sinking the well point to the desired depth, a sand filter is formed around the point by feeding coarse sand down the hole. This is referred to as "sanding in." Water flows to the well under gravity and is drawn to the surface by the vacuum in the header main.

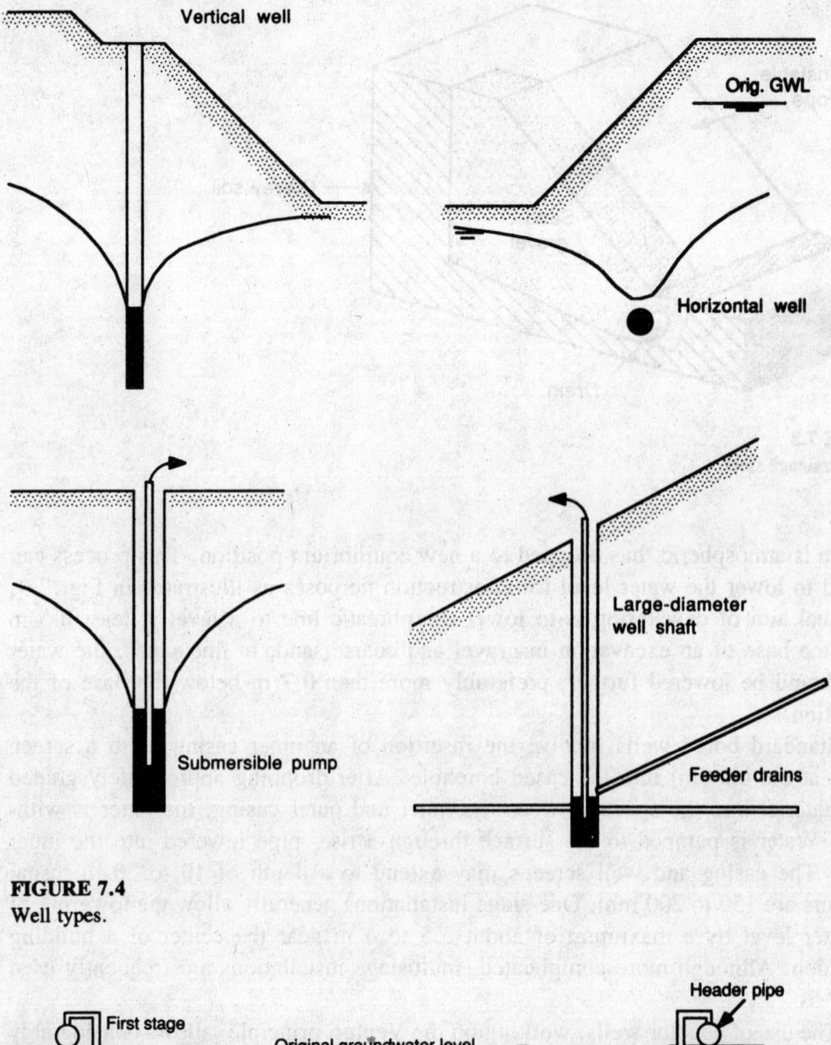

FIGURE 7.4
Well types.

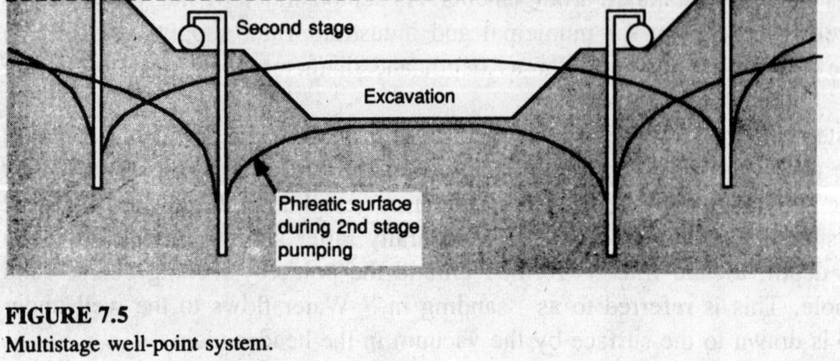

FIGURE 7.5
Multistage well-point system.

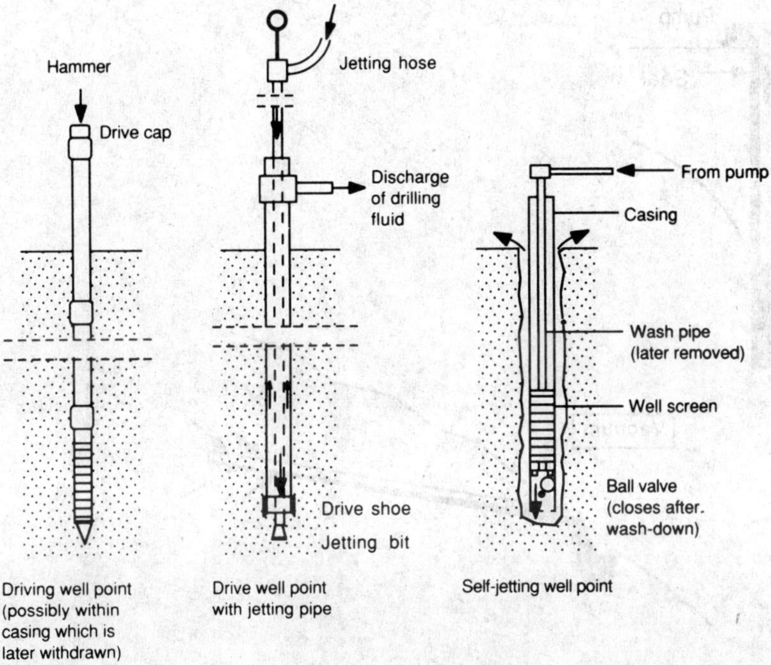

FIGURE 7.6
Types of well points.

7.2.3 Vacuum Dewatering Wells

In fine sands and silts, with permeabilities of 10^{-4} to 10^{-6} m/s, water does not flow freely under the influence of gravity, due to capillary tension. To make dewatering and stabilizing of these soils possible, a vacuum may be applied to the sealed-off filter section of the well. Seepage into the well is then increased due to the influence of the atmospheric pressure. Water inflow is generally low and wells may only require intermittent pumping out.

Vacuum action is also present in well-point systems which use a combined vacuum and centrifugal pump; the net vacuum applied at the well point is, however, only equivalent to the vacuum in the header pipe less the lift in the riser pipe. Care has to be taken that all connections in the pipe system are airtight and an effective seal is formed around the riser pipe in its upper section.

To be effective, vacuum wells have to be spaced very closely, say 1 to 2.5 m apart. The distance between rows of wells should not be more than 15 to 20 m.

Submersible pumps in combination with vacuum pumps could provide dewatering to great depth. Horizontal systems are also conceivable (Fig. 7.7).

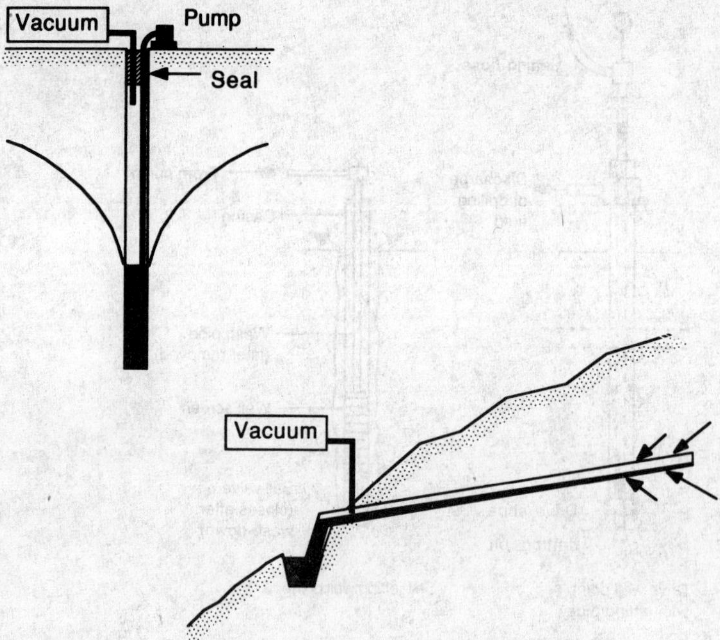

FIGURE 7.7
Vacuum dewatering wells.

7.3 FUNDAMENTAL SOIL-WATER RELATIONSHIPS

This section reviews basic phase relationships, the concept of pore pressures, the principle of effective stress, and Darcy's law. These topics are usually covered in a first course on soil mechanics and are included here for those readers who wish to briefly brush up on these concepts. At the same time it attempts to form a bridge between the vocabulary familiar to the hydrologist and that commonly used by the geotechnical engineer.

7.3.1 Phase Relationships

Soil mechanics describe unfrozen soil as a multiphase material consisting of three distinct phases: solid, liquid, and gas. In most problems these phases represent soil solids, water, and air. For the definition and derivation of basic relationships between these components, an element of soil is conveniently separated into its phases as shown in Fig. 7.8.

where V = total volume
V_s = volume of soil solids
V_w = volume of water

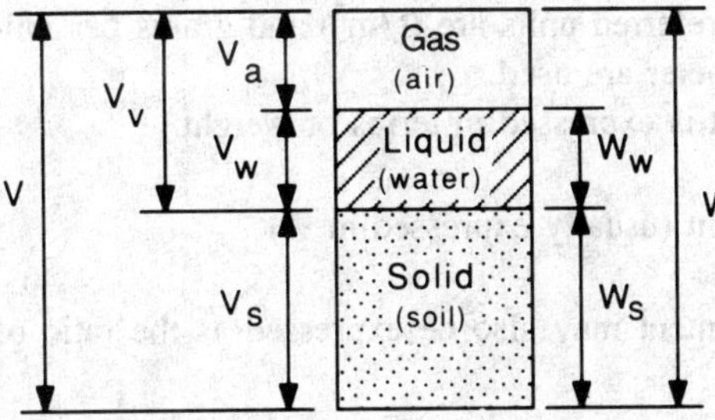

FIGURE 7.8
Soil phases—schematic representation.

V_a = volume of air
V_v = volume of voids (filled with air or water)
W = total weight
W_s = weight of soil solids
W_w = weight of water

Unit weights (kN/m^3) are defined as follows:

$$\gamma_t = \frac{W}{V} = \text{total unit weight (also referred to as bulk unit weight)}$$

$$\gamma_s = \frac{W_s}{V_s} = \text{unit weight of soil solids}$$

$$\gamma_d = \frac{W_s}{V} = \text{dry unit weight}$$

$$\gamma_w = \frac{W_w}{V_w} = \text{unit weight of water}$$

$$\gamma_b = \gamma_t - \gamma_w = \text{buoyant unit weight}$$

Specific gravities can be defined for water and soil solids, but it is the latter which is most often referred to:

$$G_s = \frac{\gamma_s}{\gamma_w} = \text{specific gravity of soil solids}$$

In this definition the unit weight of water must be measured at a temperature of 4°C. This stipulation is, however, normally not relevant, and γ_w is set equal to 9.81 kN/m^3. For most weight and volume calculations, it is recommended that $\gamma_w = 9.81$ kN/m^3 be used rather than the approximation of $\gamma_w = 10$ kN/m^3 which is usually sufficient for the computation of stresses.

For some calculations it may be preferable to work in mass units, rather than in weight units. For example, the total mass density of a soil may be expressed as

$$\rho_t = \frac{\gamma_t}{g}$$

where g is the acceleration of gravity. Preferred units are (t/m^3) and grams per milliliter; sometimes grams per cubic centimeter are used.

In soil mechanics the water content is expressed in terms of weight:

$$w = \frac{W_w}{W_s} = \text{water content (usually expressed in \%)}$$

In groundwater hydraulics, the water content may also be expressed as the ratio of volume of water over total volume.

Important volume relationships are

$$n = \frac{V_v}{V} = \text{porosity (usually expressed in \%)}$$

$$e = \frac{V_v}{V_s} = \text{void ratio}$$

$$S = \frac{V_w}{V_v} = \text{saturation (usually expressed in \%)}$$

Saturation, porosity, and water content are usually expressed in percent, while the void ratio is always a number, ranging from 0 to more than 1. A number of useful interrelationships are worth remembering:

$$G_s w = Se \tag{7.1}$$

$$n = \frac{e}{1 + e} \tag{7.2}$$

$$\gamma_t = \frac{G + Se}{1 + e}\gamma_w \tag{7.3}$$

$$\gamma_d = \frac{\gamma_t}{1 + w} \tag{7.4}$$

These expressions can be easily verified using the schematic representation of the phases of a soil element as given in Fig. 7.8, particularly if simplifying assumptions are made such as setting V_s equal to 1.

Depending on the type of deposition and stress history, soils may have a considerable range of densities, void ratios, and water contents. For typical values of the various properties of soils the reader is referred to standard soil mechanics texts. However, as a guide to the order of magnitude, the following values are reasonable assumptions for a medium dense sandy, gravelly soil:

$$G_s = 2.65$$

$$e = 0.6$$

$$n = 40\%$$

$$\gamma_d = 18 \text{ kN/m}^3$$

$$w = 20\% \text{ (saturated)}$$

$$\gamma_t = 20 \text{ kN/m}^3 \text{ (partially saturated)}$$

Note that these values do not exactly correspond to each other and that they are guidelines only. Densities could easily vary by ±20% and void ratios by ±40%. Specific gravities of soil solids generally do not vary by more than ±5%.

7.3.2 Geostatic Stresses and Pore Pressures

Geostatic stresses are stresses due to the soil's own weight in a deposit with a horizontal surface and where properties do not change appreciably in a horizontal direction. In this case we can relate the vertical stress σ_v on an element of soil at depth z simply to the weight of soil above that element:

$$\sigma_v = \gamma_t z \tag{7.5}$$

The determination of the horizontal stress is not straightforward. Mathematically the horizontal stress is obtained by multiplying the vertical stress with a factor K, the coefficient of lateral stress. The value of K depends on the stress history of the soil mass and the state of equilibrium it is in. For geostatic conditions, K is set equal to K_0, which is the coefficient of lateral stress at rest, meaning that the soil mass is neither undergoing compression nor expansion. The factor K defines the ratio of horizontal to vertical stresses in terms of effective stresses. The effective vertical stress is equal to the total vertical stress σ_v minus the pore water pressure u:

$$\sigma'_v = \sigma_v - u \tag{7.6}$$

The effective vertical stress in a saturated soil deposit can also be calculated directly by multiplying depth z with the buoyant unit weight of the soil. Effective horizontal stress and total horizontal stress is then obtained as follows:

$$\sigma'_h = K\sigma'_v \tag{7.7}$$

$$\sigma_h = \sigma'_h + u \tag{7.8}$$

Effective stresses are of particular relevance in stability problems. Wherever pore pressures increase without a corresponding increase in total stresses, such as might be due to seepage or external forces, the stability of a soil mass with frictional shear strength is reduced. In situations like this the benefits of dewatering can be clearly demonstrated.

The level at which the pore water pressure is atmospheric is referred to as the groundwater level, the water table, or the phreatic surface. In the geostatic condition, the pore pressure increases linearly with depth below the water table (Fig. 7.9). Following the laws of hydrostatics, the pore pressure is equal in all directions. As already indicated, external influences may create "excess" hydrostatic pressures, usually detrimental to stability.

Because of capillary forces, water may rise above the groundwater level. The capillary rise in soils may vary from 0.05 m in coarse gravel to more than 1 m in fine

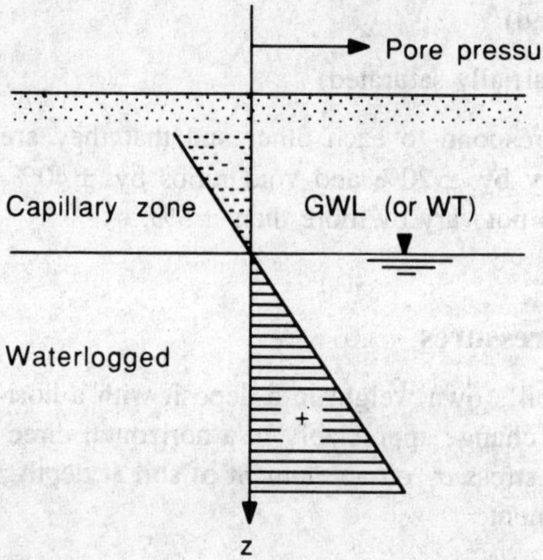

FIGURE 7.9
Pore pressure and capillary rise.

sand and to more than 3 m in silt. The pore pressure above the water table is negative. This negative pressure is referred to as tension, suction, or capillary pressure. As long as there is a continuous channel of water down to the phreatic surface, this negative pressure varies linearly with the distance from this level (Fig. 7.9). Where there is no continuity to the fully saturated zone, the magnitude of the pore water suction can be related to the water content.

The extent of the saturated zone above the water table depends on how a particular new phreatic surface evolved. The levels of saturation maintained following a fall in the groundwater level are higher than those resulting from capillary rise alone.

Depending on the type of soil and the history of water level changes, the total unit weight of soil above and below the water table may be measurably different. However, because most soils retain a significant amount of water after drainage, it is usually not appropriate to calculate stresses above the water table using dry unit weight equivalent to that of the saturated soil below.

Another aspect of importance is the conclusion that the lowering of the groundwater level will increase effective stresses below the original water table. An increase in effective stress will cause settlement of compressible strata. This is the reason why dewatering of an excavation may lead to damage of adjacent buildings. In order to avoid this problem recharging of the aquifer may be required in critical areas.

7.3.3 Drainable Pore Water

Figure 7.10 illustrates the changes which occur in the saturation of a soil when the water table is lowered from a depth z_1 to depth z_2. Because of capillary forces, the water does not drain out completely above the new water table. The ratio of the volume of water which will drain from an element of soil under gravity to its total volume is termed specific yield or phreatic storage coefficient. Typical values for coarse sands and gravels range from 0.2 to 0.3.

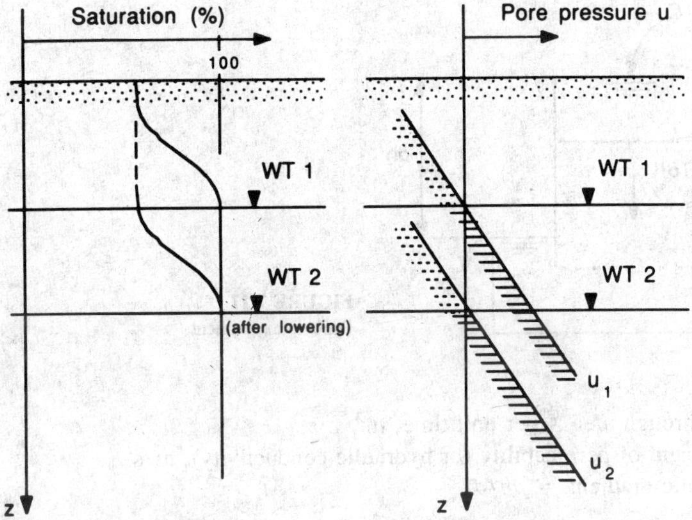

FIGURE 7.10
Saturation with changing water table.

The *specific retention* is a measure of the amount of water retained after the *specific yield* has been released. The sum of the specific yield and the specific retention must equal the porosity of the porous medium, or we can write:

$$\text{Specific retention} = \text{porosity} - \text{specific yield}$$

As pointed out earlier, a decrease in pore pressure means an increase in effective stress which leads to compression of the soil skeleton. This causes some water to be released from the soil which remains fully saturated. In soil mechanics this process is called consolidation. In groundwater hydrology the volume of water released from the soil per unit volume of aquifer per unit change in head (or water pressure) is called specific storage or *specific mass storativity*. If the specific mass storativity is multiplied by the thickness of the water-bearing layer, it becomes the *storage coefficient* (or storativity) of an aquifer.[1]

7.3.4 Darcy's Law

Figure 7.11 illustrates a simple experiment in which water is made to flow under a constant head dh through a column of sand of length dx. Darcy conducted this experiment some 130 years ago and found the following relationship:

$$Q = kiA \tag{7.9}$$

[1]The storativity of an unconfined acquifer is the same as its specific yield. For confined aquifers only the term "storativity" is used; in this case the definition of specific yield would not make sense, since the soil remains saturated. (Aquifer types are defined in Sec. 8.1.)

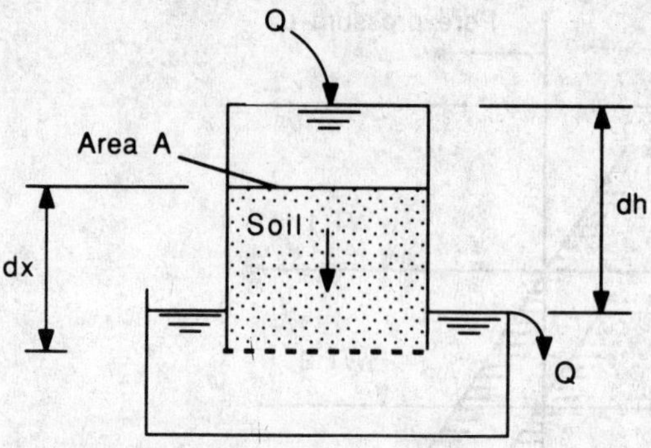

FIGURE 7.11
Darcy's experiment.

where Q = flow through area A per unit time, m^3/s
k = coefficient of permeability (or hydraulic conductivity), m/s
i = hydraulic gradient = dh/dx

Darcy's law is only valid for laminar flow. In most groundwater problems gradients are small and the motion is indeed laminar. However, this may not be the case in the vicinity of pumped wells or for water flow in very fine soils such as clays.

In the traditional soil mechanics literature, the coefficient of permeability is quoted in units of centimeters per second. Hydrologists may use all kinds of other units, such as meters per day. In these notes either centimeters per second or meters per second will be used. Typical values are:

Coarse gravel	$k = 10$ cm/s
Fine gravel	$k = 1$ cm/s
Coarse sand	$k = 10^{-1}$ cm/s
Sandy gravel	$k = 10^{-2}$ cm/s
Fine sand	$k = 10^{-3}$ cm/s
Silty sand	$k = 10^{-5}$ cm/s
Clay	$k < 10^{-7}$ cm/s

These values are meant as a guide only. The coefficient of permeability of a particular soil can easily vary by a factor of 100 if the soil is deposited or compacted at different densities. The determination of k in the field and laboratory will be discussed in Sec. 9.1.1.

The coefficient of permeability k depends on the density of the fluid and its viscosity v. The desire to have a permeability value which is independent of the fluid properties has led to the definition of the intrinsic permeability which is only related to the properties of the porous medium. With

$$k_i = \text{intrinsic permeability} \quad m^2$$

$$\eta = \text{dynamic viscosity} \quad t/m/s$$

$$\rho = \text{mass density} \quad t/m^3$$

g = acceleration of gravity m/s^2

v = kinematic viscosity m^2/s

The intrinsic permeability is calculated as follows:

$$k_i = k\frac{\eta}{\rho g} = k\frac{v}{g} \quad m^2 \tag{7.10}$$

This term is not commonly used in soil engineering but does find application in the study of pollutant flow in the ground.

In groundwater hydrology, flow through an aquifer is often expressed per unit width. The cross-sectional area A of an aquifer can be expressed as the product of aquifer height h and width b. Darcy's equation then becomes

$$Q = kihb = hkib = Tib \quad m^3/s \tag{7.11}$$

where T = transmissivity (s^{-1}) = kb.

The transmissivity term has proved useful in describing the in-plane permeability of geotextiles (Sec. 10.2).

PROBLEMS

Prefixes indicate problem type: C = calculations, B = brief answer, M = multiple choice.

Section 7.1 (Objectives and Techniques)

B7.1. Name four methods of controlling groundwater.
(*a*) ____________
(*b*) ____________
(*c*) ____________
(*d*) ____________

B7.2. Name two geotechnical processes which could be used to eliminate or control groundwater but do *not* necessarily result in a reduction of the amount of free pore water present in the soil.
(*a*) ____________
(*b*) ____________

B7.3. Name four reasons for dewatering soil or rock.
(*a*) ____________
(*b*) ____________
(*c*) ____________
(*d*) ____________

M7.4. The following method of ground modification generally results in a reduction of the amount of free water present in the soil (by water draining out of the soil or chemically combining with other molecules):
(*a*) Compaction
(*b*) Use of geomembranes
(*c*) Electroosmosis
(*d*) Soil freezing

M7.5. Lowering the groundwater level may cause
(*a*) Liquefaction.
(*b*) Settlement.
(*c*) A decrease in bearing capacity.
(*d*) Decreased effective stresses in the soil mass.

M7.6. The suitability of various groundwater lowering techniques depends significantly on
(*a*) The bearing capacity of the soil.
(*b*) The grain size distribution.
(*c*) The cohesiveness of the soil.
(*d*) The physicochemical properties of the soil.

M7.7. Dewatering of soil may
(*a*) Cause swelling.
(*b*) Reduce liquefaction potential.
(*c*) Reduce bearing capacity.
(*d*) Reduce effective unit weight.

M7.8. The *most* suitable soil for dewatering by gravity wells or slots is
(*a*) Gravel.
(*b*) Sand.
(*c*) Silt.
(*d*) Clay.

Section 7.2 (Traditional Dewatering Methods)

M7.9. Vacuum wells
(*a*) Are the same as eductor wells.
(*b*) May make dewatering in fine sands and silts possible.
(*c*) Always need submerged pumps.
(*d*) Are gravity wells using vacuum pumps.

M7.10. Vacuum wells, rather than gravity wells, may be needed for
(*a*) Coarse gravel.
(*b*) Fine gravel.
(*c*) Coarse to medium sands.
(*d*) Fine sands and silts.

B7.11. Name three types of wells which are different in hydraulic action.
(*a*) ____________
(*b*) ____________
(*c*) ____________

Section 7.3 (Fundamental Soil-Water Relationships)

C7.12. A soil has a water content of 45.2% when fully saturated (below the groundwater level). The specific gravity of the soil solids is 2.65.

Total unit weight γ_t = ____________ kN/m3
Total density ρ_t = ____________ g/cm^3
Void ratio e = ____________
Porosity n = ____________ %

Now assume that the groundwater level drops and some water drains out of the soil, reducing w to 40%. If no compression of the soil mass occurs, the new values are:

Total density ρ_t = ____________ t/m³
Total unit weight γ_t = ____________ kN/m³
Saturation S = ____________ %

C7.13. A saturated soil has a water content of 26%. The total saturated density is 2 t/m³. Determine

Dry density γ_d = ____________ t/m³
Submerged density or buoyant density ρ_b = ____________ t/m³
Specific gravity of soil solids G_s = ____________

M7.14. A soil contains 80 g of soil solids and 20 g of water. The volume of the soil solids is 30 mL. The total volume of the soil is 60 ml. Correspondingly,
(*a*) The water content is 20%.
(*b*) The void ratio is 0.5.
(*c*) The total density is 1.66 t/m³.
(*d*) The saturation is 133%.

M7.15. If the total stress in a saturated soil is 100 kPa and the effective stress is 60 kPa, the implied pore pressure is
(*a*) 0 kPa.
(*b*) 40 kPa.
(*c*) 60 kPa.
(*d*) 100 kPa.

M7.16. Lowering the groundwater level in a soil mass may cause settlement because it
(*a*) Decreases the weight of the soil.
(*b*) Causes downward seepage pressure.
(*c*) Increases effective vertical stresses in the soil.
(*d*) Makes the soil softer.

M7.17. In groundwater hydrology, the volume of water released from a unit volume of saturated soil per unit change in head (or pore water pressure) is called
(*a*) Specific storage.
(*b*) Storage coefficient.
(*c*) Conductivity.
(*d*) Permeability.

M7.18. Above the phreatic surface
(*a*) A soil is completely dry.
(*b*) A soil is always saturated.
(*c*) Pore pressures are positive (higher than air pressure).
(*d*) Pore pressures are negative (less than air pressure).

M7.19. A permeability of 10^{-3} cm/s (10^{-5} m/s) is typical for a
(*a*) Fine gravel.
(*b*) Coarse sand.
(*c*) Fine sand.
(*d*) Silty clay.

B7.20. The following notation is used to describe an element of soil: total volume = V, volume of soil solids = V_s, volume of water = V_w, weight of soil solids = W_s, and weight of water = W_w. Using these designations define:

(*a*) Water content = ______________ .
(*b*) Porosity = ______________ .
(*c*) Void ratio = ______________ .
(*d*) Saturation = ______________ .

B7.21. A sample of wet fly ash (G_s = 2.0) has a volume of 150 mL and weighs 240 g. After being dried in an oven, the sample weighs 200 grams. Determine
(*a*) Water content w = ______________%.
(*b*) Dry density ρ_d = ______________ t/m^3.
(*c*) Saturation S = ______________%.
(*d*) Void ratio e = ______________ .

B7.22. Define *one only* of the following terms:
(*a*) Specific yield (or phreatic storage coefficient)
(*b*) Specific retention
(*c*) Specific storage (or specific mass storativity)

B7.23. Darcy's law expressed in Q, k, i, and A reads:

______________ = ______________

where k = ______________ in units of ______________ .
i = ______________ in units of ______________ .
A = ______________ in units of ______________ .
Q = ______________ in units of ______________ .

CHAPTER 8

HYDRAULICS OF SLOTS AND WELLS

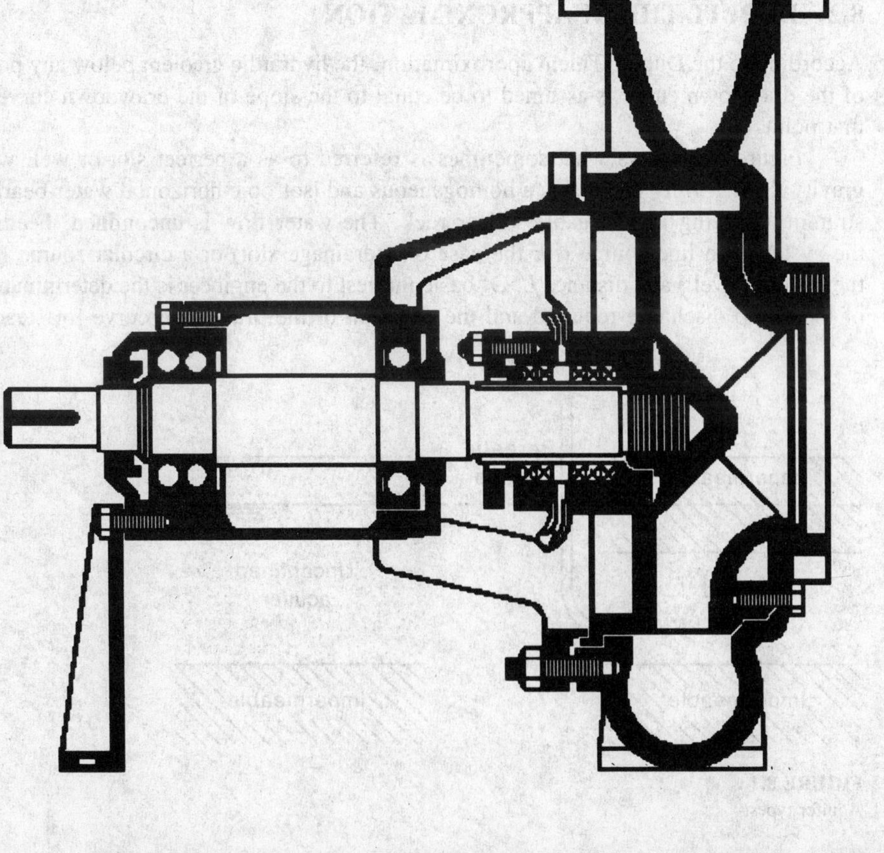

8.1 AQUIFER TYPES

A permeable soil or rock formation which stores or transmits significant amounts of water is called an *aquifer*. If it is fully saturated and confined by impervious layers at its upper and lower boundaries, it is referred to as an *artesian aquifer*. If a permeable layer is only partially filled with water, it is described as an *unconfined aquifer* (Fig. 8.1). In nature, semiconfined (or leaky) acquifers and semiunconfined aquifers may occur [for definitions see textbooks on groundwater hydrology, e.g., Hazel (1975)].

When sinking a bore, several water-bearing layers may be encountered, and therefore more than one groundwater level may be identified, possibly including temporary or permanent perched groundwater. A perched aquifer is an unconfined aquifer separated from an underlying body of groundwater by an unsaturated zone.

For the analysis of the most basic problems in the flow of water to drainage slots and wells, an aquifer is idealized to have horizontal boundaries, to consist of homogeneous and isotropic porous material, and to have infinite extent. The simplest solutions are derived for the case where the slot or well fully penetrates the aquifer. In terms of hydraulic theory, Darcy's law is assumed to be valid, and simplifying assumptions are made with respect to the hydraulic gradient.

8.2 DUPUIT-THIEM APPROXIMATION

According to the Dupuit-Thiem approximation, the hydraulic gradient below any point of the drawdown curve is assumed to be equal to the slope of the drawdown curve at that point.

Figure 8.2 shows what sometimes is referred to as a perfect slot or well with gravity flow. It fully penetrates a homogeneous and isotropic horizontal water-bearing stratum overlying impermeable soil or rock. The water flow is unconfined. Feeding the system is a line source (for the case of a drainage slot) or a circular source (for the case of a well) at a distance L. Of basic interest to the engineer is the determination of the pump discharge required and the equation of the drawdown curve for steady-

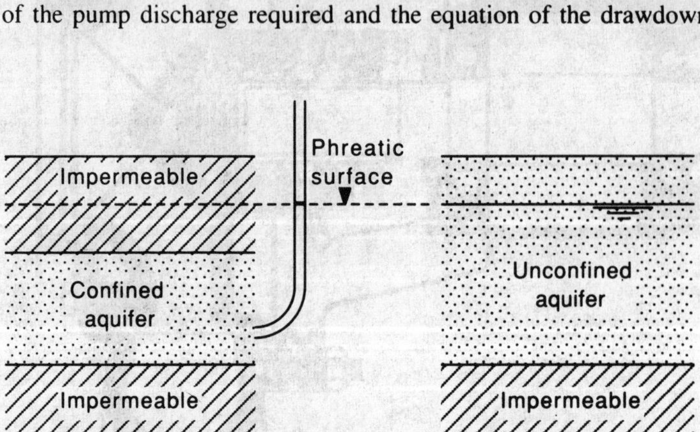

FIGURE 8.1
Aquifer types.

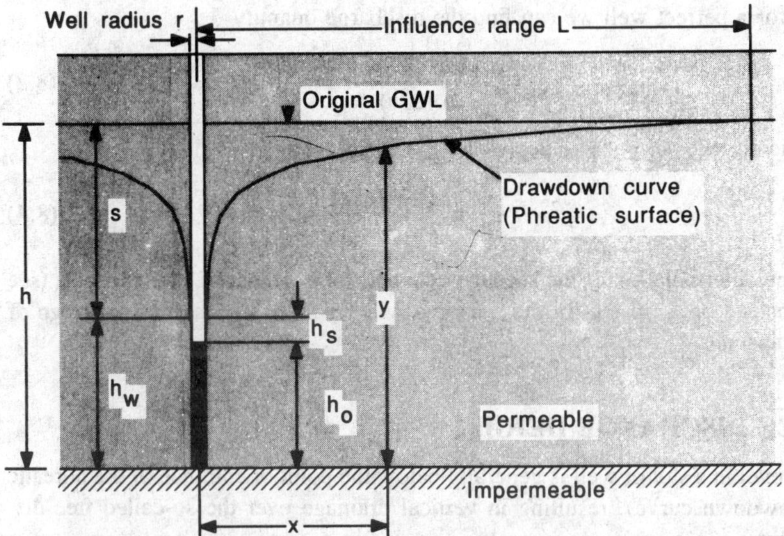

FIGURE 8.2
Perfect slot or well with gravity flow.

state seepage conditions. Their derivation makes use of Darcy's law and the Dupuit-Thiem approximation.

For a perfect slot we first find the discharge quantity q per unit length as [from Eq. (7.9)]

$$q = kiA \qquad \text{where } A = y \text{ (unit slice)}$$

$$= k\frac{dy}{dx}y$$

Now the variables are separated and both sides of the equation are integrated:

$$\int_{h_w}^{h} y\,dy = \frac{q}{k}\int_{0}^{L} dx$$

$$\frac{h^2 - h_w^2}{2} = \frac{qL}{k}$$

Solving for q (flow from one side), we find

$$q = \frac{(h^2 - h_w^2)k}{2L} \tag{8.1}$$

The total flow from both sides is equal to $2q$.

For the equation of the drawdown curve we can derive

$$h^2 - y^2 = \frac{L - x}{L}(h^2 - h_w^2) \tag{8.2}$$

Similarly, for a perfect well we can find the discharge quantity

$$Q = \frac{\pi k(h^2 - h_w^2)}{\ln(L/r)} \tag{8.3}$$

The phreatic surface (or cone of depression) is defined by

$$y^2 - h_w^2 = \frac{Q \ln(x/r)}{\pi k} \tag{8.4}$$

For the determination of the seepage quantities it is generally safe to use h_0 (see Fig. 8.2) instead of h_w in the formula, as this will result in a conservative design of the pump capacity.

8.3 FREE DISCHARGE HEIGHT

The water level in a slot or well is generally below the point of entrance of the phreatic line (or drawdown curve), resulting in vertical drainage over the so-called free discharge height

$$h_s = h_w - h_0 \tag{8.5}$$

The free discharge height depends on the drawdown in the well (or slot). The steeper the phreatic line near the well, the more significant h_s is. For rough calculations the existence of a free discharge height may be ignored.

Formulas for estimating h_s are of an empirical nature and are usually based on model tests, results of which appear to vary considerably. For slots, the distance h_s may be determined using diagrams proposed by Chapman (1956). The diagram shown in Fig. 8.3 is recommended for gravity flow to a fully penetrating slot. For wells, Kezdi (1969) gives a formula attributed to Ollos (elsewhere credited to Ehrenberger):

$$h_s = \frac{C(h - h_0)^2}{h} \tag{8.6}$$

Ollos proposed a value of $C = 0.5$. Herth and Arndts (1973) give no less than six other formulas for estimating the free discharge height in wells. Some are reproduced in App. 8B.

The ratio $(h - h_w)/(h - h_0)$ is also called the well efficiency, usually expressed in percent. It increases if the well screen has insufficient openings or is too short, if the surrounding filter zone restricts the flow, or if construction of the well contaminates the adjacent soil with fines (as sometimes happens with drilling fluids). Inefficient wells mean higher costs of pumping.

8.4 INFLUENCE RANGE

For drainage slots the influence range L should be known with reasonable accuracy, since the discharge quantity is indirectly proportional to it [Eq. (8.1)]. In the well formula [Eq. (8.3)] L (also called the radius of influence) appears in a logarithmic

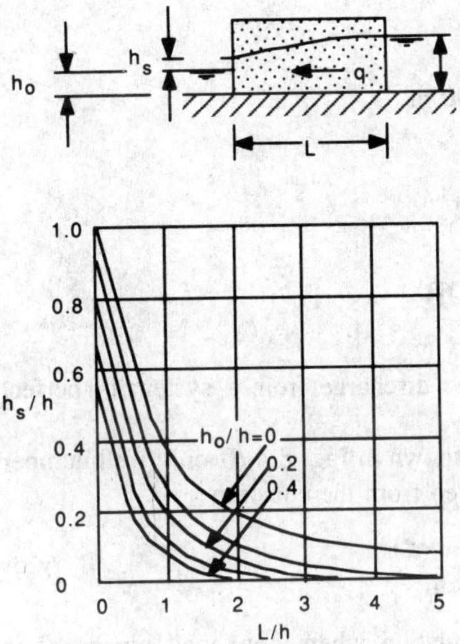

FIGURE 8.3
Free discharge surface for slots. [*Chapman (1956).*]

term, and a good assessment of the coefficient of permeability k is much more important than the value of the influence range.

Strictly speaking, pumping from an aquifer will only produce a steady-state condition of groundwater flow if the aquifer is continually recharged from somewhere. If an open watercourse or reservoir is near the dewatering point, the distance to that source will represent the distance L in the discharge formula for slots. Special formulas for wells, e.g., near a line source, are given in Apps. 8B and 8C. If no recharge occurs, the influence range will continually increase with time, although at a decreasing rate. For this nonsteady state of flow for an ordinary perfect well, Kozeny (1953) quotes a time-dependent expression for L as follows:

$$L = 1.5 \frac{\sqrt{hkt}}{n} \tag{8.7}$$

The term n represents the porosity of the soil.

In most cases it is sufficiently accurate to obtain an approximate value for L according to an empirical formula proposed by Sichardt (1928):

$$\begin{aligned} L &= C(h - h_w)\sqrt{k} \\ &= Cs\sqrt{k} \end{aligned} \tag{8.8}$$

For s in meters and k in meters per second, the value of the constant is

$$C = \begin{cases} 3000 \text{ for wells (Sichardt)} \\ 1500 \text{ to } 2000 \text{ (U.S. Corps of Engineers) for single-line well points} \end{cases}$$

The latter values given by Mansur and Kaufmann are quoted in Leonards (1962).

According to Kezdi and Marko (1969) the following values may serve as a guide for unconfined aquifers:

Coarse gravel and cobbles	$L = 500$ m
Coarse gravel	$L = 100$ to 150 m
Medium gravel	$L = 50$ m
Sand	$L = 33$ m
Fine sand	$L = 5$ to 10 m

8.5 FORCHHEIMER EQUATION FOR MULTIWELL INSTALLATIONS

Forchheimer (1930) derived a formula for the discharge from a system of perfect gravity flow wells of equal length and capacity.

Consider a point P in a field of wells as shown in Fig. 8.4. If only well number 1 is active, the water level y at P can be derived from the equation

$$Q_1 = \frac{\pi k(h^2 - y^2)}{\ln L - \ln x_1} \tag{8.9}$$

A similar equation can be written for the situation where only well number 2 is operating, and so on. For the simultaneous pumping from n equivalent wells, Forchheimer found that the total discharge quantity is equal to

$$Q = \frac{\pi k(h^2 - y^2)}{\ln L - (1/n) \ln x_1 x_2 \cdots x_n} \tag{8.10}$$

In deriving this formula it is assumed that the aquifer is thin relative to its horizontal expansion and that the rules of potential theory apply.

For a circular arrangement of wells (Fig. 8.5), the water level y at the center of the excavation can readily be calculated from a simplified version of Eq. (8.10) as follows:

$$Q = \frac{\pi k(h^2 - y^2)}{\ln L - \ln a} \tag{8.11}$$

The water level h_0 of an individual well in the circular group can be computed from Eq. (8.10) with x_1 equal to the well radius and $x_2 \cdots x_n$ the distance from well number 1 to all the other wells (Fig. 8.6).

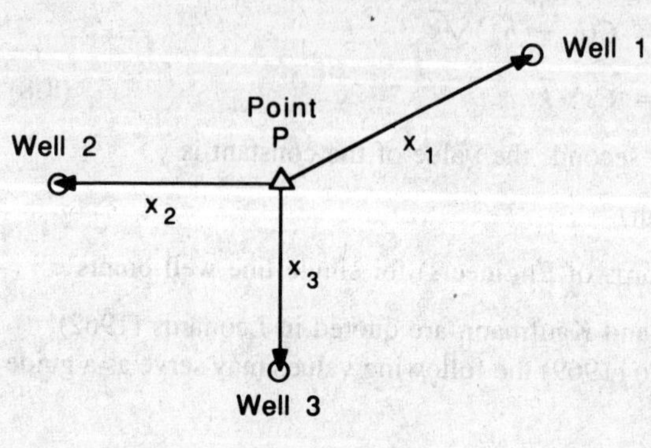

FIGURE 8.4
Multiple wells—notation.

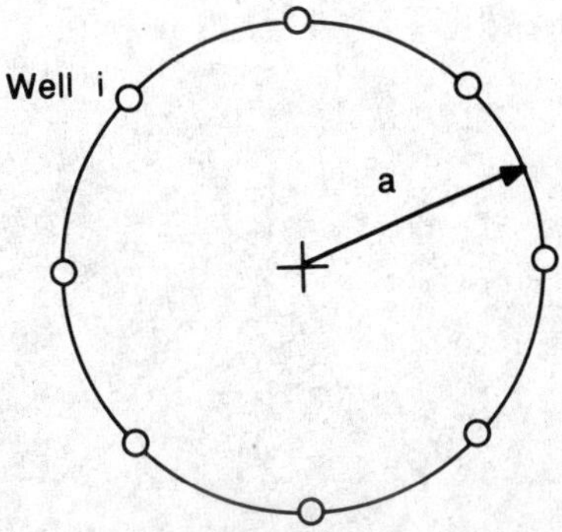

FIGURE 8.5
Circular arrangement of wells.

8.6 IMPERFECT WELLS AND OTHER CASES

Partially penetrating (imperfect) wells occur more often in practice than fully penetrating wells. According to Schröder (1966) practical rules suggest that the discharge quantity Q calculated for a perfect well is to be increased by 10 to 30% for $t = 0$ to $t > 2h$, respectively (Fig. 8.7). When the discharge from imperfect wells is estimated, the fact that vertical permeability of natural deposits is often more than 10 times smaller than horizontal permeability should also be considered. Soil layers below the well tip may therefore only contribute relatively little water. Another factor which may reduce the inflow of water from below a well is the fact that well points and sumps often only allow side entry, rather than side and bottom entry of water.

Artesian wells (Fig. 8.8) and slots have been analyzed for perfect and imperfect conditions by Kozeny, Muskat, and others, as quoted by Mansur and Kaufmann in Leonards (1962). Further references are given by Schröder (1966).

An approximate evaluation of the discharge from vacuum wells was developed by Széchy (1959).

In dewatering projects encountering inhomogeneous, nonisotropic soil deposits with uncertain recharge patterns, and where irregularly shaped excavations are needed, there is still a considerable amount of engineering judgment required in order to design

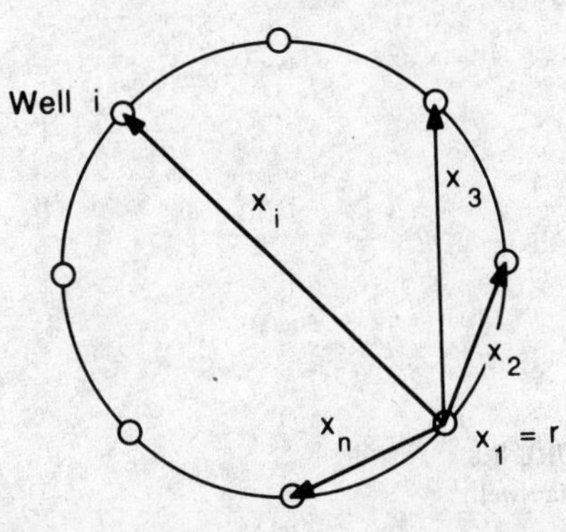

FIGURE 8.6
Water level in individual well—notation.

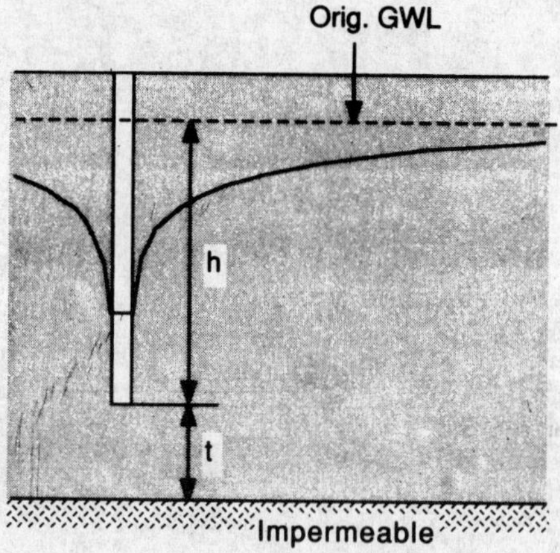

FIGURE 8.7
Partially penetrating well.

a satisfactory and economical well system, unless an extensive field testing program is undertaken.

Discharge formulas for the most common arrangements of slots and wells with gravity flow are given in Apps. 8A, 8B, and 8C for slots, single wells, and multiple wells, respectively.

8.7 DEVELOPMENT OF DRAWDOWN WITH TIME

In most dewatering projects, the major design decisions, such as the determination of the number of wells and the pump capacity required, can be made assuming a steady-state flow condition and using a constant influence range L as estimated by Sichardt's

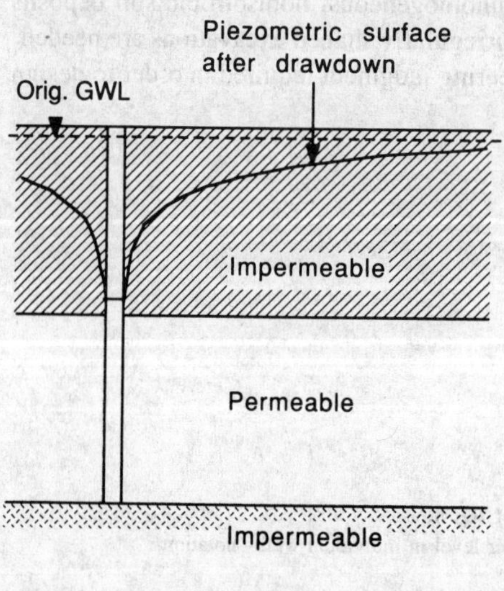

FIGURE 8.8
Artesian well.

empirical formula. Howeve⁻, if the construction schedule is tight and relatively slow draining soils are involved, it may be important to assess how long the pumps have to be operated before it is safe to commence excavating. Extra pump capacity may be required for the initial period of dewatering. The Theis formula, particularly as modified by Jacobs, provides a practical solution to this problem.

8.7.1 Theis Formula

Theis (1935) developed a nonequilibrium well formula for gravity flow to a single well, making the following assumptions:

- *"Perfect" well conditions exist*. The well fully penetrates a homogeneous, isotropic horizontal aquifer overlying an impermeable stratum.
- *The Dupuit-Thiem approximation holds*. The hydraulic gradient below any point of the drawdown curve is assumed to be equal to the slope of the drawdown curve at that point.
- *There is no recharge of the aquifer*
- *Water flows out of the pores of the soil as quickly as the drawdown of the phreatic surface occurs.*
- *The drawdown* s *is small relative to the aquifer thickness* h *so that* h − s *is approximately equal to* h. This is equivalent to assuming an artesian aquifer of thickness $h = m$.

The derivation starts with equating the water flow through a cylindrical area around the well at a distance x to the volume of water removed from the soil beyond x due to lowering of the phreatic surface (Fig. 8.9). The formula is usually written as

$$s = \frac{Q}{4\pi km} W(u) \tag{8.12}$$

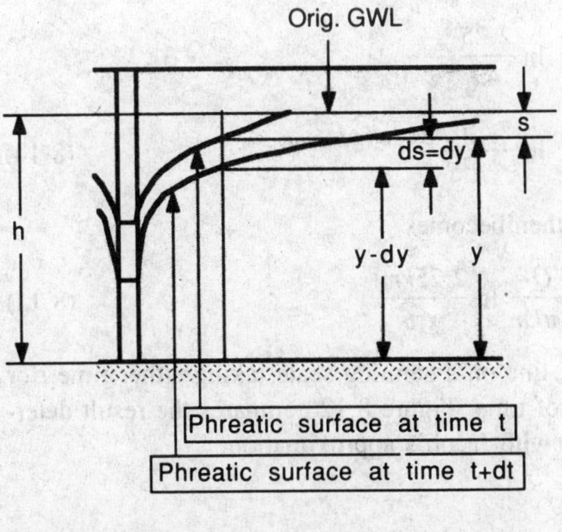

FIGURE 8.9
Lowering of phreatic surface with time.

The function $W(u)$ is the so-called well function, defined by

$$W(u) = -0.5772 - \ln u + u - \frac{u^2}{2 \times 2!} + \frac{u^3}{3 \times 3!} - \frac{u^4}{4 \times 4!} \pm \cdots \tag{8.13}$$

The variable u is given by

$$u = \frac{x^2 S}{4tkm}$$

The product of the coefficient of permeability k (also called the hydraulic conductivity) and the thickness of the aquifer m is referred to as the *transmissivity* of the aquifer. The term S represents the storage coefficient (or storativity, specific yield) of the water-bearing layer (see Sec. 7.3.3). For an unconfined aquifer it is equal to the ratio of the volume of drainable water over the total volume of the soil. Numerically the storativity must be equal to or less than the soil porosity, which, in soil mechanics, is designated with the letter n and defined as the volume of voids to the total volume of an element of soil. For confined aquifers, which remain saturated during pumping, S may be as low as 10^{-5}; for unconfined aquifers, it ranges from 0.01 to 0.3.

The function $W(u)$ is tabulated in most handbooks on hydrology and groundwater, e.g., in the book *Groundwater and Wells* [Driscoll (1986, 2d ed.)].

Figure 8.10*a* gives a numerical example of the drawdown developed at various distances from the well with time. If plotted on semilog paper, the drawdown curves plot as straight lines (Fig. 8.10*b*), except perhaps very close to their well or at distances approaching the influence range. A similar picture is obtained if the drawdown at specific locations is plotted versus log (time) (Fig. 8.11).

8.7.2 Modification by Jacob (1940)

If pumping time t is sufficiently large, the value u is small. For $u < 0.05$ the well function may be approximated by

$$\begin{aligned} W(u) &= -0.577 - \ln u \\ &= \ln \frac{2.25}{4u} \\ &= \ln \frac{2.25kmt}{x^2 S} \end{aligned} \tag{8.14}$$

The drawdown as a function of time then becomes

$$s = \frac{Q}{4\pi km} \ln \frac{2.25kmt}{x^2 S} \tag{8.15}$$

This function plots as a straight line on a semilog scale taking either time t or distance x as a variable as a function of time. Figure 8.12 compares the result determined according to the Theis formula with Jacob's approximation.

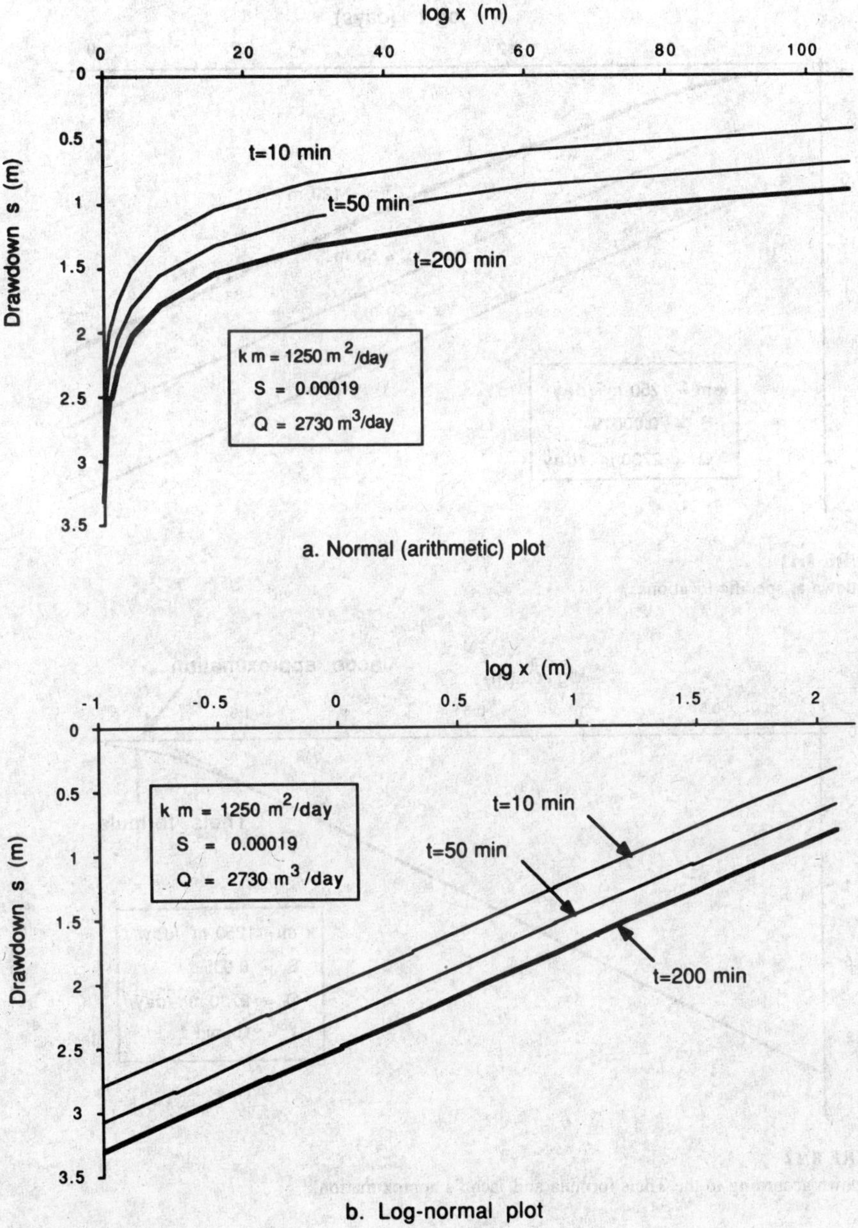

FIGURE 8.10
Drawdown at various time intervals.

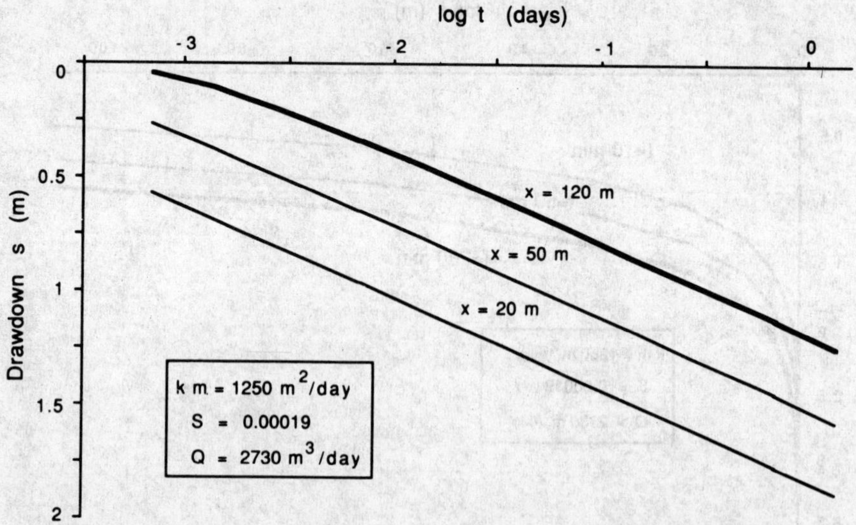

FIGURE 8.11
Drawdown at specific locations.

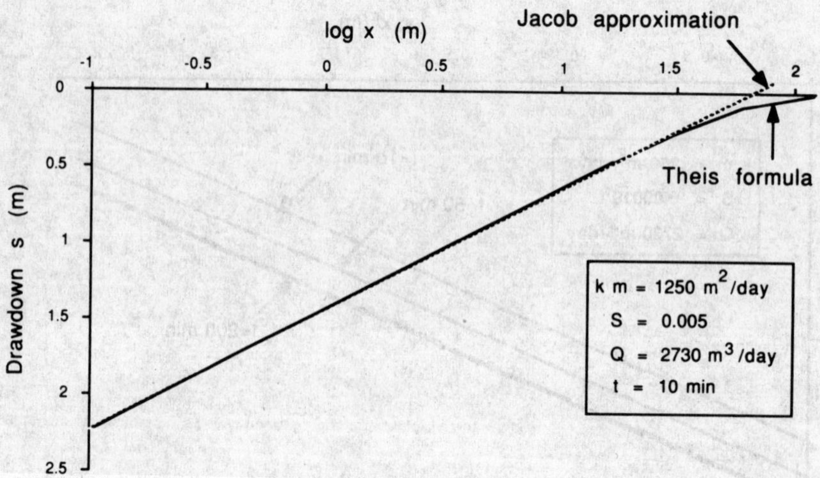

FIGURE 8.12
Drawdown according to the Theis formula and Jacob's approximation.

8.7.3 Extension to Unconfined Flow

To extend Jacob's formula from artesian to unconfined flow, we set $m = h$ and make use of the analogy

$$\underset{\text{artesian}}{2ms} = \underset{\text{unconfined}}{h^2 - y^2} \tag{8.16}$$

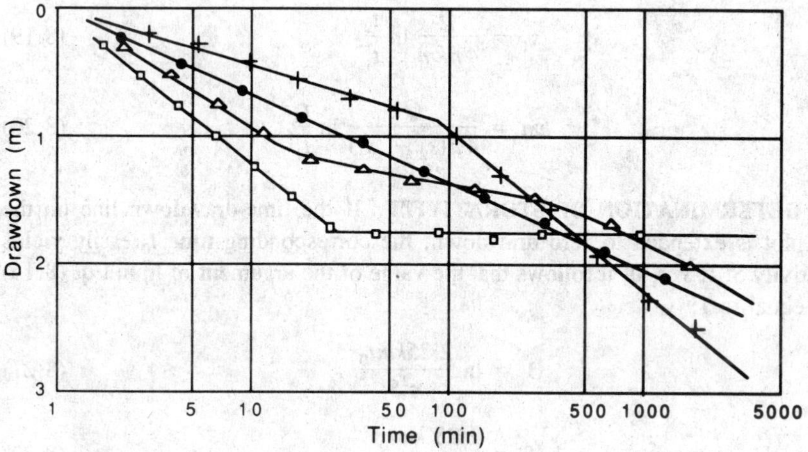

● Ideal curve
▫ Recharge of aquifer occurs within influence range
+ Impervious boundary encountered
△ Transient influence evident (other wells, tides, fluctuating recharge)

FIGURE 8.13
Time-drawdown curves encountered in the field.

which holds for fully penetrating wells. We then find

$$s = \frac{h^2 - y^2}{2h} \tag{8.17}$$

and

$$\frac{h^2 - y^2}{2h} = \frac{Q}{4\pi kh} \ln \frac{2.25kht}{x^2S} \tag{8.18}$$

8.7.4 Interpretation of Time-Drawdown Measurements

As discussed at the beginning of Sec. 8.7.1, Theis had to make a number of simplifying assumptions in order to arrive at a time-drawdown relationship. It would be rare that all these assumptions are met fully in natural geologic and hydrologic conditions. Driscoll (1986) as well as Powers and Burnett (1986) give good discussions of the effect of nonideal conditions on the results of field pumping tests. Figure 8.13 illustrates some of these cases.

8.7.4.1 DETERMINATION OF PERMEABILITY. Equation (8.15) can be used to determine the permeability k (or transmissivity $= km$) of an aquifer from time-drawdown measurements in a single observation hole, if the plot is a straight line on semilog paper, as is theoretically expected for ideal conditions. If s_1 and s_2 are the drawdowns observed at times t_1 and t_2, we can write

$$s_2 - s_1 = \frac{Q}{4\pi km} \ln \frac{t_2}{t_1} \tag{8.19}$$

and

$$km = \frac{Q}{4\pi(s_2 - s_1)} \ln \frac{t_2}{t_1} \tag{8.20}$$

8.7.4.2 DETERMINATION OF STORATIVITY. If the time-drawdown line on the semilog plot is extended to zero drawdown, the corresponding time t_0 easily yields the storativity S. If $s = 0$, it follows that the value of the argument of ln in Eq. (8.15) must be equal to 1:

$$1 = \ln \frac{2.25kmt_0}{x^2S} \tag{8.21}$$

or

$$S = \frac{2.25\ kmt_0}{x^2} \tag{8.22}$$

8.7.4.3 INTERPRETATION OF THE INFLUENCE RANGE. A value for the influence range can be calculated from Jacob's approximation by setting $s = 0$ and solving for x. We then obtain

$$L = \sqrt{\frac{2.25kmt}{S}} = 1.5\sqrt{\frac{kmt}{S}} \tag{8.23}$$

This is the same expression Kozeny (1953) gives when referring to work published in the 1930s.

Sichardt's empirical estimate of the influence range L can therefore be viewed as being associated with the development of drawdown of an unrecharged aquifer at a particular point in time. Since most natural aquifers are recharged, either underground or through rainfall, Sichardt's formula still proves to be a useful approximation.

PROBLEMS

Prefixes indicate problem type: C = calculations, M = multiple choice, B = brief answer.

Calculations

C8.1. Determine the discharge quantity q for a 100-m-long slot assuming "perfect" conditions and a free discharge height according to Chapman. The following data is given:

$$k = 2 \times 10^{-6}\ \text{m/s}$$

$$h = 4\ \text{m}$$

$$h_0 = 0.4\ \text{m}$$

C8.2. Wells are located at the corner of a square of width 8 m. The aquifer is 12 m thick. The water level in the center has to be lowered by 4 m. Determine the pump rate required. The coefficient of permeability is 0.003 m/s, and the well radius is 0.3 m. Assume h_0 equals 7.6 m for the estimate of the influence range.

C8.3. Two wells are 22.5 m apart. From each well, 25 L/s are pumped. The water level in the wells is 8.87 m above the base of the aquifer, which is 15 m thick. The well radius is 0.3 m. Estimate the permeability of the soil. Use Sichardt's formula to estimate the influence range, assuming an initial value of $k = 0.0005$ m/s. Iterate to find a more accurate value.

C8.4. Three wells are equidistant from each other. For the following data calculate the water level in the center of the triangle and in a well:

$$h = 15 \text{ m}$$

$$\text{Distance between wells} = 19.5 \text{ m}$$

$$r = 0.2 \text{ m}$$

$$Q = 47 \text{ L/s}$$

$$k = 0.0005 \text{ m/s}$$

C8.5. Two wells are spaced 10 m apart in an unconfined aquifer with a thickness of 5 m. The radius of the wells is 0.2 m, and the coefficient of permeability is 0.0004 m/s. If each well yields 0.5 L/s after steady-state conditions are reached, what is the drawdown at a point 4 m away from one well and 6 m away from the other well?

C8.6. Water is pumped from a 20-m-thick confined aquifer at a rate of 2000 m^3/day from a single well. In an observation hole at a distance of 70 m from the well, the drawdown after 10 min of pumping was 0.66 m; after 1000 min, it was 1.92 m. Calculate the coefficient of permeability (m/day) and the storativity.

C8.7. A circular excavation (diameter = 60 m) is to be made in an unconfined aquifer which is 12 m thick and underlain by an impermeable layer. The total amount of water pumped is 0.2 m^3/s. The soil has a permeability of 10^{-4} m/s, and the porosity is 0.3. Determine the time required to draw down the level at the center of the excavation by 3 m. Use the Jacob formula.

Multiple Choice

M8.8. Discharge quantities from wells are directly proportional to the
- (*a*) Diameter of the well.
- (*b*) Permeability of the soil.
- (*c*) Soil grain size.
- (*d*) Pump capacity.

M8.9. The Dupuit-Thiem approximation
- (*a*) Is used in the derivation of a simple formula for the estimation of the influence range.
- (*b*) Says that the gradient of the water flowing toward the well is the same at equal distances from the well regardless of the depth.
- (*c*) Is used to derive Darcy's law.
- (*d*) Allows us to estimate the free discharge height.

M8.10. With respect to the level at which the phreatic surface intersects the well perimeter, the water level in the well itself is
(*a*) Always higher.
(*b*) Exactly the same regardless of pumping rate and soil type.
(*c*) About the same or lower.
(*d*) Higher or lower depending on soil type.

M8.11. According to Sichardt, the influence range is *not* related to the
(*a*) Drawdown in the well.
(*b*) Configuration of wells.
(*c*) Soil permeability.
(*d*) Well radius.

M8.12. Kozeny's formula for the influence range L is
(*a*) $\sqrt{hk}/n$.
(*b*) $h5\ \sqrt{kt}/n$.
(*c*) $1.5\sqrt{hkt}/n$.
(*d*) $mv^2/2$.

M8.13. In dewatering projects, the number of wells and the pump capacity required is estimated by assuming that
(*a*) Soil behaves as water.
(*b*) Steady flow conditions exist.
(*c*) Nonsteady flow conditions exist.
(*d*) There is no recharge of the aquifer.

M8.14. For natural sedimentary soil deposits we often find that vertical and horizontal permeabilities compare as follows:
(*a*) $k_v > 10k_h$.
(*b*) $k_h > 10k_v$.
(*c*) $k_v = k_h$.
(*d*) $k_v > 100\ k_h$.

M8.15. The equation most frequently used to assess drawdown with time for gravity flow toward a well
(*a*) Was proposed by Sichardt.
(*b*) Was proposed by Forchheimer.
(*c*) Makes use of the well function.
(*d*) Assumes the aquifer is continuously recharged.

M8.16. Theis developed a formula for gravity flow to a single well for
(*a*) Nonequilibrium conditions.
(*b*) Equilibrium conditions.
(*c*) Nonisotropic soil.
(*d*) Continuous recharge conditions.

Brief Answer

B8.17. Figure P8.1 shows a perfect well with gravity flow. Name the items that are labeled a, b, c, and d.

a = ____________
b = ____________
c = ____________
d = ____________

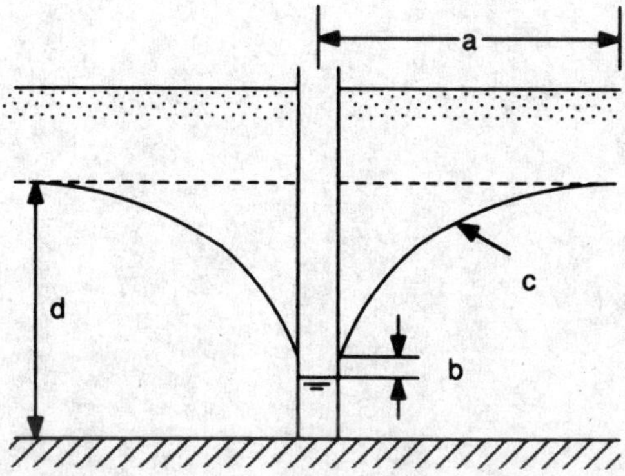

FIGURE P.8.1

B8.18. In the derivation of the equation of the drawdown curve for a slot or well, we make an approximation first proposed by Dupuit and Thiem. This means we assume ____________.

B8.19. Deriving a discharge formula for a "perfect" well or slot means we assume that four ideal or simplified conditions hold. List them.

(*a*) ____________
(*b*) ____________
(*c*) ____________
(*d*) ____________

B8.20. The designations h_s, h_w, and h_0 refer to water levels in and around the borehole. Define these terms and their interrelationship:

(*a*) h_s is the ________ .
(*b*) h_w is the ____________ .
(*c*) h_0 is the ____________ .
(*d*) h_s = ____________ .

B8.21. Identify the following formulas which are associated with the names of Darcy, Sichardt, Forchheimer, and Kozeny.

$$Q = \frac{\pi k(h^2 - y^2)}{\ln L - (1/n)\ln x_1 x_2 \cdots x_n}$$ ____________

$$Q = kiA$$ ____________

$$Q = \frac{2\pi r h_0 \sqrt{k}}{15}$$ ____________

$$Q = \frac{1.5\sqrt{hkt}}{n}$$ ____________

B8.22. Theis developed a formula for the development of drawdown with time. List two of the simplifying assumptions he made.

(*a*) ____________
(*b*) ____________

APPENDIX 8A

DRAINAGE SLOT FORMULAS

8A1 FULLY PENETRATING ("PERFECT") SLOTS

8A1.1 Unconfined Gravity Flow

For designations see Fig. A8.1.

The variable q = flow per unit length of slot.

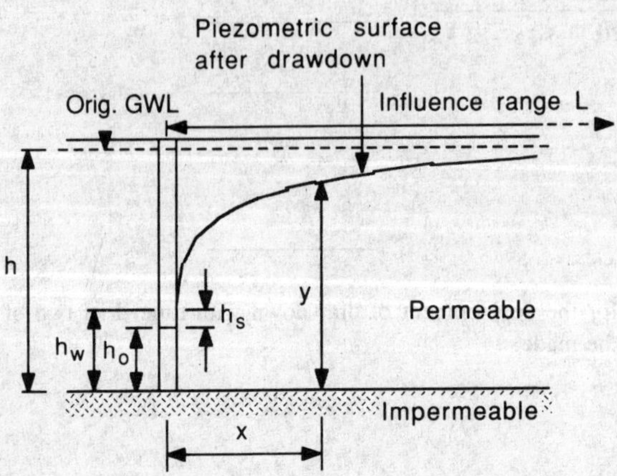

FIGURE A8.1

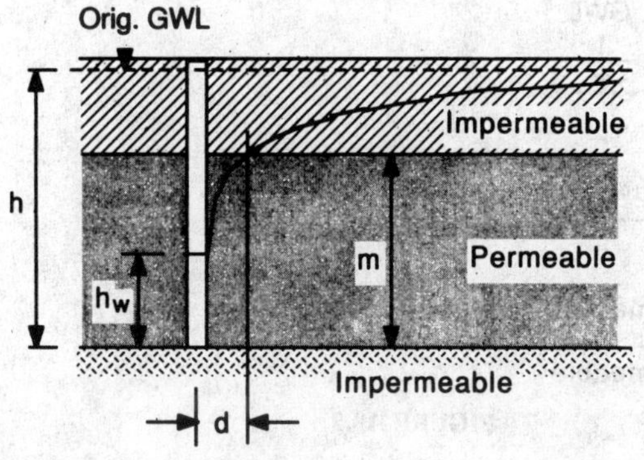

FIGURE A8.2

$$q = \frac{(h^2 - h_w^2)k}{2L} \tag{A8.1}$$

$$h^2 - y^2 = \frac{L - x}{L}(h^2 - h_w^2) \tag{A8.2}$$

The free discharge height h_s may be estimated according to Chapman (1957) from data given in Fig. 8.3.

8A1.2 Artesian Flow

For designations see Fig. A8.2.

For full artesian flow ($d = 0$, or $h_w \geq m$):

$$q = \frac{(h - h_w)km}{L} \tag{A8.3}$$

$$h - y = \frac{q(L - x)}{km} \tag{A8.4}$$

For partial artesian flow ($h_w < m$):

$$q = \frac{k[2m(h - m) + m^2 - h_w^2]}{2L} \tag{A8.5}$$

$$d = \frac{L(m^2 - h_w^2)}{2m(h - m) + m^2 - h_w^2} \tag{A8.6}$$

8A2 PARTIALLY PENETRATING SLOTS

8A2.1 Unconfined Flow, Single Slot

See Fig. A8.3.

$$q = \left[0.73 + \frac{0.27(T - t_0)}{T}\right]\frac{k}{2L}(T^2 - t_0^2) \tag{A8.7}$$

provided $L/T > 3$ [after Chapman (1957) as quoted in Leonards (1962)].

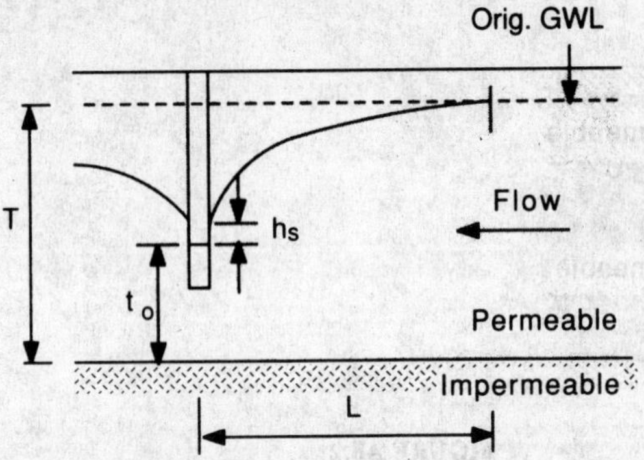

FIGURE A8.3

8A2.2 Unconfined Flow, Double Slot

See Fig. A8.4.

Residual head
$$t_d = t_0 \left[\frac{C_1 C_2}{L} (T - t_0) + 1 \right] \tag{A8.8}$$
$$(C_1 C_2)_{max} \approx 1.48$$

Factors C_1 and C_2 are from Figs. A8.5 and A8.6.

8A2.3 Artesian Flow, Single Line Source

See Fig. A8.7.

$$q = \frac{km(T - t_0)}{L + e} \tag{A8.9}$$

$$t_d = \frac{e(T - t_0)}{L + e} + t_0 \tag{A8.10}$$

Given L, m, and m_1, the distance e can be calculated from data presented in Fig. A8.8 [after Barron, as quoted by Mansur and Kaufman in Leonards (1962)].

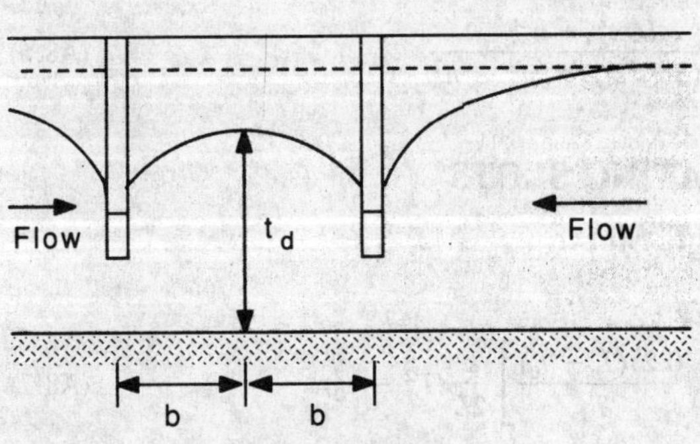

FIGURE A8.4

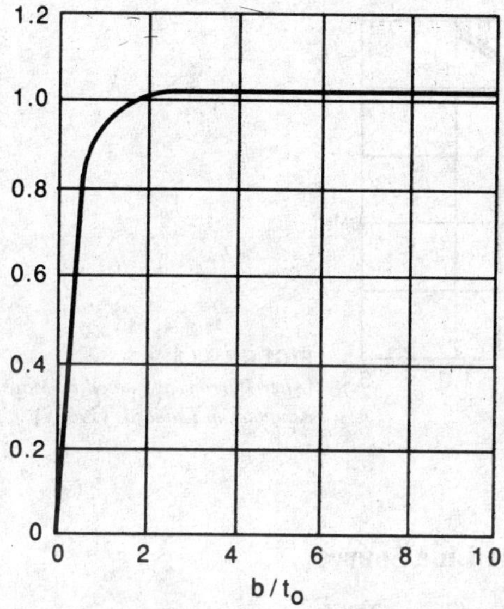

FIGURE A8.5
[*After Chapman, (1956).*]

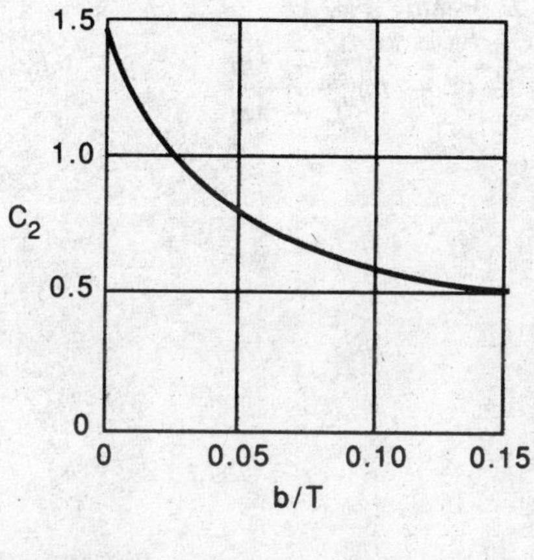

FIGURE A8.6
[*After Chapman, (1956).*]

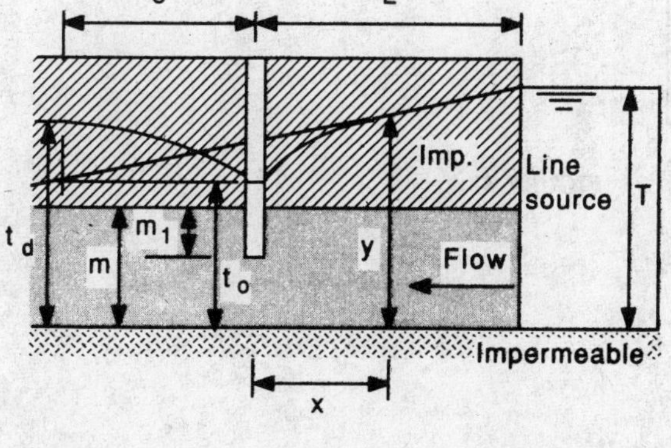

FIGURE A8.7

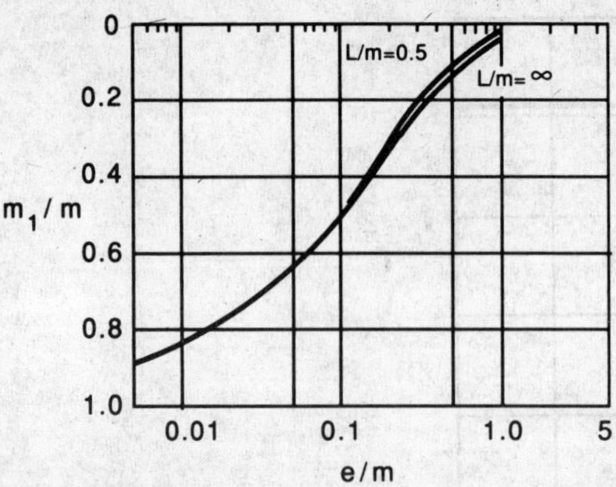

FIGURE A8.8
[*After Barron, as quoted by Mansur and Kaufman in Leonards (1962).*]

8A2.4 Artesian Flow, Double Line Source

Total flow
$$q_t = \frac{2km(T - t_0)}{L + \lambda m} \tag{A8.11}$$

$$y = t_0 + (T - t_0)\frac{x + \lambda m}{L + \lambda m} \tag{A8.12}$$

The factor λ is obtained from Fig. A8.9.

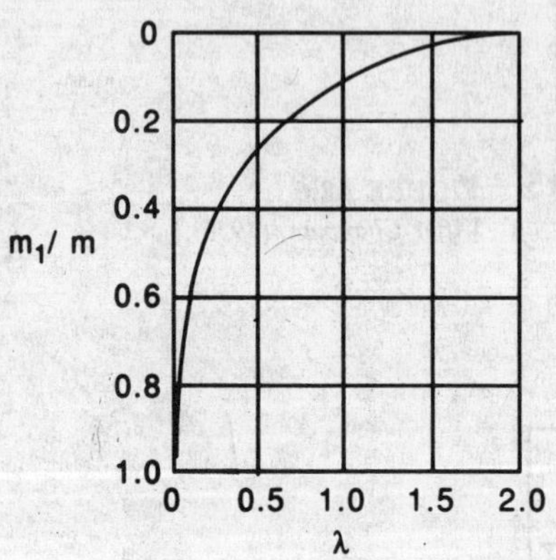

FIGURE A8.9

APPENDIX 8B

SINGLE-WELL FORMULAS

8B1 SINGLE WELL, CIRCULAR SOURCE

8B1.1 Fully Penetrating Wells

8B1.1.1 PERFECT WELL (UNCONFINED GRAVITY FLOW). For designations see Fig. B8.1.

$$Q = \frac{\pi k(h^2 - h_w^2)}{\ln (L/r)} \tag{B8.1}$$

$$y^2 - h_w^2 = \frac{Q}{\pi k} \ln (x/r) \tag{B8.2}$$

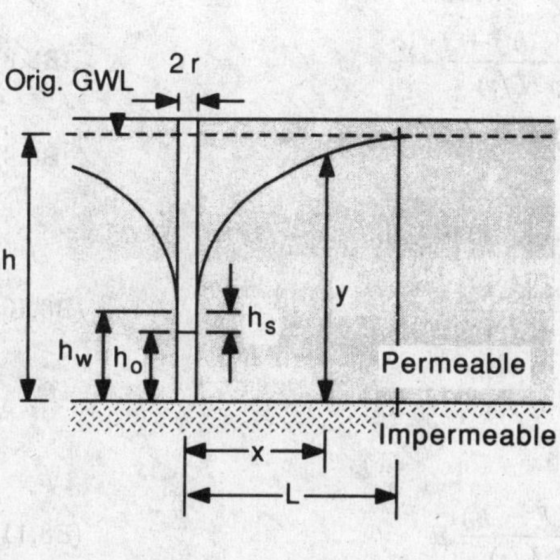

FIGURE B8.1

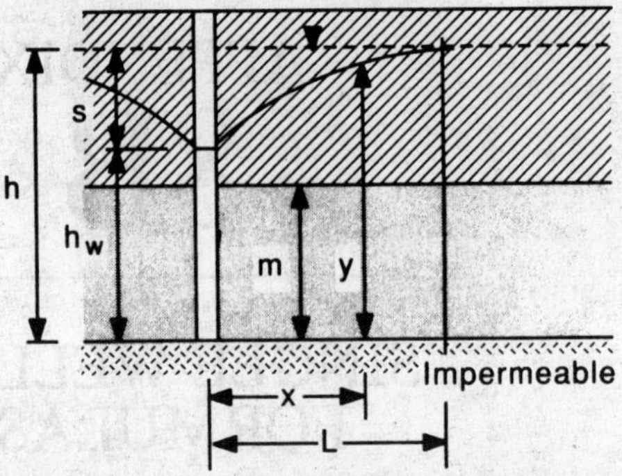

FIGURE B8.2

Note: A wide variety of formulas has been proposed for estimating the free discharge height h_s. The following expressions were presented by Herth and Arndts (1973):

$$h_s = \begin{cases} 0.5(h - h_0) & \text{(Boulton)} \quad \text{(B8.3)} \\ \dfrac{(h - h_0)^2}{2h} & \text{(Ehrenberger)} \quad \text{(B8.4)} \\ (h - h_0)\, e^{(-\alpha/\pi)} & \text{(Juhász)} \quad \text{(B8.5)} \end{cases}$$

where $\alpha = \dfrac{\sqrt{k}}{15} \dfrac{2r\pi h_0}{Q}$.

8B1.1.2 ARTESIAN WELL. See Fig. B8.2.

$$Q = \frac{2\pi kms}{\ln (L/r)} \tag{B8.6}$$

$$y - h_w = \frac{Q}{2\pi km} \ln (x/r) \tag{B8.7}$$

8B1.2 Partially Penetrating Wells

8B1.2.1 UNCONFINED GRAVITY FLOW. See Fig. B8.3.

$$Q = \frac{\pi k[T^2 - (h_w + t)^2]\alpha}{\ln (L/r)} \tag{B8.8}$$

where
$$\alpha = \sqrt{\frac{h}{T}} \sqrt[4]{\frac{2T - h}{T}} \tag{B8.9}$$

Or, after Breitenöder [as quoted by Herth and Arndts (1973)]:

$$Q = \frac{\pi k(h^2 - h_w^2)}{\ln (L/r)} \epsilon \tag{B8.10}$$

where ϵ is obtained from Fig. B8.4.

8B.1.2.2 ARTESIAN FLOW. See Fig. B8.5.

$$Q = \frac{2\pi k(T - t_0)}{\ln (L/r)} \mu \tag{B8.11}$$

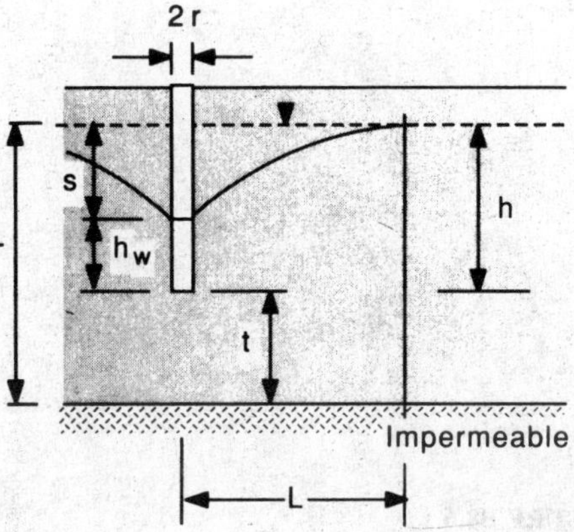

FIGURE B8.3

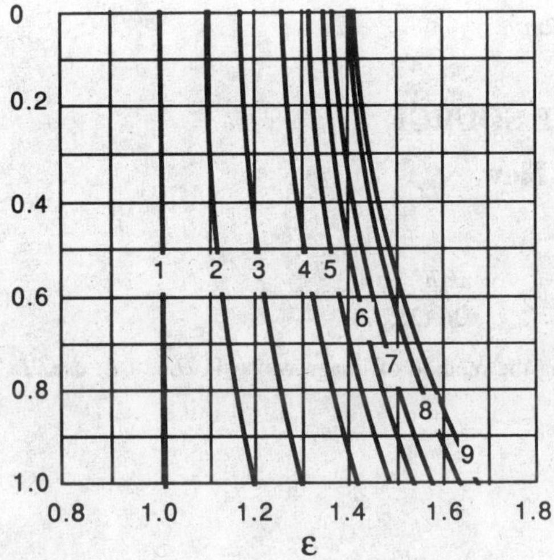

Curve no.	h/T
1	1.0
2	0.95
3	0.9
4	0.8
5	0.7
6	0.6
7	0.5
8	0.25
9	0

FIGURE B8.4

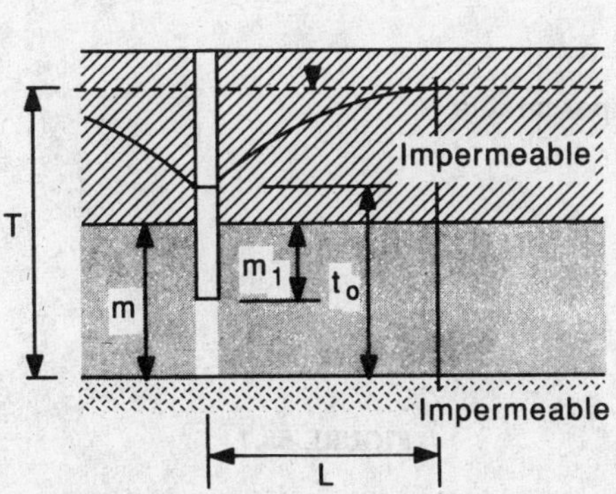

FIGURE B8.5

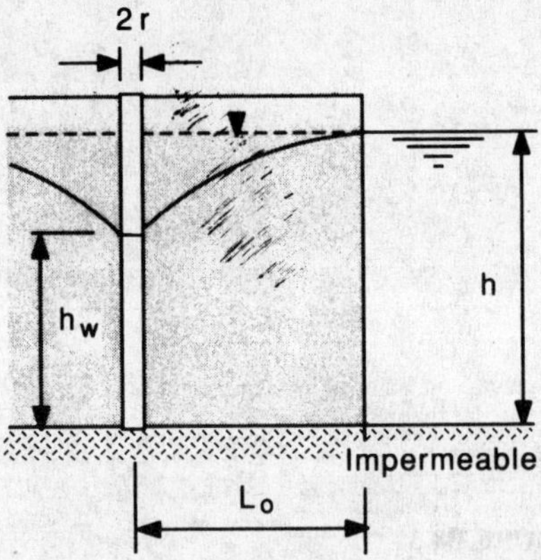

FIGURE B8.6

where $\mu = 1 + 7\sqrt{\frac{r}{2m}}\sqrt{\frac{m}{m_1}}\cos\frac{\pi m_1}{2m}$.

8B2 SINGLE WELL, LINE SOURCE

8B2.1 Unconfined Gravity Flow

For designations see Fig. B8.6.

$$Q = \frac{\pi k(h^2 - h_w^2)}{\ln (2L_0/r)} \tag{B8.12}$$

Note: This formula was derived using the method of image wells. If $2L_0 > L$, use $2L_0 = L$.

8B2.2 Artesian Flow

See Fig. B8.7.

$$Q = \frac{2\pi kms}{\ln (2L_0/r)} \tag{B8.13}$$

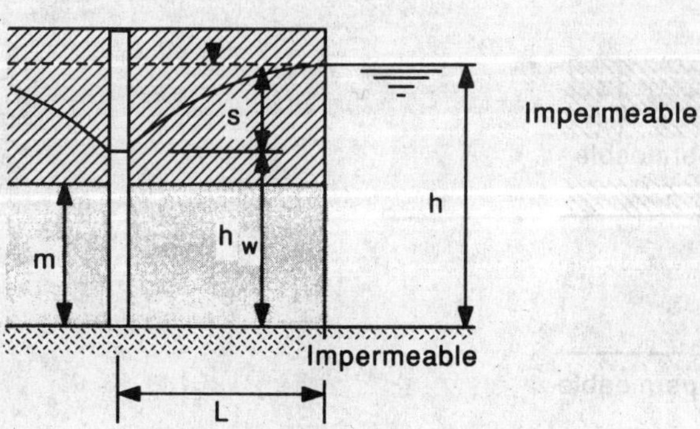

FIGURE B8.7

APPENDIX 8C

MULTIPLE-WELL FORMULAS (UNCONFINED GRAVITY FLOW)

8C1 MULTIPLE WELLS, CIRCULAR SOURCE

8C1.1 General Case

For designations see Fig. C8.1.

For well (i) only:

$$h^2 - y_i^2 = \frac{Q_i}{\pi k} \ln \frac{L_i}{x_i} \tag{C8.1}$$

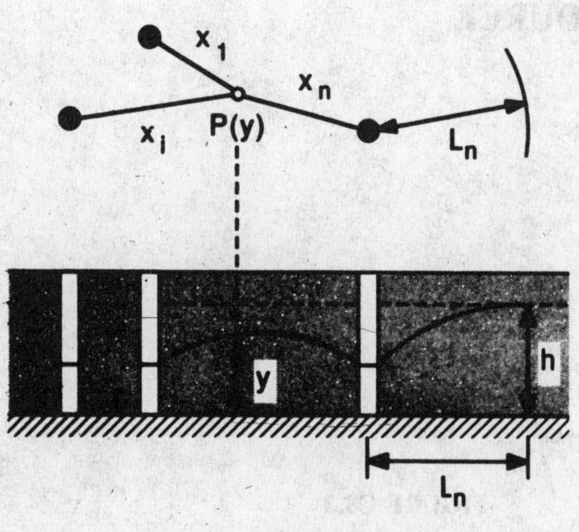

FIGURE C8.1

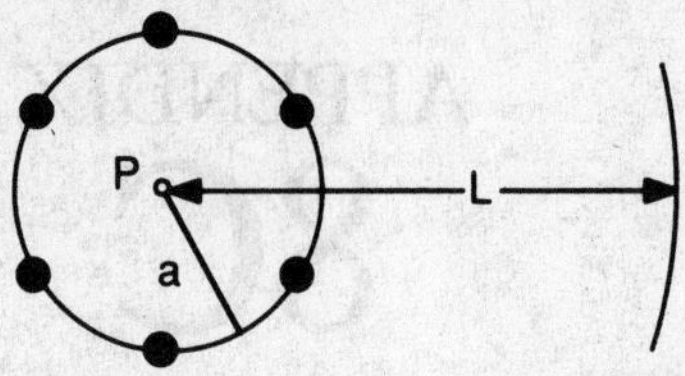

FIGURE C8.2

For n wells:

$$h^2 - y^2 = \sum_{i=1}^{n} \frac{Q_i}{k} \ln \frac{L_i}{x_i} \tag{C8.2}$$

$$Q = \sum_{i=1}^{n} Q_i \tag{C8.3}$$

Note: For artesian flow formula, replace $h^2 - y^2$ by $2ms$ as for single-well cases.

8C1.2 Circular Arrangement of Wells

See Fig. C8.2.

For n wells at distance a from point P:

$$h^2 - y^2 = \frac{Q}{\pi k} \ln \frac{L}{a} \tag{C8.4}$$

8C1.3 Rectangular Arrangement of Wells

See Fig. C8.3.

For preliminary design, replace the rectangular area by the equivalent circular area with radius a calculated as follows:

$$a = \sqrt{\frac{wl}{\pi}} \tag{C8.5}$$

8C2 MULTIPLE WELLS, LINE SOURCE

8C2.1 General Case

For designations see Fig. C8.4.

For well (i) only:

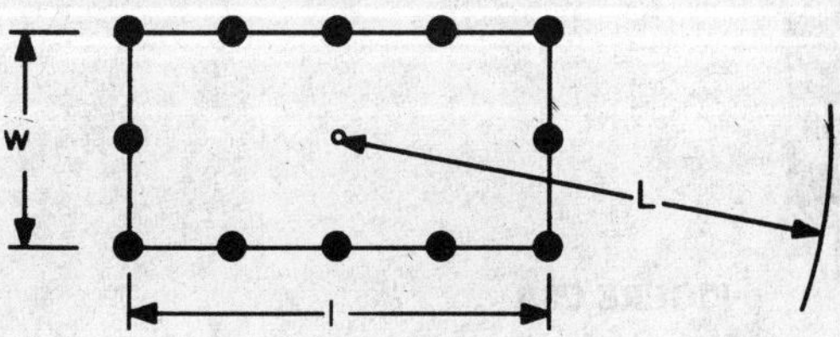

FIGURE C8.3

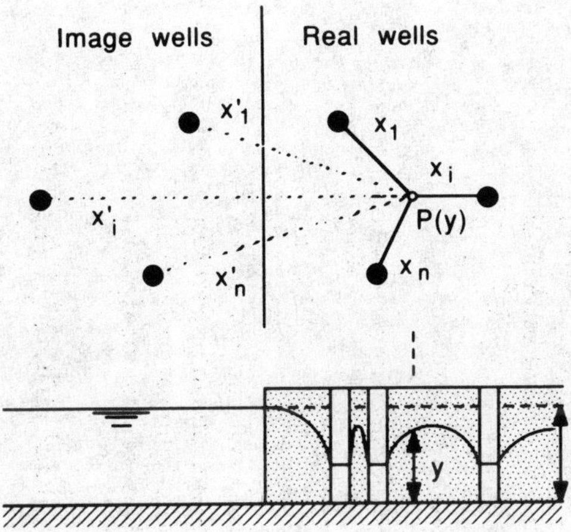

FIGURE C8.4

$$h^2 - y_i^2 = \frac{Q_i}{\pi k} \ln \frac{x_i'}{x_i} \tag{C8.6}$$

For n wells:

$$h^2 - y^2 = \sum_{i=1}^{n} \frac{Q_i}{k} \ln x_i'/x_i \tag{C8.7}$$

If $x_i' > L$, use $x_i = L$.

CHAPTER 9

DESIGN OF DEWATERING SYSTEMS (EXCAVATIONS AND SLOPES)

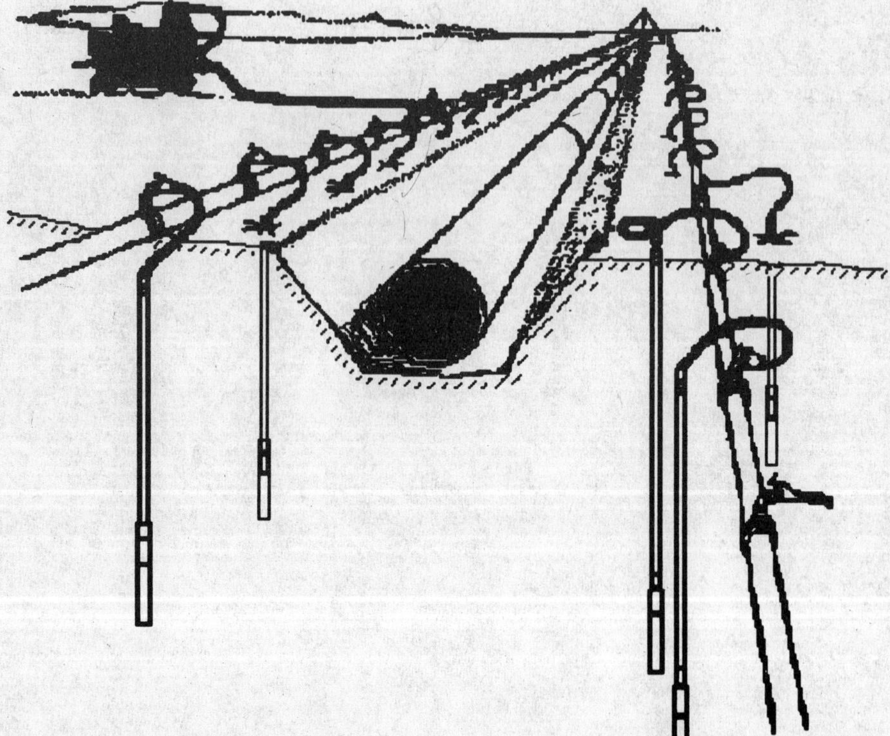

9.1 GROUND AND WELL DETERMINANTS

9.1.1 Determination of Ground Permeability

The discharge quantity from a well or slot is directly proportional to the coefficient of permeability k. For sands and gravels, the soils most suitable for dewatering by wells and open drainage systems, k may vary from 0.0001 to 0.01 m/s. A reliable estimate of k is therefore of great importance in the planning of well systems and in the prediction of the required pump capacity.

The soil properties which significantly affect the permeability are [Lambe and Whitman (1969)]:

1. Particle size
2. Void ratio
3. Mineral composition
4. Fabric (soil structure)
5. Degree of saturation

For a cohesionless soil, the void ratio is likely to be the most important determinant of its permeability; mineral composition and the fabric component of the soil structure are more important for fine-grained soils than for sands and gravels.

Because it is very difficult to obtain representative undisturbed samples of cohesionless soils, standard laboratory constant or falling head tests may not give permeability coefficients which are sufficiently reliable for design purposes. For this reason and because of the variability of natural soil deposits, field tests are highly recommended. However, if only a disturbed sample of soil is available, recourse may be taken to an empirical relationship between grain size distribution and permeability.

9.1.1.1 ESTIMATING k FROM PARTICLE SIZE. Many textbooks quote a formula attributed to Hazen (1892) who experimented with filter sands:

$$k = 100D_{10}^2 \qquad \text{cm/s} \tag{9.1}$$

where D_{10} is the diameter in centimeters corresponding to 10% passing, as read from a grain-size-distribution curve determined from a sieve analysis.

More generally the formula is written as

$$k = CD_{10}^2 \qquad \text{cm/s} \tag{9.2}$$

The constant C has been found to vary with the uniformity coefficient C_u. According to Beyer [as quoted by Schröder (1966)], C is related to C_u as follows:

$C_u = D_{60}/D_{10}$	C
1–1.9	110
2–2.9	100
3–4.9	90
5–9.9	80
10–19.9	70
> 20	60

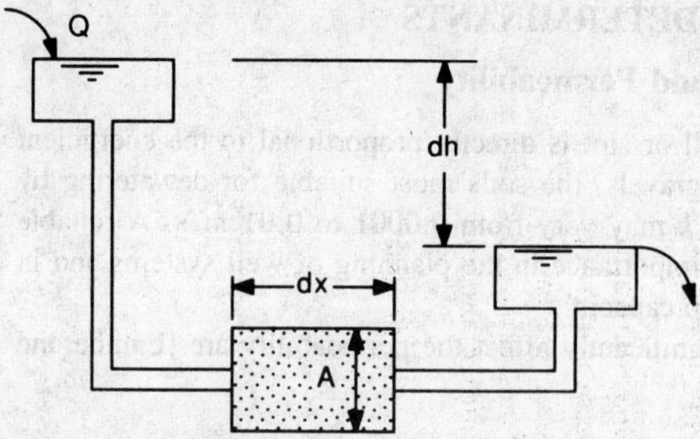

FIGURE 9.1
Constant head test.

Powers and Burnett (1986) [also see Powers (1981)] have published charts originally produced by Prugh (1959) which give the permeability as a function of D_{50} and the uniformity coefficient for either a loose, medium, or dense state of the granular soil. These correlations are said to be reliable, provided the samples tested are representative and there is no excessive stratification present.

9.1.1.2 LABORATORY PERMEABILITY TESTS. In the constant head test (Fig. 9.1) water flows from an upper reservoir through a cylindrical soil sample of length dx and cross-sectional area A to a lower reservoir. For a given difference dh in water level between the constant head reservoirs, the flow of water Q is measured. The coefficient of permeability k is calculated directly from Darcy's law [Eq. (7.9)]:

$$k = \frac{Q}{A}\frac{dx}{dh} \tag{9.3}$$

In the falling head test (Fig. 9.2) water drains from a standpipe through the soil specimen into a constant-level reservoir. At time t_1 the head of water in the standpipe

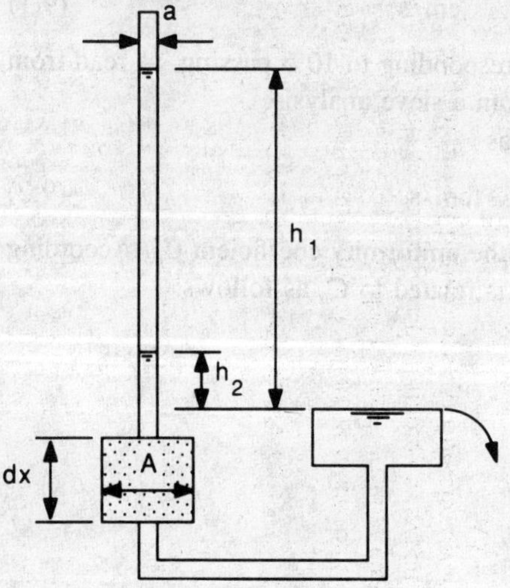

FIGURE 9.2
Falling head test.

is observed as h_1; at time t_2 it has fallen to h_2. If the cross-sectional area of the standpipe is a and that of the dx-long specimen is A, the coefficient of permeability is found to be

$$k = \frac{a\,dx}{A(t_2 - t_1)} \ln \frac{h_1}{h_2} \tag{9.4}$$

It is important to keep an account of density and saturation of the specimens tested.

9.1.1.3 FIELD PERMEABILITY TESTS. Although it seems most desirable to determine k in situ, the interpretation of field test results is not always easy because of soil disturbance, clogging of screens and filters, anisotropic conditions, and other difficulties.

Pumping test with observation holes. The most reliable field test requires the installation of a full-size pumping well plus at least two additional boreholes where the piezometric level can be observed. For steady-state conditions, the equation of the drawdown curve [Eq. (8.4)] is written in terms of the water levels y_1 and y_2 in the observation holes which are at distances x_1 and x_2 from the well. For unconfined flow the coefficient of permeability can then be calculated from

$$k = \frac{Q \ln (x_2/x_1)}{\pi(y_2^2 - y_1^2)} \tag{9.5}$$

For a confined (artesian) aquifer of thickness m the expression for k is

$$k = \frac{Q \ln (x_2/x_1)}{2\pi m(h_2 - h_1)} \tag{9.6}$$

The determination of k from measurements of the development of drawdown with time at a steady pump rate was discussed in Sec. 8.7.4.

Single-borehole tests. Formulas for single-borehole tests for a variety of geometric, soil, and water level configurations can be found in the "Design Manual" (U.S. Navy, 1962) and in Lambe and Whitman (1969) referring to Hvorslev (1949). These tests are either carried out by observing the equilibrium water level in the well for a specific pumping rate (in or out) or by recording the change in water level in the well with time after pumping has ceased (rising or falling head test).

9.1.2 Filter Criteria and Design of Well Screens.

In open drainage situations (Fig. 7.2) it is desirable that water does not exit on the slopes, in order to prevent slumping. This may be achieved by constructing a sloping filter as shown in Fig. 9.3. Cedergren (1960) provides design charts that give the desired $k(\text{filter})/k(\text{soil})$ ratio as a function of the slope S and the geometry of the filter zone.

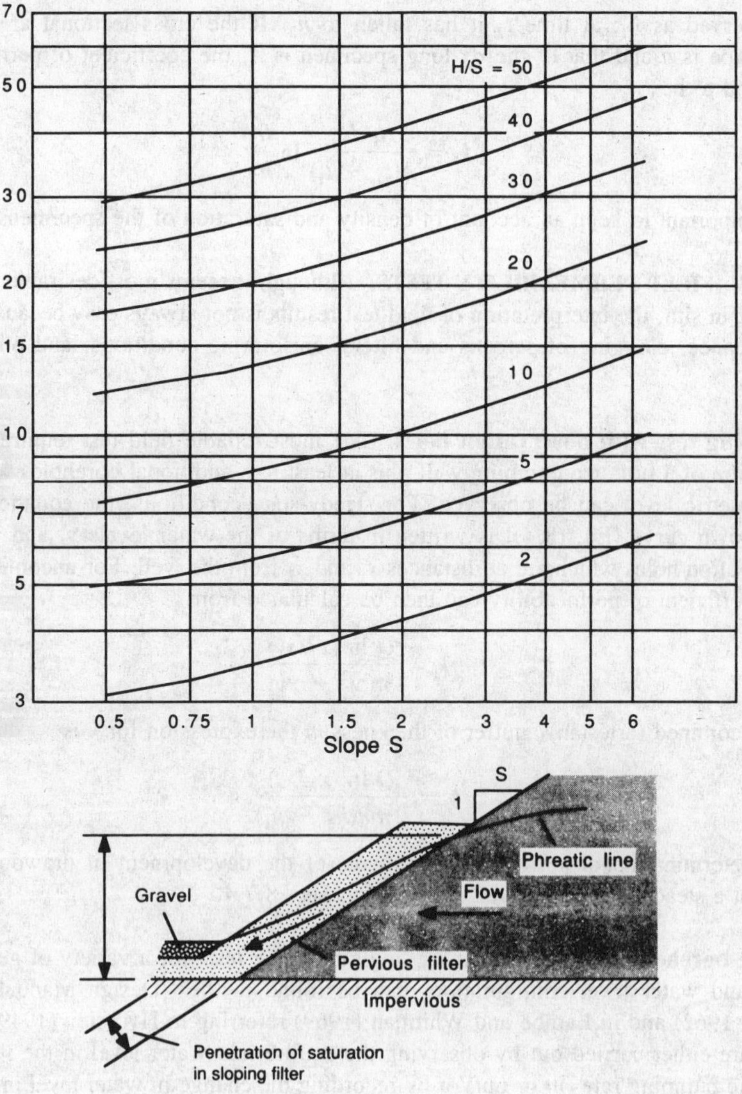

FIGURE 9.3
Sloping filter design. [*Cedergren (1967)*.]

The traditional way of preventing the migration of fine soil particles into gravel-filled drainage ditches and screened sumps is to construct one or several filter layers, usually consisting of sand with a grain size chosen according to filter criteria such as recommended by the U.S. Corps of Engineers [Eqs. (11.11), (11.12), and (11.13)]. These criteria ensure adequate flow through the filter zone, prevention of piping, and sufficient uniformity of the material. The criteria are expressed in terms of the grain size indicators D_{15}, D_{50}, and D_{85} of both the soil to be protected and the filter material.

Indicators D_{15}, D_{50}, and D_{85} represent the diameters corresponding to 15, 50, and 85% passing on a grain-size-distribution curve.

It is becoming more common to fully or partially replace granular filters by geotextiles, introduced in Chap. 11. Filter action, particularly as related to geotextiles, is discussed further in Sec. 11.2.3.

Temporary wells may consist only of perforated casing surrounded by a sand or gravel pack, chosen according to proper filter criteria. More permanent installations are likely to be fitted with commercial well screens, which have considerable advantages compared with makeshift slotted and perforated pipes. They are usually made of corrosion-resistant metal, provide continuous uninterrupted inflow over a maximum percentage of open area, have V-shaped slot openings that widen inwardly to prevent clogging, and are built strong enough to resist stresses during and after installation. A properly chosen well screen will give the well a maximum specific capacity, measured in liters per second per meter of drawdown.

9.1.3 Individual Well Capacity

The preliminary design of a dewatering system usually starts with a determination of the total quantity of water to be pumped for a circular excavation with an area equal to the one to be dewatered. In order to be able to estimate the required number of wells of a given size, knowledge of the capacity of an individual well is required. For an individual well of radius r, the discharge quantity is calculated according to

$$Q_i = 2\pi r h_w k i_e \tag{9.7}$$

where h_w can be set equal to h_0 if the free discharge height is ignored and i_e is the average entry gradient. According to empirical findings by Sichardt, the entry gradient should not exceed

$$i_{e\max} = \frac{1}{15\sqrt{k}} \tag{9.8}$$

(where k is entered in meters per second), otherwise turbulence, high head losses, and filter instability may result. This rule is recommended for well spacings larger than about 15 well diameters. The capacity of an individual well is therefore limited to

$$Q_{i\max} = 2\pi r h_0 \frac{\sqrt{k}}{15} \tag{9.9}$$

with r and h_0 in meters, k in meters per second, and Q in cubic meters per second. Note that both expressions [Eqs. (9.8) and (9.9)] are empirical rules and are not dimensionally consistent.

If $Q_{i\max}$ is plotted as a function of h_0, a straight line results as shown in Fig. 9.4. This figure also shows Q_i calculated according to Eq. (8.3) which reads (neglecting the free discharge height):

$$Q_i = \frac{\pi k(h^2 - h_0^2)}{\ln (L/r)} \tag{9.10}$$

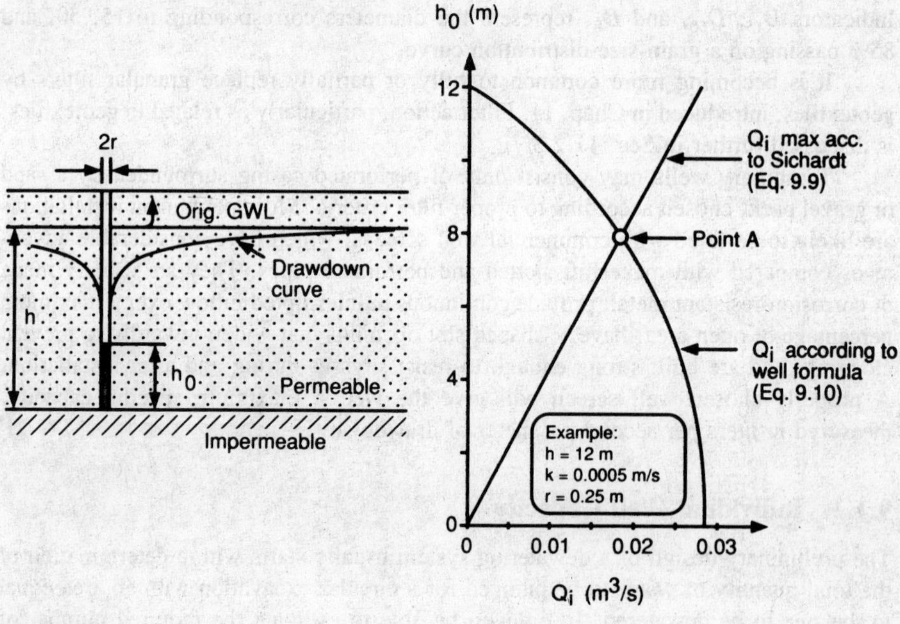

FIGURE 9.4
Well capacity.

In Fig. 9.4 this function is represented by a parabola. The point A corresponds to a desirable minimum water level in the well h_{0min} and an optimum discharge Q_{opt}, representing the individual well capacity. If Q is increased further, the maximum feasible entry gradient according to Sichardt is exceeded and the well may be pumped dry.

9.1.4 Well Diameter, Depth, and Spacing

It is of interest to note some of the guidelines published by various authors with respect to the preliminary layout of a well system. It should however be remembered that many of these design rules may have originated from engineers involved in the development of groundwater for municipal or industrial water supply, rather than for temporary dewatering projects. In addition, the technology of drilling and pumping undergoes continuous development. Nevertheless, in particular geographic areas or specific geological formations, successful traditional guidelines and techniques may be maintained.

As mentioned earlier, shallow installations using surface pumps allow a maximum suction head of about 8 m; well screens may extend the installation to a total depth of 10 to 12 m. Schröder (1966) quotes as a rule that the distance between adjacent wells should not be less than 3 to 4 m for 150-mm-diameter wells and not less than 5 to 6 m for 300- to 350-mm-diameter wells; otherwise an uneconomical system of too many wells with only marginally increased total discharge may result.

However, in well-point installations, where emphasis is on easy and quick temporary installation of reusable components, spacing is often much closer. Design charts presented by Mansur and Kaufmann (1962) consider spacings as low as 0.2 m for gravels and 1.5 m for fine sands.

For individually developed deep wells, the diameter and depth chosen will depend on the type of submersible pump to be used.

9.2 DEWATERING OF EXCAVATIONS

9.2.1 Standard Design Approach

The aims of the calculations are to determine the required pump rate and the number of wells needed in order to lower the water level by a specified amount below the base of the excavation. Calculations proceed in four steps:

Step 1: Obtain a rough guess of the total quantity of water to be pumped. We replace the actual excavation with a circular one of equal area and use an equation analogous to Eq. (8.11):

$$Q_{tot} = \frac{\pi k(h^2 - y^2)}{\ln (L/a)} \tag{9.11}$$

The values of h, y, and k are determined by the dimensions of the aquifer, the required drawdown, and the soil type. The a is the radius of the substitute circular excavation. If the actual excavation is rectangular with a length X and width Y, then

$$a = \sqrt{\frac{XY}{\pi}} \tag{9.12}$$

The L can be estimated using Eq. (8.8), but this requires an initial assumption for h_0. The assumed h_0 is checked in step 3, and the calculation is repeated if necessary.

Step 2: Estimate the number of wells needed (n). For a given well radius and the assumed wetted filter length h_0 the maximum yield of one well is calculated according to Eq. (9.9). We then find

$$n = \frac{Q_{tot}}{Q_{max}} \tag{9.13}$$

Step 3: Check original guess of h_0. Use the following version of Eq. (8.10), still referring to the circular excavation (Fig. 8.6).

$$Q_{tot} = \frac{\pi k(h^2 - h_0^2)}{\ln L - (1/n) \ln (x_1 x_2 \cdots x_n)} \tag{9.14}$$

Solving this equation for h_0, results in a new, improved value for h_0. Using this value, a new L and new Q's are computed. Steps 1 to 3 are repeated until h_0 assumed is sufficiently close to h_0 calculated in step 3.

Step 4: Return to the original excavation. Distribute the n wells around its parimeter. It is then necessary to check the water level at critical points below the excavation in order to verify whether the design requirements are satisfied. For example, the water level at the center and near the corners of the excavation should be calculated. This is done using Eq. (8.10). If the water level is too high, the pumping rate Q_{tot} has to be increased. This, in turn, will result in a reduced h_0 and an entry gradient which may be in excess of that recommended by Sichardt or, in the extreme, dry wells. If this is the case, the number of wells should also be increased and all calculations repeated until a satisfactory solution is found.

Increasing the number of wells but keeping Q_{tot} is unlikely to significantly lower the phreatic surface further, although a more even drawdown may be beneficial in critical areas.

It should be noted that because of inherent deficiencies in this method, h_0 calculated using Eq. (9.14) for individual wells around a noncircular excavation will not be a constant. To use an average value, where necessary in the calculations, appears to be a reasonable compromise.

9.2.2 Modification by Herth and Arndts

Herth and Arndts (1973) maintained that the standard approach as described in the preceding section leads to an overestimation of h_0. This is because the flow to an individual well within a group of wells located around an excavation is not equal from all directions. In particular, the flow from within the excavation area toward the wells is less than from outside the excavation. In addition, the free discharge height is not considered.

They proposed that the following modified formula be used:

$$h_0 = \sqrt{y^2 - \frac{fQ_i \ln (b/r)}{\pi k}} \tag{9.15}$$

The terms b, y, r, and h_0 are defined in Fig. 9.5.

The recommended values for the correction factor f are

$$f = \begin{cases} 1.5 & \text{for large well spacings,} \\ 2 & \text{for small well spacings } (b < 5\pi r) \end{cases}$$

The justification of the latter is that for wells closely spaced around an excavation, water flows into the wells only from one side. Thus, to calculate h_0, it is assumed that there is a $2Q_i$ flow to the well. This results in a reduced value of h_0 which should be compared with the minimum allowable h_0 calculated using Sichardt's formula [Eq. (9.9)]:

$$h_0 = \frac{15Q_{imax}}{2pr\sqrt{k}} \tag{9.16}$$

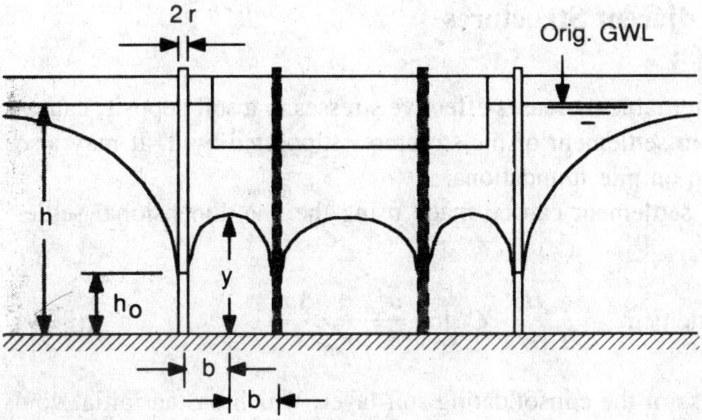

FIGURE 9.5
Multiwell system. [*Herth and Arndts (1973).*]

9.2.3 Pipelines and Pumping Plant

The velocity in suction pipes (or header pipes) with 200- to 400-mm diameters is usually kept below 1.5 to 2 m/s in order to keep friction losses within reasonable limits. Pressure lines (discharge lines) are dimensioned for velocities between 2 to 3 m/s.

The pump capacity required can be calculated using the following formula:

$$N = \frac{Qh\gamma_w}{\eta} \tag{9.17}$$

where η is the efficiency of the system, with friction losses in the pipes also being taken into account. Usually the value η is between 0.3 and 0.5.

Assuming an efficiency $\eta = 0.3$ and setting $\gamma_w = 10$ kN/m^3 (or 10 N/L) the formula simplifies to

$$N = \frac{Qh}{40} \quad \text{kW} \tag{9.18}$$

for Q in liters per second and h in meters.

The total discharge, total head, and suction lift required are the principal parameters affecting the choice of a pump. Self-priming centrifugal pumps are most common, but the water pumped has to be relatively clean. Pneumatic pumps and diaphragm pumps are able to move limited amounts of silt and sand without excessive wear. Rotary displacement pumps are used if the water pumped from sumps and wells contains more sediment. Submersible pumps are usually of the centrifugal type with one or more impellers on a vertical shaft driven by a motor.

Standby pump capacity should be available in case of breakdown or interrupted power supply or in case unexpected geological features result in increased flow into the dewatering system.

9.2.4 Settlement of Adjacent Structures and Other Side Effects

Lowering of the groundwater table increases effective stresses in a soil deposit, causes consolidation, and results in settlement of the structures supported by it. It may also cause negative skin friction on pile foundations.

A rough estimate of settlement can be made using the one-dimensional settlement formula:

$$\text{Settlement} = \frac{H}{1 + e} C_c \log \frac{\sigma'_{vo} + \Delta\sigma}{\sigma'_{vo}} \tag{9.19}$$

The H is the thickness of the consolidating soil layer, which has an initial void ratio e and a compressive index C_c. The initial average effective vertical stress in this layer is designated σ'_{vo}, and the stress increase caused by lowering the water table is $\Delta\sigma$. For a reduction in the groundwater level by Δh, the stress increase may be approximated by

$$\Delta\sigma = \Delta h \gamma_w \tag{9.20}$$

if the total unit weight of the soil does not change markedly because of dewatering.

The compressive index C_c is determined in a consolidation test or, less reliably, is estimated from empirical relationships, such as $C_c = 0.009$[liquid limit (%) − 10]. Depending on the stress history of the consolidating layer, the recompression index (often designated C_R) may give more realistic settlement values than C_c.

If the above formula indicates measurable settlement, this by itself will not necessarily mean impairment of adjacent buildings. If the consolidating layer is thick and highly impermeable, the rate of settlement may be so slow that temporary dewatering of an overlying aquifer has little effect. Also, the differential settlement experienced by a structure may be relatively small, with no effect on its function. Much will depend on the nature of the structure and the service it provides: A corrugated, ironclad, steel-framed warehouse is less susceptible to differential movement than a multistory brick building.

If settlement is likely to damage structures adjacent to the dewatered excavation, it may be possible to either provide a *cutoff* (sheet pile, slurry wall, grouted or frozen ground) or to artificially *recharge* the water-bearing layer in their vicinity in order to maintain the original groundwater level (Fig. 9.6). The cost associated with these preventative measures could, however, be more than the cost of underpinning, grouting, or repairing the affected buildings.

It should also be noted that recharging an aquifer may be more problematic than dewatering it, because

There may be practical limitations to the head of water which can be built up in the recharge well.

The ground could appear less permeable because the pumped water is contaminated with fines and clogs the walls of the recharge wells or trenches.

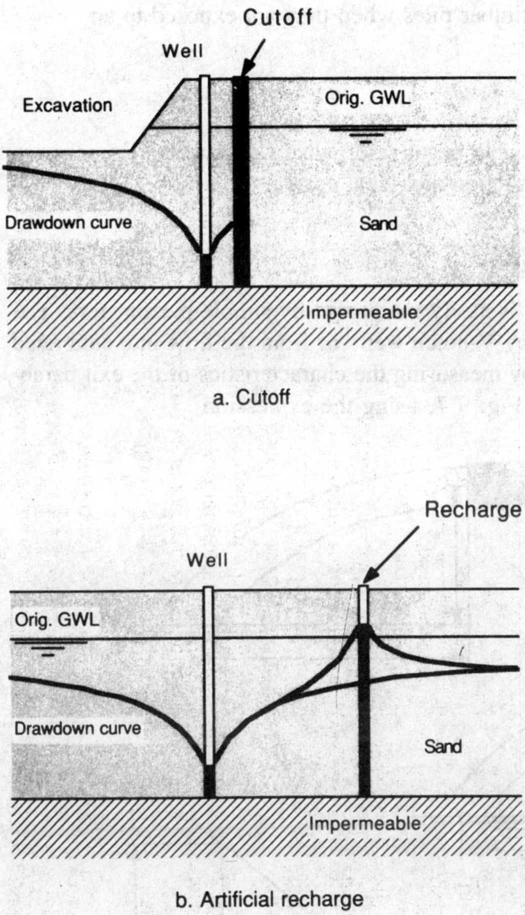

FIGURE 9.6
Preventing unwanted side effects of dewatering by cutoff or recharge.

The aquifer soil could be significantly less permeable than originally measured, if it is unsaturated prior to recharge.

Recharging the aquifer may reduce total settlements but also may increase differential settlements.

Besides settlement, other ''unwanted side effects'' [Powers (1985)] of dewatering may include the following:

Effect on nearby water supply wells

Saltwater intrusion in coastal areas (or spread of underground pollutants in industrial areas)

Degradation of old, untreated timber piles when they are exposed to air

Harmful effect on vegetation

Activation of sinkholes

Proper consideration of the possible negative effects of dewatering is necessary in order to prevent unexpected construction and legal costs.

9.2.5 Performance Evaluation

The location of the drawdown curve is easily monitored by simple piezometers. The discharge quantity can be measured by V-notch weirs or from the flow into calibrated containers. It can also be estimated by measuring the characteristics of the exit parabola of the water jet, as illustrated in Fig. 9.7, using the expression

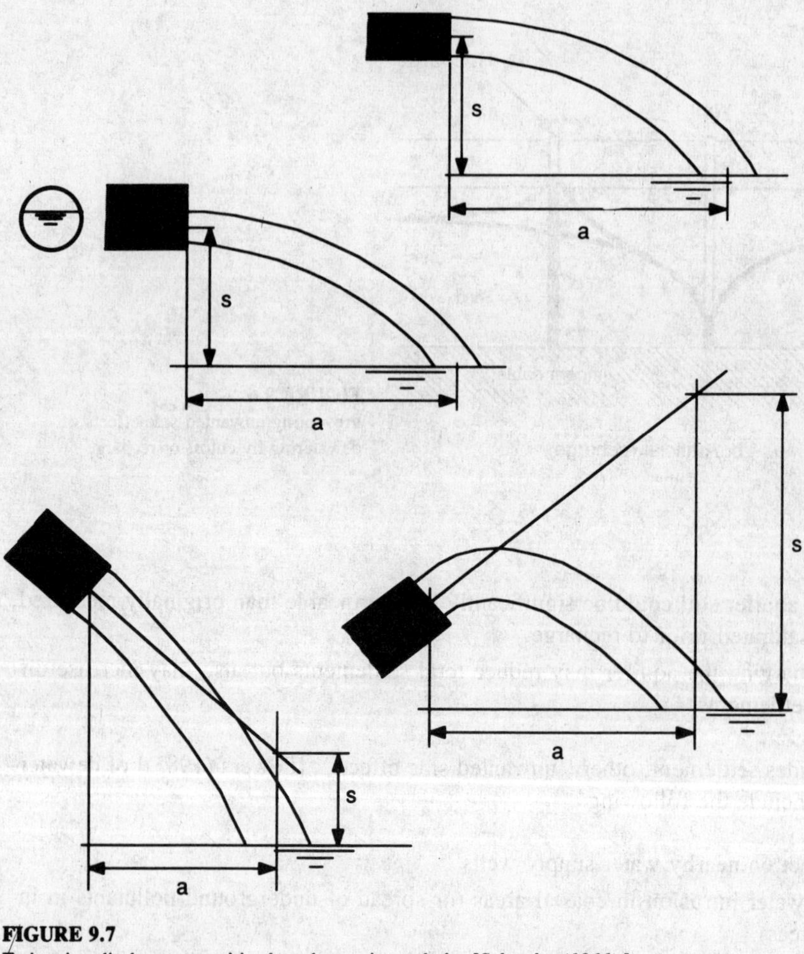

FIGURE 9.7
Estimating discharge quantities based on exit parabola. [*Schröder (1966)*.]

$$Q = 2.22F_a \frac{a}{\sqrt{s}} \tag{9.21}$$

Distances a and s in meters are as defined in Fig. 9.7. The F is the cross-sectional area of the pipe in square meters, and F_a is the cross-sectional area of the water jet at the exit point in square meters, in the case where the pipe is partially full.

Performance evaluation allows the design parameters to be checked, and the procedures to be modified during construction of multistage installations, in order to reduce costs or improve efficiency.

9.3 DRAINAGE OF SLOPES

9.3.1 Effect of Water on Slope Stability

The presence of static or flowing water in a soil or rock mass may decrease its slope stability for the following reasons:

- It reduces the shear resistance of the material by increasing its water content and/or by producing higher pore pressures.
- It increases the total weight of the material.
- It may cause seepage forces in the direction of the slope movement.
- It may cause erosion and/or piping.
- It may change the physicochemical characteristics of the soil. (This could be the case when fresh water seeps through soils originally deposited in saltwater.)
- It increases the susceptibility to liquefaction during an earthquake.

Some of these effects are difficult to assess numerically, particularly in the absence of a comprehensive field and laboratory testing program. However, the effect of reducing pore pressures on the safety factor of a slope in frictional material can be clearly shown using conventional methods of slope stability analysis.

9.3.2 Methods of Stabilizing Slopes

If a civil engineering structure is to be located in a potentially unstable area or if cutting into an existing slope endangers its stability, a variety of preventative and corrective measures can be taken.

Drainage is the most common and generally applicable method. It takes the form of surface drainage (open ditches, sloping filters, sealing of the surface) and subdrainage (vertical and horizontal drains, tunnels, stone filled trenches, etc.) as described in Chap. 7.

Drainage alone may however not be sufficient to stabilize the slope, and additional measures may have to be taken. These could include:

- Altering the geometry of the slope, i.e., flattening the slope, benching, removal of material at the head, or placing fill at the toe of the slope

- Structural reinforcement of the soil mass using retaining walls, piling, or anchors
- Improving the soil or rock properties by grouting or using other geotechnical processes

In exceptional cases, blasting may improve the stability of a slope, not only by changing the geometry but also by making jointed bedrock more permeable, thus facilitating drainage. Furthermore, blasting could result in additional consolidation and improved strength, e.g., before a road cut is constructed.

9.3.3 Analysis of Stability

There are many different methods of slope stability analysis available to the engineer. They may be categorized according to the assumptions made for the following:

- Failure geometry
- Failure law
- Type of strength or deformation parameters
- Numerical technique used

Here, in order to illustrate the effect of water on slope stability, the infinite slope analysis and the Swedish method of slices, which is based on a circular failure surface, are introduced. Both rely on the Mohr-Coulomb failure law. As it suits the evaluation of the long-term stability of the slope, the analysis is usually carried out in terms of effective stresses. If large strains and slow movement are typical for a particular stability problem, residual rather than peak shear strength should be used.

9.3.3.1 INFINITE SLOPE ANALYSIS. The infinite slope analysis is applied to problems where the likely failure plane lies parallel to the surface. This may be due to particular geological circumstances, the pattern of weathering, or because of the way the soil or rock mass was deposited or dumped. The factor of safety of an infinite slope (Fig. 9.8) is expressed as

$$F = \frac{\tau_a}{\tau_m} = \frac{\text{available shear strength}}{\text{mobilized shear strength}} \tag{9.22}$$

The available shear strength refers to the maximum possible shear stress in the potential failure plane. The mobilized shear strength is equal to the actual shear stress required for equilibrium.

For a soil of unit weight γ and with effective strength parameters c' and ϕ', we obtain for the general case

$$F = \frac{c' + (z\gamma \cos^2 \beta - u) \tan \phi'}{z\gamma \cos \beta \sin \beta} \tag{9.23}$$

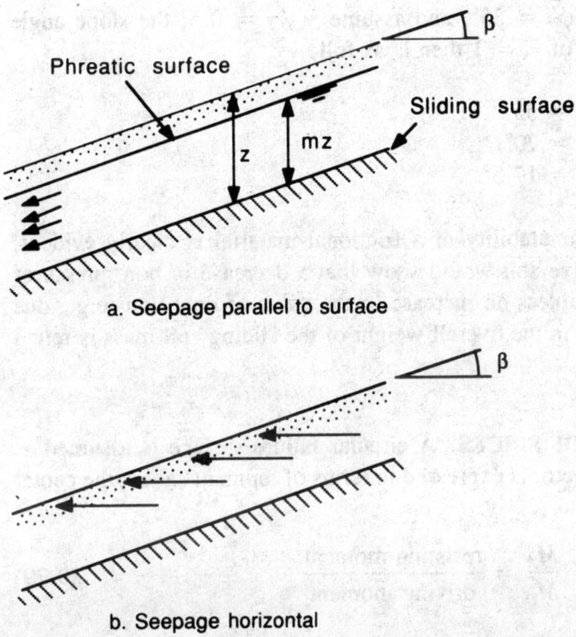

FIGURE 9.8
Infinite slope analysis.

In the special case of $c' = 0$ and $m = 1$ (water table at the surface), we obtain for seepage parallel to the slope, with γ_w representing the unit weight of water:

$$u = mz\gamma_w \cos^2 \beta \tag{9.24}$$

$$F = \left(1 - \frac{\gamma_w}{\gamma}\right) \frac{\tan \phi'}{\tan \beta} \tag{9.25}$$

$$\approx \frac{\tan \phi'}{2 \tan \beta}$$

If there is no water present above the failure surface ($m = 0$), the safety factor is approximately

$$F \approx \frac{\tan \phi'}{\tan \beta} \tag{9.26}$$

For the case of horizontal seepage

$$u = z\gamma_w \tag{9.27}$$

$$F = \left(1 - \frac{\gamma_w}{\gamma \cos^2 \beta}\right) \frac{\tan \phi'}{\tan \beta} \tag{9.28}$$

As an example, take $c' = 0$, $\phi' = 35°$, and assume $\gamma_w/\gamma = 0.5$; the slope angle equivalent to a factor of safety of $F = 1$ then is as follows:

No water present	$\beta = 35°$
Seepage parallel surface	$\beta = 20°$
Seepage horizontal	$\beta = 17.5°$

The effect of water on the stability of a frictional material is clearly evident. The analysis of a purely cohesive soil would show that a decrease in pore pressures does not increase the stability unless an increase in the value of cohesion (e.g., due to consolidation) or a reduction in the overall weight of the sliding soil mass is relied upon.

9.3.3.2 SWEDISH METHOD OF SLICES. A circular failure surface is assumed as shown in Fig. 9.9. The safety factor is expressed in terms of moments about the center of the failure circle:

$$F = \frac{M_R}{M_D} = \frac{\text{resisting moment}}{\text{driving moment}} \tag{9.29}$$

Using designations as in Fig. 9.9, the safety factor is calculated as

$$F = \frac{c'L + \tan\phi' \sum_{i=1}^{n} (W_i \cos\alpha_i - U_i)}{\sum_{i=1}^{n} W_i \sin\alpha_i}$$

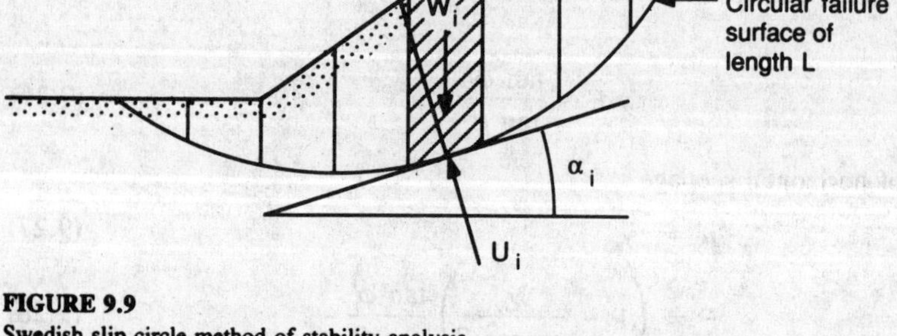

FIGURE 9.9
Swedish slip circle method of stability analysis.

where W_i is the weight of slice i and U_i is the resultant of the pore pressure acting normal to the failure arc on slice i.

The force U_i is proportional to the height of the phreatic line above the failure arc if hydrostatic conditions are assumed. The frictional resistance along the failure arc is thus greatly reduced if the water level within a sliding soil mass is increased. Provided the effective angle of internal friction is at least 10 to 20°, analysis shows a significant effect on the safety factor of a slope, even in cohesive frictional, rather than purely frictional, material.

In theory, for a purely cohesive soil, we see no improvement of stability by the relief of pore pressures alone. It should however be remembered that even highly plastic clays can have effective friction angles in excess of 0°, say 10 to 20°, as determined in a consolidated, undrained triaxial test with pore pressure measurements. Even without considering friction, drainage of cohesive soils can be seen to be beneficial for stability if the long-term reduction in water content leads to increased strength and a reduction in the total weight acting on the failure plane.

9.3.4 Design Approach

The approach to designing a drainage system for slope stabilization is basically the same as for a well system, except that topography generally allows dewatering without pumping.

As for wells, drains must be designed

- To have adequate discharge capacity
- To satisfy filter criteria

The effect of various spacings could be analyzed by drawing flow nets for idealized conditions. However, in most cases geological details such as the location of permeable layers and faults in the bedrock have an overwhelming influence on the effectiveness of the drainage system. Practical rules suggest a spacing of horizontal drains of 3 to 8 m in clayey soils and 8 to 15 m in more permeable material. Usually several drill holes are fanned out from one location in order to increase the probability of intersecting all major permeable strata and rock joints. Borehole information should be compiled so as to allow a three-dimensional assessment of the geological structure. The length of the drains depends on the particular site conditions and the location of the likely slip planes. Horizontal drains with lengths of more than 100 m have been installed as part of slope-stabilizing programs.

Vertical drains may be used in combination with a horizontal drainage system in order to intersect water-bearing layers separated by horizontal seams of impermeable clayey soil.

When attempting to stabilize an active landslide, the engineer must remember that the moving soil mass may damage the drainage system. The type of drains and their location will have to be selected accordingly; considerable maintenance work may be required in the early stages of the work.

9.3.5 Performance Evaluation

The effectiveness of the drainage system can be checked by installing piezometers and measuring the flow from the drains. Records of water levels and discharge quantities should be kept and possibly correlated with rainfall data.

Vertical and horizontal movements of the unstable soil or rock mass should be monitored by surveying surface markers and, where appropriate, borehole installations. Using a borehole inclinometer may allow identification of the actual slip surface or zone of shear failure.

9.4 CASE STUDIES

9.4.1 Dewatering of a Pumping Station Site at Cronulla, New South Wales

The Cronulla Sewage Pumping Station is located at the edge of Woolooware Bay, in an area characterized by swampy, marshy conditions. The Metropolitan Water, Sewarage & Drainage Board started construction in 1974. The station started operating in 1980.

The excavation covered an area of approximately 60 by 60 m. The deeper of the two final rings of wells reached a depth of 10 m below ground level.

The excavation was accomplished in three stages. The first stage consisted of a U-shaped layout of 150 well points, spaced 1.5 m apart and reaching a depth of 6 m. The riser pipes of 50-mm diameter were connected to a 200-mm header pipe. Three pumps were in action, each serving 50 well points. This installation was only used to initiate the excavation; it was dismantled after the next two lines of well points commenced operating. It took 5 days to lower the groundwater level to about elevation (El.) 13 m.

The second and third stages of dewatering are illustrated in Figs. 9.10 and 9.11, respectively. The contractor, Sykes Pumps Australian Pty., Ltd., used perforated plastic well points covered with cheesecloth as filter material. Lowering the water

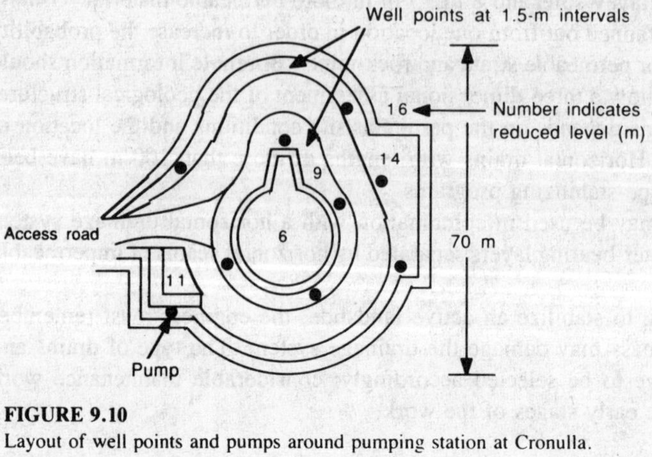

FIGURE 9.10
Layout of well points and pumps around pumping station at Cronulla.

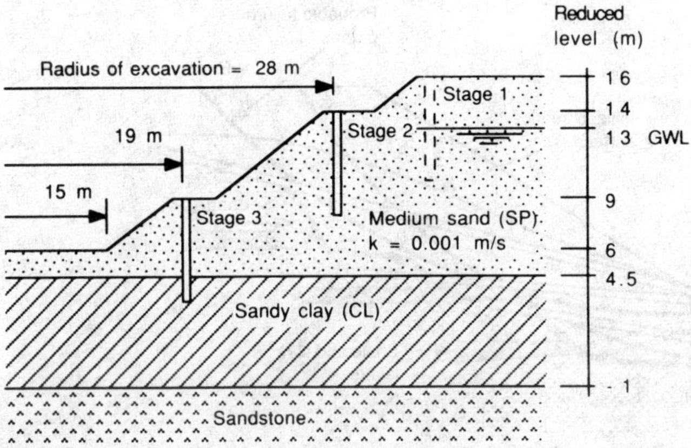

FIGURE 9.11
Schematic cross section showing levels of upper and lower well points at pumping station site at Cronulla (stages 2 and 3 of the dewatering program).

table to El. 6 m took 4 weeks, with all wells operating. After this period, every second well point was disconnected in order to prevent excess air being sucked by the vacuum-assisted centrifugal pumps; this measure increased the efficiency of the system. Five pumps were used during normal operations, when the pumping rate was on the order of 23 to 25 L/min per well point.

Dewatering was kept up for a period of about 6 months. In an attempt to prevent excessive surface erosion due to rain, embankments were treated with concrete spray or bituminous emulsions. The latter was generally considered to be the more economical of the two techniques.

9.4.2 Stabilizing Embankment at Newport, New South Wales

A major road embankment near Newport, approximately 20 m high with a slope of 35° (1.5:1), had already shown movement for several months before heavy rainfalls and a 6-m-deep cut near the toe brought about a semirotational failure, resulting in damage to the footpath and the closing of one lane of traffic. A typical cross section is shown in Fig. 9.12.

From November, 1973, to February, 1974, the subsidence at the top of the embankment reached a total of 3 m, at a rate of as much as 300 mm per week.

A geological investigation showed that there were major joint systems and several faults in the area. Water was thought to permeate through the jointed sandstone and travel along the more impermeable horizontal shale layers into the failure zone. It was decided that some 30 horizontal drains over a 100-m section of the fill should be drilled in order to reduce the pore water pressure along possible slip planes. It was estimated that reducing pore pressures in the wet contact zone above bedrock could increase the safety factor by more than 20%.

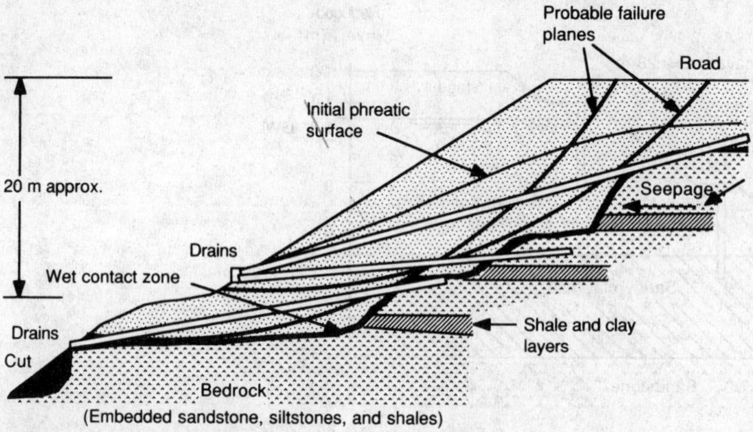

FIGURE 9.12
Embankment at Newport.

Because of the low permeability of the material in the wet contact zone, a complete relief of pore pressures did not occur. Although the rate of movement decreased so as to allow repair of the pavement, further works were thought to be necessary in order to increase the safety factor to an acceptable level. Measures considered included the building of retaining walls, placing fill at the toe of the embankment, and installing further drains.

9.4.3 Dewatering of the Morwell Open-Cut Coal Mine

This project, located in the Latrobe valley, Victoria, is described in a pamphlet by James Hardie & Co. (1973) as follows (refer to Figs. 9.13 and 9.14):

> The Morwell open cut development commenced in the early 1950's when overburden removal began. The first coal was excavated in 1960.

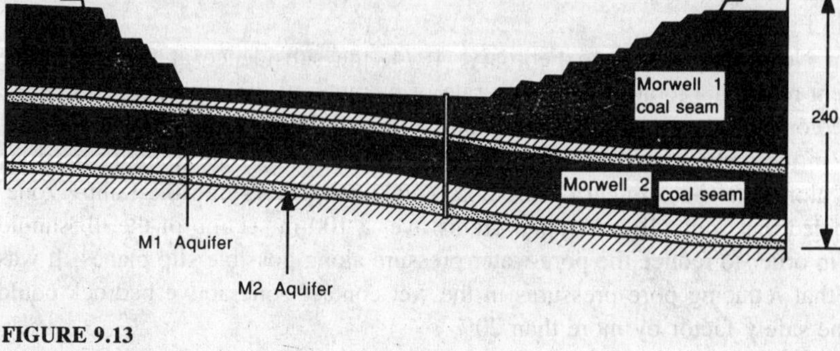

FIGURE 9.13
Morwell open cut. Diagrammatic north-south cross section.

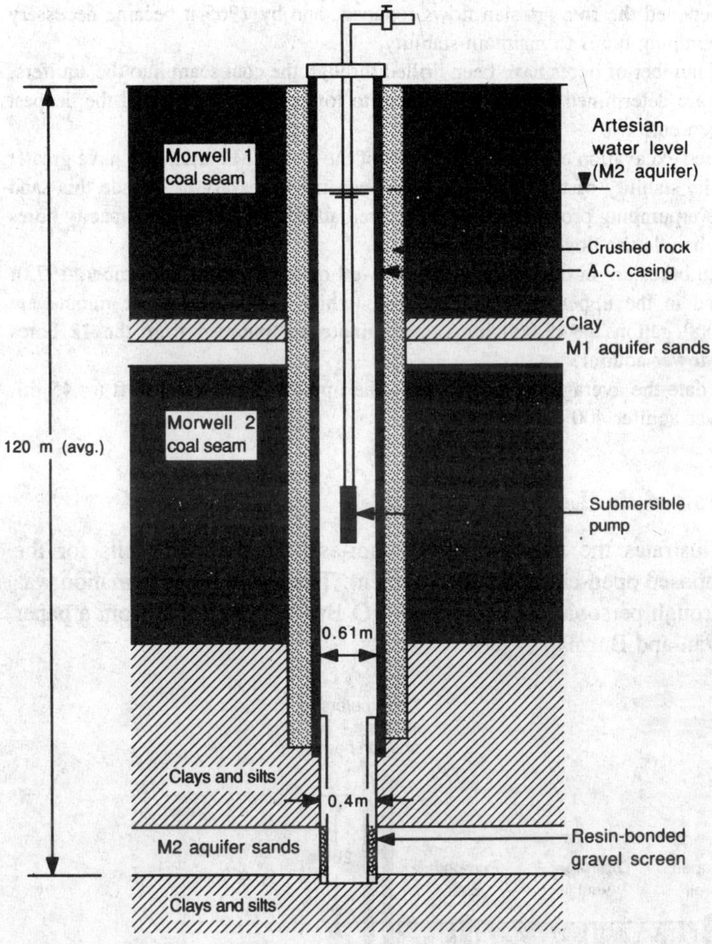

FIGURE 9.14
Diagrammatic sketch of a production bore.

The diagrammatic north-south cross-section shows the two major coal seams and the M1 and M2 aquifers, which consists of generally flat lying sand beds which underlie the coal seams.

The two aquifers contain artesian water, the original pressure levels being at the top of the M1 coal seam, immediately below the overburden.

As excavation proceeds, the weight of the coal seam is reduced and less pressure is available to equalize the pressure exerted by the artesian water. This excess pressure in the aquifers must be relieved to prevent it bursting through the now thinner coal layer above, possibly flooding the open cut and causing instability of the batters of the open cuts.

Bore pumping first commenced in 1960 from bores drilled around the perimeter of the open cut. The majority of subsequent bores were located within the open cut itself, and during the period 1960–65 the water pressures were lowered by free-flowing bores, one of which initially produced 130,000 gallons (≈500,000 liters) per hour. As the excavation deepened the free artesian flows reduced, and by 1965 it became necessary to introduce pumping bores to maintain stability.

A great number of bores have been drilled through the coal seam into the aquifers. The locations are determined by the requirement to lower pressures most at the deepest area of the open cut.

Continued excavation caused the pressures of the deeper M2 aquifer to have greater influence on the stability of the open cut, and it became necessary to include this sand layer in the bore pumping programme. This required additional and larger capacity bores and pumps to handle the flow.

Eighteen bores are in operation in the Morwell open cut (as of September, 1973). Six are located in the upper sand aquifer, from which 3000 gallons per minute are pumped. 12,000 gallons (≈45,000 liters) per minute are pumped from the 12 bores located in the lower aquifer.

At that date the average depth of bore to the upper aquifer was 150 ft (=45 m), and to the lower aquifer 400 ft (=120m).

9.4.4 Dewatering of the Lochiel Trial Pit

This case study illustrates the trial use of vacuum-assisted pumped wells for the dewatering of a proposed open-cut mine development. The following information was obtained mainly through personal communication [O'Brien (1987)] and from a paper presented by Sullivan and Burman (1986).

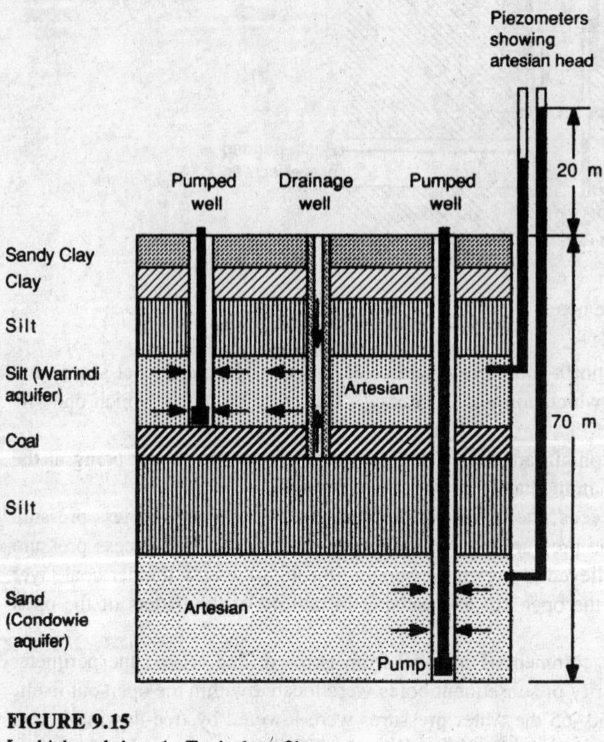

FIGURE 9.15
Lochiel coal deposit: Typical profile.

The Lochiel soft brown coal deposit, 130 km north of Adelaide, was discovered in 1982 and has since been investigated as a source of fuel for power generation in the late 1990s by the Electricity Trust of South Australia. The coal seams are embedded in fine silt and clay material. A typical profile shows 20 to 60 m of overburden above the coal zone, which is up to 20 m thick and can be divided into three main seams.

Immediately below the coal is a silt layer, which overlies fine to medium sand. A simplified profile is shown in Fig. 9.15 which also identifies the main water-bearing layers: the Warrindi aquifer and the Condowie aquifer. Additional minor aquifers may occur elsewhere in the profile. The groundwater was under artesian pressure with a hydraulic head up to 20 m above ground level. The aquifers consisted of coarse silt to medium sand and had permeabilities on the order of 10^{-5} to 10^{-4} m/s.

Open-cut mining of the Lochiel deposit obviously needs dewatering of the aquifers in order to ensure the stability of steep pit walls and a dry working area. Because of the complex geological and hydrogeological conditions a trial excavation was carried out. The trial pit was 100 m square and 29 m deep, with side slopes ranging from 45 to 60°. A 300-m-long ramp provided access to the bottom.

Eight wells were installed to pump from the Warrindi aquifer; two of these went down to the Condowie aquifer. Because of the relatively low permeability of some of the aquifers, the wells were fitted with airtight collars (Fig. 9.16) so that a vacuum

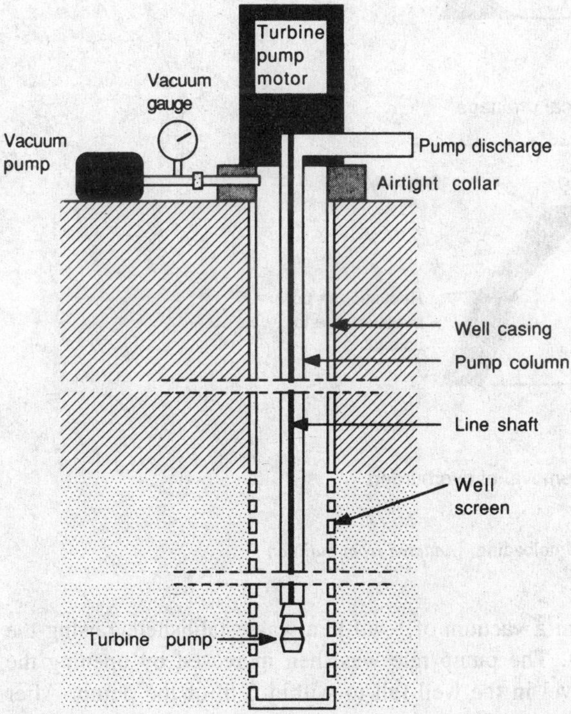

FIGURE 9.16
Lochiel trial pit: Vacuum-assisted pumped well.

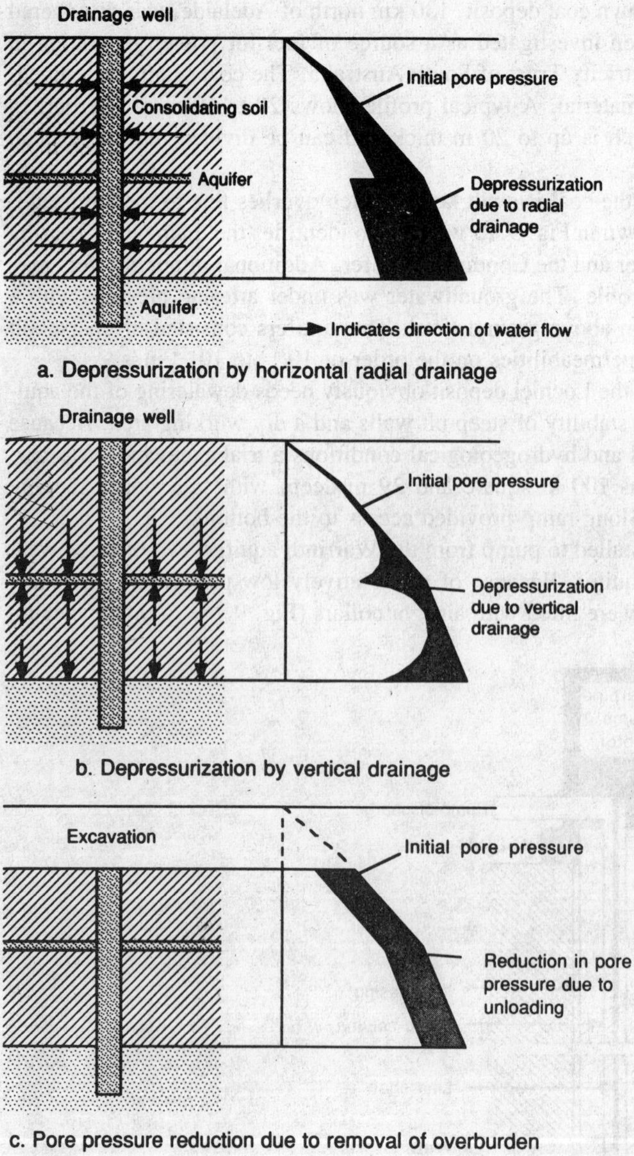

FIGURE 9.17
Pore pressure reduction by drainage and unloading. [*Adapted from Sullivan and Burman (1986)*.]

could be applied. Within 20 min a vacuum of −90 kPa was established, raising the water level in the well by 9 m. The pump rate was then increased by opening the control valve until the water level in the well fell to within 6 m of the pump. After considerable adjustment, the stable flow rate from all wells increased by 40 to 50%.

In addition to the pumped wells, some 20 sand-filled drainage wells (600-mm diameter) were installed (Fig. 9.16). They intercepted minor aquifers and made horizontal drainage possible, accelerating the consolidation of the finer-grained water-bearing layers. Water could then drain vertically downward or upward into the Warrindi aquifer, from where it was pumped to the surface.

Extensive instrumentation allowed detailed performance evaluation. The following were installed: 224 piezometers, 6 extensometers, 38 slip indicators, 10 slope inclinometers, 12 slope alarms (to warn personnel of impending slope failures), 30 surface settlement gauges, 40 survey stations, and other instruments measuring pump flow rate, vacuum pressure, and pump water levels.

The application of a vacuum to the wells was considered very successful and economical. In one particular fine-grained layer, the piezometric fall in 1 day of vacuum-assisted pumping was equal to the fall recorded over the previous 6 weeks.

The effect of dewatering was analyzed by Sullivan and Burman (1986) based on the following mechanisms, illustrated in Fig. 9.17:

a. Depressurization (lowering the pore pressure) due to horizontal drainage into the wells (horizontal consolidation)

b. Depressurization due to vertical drainage of the finer-grained layers into the more permeable aquifers (vertical consolidation)

c. Pore pressure reduction associated with unloading (change in vertical stress due to the excavation process)

The pore pressures determined for the various stages of depressurization could then be used in the stability analysis of the high walls of the planned open-pit mine. This assisted in the choice of the most economical mining method (e.g., dragline or bucketwheel excavators).

PROBLEMS

Prefixes indicate problem type: C = calculations, M = multiple choice, B = brief answer.

Calculations

C9.1. The coefficient of permeability of an 11-m-thick unconfined aquifer is to be determined by a pump-out test performed using two observation holes, 5 and 10 m from the well. For steady-state conditions at a pumping rate of 25.8 L/s the water levels in the observation holes were 3.1 and 2.4 m below the original groundwater level, respectively. Calculate the coefficient of permeability.

C9.2. Draw a diagram of h_0 versus Q for a perfect well based on the discharge formula [Eq. (9.10)] and Sichardt's formula [Eq. (9.9)] which is based on a maximum entry gradient. Calculate a new influence range for each h_0 value. Based on the diagram obtained, determine the well capacity. The well radius is 0.25 m, the aquifer has a thickness of 12 m, and the soil permeability is 0.0005 m/s.

C9.3. Wells spaced around the perimeter of a 36-m-wide circular excavation must lower the water level by a minimum of 5 m below the existing groundwater level. The unconfined water-bearing layer is 12 m thick and is located above an impermeable stratum. Other data available are

Well radius = 0.15 m

Coefficient of permeability = 10^{-4} m/s

Find the minimum number of wells and pump capacity required to achieve the dewatering target.

C9.4. A rectangular excavation is 80 m long and 40 m wide. The groundwater level is 1 m below the surface; an impermeable stratum lies at a depth of 13.5 m. The average permeability of the aquifer is 5×10^{-4} m/s. The excavation is to be 5.5 m deep. The lowered groundwater is to be at least 0.5 m below the base of the excavation.

(*a*) Find Q_{tot} for an equivalent circular excavation. You may start assuming h_0 = 4 m.

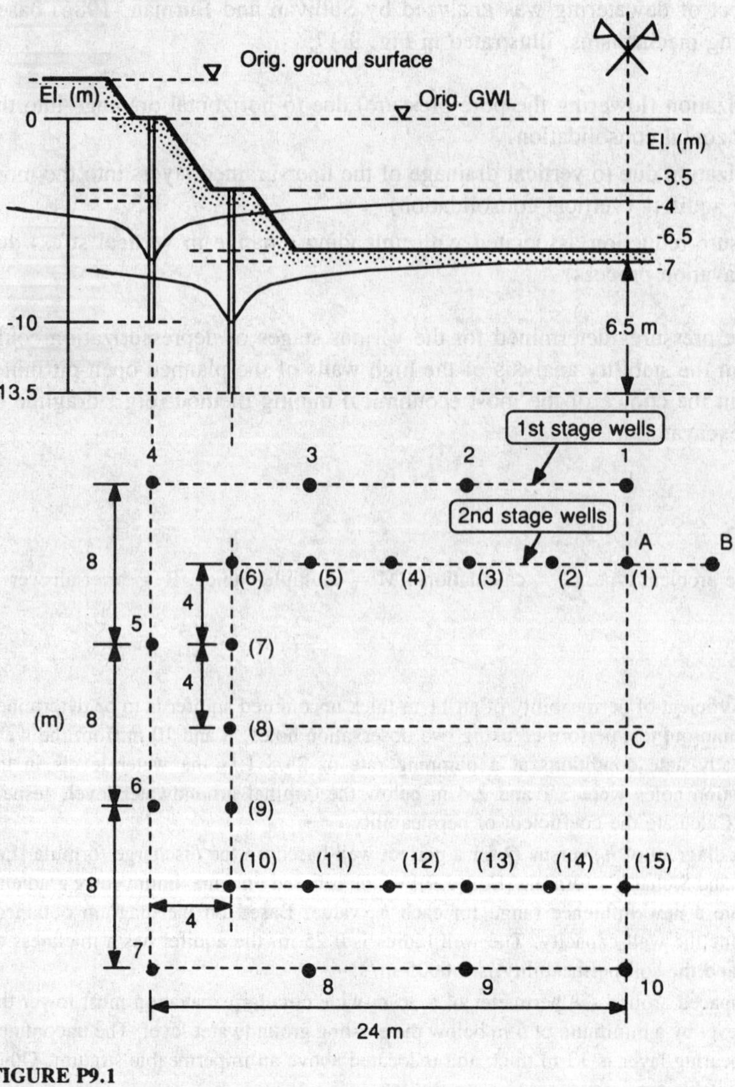

FIGURE P9.1

Layout of wells in Prob. C9.5.

(b) Estimate the number of wells required for $r = 0.10$ m. Check the value assumed for h_0 and improve it if required, revising Q_{tot}.

(c) Locate the wells around the excavation, and check the lowered water level at the most critical point using Forchheimer's equation.

C9.5. A rectangular excavation is to be dewatered in two stages as shown in Fig. P9.1. "Imperfect" conditions exist: The wells do not reach an impermeable stratum. The coefficient of permeability of the soil is 10^{-3} m/s.

(a) *First stage:* Assume 10 wells are sunk as shown. The groundwater level is to be lowered so that the second line of wells can be constructed. At the critical point A the level is to be lowered by 4 m. Proceed as follows:

Assume $s = 4$ m and estimate the influence range. (Here s is obviously underestimated, resulting in a larger Q, but this is acceptable in a preliminary design.) Determine the pump capacity required assuming perfect well conditions, and then add 20% because this is not actually the case.

Calculate the individual well capacity required. (In a real situation, this capacity, which is a function of the well radius and the wetted filter length, should now be checked. If Sichardt's formula indicates an individual well capacity less than determined above, the well diameter should be increased or more wells should be installed.)

(b) *Second stage (partial excavation):* Assume 15 wells are sunk under the protection of stage 1. The excavation is now to proceed by extending the line of wells from A to B and so on. In order to install the well at B, the original water level has to be lowered by at least 3.5 m. Determine the individual well capacity required for this step by first computing the influence range L, then Q, and then Q_i, allowing for imperfect conditions.

(c) *Second stage (final condition):* The final excavation is 40 × 16 *m*. Determine the individual well capacity required for an equivalent circular excavation with the same number of wells, given that the water level at the center (point C) has to be lowered to El. −7 m. Assume $s = 7$ m in the calculation of L.

C9.6. As shown in Fig. P9.2, there are 6 m of sand overlying a 3-m-thick clay layer. Originally the groundwater level is at the surface. A major excavation requires de-

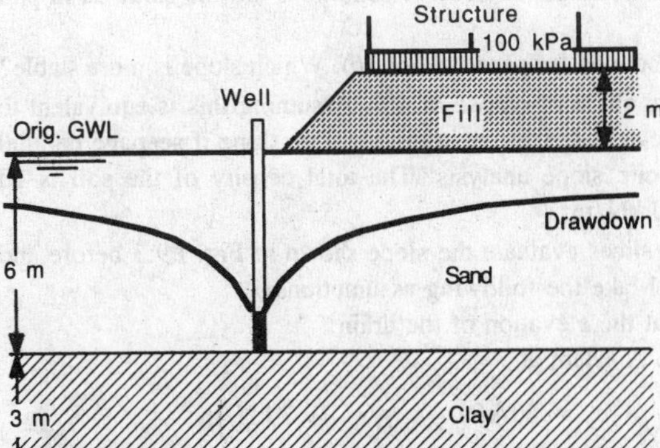

FIGURE P9.2
Settlement due to dewatering (Prob. C9.6).

watering. On an adjacent site which has been covered with 2 m of fill for a very long time, the water level is expected to be lowered by 1 to 3 m. The fill supports a raft foundation exerting 100 kPa of pressure. Using one-dimensional settlement theory, make a quick estimate of the range of settlement expected underneath the fill due to the consolidation of the clay layer during dewatering. Soil properties are as follows:

Fill: $\gamma_t = 19\ \text{kN/m}^3$

Sand: $\gamma_t = 19\ \text{kN/m}^3$

Clay: $\gamma_t = 17\ \text{kN/m}^3,\ e_0 = 1.1,\ C_c = 0.87$

Note: $S = \dfrac{H}{1 + e_0} C_c \log \dfrac{\sigma'_{vo} + \Delta\sigma}{\sigma'_{vo}}$.

C9.7. Consider the stability of an infinite slope. Assuming seepage *parallel* to the slope [Eq. (9.23)] determine the safety factor for the following conditions:

(*a*) Soil properties:

Effective cohesion = 0

Effective friction angle = 35°

Total unit weight = $18\ \text{kN/m}^3$

Unit weight of water = $10\ \text{kN/m}^3$

Geometry:

Depth of potential failure plane = 2 m

Slope angle = 30°

Water level at the surface (soil saturated)

(*b*) Same as part (*a*) but assume the phreatic surface is parallel to the slope at a depth of 1 m.

(*c*) Same as part (*a*) but assume there is *no* water above the failure plane.

C9.8. Referring to Prob. C9.7 calculate the pore pressure in the failure plane for the following conditions:

(*a*) The seepage parallel slope, water level at the surface, is as per Prob. C9.7(*a*).

(*b*) Assume there is horizontal seepage; otherwise, conditions are the same as in part (*a*).

(*c*) Compare pore pressures obtained for parts (*a*) and (*b*). Which slope is more stable?

C9.9. A cohesionless material has an angle of repose of 40°. Assuming this is equivalent to the angle of internal friction, calculate the maximum possible slope if seepage parallel to the slope occurs. Use infinite slope analysis. The total density of the soil is 20 kN/m^3 and that of water is 10 kN/m^3.

C9.10. Using the Swedish method of slices evaluate the slope shown in Fig. P9.3 before and after installation of the drain. Make the following assumptions:

(*a*) The new water table lies at the elevation of the drain.

(*b*) The soil properties remain constant at

$$\phi' = 10^\circ \qquad c' = 80\ \text{kPa} \qquad \gamma_t = 19.7\ \text{kN/m}^3$$

(*c*) Hydrostatic conditions exist (no seepage). Use eight slices of equal width. (Data adapted from E. B. Beckel (ed.), "Landslides and Engineering Practice," Chap. 9, HRB Special Report 29, 1958.)

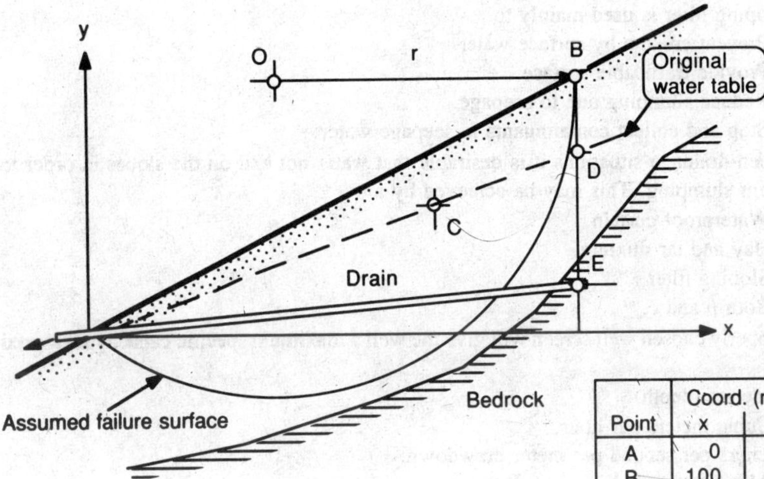

Point	Coord. (m) x	y
A	0	0
B	100	50
C	70	25
D	100	35
E	100	10
O	37.5	50

FIGURE P9.3
Geometry of slope in Prob. C9.5.

C9.11. Carry out the same analysis as in Prob. C9.10, but this time in terms of total stresses, assuming the following soil properties:
(*a*) Before drainage: $\phi = 0°$, $c = 80$ kPa, $\gamma_t = 19.7$ kN/m^3.
(*b*) After drainage: $\phi = 0°$, $c = 110$ kPa, $\gamma_t = 18.0$ kN/m^3.
Try to explain how drainage could produce changes in soil properties as inferred in this problem.

Multiple Choice

M9.12. The permeability of a particular soil is *not* significantly affected by its
(*a*) Void ratio.
(*b*) Saturation.
(*c*) Specific gravity of the soil solids.
(*d*) Fabric or structure.

M9.13. The permeability of medium to coarse sands can be determined in the laboratory using
(*a*) Constant-head-type tests.
(*b*) Specific gravity tests.
(*c*) Shear tests.
(*d*) Atterberg limits.

M9.14. The provision of filter layers in soils subject to seepage
(*a*) Cleans groundwater.
(*b*) Decreases settlement around pipes.
(*c*) Prevents migration of fine soil particles into drains or wells.
(*d*) Prevents corrosion of screens.

M9.15. A sloping filter is used mainly to
- (*a*) Prevent erosion by surface water.
- (*b*) Provide trafficable surface.
- (*c*) Reduce slumping due to seepage.
- (*d*) Stop and collect contaminants in seepage water.

M9.16. In open-drainage situations it is desirable that water not exit on the slopes in order to prevent slumping. This may be achieved by a
- (*a*) Waterproof curtain.
- (*b*) Hay and tar mixture.
- (*c*) Sloping filter.
- (*d*) Both *b* and *c*.

M9.17. A properly chosen well screen will give the well a maximum specific capacity measured in
- (*a*) Meters (depth).
- (*b*) Cubic meters per hour.
- (*c*) Liters per second per meter drawdown.
- (*d*) All of these.

M9.18. In the long term, the quantity of water which can safely be pumped from a particular individual well depends *primarily* on
- (*a*) Entry gradient of the water.
- (*b*) The number of wells in the total system.
- (*c*) The distance to adjacent wells.
- (*d*) The pump capacity.

M9.19. The capacity of an individual well does *not* depend on the
- (*a*) Diameter of the well.
- (*b*) Pump efficiency.
- (*c*) Water level in the well.
- (*d*) Maximum possible entry gradient (function of filter characteristics).

M9.20. Water may decrease the stability of a soil mass by
- (*a*) Increasing the susceptibility to liquefaction.
- (*b*) Reducing the total density.
- (*c*) Causing seepage forces opposite the direction of slope movement.
- (*d*) Preventing erosion.

M9.21. Dewatering increases slope stability in cohesionless soils mainly because it
- (*a*) Increases the friction angle of a soil.
- (*b*) Causes a change in the pH of the soil.
- (*c*) Reduces pore pressures.
- (*d*) Increases cohesion.

M9.22. Stability of an infinite slope is lowest for
- (*a*) Partially saturated soil.
- (*b*) Dry soil.
- (*c*) Seepage parallel slope.
- (*d*) Horizontal seepage.

M9.23. The practical spacing of horizontal drains in clayey soils is
- (*a*) 3 to 8 m.
- (*b*) 8 to 15 m.
- (*c*) 15 to 25 m.
- (*d*) $>$ 50 m.

M9.24. Drains installed to stabilize an active landslide require maintenance because of
(*a*) Possible damage to paint work.
(*b*) Regulations.
(*c*) Aesthetics.
(*d*) Possible movement-induced breakage and blockage.

Brief Answer

B9.25. You have to estimate the permeability of a disturbed sample of medium to coarse sand, but you are unable to conduct a permeability test. What other soil characteristics would enable you to get a value for *k*? ____________

B9.26. (*a*) A direct laboratory test to determine *k* of a relatively permeable soil is the so-called ____________ .
(*b*) Indirectly (not involving water level or flow measurement) the permeability of a sand can be estimated from its ____________ .

B9.27. The presence of water in a soil mass may decrease its slope stability because it may
(*a*) ____________ .
(*b*) ____________ .

B9.28. Besides drainage there are other methods of stabilizing a sliding soil mass, including:
(*a*) ____________ .
(*b*) ____________ .
(*c*) ____________ .

B9.29. The factor of safety of an infinite slope is usually expressed as $F = s_a/s_m$; in this formula
(*a*) s_a = ____________ .
(*b*) s_m = ____________ .

B9.30. The effectiveness of a drainage system can be checked by installing ____________ and measuring the ____________ from the drains.

CHAPTER

10

FILTRATION, DRAINAGE, AND SEEPAGE CONTROL WITH GEOSYNTHETICS

10.1 INTRODUCTION TO GEOSYNTHETICS

10.1.1 Geotextiles—Definition and Types

The term "geotextiles" appeared first in the late 1970s, describing permeable fabrics used in geotechnical engineering. Although fabrics made of natural fibers have also been used in the past, today's geotextiles are almost exclusively made of synthetic fibers (polyester, polyolefins, nylons etc.) by weaving, knitting or by bonding with partial melting, needlepunching or the addition of chemical binding agents.

Flexible sheets made of synthetic plastic or rubber as well as fabrics made "impermeable" by impregnation are referred to as geomembranes. Many engineering applications require combinations of fabrics (for strength) and solid membranes (as barriers); other products such as mats, meshes, and grids, possibly even reinforced by metal elements, may also be included in these composites. The variety of synthetic geotechnical construction materials available on the market has increased so much that the all embracing term of "geosynthetics" is gaining acceptance.

It is not intended to give a treatise of textile technology at this point, but to assist in the interpretation of information provided in subsequent sections that the following definitions are offered:

Monofilament wovens: Contains single filaments in warp (machine direction) and weft (cross-machine) direction.

Multifilament wovens: Made of multifilament yarns (strands of continuous filaments) in warp and weft directions.

Tape wovens: Split-film tapes (rather than filaments with circular or elliptical cross sections) are used.

Nonwovens: Have essentially random textile structure (e.g., common felt is a nonwoven textile). They are further categorized according to how fibers are interlocked or bonded, which is achieved by mechanical (e.g., needle punching), chemical, thermal, or solvent means.

Knitteds: The textile structure is produced by interlooping one or more ends of yarn or comparable material.

Arbitrarily selected examples of geotextiles are

Bidim, Foss Geomat, Trevira (nonwoven, needle-punched polyester)

Polyfelt, Sodoca (nonwoven, needle-punched polypropylene)

Typar (nonwoven, spun-bonded polypropylene)

Terram 700 (nonwoven, melded, 33% polyethylene, 67% polypropylene)

Geolon, Nicolon, Polytrac, Propex (woven polypropylene)

Terrafirma (woven polyester)

Terrafix 370RS, Lotrak 200 (woven-nonwoven composites)

A whole range of different geosynthetic products are now being offered on the market under more than 200 different brand names. For a more detailed review of these materials, the reader is referred to specialized textbooks such as those authored by Koerner and Welsh (1980), Rankilor (1981), Koerner (1986), and John (1987). The classification system shown in Fig. 10.1 will serve as an introductory guide.

10.1.2 Geotextile Applications

Although fabrics and membranes have been employed in civil engineering for many years, recent economic pressures and the availability of textiles specifically manufactured for engineering have led to a significant increase in their use. Geotextiles are now incorporated in many different designs, and new applications are continuously emerging.

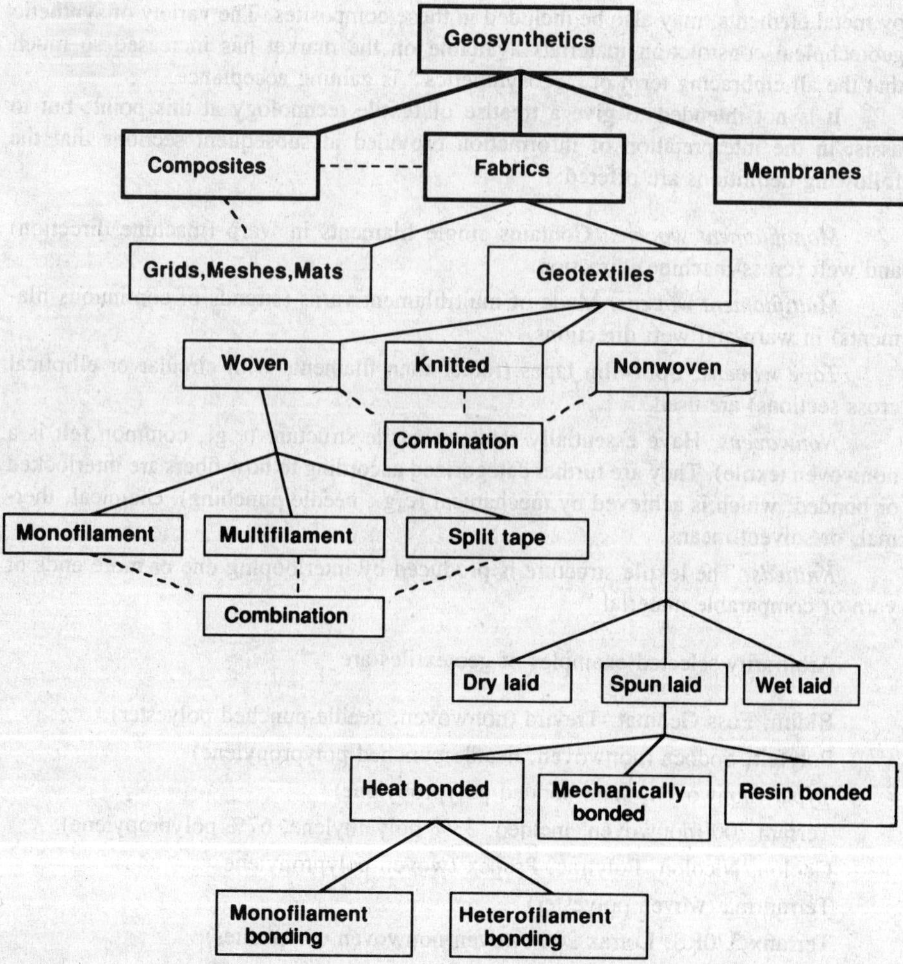

FIGURE 10.1
Classification of geosynthetics.

The main branches of engineering that use geotextiles are transportation (roads and railways), water resources, soil conservation, and public health (pollution control), as detailed in the following table:

Application	Purpose of geotextile
Pavements on soft soil	Increase bearing capacity; decrease degree of rutting
Pavement overlays	Inhibit crack propagation
Railroads	Prevent ballast contamination; distribute load on subgrade
Embankments	Improve stability; provide drainage
Retaining structures	Reinforce and protect backfill
Natural slopes	Protect slope against erosion; reinforce soil; provide drainage
Rivers, canals, reservoirs	Replace or improve traditional filter layers; erosion control
Water pollution control	Extract and collect granular pollutants; reinforce and protect geomembranes; relieve pore water or gas pressure below membranes
Shore protection	Prevent erosion and sand migration; act as filter and drainage layer
Building elements	Form soil- or soil-cement-filled bags and tubes to create columns, beams, and mats for load support and erosion and corrosion protection

Geotextiles are versatile in use, adaptable to different circumstances, and can be combined with many traditional and new building materials. The key to appropriate design lies in the understanding of the geotextile functions and the ability to relate these functions to fabric and soil properties.

10.1.3 Basic Functions

A geotextile usually fulfills one or more of the following basic functions:

1. Drainage
2. Filtration
3. Separation
4. Reinforcement

Additional related specialized functions or combinations of basic functions have been identified, such as

Containment (formwork)

Tensioned membrane (bridging of gaps, stress reduction)

Stress redistribution (cushioning)

Erosion and dispersion control

Screening or fencing

Drainage means collecting and redirecting seepage water within a soil mass or adjacent to retaining walls, culverts, and tunnel linings. Generally only nonwoven fabrics or composites have sufficient in-plane flow capacity to fulfill this function. A geotextile acts as a *filter* if it allows seepage from a water-bearing layer while preventing most soil particles from being carried away by the water flow. Drainage and filtration are hydraulic phenomena; however, it should be pointed out that both these functions may indirectly lead to a more stable soil mass, either by the process of consolidation or by preventing erosion.

Separation is achieved if the fabric prevents mixing of adjacent dissimilar soils, which may occur during construction or may be caused by repeated external loading of a soil-layer system. Most fabrics can act as separators provided they have adequate strength. Design criteria for separation therefore refer essentially to the properties as determined in standard mechanical tests. If water is present in the adjacent soil, the fabric should also be evaluated as a filter, since water flow can endanger the integrity of adjacent layers as much as mechanical action can.

In its most general sense, *reinforcement* means the inclusion of the fabric to provide tensile strength, redistribution of stresses, and/or confinement, thereby increasing the stability of a soil mass, reducing earth pressures, or decreasing deformation or susceptibility to cracking. However, special terms may describe types of reinforcement other than reinforcement by the addition of internal tensile reinforcing fabric strips or sheeting:

Fabrics are said to provide *containment* if they are used to form soil- or concrete-filled bags, tubes, or mattresses.

The fabric is seen to act as a *tensioned membrane* if it supports loads across a gap or plastic zone of soft soil.

The fabrics may be required to provide *cushioning* against localized stresses which may cause puncturing or abrasion.

If placed on the surface of a slope, the geotextile may prevent *erosion* and *dispersion* of soil due to wind, surface runoff, or wave action. Successful performance of a geotextile used for this purpose requires a combination of specific physical, mechanical, and hydraulic properties.

Table 10.1 lists the major areas of geotechnical engineering in which fabrics are used. For each area of application it is attempted to describe the role of the fabric in terms of its major functions and to identify the physical, mechanical, and hydraulic properties most relevant to those functions.

Table 10.2 gives an overview of test methods either existing or considered as standards in Australia and the United States. Many other international and national standards have been adopted or are in preparation. [For a review of American geotextile testing practice see Fluet (1987).]

Applications where filtration, drainage, and seepage control are most relevant will be discussed in this chapter. Soil reinforcement using geosynthetic sheets is treated in Chap. 19. Synthetic strips and grids are discussed in Chap. 18.

TABLE 10.1
The function of fabrics in major areas of application

Application	Major functions in order of priority*	Special consideration	Most-important properties
Primary roads and railroads	Separation Drainage Reinforcement	Repeated loading	Pore size Permeability Strength Elongation Abrasion
Retaining walls, embankments, and foundations	Reinforcement Separation	Creep	Strength Soil-fabric friction
Unpaved roads	Reinforcement Separation	Repeated loading	Strength Elongation
Erosion protection, subdrains, and seepage control	Filtration Drainage Reinforcement	Rapid changes in water level Construction stresses	Pore size Permeability Strength Abrasion
Soil drainage (accelerate consolidation, reduce water pressures)	Drainage Filtration	Clogging	Transmissivity Pore size

*Brief definition of major functions:

Filtration: The fabric prevents migration of soil particles without impeding water flow.

Drainage: The fabric collects and conveys water.

Separation: The fabric prevents mixing of adjacent dissimilar soils during construction or during repeated external loading of the soil-fabric system.

Reinforcement: The inclusion of the fabric increases the strength of a soil mass, thereby increasing stability or bearing capacity or reducing earth pressures.

Durability is of concern with respect to mechanical as well as hydraulic properties, but so far the focus of engineers and researchers has been mainly on the loss of the reinforcement function. Aging through exposure to environmental influences is further discussed in Sec. 19.1.3.

10.2 WATER FLOW THROUGH SOILS AND FABRICS

10.2.1 Permeability, Permittivity, and Transmissivity

Figure 10.2 illustrates the experiment conducted by Darcy who studied the permeability of soils (also described in Sec. 7.3.4). Water is made to flow under a constant head dh through a column of sand of length dx. From measurements of the water flow through various cross sections and under different heads, the following relationship can be deduced:

$$Q = kiA \tag{10.1}$$

TABLE 10.2
Test methods—physical, hydraulic, mechanical, and durability properties

Topic	ASTM[a]	SAA[b] method no.
General requirements, sampling, conditioning, and basic physical properties	D 4354-84	AS 3706.1
Determination of tensile properties—wide-strip method	D 4595-86	AS 3706.2
Determination of tearing strength—trapezoidal method	D 4533-85	AS 3706.3
Test method for breaking load and elongation (grab test)	D 4632-86	
Determination of indirect biaxial tensile strength—CBR plunger method	—	AS 3706.4
Determination of puncture-resistance—drop cone method	—	As 3706.5
Determination of seam strength	D 4884-89	AS 3706.6[c]
Determination of pore size distribution—dry-sieving method	D 4751-87	AS 3706.7
Determination of pore size distribution—wet-sieving method	—	AS 3706.8[c]
Determination of permittivity	D 4491-85	AS 3706.9
Determination of transmissivity	D 4716-87	AS 3706.10[c]
Determination of durability—general requirements	—[c]	AS 3706.11
Determination of durability—resistance to degradation by ultraviolet light and heat	D 4355-84	AS 3706.12
Determination of durability—resistance to degradation by hydrocarbons or chemical reagents	—[c]	AS 3706.13
Determination of durability—resistance to degradation by biological attack	—[c]	AS 3706.14

[a]ASTM—American Society for Testing and Materials. (Some 14 additional standards have been introduced or are being drafted by this organization.)

[b]SAA—Standards Association of Australia.

[c]In course of preparation.

where Q = flow through area A per unit time, m^3/s
A = cross-sectional area, m^2
k = coefficient of permeability, m/s (also called hydraulic conductivity)
$i = dh/dx$ = hydraulic gradient (In the mathematics of hydraulics, this term is negative; nevertheless, in soil mechanics computations it is usually taken as a positive value.)

The same experiment could be carried out with one or more horizontal layers of fabric replacing the sand. In this case the term dx would represent the fabric sample thickness. There is, however, a problem with measuring a representative fabric thickness: Woven geotextiles are very thin, and nonwovens are distinctly compressible. It is therefore desirable to indicate the permeability of fabrics unaffected by the variability of thickness measurements. This has led to the definition of the permittivity ψ. Let us rewrite Darcy's law [Eq. (10.1)] and define ψ:

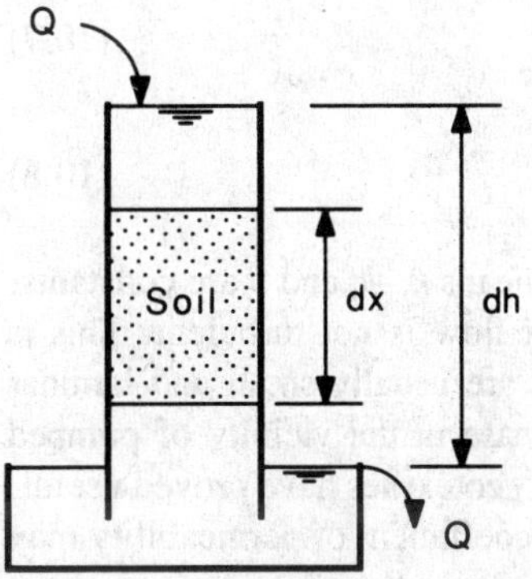

FIGURE 10.2
Darcy's experiment.

$$Q = \frac{k}{dx} dh\, A = \psi\, dh\, A \tag{10.2}$$

where
$$\psi = \frac{k}{dx} = \text{permittivity } (\text{s}^{-1}) \tag{10.3}$$

For water flow normal to the plane of geotextiles the permittivity is simply the coefficient of permeability divided by the fabric thickness. It is the preferred measure of water flow capacity across the geotextile plane.

In drainage applications it is important to know at which rate water can be transmitted within a strip of geotextile. Referring back to Fig. 10.2, we now consider the cross-sectional area A to be defined by the width and thickness of a fabric strip:

$$A = ab$$

where a is the width of the strip of fabric and b is the thickness of the fabric, both in meters.

Equation (10.1) can then be rewritten in such a way that we can define transmissivity θ:

$$Q = (kb)ia = \theta ia \tag{10.4}$$

with
$$\theta = kb = \text{transmissivity } (\text{m}^2/\text{s}) \tag{10.5}$$

The transmissivity is the quantity of water which flows within the plane of an aquifer of unit width under a unit gradient. It is the preferred measure of the in-plane water flow capacity of a geotextile. It is only relevant for relatively thick nonwoven or composite fabrics.

Equations (10.1), (10.2), and (10.4) express the flow rate in terms of the coefficient of permeability, permittivity, and transmissivity of a water-bearing layer. Alternatively, we could express the latter in terms of quantities measurable in an experiment, such as flow rate, cross section, and gradient:

$$k = \frac{Q\, dx}{A\, dh} \quad \text{m/s} \tag{10.6}$$

$$\psi = \frac{Q}{A\,dh} \qquad \text{s}^{-1} \tag{10.7}$$

$$\theta = \frac{Q}{ia} \qquad \text{m}^2/\text{s} \tag{10.8}$$

Darcy's law is only valid for laminar flow. This means k, ψ, and θ are constants, that is, independent of the gradient, only if the water flow is not turbulent. This is true for most groundwater problems, where gradients are usually small and laminar flow conditions exist. However, this may not be the case in the vicinity of pumped wells, erosion control structures, and other areas where geotextiles have proved useful. In situations like this, one either has to accept that the coefficient of permeability may vary with certain limits or one has to introduce an alternative flow law, such as proposed by the Delft Hydraulics Laboratory [Ogink (1975)]:

$$Q^n = k_f A \frac{dh}{dx} \tag{10.9}$$

where k_f is the generalized coefficient of permeability and n is a factor ranging between 1 and 2 (purely laminar flow: $n = 1$; purely turbulent flow: $n = 2$).

10.2.2 Factors Affecting Permeability

Theoretical analysis of flow through porous media shows that the coefficient of permeability k is related to the following properties:

Unit weight and viscosity of the permeant

Void ratio or porosity

Effective particle or pore size

Particle or pore shape

Experiments show that, in soils, saturation also has an important influence on permeability.

Typical values for soils are (order of magnitude only)

Coarse gravel	$k = 10^{-1}$ m/s
Fine gravel	$k = 10^{-2}$ m/s
Coarse sand	$k = 10^{-3}$ m/s
Fine sand	$k = 10^{-5}$ m/s
Silt	$k = 10^{-8}$ m/s
Clay	$k < 10^{-9}$ m/s

The preceding values are meant as a guide only. The permeability of an individual sand could easily vary by a factor of 10 depending on its relative density. The finer the soil, the larger the possible variation. Changing the density of a silt could change its permeability by a factor of 100. The very slow movement of water in clays could vary by a factor of 1000 depending on its void ratio and soil structure (flocculated or dispersed). Similar to manufactured geotextile, natural sedimentary soil deposits can be anisotropic with respect to permeability. Horizontal permeability could easily be 10 times the vertical one.

The coefficient of permeability of geotextiles quoted in the literature is usually the permeability across the fabric, for flow normal to its plane. To prevent misunderstanding, it is desirable to designate it k_n. Similarly, the in-plane permeability should be referred to as k_p if it is not expressed in terms of transmissivity. Typical values for the most common geotextiles and geocomposites fall into the following ranges:

Nonwovens:	$k_n = 4 \times 10^{-3}$	to	10^{-4}	m/s
	$\psi = 4$	to	4×10^{-2}	s^{-1}
	$\theta = 5 \times 10^{-4}$	to	10^{-6}	m^2/s
	(Depends greatly on normal stress acting on the geotextile)			
	$\frac{k_p}{k_n} = 5$ to 250			
Wovens:	$k_n = 10^{-4}$	to	10^{-5}	m/s
	$\psi = 2 \times 10^{-1}$	to	3×10^{-2}	s^{-1}
Geocomposite strip and sheet drains:	$\theta = 10^{-3}$	to	100	m^2/s

Considerable theoretical and experimental work is currently going on in an effort to study the basic principles of flow across and within the plane of a geotextile and the influence of adjacent soils on the geotextile's permeability. Factors identified as important include

Fiber type, size, and orientation
Void ratio or porosity
Confining pressure
Repeated loading
Contamination
Aging (time)
Nonlaminar flow

When dry, some fabrics exhibit resistance to wetting. In such cases initial permeability is low but rises until the fabric reaches saturation. Permeability may also be reduced through air bubbles trapped in the geotextile. This is the reason why testing standards usually require careful saturation of fabric specimens before they are subjected to water flow. In addition, permeability measurements will be more consistent with the use of deaired water rather than tap water.

It is hoped that mathematical relationships will be developed which will assist in estimating the permeability for specific service conditions based on that measured under standard laboratory conditions.

10.2.3 Laboratory Flow Tests

10.2.3.1 NORMAL FLOW (PERPENDICULAR TO FABRIC). Most common is the constant-head-type test as originally used for Darcy's experiments with soil. Figure 10.3 shows a typical test arrangement where water flows across one or several layers of geotextile of total thickness dx and cross-sectional area A. The flow rate Q can be controlled by adjusting the upper constant head tank or by a valve in the feeder line originating from a high-capacity source of pressurized water. The head loss across the fabric is measured using standpipes. Most relevant to practical problems is a dh within the range of 0 to 500 mm. The coefficient of permeability for flow normal to the fabric, designated with a subscript n, is calculated directly from Darcy's law:

$$k_n = \frac{Q\,dx}{A\,dh} \quad \text{m/s} \tag{10.6}$$

Alternatively, a falling head test could be performed (Fig. 10.4), where water drains from a vertical tube through the fabric specimen into a constant-level reservoir. At time t_1 the head of water in the tube is observed as h_1; at time t_2 it has fallen to h_2. If the cross-sectional area of the vertical tube is A_0, the coefficient of permeability is found to be

$$k_n = \frac{A_0\,dx}{A(t_2 - t_1)} \ln \frac{h_1}{h_2} \tag{10.10}$$

The falling head test has the disadvantage that the possible variation of the permeability with a change in head is not readily apparent. In addition, the high flow rate through relatively open fabrics makes it difficult to achieve good accuracy.

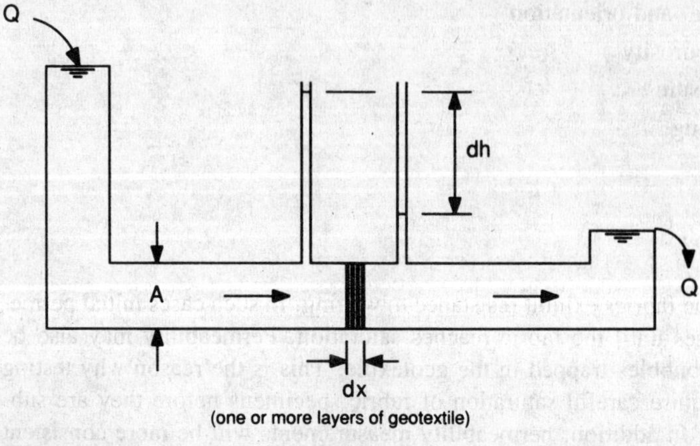

FIGURE 10.3
Constant head test.

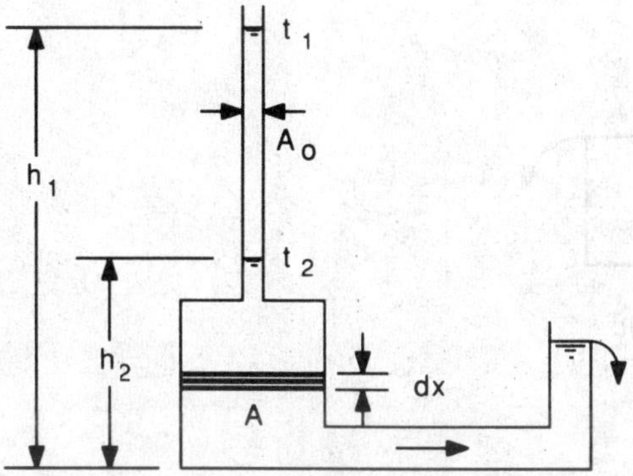

FIGURE 10.4
Falling head test.

As indicated earlier, it is becoming more common to quote the water flow capacity across the fabric in terms of the permittivity $\psi = k_n/dx$ rather than the coefficient of permeability.

10.2.3.2 IN-PLANE FLOW. The constant and falling head types of permeameters can be adopted to measure the in-plane permeability k_p of a fabric. It is now commonly reported in the form of the transmissivity $\theta = k_p b$, where b is the thickness of the fabric. The transmissivity varies with the thickness and density of the fabric, which in turn is a function of the compressive stress applied to the fabric. Various types of apparatuses have been developed to apply a load to the fabric and at the same time measure the flow along the plane of it.

Figure 10.5*a* shows the principle of an apparatus developed by ICI Fibres. A rectangular sample is encapsulated in a flexible impermeable membrane and placed horizontally between two 50-mm-thick sand layers. This complete sandwich is loaded via a rigid plate. Using different reservoir levels and applied loads, the effect of different gradients and compressive stresses can be studied. Similar devices using rectangular specimens with planar flow have been developed by other researchers, mainly differing in the way the compressive load is applied.

Figure 10.5*b* illustrates a device developed by the Hydraulic Engineering Institute in Bucharest. It measures the lengthwise permeability of a cylindrical sample of geotextile. Lateral pressure is applied via an elastic impermeable membrane. This type of apparatus has some resemblance to triaxial testing equipment used in a soil laboratory.

Figure 10.5*c* shows an apparatus such as is used at Queen's University, Kingston, Ontario. A circular disk of fabric is placed between two metal plates. Water flow is radial from the center outward. RILEM recommends a similar test and suggests a hydraulic head to 500 mm and confining pressures up to 500 kPa. In addition, RILEM (The International Union of Testing and Research Laboratories for Materials and Structures) suggests that a sheet of elastic material of given hardness and thickness be placed between the geotextile and the metal plates. Some types of soil consolidation test equipment could be adapted for this kind of test.

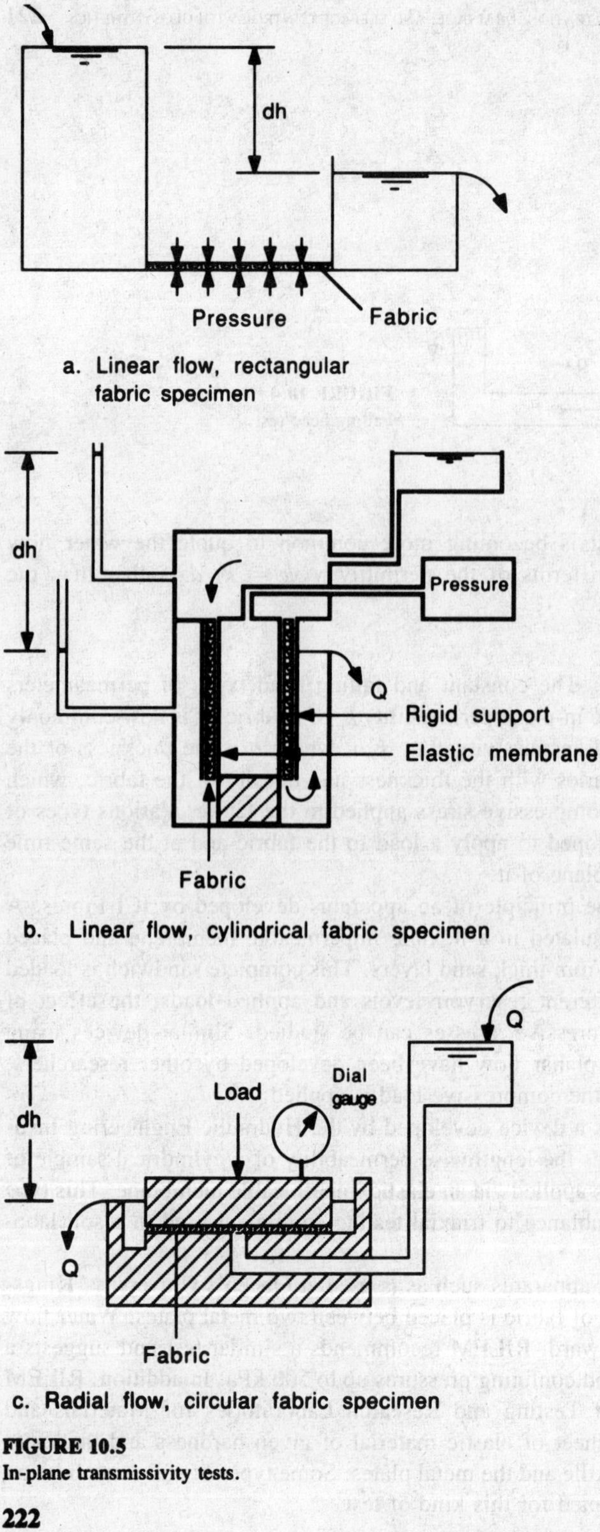

FIGURE 10.5
In-plane transmissivity tests.

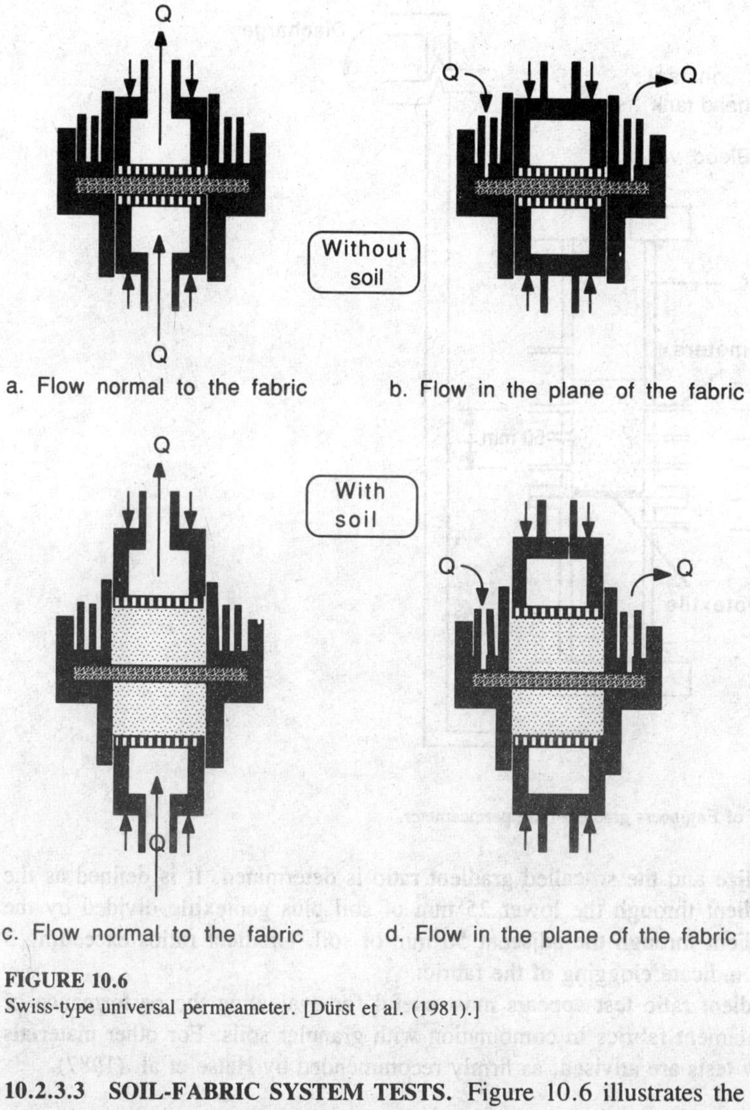

FIGURE 10.6
Swiss-type universal permeameter. [Dürst et al. (1981).]

10.2.3.3 SOIL-FABRIC SYSTEM TESTS. Figure 10.6 illustrates the use of an apparatus which was built at the Swiss Federal Institute of Technology and which can be used to determine the permittivity and transmissivity of a geotextile alone and when sandwiched between two layers of soil. The latter is of importance with respect to the long-term behavior of fabrics buried in soil and subjected to water flow. This apparatus can be used to study the filtration characteristics and clogging susceptibility.

The permeameter shown in Fig. 10.7, which was developed by the U.S. Army Corps of Engineers, is used to provide data to develop filter criteria for the prevention of piping and erosion and to evaluate clogging resistance of geotextiles. It is essentially a constant-head-type apparatus which allows the measurement of the head loss along a soil-geotextile system. After the test is run for some hours (or days), the piezometer

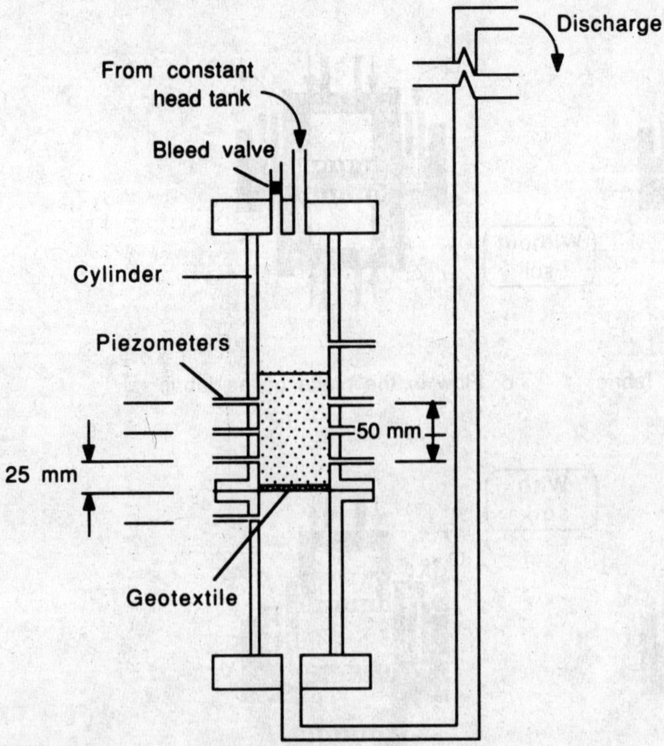

FIGURE 10.7
U.S. Army Corps of Engineers gradient ratio permeameter.

readings stabilize and the so-called gradient ratio is determined. It is defined as the hydraulic gradient through the lower 25 mm of soil plus geotextile divided by the hydraulic gradient through the adjacent 50 mm of soil. Gradient ratios exceeding 3 were found to indicate clogging of the fabric.

The gradient ratio test appears most useful for evaluating the performance of woven monofilament fabrics in combination with granular soils. For other materials long-term flow tests are advised, as firmly recommended by Halse et al. (1987).

10.2.3.4 CAPILLARY TESTS. Queen's University and others have carried out tests to determine the capillary rise in geotextiles. The type of apparatus used is shown in Fig. 10.8. The adjustable crossbar is set, e.g., at an initial height of 250 mm, and the dry geotextile is draped over it. One end of the fabric strip is immersed in the graduated reservoir filled with dyed water; the other is connected to an evaporation dish set at a level below the reservoir. The crossbar is lowered daily until flow is noticed. The height at which siphoning occurs is recorded.

Gamski and Rigo (1982) think that in particular circumstances it would be useful to have a geotextile drain perform as a siphon, especially since it can reestablish itself after drying out. However, the main merit of the capillary test may be as an indicator of pore size, but so far not much data is available.

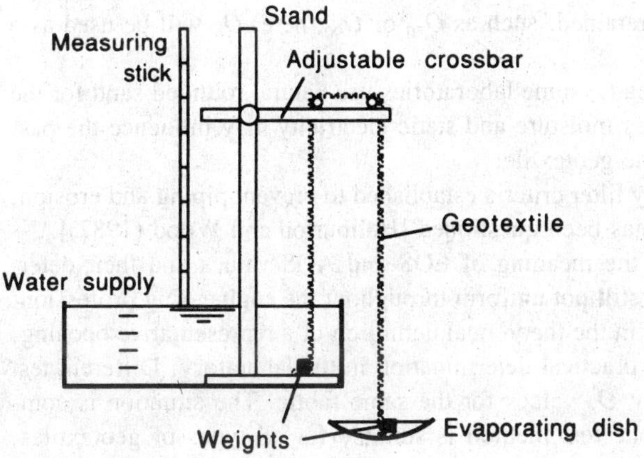

FIGURE 10.8
Capillarity test (Queens University, Kingston).

10.3 POROMETRY

Porometry is the measurement of sizes of pores and their distribution. This has proved to be a difficult problem because of the variety of fabrics used for geotechnical purposes.

10.3.1 Percent Open Area

The image of the geotextile is projected onto a screen, and the outline of the open area is traced. The ratio of the area of the openings to the total area of geotextile is expressed in percent and termed the "percent open area." This method is not really suitable for nonwoven textiles. Haliburton and Wood (1982) found that for woven fabrics the percent open area is directly related to the clogging potential, at least for the specific soil types that they tested.

10.3.2 Equivalent Opening Size and Apparent Opening Size—Dry Sieving

One way the pore size distribution of geotextiles can be measured is by sieving glass beads of known size through a screen made of the fabric being tested. The percent of glass beads retained on the sieve after 20 min or more of shaking is determined and can then be plotted versus the particle size used. The test is conducted by using successively coarser beads until 5% or less are passing the geotextile. The particle size corresponding to 5% passing is termed the *equivalent opening size (EOS)*. It is reported in microns or as the U.S. sieve number having openings closest in size to that particle size. It is also designated as O_{95}. A value of $O_{95} = 0.3$ mm means that 95% of particles with a diameter of 0.3 mm are retained on the fabric after shaking for 20 min. The *apparent opening size (AOS)* is equivalent to the EOS but is also quoted for other percentages retained, such as O_{50} or O_{90}; here, O_n will be used as a general designation.

Instead of using glass beads, some laboratories use natural rounded sand for the test, because in humid climates moisture and static electricity may influence the passage of glass beads through the geotextile.

The EOS is used in many filter criteria established to prevent piping and erosion, but its relevance to clogging has been questioned [Haliburton and Wood (1982)].

It should be noted that the meaning of EOS and AOS values and their determination in the laboratory are still not uniform throughout the engineering profession. There are not only differences in the theoretical definition of a representative opening, but variations also exist in its practical determination in the laboratory. Different test methods will produce different O_n values for the same fabric. The situation is complicated by the fact that no one test method is suitable for all types of geotextiles, ranging from coarse wovens to thick nonwovens. As a consequence, for example, filter criteria developed in different countries may not be directly comparable, even if they appear to refer to the same O_n value.

Confusion also arises because the industrial and chemical filtration expert may define O_n in terms of numbers of openings. For instance, according to Rankilor (1981), O_{50} means half the number of the pores in a geotextile is smaller than O_{50} and the other half is larger.

10.3.3 Wet Sieving

In order to evaluate the soil retention capacity of geotextiles under tidal conditions, some organizations are using wet-sieving-type tests. In one particular procedure, favored in France, a basket with a fabric bottom and holding a well-graded standard soil is repeatedly submerged in water, maybe 2000 times or more. The soil fraction which escapes through the textile during this procedure is analyzed by sieving and sedimentation. The particle size equivalent to 98% passing of this fraction is then designated as D_{98} and termed the *maximum effective pore size*.

An alternative procedure is followed at the Franzius Institute of the University of Hannover and is likely to become a German Standard. A well-graded, slightly silty sand serves as a test soil. While the sample fabric and soil is vibrated for 15 min, water is sprayed on top of it. The grain size distribution of the original soil as well as that of the portion which passes through the fabric are determined with a set of sieves as follows (mm opening):

2.50; 1.18; 0.71; 0.50; 0.40; 0.20; 0.10; 0.063; 0.045 mm

For each size fraction, defined by an upper bound D_u (mm) and a lower bound D_l (mm), the original total mass M_t and the mass M_p which passed through the fabric are recorded. The largest fraction for which

$$\frac{M_p}{M_t} 100 \geq 10\% \tag{10.11}$$

is chosen to calculate the *effective opening diameter* D_{eff} according to

$$D_{eff} = D_1 + (D_u - D_1)\frac{M_p}{M_t} \tag{10.12}$$

10.4 FILTER ACTION

Where water exists from an earth structure or flows from a finer to a coarser soil layer, there is a danger that fine soil particles will be carried along with the water, leading to piping (internal erosion), external erosion, instability (due to a buildup in pore pressures or a change in geometry), and/or silting up of the drainage pipe, trench, or basin. Preventing damage of this kind has traditionally been achieved by providing one or more graded filter layers, increasing in grain size (and permeability) in the direction of the water flow toward the seepage collection system. A natural filter zone of sizable proportions may also act as a drainage layer, meaning there is flow into the filter as well as along the plane of the filter. This would be the case for a so-called sloping filter protecting the downstream surface of a water-retaining dam. Typical situations requiring filter zones are illustrated in Fig. 10.9.

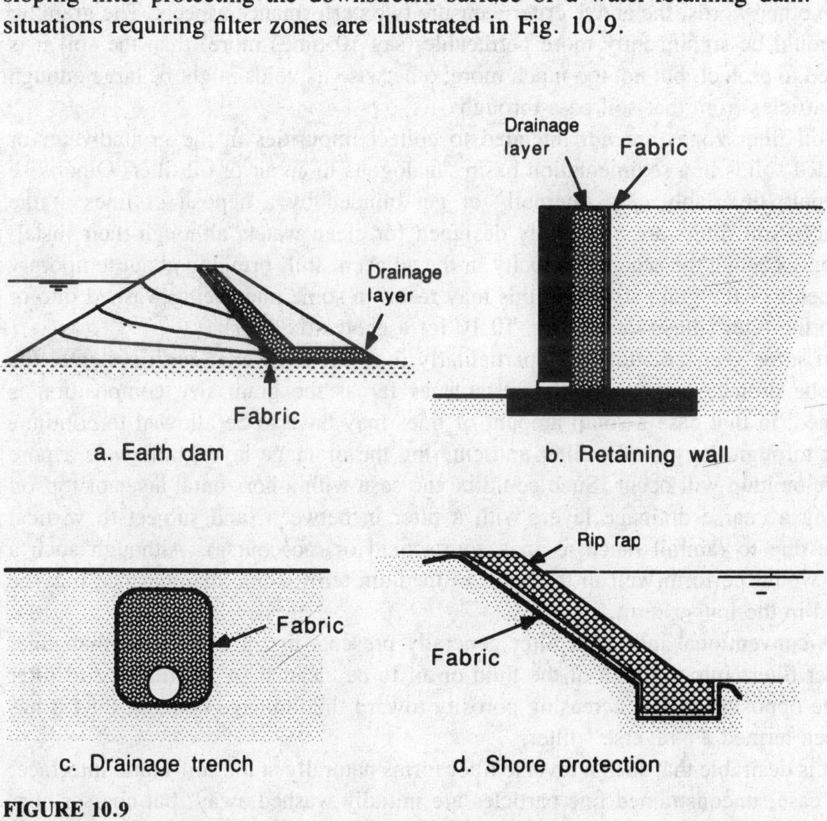

FIGURE 10.9
Typical filter applications.

10.4.1 Conventional Granular Filters

Well-accepted filter criteria for *granular materials* include the guidelines from the Corps of Engineers at Vicksburg, which require the following grain size relationships [e.g., as quoted by Lambe and Whitman (1969)]:

Piping (or retention) criterion: $$\frac{D_{15}\text{ filter}}{D_{85}\text{ soil}} < 5 \qquad (10.13)$$

Permeability criterion: $$4 < \frac{D_{15}\text{ filter}}{D_{15}\text{ soil}} < 20 \qquad (10.14)$$

Uniformity criterion: $$\frac{D_{50}\text{ filter}}{D_{50}\text{ soil}} < 25 \qquad (10.15)$$

Diameters D_{15}, D_{50}, and D_{85} correspond to 15, 50, and 85% passing, as read off a grain size distribution diagram. If these relationships are satisfied, ongoing migration of soil fines into the filter and drainage layer is prevented without impeding water flow.

In other words, the above criteria ensure two performance aspects: The granular filter should be significantly more permeable (say 10 times more) than the soil it is supposed to protect, but not too much more; otherwise its voids might be large enough to let particles from that soil pass through.

Soil filter zones are *not* intended to collect impurities in the groundwater or suspended solids in a sedimentation basin, analogous to an air or oil filter. Otherwise they would invariably clog internally or get blinded by a deposit of fines at the boundary. Soil filters are essentially designed for clean water, although their installation may change the seepage velocity in the adjacent soil, bringing about temporary adjustments of the grain structure; this may result in some fines being washed into or through the filter (illustrated in Fig. 10.10 for a geotextile filter).

In some seepage situations, particularly in artificial layered earth systems, the soil to be protected is inherently unstable as far as the grain size composition is concerned; in that case a small amount of fines may have to be allowed to continue passing through the granular filter and entering the drainage layer; otherwise a pore pressure buildup will occur. Such could be the case with a horizontal layer of topsoil overlying a coarse drainage layer, with a filter in between, and subject to vertical seepage due to rainfall (such as in a sports field or racecourse). Although such a system would perform well in the short to medium term, some maintenance may be required in the longer term.

A conventional industrial filter generally presents first a coarse and then finer and finer filters into the path of the fluid or air to be cleaned. A multilayer soil filter does the opposite: It has increasing porosity toward the drainage conduit; thus it has also been termed a ''reverse'' filter.

It is desirable that such a reverse filter forms naturally at the soil-fabric interface. In that case, unconstrained fine particles are initially washed away, but coarser constituents eventually form bridges over fabric openings and voids created in the soil

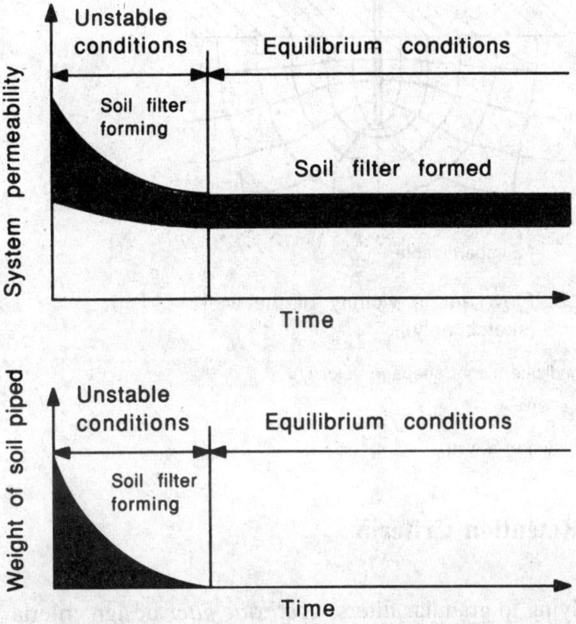

FIGURE 10.10
System permeability and piping during soil filter zone formation. [*After Lawson (1986).*]

matrix. In the field, this adjustment may take days or weeks but leads to steady-state seepage conditions.

The opposite of reverse filter formation is the deposition of a filter cake. This has the following possible outcomes:

(*a*) The seepage diminishes until no further soil particles are dislocated. This represents a new equilibrium state, but by that time the drainage system may no longer be effective.

(*b*) Pore pressures increase, leading to instability: loss of bearing capacity, slope failures, increased lateral pressures.

(*c*) Seepage is redirected, possibly causing flooding, erosion, or stability problems elsewhere.

Fine cohesionless silty and sandy soils are most susceptible to internal erosion when subjected to increased seepage gradients due to the construction of a drainage system or a well (see Fig. 10.11); they are also the types of soil most likely to block downstream drainage pipes. In addition, fine uniform sands or silts and gap-graded soils need protection by filters when subjected to seepage. The same applies to dispersive silty and clayey soils, although the seepage quantities involved may be considerably smaller, unless cracks due to shrinkage or differential settlement open up a drainage path. Clay and coarse sand particles are more resistant to dislocation than silt, clay because of cohesion and coarse sand because of size.

FIGURE 10.11
Increase in seepage velocity after installation of a well.

10.4.2 Permeability and Retention Criteria for Geotextiles

Similar to the formulations applying to granular filters, *geotextile filter* design criteria also relate to adequate filter flow capacity (permeability criterion) and prevention of continuing soil particle migration or ''piping'' (retention criterion). Several different criteria have been proposed, as discussed by Lawson (1982), Hoare (1982), and others.

Two recent sets of filter design guidelines are those by Giroud (1982) and Lawson (1986), both presented in Table 10.3.

Giroud's equations are based on theoretical considerations and engineering judgment. His permeability criterion indicates that a geotextile needs to have only one-tenth of the permeability of the soil and will still not cause excessive pore pressure buildup in the soil (it is of course much thinner than a granular filter layer, thus causing only a small head loss). The R_G factor in the piping criterion is a function of the relative density I_D and the uniformity coefficient C'_u as shown in Table 10.4.

Lawson (1986) generalized the relationship between O_{90} (apparent opening size, meaning 90% of particles of this size are retained on the fabric after a specified time of shaking) and the grain size distribution of the soil, represented by the diameter D_n, which is the particle size below which lie n percent of the whole sample. Given O_{90},

TABLE 10.3
Filter design guidelines

Author	Permeability criterion	Retention criterion	For constants see
Giroud (1982)	$k_n > 0.1\ k_{soil}$	$O_{95} \leq R_G D_{50}$	Table 10.4
Lawson (1986)	$O_{90} \geq CD_n$	$O_{90} \leq CD_n$	Fig. 10.12 (for specific soil type)

TABLE 10.4
Factor R_G for retention criterion according to Giroud (1982)

Relative density I_D	Linear soil uniformity coefficient C'_u*	
	$1 < C'_u < 3$	$C'_u > 3$
< 35% (loose)	$R_G = C'_u$	$R_G = 9/C'_u$
$35\% < I_D < 65\%$	$R_G = 1.5C'_u$	$R_G = 13.5/C'_u$
$I_D > 65\%$ (dense)	$R_G = 2C'_u$	$R_G = 18/C'_u$

*$C'_u = (D'_{100}/D'_0)^{0.5}$ where D' values are found by linear extrapolation of the tangent to the central portion of the grain size distribution curve.

the factor C can be determined from Fig. 10.12 for different percentile values of n and particular hydraulic performance standards, delineated by the permeability limit and piping (retention) limit. The relationships implied by this figure were found to be valid for results of laboratory experiments on well-graded residual soils from Hong Kong.

Lawson (1986) defines the four regions in Fig. 10.12 as follows:

Region 1. This is where it is possible for the permeability of the geotextile to be less than the base soil (the soil to be protected). (*Note:* This may result in reduced capacity of the drainage system and a buildup of pore pressures in the soil.)

Region 2. Zero soil is piped through the filter, and the permeability of the soil-geotextile system attains equilibrium.

Region 3. An initial amount of soil is piped through the system, and then the permeability of the soil-geotextile system attains equilibrium (Fig. 10.10).

Region 4. Continual uncontrolled piping occurs through the geotextile.

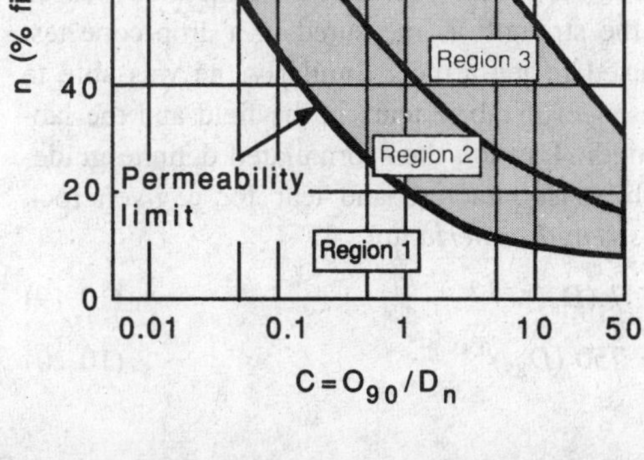

FIGURE 10.12
Values of C for different base soil particle percentiles n for well-graded residual soils (completely decomposed granite). [*According to Lawson (1986).*]

A study of other filter design criteria indicates that the permeability requirement is often related to D_{15} as a representative grain diameter while piping limits are usually related to the average diameter D_{50}, D_{85}, or D_{90}. If detailed soil-fabric filter performance data is lacking, the following criteria could be considered:

Retention criterion: $$O_{90} \leq D_{85} \tag{10.16}$$

Permeability criterion: $$O_{90} > D_{15} \tag{10.17}$$

10.4.3 Clogging Criterion Based on Gradient Ratio Tests

Poorly graded soils, such as gap-graded soils, have been identified as problem materials as far as filter protection is concerned, because seepage water may bring with it enough soil particles to clog the filter, thus making it less permeable.

Laboratory permeameter tests on specific soil-geotextile systems, such as developed by the U.S. Corps of Engineers, allow measurement of the gradient increase near the clogging fabric. This value divided by the normal gradient in the soil (after flow has stabilized) is termed the gradient ratio (see Sec. 10.2.3 and Fig. 10.7). The *clogging criterion* proposes that the gradient ratio GR should be less than 3 in such a filtration test in order to ensure satisfactory performance in the field:

$$\text{GR} < 3 \tag{10.18}$$

Rather than adopting a limiting GR value as a design criterion, it may be more appropriate to observe the long-term performance of a soil-filter system in a laboratory permeameter. As illustrated in Fig. 10.13, a geotextile should be acceptable as a filter for a particular soil if, for a given overall gradient, the flow rate, system permeability, and gradient ratio stabilize after an initial period of adjustment, regardless of the absolute GR value. Interpretation of long-term filter performance data is complicated by the fact that the textile may undergo chemical degradation and clogging due to biological growth rather than particle movement [GRI (1988)].

10.4.4 Strength Criteria for Filters in Slope Protection

After investigating the performance of a large number of erosion control structures, Lawson (1982) found a correlation between the occurrence of fabric punctures (caused mainly by construction processes) and the strength as measured in a drop cone test for particular characteristic rock sizes used in the armor. Similarly, he was able to establish a relationship between the presence of fabric tears in the field and the laboratory-determined trapezoidal tear strength. Lawson then formulated definite guidelines for the selection of fabrics which resist puncture and tear for a given rock diameter. Expressed in formulas, these *strength criteria* are

$$P_c\sqrt{H} < 7 \cdot (D_{85})^{-0.9} \tag{10.19}$$

$$F_T > 750\ (D_{85})^{0.45} \tag{10.20}$$

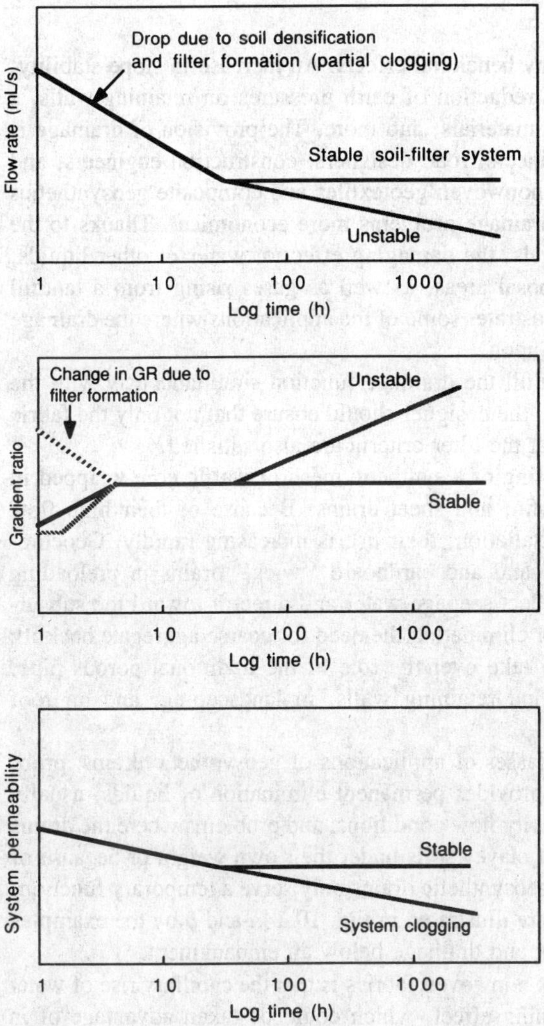

FIGURE 10.13
Filter formation and clogging as observed in soil-fabric systems during a constant head permeameter test (schematic only).

where P_c = cone penetration value, mm
H = height of rock placement, m
D_{85} = characteristic rock diameter, m
F_T = trapezoidal tear strength, N

In situations where severe dynamic loading by waves and currents are expected, Heerten and Wittmann (1985) recommend the use of composite geotextiles where the filter fabric is connected to a thick, high-porosity roughness layer. The purpose of this layer is to prevent any downslope soil particle movement below the fabric, which may lead to benching of the embankment and fabric distortion.

10.5 DRAINAGE

Dewatering of a soil mass has many beneficial effects: An increase in slope stability, an increase in bearing capacity, a reduction of earth pressures on retaining walls, a reduction of the weight of borrow materials, and more. The provision of drainage is a central task of geotechnical engineers, road designers, construction engineers, and builders. The availability of thick nonwoven geotextiles and composite geosynthetics have made the solution of many drainage problems more economical. Thanks to the transmissivity of these new materials, the damaging effect of water or other liquids, such as leachates from waste disposal areas, as well as gases rising from a landfill may be controlled. Figure 10.14 illustrates some of the applications where the drainage function of geosynthetics is relied upon.

Thick nonwovens usually fulfill the drainage function simultaneously with the filter action. Where this is the case, the designer should ensure that not only the fabric has adequate transmissivity but that the filter criteria are also satisfied.

Geocomposite drains, consisting of a synthetic mesh or waffle core wrapped in geotextile, are available as strip, fin, and sheet drains. Because of their high flow capacity and easy, economical installation, their use is increasing rapidly. Geocomposite strip drains have replaced sand and cardboard "wick" drains in preloading projects (Fig. 11.2). Fin drains collect seepage water and direct it toward the subsurface drainage structure, reducing or eliminating the need for coarse aggregate backfill; high-capacity fin drains may even take over the role of the traditional porous pipe. Sheet drains find application behind retaining walls, in landscaping, and on roof gardens.

One could distinguish two classes of applications of geosynthetic drains: problems where the drainage material provides permanent elimination of liquids, usually analyzed assuming steady-state gravity flow conditions; and problems where the drains serve to accelerate consolidation of clayey soils under their own weight or because of a surcharge. In the latter case, the geosynthetic drains only serve a temporary function. These two categories of problems are illustrated in Fig. 10.14*a* and *b* by the examples of drainage behind a retaining wall and drainage below an embankment.

An interesting aspect of thick nonwoven fabrics is that the capillary rise of water within them may lead to a siphoning effect, which could be taken advantage of in some drainage situations. On the other hand, a geosynthetic with large voids could serve as a break against capillary rise, e.g., in order to prevent frost heave, saltwater intrusion, or expansive soil problems.

In all design situations it should be remembered that nonwoven fabrics and composite synthetic drains are compressible. Their transmissivity is therefore significantly affected by the normal pressure they are subjected to. Under medium to high pressures, long-term creep may become significant and should be considered by the designer.

Approximate required flow rates under a hydraulic gradient of 1 and typical normal pressures for a number of common drainage applications are presented in Fig. 10.15 [according to Koerner et al., (1986)].

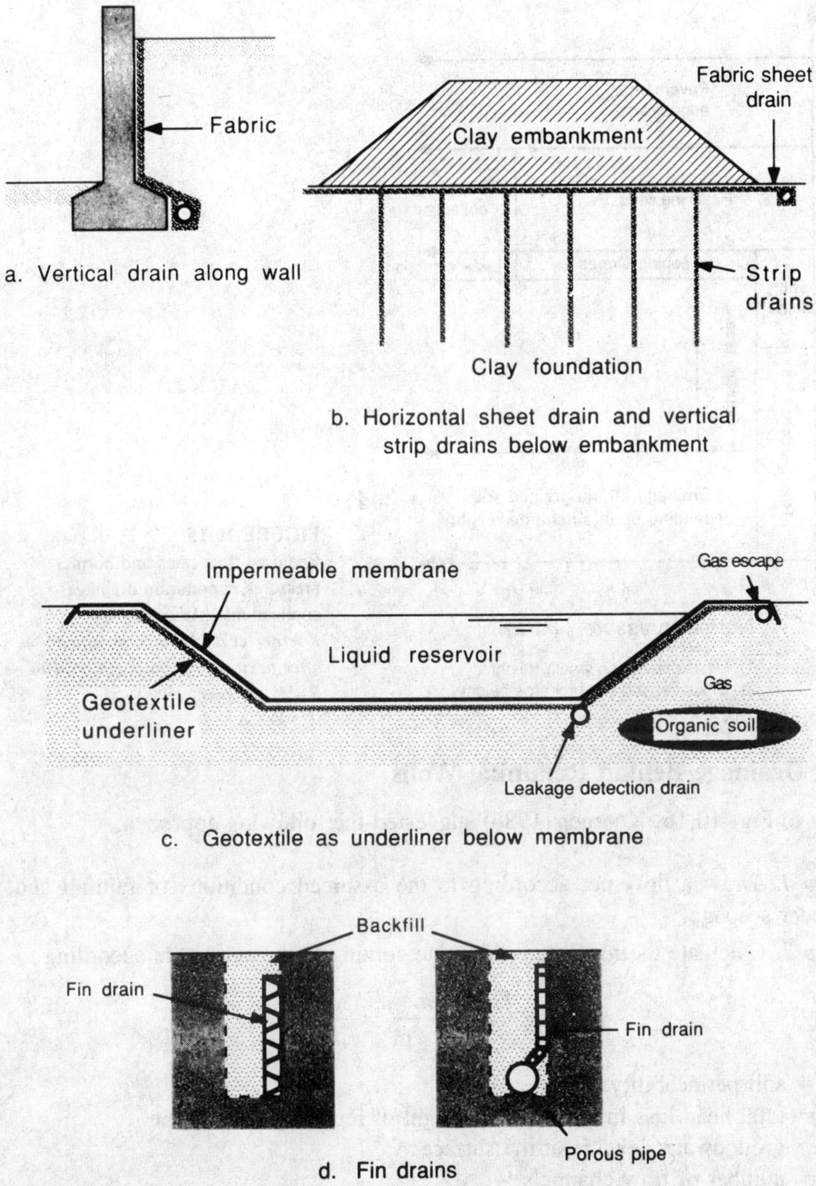

FIGURE 10.14
Typical drainage applications of geotextiles (or geocomposites).

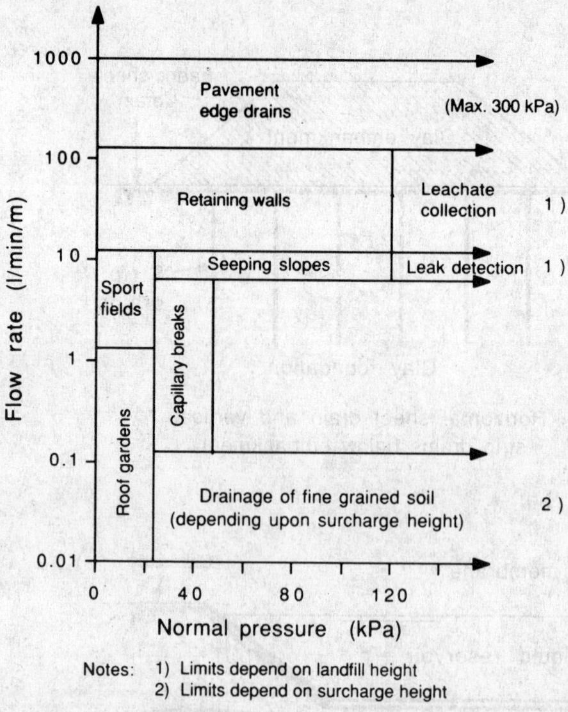

FIGURE 10.15
Required flow rates and normal pressures in common drainage applications. [*Adapted from Koerner et al. (1986), modified after personal communication with R. M. Koerner, 1988*]

10.5.1 Drainage Behind Retaining Walls

Referring to Fig. 10.16, Koerner (1986) suggested the following approach:

Step 1. Draw a flow net according to the assumed conditions of rainfall and groundwater seepage.

Step 2. Calculate the maximum flow rate coming to the geotextile according to

$$q = kh\frac{F}{N} \qquad \frac{\text{m}^3}{\text{s}\cdot\text{m}} = \frac{\text{m}^2}{\text{s}} \tag{10.21}$$

where k = soil permeability, m/s
h = total head loss in the flow = height of retaining wall, if the groundwater level is at the surface
F = number of flow channels
N = number of head drops

Step 3. Determine the flow gradient i within the geotextile. For a vertical drain, $i = 1$.

Step 4. Calculate the required transmissivity.

$$\theta = \frac{q}{i} \qquad \frac{\text{m}^2}{\text{s}} \tag{10.22}$$

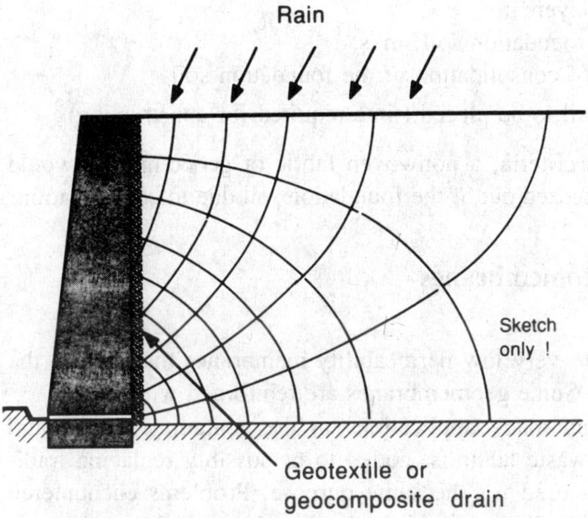

FIGURE 10.16
Geotextile drainage behind retaining wall.

Step 5. Select a geotextile or composite drain which will safely provide the required transmissivity at the maximum earth pressures expected.

10.5.2 Drainage Below Embankment

Giroud (1983) applied consolidation theory to the problem shown in Fig. 10.17 which represents impervious surcharge fill placed on a compressible and saturated clay foundation. With adequate transmissivity, a geotextile may then serve as a drainage layer for conveying water expelled from the soil by the consolidation process. After Giroud (1983), the required geotextile transmissivity θ_{req} (m^2/s) can be expressed in the following terms:

$$\theta_{req} = \frac{B^2 k}{\sqrt{c_v t}} \tag{10.23}$$

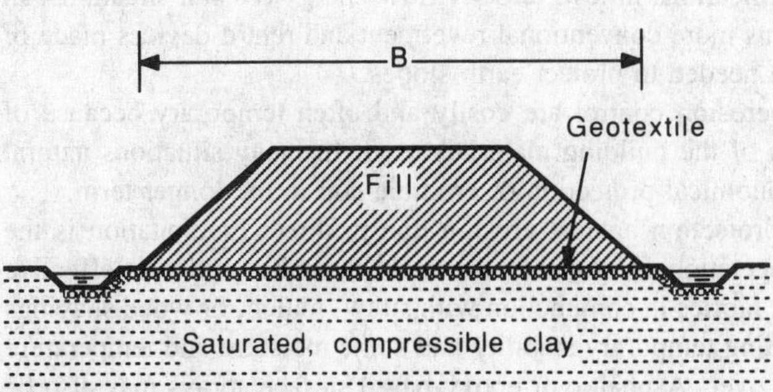

FIGURE 10.17
Geotextile drainage below surface fill.

where B = width of surcharge layer, m
k = permeability of the foundation soil, m/s
c_v = vertical coefficient of consolidation of the foundation soil
t = time for surcharge fill to be placed (time required for construction)

Selected according to this criteria, a nonwoven fabric or geocomposite would be able to carry all the water squeezed out of the foundation soil due to consolidation.

10.5.3 Drainage Below Geomembranes in Waste Containment

By definition, geomembranes are very low permeability membranes that control the flow of liquids in soil and rock. Some geomembranes are reinforced with fabrics.

Geomembranes have found many applications in the containment of liquid wastes and seepage from solid-waste landfills, added to or possibly replacing traditional layers of compacted clay used for the same purpose. Problems encountered with accidental puncturing and the formation of gas bubbles beneath the membranes have led to the design of multiple-layer barriers incorporating geotextiles. In these applications geotextiles help to protect the membrane from localized high stresses and uneven settlement and possible cracks in the foundation soil. Furthermore, because of their in-plane drainage capacity (transmissivity) they can prevent the accumulation of gas or water under pressure below the membranes. This drainage capacity may be enhanced by strip drains with higher flow capacity than fabrics alone.

10.6 SCREENS AND BAFFLES FOR THE CONTROL OF SEDIMENT TRANSPORT BY WIND AND WATER

As discussed by Ingles (1983), the fundamental processes of particulate transport are known. Whether erosion, transportation, or deposition takes place depends on particle size and velocity of wind or water. In order to control erosive action, basically two approaches are possible: reduce velocity by baffles and barriers, or increase the particle size by agglomeration or containment.

Geotextiles have applications in low-velocity, low-energy erosion situations. In high-energy flow conditions more conventional revetment and retard devices made of concrete and steel may be needed to protect earth slopes.

Artificial means of erosion control are costly and often temporary because of corrosion and degradation of the building materials used. In many situations natural vegetation is the most economical protection of erodible soil in the longer term.

Particularly where protection against erosion due to direct precipitation is the objective, establishing vegetation is probably the most desirable and cost-effective measure. Open-mesh-type fabrics in conjunction with straw, mulch, or wood shavings and seeding help in providing temporary stability and some moisture and temperature control until growth takes over. Asphaltic or epoxy-based surface sprays may also be effective in combination with geonets and geogrids. These can be anchored to the base soil with pegs. The success of these measures depends to a great deal on local knowledge and experience. Some problems may be encountered with follow-up grass cutting and with loose pegs causing foot injuries to people stepping on them in revegetation zones near recreational areas.

Open-mesh-type geotextiles can be used to make sand fences in dune management. Denser fabrics can act as a silt curtain preventing floating and suspended matter from flowing downstream. In these cases, the geotextile basically serves to reduce wind or water velocity, leading to deposition, or represents a screen collecting particles. As it becomes more common for construction management to adopt site erosion control measures in order to prevent the associated siltation, the importance of designing silt fences which are able to resist pressures due to water and accumulated sediments will increase. Koerner (1986) gives guidelines and experiments which can assist in the solution of these problems.

PROBLEMS

Prefixes indicate problem type: C = calculations, M = multiple choice, B = brief answer, D = discussion.

Multiple Choice

M10.1. Geotextiles, geogrids, geomembranes, and geonets, etc., have one thing in common; namely, they almost exclusively
(*a*) Act as filters.
(*b*) Provide reinforcement.
(*c*) Are made of synthetic materials.
(*d*) Are woven or needle-punched nonwoven materials.

M10.2. By definition,
(*a*) Geotextiles are virtually impermeable.
(*b*) Geomembranes are virtually impermeable.
(*c*) Geogrids are very rigid.
(*d*) Composites are made from a mixture of polymers.

M10.3. The main purpose of geotextiles in river, canal, and reservoir works is to
(*a*) Increase bearing capacity.
(*b*) Distribute load.
(*c*) Replace or improve traditional filter layers.
(*d*) Relieve pore pressures.

M10.4. Major functions of geotextiles in road and railroad applications are
(*a*) Filtration, drainage, and reinforcement.
(*b*) Drainage and filtration.
(*c*) Separation and filtration.
(*d*) Separation, drainage, and reinforcement.

M10.5. The filtration function is defined by which of the following statements:
(*a*) The fabric prevents migration of soil particles without impeding water flow in the long term.
(*b*) The fabric collects and conveys water.

(*c*) The fabric collects contaminants in the seepage water, which may otherwise clog the drainage system.
(*d*) The fabric prevents coarse drainage material to be punched into the adjacent finer soil during construction.

M10.6. Which one of the following statements is *wrong*?
(*a*) Transmissivity has units of m^2/s.
(*b*) Permeability has units of m/s.
(*c*) Permittivity has units of s^{-1}.
(*d*) Hydraulic conductivity has units of m^3/s.

M10.7. Which one of the following statements is *wrong*?
(*a*) The hydraulic gradient is a measure of head loss per length of flow path.
(*b*) Equipotential lines are lines of equal elevation head.
(*c*) Flow from one point to another is caused by a difference in total head.
(*d*) Flow through an area per unit time is proportional to the coefficient of permeability and the gradient.

M10.8. A typical permeability of a fine sand is
(*a*) 10^{-1} m/s.
(*b*) 10^{-2} m/s.
(*c*) 10^{-5} m/s.
(*d*) 10^{-8} m/s.

M10.9. An apparent opening size (AOS) of $O_{95} = 0.3$ mm means
(*a*) Using the fabric as a sieve, 95% of particles of a uniform sand with a diameter of 0.3 mm are retained after a standard shaking period.
(*b*) Using the fabric as a sieve, 95% of particles of a uniform sand with a diameter of 0.3 mm pass through after a standard shaking period.
(*c*) 95% of the fabric pores have an equivalent diameter of 0.3 mm.
(*d*) The fabric openings apparent in microscopic evaluation are 0.3 mm wide.

M10.10. In respect to geotextile filter action, which one of the following statements is *wrong*?
(*a*) Permeability criteria ensure that the fabric has adequate flow capacity.
(*b*) Piping (or retention) criteria ensure that continuing soil particle migration is prevented.
(*c*) Giroud expressed the permeability criterion as $k_n < 0.1k_{soil}$.
(*d*) Piping limits are usually related to D_{50}, D_{85}, or D_{90}.

M10.11. A nonwoven fabric or drainage composite behind a concrete retaining wall is beneficial mainly because it
(*a*) Reduces compaction requirements for the backfill.
(*b*) Prevents discoloration of the concrete.
(*c*) Reduces lateral pressures on the wall.
(*d*) Reduces the plasticity of the backfill.

M10.12. A nonwoven fabric or drainage composite below an embankment on a soft clayey ground
(*a*) Acts principally as a separator.
(*b*) Acts principally as a reinforcement layer (increase in bearing capacity).
(*c*) Accelerates consolidation and subsequent gain in strength.
(*d*) Causes compaction of the ground and thus increases its bearing capacity.

M10.13. With respect to the gradient ratio (GR) test, which one of the following statements is *wrong*?

(*a*) GR criteria have been established to prevent piping.
(*b*) The more soil particles accumulate on or in the fabric during the test, the higher the GR values.
(*c*) GR criteria are most relevant for granular soils protected by woven monofilament fabrics.
(*d*) The use of the GR test for a nonwoven fabric adjacent to a clayey silt is not recommended.

Brief Answer

B10.14. Name four polymer types used to make geosynthetics:
(*a*) ____________
(*b*) ____________
(*c*) ____________
(*d*) ____________

B10.15. What type of geotextiles are Bidim, Polyfelt, and Typar (woven, knitted, nonwoven, composite, stitched)? ____________

B10.16. Name two hydraulic properties of geotextiles which are important in geotechnical engineering:
(*a*) ____________
(*b*) ____________

B10.17. Name four basic functions of geotextiles:
(*a*) ____________
(*b*) ____________
(*c*) ____________
(*d*) ____________

B10.18. If a geotextile successfully acts as a filter, its openings must be
(*a*) Large enough so as not to ____________ .
(*b*) Small enough so as to ____________ .

Calculations

C10.19. In a laboratory test, nominal thickness, flow of water normal through an 80-mm-diameter fabric specimen, and head loss across the fabric were measured as follows:

Fabric type	Nominal thickness, mm	Head loss across fabric, mm	Flow, L/s
Woven	0.53	682	0.278
Nonwoven	1.64	6	0.214

For each fabric, calculate
(*a*) Permittivity (s^{-1}).
(*b*) Coefficient of permeability (m/s).

C10.20. In a radial transmissivity device, water flows from the central area ($r_0 = 8$ mm) to the outside perimeter ($r_1 = 62.5$ mm) of the fabric specimen. The following measurements were taken with a constant head loss $dh = 1.59$ m:

Pressure, kPa	Flow Q, L/s
20	0.00925
50	0.00550
100	0.00337
200	0.00142
500	0.00039

Calculate the transmissivity (m²/s) of this nonwoven geotextile according to

$$\theta = \frac{Q}{2\pi\, dh} \ln \frac{r_1}{r_0}$$

Plot θ versus the applied pressure. (*Note:* The same nonwoven fabric is involved here as in Prob. C10.19.)

C10.21. A dry sieving test with a fabric gave the following data:

Average particle diameter of fraction, mm	Percent retained
0.09	34
0.115	63
0.14	79
0.18	89
0.255	98

Plot the result on log-probability paper (Fig. P10.1), and estimate the equivalent opening size (or O_{95}).

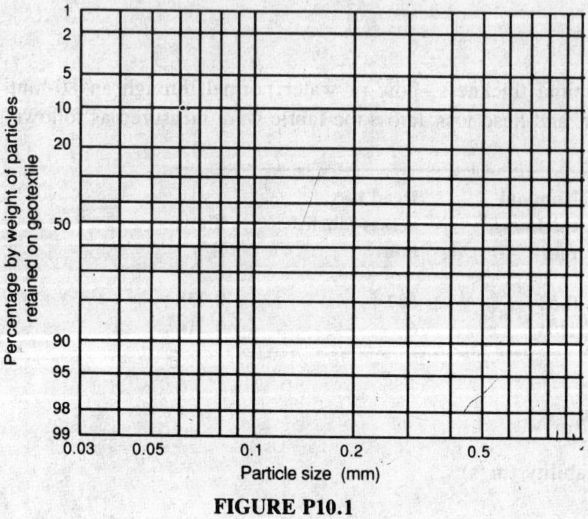

FIGURE P10.1

C10.22. Wet sieving through a fabric according to the German method described in Sec. 10.3.3 yielded the following results:

Sieve opening, mm	Total mass of fraction M_t, g	Mass of soil fraction which passed through the fabric M_p, g
0.40	23.93	0
0.20	101.99	0.20
0.10	70.39	1.45
0.063	15.13	8.02
0.045	12.22	11.83
< 0.045	1.49	1.49

Determine the effective opening diameter D_{eff}.

C10.23. A loose silty sand has the following grain size characteristics: $D_{15} = 0.08$ mm, $D_{50} = 0.45$ mm, $D_{85} = 0.9$ mm. By extrapolation of the central portion of the grain size distribution curve, $D_{100} = 1$ mm and $D_0 = 0.1$ were found.

(*a*) Determine maximum diameters D_{15} and D_{50} of a filter material satisfying conventional granular filter criteria.

(*b*) Determine the maximum O_{95} value of a filter fabric which would satisfy Giroud's retention criteria.

C10.24. Riprap is to be placed on a geotextile covered slope. It is estimated that the individual rock pieces used will have a diameter of 0.5 m and may fall from a maximum height of 2 m. According to Lawson's criteria, determine the required fabric strength in terms of the

(*a*) Maximum allowable cone penetration value (mm).

(*b*) Minimum required trapezoidal tear strength (N).

C10.25. Consider a 4-mm-high vertical retaining wall with horizontal backfill of permeability $k = 0.0005$ m/s. A geosynthetic drainage layer is placed adjacent to the wall leading to a horizontal drain at its base.

(*a*) Sketch a flow net with five flow channels for the condition that rainfall completely saturates the backfill.

(*b*) Estimate the maximum flow rate into the geosynthetic drainage layer ($m^3/s \cdot m$).

C10.26. A geosynthetic is to provide drainage behind a retaining wall with a vertical backface. The estimated flow into the drain is 0.002 m^3/s.

(*a*) Determine the required transmissivity of the geosynthetic.

(*b*) Would an ordinary single layer of nonwoven fabric be adequate?

C10.27. An area to be consolidated by preloading with a clay soil is 20 m wide. The subsoil has a coefficient of consolidation of 0.0004 cm^2/s and a permeability of 6×10^{-8} cm/s. The surcharge is to be applied in a 10-day construction period. A fabric is to be placed on top of the subsoil so that water expelled from it can drain away freely and the presence of the impervious fill does not impede the consolidation process.

(*a*) What transmissivity is required from the geotextile according to Giroud's recommendation?

(*b*) Is a thick nonwoven fabric likely to be adequate?

Discussion

D10.28. What is the difference between the function of an industrial filter and a soil filter zone?

D10.29. What are the consequences of the clogging of a soil filter zone?

D10.30. Describe the process and consequences of "piping" in soil mechanics.

D10.31. In a gradient-ratio-type test water flows through soil and then through a geotextile under a constant head. It was noticed that on the first day the flow increased by about 10%, and then decreased for about 2 weeks, and from there on remained steady at 30% of the initial flow rate. Try to explain this behavior.

D10.32. What factors would affect the durability of geotextile filter or drainage layers embedded in soils?

D10.33. Why would there be a need for strength criteria for fabrics used in hydraulic applications?

CHAPTER

11

PRELOADING AND THE USE OF VERTICAL DRAINS

11.1 PURPOSE OF PRELOADING AND VERTICAL DRAINS

Preloading or *precompression* increases the bearing capacity and reduces the compressibility of weak ground by forcing loose cohesionless soils to densify or clayey, silty soils to consolidate. It is achieved by placing a temporary surcharge on the ground prior to the construction of the planned structure. It is a method of preempting potentially damaging settlements on soft soil. A similar strategy is often employed in the construction of liquid storage tanks, which are test-loaded with water before being used to store dangerous chemicals.

In the case of a building, the surcharge would normally be equivalent or higher than the expected bearing pressure. An embankment could be overbuilt prior to final shaping and commissioning in order to more rapidly achieve the level of settlements expected for the actual design.

The surcharge generally consists of earth fill. In cohesionless sands and gravels, lowering the groundwater level may provide an alternative means of temporarily increasing effective vertical stresses.

Although it can be applied to all types of soil, the preloading technique is most advantageous for modifying soft cohesive ground. The process can be speeded up with vertical drains and, in the case of relatively impermeable fill, with a horizontal drainage layer at the original ground surface.

Preloading, with or without vertical drains, is only effective in causing substantial preemptive settlement if the total applied load significantly exceeds the preconsolidation pressure of the foundation material (see Sec. 11.3.1).

Vertical drains are installed in order to accelerate settlement and gain in strength of soft cohesive soil. Without installing vertical drains, bearing failures may occur during placement of the fill and settlement of clay soils may extend over many years. Because highly efficient drain installation methods have been developed, preloading combined with vertical drains has become an economic alternative to the installation of deep foundations or other methods of ground improvement. Vertical drains are also used to advantage in the construction of permanent fills, such as highway embankments on soft ground. Basic design principles are the same, whether the surcharge is permanent or only temporary.

Vertial drains accelerate primary consolidation only, because significant water movement is associated with it. Secondary consolidation causes only very small amounts of water to drain from the soil; secondary settlement, therefore, is not speeded up by vertical drains (see Sec. 11.4.5).

As outlined in Chap. 19, high-strength geosynthetic sheet reinforcement in combination with vertical drains may make it possible to construct embankments on very soft soil, which may have a shear strength of less than 5 kPa.

Only relatively impermeable soils [those with a vertical coefficient of consolidation $c_v < 3 \times 10^{-7}$ m^2/s, according to Rowe (1968)] potentially benefit from vertial drains. Soils which are more permeable will usually consolidate under a surcharge at an acceptable rate on their own.

Vertical drains are particularly effective where a clay deposit contains many thin horizontal sand or silt lenses (so-called microlayers). However if these microlayers are continuous in a horizontal direction, only little may be gained from vertical drains under a surcharge of limited extent, since rapid drainage of the foundation material then occurs, whether the drains are placed or not.

The major beneficial effects of preloading and vertical drains are illustrated in Fig. 11.1: Preloading reduces total and differential settlement and may allow for economies in the choice of the foundation system. Vertical drains speed up the settlement process but do not reduce the amount of deformation under a given load.

This chapter looks at ways of providing vertical drainage and reviews the principles of consolidation relevant to the placement of fill on soft ground. This leads to a guide for optimizing the spacing of drains (given time constraints and allowable deformation) and for predicting corresponding settlement and pore pressure dissipation. Subsequently, additional information is provided which will assist in determining the soil parameters needed for analysis and the monitoring of the consolidation process during construction.

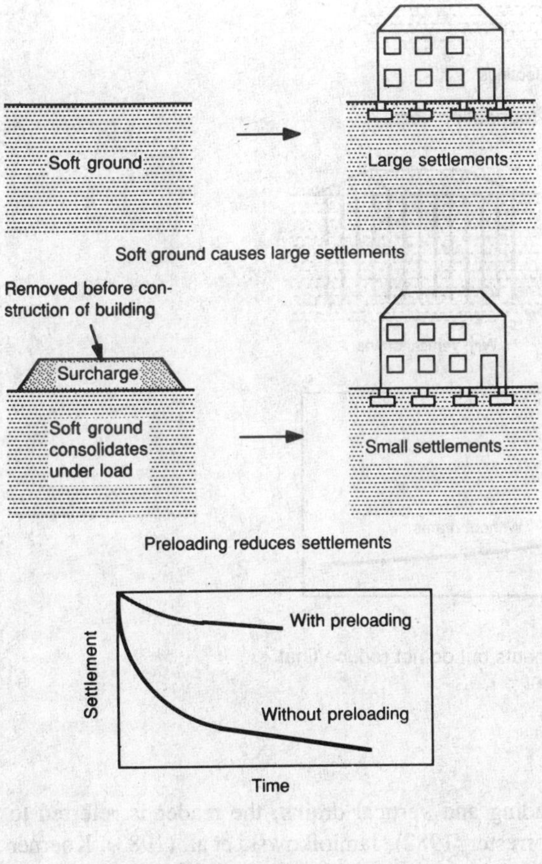

a. Preloading reduces settlements

FIGURE 11.1
Beneficial effects of preloading.

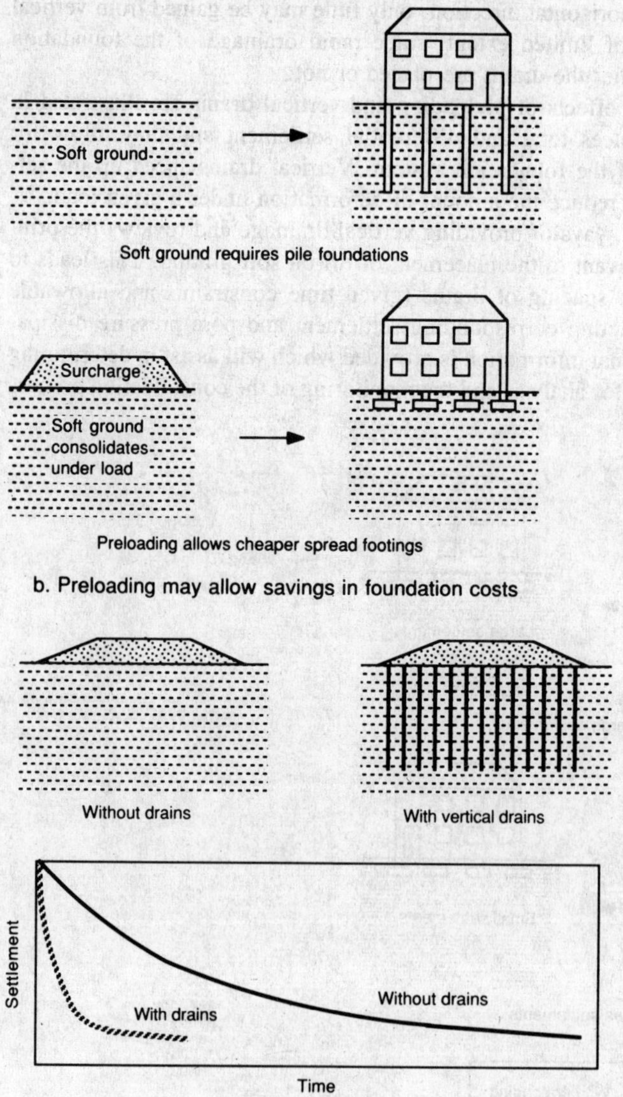

b. Preloading may allow savings in foundation costs

c. Vertical drains accelerate settlements but do not reduce final movement

FIGURE 11.1
(Continued)

For further details on preloading and vertical drains, the reader is referred to Johnson (1970), Hansbo (1979), Forrester (1982), Jamiolkowski et al. (1983), Koerner (1986), Rixner et al. (1986), and Holtz (1987). An informative group of papers was republished in 1982 by Thomas Telford, Ltd., under the title "Vertical Drains" (originally published as a Symposium in Print in Géotechnique, March 1981).

11.2 METHODS OF PROVIDING VERTICAL DRAINAGE

11.2.1 Cylindrical Sand Drains

In early applications of vertical drains, sand drains consisted simply of boreholes filled with sand. The holes may have been formed by driving, jetting, or augering and would typically have had diameters of 200 to 450 mm and would be spaced 1.5 to 6 m apart. To facilitate construction, minimize sand wastage, and ensure continuity of the drain, the sand may be prepacked in a fabric sock, such as the sandwick drain, which has a diameter of 65 mm [Dastidar et al. (1969)].

A large-diameter sand, or rather gravel drain (or gravel column!), in a fine-grained soil not only enables rapid consolidation of the surrounding material but may also provide vertical compressive reinforcement and conceivably could transfer surface loads to a bearing stratum at depth. The higher the reinforcing effect, however, the lower will be the surcharge-induced consolidating stresses in the foundation soil. Large-diameter gravel columns would therefore defeat the purpose of preloading and are most appropriately discussed under pile foundations.

11.2.2 Geosynthetic Drains

Most synthetic drains are of a strip (or band) shape, although circular plastic drainage pipes wrapped in a geotextile could also serve as vertical drains.

The first strip drain was developed by the Swedish Geotechnical Institute; see Kjellman (1948). It was made of cardboard with internal ducts. This type was later superseded by thin fluted PVC drains. Today, there are said to be more than 50 different makes of drains on the market, mostly of composite construction: a corrugated or studded inner core wrapped in a filter fabric, normally a nonwoven geotextile (Fig. 11.2). Similar to the original Kjellman wick drain, strip drains are generally about 100 mm wide and 2 to 6 mm thick. Details of some of the strip drains available are given in Table 11.1.

TABLE 11.1
Details of some selected strip drains*

Type	Core material	Filter material	Dimensions,† mm
Kjellman	Paper	Paper	100 × 3
PVC	PVC	None	100 × 2
Geodrain	PE	Cellulose	95 × 4
Mebradrain	PP	PP or PES	95 × 3
Alidrain	PE	PES	100 × 6
Colbond	PES	PES	100 × 6
Hitek	PE	PP	100 × 6

*Compare Koerner (1986), McGown and Hughes (1981), and Gambin (1987). PE = polyethylene, PVC = polyvinyl chloride, PP = polypropylene, PES = polyester.

†Dimensions may vary with different models of the same drain.

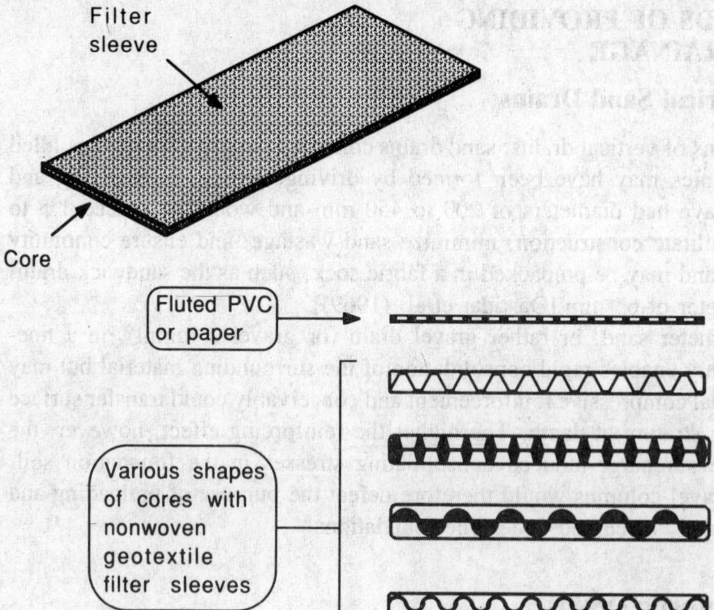

FIGURE 11.2
Typical core shapes of strip drains.

Advantages of synthetic drains are

Easy, rapid installation is possible.

Made of uniform material, easily stored and transported.

Equipment needed is lighter than the rigs required for equivalent sand drains.

Tensile strength of the strips helps to preserve continuity.

Low costs (site treatment may be possible for only one-fourth the costs of traditional sand drains).

In a typical installation of 15-m-long drains, spaced 1 to 2.5 m apart, a placement rate of up to 375 m/h is possible.

Because radial consolidation theory assumes the drain has a circular cross section, an equivalent diameter has to be calculated for strip drains. Two different approaches are possible. An equivalent sand drain diameter can be found based on equal void area or equal circumference.

Equal void area [as mentioned by Koerner (1986)]:

$$\text{Equivalent diameter } d_e = \frac{\sqrt{4Btn_d/\pi}}{n_s} \tag{11.1}$$

where B = width of strip
t = thickness of strip
n_d = void area/total cross-sectional area of strip
n_s = porosity of sand drain

Equal circumference:

$$\text{Equivalent diameter } d_e = \frac{2(B + t)}{\pi} \tag{11.2}$$

Equation (11.2) [preferred by Hansbo (1979)] is usually more conservative than Eq. (11.1), which means it results in a smaller equivalent diameter.

11.3 PRELOADING WITHOUT VERTICAL DRAINS

11.3.1 Simple Preloading of a Building Site

"Simple" preloading in this context means that a surcharge equal to a future site load is applied; when consolidation of the foundation soil is practically complete (say 90% complete), the surcharge is removed and the new building is erected.

11.3.1.1 SETTLEMENT CALCULATIONS (VERTICAL CONSOLIDATION). Figure 11.3 illustrates the case where a soft, normally consolidated foundation is preloaded with earth fill exerting a pressure equivalent to a future building with a raft foundation. The changes in vertical stress and strain of an element of soil at depth z are shown in the form of a (slightly idealized) e–log p diagram as obtained in a one-dimensional consolidation test (also called an odometer test).

Let us assume that the soil element chosen is representative for the whole layer. (For improved accuracy the consolidating soil should be divided into several thinner layers.) The final settlement under the surcharge then is

$$s_{\mathrm{Sf}} = \frac{H}{1 + e_0} C_c \log \frac{p_0 + \Delta p}{p_0} \tag{11.3}$$

where s_{Sf} = final settlement due to surcharge, m
H = thickness of consolidating layer, m
e_0 = initial void ratio of representative element of soil
= wG_s (for a saturated soil, where w = water content and G_s = specific gravity of soil solids)
C_c = compressive index (slope of e–log p curve in primary consolidation)
p_0 = initial vertical pressure, kPa
Δp = stress increase due to surcharge, kPa; for one-dimensional conditions it is equal to surcharge and is constant with depth

After say 90% of this settlement has taken place, the surcharge is removed and construction of the building begins. (If considerable time elapses during that construction phase, a small amount of swelling may take place.)

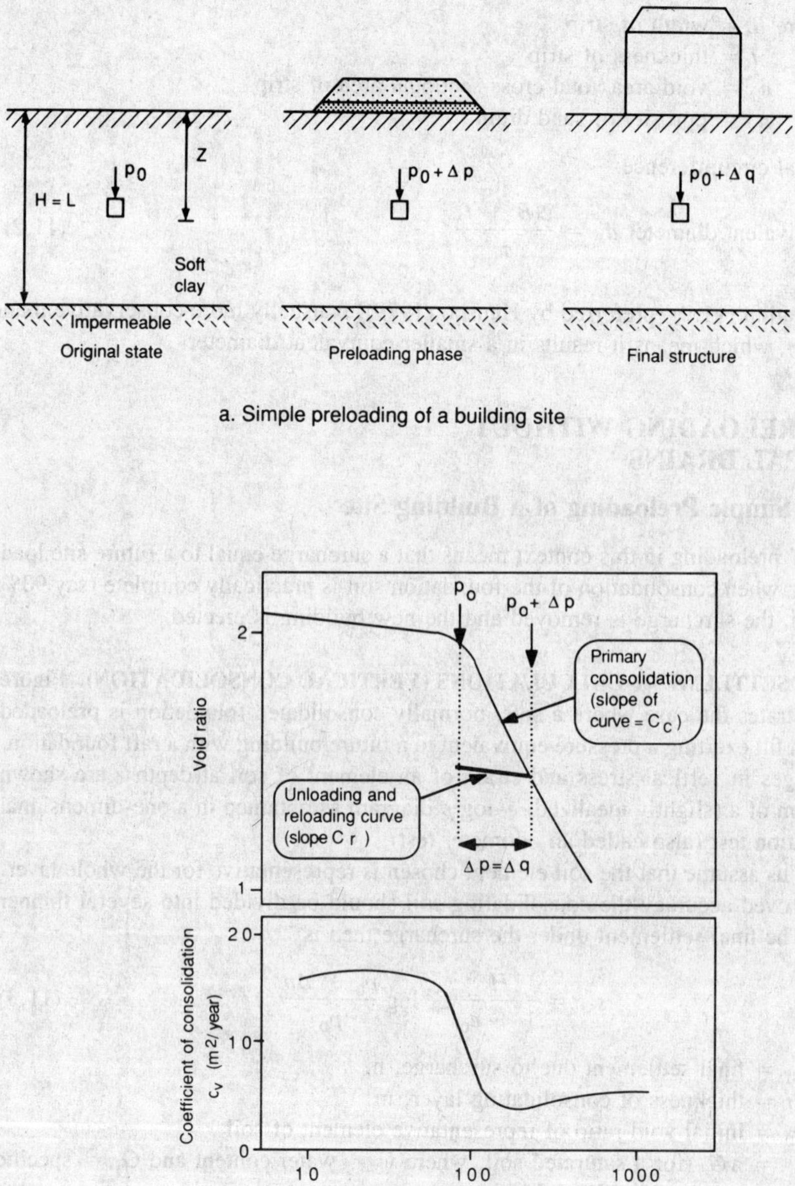

FIGURE 11.3
Preloading a building site.

Upon reloading of the foundation soil, the final settlement s_{Bf} of the building (exerting a pressure Δq) could be expressed in terms of the reloading index C_r, which is equal to the slope of the reloading curve on the e–log p diagram (Fig. 11.3):

$$s_{Bf} = \frac{H}{1 + e_0} C_r \log \frac{p_0 + \Delta q}{p_0} \qquad (11.4)$$

In Fig. 11.3 the e–log p curve was assumed to show a distinct break at the original overburden pressure of the normally consolidated soil element; this means the overburden pressure was assumed to be equal to the preconsolidation pressure. For soil samples obtained in the field this section of the e–log p diagram is generally curved, and the preconsolidation pressure p_c has to be deduced using Casagrande's graphical procedure (illustrated in most standard soil mechanics texts). Adding a further complication, the present overburden pressure p_0, calculated from bore-log information, may turn out to be less than the graphically determined preconsolidation pressure, even if the soil is truly normally consolidated. This is because of the secondary consolidation since the geological formation of the deposit, as illustrated in Fig. 11.4 [compare Bjerrum (1972)].

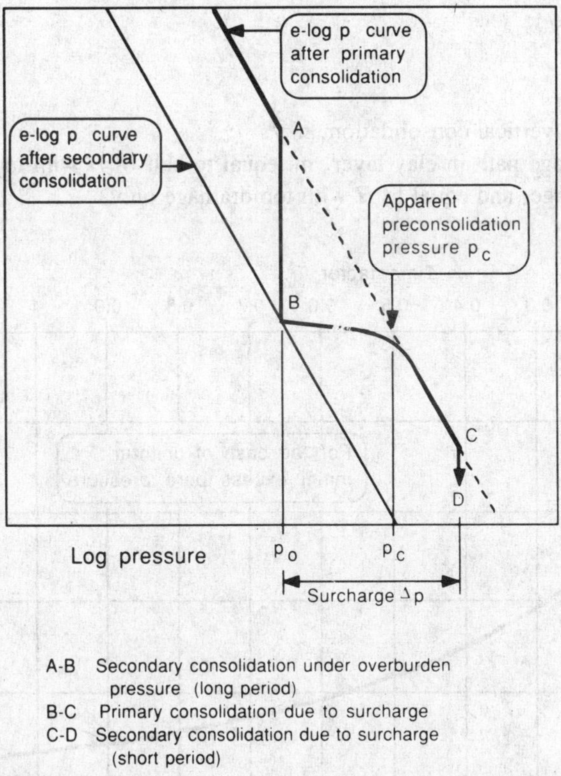

FIGURE 11.4
Compression of soil deposit under its own weight and during subsequent surcharging.

The above considerations are crucial for meaningful preloading. In order for this construction procedure to cause significant preemptive settlement, the preloading stresses must be in the range of primary consolidation. In other words, the pressures imposed by the surcharge must exceed $p_c - p_0$; the greater the excess, the more effective preloading will be. The magnitude of the surcharge is of course limited by the maximum support provided by the foundation soil, which can be estimated using a bearing-capacity formula or a slip-circle analysis.

11.3.1.2 RATE OF SETTLEMENT (VERTICAL CONSOLIDATION). In relation to the final settlement s_f, the settlement s_t at time t can be expressed as

$$s_t = U_v s_f \tag{11.5}$$

where U_v is the average consolidation ratio (vertical consolidation).

For constant initial hydrostatic excess pressure (equal to the surcharge for one-dimensional consolidation), Terzaghi (1943) derived the following value [as presented by Taylor (1948)]:

$$U_v = 1 - \sum_{m=0}^{m=\infty} \frac{2}{M^2} e^{-M^2 T_v} \text{ with } m = 0, 1, 2, 3 \ldots \tag{11.6}$$

where $M = (2m + 1)\pi/2$
$T_v = c_v t/L^2$
t = time, s
c_v = coefficient of vertical consolidation, m^2/s
L = longest drainage path in clay layer, m; equal to half of H with top and bottom drainage, and equal to H with top drainage only

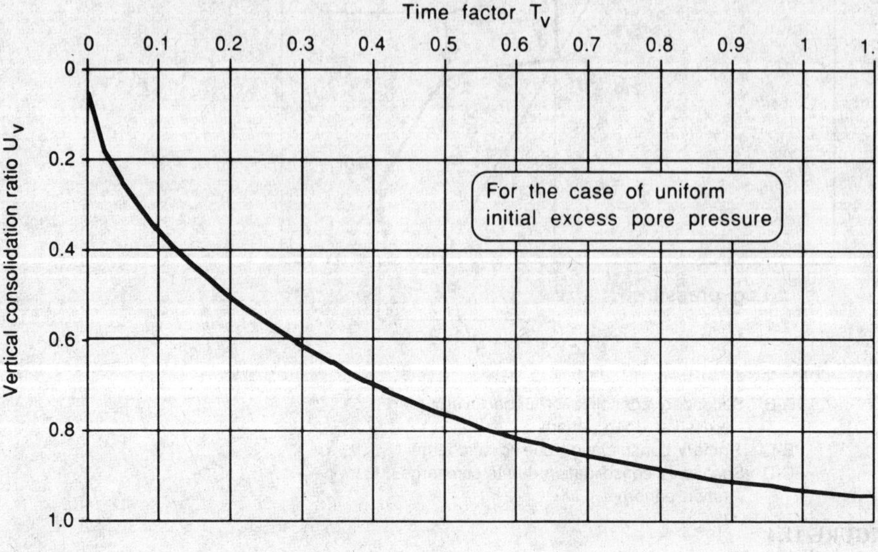

FIGURE 11.5
Average vertical consolidation ratio U_v (= percent vertical consolidation).

Values of U_v as a function of T_v are given in Fig. 11.5.

The coefficient of consolidation c_v is not a soil constant but decreases with increasing vertical stress as indicated in Fig. 11.3. Particularly around the preconsolidation pressure, c_v could drop sharply, say by a factor of one-half, although no general guidelines can be given. Soil disturbance or remolding (more relevant if vertical drains are installed) can similarly lower the c_v value, which means more time is required to reach a certain percentage of the final settlement.

11.3.2 Consolidation under a Structural Fill with Surcharge

Assume a highway embankment is placed on highly compressible soil. The expected amount of final settlement (or a given percentage thereof) can be reached quicker if an additional surcharge is placed on top of the embankment (Fig. 11.6).

One possible question posed by such a construction sequence is, "How much extra surcharge q (kPa) has to be applied for a given time t in order to produce x% of the final settlement s_{pf} under the embankment load p alone?" Calculations would proceed as follows:

Step 1: Calculate s_{pf} and s_{p+q}(final) according to Eq. (11.3) for an assumed q; these are the final settlements due to embankment loading alone and embankment plus surcharge, respectively.

Step 2: $s_{p+q}(t) = xs_{pf}/100$ (settlement under $p + q$ at time t).

Step 3: $U_v = s_{p+q}(t)/s_{p+q}(\text{final})$.

Step 4: Find T_v for given U_v (Fig. 11.5).

Step 5: $t_{calc} = T_v L^2/c_v$.

Step 6: If $t_{calc} > t$, increase q; if $t_{calc} < t$, decrease q; repeat calculations until $t_{calc} = t$.

Ideally, no excess pore pressures should remain after removal of the extra surcharge. However, even with $x = 100\%$, excess pore pressures may remain in the clay furthest from the surface or drainage layer, e.g., at time t_1 for example shown in Fig. 11.6; leaving the extra surcharge on until time t_2 would be desirable.

Another basic problem is to assess whether the soft soil is capable of supporting the fill. A simple bearing-capacity analysis using undrained shear strength c_u (also designated s_u) and the equation $q_{b.c.} = c_u N_c$ can be used as a guide (see App. 18B). Alternatively, a slip-circle analysis could be carried out (see App. 18A).

If the foundation soil is too weak, it has to be loaded in stages or reinforced by a high-strength geotextile (Chap. 19).

Loading in stages relies on the fact that a soil's shear strength increases as it consolidates. The Mohr envelope for a consolidated-undrained (CU) test gives an indication of the possible improvement with each stage of loading: $\Delta s_u = \Delta\sigma_c' \tan \phi_{cu}$, where $\Delta\sigma_c$ represents the increase in vertical effective stress and ϕ_{cu} the CU friction angle.

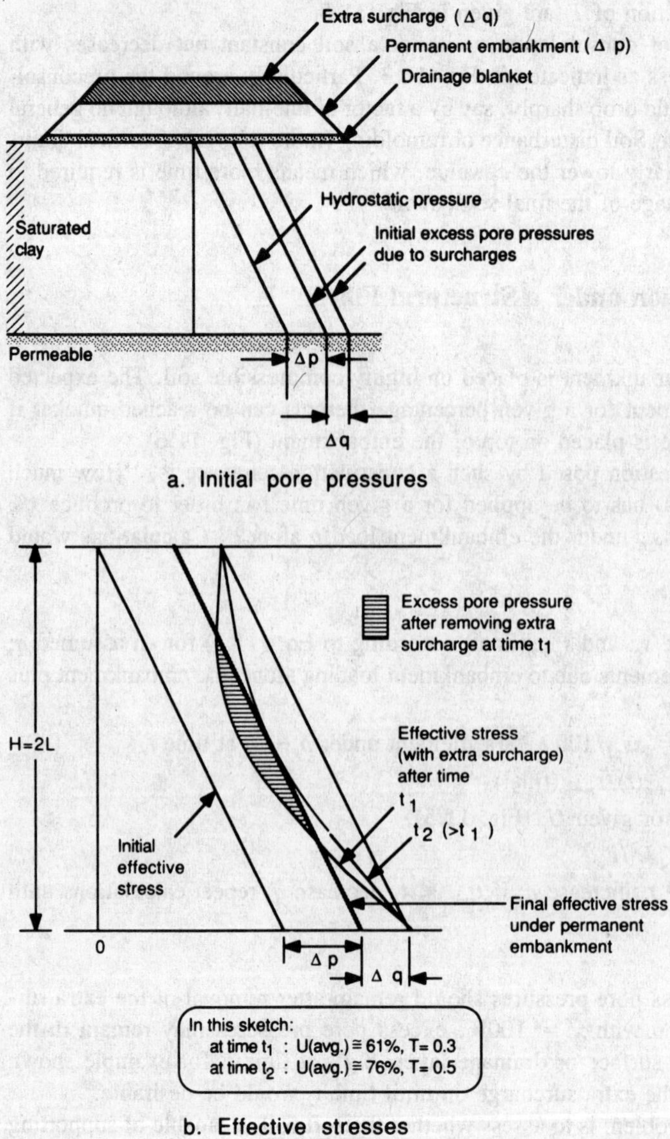

FIGURE 11.6
Consolidation under embankment with surcharge.

For one-dimensional consolidation, the logarithm of strength runs approximately parallel to the virgin compression line on an *e*-log (pressure, strength) diagram [Fig. 11.7; compare Bjerrum (1972) and Taylor (1948)]. Given, for instance, the undrained strength at the preconsolidation pressure, the possible increase in shear strength can be estimated.

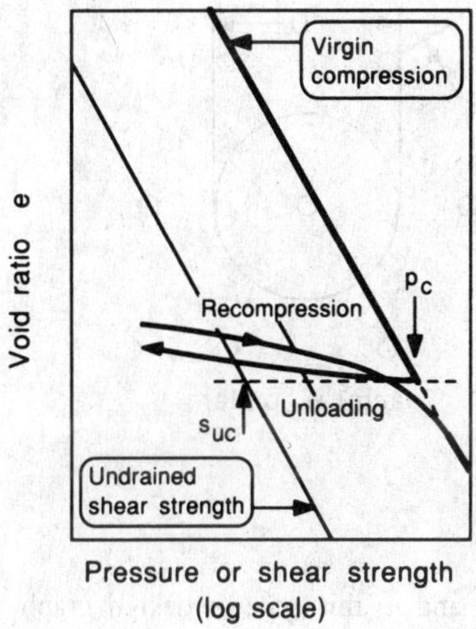

p_c = preconsolidation pressure for reloaded soil

s_{uc} = undrained shear strength of preconsolidated soil

FIGURE 11.7
Plot of void ratio versus pressure and strength.

Predicting strength gain after partial consolidation and for particular stress histories can present a very complex problem requiring advanced methods of analysis [e.g., Ladd and Foott (1974)].

11.4 PRELOADING WITH VERTICAL DRAINS

11.4.1 Radial Consolidation

The solution for radial water flow toward the central drain of a cylinder of soil undergoing one-dimensional strain goes back to Rendulic (1935). The result is generally expressed in terms of the average consolidation ratio for radial drainage U_r:

$$U_r = 1 - \exp \frac{-8T_r}{\alpha} \tag{11.7}$$

where U_r = average consolidation ratio for radial drainage = average degree of consolidation for radial drainage

$T_r = c_h t/D^2$ = time factor for radial drainage

c_h = coefficient of horizontal (or radial) drainage

t = time elapsed since application of the surcharge

D = equivalent diameter of cylinder of soil around drain (see Fig. 11.8) = $1.06s$ for triangular pattern of drains (s = spacing) = $1.13s$ for square grid pattern of drains

$n = D/d$

d = drain diameter (or equivalent diameter for strip drains)

$\alpha = n^2 \ln n/(n^2 - 1) - (3n^2 - 1)/4n^2$

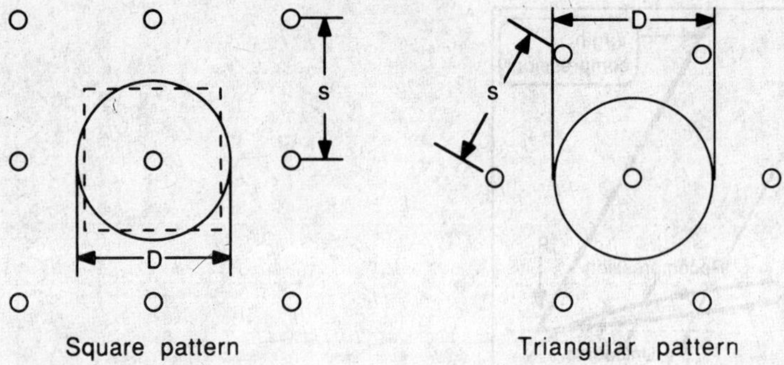

FIGURE 11.8
Vertical drain patterns.

Figure 11.9 gives U_r as a function of T_r and n; this type of design graph was first presented by Barron (1948).

11.4.2 Combined Vertical and Radial Consolidation

It would not be unusual in practice that the drain spacing giving a specified percent consolidation over a fixed time period would be based on considering radial consolidation only. However, Carillo (1942) showed how an average degree of consolidation U_{vr} for combined vertical and radial water flow can be calculated.

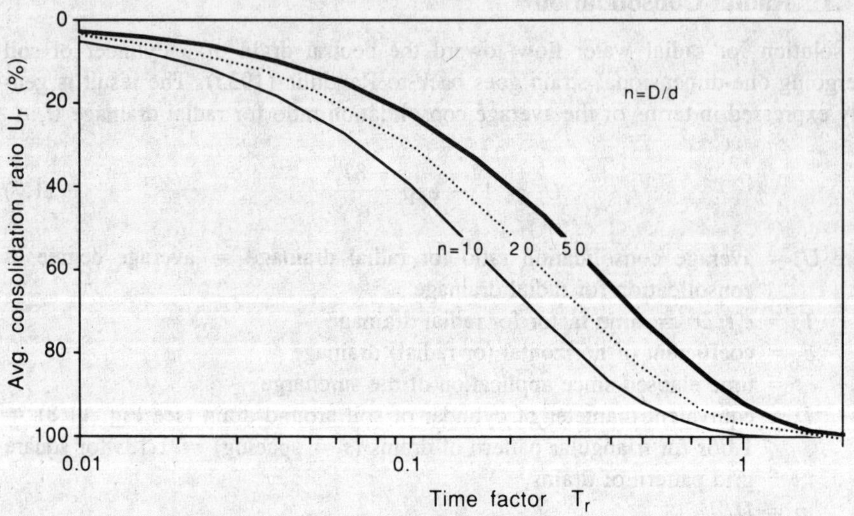

FIGURE 11.9
Average consolidation ratio U_r versus time factor for radial consolidation.

$$1 - U_{vr} = (1 - U_v)(1 - U_r) \tag{11.8}$$

A typical practical problem would be the following: A highway embankment is to be built on soft clayey ground subject to large settlement. Vertical drains are to be used in order to produce say 90% of the expected final primary settlement in a given time t. How close do the drains have to be spaced in order to achieve this objective? Assuming all relevant soil properties as well as the drain type (and its equivalent diameter) are given, one would proceed as follows:

Step 1: Calculate T_v given c_v, L, and t; then determine U_v [Eq. (11.6) or Fig. 11.5].

Step 2: Set $U_{vr} = 0.9$ (must be $> U_v$).

Setp 3: Find U_r from Eq. (11.8).

Step 4: Rewrite Eq. (11.7) in the form

$$D^2\alpha = \frac{-8c_h t}{\ln(1 - U_r)}$$

and solve for D by successive approximation.

Step 5: Compute spacing s for triangular or square pattern (Fig. 11.7).

11.4.3 Effect of Smear and Drain Resistance

Disturbance of the soil adjacent to the drain is likely to decrease its permeability and thus slow down the consolidation process. This effect is described as "smear." The smear effect is believed to increase with increasing drain diameter. It is also known to depend on the method of installation: Dynamic driving of the mandrel (or lance), for instance, creates more disturbance of the soil than static pushing. The size of the mandrel or anchor plate (for strip drains) also has an influence on the extent of the disturbed zone.

In addition, the drains (or wells) present resistance to the water flowing in them. The longer the drainage path within the drain, the slower will be the consolidation and gain in strength in the zone furthest from a permeable layer adjoining the soil being treated.

Hansbo et al. (1981) have proposed that a modified factor α is used in Eq. (11.7) which is rewritten as follows (see Fig.11.10):

$$U_{rz} = 1 - \exp\frac{-8T_r}{\alpha_s} \tag{11.9}$$

with U_{rz} = average degree of consolidation at depth z for radial drainage and

$$\alpha_s = \ln\frac{n}{m} + \frac{k_c}{k_c'}\ln m - \frac{3}{4} + \pi z(2L - z)\frac{k_c}{q_w}$$

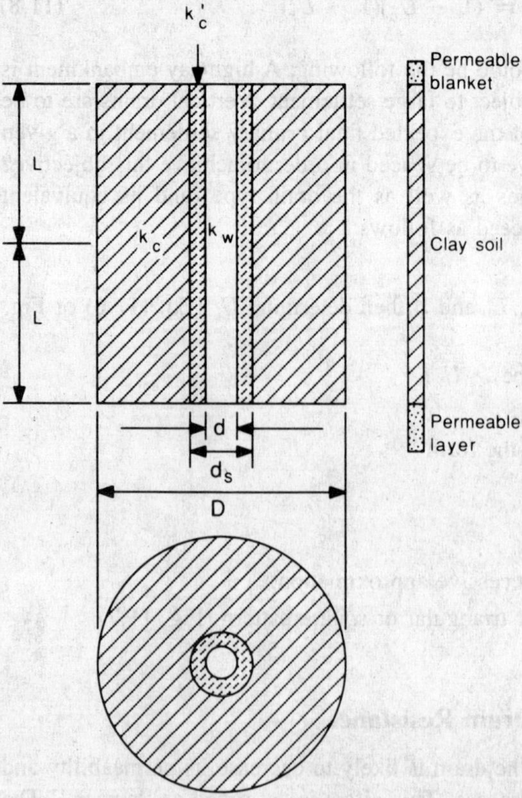

FIGURE 11.10
Notation for evaluating smear effect and well resistance.

where $n = D/d$
D = equivalent diameter or cylinder of soil around drain (see Fig. 11.8)
d = drain diameter (or equivalent diameter for strip)
$m = d_s/d$
d_s = diameter of the smeared zone
k_c = horizontal permeability of the undisturbed consolidating soil
k_c' = horizontal permeability of the smeared soil
L = longest drainage path along vertical drain
$q_w = k_w A_w = k_w \pi d^2/4$ = discharge capacity of the drain well (If there is little or no drain resistance, q_w is very large and U_{rs} is constant with depth.)
k_w = axial permeability of the drain well
A_w = cross-sectional area of drain

Just how significant the smear effect and drain resistance are is still a matter of debate. Modern geotechnically designed synthetic drains have a relatively large core (at least compared with the early paper and wick drains!). If these drains are properly installed by specialized equipment, the consolidation process should not be unduly slowed down. The overall effect of nonideal flow conditions is probably less significant than the uncertainties involved in choosing a representative value of c_h for the soil deposit being dewatered by vertical drains.

11.4.4 Secondary Consolidation

Secondary consolidation causes that part of the settlement which takes place after hydrostatic excess pore pressures have fully dissipated, or, in other words, are immeasurably small. It can be attributed to time-dependent creep of the soil "skeleton."

Vertical drains have virtually no effect on secondary settlement. This is the reason why soils subject to extreme magnitudes of secondary settlement, such as peats and organic soils, may not benefit greatly from vertical drains.

Nevertheless, secondary settlement is mentioned here because it could be very significant in the prediction of the settlement of structures built on preloaded soil: The main component of long-term settlement of these structures will be of the secondary kind!

Secondary settlement produced from time t_0 to t can be expressed as

$$s_s = c_\alpha H \log \frac{t}{t_0} \qquad (11.10)$$

where H is the layer thickness and c_α is the parameter representing the vertical strain per log cycle increase in time during secondary consolidation (also simply defined as the rate of secondary consolidation).

Precompressed clays have a lower rate of secondary consolidation than normally consolidated clays. After being subjected to surcharging, a clay layer is in an overconsolidated state, and as a result the c_α value is reduced to say 50% or even 10% of its original value. For example, a normally consolidated clay may have a value of $c_\alpha = 0.01$; after preloading, c_α could be reduced to the range of 0.005 to 0.001.

11.5 ASSESSMENT OF GROUND CONDITIONS

Preloading projects require detailed site investigations that likely involve boring, penetration testing (including piezocone), sampling, and laboratory experiments. Particular emphasis must be placed on

Stratification (presence of permeable microlayers)

Drainage conditions (above and below the consolidating clay layer)

Consolidation and permeability characteristics (in vertical and horizontal directions)

Strength and stress-strain characteristics (stress history)

The determination of c_h (the horizontal coefficient of consolidation), needed for the design of vertical drains, is not part of routine foundation investigations. Table 11.2 lists several laboratory and field procedures currently used in evaluating c_h. At

TABLE 11.2
Determination of c_h, the horizontal coefficient of consolidation

Where done	Test	Comments; references
Laboratory	Std. consolidation or triaxial test	Modify equipment to provide radial drainage or cut specimen appropriately. Escario and Uriel (1961), Marsh (1963)
Laboratory	Special (Rowe cell)	Rowe and Barden (1966)
In situ	Piezocone	Torstensson (1975)
In situ	Permeability test	Constant or falling head tests yield permeability, from which c_v and c_h can be calculated. Wilkinson (1968)
In situ	Consolidation	Clarke et al. (1979)
In situ	Trial embankment	Interpretation of piezometer readings: Johnson (1970). Interpretation of field settlement records: Asaoka (1978), Magnan and Deroy (1980), Magnan and Mieussens (1980)

least one reference is given for each method, but only one method will be discussed further, namely the Asaoka procedure (1978) for the interpretation of settlement records from a trial embankment.

11.5.1 Trial Embankments

Building a trial embankment in order to determine the feasibility of preloading and vertical drains avoids uncertainties associated with soil disturbance during sampling and implicitly takes into account nonideal conditions created by drain installation equipment. In addition it may be possible to directly evaluate various types of drains or spacing patterns. A trial embankment can also give guidance with respect to the bearing capacity of the ground: One section of the trial fill could be deliberately overbuilt in order to induce shear failure.

Good knowledge of site conditions is still required in order to know how representative the chosen test location is for the whole project. According to Forrester (1982) a minimum of 6 months should be allowed for monitoring of the trial embankment; otherwise it may not be worth constructing.

Based on Forrester's report (1982), the following recommendations for trial embankments are highlighted:

The location of the trial embankment should be chosen so that it forms part of the final earth structure, thereby reducing extra costs.

The shape of the trial embankment should allow for approximation of the strain and drainage conditions of the final structure.

In the case of a land reclamation project, the trial fill should be placed over a wide enough area to produce a reasonably linear stress increase with depth; this means the width across the top of the embankment should be at least equal to the depth of the soft soil below. Where this is not feasible or possible, the analysis of test results and the prediction of performance of future, different sized structures may become rather complex.

The height of the trial embankment should ideally be equal to that of the planned embankment because the soil's permeability and coefficient of consolidation are likely to decrease with the applied load.

A bearing-capacity or slip-circle analysis should ensure that the trial embankment will stand up.

Instrumentation should include settlement plates and gauges, leveling points, alignment stakes, and piezometers with a rapid response.

How many variables should be investigated depends on the time available and the limitations imposed on costs. Determining the effect of different spacings of vertical drains forms part of most trial embankment exercises.

11.5.2 Asaoka (1978) Method of Determining c_h

This procedure requires settlement readings taken at constant time intervals Δt, or equivalent values interpolated from a time-settlement curve. On the so-called Asaoka diagram, the settlement reading s_n (on the y axis) is plotted versus the preceding settlement s_{n-1} (on the x axis) as illustrated in Fig. 11.11. A line is drawn through the points plotted, extrapolated, and intersected with the 45° line. This intersection point represents the final settlement s_f.

In plotting real data, it is not unusual that the first one or two points on the Asaoka line do not coincide with the line drawn through subsequent points. This could be due to the effect of the length of the construction period, nonuniform initial pore

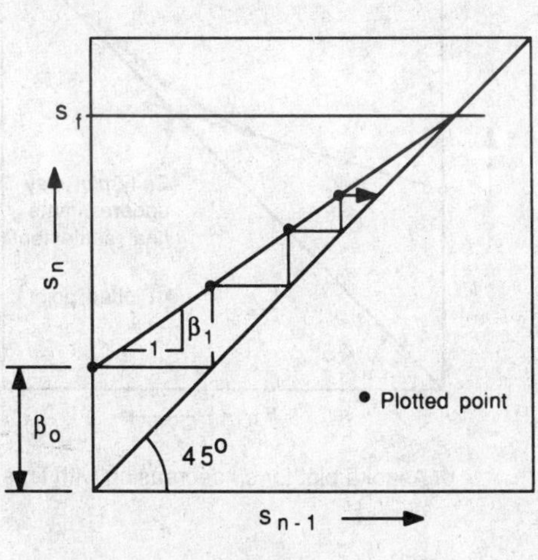

Note: The first one or two points plotted may not coincide with the straight line drawn through subsequent points, as explained in text (Section 11.5.2).

FIGURE 11.11
Asaoka diagram.

pressure increase with depth (e.g., due to loading over a limited area), inhomogeneous ground, and other reasons. Irregular early points are generally ignored in the standard analysis.

The Asaoka method is based on the fact that one-dimensional consolidation settlement at times 0, Δt, $2\Delta t$, $3\Delta t$, etc., can be expressed mathematically in the general form

$$s_n = \beta_0 + \sum_{i=1}^{i=w} \beta_n s_{n-i} \tag{11.11}$$

where β_0 is a settlement value (m) and β_n is a dimensionless ratio.

For $w = 1$, a first-order approximation is obtained:

$$s_n = \beta_0 + \beta_1 s_{n-1} \tag{11.12}$$

which takes the primary consolidation settlement into account. It is apparent that β_1 is the slope of the line drawn through the settlement points in the Asaoka diagram. The overall effective value of the horizontal coefficient of consolidation c_h' can be calculated from the following equality:

$$-\frac{\ln \beta_1}{\Delta t} = \frac{8c_h'}{D^2\alpha} + \frac{\pi^2 c_v}{4H^2} \tag{11.13}$$

where D, α, and H are defined as in Eq. (11.7). The coefficient of vertical consolidation c_v could be obtained in a standard laboratory test. The last term in Eq. (11.13) is small compared to the term containing c_h', and therefore any error in c_v will have only a little effect.

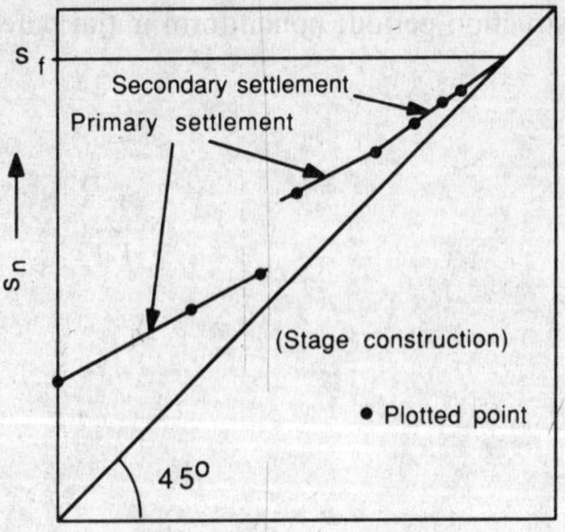

a. Asaoka plot for stage construction and secondary settlement

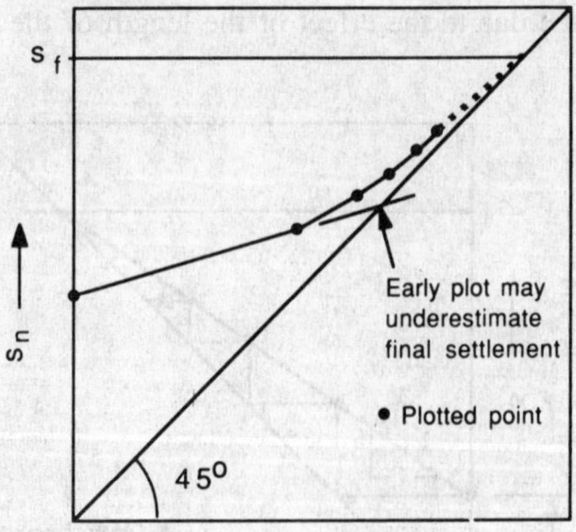

b. Asaoka plot for c_v decreasing with time

FIGURE 11.12

Asaoka plots for the cases of stage loading and secondary settlement, and a c_v decreasing with time (schematic diagrams only).

Figure 11.12 shows a few variations in the Asaoka diagram. Stage construction results in a series of more or less parallel straight lines on the Asaoka plot. With a constant embankment load, a break in the slope of the plot indicates the end of primary settlement. The continuation represents secondary settlement; the final settlement at the intersection with the 45° line will then also include secondary settlement. A curved line of primary settlement points is likely to be an indication of a c_h' value decreasing as the effective stress in the soil increases. An attempt to use a higher-order version of Eq. (11.11) ($w > 1$) for predicting the final settlement could then be made.

Settlement s_t at time t can be calculated as a fraction of the final settlement s_f from

$$\frac{s_t}{s_f} = 1 - \frac{8}{\pi^2} \exp\left[-\left(\frac{8c_h'}{D^2\alpha} + \frac{\pi^2 c_v}{4H^2} \right) t \right] \tag{11.14}$$

For preloading without vertical drains one can set $c_h' = 0$, and Eq. (11.13) yields c_v. The settlement at a particular time can be predicted from Eq. (11.14).

11.6 INSTRUMENTATION AND PERFORMANCE MONITORING

Monitoring of the behavior of embankments on soft ground is essential in order to prevent sudden failures, to recognize changes in the rate of consolidation (e.g., so that the construction schedule can be adjusted accordingly), and to determine or verify design parameters. Performance evaluation will also help to improve settlement predictions and construction efficiencies in future projects.

It is not intended to give details of instruments used to measure the deformation of embankments and fills and pore pressures developed in the foundation. For this the reader is referred to Hanna (1973) and specialist articles on preloading [e.g., Magnan and Mieussens (1980)]. This section merely gives an idea of the minimum requirements for the geotechnical monitoring of a preloading project.

Figure 11.13 shows the layout of the basic instrumentation of an embankment built on ground modified with vertical drains. Included are

- Settlement plates, gauges, and leveling points
- Piezometers
- Alignment stakes

More detailed guidelines are given by Magnan and Mieussens (1980), who suggest a number of different configurations, depending on site conditions and the performance aspects to be monitored.

11.6.1 Settlement Observations

Settlement gauges, e.g., of the hydraulic type, measure the long-term settlement at the original ground surface, covered with a permeable blanket during construction;

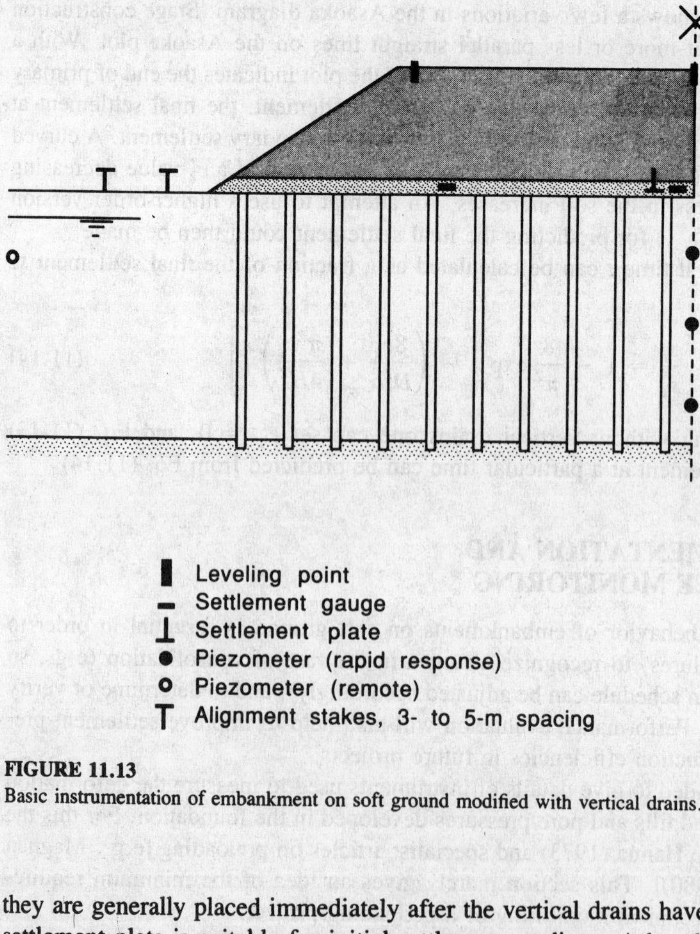

FIGURE 11.13
Basic instrumentation of embankment on soft ground modified with vertical drains.

they are generally placed immediately after the vertical drains have been installed. A settlement plate is suitable for initial settlement readings. A benchmark is set up on stable ground at a reasonable distance from the fill. Leveling points are placed at the end of construction on top of the embankment.

Settlement readings are then taken at regular intervals, say weekly to monthly, where vertical drains are installed. For preloading without drains, settlement measurements would not have to be taken that frequently. Settlement readings are then interpreted using the Asaoka procedure.

11.6.2 Piezometer Readings

Knowledge of pore pressures is necessary in order to do a stability analysis in terms of effective stresses, as will be required when the foundation soil is undergoing consolidation under the surcharge. For monitoring the overall consolidation process, settlement records appear to be a more reliable measure than piezometer readings.

More than one piezometer may be installed in a hole, if they are separated from each other by virtually impermeable bentonite plugs. A remote piezometer serves to record the natural groundwater level.

Piezometers are generally installed either during or immediately after the installation of the drains.

11.6.3 Alignment Stakes

Alignment stakes are set out parallel to the embankment slope at the beginning of the placement of the fill. They provide a simple means of observing the lateral displacement of the foundation soil during construction. They could give early warning of an impending bearing failure, even if only monitored visually. Their rate of movement is likely to be related to a change in pore pressure under the embankment.

A more sophisticated alternative to alignment stakes would be to measure the lateral deformation in a borehole using an inclinometer.

PROBLEMS

Prefixes indicate problem type: C = calculations, M = multiple choice, B = brief answer, D = discussion.

Calculations

C11.1. A soft soil deposit has an undrained strength of 10 kPa. How much gravel fill ($\gamma = 20\ \text{kN/m}^3$) can be placed on top of it without causing bearing failure ($N_c = 5.7$)

(*a*) If the groundwater level is at the surface?

(*b*) If there is 1 m of water above the original ground surface?

C11.2. The sand considered for use in sand drains has a porosity of 0.25. The alternative geocomposite drain measures 100 × 5 mm and has a void area of 95%. Calculate the equivalent sand drain diameter for a synthetic drain for the following conditions:

(*a*) Equal void area.

(*b*) Equal circumference.

C11.3. A 4-m-thick horizontal clay layer lies below 5 m of sand and above impermeable bedrock. Calculate the settlement under a 100-kPa surcharge of large extent. Soil properties and stresses at the midpoint of the clay layer, thought to be representative of all the clay, are as follows:

$$\begin{aligned} \text{Initial effective stress} &= 70\ \text{kPa} \\ \text{Initial void ratio} &= 1.8 \\ \text{Compressive index} &= 0.8 \\ \text{Coefficient of consolidation} &= 10^{-3}\ \text{cm}^2/\text{s} \end{aligned}$$

Calculate

(*a*) The expected final settlement.

(*b*) The time required for 90% of the consolidation to be complete.

C11.4. A land reclamation project requires 3 m of sand-gravel fill (unit weight = 18 kN/m^3) to be placed on a deposit with the following profile (and average soil properties): Groundwater level at the surface.

0 to 6 m: Soft silty clay $w = 65\%$
$G_s = 2.7$
$C_c = 1.0$
$c_v = 1\ \text{m}^2/\text{year}$
$c_u = 20$ kPa

> 6 m: Dense shale

(a) Estimate the final settlement s_f under the proposed fill.
(b) Calculate the time t_{f90} required for 90% of the settlement s_f to take place.
(c) How much extra surcharge is required to produce a settlement equivalent to 90% s_f in half of the time t_{f90}?
(d) Is there any excess pore pressure left after removing the extra surcharge at time $t = t_{f90}/2$?
(e) Could the fill and the extra surcharge be placed all at once without causing a bearing failure?

C11.5. Given are the following details pertaining to a preloading project:

Surcharge $\Delta p = 80$ kPa
Consolidating soil $w = 50\%$
$w_1 = 71\%$ (liquid limit)
$I_p = 47\%$ (plasticity index)
$\gamma = 1.7\ \text{t/m}^3$
$C_c = 0.7$
$c_v = 0.6\ \text{m}^2/\text{year}$
$c_h = 1.2\ \text{m}^2/\text{year}$
$L = 20$ m (longest drainage path)
Sand drains $d = 0.10$ m
$s = 2$ m (triangular spacing)

(a) Determine the average consolidation ratio U_{vr} (combined vertical and radial drainage) after 10 months of preloading.
(b) Calculate the settlement obtained after 10 months with and without drains.

C11.6. For the data given in Prob. C11.5, but for radial drainage only, calculate U_{rz} at depth $z = 10$ m,
(a) With zero well resistance.
(b) With an axial drain permeability $k_w = 1000$ m/year.
(c) Zero well resistance, but a smear effect expressed by $k_c'/k_c = 0.3$.

C11.7. Assume primary settlement of a 5-m-thick layer of clay is complete after 1 year. A one-dimensional consolidation test indicated $c_\alpha = 0.005$ for the appropriate stress change. Estimate the secondary settlement for the following 20 years.

C11.8. A 6-m-high embankment was built in stages on a 25-m-thick deposit of compressible clay (top and bottom drainage possible). The initial cover was 1.5 m, followed 100 days later by an additional 2.5 m of fill. The embankment was completed over a short period approximately 400 days after initial construction. The following settlements were recorded:

Time t, days	Settlement, cm
0	0
100	16
200	50
300	60
400	90
500	150
600	170
700	180

Using the Asaoka procedure:

(*a*) Estimate the final settlement of the completed embankment.

(*b*) Backfigure the coefficient of vertical consolidation (m^2/s) for the last two stages of construction.

[Problem adapted from Magnan and Deroy (1980) p. 50.]

C11.9. A 6-m-high fill was placed in 75 days on a 15-m-deep deposit of soft clay overlying impermeable bedrock. Cardboard drains with an equivalent diameter of 5 cm were spaced in a triangular pattern with a spacing of 3 m. Settlements measured were

Time t, days	Settlement, cm
0	0
50	35
100	54
150	61
200	65
250	68
300	70
350	72
400	74
450	75
500	75.5

Use the Asaoka procedure to

(*a*) Estimate the final settlement.

(*b*) Calculate the coefficient of horizontal consolidation, assuming the vertical consolidation coefficient is 10^{-6} m^2/s.

[Problem adapted from Magnan and Deroy (1980) p. 51.]

Multiple Choice

M11.10. (Choose the *incorrect* statement.) Preloading of a site prior to building the permanent structure

(*a*) Reduces the settlement of the future structure.

(*b*) Increases the bearing capacity of the ground.

(*c*) Improves the engineering behavior of clays but not of coarse-grained soils.

(*d*) May not be cheaper than the use of (alternative) deep foundations.

M11.11. (Choose the *incorrect* statement.) Small-diameter vertical drains under a permanent embankment

(*a*) Accelerate settlements.

(*b*) Are not likely to be beneficial for peats and other highly organic soils.

(*c*) Are not effective if horizontal silt and sand microlayers in the ground are discontinuous.

(*d*) Do not significantly reduce settlement under a surcharge.

M11.12. (Choose the *incorrect* statement.) With respect to vertical drains, it can be said that

(*a*) Soil "smear" along the drain reduces the rate of consolidation.

(*b*) Using a coarse rather than very fine sand for vertical drains increases the consolidation ratio after a given preloading time.

(*c*) Using equal circumference composite synthetic drains rather than sand drains reduces the final settlement of an embankment.

(*d*) The longer the vertical drains, the more significant is the effect of "well resistance."

M11.13. (Choose the *incorrect* statement.) In monitoring the performance of a trial embankment

(*a*) Settlement records appear to be more reliable than piezometer readings for assessing the overall degree of consolidation.

(*b*) Alignment stakes alongside the fill are placed to facilitate the construction process.

(*c*) Inclinometers can be used to measure lateral deformation.

(*d*) It is difficult to prevent damage to settlement-measuring devices and standpipes protruding to the surface of the fill.

Brief Answer

B11.14. Choose a name, either Rendulic, Terzaghi, Barron, or Kjellman, to complete each of the following statements:

(*a*) A basic solution for radial consolidation was presented by ____________ in 1935.

(*b*) The first strip drain was developed by ____________ .

(*c*) The first extensive work on vertical drains was published by ____________ in 1948.

(*d*) One-dimensional vertical consolidation theory was first published in a book by ____________ .

B11.15. A trial embankment is to be built for the purpose of assessing the feasibility of preloading with vertical drains. Give at least one recommendation with respect to

(*a*) Location of trial embankment.

(*b*) Shape of fill.

(*c*) Stability analysis required.

(*d*) Instruments required.

B11.16. List four different types of instruments relied upon in the monitoring of the performance of a trial embankment.

Discussion

D11.17. Is the preloading method really a hydraulic modification of the ground or should it be discussed as a mechanical modification?

D11.18. Under what conditions are vertical drains likely to be an economic means of improving soft ground?

D11.19. How could the stress history of a soil deposit affect its suitability for preloading with vertical drains?

D11.20. How can you measure the settlement of the original ground surface during and after the placing of say 5 m of fill?

D11.21. After reviewing technical and promotional literature on piezometers, discuss the merits of the different types available.

CHAPTER 12

ELECTROKINETIC DEWATERING AND STABILIZATION

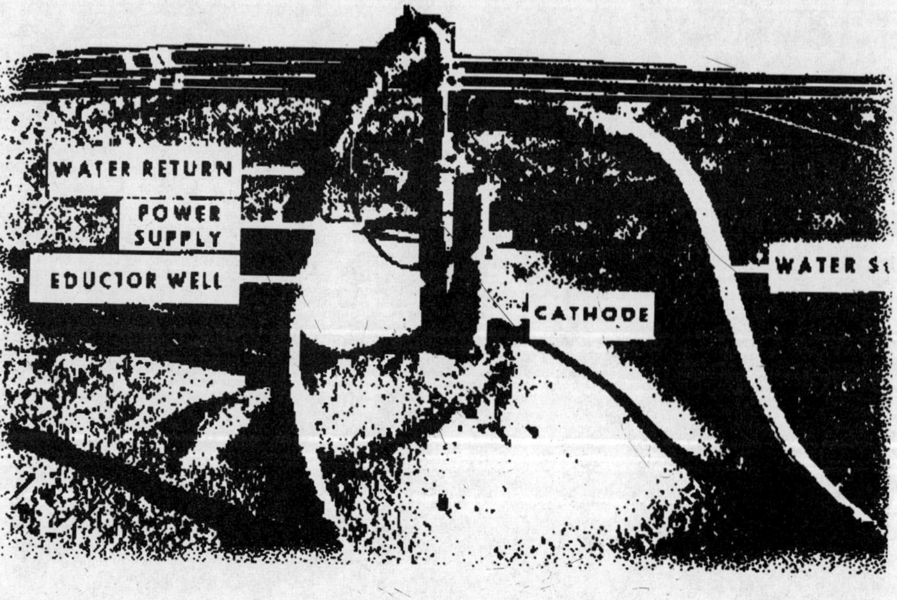

The principal effect of applying an electric potential to a saturated soil mass is to cause pore water flow toward the cathode (the negative terminal). If the water collected at the cathode is removed, the net result is a lower soil moisture content, a correspondingly increased strength, and reduced compressibility. This process is equivalent to that of consolidation. Indeed, consolidation theory as used to analyze the densification (or dewatering) of fine-grained soils under a surcharge can be adapted to predict the effect of electroosmosis.

Because dewatering is its main effect, electrokinetic consolidation is in this text arbitrarily classified as a hydraulic method of ground modification. Actually, additional phenomena may take place which cause changes to the soil's structure as well as to its chemical composition; these changes could result in a strength increase above that expected from a change in water content alone. For example, inadvertent release of chemicals from the electrodes may harden the soil surrounding them. This effect can be enhanced by injecting additives into the soil and thus making a primarily hydraulic method of ground modification into a physicochemical technique resembling grouting.

Electroosmosis has fascinated many eminent engineers and scientists. For a few it has been or still is a life-long interest, sometimes delivering the technical solution for the seemingly impossible, at other times disappointing in its results. Ever since Leo Casagrande registered his German patent (1935) and brought electroosmosis to the attention of the civil engineering profession (1949, 1952, 1981), electrokinetic phenomena have received sporadic attention by researchers and practitioners in Switzerland [Schaad and Haefeli (1947); Jaecklin (1968)], Japan [Mise (1961)], Austria [Veder (1963, 1973)], Australia–New Zealand [Nettleton (1963); Gladwell (1965); Johnston (1977, 1978)], Scandinavia [Bjerrum (1967); Hansbo (1970); Eggestad (1983)], and in the United States [Vey (1949); Fetzer (1967); Gray and Mitchell (1967); Wan and Mitchell (1976); Banerjee and Mitchell (1980)]. This list is by no means a complete bibliography; in particular it does not give credit to work done in Eastern Europe.

12.1 BASIC ELECTROKINETIC PHENOMENA

Electrokinetics of soil-water systems is concerned with the relative movement of the solid versus the liquid phase under the influence of an electric field. Basic electrical phenomena in soils include electroosmosis (or electroconsolidation), flow potential (or streaming potential), electrophoresis (also called cataphoresis), electric effects in soil contact zones, and electrostabilization (or electrohardening).

12.1.1 Electroosmosis

Reuss (1809, in Russia) was the first to observe water flow in soils as a result of a direct electric current passing through it. In analogy to similar, already known phenomena associated with the flow through capillaries and membranes, it was called electroosmosis in soils. Reuss' experiment is shown in Fig. 12.1. Two glass tubes

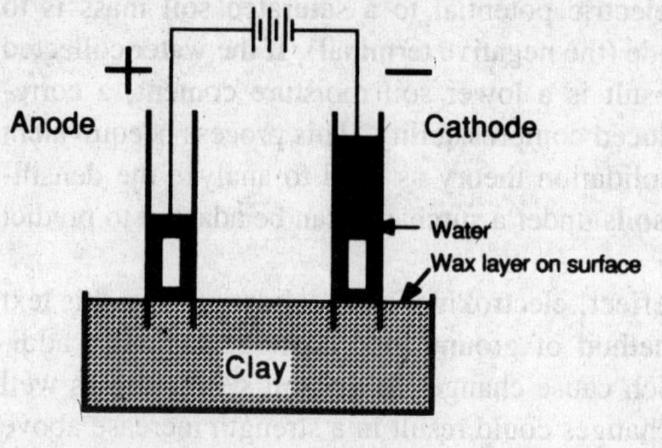

FIGURE 12.1
Schematic arrangement of Reuss' experiment. [*Reuss (1809).*]

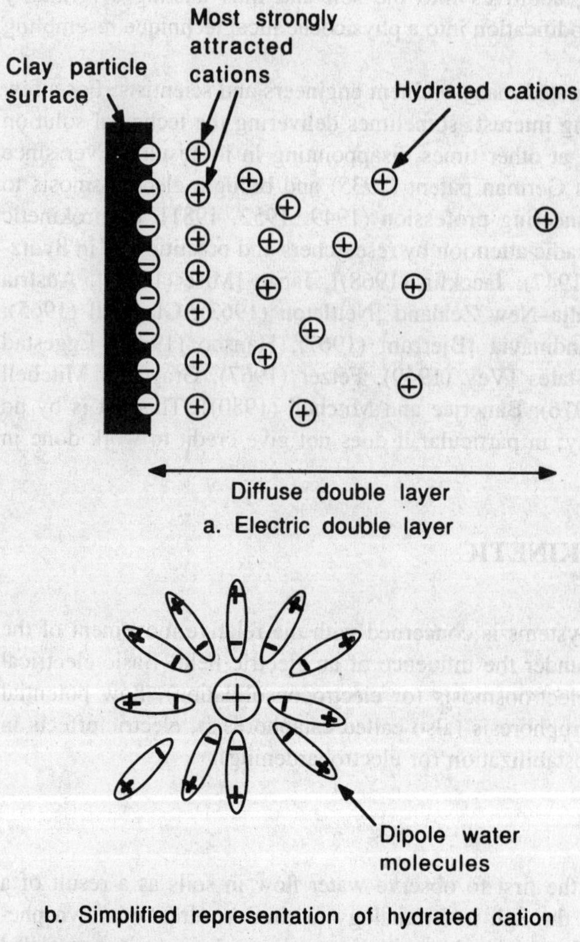

FIGURE 12.2
Model of electric double layer and cation hydration.

with electrodes are pushed into clayey soil. Upon the application of an electric potential, water is observed to flow in the direction of conventional current towards the cathode, raising the water level in the surrounding glass tube.

The explanation of this phenomena lies in the electrochemical nature of soil particle surfaces and the pore water. According to colloidal theory, a soil particle (particularly a clay particle), when suspended in water, has a negatively charged surface. Surrounding the particle is a so-called diffuse double layer of positive ions, such as Na^+, K^+, or Ca^{++}. Most rigidly attached to the surface is the first layer of positive ions; with greater distance from the solid particle, the attractive force diminishes, giving rise to an increasingly diffuse ionic atmosphere (the second layer), where the ions are relatively free to move (Fig. 12.2*a*). Beyond the double layer, the ion concentration is equal to that of the "free" pore water. Because of the polar nature of water molecules they are oriented around cations as shown in Fig. 12.2*b*. In the presence of water, the radius of an ion may thus increase to several times its original, nonhydrated dimension.

Applying an electrical potential to the saturated soil causes the hydrated positive ions to move toward the negative electrode (the cathode), dragging free water with them. The movement is primarily generated in the diffuse double layer, also called soil moisture film (or "boundary" water film), where the cations dominate. The higher the soil particle surface area, the more soil moisture film transfer will occur. Other important determinants are the magnitude of the electrical potential applied and the viscosity of the pore fluid.

Soil moisture film movement can also be caused by temperature differences, even in unsaturated soils, leading to the term "thermoosmosis." This process helps to explain certain phenomena associated with freeze-thaw cycles in soils.

12.1.2 Flow Potential

If a hydraulic or thermal gradient forces water to flow through a porous soil system, the electrically charged moisture film adhering to the soil particles is moved with it, inducing an electric current. The small but measurable electric potential generated is called the flow potential or streaming potential. This process is the reverse of electroosmosis and was first described by Quincke (1861). The flow potential may itself give rise to an electroosmotic backflow (in a direction opposite to the net fluid flow) involving the cations closest to the wall of the soil capillaries.

Although it is an interesting soil phenomena, the flow potential does not seem to have any practical geotechnical applications beyond research. But who knows? It may provide an explanation for believers in water divining!

12.1.3 Electrophoresis

If electrodes are immersed in a soil suspension rather than in a coherent soil mass, freely moving negatively charged clay particles will move toward the anode when a potential is applied. Conversely, free positively charged ions will be drawn toward the cathode. Measurement of the velocity reached by the clay particles gives infor-

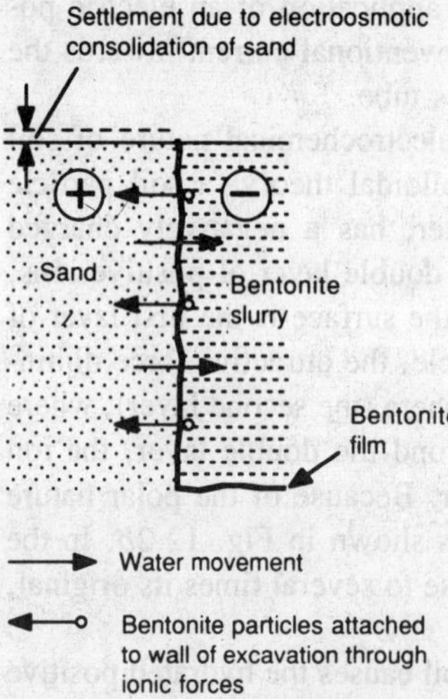

FIGURE 12.3
Electrokinetics of a bentonite slurry wall. [*After Veder (1963).*]

mation on the electrokinetic potential of the double layer, the so-called zeta (ζ) potential.

The reverse process is also recognized: Electrically charged particles moving in a solution create an electrophoretic potential.

Electrophoresis is thought to play a role in the supporting effect of bentonite in a slurry wall [Veder (1963)]. Bentonite in the slurry is negatively charged relative to adjacent sand (Fig. 12.3). The bentonite particles are attracted to the sand, forming a clayey boundary film; at the same time pore water is drawn from the sand into the slurry. With theoretical and experimental evidence, Veder argues that the electrokinetic lateral pressure exerted by the bentonite is on the order of 10 kPa!

12.1.4 Electric Potentials between Natural Soil Layers

Veder (1963, 1973) has repeatedly pointed to the significance of electrical potentials (say 10 to 40 mV) between soil layers in natural deposits. The resultant electroosmotic flow may create high pore pressures in the contact zone, leading to land instability or foundation failure. It is speculated that these potentials are maintained through continuing physicochemical processes in the soil, possibly even caused by the actions of microorganisms.

Veder's theory was prompted by observation of the astounding stabilizing effect of driving a few metal rods (15- to 25-mm diameter, at 3- to 8-m intervals) through unstable soil in several active landslides. Neither the consideration of reinforcement action nor drainage along the rods appeared to explain their beneficial effect. Veder believed that insertion of metal rods provides a short circuit between the soil layers, gradually reducing the potential between them. This stops electroosmotic flow and relieves excess pore pressures in the contact zone.

Veder (1973) further speculated that the success of soil nailing and even rock bolting (as used in the new Austrian method of tunnel construction) may be partly due to eliminating electroosmotic phenomena.

12.1.5 Electrostabilization and Electrohardening

Electrostabilization is really just another name for electroosmosis and variations thereof. It emphasizes the increase in strength of the soil, while for many engineers electroosmosis is simply considered a dewatering technique like well points.

Experimenting with a medium plastic clay, Jaecklin (1968) found, like others before him, that the strength increase achieved by electroosmosis was considerably higher than for a corresponding decrease in moisture content achieved by standard consolidation. This appears to be due to physicochemical changes in the soil, amplified by the release of cations at the corroding anode.

Jaecklin also reported that as the water content decreased by electroosmotic flow, both Atterberg limits (plastic limit w_p and liquid limit w_L) went up, and even the plasticity index increased; but notably the liquidity index [(natural water content $- w_p)/(w_L - w_p)$] fell markedly, which means the stiffness of the soil increased. Jaecklin's results and others are further discussed in Sec. 12.3.2.

The stabilizing (strengthening) effect of electroosmosis can be enhanced by the appropriate choice of the electrode material (e.g., aluminum instead of steel) or the addition of sodium silicate or calcium chloride at the anode. Engineers may describe this geotechnical process as *electrochemical hardening* or electrogrouting, rather than electroosmosis.

12.2 THEORETICAL CONCEPTS AND LABORATORY MEASUREMENTS OF ELECTROOSMOSIS

12.2.1 Basic Equations

The differential equation describing consolidation involving linear drainage is

$$\frac{\partial u}{\partial t} = c_v \frac{\partial^2 u}{\partial x^2} \tag{12.1}$$

where u = pore pressure, kPa
t = time, s
x = coordinate as defined in Fig. 12.4, m
c_v = coefficient of consolidation, m^2/s

After Schaad and Haefeli (1947), the flow velocities due to the hydraulic and the electrical potential gradient are combined in

$$-v = \frac{k_h}{\gamma_w}\frac{\partial u}{\partial x} + k_e \frac{\partial E}{\partial x} \tag{12.2}$$

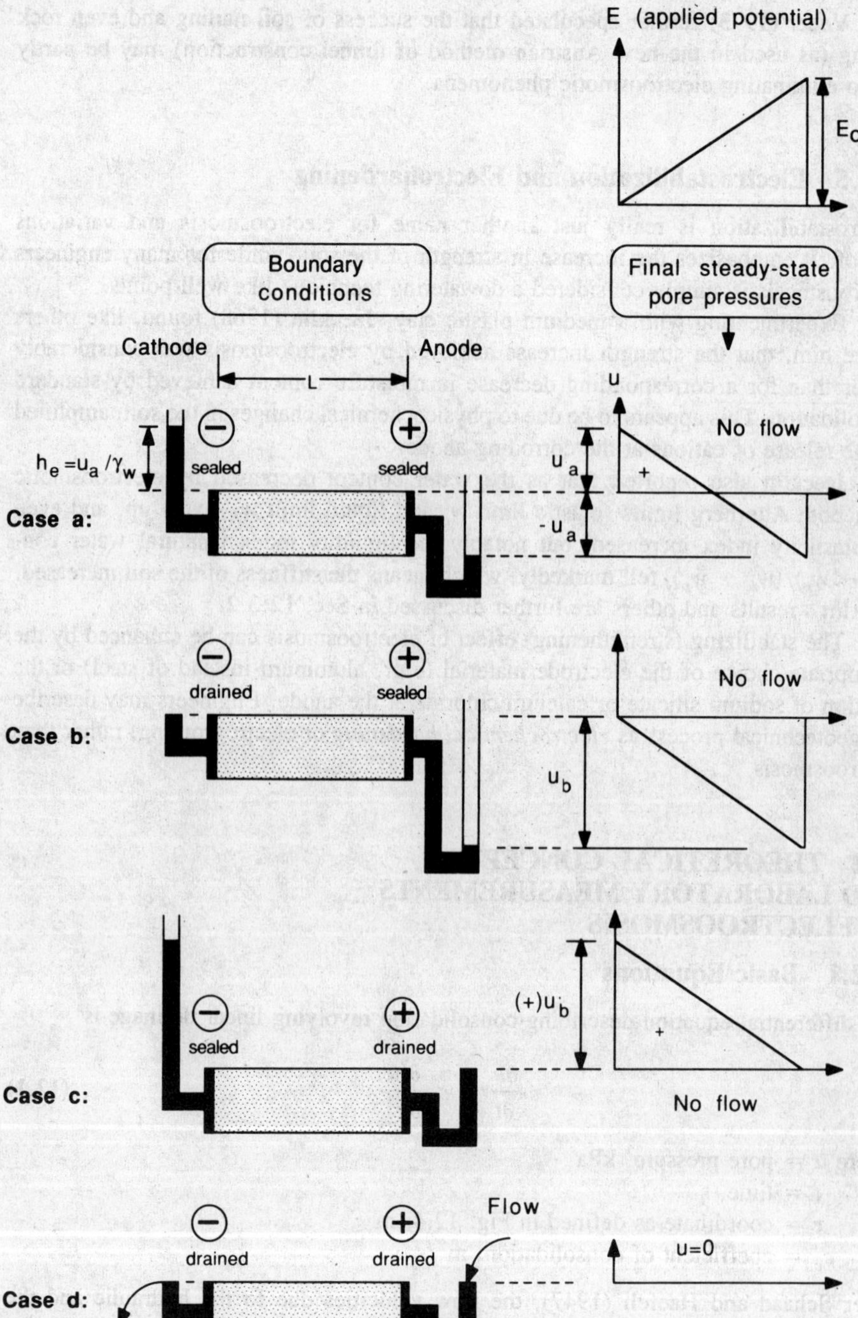

FIGURE 12.4
Final steady state for various boundary conditions.

where v = flow velocity, m/s
k_h = coefficient of hydraulic permeability, m/s
γ_w = unit weight of water, kN/m^3
E = electrical potential, V
k_e = coefficient of electroosmotic permeability, (m/s)/(V/m) = m^2/(s · V)

Note that in Eq. (12.2), the sign convention is as generally used in hydraulics, where a gradient producing flow is negative; this is opposite to the common practice in soil mechanics.

Ohm's law is written as

$$-i = \frac{\partial E}{\rho\, \partial x} \tag{12.3}$$

where i is the electric current density (A/m^2) and ρ is the resistivity of soil ($\Omega \cdot$ m).

The condition that the electric potential varies linearly is expressed by

$$\frac{\partial^2 E}{\partial x^2} = 0 \tag{12.4}$$

12.2.2 Boundary Conditions

Figure 12.4 depicts the four basic possible boundary conditions and the corresponding final steady-state pore pressures for a linear potential gradient. Case a is possible in the field, but in most practical applications conditions are as for case b, where maximum "drying out" is achieved at the anode and water appearing at the cathode is drained away. Cases c and d are mostly relevant for laboratory conditions, but case d also corresponds to the situation where a constant groundwater level is maintained.

Magnitudes of the maximum pore water pressures (or tensions) for the most important cases are (see Fig. 12.4):

Case a: Anode sealed and cathode sealed (no drainage or water access). For the final steady-state condition ($t = \infty$) under an applied potential difference E_0 the pore pressure distribution is linear and defined by

$$u_a = 0.5\,\frac{k_e}{k_h}\,E_0\gamma_w \tag{12.5}$$

Case b: Anode sealed and cathode drained. The pore pressure at the anode is found to be

$$u_b = -\,\frac{k_e}{k_h}\,E_0\gamma_w \tag{12.6}$$

Because the cathode is drained, the pore pressure is zero at the cathode.

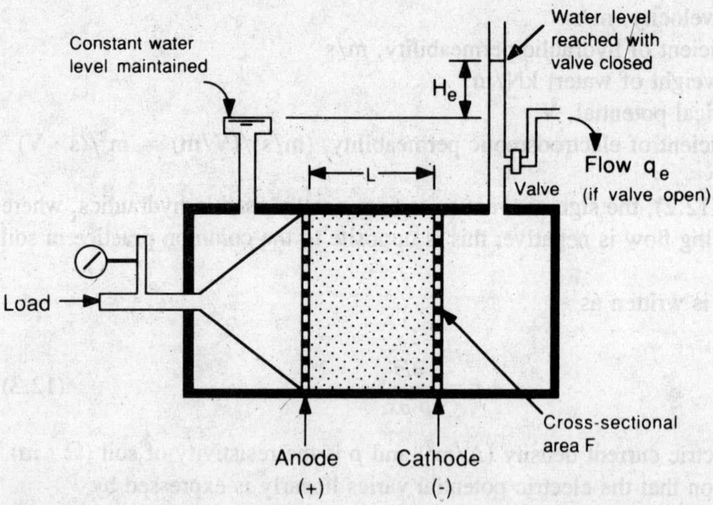

FIGURE 12.5
Electroodometer. [*After Schaad and Haefeli (1947).*]

12.2.3 Interpretation of Laboratory Tests

12.2.3.1 STEADY-STATE METHODS. Equations (12.2) and (12.6) can be used to interpret the final steady state attained in an electroodometer such as is used by Schaad and Haefeli (1947) and shown schematically in Fig. 12.5.

When the electric potential is applied with the valve *open*, and equal hydraulic head is maintained at the inflow as well as at the outflow ($\partial u/\partial x = 0$), then the maximum flow reached is q_e:

$$q_e = vF = k_e F \frac{E_0}{L} \quad \text{m}^3/\text{s} \tag{12.7}$$

where F is the cross-sectional area of the soil specimen (equal to the electrode area) and L is the length of the specimen. Measuring q_e for a given geometry and applied potential will thus allow k_e to be calculated.

If the valve is *closed* when the electric potential is turned on, an electroosmotic excess pore pressure u_e [= u_b from Eq. (12.7)] will develop at the cathode, raising the water level in the standpipe to a final excess head H_e, at which time the flow through the soil ceases:

$$H_e = \frac{u_e}{\gamma_w} = \frac{k_e}{k_h} E_0 \tag{12.8}$$

Equation (12.8) yields k_h for known k_e, given E_0 and measured H_e.

Other types of combined electroosmotic and static odometers are shown in Fig. 12.6. Type *a* was used in some tests by Jaecklin (1968). Type *b* is more common and was the one used by Morris et al. (1985). Some of the test results obtained by both are further discussed in Sec. 12.3.3.

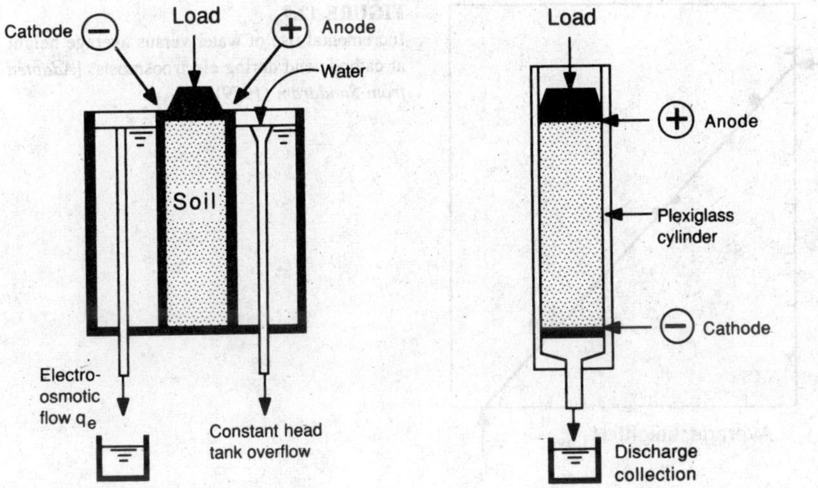

a. Anode and cathode drained at constant head (after Jaecklin, 1968)

b. Anode sealed, cathode drained (after Morris et al., 1985)

FIGURE 12.6
Combined electroosmotic and static odometers.

12.2.3.2 RATE-OF-RISE METHOD. Sundaram (1979) pointed out that if the rise in hydraulic head to H_e is monitored at regular time intervals (test arrangement as in Fig. 12.5 with valve closed), then it is possible to calculate, at least approximately, the electroosmotic permeability k_e as well as the hydraulic permeability k_h. In addition, the final equilibrium value H_e can be predicted without waiting until a steady state is achieved. The procedure is as follows:

Step 1: Take a number of standpipe readings at regular time intervals Δt (typically 4 to 5 values at hourly intervals). Produce a table of values including

Time	Reading of head	Rise over time Δt	Average head over time Δt
$\vdots$	$\vdots$	$\vdots$	$\vdots$
t_i	H_i	$\Delta H_i = H_i - H_{i-1}$	$H_{i,\text{avg}} = (H_i + H_{i-1})/2$
$t_{i+1} = t_i + \Delta t$	H_{i+1}	$\Delta H_{i+1} = H_{i+1} - H_i$	$H_{i+1,\text{avg}} = (H_{i+1} + H_i)/2$
$\vdots$	$\vdots$	$\vdots$	$\vdots$
t_{final}	H_e		

Step 2: Plot the incremental rise ΔH_i in the standpipe versus the average head $H_{i,\text{avg}}$ over the time interval Δt_i as shown in Fig. 12.7.

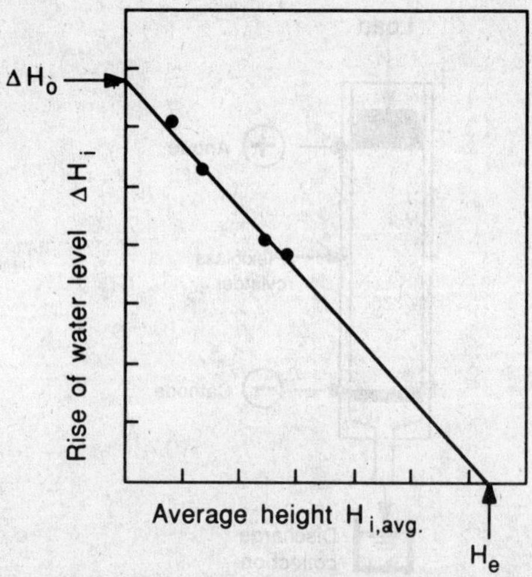

FIGURE 12.7
Incremental rise of water versus average height at cathode end during electroosmosis. [*Adapted from Sundaram (1979).*]

Step 3: Draw a straight line through the plotted points. Extrapolate to find ΔH_0 (corresponds to $H_i = 0$) and H_e (corresponds to $\Delta H_i = 0$).

Step 4: Find $$k_h = \frac{\Delta H_0 aL}{FH_e \Delta t} \tag{12.9}$$

and $$k_e = \frac{H_e k_h}{E_0} \tag{12.10}$$

where F = cross-sectional area of soil (area of electrodes)
a = cross-sectional area of standpipe
L = length of soil specimen.

12.2.4 Transient Pore Pressures

12.2.4.1 ESRIG SOLUTION. As a solution of Eq. (12.1) for boundary conditions as described in case *b* (sealed anode, drained cathode), Esrig (1968) gave the following expression for the pore pressure at distance *x* from the cathode at time *t*:

$$u(x, T) = -\frac{k_e}{k_h} E_0 \gamma_w \left[\frac{x}{L} - \frac{2}{\pi^2} f(T)\right] \tag{12.11}$$

where $$f(T) = \sum_{n=0}^{n=\infty} \frac{(-1)^n}{m^2} \sin\left(\frac{m\pi x}{L}\right) e^{-m^2\pi^2 T}$$

$$m = n + \tfrac{1}{2} \qquad n = 0, 1, 2, \ldots$$

$$T = \frac{c_v t}{L^2}$$

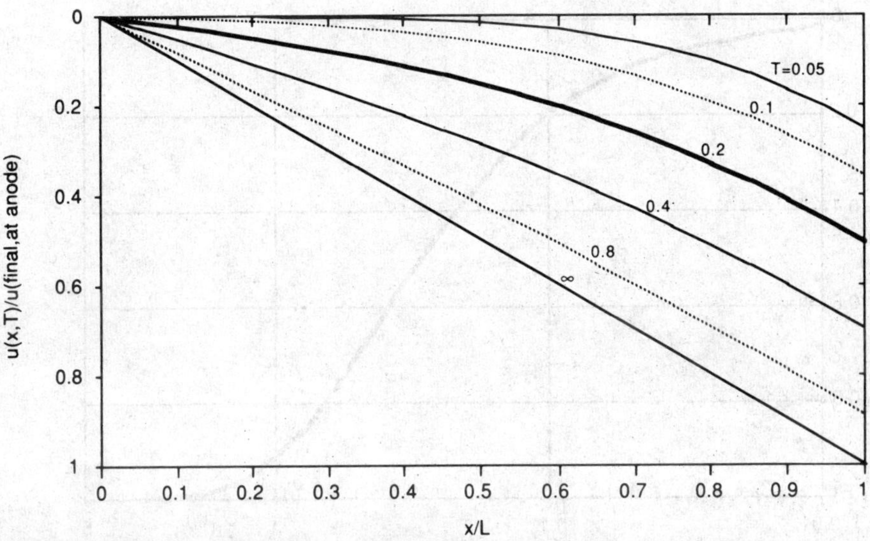

Note: u(final, anode)= $-\frac{k_e}{k_h} E_o \gamma_w$

x= distance from cathode

L= distance between anode and cathode

FIGURE 12.8
Pore pressure isochrones (Esrig solution).

Isochrones of pore pressure for various time factors T are given in Fig. 12.8 as a function of the distance from the cathode.

The average degree of consolidation $\overline{U}$ over the distance L can be calculated according to

$$\overline{U} = 1 - \frac{4}{\pi^3} \sum_{n=0}^{n=\infty} \frac{(-1)^n}{m^3} e^{-m^2\pi^2 T} \tag{12.12}$$

and is given graphically in Fig. 12.9.

12.2.4.2 MODIFICATION BY JOHNSTON AND BUTTERFIELD (1977). The foregoing solution for pore pressures during electroosmotic consolidation is based on a constant initial electrical gradient and constant soil parameters, such as hydraulic and electroosmotic permeabilities and resistivity.

Johnston and Butterfield (1977) [also see Johnston (1978)] modified the above analysis so that the results more closely predicted their experimental results. This was achieved by changing the boundary conditions so that the initial electrical gradient at the sealed boundary is infinite and decreases in a consistent way to a uniform gradient at the completion of consolidation. The revised expressions corresponding to Eqs. (12.11) and (12.12) become

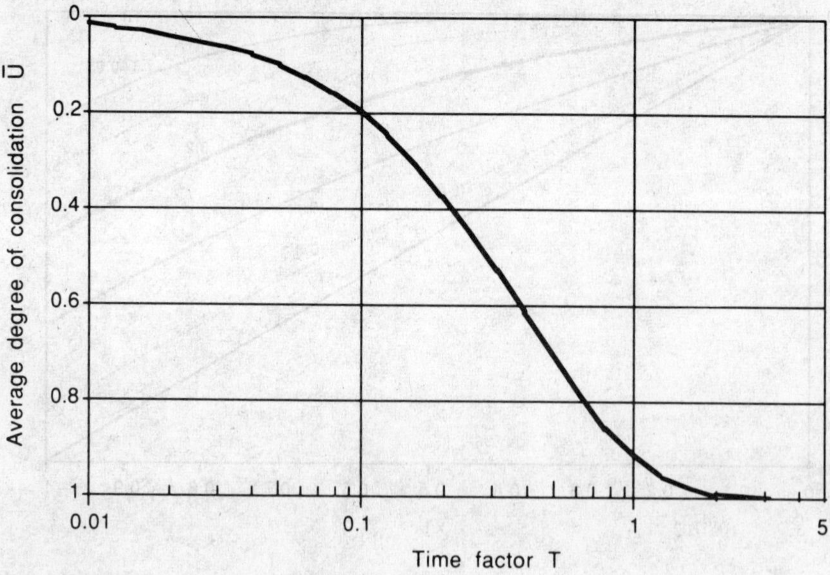

FIGURE 12.9
Average consolidation ratio (Esrig solution).

$$u(x, T) = -\frac{k_e}{k_h} E_0 \gamma_w \left[\frac{x}{L} + \frac{2}{\pi} f'(T) \right] \tag{12.13}$$

where $$f'(T) = \sum_{n=1}^{n=\infty} \frac{(-1)^n}{n} \sin\left(\frac{n\pi x}{L}\right) e^{-n^2\pi^2 T} \qquad n = 1, 2,, 3, \ldots$$

and $$\overline{U} = 1 - \frac{8}{\pi^2} \sum_{n=1}^{n=\infty} \frac{1}{n^2} e^{-n^2\pi^2 T} \qquad n \text{ odd} \tag{12.14}$$

Results for Eqs. (12.13) and (12.14) are shown graphically in Figs. 12.10 and 12.11.

The rapid drop in the electric potential around a sealed anode may not only be due to the reduction in moisture content but also to the development of shrinkage cracks and gas formation leading to soil separation from the electrodes [e.g., see Jaecklin (1968) and Johnston (1977)]. This may be overcome by appropriate design of the electrodes or the addition of salt solutions.

12.2.4.3 WAN AND MITCHELL'S (1976) EXTENDED SOLUTION. Wan and Mitchell (1976) solved Eq. (12.1) for the condition where there is a linearly variable initial pore pressure in the soil prior to activating an electric potential. This solution has a direct application for the case where, after completion of a first stage of electroosmotic consolidation, the electrodes and drainage conditions are reversed and a second stage is commenced.

For a situation as defined by case *b* (Fig. 12.4), which corresponds to the normal field conditions, the final pore water tension is highest at the anode and decreases

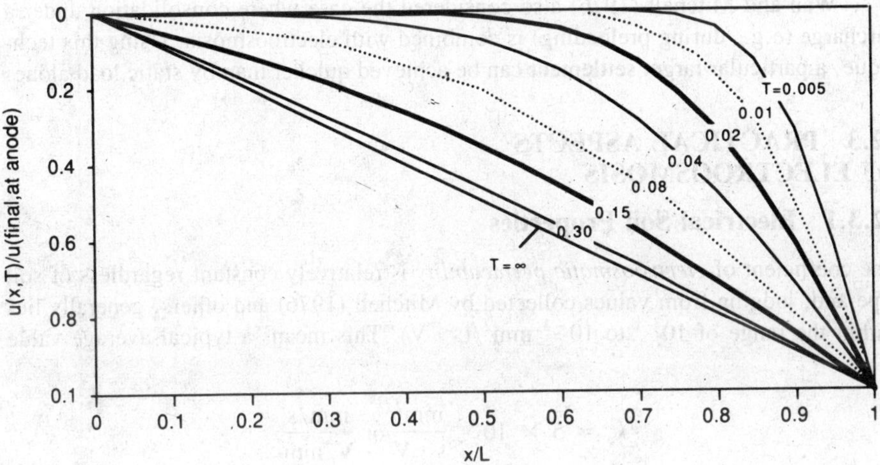

FIGURE 12.10
Pore pressure isochrones (Johnston and Butterfield solution).

linearly to zero at the cathode. As a result, consolidation is a maximum at the anode but is minimal near the cathode. By reversing the electrodes, sealing the old cathode, and draining water off at the new cathode, the water content near the old cathode can be reduced much further. Although there will be a rebound of pore pressures at the original anode, overall consolidation and gain in strength will be higher and more uniform.

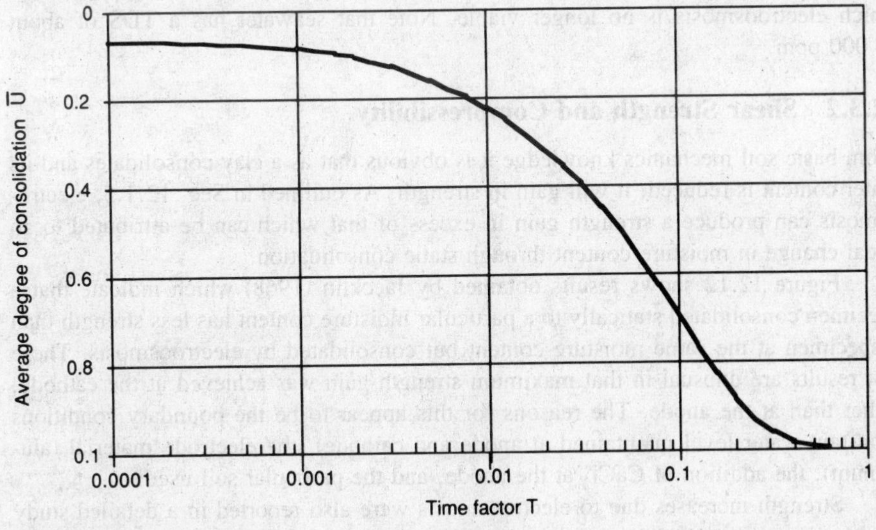

FIGURE 12.11
Average consolidation ratio (Johnston and Butterfield solution).

Wan and Mitchell (1976) also considered the case where consolidation under a surcharge (e.g., during preloading) is combined with electroosmosis. Using this technique, a particular target settlement can be achieved quicker than by static load alone.

12.3 PRACTICAL ASPECTS OF ELECTROOSMOSIS

12.3.1 Electrical Soil Properties

The coefficient of *electroosmotic permeability* is relatively constant regardless of soil type and, judging from values collected by Mitchell (1976) and others, generally lies within the range of 10^{-4} to 10^{-3} $mm^2/(s \cdot V)$. This means a typical average value is

$$k_e = 5 \times 10^{-3} \frac{mm^2}{s \cdot V} \text{ or } \frac{mm/s}{V/mm}$$

Hydraulic permeability k_h varies considerably more with soil type and consistency than does k_e. The less the hydraulic permeability, the stronger will be the electroosmotic effect on pore pressures; the more impermeable the soil, however, the longer it will take to develop these pore pressures.

Resistivity ρ of moist soils varies, according to the geophysicists, normally from about 5 to 500 $\Omega \cdot m$. Soils treated with electroosmosis are likely to be at the lower end of this range, approaching, but hopefully not too closely, the resistivity of salty water (2 to 20 $\Omega \cdot m$).

Resistivity is largely determined by the ionic content of the pore water. According to Johnston (1978), there appears to be an upper limit of total dissolved salt content (TDS), somewhere between 6000 to 14 000 parts per million (ppm) above which electroosmosis is no longer viable. Note that seawater has a TDS of about 35 000 ppm.

12.3.2 Shear Strength and Compressibility

From basic soil mechanics knowledge it is obvious that as a clay consolidates and its water content is reduced, it will gain in strength. As outlined in Sec. 12.1.5, electroosmosis can produce a strength gain in excess of that which can be attributed to an equal change in moisture content through static consolidation.

Figure 12.12 shows results obtained by Jaecklin (1968) which indicate that a specimen consolidated statically to a particular moisture content has less strength than a specimen at the same moisture content but consolidated by electroosmosis. These test results are unusual in that maximum strength gain was achieved at the cathode, rather than at the anode. The reasons for this appear to be the boundary conditions (constant water level maintained at anode and cathode), the electrode material (aluminum), the addition of $CaCl_2$ at the anode, and the particular soil used.

Strength increases due to electroosmosis were also reported in a detailed study by Morris et al. (1985), who tested a low-plastic sensitive clay obtained in tube samples from boreholes. An apparatus of type *b* in Fig. 12.6 was used with stainless-steel electrodes (anode sealed, cathode drained); these conditions resulted in the high-

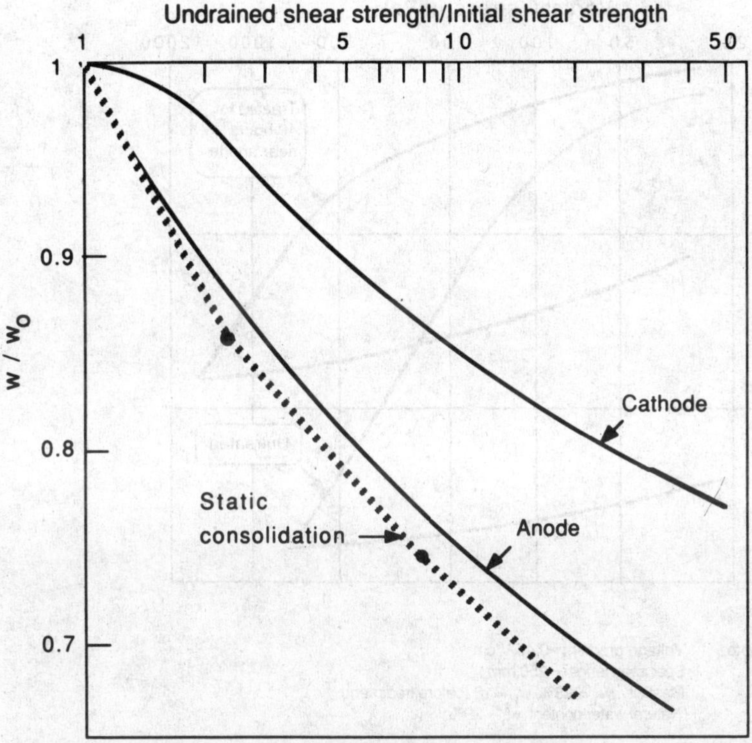

Note: w = water content, w_0 = initial water content
Test apparatus as Type (a) in Figure 12.6
(anode and cathode drained!)
Aluminium electrodes
$CaCl_2$ solution added at anode
Distance between electrodes = 50 mm, Voltage 2.5 to 20 V
Initial water content = 26.6%
Initial shear strength = 20 kPa (Vane shear apparatus)
Initial plasticity: w_P = 16.8, w_L = 36.4
Final plasticity (cathode): w_P = 22.7, w_L = 51.0

FIGURE 12.12
Relative water contents and corresponding relative shear strength after static consolidation or electroosmosis. [*After Jaecklin (1968).*]

est strength near the anode. Consolidated undrained triaxial tests showed a cohesion component added to the Mohr envelope obtained for static consolidation if electroosmosis was activated at the same time. Major improvement due to electroosmotic treatment was also found in the resistance to cyclic loading of this soil. Atterberg limits, when plotted on the plasticity chart, indicated a shift upward along the *A* line after electroosmosis, generally decreasing the liquidity index.

Figure 12.13 shows two static consolidation test results, one for a sample treated with electroosmosis and the other for an untreated sample [selected results from Morris et al. (1985)]. Electrical treatment of the soil resulted in increased stiffness and an apparently increased preconsolidation load.

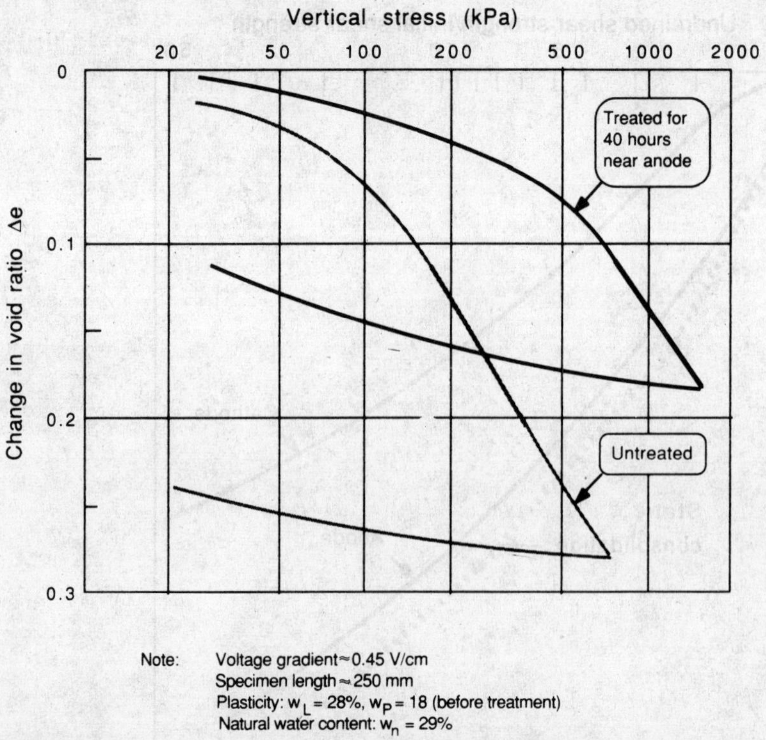

FIGURE 12.13
Effect of electroosmosis on one-dimensional compression. [*Selected results from Morris et al. (1985).*]

12.3.3 Settlement Predictions

For boundary conditions as described in case *b* (Fig. 12.4), the negative pore pressure developed results in an increase in effective stress $\Delta\sigma'$ of equal magnitude:

$$\Delta\sigma' = -\Delta u \tag{12.15}$$

and settlement for a layer of soil of thickness H can then be calculated according to usual formulations, such as

$$s = \frac{H}{1 + e_0} C_c \log \frac{\sigma_0' + \Delta\sigma'}{\sigma_0'} \tag{12.16}$$

where e_0 is the initial void ratio and C_c is the compressive index. A complication is that the settlement characteristics of a soil may change during electroosmosis because of structural changes caused by the current.

Corresponding to the distribution of pore pressures reached in the final steady state, settlement will also vary from a maximum at the sealed anode to a minimum near the cathode.

It appears that basic theory is likely to greatly overestimate pore pressure changes and the resulting settlement. Recognition of discrepancies between theory and field data prompted the modifications proposed by Johnston and Butterfield (1977) mentioned earlier. In addition there is also the school of thought [Johnston (1978)] which says that the magnitude of negative pore pressures developed cannot exceed 100 kPa, otherwise cavitation will occur. If this is accepted, the maximum increase in effective stress due to electroosmosis would also be limited to 100 kPa.

The time-settlement calculations can proceed as for ordinary consolidation, except that the relationships for pore pressures, consolidation ratios, and time factors as set out in Sec. 12.2.4 must be used.

12.3.4 Electrode Types and Layout

Many different metallic structural elements can serve as electrodes. Common ones are rebars, gas pipes, scrap-sheet piles, well points, aluminum rods, and other types. Most effective for dewatering are drained cathodes, thus the preference for perforated pipes or steel well points for this task. It has been pointed out by Johnston (1978) that even if the cathode consists of a solid bar, a drainage path quickly forms along its side, because of the buildup of excess hydrostatic pressure and the likely release of hydrogen due to electrolytic action. At the surface, water is collected in drainage channels leading to a pump sump where necessary.

The anode is usually a solid bar, unless provision is to be made for the addition of a chemical agent in order to enhance the hardening effect. Anodes, whether steel or aluminum, corrode quite rapidly during electroosmosis, as the chemical reactions producing hardening proceed.

A review of the relevant literature indicates that electrodes have a length of 2 to 15 m, or more. The spacing within each row of anodes or cathodes could be as little as 1 m. Typically, the distance between rows of anodes and cathodes is 2 to 5 m.

Based on an analysis of electrical flows for different electrode layouts, Mitchell (1981) pointed out that a hexagonal arrangement of anodes around a central cathode is more efficient than linear (rows) or square patterns. A hexagonal arrangement was earlier described by Mise (1961).

12.3.5 Power Supply and Consumption

The current per electrode between rows of anodes and cathodes at a distance L could be estimated approximately from

$$I = \frac{E_0 F}{L\rho} \tag{12.17}$$

where E_0/L = potential gradient applied, V/m

F = cross-sectional area of soil subject to current flow between two electrodes ($\approx$ depth $\times$ distance between anodes or cathodes)

ρ = resistivity, $\Omega \cdot$ m

Equation (12.17) combined with Eq. (12.7) gives

$$q_e = I\rho k_e = Ik_i \tag{12.18}$$

where k_i is the amount of water transferred per unit charge passed [$m^3/(s \cdot A)$].

The power consumption can then be formulated as

$$P_e = E_0 I = \frac{E_0 q_e}{k_i} \quad \text{W} \tag{12.19}$$

As indicated by data presented by Mitchell (1976), the value k_i could easily vary by a factor of 10 for different moisture contents of a soil within its plastic range. Typical values of k_i for clays range from 10^{-5} to 10^{-8} $m^3/(s \cdot A)$. Since k_e is relatively constant for a particular soil, variations in the electroosmotic flow rate can be mainly attributed to variations in resistivity.

The time required to apply a particular potential gradient will depend on the degree of consolidation (or settlement) desired. If this time is calculated for say a consolidation ratio of 80%, then, at least theoretically, the total power consumption could be calculated. The problem lies with the fact that the soil's permeabilities as well as its resistivity could decrease drastically as electroosmosis takes place, due to changes in water content, shrinkage cracks, gas formation at the electrodes, and electrochemical changes, including corrosion of the electrodes.

In the absence of results of full-scale field trials, power consumption could be estimated simply based on past experience with a variety of projects, as outlined in the following paragraphs. The guidelines given are, however, not very reliable. Electroosmosis seems to confirm the saying, "Exceptions prove the rule!"

The typical potential applied in the field is in the range of 50 to 100 V. The current may be 20 to 200 A or more. The higher the current, the bigger are the problems with power output, cabling, heat generation, and safety. The potential gradient applied seems to be below 50 V/m in most past projects.

The energy consumption on dewatering projects has sometimes been expressed in terms of power used per cubic meter of soil. Values on the order of 30 kW · h/m^3 have been quoted for a number of successful projects.

The economics of electroosmosis is obviously significantly affected by the cost of the electrical energy used. This could vary significantly depending on whether the power is generated on site or comes from a nearby power station.

12.4 TYPICAL APPLICATIONS OF ELECTROOSMOSIS

12.4.1 Increasing Strength of Sensitive Clay Prior to Excavation

A well-documented case of successful application of electroosmosis is the consolidation of the sensitive clay at Ås in Norway [Bjerrum et al. (1967)]. This project is also referred to in the books by Mitchell (1976) and Lee et al. (1983). More than

2000 m^3 of soil were treated over a period of 120 days. A total of 186 steel reinforcing rods, 19 mm in diameter and about 10 m long, served as electrodes. The spacing between the 10 rows of electrodes was 2 m, and the applied potential was 40 V.

The soil properties reported were

$$k_h = 2 \times 10^{-10}\ \text{m/s} \qquad k_e = 2 \times 10^{-9}\ \text{m}^2/(\text{s} \cdot \text{V})$$

$$c_v = 0.5 \times 10^{-7}\ \text{m}^2/\text{s} \qquad C_c = 0.4$$

$$e_0 = 0.86 \qquad \rho = 47\ \Omega \cdot \text{m}$$

The soil had a natural water content of 31% which was reduced by some 4% by the treatment. At the same time, however, the liquid limit increased from 19 up to 29% (average about 25%) and the plastic limit rose by a couple of percent. Average undrained shear strength increased from about 10 to 60 kPa, and remolded strength increased from virtually 0 to about 30 kPa. Maximum settlements observed were 300 mm after 50 days and 500 mm after 120 days.

Another Norwegian project employing electroosmosis is described by Eggestad (1983). The objective was to stabilize an excavation in soft sensitive marine clay near Oslo. Original soil properties were

$$w_n = 32 \text{ to } 45\% \qquad w_L = 25 \text{ to } 27\% \qquad w_p = 19 \text{ to } 22\%$$

Undrained shear strength:	4 to 15 kPa
Sensitivity:	> 60
Salinity:	1.0 equiv. gram of NaCl per liter
Unit weight:	19 to 19.6 kN/m^3
Clay content (> 2μm):	33 to 47%
Electroosmotic permeability:	$k_e = 6 \times 10^{-5}$ to 9×10^{-5} cm/s
Resistivity:	$\rho = 30$ to $42\ \Omega \cdot$ m

After a pilot program proved the feasibility of electroosmosis, a total area of 2500 m^2 of ground was treated by rows of electrodes 1 m apart. Reinforcing rods, 20 mm in diameter and 6 m long, served as electrodes. Generators normally used for welding supplied a 50-V potential. The polarity of the system was reversed after 39 days. After a total treatment of about 90 days, the strength of the clay had more than doubled. Settlement observed was on the order of 300 to 400 mm, suggesting a reduction in water content of only about 3%, whence the gain in strength must have been partly due to physicochemical changes near the electrodes. Analysis of the cation exchange complex of the clay indicated a decrease in adsorbed Na^+ and an increase in Fe^{++}. The pore water showed a decrease in Na^+ and a pronounced increase in Ca^{++}.

Eggestad noted that the applied potential dropped rapidly close to the anode, giving a gradient of only about 10 V/m over most of the distance between opposite electrodes. This corresponds to the observation that the water extracted was about 5 times less than expected from the water flow equation [Eq. (12.7)]. Total power consumption was 7 kW · h/m^3 of stabilized clay.

12.4.2 Improvement of Foundations Below Dam

A major project of electroosmotic soil improvement was described by Fetzer (1967). The objective was to stabilize a 18-m-thick clay layer below the 24-m-high West Branch Dam. A total of 580 000 m^3 of soil was treated by the installation of 990 cathodes (64-mm steel pipes) and 660 anodes (64-mm steel pipes), all 40 m long. A potential of 100 to 150 V was applied over an anode-cathode distance of 6 m, resulting in a current of some 40 A. The treatment lasted 10 to 12 months.

12.4.3 Friction Pile Stabilization

As discussed by Johnston (1978), electroosmosis can be used to alter the side resistance of steel piles in clay. If the pile acts as an anode, negative pore pressures develop instantly adjacent to the pile, increasing side resistance and thus pile capacity. If the pile becomes a negative electrode (a cathode), then the opposite occurs: High pore pressures develop which reduce side friction. This can be used to advantage when driving piles or when it is necessary to reduce negative skin friction; the latter may, for example, develop along an end-bearing pile embedded in soil consolidating under a surcharge.

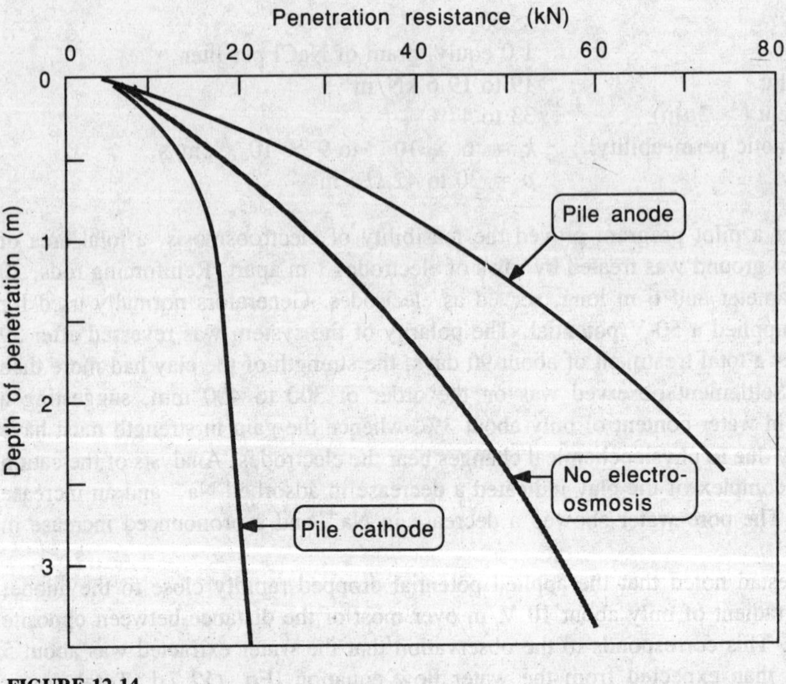

FIGURE 12.14
Penetration resistance of a model pile for tests with and without electroosmosis. [*After Johnston (1978); only average results plotted.*]

The principle of short-term electroosmotic effects is well demonstrated by the results of experiments with a model pile shown in Fig. 12.14 [after Butterfield and Johnston (1980)]. Penetration resistance is increased if the pile acts as an anode and is decreased if the pile is a cathode.

Short-term changes of pore pressure along the pile may disappear altogether when the electrical potential is removed. In order to obtain permanent changes to the pile behavior, electroosmosis should be applied over a period of days or weeks. Under these conditions, irreversible physicochemical changes in the soil have been observed to permanently increase the penetration resistance of piles. The increase in bearing capacity has also been explained by stronger bonding of the clay to the pile, bringing about an effective increase in its diameter [e.g., Spangler and King (1949)]. Electroosmotic treatment can result in a doubling of the capacity of a steel pile, as reported by Soderman and Milligan (1961).

12.4.4 Field Determination of Consolidation Characteristics of a Soil

Banerjee and Mitchell (1980) have developed a technique for the determination of the coefficient of consolidation and the coefficient of volume change of a soil in situ by electroosmosis. The probe consists of a hollow cylindrical tube with two strip ring electrodes built in. Transducers can measure the pore pressure buildup on the application of an electric potential if both electrodes are sealed. If the electrodes are drained, a flow of water occurs between them; the electroosmotic permeability is determined from the measurement of this flow.

PROBLEMS

Prefixes indicate problem type: C = calculations, M = multiple choice, B = brief answer, D = discussion.

Calculations

C12.1. An experiment is set up as in Fig. 12.5. A potential of 4 V was applied to a 200-mm-long specimen with a cross-sectional area of 0.005 m^2. With the valve open, a flow of 43.2 mL/day was recorded. With the valve closed, a water rise in the standpipe of 200 mm was observed. Determine k_e and k_h.

C12.2. A static consolidation test indicated a hydraulic permeability of 6×10^{-9} m/s. In an electroosmotic odometer a 100-mm-long specimen with a cross-sectional area of 0.003 m^2 was set up, with both electrodes drained. Under a voltage difference of 2 V, a flow of 1 mL/day was recorded. Determine the electroosmotic permeability, and calculate the theoretical rise in a standpipe at the cathode, if that was sealed.

C12.3. Sundaram (1979) gave the following results for an electroodometer test on kaolinite:

Cross-sectional area	$F = 12.56$ cm^2
Length of sample	$L = 9.7$ cm

Cross-sectional area of standpipe $a = 0.635\ cm^2$
Voltage applied $E_0 = 9.7\ V$

The average height H_{avg} of the water in the standpipe and height increments ΔH over time intervals $\Delta t = 60$ min were recorded:

H_{avg}, cm	ΔH, cm
15.9	1.6
27.18	1.35
47.52	1.05
57.00	1.00

Determine the electroosmotic and hydraulic permeability.

C12.4. In the laboratory, a 30-cm-long soil sample, 10 cm in diameter, is subjected to a potential difference of 5 V, and a current of 2.6 mA was observed. What is the resistivity of the soil?

C12.5. Estimate the power (in watts) required to set up a potential gradient of 20 V/m between two rows of opposite electrodes placed on a square grid 2 m apart. The soil has a resistivity of 50 Ω · m.

C12.6. The following problem attempts to apply basic theory to an actual case study, not necessarily with great success. Particularly, answers to parts (*a*) and (*b*) are greatly in excess of observations. Typical properties for the Ås clay in Norway are [Bjerrum et al. (1967)]:

$$k_h = 2 \times 10^{-10}\ \text{m/s} \qquad k_e = 2 \times 10^{-9}\ \text{m}^2/(\text{s} \cdot \text{V})$$

$$c_v = 0.5 \times 10^{-7}\ \text{m}^2/\text{s} \qquad C_c = 0.4$$

$$e_0 = 0.86 \qquad \rho = 47\ \Omega \cdot \text{m}$$

An electroosmotic dewatering system is installed. The electrodes are embedded in 10 m of soil. The anode-cathode distance is 2 m. A 40-V potential is applied.

(*a*) Calculate the average final (negative) excess pore pressure expected after electroosmosis has been applied and calculate the corresponding settlement.

(*b*) How long does it take for 50% of the expected final settlement to take place, according to the Esrig solution?

(*c*) What is the (excess) negative pore pressure midpoint between two electrodes for conditions defined by part (*b*)?

(*d*) Approximately how much current is initially flowing in a 1-m-wide, 10-m-deep, and 2-m-long slice of soil between the row of electrodes? How much power is used per hour?

(*e*) Revise your settlement estimate of part (*a*) by assuming that the voltage applied is only 75% effective and that the negative pore pressure cannot be less than 100 kPa.

(*f*) Repeat the estimate of part (*b*) but use an average consolidation ratio according to Johnston and Butterfield.

Multiple Choice

M12.7. If an electric potential is applied to two sealed electrodes embedded in a saturated soil,

(*a*) A permanent water flow from the cathode to the anode is established.

(*b*) A streaming potential is set up.
(*c*) Pore pressures at the anode are reduced.
(*d*) Calcium chloride is released at the anode.

M12.8. The reverse process of electrophoresis is
(*a*) Cataphoresis.
(*b*) Electroosmosis.
(*c*) Electrophoretic potential.
(*d*) Flow potential.

M12.9. According to Veder, some slopes become unstable because of natural electric potentials between soil layers. This is due to
(*a*) Downslope electric forces.
(*b*) Seepage forces parallel to the slope.
(*c*) Pore pressure buildup in the contact zone.
(*d*) Increase of soil unit weight.

M12.10. Which one of the following four test setups would result in the maximum pore water tension at the anode?
(*a*) Anode sealed, cathode sealed.
(*b*) Anode drained, cathode sealed.
(*c*) Anode drained, cathode drained.
(*d*) Anode sealed, cathode drained.

M12.11. The resistivity of moist soils is most likely in the range of
(*a*) 5 to 500 Ω.
(*b*) 5 to 500 kΩ.
(*c*) 5 to 500 Ω · m.
(*d*) 5 to 500 Ω/m^2.

M12.12. After electroosmotic treatment of clayey soils, it has been observed that
(*a*) The liquid limit w_L goes down.
(*b*) The plasticity index I_p decreases.
(*c*) The liquidity index I_L decreases.
(*d*) The resistivity ρ decreases.

M12.13. The liquidity index I_L is defined as
(*a*) (Natural water content $- w_p)/(w_L - w_p)$.
(*b*) $(w_L -$ natural water content$)/(w_L - w_p)$.
(*c*) The slope of the straight line through plots of water content versus blows in the liquid limit test.
(*d*) Plasticity index/% clay.

Brief Answer

B12.14. A "Who's Who" in electroosmosis (EO) contains names such as Casagrande, Haefeli, Mitchell, Quincke, Reuss, Schaad, and Veder. For each statement, choose the name(s) described.
(*a*) Discovered EO in soils: ____________ .
(*b*) First described electrophoresis: ____________ .
(*c*) Registered patent on EO in 1935: ____________ .
(*d*) Carried out tests in the 1940s in Zurich: ____________ and ____________ .
(*e*) Californian professor with a long involvement in EO research: ____________ .
(*f*) Austrian professor who recognized electric potential between natural soil layers:

B12.15. Add the missing word in the following statements.

(*a*) ____________ of soil-water systems is concerned with the relative movement of the solid versus the liquid phase under the influence of an electric field.

(*b*) The process of water flowing in soils as a result of a direct electric current passing through it is called ____________ in soils.

(*c*) ____________ is the reverse process of electroosmosis.

(*d*) The phenomenon of clay particles moving in a suspension of water under the influence of an electric field is called ____________

B12.16. List four typical applications of electroosmosis in geotechnical engineering.

(*a*) ____________

(*b*) ____________

(*c*) ____________

(*d*) ____________

Discussion

D12.17. Explain the role of the "double layer" around clay particles in the process of electro-osmosis.

D12.18. Let us say that the moisture content of two identical samples of clay was reduced by the same amount; one specimen was subjected to static consolidation, the other to electroosmosis. Which specimen is likely to show a higher strength increase, and why is there a difference?

D12.19. What is the purpose of reversing electrodes (anode ↔ cathode) during electroosmotic dewatering?

PART IV

PHYSICAL AND CHEMICAL MODIFICATION

CHAPTER 13

MODIFICATION BY ADMIXTURES

13.1 TERMINOLOGY, CONSTRUCTION TECHNIQUES, AND TYPICAL USES

This chapter deals with modifying soils by the mechanical addition of granular materials or chemical compounds such as cement, lime, bitumen, and calcium chloride. The purpose of mixing these additives with the ground is to

Increase strength
Reduce deformability
Provide volume stability (control shrinking and swelling)
Reduce permeability
Reduce erodibility
Increase durability (inhibit degradation of aggregates)
Control variability

Improving the engineering properties of a soil by admixtures is often simply referred to as *soil stabilization,* particularly in roadworks. It must also be pointed out that the expression *modifying* soils may have a special meaning for road engineers; if a subgrade is described as "cement-modified," this may imply that only a small amount of cement was added, say less than 3% by weight. In this book, however, no such distinction is made between soil stabilization and modification, and the latter term is used in its most general sense.

On specific civil or mining engineering projects, soil additives may help to

Enhance subgrade or subbase properties in order to reduce the required overall pavement thickness
Improve trafficability on construction sites
Prepare the ground for shallow foundations
Stabilize slopes by improving the soil's shear strength and/or by changing the failure geometry
Reduce erosion by surface runoff or internal seepage (piping)
Construct embankments
Form load-bearing columns (in situ)
Improve the workability of borrow materials
Reduce traffic-generated dust
Contain hazardous wastes
Rehabilitate polluted or mined ground

Different ways of mixing additives with soil are illustrated in Fig. 13.1.

Traditional *surface stabilization* begins with excavating and breaking up (comminution) of the soil. Then the stabilizer is added, and water is added if necessary. Soil and additives are mixed thoroughly, compacted, and allowed to cure (Fig. 13.1*a*). Most commonly, mix-in-place equipment is used, with a processing depth from 150

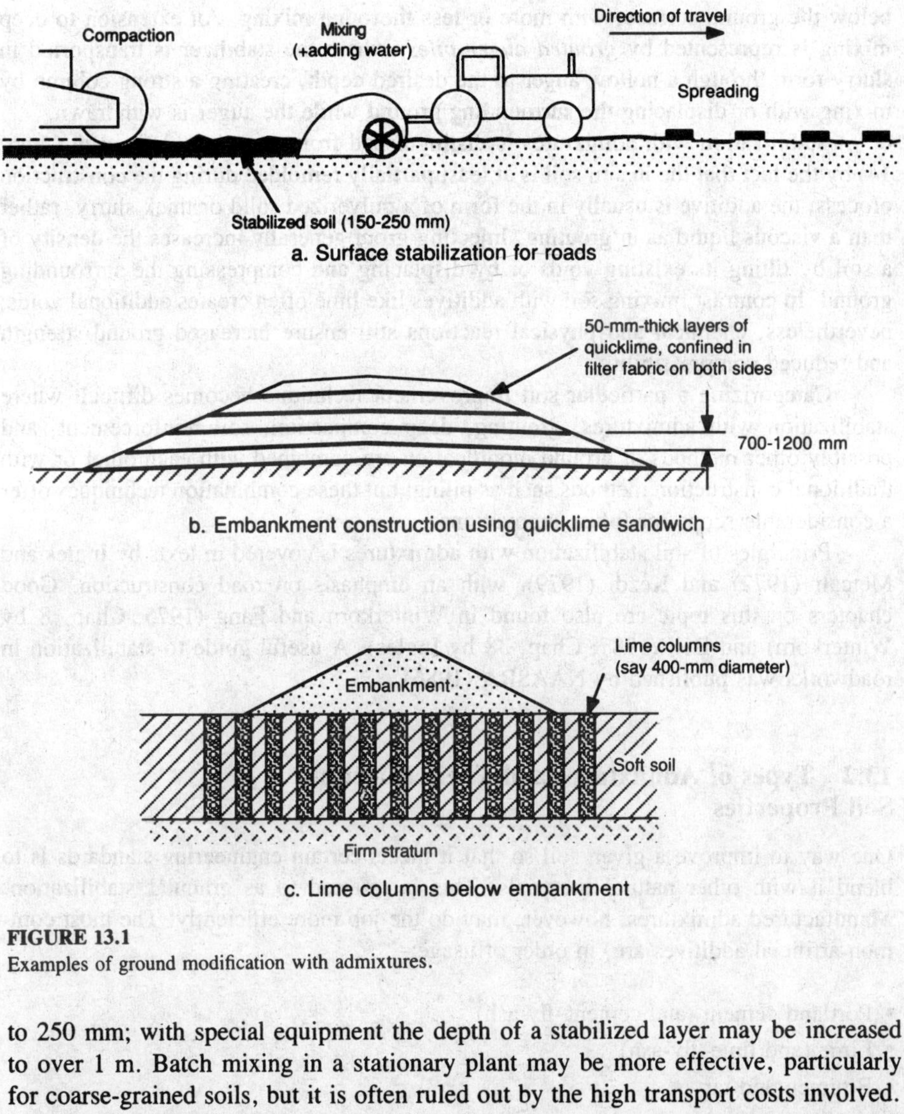

FIGURE 13.1
Examples of ground modification with admixtures.

to 250 mm; with special equipment the depth of a stabilized layer may be increased to over 1 m. Batch mixing in a stationary plant may be more effective, particularly for coarse-grained soils, but it is often ruled out by the high transport costs involved.

Additives such as lime can be beneficial without thorough mixing, although more lime may be required to achieve the desired effect. Yamanouchi et al. (1982) described a construction technique, where 50-mm-thick layers of quicklime were confined by filter fabric and alternated with 0.7- to 1-m lifts of compacted cohesive soil (called *sandwich construction*, see Fig. 13.1*b*).

Similarly, without mechanical intermixing, vertical *lime columns* can be formed which strengthen the surrounding soil, enabling a load transfer to a firm stratum below (Fig. 13.1*c*). This technique can also be used to stabilize an active landslide.

The term *deep mixing* is applied to techniques where piles, walls, or foundation blocks are formed by introducing lime, cement, slag, and other additives into the soil below the ground surface, with more or less thorough mixing. An extension to deep mixing is represented by *grouted auger piles,* where the stabilizer is transported in slurry form through a hollow auger to the desired depth, creating a strong column by mixing with or displacing the surrounding ground while the auger is withdrawn.

Stabilization with admixtures is distinguished from *grouting* (defined in Chap. 14) by the fact that the in situ soil is at least partially remolded during the construction process; the additive is usually in the form of a pulverized solid or thick slurry, rather than a viscous liquid as in grouting. Injecting grout generally increases the density of a soil by filling its existing voids or by displacing and compressing the surrounding ground. In contrast, mixing soil with additives like lime often creates additional voids; nevertheless, chemical and physical reactions still ensure increased ground strength and reduced compressibility.

Categorizing a particular soil improvement technique becomes difficult where stabilization with admixtures, grouting, deep compaction, soil reinforcement, and possibly other methods of ground modification are combined with each other or with traditional construction methods such as piling; but these combination techniques offer a considerable scope for future innovations.

Principles of soil stabilization with admixtures is covered in texts by Ingles and Metcalf (1972) and Kezdi (1979), with an emphasis on road construction. Good chapters on this topic are also found in Winterkorn and Fang (1975; Chap. 8 by Winterkorn) and Bell (1987; Chap. 38 by Ingles). A useful guide to stabilization in roadworks was published by NAASRA (1986).

13.2 Types of Admixtures and Their Effect on Soil Properties

One way to improve a given soil so that it meets certain engineering standards is to blend it with other natural materials. This is referred to as granular stabilization. Manufactured admixtures, however, may do the job more efficiently. The most common artificial additives are, in order of usage:

- Portland cement (and cement–fly-ash)
- Lime (and lime–fly-ash)
- Bitumen and tar

The reason for their popularity is that they are applicable to a considerable range of soil types, they are widely available, their costs are relatively low, and they are environmentally acceptable.

Uses for other chemicals, in particular calcium chloride ($CaCl_2$), have been found in earth construction, as dust palliatives on unpaved roads, or for other purposes. Many other chemicals, either designed in the laboratory or available as industrial-waste products, have been tried, with more or less success.

Hoshiya and Mandal (1984) found that adding 0.5% by weight of metallic powder (aluminum and cast iron) to a silty clay increased the clay's cohesion. They interpreted this effect as reinforcement by inclusions. The question is whether their tests would not be more appropriately described as stabilization by granular admixtures, perhaps with secondary effects of a physicochemical nature.

13.2.1 Granular Admixtures

The strength of predominantly coarse-grained materials is largely related to their density, which in turn depends significantly on the particle size distribution. Maximum density is obtained in soils which have a particle size distribution which can be approximated by the following expression [NAASRA (1986)]:

$$p = 100\left(\frac{d}{D}\right)^n \tag{13.1}$$

where p = percentage passing the sieve with aperture d
D = maximum particle size
n = exponent

For most soils used in pavement construction, densest particle packing is achieved when n is in the range of 0.45 to 0.5. Such a soil would be described by a geotechnical engineer as "well-graded," meaning there is a broad range of particle sizes present. This is in contrast to poorly graded (uniformly sized) soils which also show poor compactibility. Note that to a geologist, a well-graded soil is known as a "poorly sorted" material.

Where a borrow area contains poorly graded but variable soils, mixing material from different locations may bring an improvement in the grain size distribution and thus in density and related properties. Alternatively, selected materials may be imported and mixed with the on-site soil in order to gain higher strength and lower compressibility.

Fine-grained base and subbase materials with excessive plasticity may be improved by the addition of nonplastic coarse fractions. The benefits of such measures could be evaluated by strength tests (CBR, Texas Triaxial, etc.) and measurements of shrinkage and swelling (linear shrinkage).

Modification with granular additives has also been called "mechanical stabilization," but in this book "mechanical" infers the application of short-term external forces, as in compaction, not the use of admixtures. Mixing granular materials for achieving maximum density and some cohesiveness was important in the early days of building unsealed roads but is more or less being phased out on economic grounds.

Granular admixtures may still have some importance in efforts to use waste materials for pavement construction. By-products of industrial processes, such as slag and fly ash, often have a very limited range in grain sizes. This may make them unacceptable as high-quality road materials according to existing standards. Combinations of granular waste products from different sources, or intermixing with natural materials, could offer an economic and environmentally welcome answer to both the growing scarcity of good aggregates and the problem of disposing of large quantities of granular waste products.

13.2.2 Portland Cement

The most commonly used additive for soil stabilization is ordinary Portland cement. Problems may be encountered with soils which contain excessive amounts of organic matter, sulfates, or salts. Whether sulfate-resisting cements are of benefit or not is disputed [Sherwood (1962)]. Additives such as retarding agents could be appropriate where a delay in compacting after mixing is expected. Lime addition assists comminution and mixing with cement, while calcium chloride has been used in the presence of organic matter. Fly ash may partially replace cement and act as a filler.

Soils with cement admixtures are generally termed *cement-stabilized* or simply *cement-treated*. Mitchell (1976) reserves the name *soil-cement* for hardened mixtures of Portland cement, soil, and water that contain sufficient cement to pass the ASTM-PCA durability (wet-dry and freeze-thaw) tests.

Road engineers may arbitrarily distinguish between cement modification and cement stabilization, depending on the cement content or the degree of strength achieved, say less or more than 0.8 MPa in unconfined compression (after 7 days moist curing).

Gravels, crushed rock, or coarse sands with enough cement to achieve a stress-strain modulus in the range of 2000 to 20 000 MPa are described as *cement-bound* and may serve as subbase or base course materials.

13.2.2.1 SOIL-CEMENT-WATER REACTIONS. The reaction of cement and water forms cementitious calcium silicate and aluminate hydrates, which bind soil particles together. The hydration releases $Ca(OH)_2$, slaked lime, which in turn may react with components of the soil, such as clay minerals. While hydration occurs immediately upon contact of cement and water, secondary reactions are slower and may go on for many months, similar to soil-lime interaction.

Because the primary reaction (hydration) is independent of the soil type, cement stabilization is effective for a wide range of soils. Difficulties are usually only encountered with highly organic soils or coarse gravels. If the latter need stabilization at all, additional granular admixtures may assist. With fine-grained soils, limits of applicability may be imposed by the difficulty of mixing, particularly with wet highly plastic clays (liquid limit > 40%); in such a situation, treatment with quicklime may be more advantageous.

13.2.2.2 ENGINEERING BENEFITS OF CEMENT STABILIZATION. Major gains in the cement treatment of soils are

- Increased strength and stiffness
- Better volume stability (less moisture sensitivity, control of frost heave)
- Increased durability

Before discussing the effect of stabilization, it is important to realize that its results depend to a large extent on the degree of mixing and compaction achieved in the specimen tested. The mixing efficiency can be evaluated by the ratio E_m as follows:

$$E_m = \frac{S_f}{S_L} \tag{13.2}$$

where S_f is the strength of the soil mixed in the field and compacted in the laboratory and S_L is the strength of soil mixed and compacted in the laboratory.

Good construction procedures may result in efficiencies in excess of 80%. Typical cement contents in soil stabilization range from 2 to 10%. The lower limit is dictated by the difficulties in evenly mixing small quantities of cement with soil. The upper limit is related to the problem of shrinkage cracking.

Density and plasticity. Kezdi (1979) reports that cement treatment may slightly increase the Proctor maximum dry density of sands and highly plastic clays, but that of silts may be decreased; small changes in the optimum moisture content also occur.

Cement reduces the plasticity index of a cohesive soil. Whether this is mainly due to an increase in the plastic limit or a reduction of the liquid limit depends on the type of soil.

Strength. The strength of cement-stabilized cohesionless soil increases with higher densities. For cohesionless soils with and without cement, water content and method of compaction are also important. Other factors, such as the time elapsed between mixing and compaction, length of curing, temperature, humidity, and specimen size should also be considered when comparing laboratory test results.

A most comprehensive review of the strength properties of cement-stabilized soils was presented by Mitchell (1976). The following comments on strength are largely drawn from his paper.

As illustrated schematically in Fig. 13.2, the unconfined compressive strength q_u is generally described as increasing linearly with the cement content percent C. This increase is more pronounced for coarse-grained soils than for silts and clays Like q_u, other strength parameters such as the cohesion intercept and the friction angle

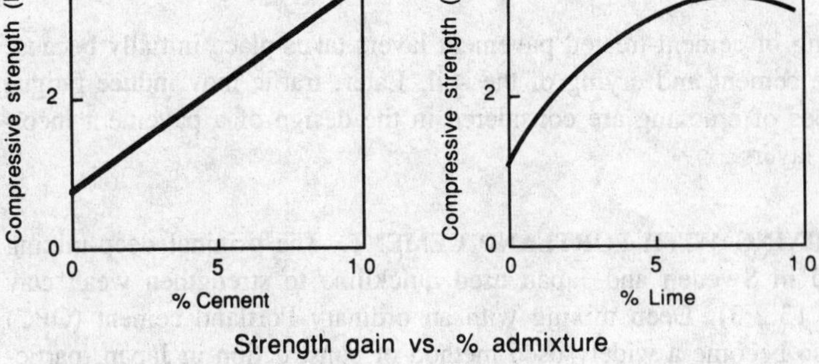

FIGURE 13.2

Gain in strength versus percent of admixture—typical characteristics for cement and lime.

increase with C and curing time. Mitchell (1976) gave the following relationship between q_u and the curing time:

$$q_u(t) = q_u(t_0) + K \log \frac{t}{t_0} \tag{13.3}$$

where $q_u(t)$ = unconfined compressive strength at t days, kPa
$q_u(t_0)$ = unconfined compressive strength at t_0 days, kPa
K = $480C$ for granular soils and $70C$ for fine-grained soils
C = cement content, % by mass

According to Mitchell, the flexural strength is on the order of one-fifth to one-third of the unconfined compressive strength. Extended to tension ($\sigma < 0$) and to large normal stresses ($\sigma > 70$) the failure envelope of compacted cement-treated soils plotted on a Mohr diagram appear curved; data presented by Mitchell suggests that a modified Griffith crack theory more accurately reflects the behavior of these soils than the traditional Mohr-Coulomb theory.

A relationship between the unconfined compressive strength and the California bearing ratio has also been suggested. According to NAASRA (1986) there are, however, limitations to the use of CBRs (as well as Texas triaxial tests) in pavement design with stabilized materials; they are said to be inapplicable to cement-bound layers. Also, in well-graded granular material, small additions of cement may give rise to inordinate increases in the CBR, which are inappropriate for design. As far as subgrades are concerned, NAASRA (1987) suggests that stabilized soils should not be assigned a CBR greater than 15.

Elastic properties are of particular interest to pavement design with cement-bound subbases and base courses. Typically the modulus of a granular soil can be increased through cement treatment from say 200 to 2000 MPa or 400 to 20 000 MPa. Poisson's ratio is in the range of 0.1 to 0.35, with coarser soils likely to lie in the lower part of this range of values.

Swelling and shrinkage. Even small additions of cement to an expansive subgrade soil significantly reduce shrinkage and swell, generally below 1%. Cement also provides stability against freeze-thaw cycles and repeated wetting and drying.

Cracking. Cracking of cement-treated pavement layers takes place initially because of hydration of the cement and drying of the soil. Later, traffic may induce fatigue cracking. Both types of cracking are considered in the design of a pavement incorporating stabilized layers.

13.2.2.3 DEEP MIXING WITH PORTLAND CEMENT. The original deep-mixing methods developed in Sweden and Japan used quicklime to strengthen weak clay deposits (see Sec. 13.2.3). Deep mixing with an ordinary Portland cement (OPC) slurry has since also become a widely used method of construction in Japan, particularly in marine clays.

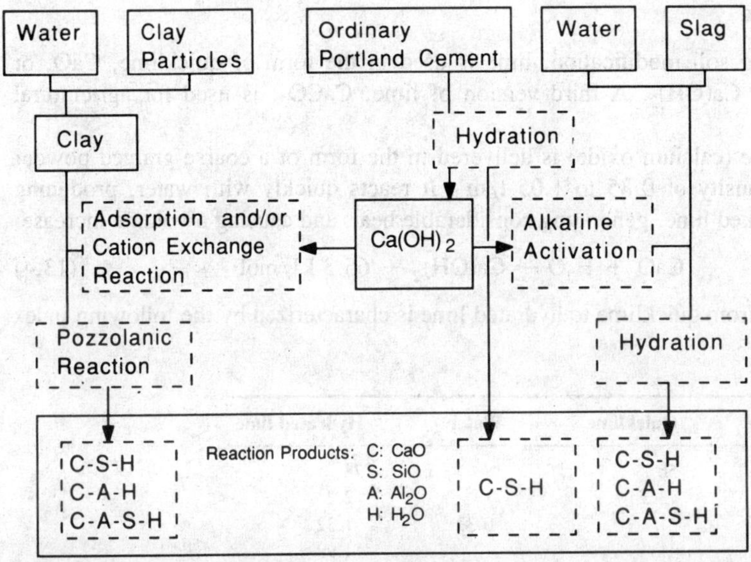

FIGURE 13.3
Chemical reaction between soil and hardening agents. [*After Saitoh et al. (1985).*]

The effect of adding OPC and blast furnace slag to several marine clays and other soils was investigated by Saitoh et al. (1985), who identified the following reactions between soil, cement, and slag, shown diagrammatically in Fig. 13.3:

(*a*) Hydration of the OPC, producing $Ca(OH)_2$. (The calcium hydroxide generated equals up to 25% of the weight of the cement.)
(*b*) Adsorption of the $Ca(OH)_2$ by the clay (or a cation exchange reaction).
(*c*) If and when the clay is saturated with $Ca(OH)_2$, a pozzolanic reaction between these two components occurs.

The end result is that an improved soil is obtained which contains hardened cement particles and also soil particles hardened by their pozzolanic reaction with products of the hydration process.

The success of deep mixing varies with the type of soil being treated, more specifically on their calcium hydroxide adsorption capacity and their pozzolanic reactivity. Some soils show a marked increase in strength once enough $Ca(OH)_2$ is available to cause a pozzolanic reaction. Saitoh et al. (1985) recommend solidification tests using slaked lime in order to investigate the pozzolanic reactivity of a clay.

Saitoh et al. (1985) also concluded that from the point of view of its wide applicability with respect to soil types, deep mixing with OPC, blended with blast furnace slag, is more advantageous than quicklime and slaked lime.

13.2.3 Lime

For engineering soil modification, lime is used in the form of quicklime, CaO, or hydrated lime, $Ca(OH)_2$. A third version of lime, $CaCO_3$, is used for agricultural purposes only.

Quicklime (calcium oxide) is delivered in the form of a coarse-grained powder with a bulk density of 0.85 to 1.05 t/m^3. It reacts quickly with water, producing hydrated or slaked lime, generating considerable heat, and causing a volume increase:

$$CaO + H_2O \rightarrow Ca(OH)_2 + 65.3 \text{ kJ/mol} \tag{13.4}$$

The transition from quicklime to hydrated lime is characterized by the following index properties:

	Quicklime	Water	Hydrated lime
Molecular weight	56	18	78
Specific gravity	3.3	1	2.2
Relative weight	1	0.32	1.32
Relative volume	1		1.99

On a construction site, the slaking process typically creates clouds of steam. Quicklime must be handled with care; it will burn exposed skin when moisture is present. It may also cause corrosion of equipment.

Slaked lime is used in the form of a fine powder with a bulk density or 0.45 to 0.6 t/m^3 or as a slurry with a water content of 80 to 100%.

Quicklime is more cost-effective than slaked lime in terms of handling and transport. As the figures above illustrate, it could be said that a truckload of quicklime has 25% more lime available for the reaction with soil than slaked lime!

Impurities such as silica, alumina, or carbonates may reduce the reactivity of commercial lime but are not harmful. Dolomitic lime, which contains significant amounts of magnesium oxide, is not as effective as calcium lime.

13.2.3.1 SOIL-LIME REACTIONS. Short-term reactions include hydration (for quicklime) and flocculation (ion exchange). Longer-term reactions are cementation and carbonation.

Hydration: Quicklime will immediately react with the water in the soil. This drying action is particularly beneficial in the treatment of moist clays. In the placement of lime columns and layers, the heat generation and expansion of the lime further enhance the consolidation effect.

Flocculation: When lime is mixed with clay, sodium and other cations adsorbed to the clay mineral surfaces are exchanged with calcium. This change in the cation exchange complex affects the way the structural components of the clay minerals are connected together. Lime causes clay to coagulate, aggregate, or flocculate. The clay's plasticity (measured in terms of Atterberg limits) is reduced, making it more easily workable and potentially increasing its strength and stiffness.

Cementation: This second stage of clay-lime reaction removes silica from the clay mineral lattice to form products not unlike those of cement hydration [see Ingles (1970) and Diamond and Kinter (1965)]. The structure of the stabilized clay was described by Herzog (1967) as an assembly of hard-skinned, lime-poor lumps of clay embedded in a lime-rich, fine-grained soil matrix.

Cementation is the main contributor to the strength of the stabilized soil. The higher the surface area of the soil, the more effective is this process; note that lime is not suitable for improving clean sands or gravels. Cementation is, however, limited by the amount of available silica. Increasing the quantity of lime added will increase strength only up to the point where all the silica of the clay is used up; adding too much lime can actually be counterproductive. This is in contrast to stabilization with cement, where strength continues to improve with the amount of admixture (see Fig. 13.2).

Cementation on the surface of clay lumps causes a rapid initial strength gain, but further diffusion of the lime in the soil will bring about continued improvement in the longer term, measured in weeks or months.

Carbonation: Reaction of lime with carbon dioxide in the open air or in voids of the ground forms a relatively weak cementing agent. This may be beneficial where lime is plentiful; the $CaCO_3$ formed will not react any further with the soil.

13.2.3.2 EVALUATION OF SOIL REACTIVITY. Lime is primarily used for the treatment of clayey soils. It is not very effective for cohesionless soils unless other materials are also added, such as fly ash, furnace slag, or other pozzolans.

A soil where lime treatment leads to cementation is termed a "reactive" soil. According to Dunlop (1977), a reactive soil will give CBRs in excess of 100 or unconfined strengths higher than 0.86 MPa after lime treatment. The most reactive clays are those containing minerals belonging to the montmorillonite group, distinguished by a three-layer primary element in their crystalline structure and a high base exchange capacity. Less reactive are illites, kaolonites, and chlorites. Despite this knowledge, it is still prudent to carry out conventional laboratory soil testing in order to evaluate the reactivity of a particular material, rather than rely on physicochemical index properties, such as surface area and exchange capacity.

A quick indication of the soil response to lime can be gained from measuring the penetration resistance of a compacted sample of the original and treated soil with a Proctor needle. This may, however, not be very helpful in identifying soils with relatively low reactivity [Dunlop (1977)].

Another test consists of submerging a specimen of treated and untreated soil in water. If the original soil disintegrates, but the lime additive helps to maintain its shape, then some cementation must occur and the soil is at least partially reactive.

Practical lime admixtures range from 2 to 8%. Optimal and/or most-economical mixes should be determined using the appropriate performance-related laboratory procedures, such as the CBR or Texas triaxial tests. Moist curing for 28 days before testing is desirable, but in industry CBR samples may only be cured for 1 to 4 days prior to soaking and testing.

13.2.3.3 ENGINEERING BENEFITS OF LIME STABILIZATION. The main benefits of lime stabilization of clays are improved workability, increased strength, and volume stability. Probably of lesser practical significance is a possible increase in permeability.

The following paragraphs describe typical changes in soil properties initiated by lime. There are exceptions to the rule, particularly where organic materials or chemicals such as sulfates are present.

Workability is improved because flocculation makes the clay more friable; this assists comminution for effective mixing and compaction. The plasticity of the soil decreases, mainly because of an increase in the plastic limit; the liquid limit may increase or decrease, depending on the type of soil (Fig. 13.4). The immediate gain in strength upon mixing is advantageous where there is a need to quickly improve the trafficability of soft clayey ground. Lime makes the soil firmer and more water-resistant.

Lime increases the optimum water content for compaction, which is an advantage when dealing with wet soil. Flocculation and cementation will make the soil more difficult to compact; therefore, the maximum dry density achieved with a particular compactive effort is reduced (Fig. 13.5). The compaction curve for lime-treated clay is generally flatter, which makes moisture control less critical and reduces the *variability* of the density produced.

In the first few hours after mixing, lime additives cause a steady increase in strength, but at a slower rate than cement. The need for compaction immediately after mixing is therefore less critical for lime than cement (Fig. 13.6).

Lime increases the *strength* of a clayey soil, typically demonstrated in terms of the unconfined compressive strength or CBR test results. Care has to be taken in comparing strength values of different specimens because of the many variables involved: size and height/diameter ratio, density or compactive effort, water content,

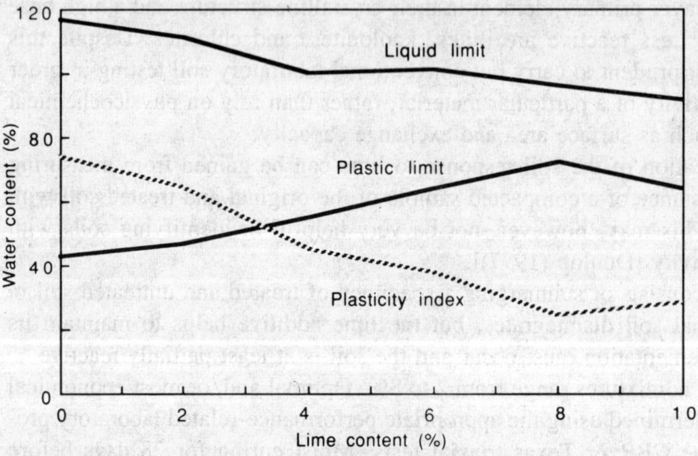

FIGURE 13.4
Effect of lime on the plasticity of a black soil (Inverell, Australia).

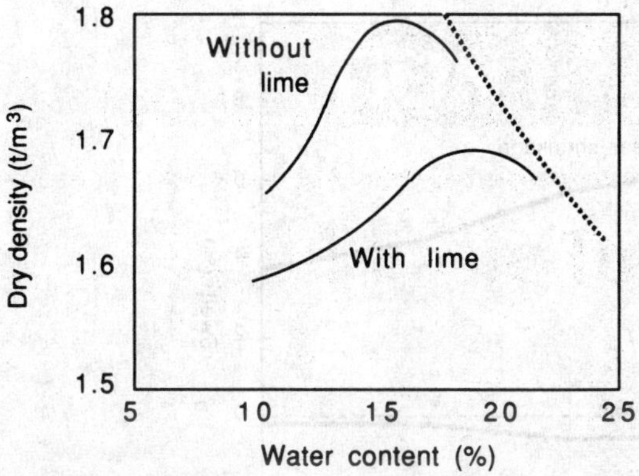

FIGURE 13.5
Typical effect of lime on the compaction curve of a low plastic clay.

curing time before and after compaction, temperature and humidity during curing, type and amount of lime, and more.

Related to strength is improved durability under traffic or resistance to the action of water, wind, and freeze-thaw cycles. Low-volume unsealed country roads on clay ground can be greatly improved by stabilizing the top 150 to 200 mm with lime; wheel path rutting and potholing is reduced, resulting in lower maintenance costs.

Thomson (1968) showed that the unconfined compressive strength (UCS) is a good indicator of other strength parameters and stress-strain moduli. Like the UCS, both the friction angle and the cohesion intercept increase with the lime content percent *L*. Tensile strength also improves with *L*, except at very low percentages, where the soil is dehydrated and crumbles but not enough lime is available to start the cementation process.

Improved *volume stability* means reduced shrinkage and swell upon drying and wetting, respectively. A typical result of a shrinkage and swelling test is shown in Fig. 13.7. It is not always straightforward to obtain comparable initial conditions for volume change tests. Linear shrinkage, e.g., is generally measured from an initial state equivalent to the liquid limit. Lime changes the Atterberg limits, and it then becomes debatable at which initial water content the soil should be prepared in order

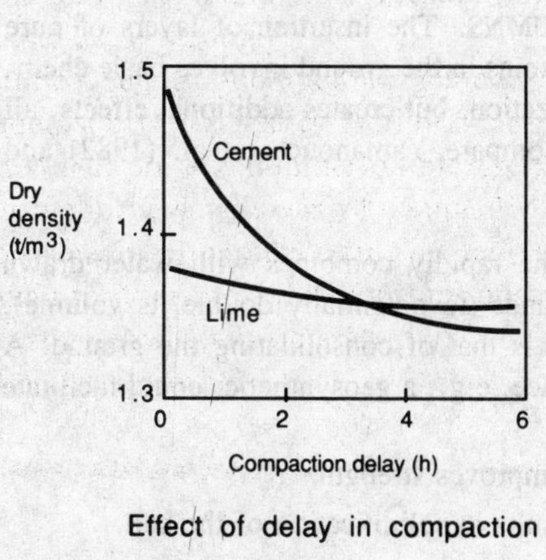

FIGURE 13.6
Effect of compaction delay in cement and lime stabilization—schematic only. [*Compare Ingles and Metcalf (1972).*]

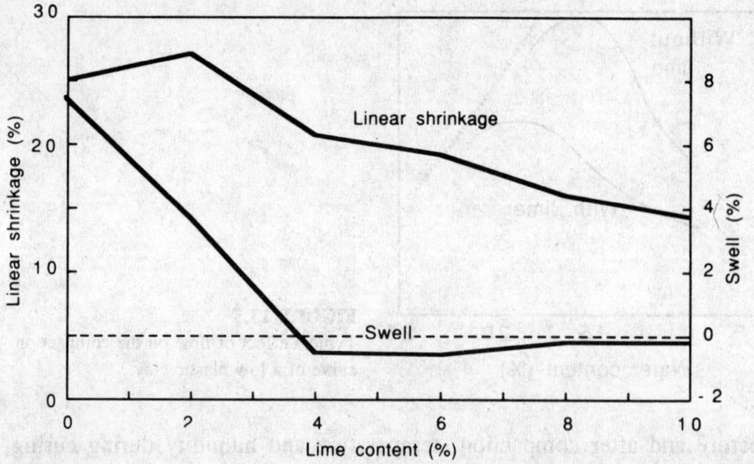

FIGURE 13.7
Effect of lime on the linear shrinkage and swelling of a compacted black soil (Inverell, Australia).

to properly reflect the effect of various amounts of admixture; note that the linear shrinkage test is an index test and does not simulate field conditions.

An increase in *permeability* could mean better drainage and less pore pressure buildup under load. Too high a permeability may, however, lead to softening of the unstabilized soil below. Lime is one of the few admixtures which has the potential to increase the permeability of a clayey soil, while there are many techniques of ground modification available which can reduce the permeability of ground, such as bituminous admixtures, chemical grouts, and geomembranes. Lime makes soil pores larger (permeability is increased!),and thus *frost* penetration is generally reduced.

Brandl (1981) indicated that the frost heave of a soil may be larger if it is lime-treated, if the soil is frozen within 1 month of compaction; but frost heave is less if the stabilized soil is allowed to cure for say 3 months or more before freezing.

13.2.3.4 QUICKLIME LAYERS AND COLUMNS. The insertion of layers of pure quicklime in embankments and pure lime columns in the ground involves basic chemical reactions, as described for surface stabilization, but creates additional effects, all of which may be summarized as follows [compare Yamanouchi et al. (1982) and Kitsugi and Azakami (1982)]:

1. *Dehydration and consolidation*. Quicklime rapidly combines with water drawn from the surrounding soil. The lime expands (to potentially double its volume), and heat is generated. The overall effect is that of consolidating the ground. A filter-drainage layer at the lime-soil interface, e.g., a geosynthetic, could facilitate the consolidation process.
2. *Reduction of soil plasticity*. This in turn improves strength.
3. *Longer-term pozzolanic reaction*. This results in solidification of the soil.

4. *Longer-term reaction of the lime with carbon dioxide contained in air voids in the soil.* In embankments this may increase the rigidity of the lime layer or column to provide a kind of structural reinforcement.

The drying out of the ground with lime columns has been compared with the effect of preloading. In both cases, settlement under future buildings will be reduced.

Water content reduction due to lime hydration. Kitsugi and Azakami (1982) indicated that the reduction in moisture content Δw in the soil surrounding quicklime piles could vary from a few percent to as high as 30%, decreasing with distance from the lime, but noticeable as far as 1 m away. Their field measurements indicate typical moisture reductions from 50 to 40% or from 100 to 80%, associated with a two- or threefold strength increase. The equivalent preload would have been on the order of 100 kPa.

Given the original water content w_0 of a soil and the ratio a_s of the mass of lime added per mass of soil solids, and knowing that 1 kg of CaO absorbs 0.32 kg of water through hydration, the water content w' after treatment can be calculated from

$$w' = \frac{M_w - 0.32a_sM_s}{M_s + 1.32a_sM_s} = \frac{w_0 - 0.32a_s}{1 + 1.32a_s} \tag{13.5}$$

and the reduction in water content Δw is

$$\Delta w = w_0 - w' = w_0 - \frac{w_0 - 0.32a_s}{1 + 1.32a_s} \tag{13.6}$$

where M_w = mass of water
M_s = mass of soil solids
Δw = average reduction of soil water content (ratio)
w_0 = original soil water content
w' = water content of the soil-lime system after slaking
a_s = mass ratio of lime to soil

According to the above expression, an initial water content of 50% would be reduced to 41% if the ground has a lime content equivalent to 10% of the dry mass of the soil.

Water loss due to heat generation. The heat produced in the hydration of lime accelerates the development of cementation, and thus strength, but it also has a drying out effect by evaporating water. The quantity of heat given off by 1 kg of quicklime during slaking is

$$C_s = \frac{C_m}{M_m} M = \frac{65.3}{56} 1000 = 1166 \frac{\text{kJ}}{\text{kg}} \tag{13.7}$$

where C_m = molar heat capacity of quicklime
M_m = molecular mass, g
M = 1000 g

As a theoretical exercise, we could calculate the amount of water ΔM_w which could be heated from 20 to 100°C and then evaporated by the energy released by slaking:

$$\Delta M_w = \frac{C_s a_s M_s}{L_w + \Delta T \; C_w} \tag{13.8}$$

$$= \frac{1166}{2260 + 80 \times 4.2} a_s M_s = 0.45 a_s M_s$$

where L_w = heat of vaporization, kJ/kg
C_w = heat capacity of water, kJ/(kg · °C); (see Chap. 15 for a definition of these terms.)

$$\Delta w = \frac{\Delta M_w}{M_s} = 0.45 \; a_s \tag{13.9}$$

For a_s = 10%, the maximum theoretical reduction in water content through evaporation would therefore be 4.5%.

The drying or consolidation process is further enhanced by capillary water absorption by the hydrated lime and the compressive stress exerted by the expanding lime layer or column, but both of these phenomena are more difficult to assess.

13.2.3.5 LIME-SOIL COLUMNS CREATED BY DEEP MIXING. Pure unslaked lime columns were developed in Japan in the 1960s and are now widely used there. Concurrently, a lime-soil piling technique was perfected in Sweden [Broms and Bowman (1978); Broms (1982)]; this construction method is most appropriately described as a deep mixing technique, although the piles produced are often described as lime columns.

Kujala, Halkola, and Lahtinen (1985) reported on the experience gained on seven projects of deep mixing with quicklime. In situ shear strength was measured using a specially developed penetrometer and vane-bore apparatus. Stress-strain moduli were determined from screw-plate compressometer tests and backfigured from observed settlement. Strength as well as moduli increased significantly. It was noticed that if the strain exceeded 1%, significant creep occurred; this was attributed to the compression of air- and water-filled voids and to the fracturing of the microstructure of the lime-soil column.

When compared with laboratory results, field measurements showed larger gains in performance. This was attributed to the effect of the heat generated in the field, which reached a maximum value of 240°C; increased ground temperature was still evident some 2 weeks after construction. (In pure lime columns, temperatures go even higher, into the range of 300 to 400°C.)

The shear strength of the lime-soil columns was still increasing 1000 days after their installation. In the Helsinki area the ground improvement was consistent provided the sulphur content was less than 0.1% and organic matter (humus) did not exceed 2%.

13.2.4 Calcium Chloride

Calcium chloride ($CaCl_2$) has been used in highway construction and maintenance since early this century. It becomes available as a by-product from making sodium carbonate or from other chemical processes. In Australia it is obtained by reacting lime or limestone with hydrochloric acid, which is a by-product in the manufacture of fluorocarbons; it is available under the trade name PACWET as a 29% solution.

13.2.4.1 PHYSICAL PROPERTIES. Calcium chloride is an inorganic salt with a number of physical properties which can be of advantage in geotechnical engineering.

Hygroscopicity: Calcium chloride is hygroscopic; that means it attracts and absorbs moisture from the atmosphere. This is a function of the relative humidity and the temperature. At 25°C and 90% humidity, 1 kg of $CaCl_2$ takes up more than 4 kg of water.

Deliquescence: A deliquescent substance is one that liquifies in moisture of its own absorption. Again, temperature and humidity determine this behavior.

Solubility: Calcium chloride is highly soluble; 59.5 g dissolve in 100 mL of water at 0°C, more at higher temperatures. According to Slesser (1943), the solubility of a substance in water is a major factor in determining the extent to which the vapor pressure, density, surface tension, and freezing point of the water can be altered by the addition of the substance.

Vapor pressure: This is the tendency shown by a substance to pass from the liquid or solid state into the gaseous state. For the same humidity and temperature, the vapor pressure of a calcium chloride solution is always lower than that of water [Burggraf (1933)]. Adding calcium chloride thus results in a lowering of the rate of evaporation.

Surface tension: Calcium chloride has higher surface tension than water.

Freezing point: A $CaCl_2$ solution has a lower freezing point than water. Complete freezing occurs at −51°C. At lesser subzero temperatures only pure water freezes out, increasing the concentration of the remaining solution.

13.2.4.2 EFFECT ON SOIL PROPERTIES. Calcium chloride has several physicochemical effects on the fine-grained component of a soil. If, e.g., sodium ions (Na^+) are present in the exchangeable cation complex around the negatively charged clay platelets and they are replaced by Ca^{++}, the thickness of the diffuse double layer (see Sec. 12.1.1) is reduced. This could mean lower plasticity and increased strength. It has also been suggested that $CaCl_2$ reduces intergranular repulsion and strengthens molecular bonds between particles, at least in some types of soil.

Figure 13.8 shows the results of compaction tests on a gravelly clay with and without $CaCl_2$. In this example, calcium chloride aided compaction, an effect which is variously attributed to increased surface tension of the moisture films, a lubrication effect, and/or increased density of the pore water. However, in some soils, compaction becomes more difficult after the addition of $CaCl_2$.

Kezdi (1979) states that $CaCl_2$ generally decreases strength. Others [Thornburn and Mura (1969)] report increased CBRs after calcium chloride treatment.

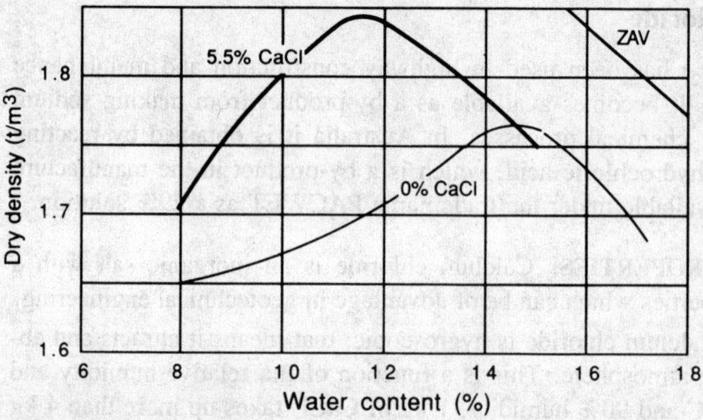

FIGURE 13.8
Compaction curves of a gravelly clay with and without calcium chloride. *(Courtesy Pacific Chemical Industries Pty. Ltd., 1983.)*

There is agreement, however, on the ability of $CaCl_2$ to reduce evaporative water losses from soils. This characteristic facilitates moisture control during construction and helps in the control of dust generated on unpaved roads.

13.2.4.3 ENGINEERING USES. During the period between 1930 and 1945 more papers were written about the use of calcium chloride in geotechnical engineering than at any other time before or since [Slate and Johnson (1958)]. Interest then waned, at least partly due to concerns about corrosion to motor vehicles, damage to vegetation along a treated road, and the salt's possible leaching out or migration into lower soil layers. The concern about corrosion does not seem to have arisen from soil stabilization work but from uses of $CaCl_2$ in melting snow and ice on highways.

Today, $CaCl_2$ still has a variety of uses, but it is probably most appreciated as a dust palliative on highly trafficked unpaved roads, such as haul roads in mining and on large earth-moving projects. Dust causes problems of poor visibility, increased vehicle wear, and driver fatigue and adversely affects the environment. If the road surface is kept moist, less dust is created; fines which provide cohesion to the road surface are not lost, reducing maintenance requirements. Calcium chloride helps to keep up a higher moisture content for a longer period than would exist in untreated soil, as the more than 50-year-old records of Burggraf (1933) show (Fig. 13.9).

In cold regions, $CaCl_2$ can assist in thawing frozen soils and thus make their compaction possible. The strength reduction after repeated freeze-thaw cycles is less with $CaCl_2$ treated soils [Kezdi (1979)].

Calcium chloride as a secondary additive can also bring benefits in cement or lime stabilization by increasing early strength values. Only 0.5 to 1.5% of $CaCl_2$ may be needed for this purpose.

Miscellaneous geotechnical uses of calcium chloride are found in electrokinetic stabilization (see Sec. 12.1.5) and grouting (Joosten process, Sec. 14.3).

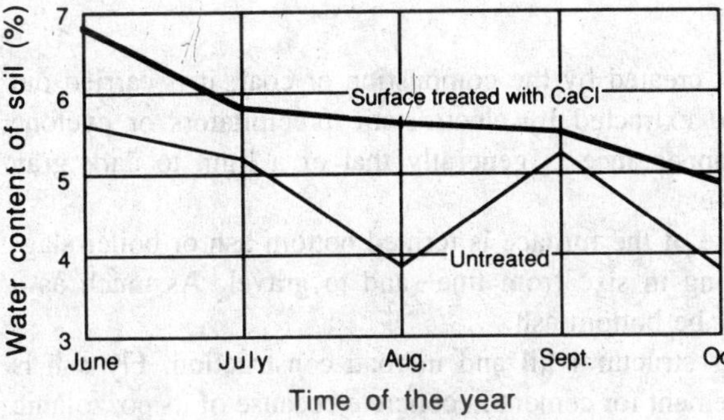

FIGURE 13.9
Moisture loss from a sandy clay. [*Burggraf (1933)*.]

13.2.5 Other Chemical Admixtures

Sodium chloride (NaCl) has similar properties to $CaCl_2$ and is the cheaper product of the two, but its overall benefits are generally rated lower. Brandl (1981) maintains that $CaCl_2$ as well as sodium chloride reduce the unconfined compressive strength and with it almost all other important engineering properties.

According to Brandl (1981), *sodium hydroxide* (NaOH) and hydrated lime ($Ca(OH)_2$) cause similar changes to a fine-grained soil, such as reducing its plasticity and increasing its strength through long-term reactions. Ingles (1970) found sodium hydroxide particularly effective in aluminous soils but warned of its caustic nature. As a solution in water, sodium hydroxide may assist compaction, increasing the density achieved with a particular compactive effort.

Gypsum ($CaSO_4 \cdot 2H_2O$) alone is not effective as a soil stabilizer for engineering purposes, although it has agricultural uses in treating topsoil. However, Kujala (1983) showed that cheap waste gypsum can enhance the stabilizing properties of unslaked lime by accelerating its reaction with the soil in deep mixing. This process has been called *gypsum-lime* stabilization.

Sodium silicate is known to reduce the plasticity index of a clay. NAASRA (1986) indicates that additions up to 2% by mass have been considered effective in road engineering, but the advantages are small.

Lignosulfonate derivatives, such as chrome-lignin are by-products of the paper industry (lignin is the natural cement in wood). They can be used as a binder in road materials. As a dispersing agent, they can make a clayey soil less permeable and reduce moisture evaporation. Concern has been expressed about the toxic nature of hexavalent chromium which is used as a catalyst in lignosulfonate stabilizers.

Most chemicals used in grouting (Chap. 14) are also suitable for surface stabilization, but their cost is likely to be prohibitive. These include aniline-furfural, calcium acrylate, lignosulfonate resins, and many proprietary products of unknown composition.

13.2.6 Fly Ash

Fly ash is a solid waste product created by the combustion of coal; it is carried out of the boiler by flue gases and extracted by electrostatic precipitators or cyclone separators and filter bags. Its appearance is generally that of a light to dark gray powder of predominantly silt size.

Ash removed from the base of the furnace is termed bottom ash or boiler slag. It is coarser than fly ash, ranging in size from fine sand to gravel. As much as a quarter of the ash produced may be bottom ash.

Bottom ash serves well as structural fill and in road construction. Fly ash is regularly used as a partial replacement for cement in concrete because of its pozzolanic properties; it is also the form of ash which has the greatest potential for use in ground modification.

In 1986 some 65 to 70 million metric tons of fly ash were produced in the United States alone. Only 15 to 20% of this massive amount was used constructively; less than half of that was used in the manufacture of concrete. The rest is pumped in slurry form into lagoons or is conditioned by the addition of 10 to 15% water and disposed of as more or less engineered landfills.

Making more productive use of fly ash would have considerable environmental benefits, reducing land, air, and water pollution. Increased use as a partial cement or lime replacement would also represent a savings in energy (fly ash has been called a high-energy waste material).

Besides using fly ash alone as a structural fill material, scope exists for employing techniques of ground modification to find more medium- to high-volume applications in the following ways:

Add cement or lime to stabilize the fly ash
Stabilize soils with cement–lime–fly-ash mixes
Use fly ash in the containment of toxic wastes (see Sec. 13.4)

The Electric Power Research Institute has produced a comprehensive design manual for the use of fly ash in structural fills and highway embankments and for subgrade stabilization and land reclamation [EPRI (1986)]. Another good source of information are the proceedings of conferences organized by the American Coal Ash Association, which provide a regular update in fly ash technology.

13.2.6.1 PROPERTIES OF FLY ASH

Chemical composition and reactivity. A microscopic view of fly ash reveals mainly glassy spheres with some crystalline and carbonaceous matter. The principal chemical constituents are silica (SiO_2), alumina (Al_2O_3), ferric oxide (Fe_2O_3), and calcium oxide (CaO). Other components are magnesium oxide (MgO), titanium oxide (TiO_2), alkalies (Na_2O and K_2O), sulphur trioxide (SO_3), phosphorous oxide (P_2O_5), and carbon (related to the "loss-on-ignition"). Water added to fly ash usually creates an alkaline solution, with a pH in the range from 6 to 11.

Fly ash is a heterogeneous material. Factors affecting the physical, chemical, and engineering properties of fly ash include:

Coal type and purity
Degree of pulverization
Boiler type and operation
Collection and stockpiling methods

There is no single chemical or physical property which gives a reliable indication of the pozzolanic reactivity of fly ash. Cementitious calcium silicate and calcium aluminosilicate hydrates are formed when the glassy components of the fly ash ($3Al_2O_3 \cdot SiO_2$ or "mullite") react with water and lime. Critical to the pozzolanity of fly ash are conditions such as

Amount of silica and alumina in the fly ash
Presence of moisture and lime
Fineness of the fly ash (surface area)
Low carbon content

The degree of self-hardening of ash is also highly dependent on the ash's density, temperature, and age.

ASTMC618 distinguishes between class F and class C fly ash. Class F fly ash is normally produced from burning anthracite or bituminous coal; it has pozzolanic properties, which means that it will react with lime to form cementitious compounds. Class C fly ash is normally produced from burning subbituminous or lignite coal; in addition to being pozzolanic, it has cementitious properties of its own.

Engineering properties. The specific gravity of the ash particle ranges from 1.9 to 2.5, which is below that normally measured for soil solids. Some of the ash particles may actually float if they consist of hollow glass spheres (cenospheres); these have numerous industrial applications.

The average grain size D_{50} of fly ash is likely to be in the range of 0.02 to 0.06 mm. Fly ash is nonplastic and, in a dry state as collected, completely cohesionless. This lack of cohesion makes nonhardening fly ash highly erodible. In a moist, unsaturated state, surface tension of the pore water gives fly ash an apparent cohesion; if and when pozzolanic reaction occurs, considerable unconfined compressive strength is observed, increasing with age. The friction angle as measured in consolidated drained triaxial tests is typically on the order of 30°, but values as low as 20° and as high as 40° have been reported.

As a guide, compacted ash may have a dry density anywhere between 1.2 and 1.9 t/m^3 and a corresponding optimum moisture content ranging from 30 down to 15%; however, more extreme values are also reported in the literature, such as γ_{dmax} = 0.7 t/m^3 and w_{opt} = 60%. Low compacted density points to a potential advantage in the use of fly ash as backfill or embankment material: Low unit weight means low overburden pressures and, combined with a high friction angle, also low earth pressures.

EPRI (1986) reports that the compression index C_c of fly ash can range from 0.05 to 0.37 for initial loading. In recompression, these values are much lower: 0.006 to 0.04. The compressibility of compacted ash must rate as small when compared with clayey soils.

Compacted dry fly ash may swell upon wetting if subjected to vertical pressures less than those equivalent to 0.5 to 1 m of fly ash fill. Goelen (1982) reported 11 to 14.5% free swell for a particular ash tested.

The permeability of a fly ash compacted to standard maximum dry density depends on the coal type it is derived from [EPRI (1986)]:

Coal type	Permeability of fly ash, cm/s
Bituminous	10^{-4} to 10^{-7}
Subbituminous	10^{-5} to 3×10^{-6}
Lignite	9×10^{-6} to 10^{-7}

Considerable capillary rise of water in fly ash fills can occur—on the order of 2 m, and possibly more.

Fly ash is classed as a frost-susceptible material, which is a major drawback in its use for road construction in regions with cold winters.

Negative environmental impacts from a fly ash fill are unlikely, but a study has to be made of the chemical composition of its leachate; its corrosivity on buried pipes, culverts, or other structural elements; and its radioactivity (radium-226).

13.2.6.2 FLY ASH STABILIZED WITH LIME, CEMENT, AND/OR AGGREGATE. The use of mixtures of lime (L) or cement (C) and fly ash (F) with aggregate (A) giving LFA, CFA, or LCFA bases or subbases for pavements is relatively well established in most countries. Guidelines for design and construction were given by Barenberg (1974) and others. Many local authorities have published criteria for the incorporation of pozzolanic materials with cement or lime in aggregate layers, either rated as bound or unbound layers, depending, e.g., on whether their indirect tensile strength is above or below 80 kPa [NAASRA (1986)].

To build a subbase or base course with lime- or cement-stabilized ash alone is not yet common, but this is one of the high-volume ash applications being promoted by ash producers.

Referring to British and American experience, EPRI (1986) quoted the following criteria as part of their design recommendation for a cement-stabilized fly ash base course:

Minimum strength: The 7-day unconfined compressive strength of the mix, when cured under moist conditions at 21 ± 2°C, must exceed 2.8 to 3.1 MPa for cylindrical specimens having a length to diameter ratio of 2:1. If cubical specimens are used, a strength of 3.5 MPa is recommended.

Maximum strength: An upper limit of strength of 5.5 MPa is advised to avoid distinct cracking which may reflect through the asphalt surface layer.

Aging criteria: The unconfined comrpessive strength of the mix is observed to increase with time.

Similar guidelines hold for lime-stabilized fly ash base courses, except that the design criteria refer to the 28-day, rather than the 7-day strength, because of the slower rate of cementation. The minimum strength required is also correspondingly higher (3.7 to 4.1 MPa). In some areas, standard strength tests must be complemented by the evaluation of durability, such as through freeze-thaw tests.

Figure 13.10 shows the compaction and strength characteristics of compacted Australian fly ash with the addition of cement or lime. For this (class F) fly ash, lime was ineffective as a stabilizer; considerable amounts of cement were needed to achieve strengths as would be required in a base course. In these tests, the ash-cement combinations were compacted within 5 to 10 min after the addition of water. Ash-lime-water mixtures were allowed to cure overnight before compaction.

13.2.6.3 SOILS MODIFIED WITH FLY ASH AND CEMENT OR LIME. For cohesionless soils or soils with very low plasticity (plasticity index < 10), cement will be more effective than lime, either alone or when combined with fly ash. For more plastic soils, either cement or lime may be added with fly ash. Only a soils testing program can indicate optimal mixes and relative economies. Fly ash could also serve as a filler in the bituminous stabilization of coarse-grained materials.

Figure 13.11 demonstrates the effect of fly ash on the density and strength of a cement-stabilized sand. The sand in question is of medium grain size ($D_{50} = 0.3$ mm), is fairly uniform (USCS classification SP), and is from the Woy Woy area, New South Wales. The Miniature Harvard Compaction Test was used in these experiments; this procedure allows easy preparation of a large number of specimens, but the density results may not be equivalent to Proctor compaction and the strength values may be affected by the small size of the specimen. The class F fly ash added acted primarily as a filler, enhancing the binding effect of the cement. All the materials were mixed in a dry state. Both, the density as well as the unconfined compressive strength showed maximum values when the mix was proportioned at around 20% fly ash to 80% sand.

Stabilization of a sandy road base with a fly-ash–cement mix, rather than cement alone, creates a less-permeable stiffer layer. This may result in reduced long-term maintenance. Initial financial benefits depend on local material and transport costs.

It has also been demonstrated that cement–fly-ash–sand or cement–fly-ash–gravel mixtures shrink less than soil-cement mixtures [Natt and Joshi (1984)]. Greater shrinkage is observed in these combinations if the cement is replaced by lime.

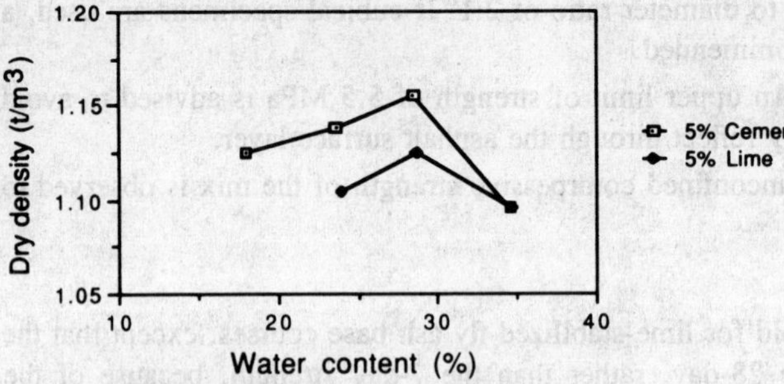

a. Compaction curves of fly ash with 5% additive

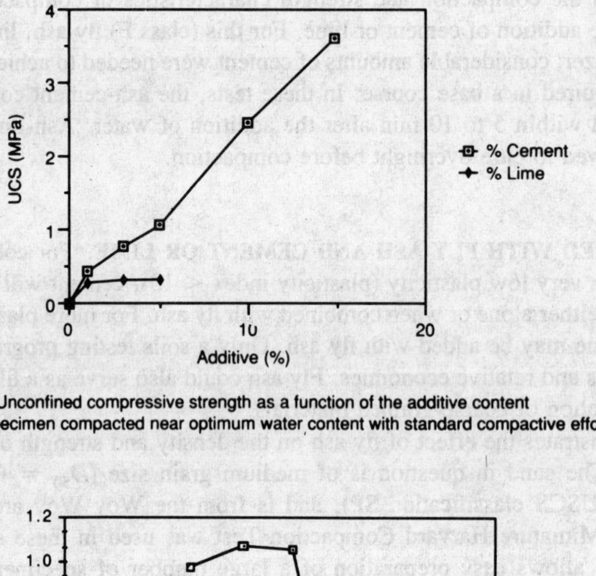

b. Unconfined compressive strength as a function of the additive content (specimen compacted near optimum water content with standard compactive effort).

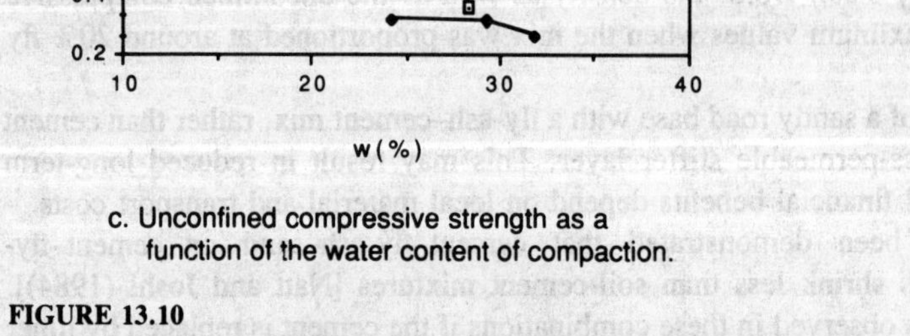

c. Unconfined compressive strength as a function of the water content of compaction.

FIGURE 13.10

Compaction and strength characteristics of (class F) fly ash from Vales Point power station (New South Wales).

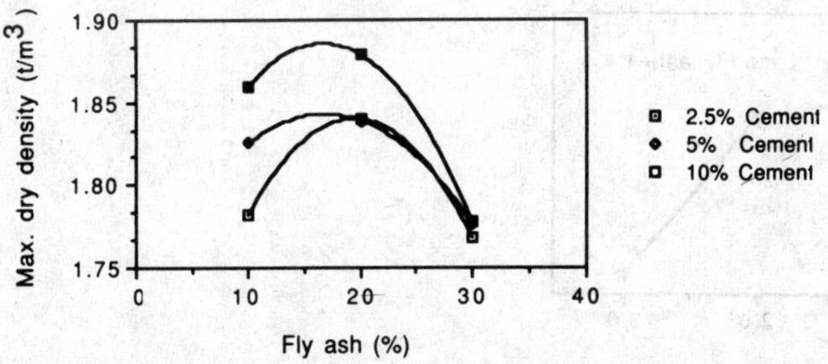

a. Maximum dry density of sand–fly-ash–cement mixes
10% fly-ash corresponds to 90% sand, while % cement is expressed
terms of the fly-ash–soil mix)

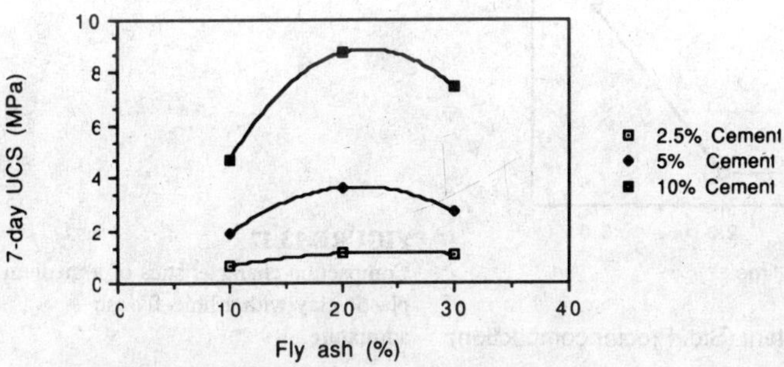

b. Unconfined compressive strength

FIGURE 13.11
Density and unconfined compressive strength of a medium sand–fly-ash–cement mix compacted to maximum dry density in a Miniature Harvard Compaction test mold.

Figure 13.12 presents some test results obtained with an inorganic clay of intermediate plasticity ($w_L = 45\%$, $I_p = 22\%$) from Gosford, New South Wales. The soil was air-dried and then broken down into small crumbs. The soil was premoistened for 24 h before lime, fly ash, and additional water were added immediately before compaction.

As Fig. 13.12 shows, lime and fly ash reduce the maximum dry density of clay. The corresponding optimum water content tends to increase, although results at low lime percentages ($\leq 2\%$) can be inconsistent. The unconfined compressive strength of this clay rose with the addition of fly ash to the lime (Fig. 13.13): indications are that additional strength gains could have been achieved if the lime–fly-ash content would have been increased further. In these experiments the lime–fly-ash mix formed a coating around the soil crumbs, which was visible as a light-colored matrix in the compacted specimen.

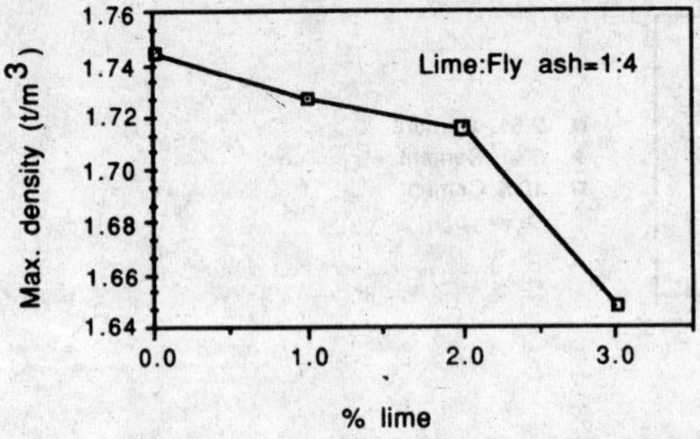

a. Maximum dry density (Std. Proctor compaction)

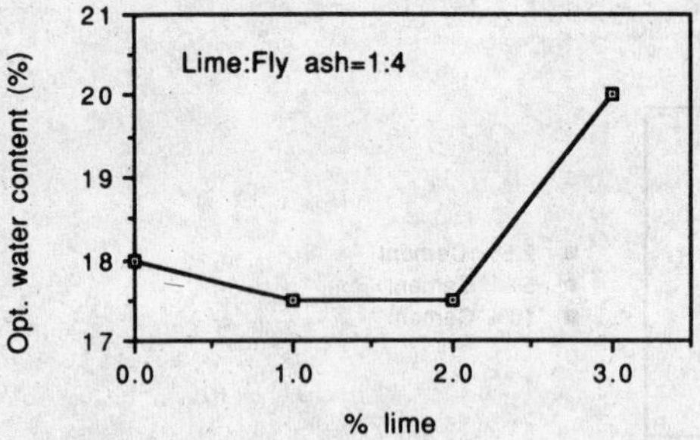

b. Optimum water content (Std. Proctor compaction)

FIGURE 13.12
Compaction characteristics of a medium plastic clay with a lime–fly-ash admixture.

13.2.7 Bitumen and Tar

13.2.7.1 TERMINOLOGY. *Bitumen* is a product obtained after processing the residue that remains after the distillation or evaporation of crude petroleum. *Tar* is the result of destructive distillation of coal or other carbonaceous material. *Asphalt* consists of

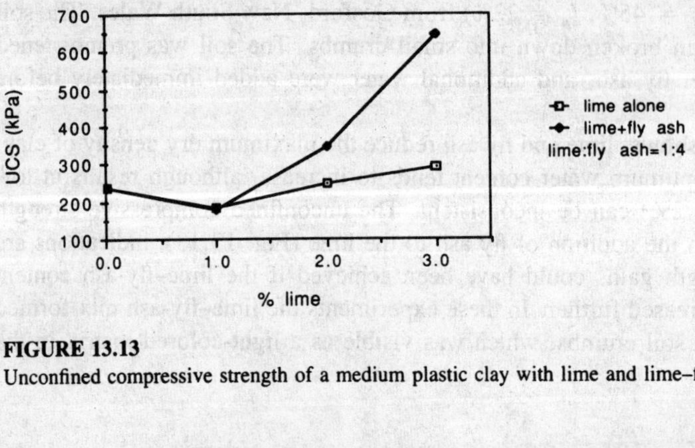

FIGURE 13.13
Unconfined compressive strength of a medium plastic clay with lime and lime–fly-ash additive.

inert mineral particles impregnated or cemented by bitumen. Nevertheless, the term "bitumen" is often taken to include both tar and asphalt. The use of all three materials in ground modification is collectively called *bituminous stabilization*. Its product is variously referred to as

Soil-bitumen (or soil-asphalt): A waterproofed cohesive soil (4 to 7% bitumen)

Sand-bitumen (or sand-asphalt): Sand particles cemented together by a bituminous admixture (maximum strength may be achieved with only 3 to 4%, but as much as 10% has been added in practice)

Waterproofed granular soil: A well-graded soil with some low plastic fines, waterproofed with small amounts of bitumen (1 to 2%)

Oiled earth: A bituminous emulsion or cutback is sprayed on the soil surface, providing water and abrasion resistance (4 to 5 L/m^2)

Bitumen is usually mixed into the soil in the form of an emulsion or cutback or as hot bitumen; the latter is added to the soil by the foaming or high-impact process.

In a bitumen *emulsion* small droplets (4 to 10 μm) are dispersed in water and are prevented from coagulation by chemical emulsifiers. Because most naturally occurring aggregates have negatively charged particle surfaces, *cationic* emulsions have more universal application than anionic emulsions. In a *cutback*, the viscosity of the bitumin is temporarily reduced by a volatile solvent, or "cutting oil," which evaporates after placement; naphtha, kerosene, and diesel oil can be used for this purpose, here listed in order of decreasing rate of curing. In the *foaming* process, steam is blown through hot bitumen or tar through specially designed nozzles, forming thin-film bubbles with excellent coating ability. In the *high-impact* process, the bitumen is dispersed through atomizing jets onto the soil.

Bituminous stabilization is generally used to

Waterproof soils or at least reduce water absorption

Add cohesion to granular soils

Most suitable for bituminous admixtures are sandy gravels, sands, clayey and silty sands, and fine crushed rock; but even highly plastic clays can be treated successfully, although higher quantities of bitumen may be required. Cohesive soils should be dry enough to be broken up readily; pretreatment with lime could be advantageous, particularly in acidic soils. In coarser soils, lime, fly ash, or other mineral fillers could reduce the optimal amount of bitumen needed for stabilization; lime also improves particle coating.

Bitumen is not as common as the other major stabilizers, lime and cement, mainly because of its relatively high costs. One should also be aware that considerable expertise is required in controlling temperatures (and viscosity!) in hot applications, in choosing correct proportions and mixing time for emulsions and cutbacks, and in optimizing curing rates and conditions.

13.2.7.2 ENGINEERING CHARACTERISTICS OF BITUMEN-STABILIZED SOILS. Of special interest are properties with respect to compaction, strength, and water absorption.

Kezdi (1979) as well as Ingles and Metcalf (1972) quoted results indicating that the *maximum dry density* achieved with a constant compactive effort falls with increasing bitumen content. Figure 13.14*a* illustrates that this is not necessarily true, at least not for tar-stabilized soils. The data plotted was obtained by Giffen et al. (1978)

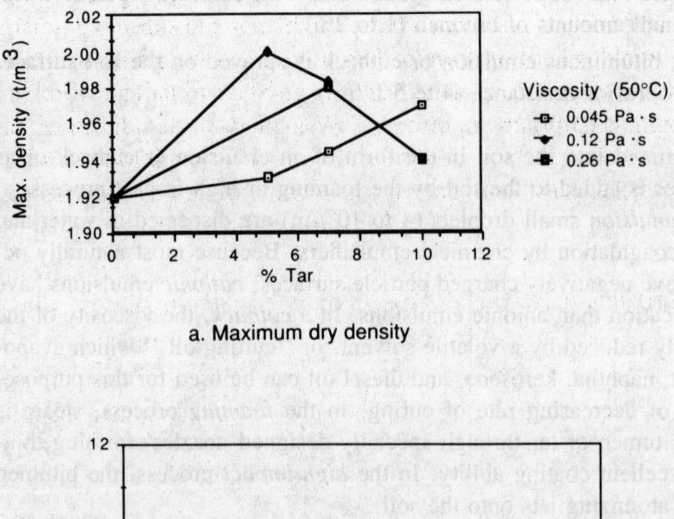

a. Maximum dry density

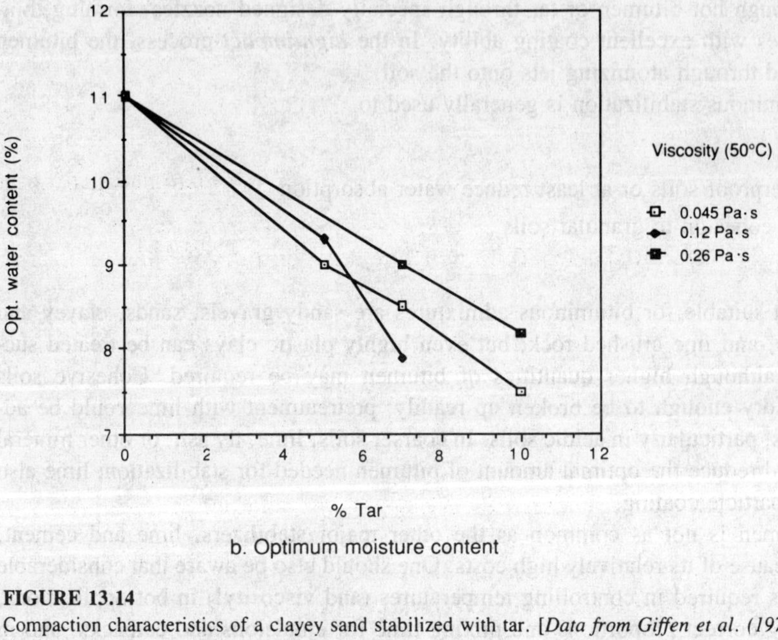

b. Optimum moisture content

FIGURE 13.14

Compaction characteristics of a clayey sand stabilized with tar. [*Data from Giffen et al. (1978).*]

who set out to prove that a so-called nonstandard gravel could be suitable as a base course, if stabilized with tar.

Figure 13.14*b* shows that the optimum amount of water necessary for reaching maximum density decreases with increasing tar content. Note, however, that the corresponding total amount of liquid added (water and tar) actually rises.

The *strength* of compacted bitumen-stabilized soil is most commonly measured in an unconfined compression test; alternatively, the California bearing ratio can be determined. Other strength values are also reported in the literature, such as the Hubbard field test and the Florida bearing value [see Ingles and Metcalf (1972)]. It is apparent that asphaltic pavement engineers may favor standards of compaction and testing other than those carried out routinely in general geotechnical engineering. The difference between bitumen-stabilized soils and asphaltic pavement surface layers is not very distinct, and an overlap between these two areas of expertise exists. A similar situation exists with respect to concrete technology and cement stabilization.

Figures 13.15 and 13.16 again refer to work carried out by Giffen et al. (1978). They show the degree of *water absorption* and the corresponding loss of strength as a function of the soaking period. In these experiments, the unstabilized soil collapsed when immersed in water.

Figure 13.16 demonstrates a typical finding in bituminous stabilization: Initially there is an increase in strength with the quantity of binder added until maximum stability is reached; thereafter increasing the bitumen content reduces strength. Unsoaked, the strength of bitumen-stabilized soils generally peaks at a lower binder content than with a soaked strength. As the bitumen content increases past the peak strength, the difference between unsoaked and soaked strength diminishes.

The effectiveness of bitumen in imparting cohesion and waterproofing significantly depends on the nature of the soil. As an example, the use of emulsions is said to be best for well-graded sands with a fines content in the range of 8 to 20%. Increasing fines content within this range means improved dry strength but reduced water resistance. Too many fines could present problems with mixing stability and uniformity. Lack of fines, combined with too rigorous mixing, could also result in an unstable mixture, causing a loss of adhesion or "stripping."

13.3 Admixtures in Pavement Design

The aim of pavement design is to provide the most economic support system for the traffic anticipated, considering the characteristics of the subgrade and the available construction materials. Not only the cost of initial construction but also longer-term maintenance costs have to be considered. A good design will also take into account the planned road cross section and the local environment, particularly expected changes in temperature and moisture content of the ground.

A comprehensive coverage of the principles of pavement design is clearly outside the scope of this book. This section is merely intended to give an idea of how stabilized layers can be incorporated in the pavement structure and how their effect can be analyzed. The discussion is restricted to the design of flexible pavements.

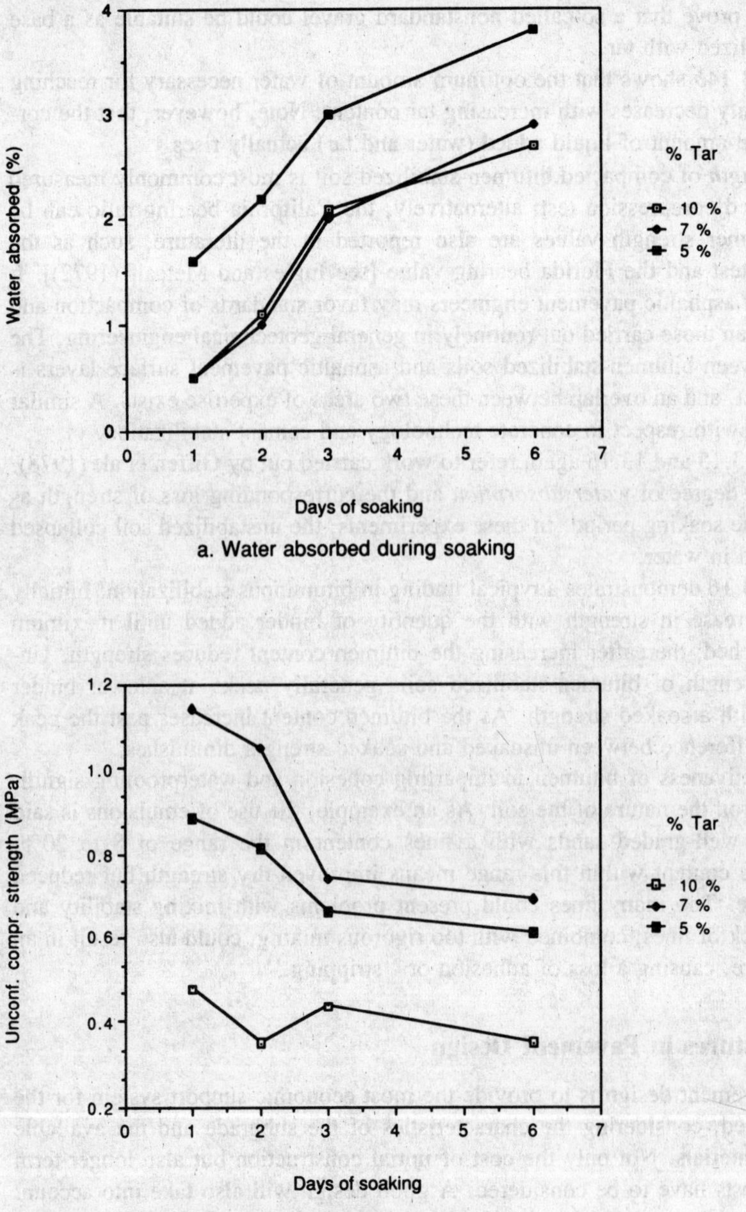

FIGURE 13.15

Water absorption and soaked strength of a clayey sand stabilized with tar of a viscosity of 0.045 Pa·s. [*Data from Giffen et al. (1978).*]

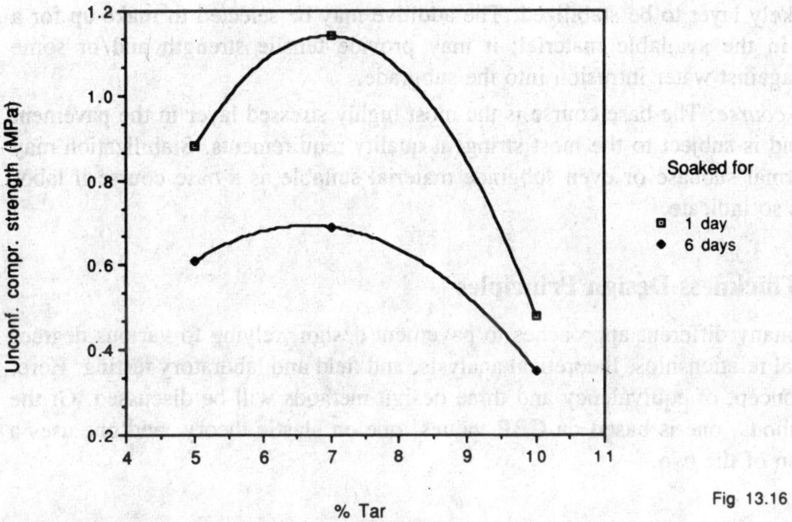

FIGURE 13.16
Soaked unconfined compressive strength of a tar-stabilized clayey sand. [*Data from Giffen et al. (1978).*]

13.3.1 The Pavement Structure

Figure 13.17 shows the typical structure of a flexible pavement. The pavement thickness includes all layers above the subgrade. Base, subbase, and subgrade may all contain stabilizing additives, such as cement, lime, and bitumen.

Subgrade: Admixtures may simply be used to improve the trafficability of the subgrade during the initial stages of construction. If the subgrade properties are significantly improved over say 100 to 150 mm in depth, this layer becomes a subbase and is part of the total pavement structure.

Subbase: A subbase generally consists of imported material, better in quality than the subgrade but not meeting the requirements for a base course. The subbase is

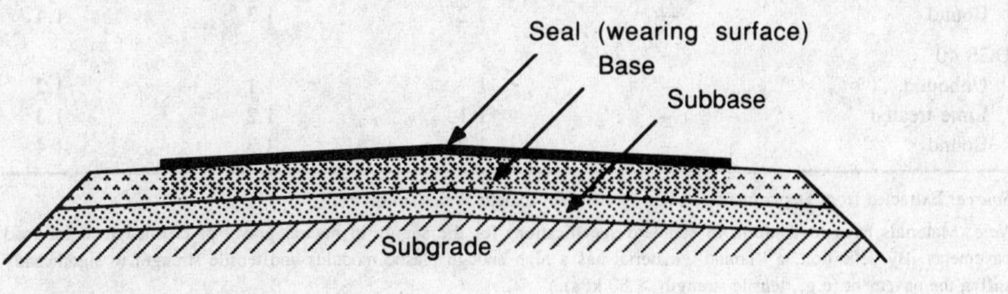

FIGURE 13.17
Typical cross section of a flexible pavement.

the most likely layer to be stabilized. The additive may be selected to make up for a deficiency in the available material; it may provide tensile strength and/or some protection against water intrusion into the subgrade.

Base course: The base course is the most highly stressed layer in the pavement structure and is subject to the most stringent quality requirements. Stabilization may make a normal subbase or even subgrade material suitable as a base course if laboratory tests so indicate.

13.3.2 Thickness Design Principles

There are many different approaches to pavement design, relying to various degrees on empirical relationships, theoretical analysis, and field and laboratory testing. Here, only the concept of equivalency and three design methods will be discussed. Of the design methods, one is based on CBR values, one on elastic theory, and one uses a combination of the two.

TABLE 13.1
Pavement layer equivalencies

	Depth below surface, mm			
Material classification	**0 to 100**	**100 to 200**	**200 to 300**	**300+**
NGB:				
Unbound	1.0	1.05	1.1	1.15
Bound	1	1.1	1.2	1.25
NGS:				
Unbound	—	1	1	1
Bound	—	1.1	1.15	1.2
DGB 20:				
Unbound	1	1.1	1.2	1.3
Lime-treated	1	1.15	1.3	1.45
Bound	1	1.2	1.4	1.6
DGS 20:				
Unbound	—	1	1.1	1.2
Lime-treated	—	1.1	1.2	1.3
Bound	—	1.2	1.3	1.4
DGS 40:				
Unbound	—	1	1.1	1.2
Lime-treated	—	1.1	1.2	1.3
Bound	—	1.2	1.3	1.4

Source: Extracted from Department of Main Roads, New South Wales, Form 76.

Note: Materials have to conform to standard specifications for the supply of natural gravel or crushed rock for road pavements. By definition, a "bound" material has a high enough elastic modulus and tensile strength to significantly stiffen the pavement (e.g., tensile strength > 80 kPa).

Abbreviations: NGB = natural gravel base; NGS = natural gravel subbase; DGB 20 = dense graded base, 20 mm nominal size; DGS 20 = dense graded subbase, 20 mm nominal size; DGS 40 = dense graded subbase, 40 mm nominal size.

13.3.2.1 LAYER EQUIVALENCIES. According to the equivalence concept, granular materials which already satisfy base course or subbase criteria, but are nevertheless treated with cement or lime, allow a reduction in thickness of the particular pavement layer. For example, if a 200-mm-thick natural gravel base is cement-bound, it is considered to be equivalent to a 220-mm-thick gravel base. Table 13.1 gives an extract of pavement layer equivalences according to the New South Wales Department of Main Roads (1983). This table serves as an illustration only; actual use in design would require detailed knowledge of relevant material specifications.

13.3.2.2 BASIC CBR DESIGN. A conventional basic CBR design chart (Fig. 13.18) gives the aggregate cover required as a function of the CBR of the subgrade and the design traffic expressed in ESAs (equivalent standard axle loads). The granular material is assumed to have a CBR in excess of 80 and has either a bituminous seal or asphalt cover less than 25 mm thick.

This design procedure can be extended to a multilayer pavement, according to the following principles [Metcalf (1979)]:

(*a*) The CBR of the subgrade determines the minimum total thickness of pavement required.
(*b*) The CBR of any pavement layer determines the thickness required above that layer.
(*c*) The minimum CBR to carry traffic with only a bituminous seal coat surfacing is 80.

This means that stabilized layers may form part of the pavement structure, but the total thickness would never be less than that required for standard base course material. It should also be noted that the CBR method basically caters for the case where the strength of the pavement layers always increases toward the surface.

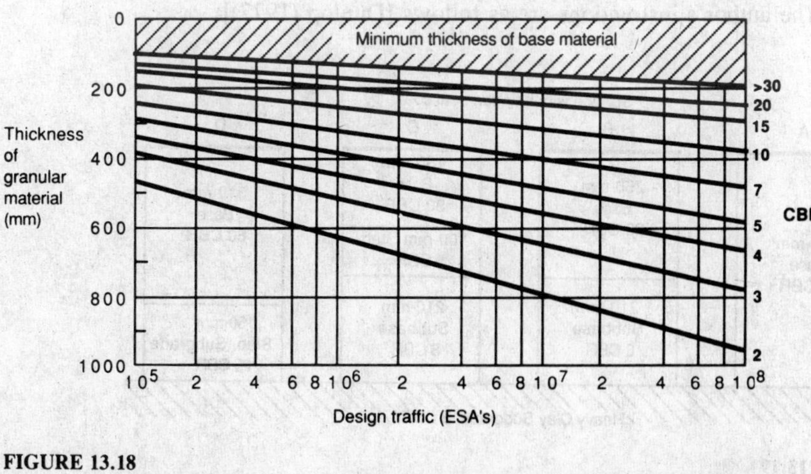

FIGURE 13.18
CBR design chart for granular pavements with thin bituminous surfacing. [*NAASRA (1987).*]

According to NAASRA (1987), a subgrade material which has been stabilized should not generally be assigned a CBR greater than 15. This is because the CBR may overestimate the in-service strength of a highly cemented granular soil; under the influence of large traffic-induced strain, cemented bonds are likely to break, a condition which is not simulated in the CBR test.

Examples of alternative CBR-based pavement designs is given in Fig. 13.19 [after Metcalf (1979)]. Note that the total thickness of all options is the same.

Shrinkage and fatigue cracking of a highly stabilized subbase or base course (e.g., 9% cement) may be reflected in the asphaltic surface layer. In order to prevent this type of damage, an unbound granular layer may be used to separate the asphalt from the soil-cement layer. This pavement structure is also described as "upside-down" or "sandwich" construction and requires special design considerations (see discussion in Sec. 13.4.2.4) and/or construction techniques, such as deliberate pre-cracking by rollers or traffic.

13.3.2.3 SIMPLIFIED DESIGN OF A LIME-MODIFIED SUBGRADE. Dunlop (1977) proposed a simple design method for assessing the effect of treating a soft subgrade (CBR < 5) with small quantities of lime. In Dunlop's terminology, "lime-modified" means that mainly the flocculating rather than cementing action of the lime is called upon, resulting in a tensile strength of less than 80 kPa (after 14 days curing at 20°C). As a consequence, undesirable effects such as excessive shrinkage and fatigue cracking are minimized.

The design is based on the elastic analysis of a multilayer system but allows the properties of the natural subgrade and the lime-modified soil to be represented in terms of CBRs. It thus represents a combination of a CBR and a mechanistic approach. Calculations were carried out using the program BISTRO by Shell [Peutz et al. (1967)]. The limiting performance criteria underlying the design is either the vertical strain at the top of the subgrade layer or the tensile strength in the lime-modified layer, whichever is the more critical. Dunlop's design chart is reproduced in Fig. 13.20. The author's instructions are as follows [Dunlop (1977)]:

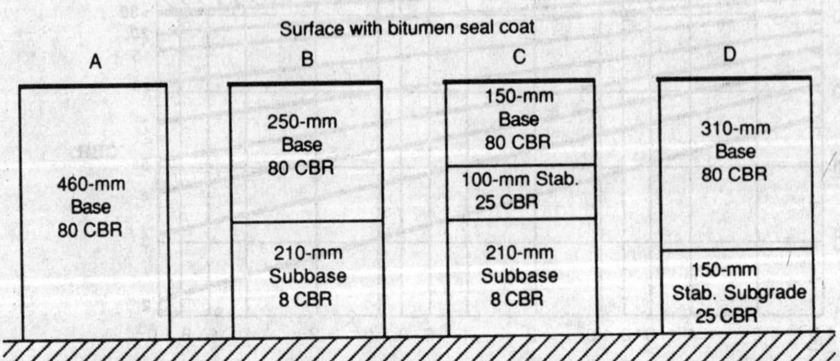

FIGURE 13.19
Examples of California bearing ratio pavement design. [*Metcalf (1979)*.]

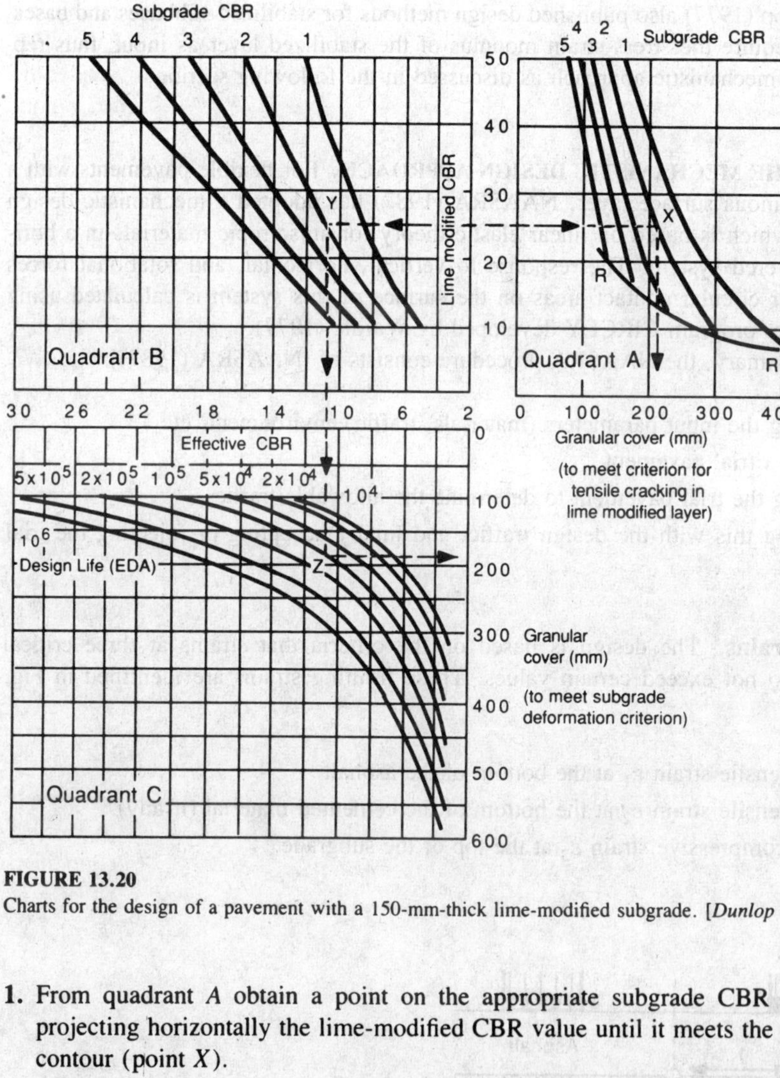

FIGURE 13.20
Charts for the design of a pavement with a 150-mm-thick lime-modified subgrade. [*Dunlop (1977).*]

1. From quadrant *A* obtain a point on the appropriate subgrade CBR contour by projecting horizontally the lime-modified CBR value until it meets the appropriate contour (point *X*).
2. Project this point down vertically to obtain the granular cover required to prevent tensile cracking.
3. If the projected point falls below curve *R-R*, then tensile cracking is not critical.
4. Now check on compressive strain in the subgrade by projecting the lime-modified CBR horizontally into quadrant *B* to cut the appropriate subgrade CBR contour (point *V*).
5. Drop this point down vertically to meet the design loading curve in quadrant *C* at point *Z*.
6. Project this point horizontally to obtain the minimum granular cover to prevent deformation in the subgrade.

Dunlop (1977) also published design methods for stabilized subbases and bases, but these require the stress-strain modulus of the stabilized layer as input, thus representing a mechanistic approach as discussed in the following section.

13.3.2.4 THE MECHANISTIC DESIGN APPROACH. For flexible pavements with a thick bituminous surface layer, NAASRA (1987) has adopted a mechanistic design procedure which is based on linear elastic theory for anisotropic materials in a horizontally layered system. The response to vertical, horizontal, and rotational forces applied over circular contact areas on the surface of this system is calculated using the computer program CIRCLY developed by Wardle (1977).

In summary, the NAASRA procedure consists of [NAASRA (1987)]:

- Evaluating the input parameters (materials, traffic, environment, etc.)
- Selecting a trial pavement
- Analyzing the trial pavement to determine the allowable traffic
- Comparing this with the design traffic, and finally accepting or rejecting the trial pavement.

Critical strains. The design is based on the criteria that strains at three critical locations do not exceed certain values. These limiting strains are identified in Fig. 13.21 as

The tensile strain ε_1 at the bottom of the asphalt

The tensile strain ε_2 at the bottom of the cemented material (if any)

The compressive strain ε_3 at the top of the subgrade.

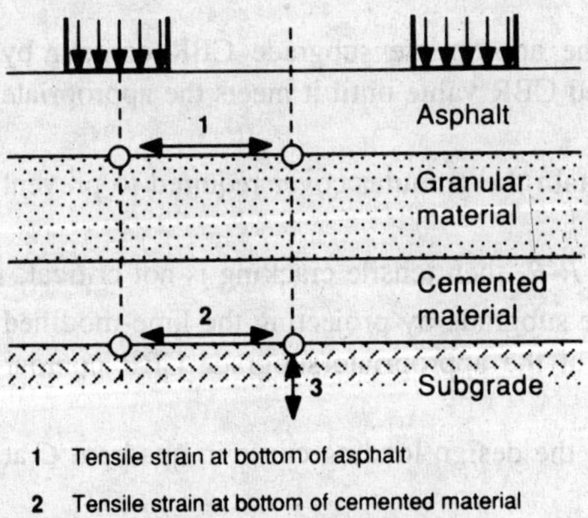

FIGURE 13.21
Pavement model for the NAASRA mechanistic design procedure showing locations of critical strains due to applied loading. [*NAASRA (1987)*.]

Limiting strains are related to the allowable number of load repetitions N before unacceptable rutting or fatigue cracking occurs. The mathematical relationships used are based on experience. They are

Asphalt fatigue criteria:

$$N = \left[\frac{6918(0.856V_B + 1.08)}{S_{\text{mix}}^{0.36}\varepsilon_1}\right]^5 \tag{13.10}$$

$$= \left(\frac{4509}{\varepsilon_1}\right)^5 \quad \text{for } E_{\text{asphalt}} = 2800 \text{ MPa} \tag{13.11}$$

Fatigue criteria for cemented materials (valid for a typical modulus of 2000 MPa):

$$N = \left(\frac{280}{\varepsilon_2}\right)^{18} \tag{13.12}$$

Subgrade failure criteria:

$$N = \left(\frac{8511}{\varepsilon_3}\right)^{7.14} \tag{13.13}$$

where $\varepsilon_1,\varepsilon_2,\varepsilon_3$ = critical strains as defined above (in microstrains)
N = allowable number of load repetitions
V_B = percentage by volume of bitumen in the asphalt
S_{mix} = stress-strain modulus of mix, MPa

In unbound granular layers, no tensile stresses exist; compressive stresses do not create a problem either. These layers are therefore not critical in terms of strain.

Material input parameters. All materials are assumed to be elastic and either isotropic or anisotropic. For the latter case, three stress-strain moduli (vertical, horizontal, and shear) and two Poisson's ratios are required.

Characterization of a *subgrade* begins with estimating the vertical modulus E_v from its CBR, based on an empirical relationship such as

$$E_v = 10 \text{ CBR} \qquad \text{MPa} \tag{13.14}$$

The NAASRA guide suggests that the horizontal modulus E_h be set equal to half the vertical modulus. Poisson's ratio ν of a cohesive subgrade can be set at 0.45, and for a cohesionless soil at 0.35. The shear modulus is equal to $E_v/(1 + \nu)$.

The modulus of an *unbound granular* layer not only depends on the material itself and the prevailing stress level but also on the stiffness of underlying layers. In the laboratory it could be measured in a triaxial test under repeated loading. In the field it could be backfigured from deflection measurements. In the analysis, sublayering may be needed to improve computational accuracy. Typical presumptive elastic parameters are given in Table 13.2.

TABLE 13.2
Typical presumptive elastic parameters for pavement layers

Pavement layer	Typical modulus, MPa	Typical Poisson's ratio
Unbound granular material (modified compaction):		
High-quality crushed rock	500	0.35
Base-quality gravel	400	0.35
Subbase gravel	300	0.35
Cemented material:		
Crushed rock (2 to 3% cement)	5 000	0.2
Base-quality natural gravel, 4–5% cement	5 000	0.2
Subbase-quality natural gravel, 4–5% cement	2 000	0.2
Asphalt (values typical for New South Wales) temperature:		
10°C	115 000	0.4
25°C	3 500	0.4
40°C	620	0.4

Source: Extract of data presented by NAASRA (1987).

Layers of *cemented* materials are assumed to be isotropic and uniform. In the laboratory, flexural testing of a simple beam (50- by 50-mm cross section) under repeated loading is considered to give the most representative values, when the results are interpreted using elastic theory. Direct tension testing or triaxial testing could also be used. Typical values are given in Table 13.2.

Elastic characterization of *asphalt* is rather difficult. It depends on many factors, such as bitumen class and content, air voids, aggregate type, temperature, rate of loading, and age. Flexure tests and indirect tension tests are most commonly used, in the laboratory, to determine moduli. Asphalt is considered isotropic; typical presumptive values for New South Wales, Australia, are given in Table 13.2.

Examples of design charts. The CIRCLY program can be run repeatedly to develop design charts for particular conditions in regard to traffic loadings, axle load distributions, material characterization, and basic pavement structure.

Three design graphs have been selected from the NAASRA pavement design guide [NAASRA (1987)] as an illustration for the mechanistic design approach of a pavement structure with and without a stabilized layer. They use typical presumptive elastic properties as described earlier but are only valid for a subgrade with a CBR of 3. They allow the design of three alternative structures above this subgrade:

Figure 13.22: An asphalt layer (E = 2800 MPa) above an unbound granular material

Figure 13.23: An asphalt layer above cemented material with a modulus of 2000 MPa

Figure 13.24: Asphalt above a 100-mm-thick unbound granular layer, separated from the subgrade by a cemented material with a modulus of 2000 MPa

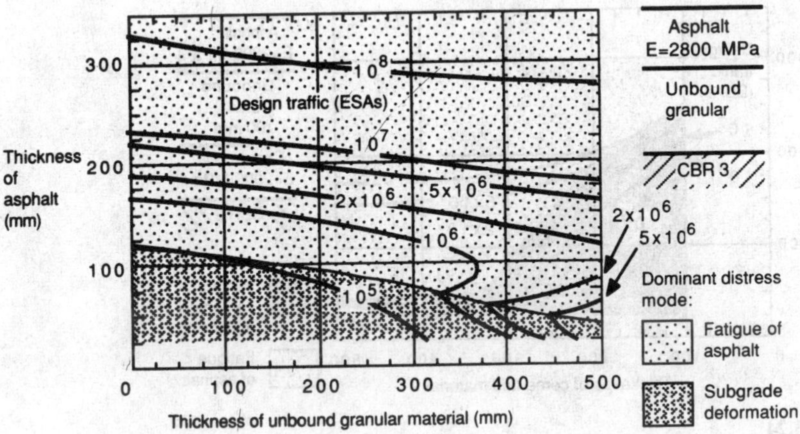

NOTE 1. Allowance to be made for construction tolerances.
2. For an explanation why more than one asphalt thickness is satisfactory, refer to NAASRA (1987).

FIGURE 13.22
NAASRA design chart no. EC 2. [*NAASRA (1987).*]

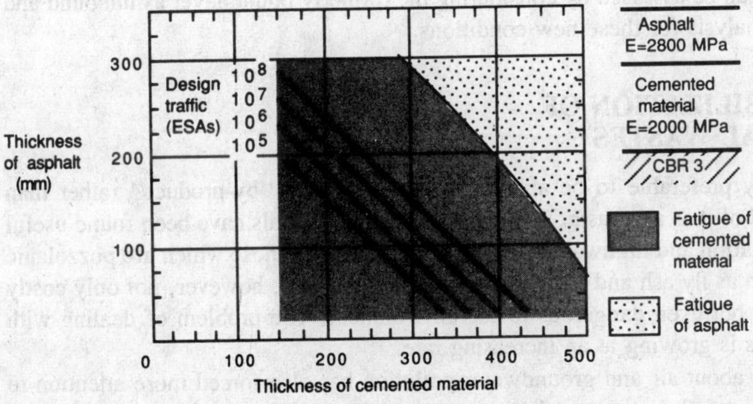

Note 1. Allowance to be made for construction tolerances.
2. For pavements where the cover over the cemented material exceeds 100 mm, the second phase of life of the pavement after the cracking of the cemented material may be considered by assuming that the cemented layer has become unbound.
3. For designs with asphalt thickness <100 mm, the upper 150 mm of subgrade should consist of material of CBR > 15 to provide resistance to infiltration through shrinkage cracks.

FIGURE 13.23
NAASRA design chart no. EC 9. [*NAASRA (1987).*]

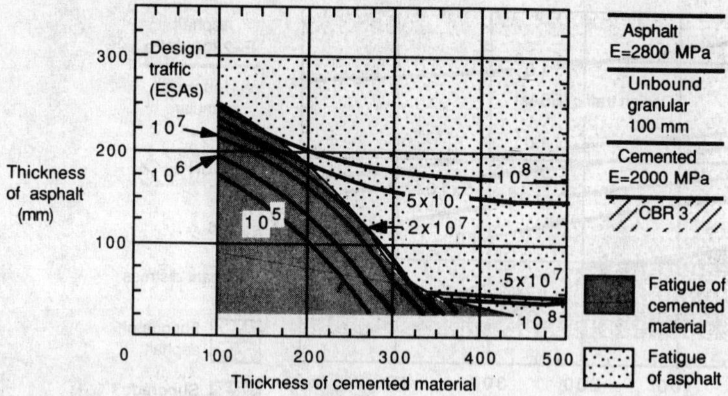

FIGURE 13.24
NAASRA design chart no. EC 23. [*NAASRA (1987).*]

For a given traffic loading, these graphs allow various suitable thickness combinations of asphaltic surface course and cemented base or subbase to be determined. Economic consideration would indicate which one would be the most desirable.

It should be noted that even if the cemented layers exceed the allowable strain and start cracking, the pavement may remain serviceable for some time. This post-cracking life can be assessed by considering the formerly bound layer as unbound and redoing the analysis for these new conditions.

13.4 STABILIZATION OF INDUSTRIAL WASTES

It is obviously preferable to be able to sell an industrial "by-product" rather than have to get rid of it as a "waste." Many solid waste materials have been found useful in land reclamation and highway construction, particularly those which are pozzolanic in nature, such as fly ash and blast furnace slag. Others are, however, not only costly to dispose of, but even dangerous to the environment. The problem of dealing with waste products is growing at an increasing rate.

Concern about air and groundwater pollution has also forced more attention to the purification of flue gases and treatment of effluents emanating from industrial plants. A variety of granular materials, clays, and liquids are used in the neutralization and detoxification processes; once spent, they are usually disposed of in sludge lagoons or landfills.

Regulatory agencies today place strict requirements on the design of new disposal sites. Many existing waste deposits have been reassessed in recent times and found to be hazardous to the environment. Pressure is therefore on to find ways of either utilizing industrial waste in engineering construction or developing methods of stabilizing it so that it presents no danger to groundwater resources. Ideally, a disposal site should be stable enough to have the potential to be redeveloped, when it is either full or when the industrial activity has ceased.

The following brief case studies are meant to indicate how the principles of ground modification can be applied to waste stabilization. They illustrate the experimental approach to optimizing the mixtures of waste, natural materials, and chemicals and the need for extensive laboratory testing and evaluation of performance in the field.

13.4.1 Lime and Fly Ash Stabilization of Calcium Sulfate

Usmen and Moulton (1984) reported on the performance of an experimental base course section built with coarse and fine waste calcium sulfate, lime, and fly ash. Waste sulfates are generated in hydrofluoric acid and phosphoric acid production, fluidized bed combustion, desulfurization of flue gases in power stations by scrubbers, and the neutralization of acid mine drainage with lime; it is usually produced in slurry form but solidifies in sedimentation ponds. Four 30-m-long test sections were built, with the following material combinations:

	Mix no.			
Material, %	**1**	**2**	**3**	**4**
Coarse sulfate			65	75
Fine sulfate	30	50		
Lime	5	5	5	5
Fly ash	65	45	30	20

The particle size of the coarse sulfate was that of fine gravel, and the fine sulfate was that of fine sand. Mix 4 showed the highest density in the laboratory, but in the field, better compaction (> 97% std. Proctor) was recorded for mixes 2 and 3. Mix 2 showed the highest unconfined compressive strengths: 3, 8.3, and 7.6 MPa after curing for 7, 28, and 56 days, respectively. Mix 2 also showed the highest freeze-thaw durability and, on the test section, the least deflection under load. Continued measurements of the compressive strength of field samples indicated strength losses during the cold seasons and gains during the warmer and more humid months. Overall, the strength of mixes 1 and 2 increased over the 4-year observation period on the order of 50%.

13.4.2 Mechanical Stabilization and Fixation of Petroleum Wastes

Wastes from petroleum refining accumulate in the form of liquids or sludge which are normally placed in lagoons for sedimentation and drying by evaporation. What is left in the long term is a soft deposit of contaminated soil, unsuitable for redevelopment. Because of the danger of a toxic leachate, the deposit may have to be completely encapsulated.

Martin et al. (1985) investigated the suitability of hydrated lime and fly ash to stabilize a mixture of three waste products: an acid hydrocarbon sludge, spent attapulgite clay and small quantities of catalyst fines, and a processed aluminum oxide. The objective of this exercise was twofold:

Immobilize the contaminants in a stable and durable matrix of solids so that any leachate produced through infiltration or consolidation is tolerable

Modify the mechanical properties of the waste so that it is volumetrically stable and able to support loads

The mixtures prepared had sludge/clay/fly-ash/lime/fines ratios (by weight) as follows: 1/1/0.75 to 1/0 to 0.5/0 to 0.2. Curing conditions were found to be of great significance; test variables included confinement (freestanding or in tubes), boundary shape, and humidity and moisture content.

Unconfined compression tests on compacted specimens served as indicators of mechanical strength and deformability. Consolidation tests yielded compressibility and permeability data. Volume stability was observed over a 5-month period in terms of moisture losses and shrinkage in a ventilated room. Leachate quality was evaluated in permeameter tests; distilled water was the permeant, and the leachate was examined by light transmittance and pH measurements.

Martin et al.'s (1985) study demonstrated the relevance of the principles and techniques of ground modification to waste disposal problems. But it also raised many new issues where the engineer is unable to draw on past experience or on a well-established database. An example of these is the difficulty of predicting leachate quality from a deposit containing a variety of waste products which interact with each other as well as with stabilizing additives.

PROBLEMS

Prefixes indicate problem type: C = calculations, M = multiple choice, B = brief answer, D = discussion.

Calculations

C13.1. Given that the maximum grain size of a soil is 5 mm, calculate the diameters D_{10}, D_{30}, and D_{60} (corresponding to 10, 30, and 60% passing) so that maximum density is achieved. Check the USCS classification of such a soil. Use Eq. (13.1) with $n = 0.5$.

C13.2. A soil thoroughly mixed and then compacted in the laboratory had a strength of 2 MPa. The same soil mixed in the field, but otherwise equally formed into a specimen in the laboratory, had a strength of 1.6 MPa. What is the mixing efficiency?

C13.3. A granular soil with a cement content of 10% has a strength of 6 MPa after 10 days of curing. What strength can be expected after 1 year?

C13.4. If the unconfined compressive strength of a cement-treated fine-grained soil is 2 MPa, in what range do you expect its flexural strength to be? Give the answer in kilopascals.

C13.5. Calculate the following changes of quicklime through hydration:

(*a*) Mass of 1 kg of quicklime after hydration.

(*b*) Volume of 1 m^3 of solid quicklime after hydration.

C13.6. (*a*) Estimate the reduction in water content due to hydration of a saturated soil treated with quicklime, given the following information:

Initial water content = 60%
Lime is 90% pure quicklime
Amount of lime added is equal to 15% of the dry soil

(*b*) Calculate the loss of water in the soil if all the heat generated during hydration could be used up for this purpose.

C13.7. How much cement would have to be added to Vales Point fly ash (Fig. 13.10) to make it suitable as a base course according to recommendations by the EPRI?

C13.8. A basic CBR method is to be used for the design of a granular pavement with a thin bituminous surface.

(*a*) Determine the thickness required on a subgrade with a CBR of 3 if the design traffic is 10^6 ESAs.

(*b*) Would the total thickness required be reduced if a 100-mm-thick stabilized subbase is part of the design?

C13.9. Using the Dunlop design chart (Fig. 13.20), indicate what granular cover is required above the 150-mm lime-modified layer, given the following conditions:

Untreated soil:	CBR = 3
Lime-modified soil:	CBR = 20
Design traffic:	10^6 EDA

C13.10. Calculate the limiting strains (in microstrain) assumed in the NAASRA mechanistic design in the asphalt (E = 2800 MPa), cemented material, and subgrade, for $N = 10^5$.

C13.11. Estimate the stress-strain modulus equivalent to a CBR of 3.

C13.12. Design three alternative pavements for conditions covered by Figs. 13.22, 13.23, and 13.24. Assume a design traffic of $N = 10^6$ ESAs and a standard asphalt thickness of 150 mm. Indicate the type of distress expected at the end of the design life.

Multiple Choice

M13.13. In older books on road engineering, ''mechanical'' stabilization meant

(*a*) Changing the grading of a soil in order to get maximum density.

(*b*) Compacting using a heavy tamper.

(*c*) Strengthening by reinforcing inclusions (steel strips, fabrics).

(*d*) Providing retaining walls to unstable slopes.

M13.14. In deep mixing of Portland cement with clay, three reactions take place: pozzolanic reaction (P), adsorption (A), and hydration (H). These reactions occur in the following sequence:

(*a*) PAH

(*b*) AHP

(*c*) HAP

(*d*) HPA

M13.15. With respect to cement stabilization of pavement layers, which one of the following statements is definitely *not* correct:

(*a*) Letting traffic in straight away or within 24 h after stabilization (and initial compaction) causes desirable closely spaced precracking.
(*b*) The lower the temperature, the higher the strength of the stabilized soil.
(*c*) Organic matter is deleterious.
(*d*) Seawater is not necessarily bad.

M13.16. A type of lime which is *not* suitable for soil stabilization is

(*a*) $Ca(OH)_2$
(*b*) CaO
(*c*) $CaCO_3$
(*d*) CaO + MgO

M13.17. In lime stabilization, short-term reactions are

(*a*) Hydration and carbonation.
(*b*) Hydration and cementation.
(*c*) Flocculation and hydration.
(*d*) Carbonation and cementation.

M13.18. Which one of the following effects is *not* true for a lime treatment of a cohesive soil:

(*a*) The plastic limit is increased.
(*b*) The plasticity index is decreased.
(*c*) The liquid limit may decrease or increase.
(*d*) The strength and maximum density both increase.

M13.19. One of the few admixtures which may increase the permeability of a soil is

(*a*) Cement.
(*b*) Lime.
(*c*) Bitumen.
(*d*) Calcium chloride.

M13.20. Which one of the following admixtures is likely to show the most linear increase in strength with amount added?

(*a*) Cement
(*b*) Lime
(*c*) Bitumen
(*d*) Calcium chloride

M13.21. A deliquescent substance

(*a*) Forms bubbles in contact with water.
(*b*) Absorbs water.
(*c*) Dissolves in water.
(*d*) Liquifies in its own absorbed moisture.

M13.22. Of the following additives, which one could be used for all the following purposes: to compact frozen soils, to melt ice, to reduce dust, to maintain the moisture in the soil, and for electrokinetic stabilization?

(*a*) Sodium silicate
(*b*) Asphaltic emulsion
(*c*) Calcium carbonate
(*d*) Calcium chloride

M13.23. It is possible that by the year 2000, the annual production of fly ash in the United States of America will be

(*a*) 100 000 t.

(*b*) 1 000 000 t.
(*c*) 10 000 000 t.
(*d*) 100 000 000 t

M13.24. If sufficient cement or lime is added to a subbase-quality gravel, it is rated as a cemented (or bound) material. Typically its stress-strain modulus would change from
(*a*) 0 to 300 MPa.
(*b*) 300 to 2000 MPa.
(*c*) 2000 to 300 MPa.
(*d*) 20 000 to 300 MPa.

M13.25. Assume a dense base-quality graded gravel (20 mm nominal size) is treated with cement so that it rates as a bound material. It is placed in a layer at a depth > 300 mm. A simplified approach suggests that the equivalent layer of untreated material would have to be thicker by
(*a*) 10%.
(*b*) 20%.
(*c*) 40%.
(*d*) 60%.

M13.26. Granular base course material which can carry traffic with only a bituminous seal must have a CBR of
(*a*) > 2000.
(*b*) > 80.
(*c*) > 30.
(*d*) < 30.

M13.27. Dunlop (1977) defines a "lime-modified" soil as one whose tensile strength is
(*a*) < 80 kPa.
(*b*) > 80 kPa.
(*c*) > 2 MPa.
(*d*) < 2 MPa.

M13.28. The NAASRA mechanistic design is based on the following performance criteria
(*a*) The tensile strength of the asphalt and the compressive strength of the base course and subgrade.
(*b*) The strength of the subgrade only.
(*c*) The surface deflection under traffic.
(*d*) The critical horizontal tensile strains in the asphalt and the cemented layer (if any) and the vertical compressive strain in the subgrade.

Brief Answer

B13.29. Soil stabilization by lime, cement, and bitumen has the following objectives:
(*a*) Increase ______________ .
(*b*) Increase ______________ .
(*c*) Reduce ______________ .
(*d*) Reduce ______________ .
(*e*) Control ______________ .
(*f*) Control ______________ .

B13.30. Soil additives help to solve many construction or design problems in civil and mining engineering. Give a specific example with respect to
(*a*) Trafficability.

(*b*) Workability.
(*c*) Hazardous wastes.
(*d*) Dust.

B13.31. In some books and standards, stabilization with granular admixtures is called ______________.

B13.32. In soil stabilization work, the practical limits of cement addition are generally set between 2 and 10%.
(*a*) The reason for setting a lower limit of cement addition is the problem of ______________.
(*b*) The reason for setting an upper limit of cement addition is the problem of ______________.

B13.33. Choose the correct word (decreases, increases) to complete the following statements:
(*a*) If cement is added to a clay, the plasticity index ______________.
(*b*) If cement is added to a clay, the linear shrinkage ______________.
(*c*) If lime is added to a clay, the swelling pressure ______________.
(*d*) If lime is added to a clay, the compressive strength ______________.

B13.34. Standard laboratory compaction tests on the same clay, alone and then lime-stabilized, would most likely indicate that
(*a*) With lime, the optimum water content ______________ (decreases, increases).
(*b*) With lime, the compaction curve is ______________ (more peaked, flatter).

B13.35. List four factors which affect the engineering properties of the ash produced at different power stations using the same coal.
(*a*) ______________
(*b*) ______________
(*c*) ______________
(*d*) ______________

B13.36. Identify the following materials:
(*a*) A dark viscous liquid obtained by dry distillation of wood, coal, etc.: ______________.
(*b*) A black organic material, either occurring naturally or produced by destructive distillation (pyrogenic) of organic substances, such as pitch or the residue after distillation of petrol: ______________.

B13.37. Give two main reasons for bituminous stabilization:
(*a*) ______________
(*b*) ______________

B13.38. If the data of Fig. 13.14*b* was replotted in order to show the variation of w_{opt} versus the total amount w_{tot} of liquid present (tar + water), would the optimum water content increase or decrease with w_{tot}?

B13.39. Indicate two possible objectives of waste stabilization using concepts of ground modification:
(*a*) ______________
(*b*) ______________

Discussion

D13.40. Discuss various meanings of the words stabilization, modification, consolidation, and mechanical stabilization in civil engineering literature.

D13.41. Explain the usage of the terms cement-treated, cement-bound, and soil-cement in some of the geotechnical literature.

D13.42. How can the lime-reactivity of a soil be assessed? Are there any quick indicator-type tests?

D13.43. Compare the techniques of pure lime columns and lime-soil columns created by deep mixing.

D13.44. Discuss the need for constructive uses of industrial waste materials such as fly ash and slag.

D13.45. The addition of fly ash may significantly increase the maximum density of a sand, but not that of a clay. How do you explain that behavior?

D13.46. The strength of a bitumen–granular-soil mix shows a peak for a particular bitumen content. Why doesn't strength continue to increase with the amount of stabilizer added?

D13.47. Describe design and construction measures which could prevent or reduce reflection cracking (the cracking of the asphaltic surface layer which is caused by shrinkage or fatigue cracking of an underlying cemented layer).

D13.48. How would you evaluate the suitability of a granular industrial-waste product for road construction? List desirable and undesirable properties, and speculate how conventional additives may be used to overcome shortcomings.

D13.49. What properties of fly ash could make it suitable for the immobilization of liquid or semiliquid wastes, perhaps in combination with other materials?

CHAPTER 14

MODIFICATION AT DEPTH BY GROUTING

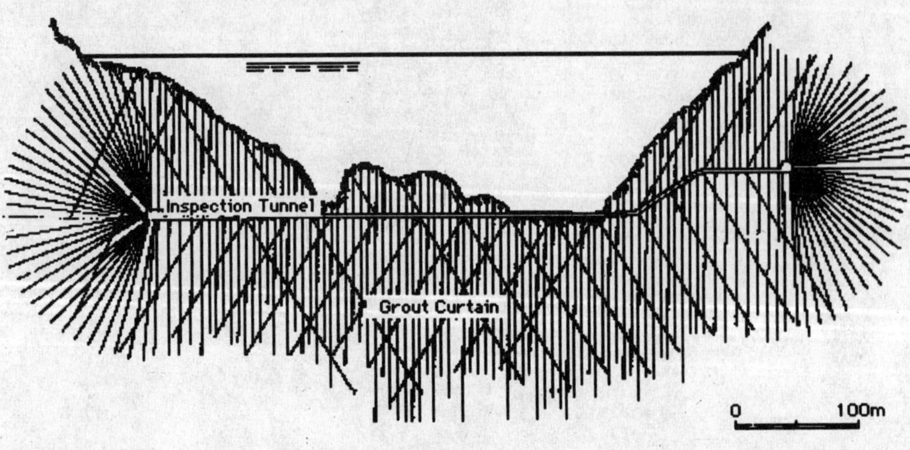

14.1 INTRODUCTION TO GROUTING

14.1.1 Definition and Purpose of Grouting

Grouting is defined as the injection of fluidized materials into voids of the ground or spaces between the ground and adjacent structures, generally through boreholes and under pressure. Many grouts are designed to set (gel or harden) after injection, either instantly or over a period of time.

The main objectives of grouting are to produce a stronger, denser, and/or less-permeable soil or rock; it may also simply serve to fill voids which are otherwise inaccessible and may prevent adequate stress transfer within the ground or from a structure to the ground.

14.1.2 Categories of Grouting

In this introduction, grouting techniques are classified according to the method used to introduce the grout into the ground. However, other criteria could be used to differentiate grouting methods, such as the type of grout material injected, the typical applications, the layout of injection points, and the sequence of construction.

Distinguished by the mode of entry into the soil or rock, the basic categories of grouting are (as shown in Fig. 14.1):

- Penetration grouting (intrusion, permeation)
- Displacement grouting
- Compaction grouting (including slab-jacking)
- Grouting of voids
- Jet grouting (replacement)
- Special grouting applications and techniques, including electrogrouting

Penetration grouting[1] describes the process of filling joints or fractures in rock or pore spaces in soil with a grout without disturbing the formation. More specifically, permeation grouting refers to the replacement of water in voids between soil particles with a grout fluid at low injection pressure so as to prevent fracturing.

Displacement grouting is the injection of grout into a formation in such a manner as to move the formation; it may be *controlled*, as in compaction grouting (see below), or *uncontrolled*, as in high-pressure soil or rock grouting which leads to splitting of the ground, also called hydrofracture.

In *compaction grouting* a very stiff (say 25-mm slump) mortar is injected into loose soils, forming grout bulbs which displace and densify the surrounding ground,

[1]The Committee on Grouting of the Geotechnical Engineering Division of the American Society of Civil Engineers has published a preliminary glossary of terms relating to grouting (1980). Where possible, their recommendations will be followed.

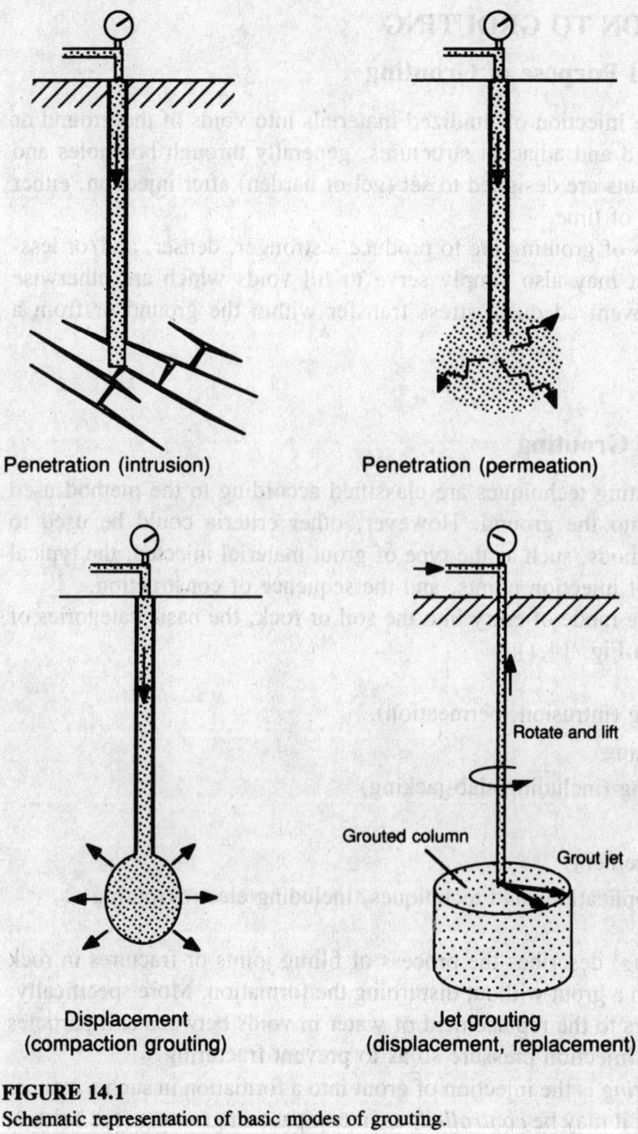

FIGURE 14.1
Schematic representation of basic modes of grouting.

without penetrating the soil pores. With slightly more fluid grout, thick fissures rather than bulbs may form; this is sometimes referred to as "squeeze grouting."

A technique similar to compaction grouting may be employed for the lifting and leveling of heavy structures [King and Bindhoff (1982)]. A special application is *slab-jacking* (or mud-jacking), where grout is injected under a concrete slab in order to raise it to a specified grade.

Grout may also be used simply to *fill voids*, such as may develop below the joints in a concrete pavement through pumping. Special terms have evolved for the grouting behind the lining of a tunnel due to overbreak: *Backpack grouting*, contact grouting, or more specifically crown grouting are found in the relevant literature, in addition to interface and gap grouting.

A special technique related to backpack grouting can be used for high-pressure shafts and tunnels. Described as *prestress grouting* [Harris (1982)], it involves simultaneous injection of grout through a multipoint system into the space between the tunnel lining and the rock; this creates a balanced confining stress in excess of the expected hydraulic pressure within the tunnel.

Jet grouting is a technique where high-speed water jets emanating from a drill bit cut into alluvial soils; as the drill bit is withdrawn, grout is pumped through horizontal nozzles and mixes with or displaces the soil. The original foundation material is thus replaced with a stronger and/or more impermeable grout-soil mixture. Jet grouting may be used to form cutoff walls, do underpinning, or form deep foundations similar to grouted auger piles.

Electrogrouting is a term used for promoting electrochemical hardening during electroosmosis by adding chemicals, such as sodium silicate or calcium chloride, at the anode (see Chap. 12). Under the influence of the electric field, these chemicals permeate the ground, flowing in the direction of the cathode, while the anode becomes a grout injection pipe.

14.1.3 The Art of Grouting

An overview of typical grouting applications is given in Fig. 14.2. There is no doubt that the largest quantities of grout (usually cement-based) are used in creating more or less impervious curtains below dams in order to reduce water losses, uplift pressures, and reduce the potential for hydraulic fracturing (leading to piping failure) under operating conditions. Foundation grouting for increased stability and reduced compressibility is probably next in importance. It may serve a permanent or temporary function: Only a temporary increase in strength may be required for surface or underground excavations. In some cases, grouting may provide an elegant, if not the only way, of repairing existing structures or making up for inaccurate or imperfect construction procedures.

For many engineers, grouting is still considered an art rather than a science. Its successful application requires a great deal of experience, thorough knowledge of geological conditions, and an awareness of equipment capabilities and limitations. Houlsby (1982) went even further in saying that "grouting requires an intuitive perception of just what the liquid grout does as it flows through the open joints and cracks hidden away down there underground."

This chapter concentrates on the engineering principles involved in grouting; those interested in further information, particularly practical construction aspects, are well advised to refer to the books by Cambefort (1967), Bowen (1981), Karol (1983), and Ewert (1985), in addition to conference proceeding, such as those edited by Baker (1982, 1985). Specialist chapters found in more general texts or reference books are also helpful [Winterkorn and Fang (1975); Koerner (1984); Bell (1987)] in addition to course notes [e.g., Tomiolo (1982); Redaelli (1982)].

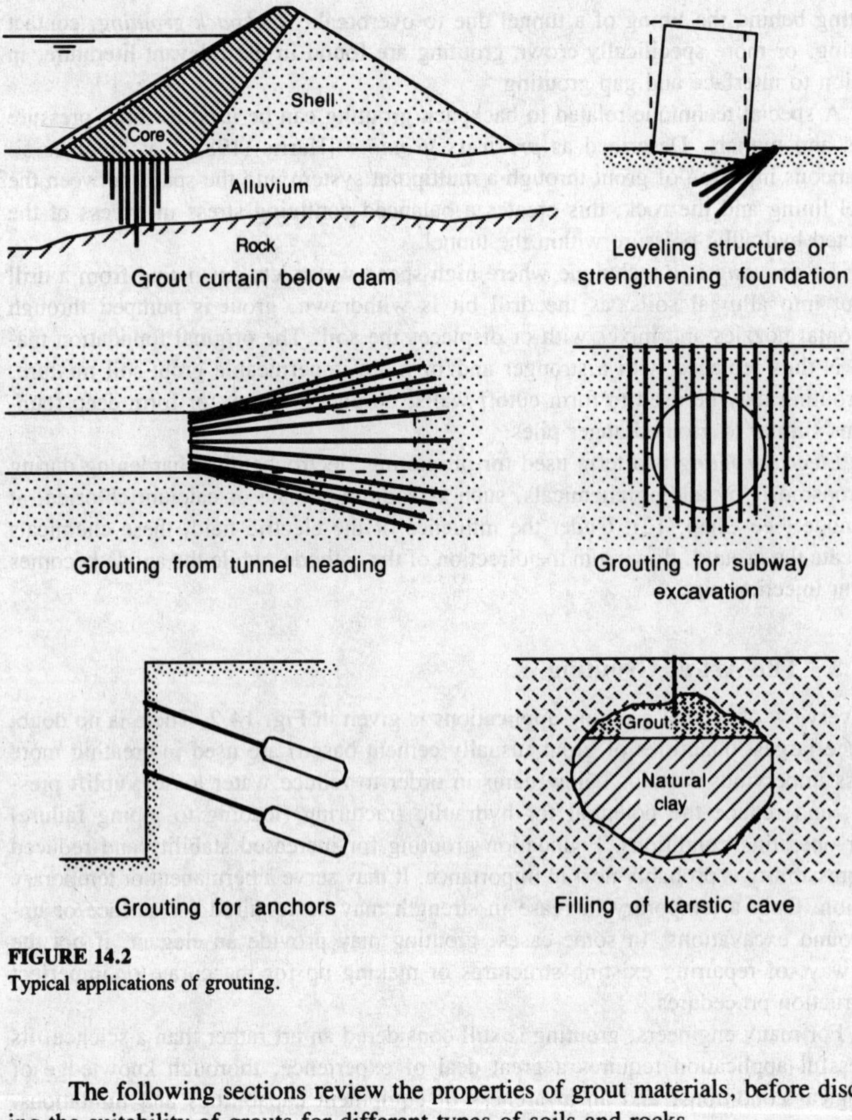

FIGURE 14.2
Typical applications of grouting.

The following sections review the properties of grout materials, before discussing the suitability of grouts for different types of soils and rocks.

14.2 GROUT MATERIALS

14.2.1 Classification of Grout Materials

Three basic types of grout are differentiated according to composition:

Suspensions: Small particles of solids are distributed in a liquid dispersion medium. Examples: cement and clay in water.

Emulsions: A two-phase system containing minute (colloidal) droplets of liquid in a disperse phase. Example: bitumen and water. Also in this category are foams, created by emulsifying a gas into the grout material, which could be cement or an organic chemical. Foaming agents, such as additives which increase surface tension, assist in forming bubbles by agitation; alternatively, they may induce gas-forming chemical reactions.

Solutions: Liquid homogeneous molecular mixtures of two or more substances. Examples: sodium silicate, organic resins, and a wide variety of other so-called chemical grouts. A difference may be made between colloidal solutions (e.g., silica or lignochrome gels) and pure solutions (e.g., phenolic and acrylic resins, aminoplasts). (*Note:* A cement grout is not commonly classified as a chemical grout; it is usually classified as a suspension grout.)

Principal types of grouts as distinguished by Cambefort (1987) are listed in Table 14.1, which also gives an indication of appropriate uses, construction controls, and relative costs.

Cambefort (1977) characterizes *foam grouts* by the following parameters:

Expansion coefficient $$e_g = \frac{\text{volume of gas}}{\text{volume of liquid}}$$

Bulking coefficient $$C_b = \frac{\text{total volume}}{\text{volume of liquid}} = 1 + e_g$$

Air ratio $$n_g = \frac{\text{volume of gas}}{\text{total volume}} = \frac{e_g}{1 + e_g}$$

TABLE 14.1
Principal types of grout

	State							
	Suspensions			Liquids			Aerated emulsions	
	Unstable		Stable	Chemical products				
Grout type	Cement	Bentonite + cement	Deflocculated bentonite	Sodium silicate hard gels	Sodium silicate diluted gels	Organic resins	Cement foams	Organic foams
Range of uses	Fissures	Sands and gravels, k m/s					Cavities	High water flows
		$>5 \times 10^{-4}$	$>10^{-4}$	$>10^{-4}$	$>10^{-5}$	$>10^{-6}$		
Grouting control	Refusal pressure	Limited quantities					Filling	
Relative cost for the products to fill 1-m³ voids	4.2 (deposit with γ_d = 1.5)	1 (cement 200 kg; betonite 30 kg)	0.8–1	6	2–4	10–500	1.2	10

Source: After Cambefort (1987).

The relationship between the expansion coefficient e_g and the air ratio n_g is analogous to that of the void ratio and porosity in a soil.

The rheological properties and ground penetrability of foams are not only related to the expansion coefficient, which could vary from less than 3 for cement-based foams up to 50 for organic foams [Cambefort (1977)], but also on the bubble size distribution. The latter can be controlled, to a degree, by the choice of foaming agent and the method of bubble formation.

With *chemical grouts,* toxicity and permanency have also become an issue. Table 14.2 [after Greenwood and Thomson (1984)] lists several groups of chemical grouts, their principal characteristics, and problems involved. Examples of commercial chemical grouts are given in Table 14.3.

Permanency refers to the resistance against mechanical deterioration due to freeze-thaw or wetting and drying cycles and chemical degradation by reactions with the groundwater or soil constituents. Permanency may also be threatened by hydraulic displacement or erosion by seepage through residual void space left after inadequate grouting or shrinking and consolidation of the grout itself.

Toxicity refers to health hazards in handling the grout and its effects on the quality of the groundwater it is in contact with. Unfortunately, the high-strength,

TABLE 14.2
Chemical grouts

Chemical grout	Initial viscosity,* 10^{-3} Pa · s	Gel time, min	Strength in coarse sand, MPa	Risk and toxicity†	Remarks
Silicate	1.5–40	1–200	0.7–3.0	Household chemicals	Only one or two stable gels of high penetrability
Lignochromes	2.5–20	5–120	1.0–1.75	Dermatitis risk	Hexavalent chromium is an accumulative pollutant; needs clarification to remove particles
Phenolic resins	1.5–10	5–60	1.0–3.0	Respiratory irritant; caustic	Poor gel time control with high strengths; some need clarification
Acrylic resins	1.3–10	1–200	1.0–3.0	AM-9 neurotoxic (banned in Japan)	Latest forms less toxic than AM-9 and not neurotoxic
Aminoplasts	6.0–30	40–300	1.0–3.5	Respiratory irritant to users when pure	Very viscous unless pure
Polyurethane	19.0–150	Reacts instantly with water	0.8–1.0	Irritant; toxic gases when burned—banned in mines	Gaseous foam expands fluid

Note: Tabulated figures are indicative for common formulations and cannot be used without reservation.

*10^{-3} Pa · s = 1 cP = 0.01 dyne · s/cm^2; water has a viscosity of about 1 cP at 20°C.

†All gelling chemicals are toxic if mishandled.

Source: After Greenwood and Thomson (1984).

TABLE 14.3
Examples of commercial chemical grouts

Chemical grout type	Commercial name	Reference (see general references below)
Silicate[a]	SIROC	Karol (1983)
	Hardener 600	"
	Clean Rock	Hoshiya et al. (1982)
	Glyoxal	Graf et al. (1982)
	Modified Earthfirm	"
	Silicate Bicarbonate	"
	Geloc-3	Zeigler and Wirth (1982)
	Monodur	Stetzler (1982)
Lignochrome	Blox-All	Karol (1982)
Phenolic resins	Rocagil	Karol (1982)
	Geoseal	"
	Terranier	"
Acrylic resins[b]	AM-9 (or Q-Seal, PWG)[c]	Clarke (1982) and Karol (1982)
	AC-400[d]	Clarke (1982)
	Rocagil BT	Karol (1982)
	Injectite 80	Berry (1982)
Aminoplasts	Herculox	Karol (1982)
	Rocagil[e]	"
	Diarock	"
	Cyanaloc 62	"
Polyurethane	TACSS[f]	Karol (1982)
	CR 250	"
	CR 260	"

General references: Primarily Karol (1983) and papers presented at the 1982 Conference on Grouting in Geotechnical Engineering, New Orleans, including two papers by Karol [proceedings edited by Baker (1982)].

[a]Most widely used type of chemical grout.

[b]Second most widely used type of chemical grout.

[c]No longer marketed in most countries since 1978.

[d]Low toxic replacement of AM-9, introduced in 1980.

[e]Rocagil products cover a range of chemical grouts.

[f]TACSS = "Takenaka Aqua-reactive Chemical Stabilization System."

highly permanent grouts seem to present the greatest risk with respect to handling, groundwater pollution, and corrosion.

14.2.2 Rheology of Grouts

Rheology is the study of flowage of materials. In this section, after reviewing basic rheological properties relevant to grouting, commonly used models of rheological behavior are introduced and applied to tubular flow. Penetrability of porous and fissured materials will be discussed further under the headings of permeation of grouting of soils (Sec. 14.4) and rock joints and fissures (Sec. 14.5).

14.2.2.1 BASIC RHEOLOGICAL PROPERTIES. Important basic characteristics of grouts are stability, setting time, and viscosity. Further engineering properties which are important include density, particle size (of suspensions), and the ultimate strength of grouted soils, but their determination does not need further explanation.

Stability. A grout is referred to as *stable,* if its particles remain in suspension of solution until it has reached the destination in the ground. If sedimentation occurs as soon as the grout is no longer agitated by the mixer or through turbulence in the grout pipes, it would be considered unstable. The breaking of an emulsion and the exudation of liquids from colloidal gels (syneresis) could also be considered a sedimentation process.

Rather than talking of sedimentation (or "settling out"), the grouting engineer may describe the separation of solids from the liquid (or vice versa) as "bleeding." The settling out of solid particles from a suspension while the liquid component travels further into the soil or rock mass is also referred to as "filtration," although the process is not just related to the relative sizes of the particles and the voids, but more importantly to the flow velocity of the grout.

Settling of particles out of suspension when the grout becomes stationary results in part of the voids containing water rather than grout. The same phenomenon, but on a larger scale, occurs in filling natural or artificial underground openings. Redrilling and additional grouting may then be necessary for achieving maximum strength and minimum permeability.

An indication of the stability of a suspension can be gained from a simple laboratory test where the grout is thoroughly mixed in a graduated cylinder and then left standing. The column of sediment-free water found after the particles have settled out, expressed in percent of the original sample height, is called the bleeding value. Pure cement grouts have high bleeding values and are therefore rated as unstable. The addition of bentonite improves the stability of cement grouts, as does high-speed mixing.

Setting time. Setting time is the time required for the grout to harden. Cement-based grouts normally set within 4 to 24 h, depending on the additives used. Setting or *gel time* can be critical for chemical grouts, which can set very rapidly, possibly within minutes.

Viscosity. Viscosity μ is defined by Newton's law of viscosity as the proportionality factor relating the shear resistance τ in a fluid to the velocity gradient dv/dz, which represents the rate at which one layer of fluid moves relative to an adjacent layer.

$$\tau = \mu \frac{dv}{dz} \tag{14.1}$$

The above equation holds for laminar flow; this means fluid particles follow smooth streamlines which are not disturbed through eddy currents. The SI unit for viscosity is the pascal-second (Pa · s). Traditionally the unit of centipoise (cP) has been used (1 cP = 10^{-3} Pa · s, which is the viscosity of water at about 20°C).

The viscosity μ is also called the dynamic viscosity or absolute viscosity. When divided by the mass density ρ, it becomes the kinematic viscosity v:

$$v = \frac{\mu}{\rho} = \frac{\mu g}{\gamma} \tag{14.2}$$

where $g = 9.81 \text{ m/s}^2$ (acceleration due to gravity). The units of kinematic viscosity are square meters per second; another unit name which may be used is the centistoke (cSt) where $10^{-6} \text{ m}^2/\text{s} = 1 \text{ cSt}$.

For turbulent flow, Eq. (14.1) has to be modified by including the so-called dynamic eddy viscosity η:

$$\tau = (\mu + \eta)\frac{dv}{dz} \tag{14.3}$$

Turbulent flow is important as far as maintaining the stability of the grout during pumping is concerned; when it comes to evaluating the extent of grout penetration into the ground, laminar conditions are generally assumed.

The transition from laminar to turbulent flow is usually described in terms of the dimensionless Reynolds number R_e, defined as

$$R_e = \frac{\rho v L}{\mu} \tag{14.4}$$

where v is the flow velocity and L is a characteristic length, e.g., the pipe diameter. A low Reynolds number means the viscous forces dominate over the inertia forces and laminar flow occurs.

Viscosity of a fluid can be measured directly or indirectly. A direct measurement is obtained with a concentric-cylinder viscometer, shown in Fig. 14.3. The inner

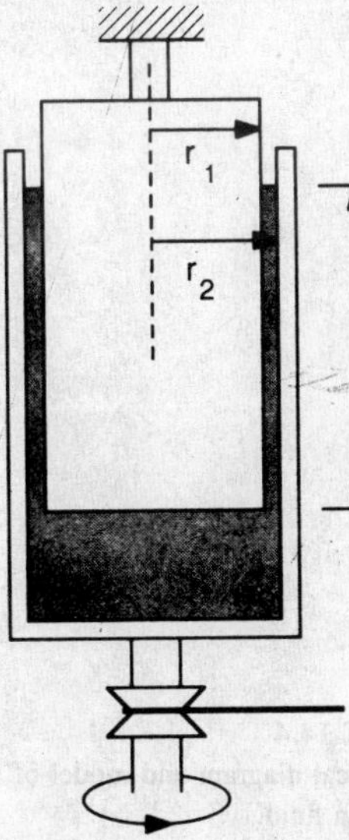

FIGURE 14.3
Measurement of viscosity.

cylinder is stationary, the outer rotates. Because of the shear resistance in the fluid between the cylinders, a torque is exerted on the inner cylinder. Neglecting the effect of the fluid below the bottom of the inner cylinder, the viscosity can be determined from

$$\mu = \frac{15Tb}{\pi^2 r_1^2 r_2 hn} \tag{14.5}$$

where T = torque, Nm
$b = r_2 - r_1$ = space between cylinders, m
r_1 = radius of inner cylinder, m
r_2 = inside radius of outer cylinder, m
h = height of inner cylinder, m
n = speed of rotation, r/min

On a construction site, a check of the viscosity of a grout can be obtained by measuring the time required for a certain amount of fluid to flow through a standardized funnel. Most popular is the Marsh cone shown in Fig. 14.6; the time recorded is usually for the outflow of the first 1000 mL. Other test equipment use cones with wider and shorter funnels and with larger openings, or measure the travel distance of the grout along a trough under standard conditions. As pointed out by Lombardi

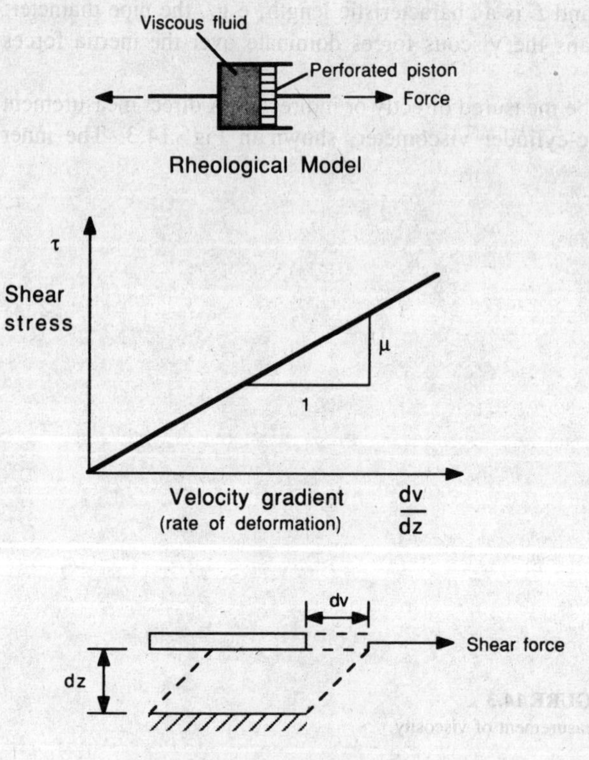

FIGURE 14.4 Rheological diagram and model of a newtonian fluid.

(1985), the Marsh cone measures a combination of rheological properties, rather than viscosity alone, and its results should be termed "apparent viscosity."

The viscosity of "evolutive" grouts, like silica gels, increases gradually until they set. Acrylic and other resins and other "nonevolutive" grouts show constant viscosity until they set almost instantaneously; their setting time, usually measured in minutes, is controlled by the type of catalyst present and the temperature.

Ideally, a grout should have low viscosity, a controllable setting time, and high strength once it is in the ground. In addition it should be nontoxic, permanent, and cheap.

14.2.2.2 RHEOLOGICAL MODELS. In a *newtonian* fluid the shear stress τ varies proportionally to the velocity gradient dv/dz, as expressed in Eq. (14.1) and shown in Fig. 14.4. The mechanical model corresponding to a newtonian fluid is a dashpot. (A rheological model is called linear if there is a linear relationship between stresses and strains and their derivatives with respect to time. A fluid which has a nonlinear relationship between τ and dv/dz is classified as *non-newtonian*.

Figure 14.5 illustrates the behavior of a material which has an initial yield stress which must be overcome before continuous deformation takes place. The mechanical model is that of a dashpot and a frictional resistance element in parallel, and substances which follow its rules are called *Bingham* bodies. Strictly speaking, a Bingham body is not a fluid, but rather a visco-plastic solid; nevertheless the term "Bingham fluid"

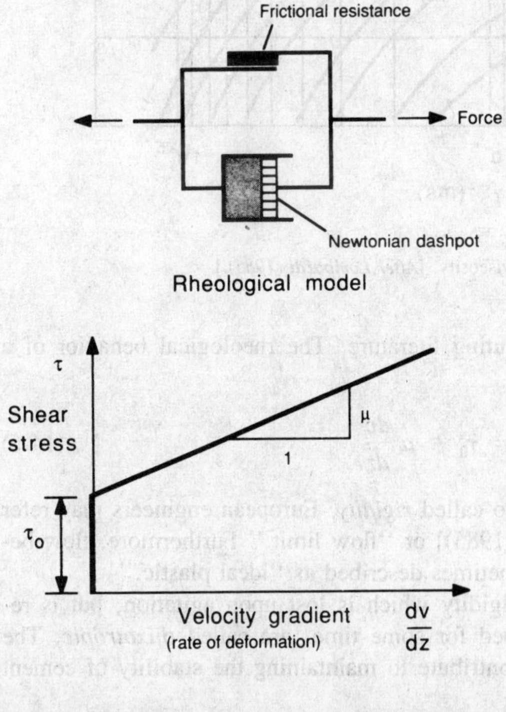

FIGURE 14.5
Rheological diagram and model of a Bingham body.

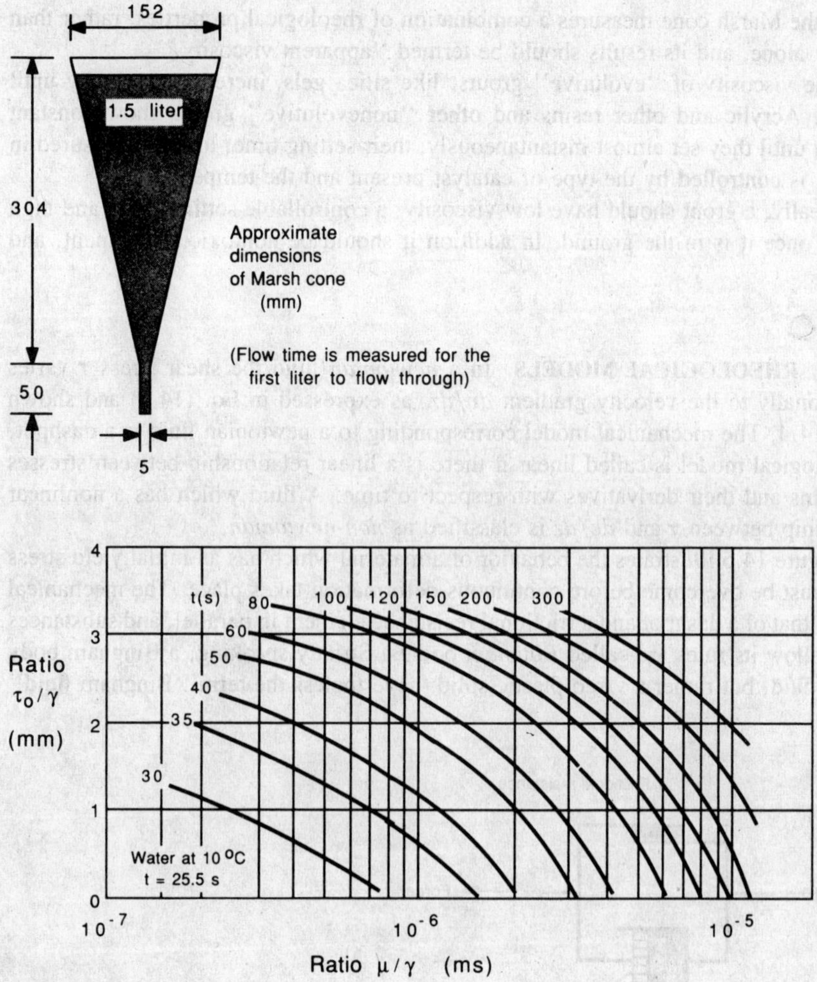

FIGURE 14.6
Marsh cone flow time versus yield stress and viscosity. [*After Lombardi (1985).*]

is used quite extensively in the grouting literature. The rheological behavior of a Bingham body is expressed by

$$\tau = \tau_0 + \mu \frac{dv}{dz} \tag{14.6}$$

The initial yield stress τ_0 is also called *rigidity;* European engineers may refer to it as ''cohesion'' [see Lombardi (1985)] or ''flow limit.'' Furthermore, flow behavior according to Eq. (14.6) is sometimes described as ''ideal plastic.''

Suspensions which exhibit a rigidity which is lost upon agitation, but is recovered after they are left undisturbed for some time, are called *thixotropic*. The thixotropic properties of bentonite contribute to maintaining the stability of cement grouts. As such grouts become stationary, they tend to settle out, "consolidate," or "bleed" water, a process which is slowed down or inhibited by thixotropy until the cement hardens.

Water behaves as a newtonian fluid. Clay particles in suspension represent a non-newtonian substance. A clay or cement grout may be approximately treated as a Bingham body. Bentonite clay is a typical thixotropic material.

For Bingham grouts, Lombardi (1985) proposed a simple plate "cohesion" meter. A thin steel plate with rough surfaces is immersed in the grout. The initial yield stress or "cohesion" τ_0 can be determined from the amount of grout sticking to the plate. The thickness of the grout cover (computed from the total weight of grout and its unit weight γ) is equal to τ_0/γ. Given the flow time from a Marsh cone and τ_0, Lombardi's graph (Fig. 14.6) allows the true viscosity to be determined.

14.2.2.3 VISCOUS AND VISCO-PLASTIC FLOW IN PIPES. The difference in behavior of newtonian fluids and Bingham grouts can be well demonstrated for the case of tubular flow.

A *newtonian liquid* starts to flow as soon as there is a hydraulic gradient i, defined as the total head loss Δh over the distance Δx in the direction of the flow.

$$i = \frac{\Delta h}{\Delta x} \tag{14.7}$$

The flow rate Q, the velocity variation $v(z)$ with distance from the center, and the maximum velocity v_{max} at the center of the tube ($z = 0$, see Fig. 14.7a) are then given by

$$Q = \frac{\gamma i \pi r^4}{8\mu} \tag{14.8}$$

$$v(z) = \frac{\gamma i}{4\mu}(r^2 - z^2) \tag{14.9}$$

$$v_{max} = \frac{\gamma i}{4\mu} r^2 \tag{14.10}$$

where r is the radius of the tube and γ is the unit weight of the liquid. Equation (14.8) is equivalent to Poiseuille's law, also known as the Hagen-Poiseuille equation.

Obviously, grout flow in soil or rock is much more complex than viscous flow in a pipe. Nevertheless some fundamental conclusions can be drawn which hold for laminar flow:

The rate of grout intake is proportional to the hydraulic gradient (or pressure) applied.

The rate of grout intake is inversely proportional to the viscosity of the grout.

For a given pressure and viscosity the grout flow increases with the fourth power of the radius of a void! (For soils, void size is directly related to grain size.)

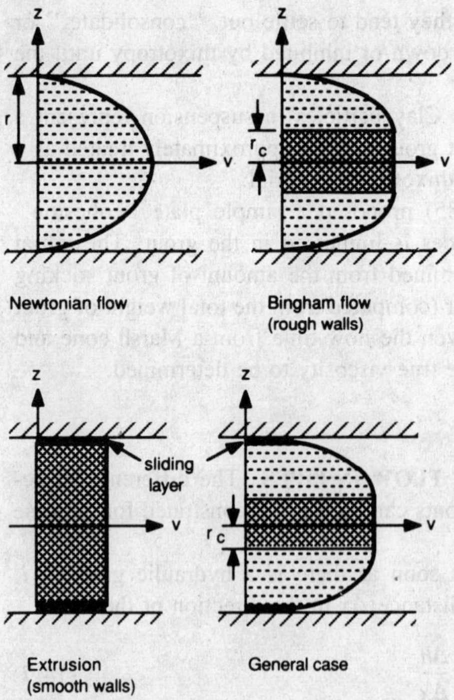

FIGURE 14.7
Velocity profiles for viscous and visco-plastic flow in pipes. [*After Lombardi (1985).*]

In the visco-plastic flow of a *Bingham substance* a rigid core of diameter $2r_c$ is pushed through the center of the pipe (Fig. 14.7*b*). The diameter of this core depends on the yield stress τ_0. A core element of length Δx is subjected to a pressure equal to $\Delta h\ \gamma$ over a cross-sectional area πr_c^2. Movement is resisted by the shear stress on the cylindrical surface of the core, equal to $2\pi r_c \tau_0\ \Delta x$. Therefore,

$$\Delta h\ \gamma\ \pi r_c^2 = 2\pi r_c \tau_0\ \Delta x$$

and

$$r_c = \frac{2\tau_0}{i\gamma} \qquad \text{where } i = \frac{\Delta h}{\Delta x} \tag{14.11}$$

If $r_c > r$, no movement can take place. The minimum gradient i_{min} required to cause flow corresponds to the condition where the initial yield stress is just reached at the boundaries of the pipe:

$$i_{min} = \frac{2\tau_0}{r\gamma} \tag{14.12}$$

The above equation can also be used to calculate the maximum distance Δx_{max} which a Bingham grout would travel in a tube of radius r under an applied head Δh (before the gradient $\Delta h/\Delta x$ is less than i_{min}):

$$\Delta x_{max} = \frac{\Delta h \gamma r}{2\tau_0} \tag{14.13}$$

The analysis further yields [after Lombardi (1985)]:

$$Q = \frac{\gamma i \pi r^4}{8\mu}\left[1 - \frac{4r_c}{3r} + \frac{(r_c/r)^4}{3}\right] \tag{14.14}$$

$$v(z) = \begin{cases} v_{\max} = \dfrac{\gamma i}{4\mu} r^2 \left(1 - \dfrac{r_c}{r}\right)^2 & \text{for } z < r_c \\ \dfrac{\gamma i}{4\mu} r^2 \left[\left(1 - \dfrac{r_c}{r}\right)^2 - \dfrac{r_c^2}{r^2}\left(1 - \dfrac{z}{r_c}\right)^2\right] & \text{for } z > r_c \end{cases} \tag{14.15}$$

These expressions are valid for flow between rough boundaries, which means the shear stress along the wall of the pipe is larger than τ_0.

If the pipe walls are very smooth, pure extrusion may occur, characterized by a constant velocity profile. This case is discussed by Lombardi (1985), who also analyzed a general case of visco-plastic flow, where the shear stress in a smooth boundary layer of specified thickness is reduced by a given factor α, according to

$$\tau = \alpha\left(\tau_0 + \mu \frac{dv}{dz}\right) \qquad \alpha < 1 \tag{14.16}$$

The velocity profiles for newtonian and Bingham flows are compared in Fig. 14.7, which also illustrates the limiting case of extrusion and general visco-elastic flow [after Lombardi (1985)]. The following conclusions can be drawn:

For newtonian fluids, the flow velocity has a parabolic distribution with a maximum value in the center. Flow starts as soon as there is a hydraulic gradient.

Bingham flow in pipes is characterized by a stiff core moving along the center of the cross section. A minimum pressure, related to the rigidity (initial yield stress) of the substance, is required to induce flow.

For equal conditions and viscosities, a newtonian grout travels quicker than a Bingham grout.

Generally, as grout moves radially away from an injection point, its velocity reduces. Grout movement stops when either the setting time is reached or, for Bingham fluids, when the hydraulic gradient reduces below the minimum value required.

Viscous and visco-plastic flow for boundary conditions other than those of tubular flow, and more appropriate to the grouting of soils and rocks, will be introduced in Secs. 14.3 and 14.4.

14.3 GROUTING TECHNIQUES AND CONTROLS

According to a historical review by Bowen (1981), grouting in civil engineering dates back some 300 years. At that time injected clay, lime, and cement started to be used

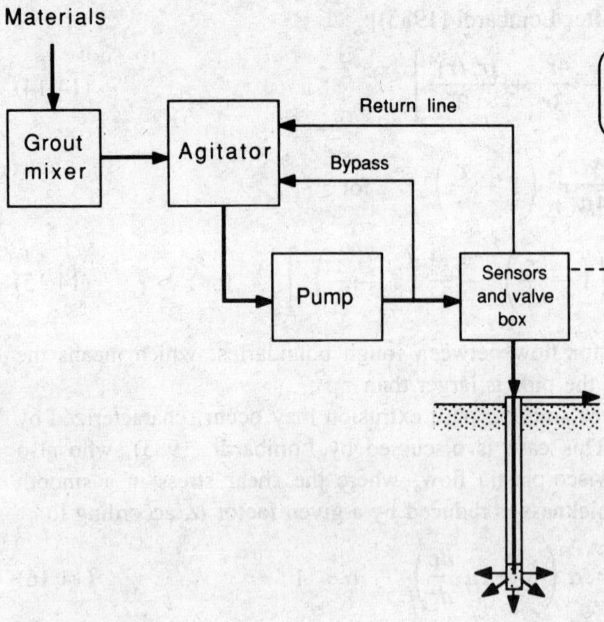

FIGURE 14.8
Equipment used in modern cement grouting.

to repair masonry walls, fill cracks in load-bearing structures, and seal off water flow in rock fissures.

Initially grout consisted only of a single or multicomponent material, mixed with a reactant before injection (''one-shot'' technique), designed to set within a few minutes or hours. Improvement in the grouting methods depended to a large extent on the development of an efficient mixing and pumping technology. H. J. Joosten, around 1925, invented a two-shot grouting technique suitable for stabilizing soils as small-grained as fine sands [see Joosten (1953)]. He pumped sodium silicate and calcium chloride into the ground from two separate pipes; these chemicals reacted instantly in the soil when mixed.

Figure 14.8 shows schematically the equipment employed in modern cement grouting. On large-scale projects, computers control mixing and injection of grout and record pressures and quantities pushed into the voids or fissures of the ground.

Grout could be injected as drilling proceeds, but not much control could be exerted in such an approach. It is preferable to organize drilling and grouting as separate phases, e.g., with grouting commencing once the borehole is completed, or alternating with drilling in stages. In order to treat a particular ground stratum, the corresponding length of borehole is isolated by expanding rubber ''packers'' built into the drilling rods; grout is then only allowed to flow into the soil or rock from between two packers or, if a single packer is used, between it and the bottom of the hole.

Packers in contact with the ground are only feasible in rock grouting. For stage grouting in soils, the sleeve tube (or *tube à manchette*) has been developed (Fig. 14.9). The procedure is as follows:

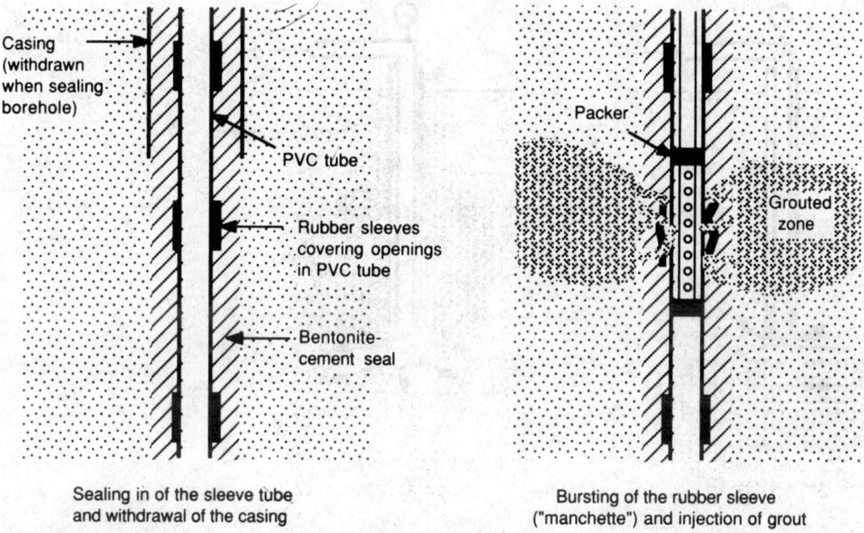

FIGURE 14.9
Sleeve tube grouting (*tube à manchette*).

(***a***) The hole is drilled and cased.
(***b***) A steel or plastic tube, slotted at regular intervals, is inserted. The vertical slots are covered with a rubber sleeve.
(***c***) As the casing is withdrawn, the space between the sleeve tube and the borehole wall is sealed with a cement-bentonite grout.
(***d***) After the seal has set, the grouting tube is inserted. Grout exits between two packers allowing injection through selected slots. With increasing pressure, the rubber sleeve bursts and grout flows into the soil.

With the sleeve tube technique, grouting can be repeated in the same hole, e.g., using different viscosity grouts or different chemicals in a planned sequence. Flexibility is important where permeabilities of the ground vary significantly from point to point.

Grouting in stages may proceed in a descending (downhole, downstage) or ascending (upstage) direction (Fig. 14.10). In the descending method, impregnation of the ground occurs in advance of the borehole, which could be advantageous in loose soil or rock. In the ascending technique grouting follows drilling as a separate phase; a benefit would be that water pressure testing is possible immediately prior to grouting, allowing for a choice of the most suitable grout type, pressure, and quantity of grout for that particular stratum.

Experience and intuitive judgment, if not art, are used in deciding on the pattern of boreholes, the sequence of grouting holes within a group, the stages of injection along a single hole, the pressures, viscosities, and quantities to be used in each phase of the operation, etc.—all aimed at minimum wastage of grout, least damage to the ground, and maximum gain in strength or reduction in seepage.

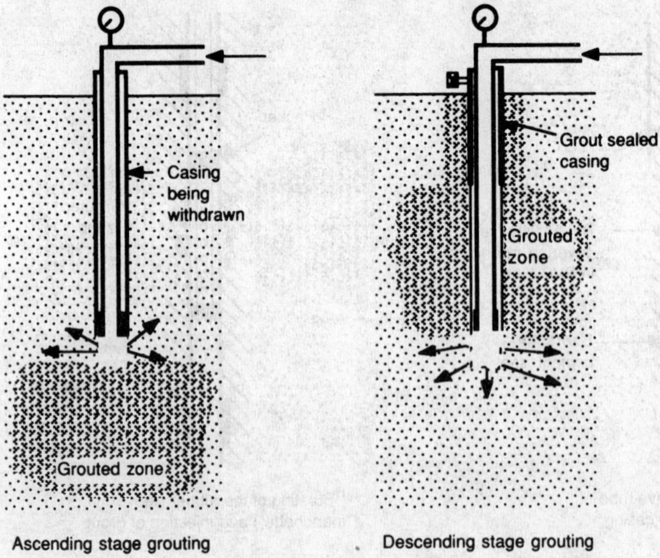

FIGURE 14.10
Descending and ascending stage grouting.

Some engineers say that the success of grouting can only be assessed after the job is completed, when seepage measurements or settlement observations for the structure under operating conditions are available. However, some controls are possible, if not vital during the actual grouting process or at intervals during construction. They may include:

Monitoring the grout taken as a function of pressure

Observing ground heave

Recording piezometer levels

Digging inspection pits

Retrieving core samples for examination and laboratory testing

Photographing walls of boreholes for visual inspection (particularly effective if grout has been dyed)

Conducting pressuremeter tests, possibly penetration tests

Conducting borehole permeability tests

Obviously the better the site investigation before grouting, the better is the chance of selecting a successful grouting strategy.

14.4 PERMEATION GROUTING OF SOILS

Grout permeation through soil is generally related to the grout's permeability, measured in terms of the coefficient of permeability k according to Darcy's law [see

Eq. (7.9)]. For a particular fluid, k is primarily a function of the void ratio (or corresponding porosity or density), but particle size distribution, soil structure, saturation, and other factors also influence its value.

If soil voids are represented by a system of tubes with equal permeability, then the formulas presented in Sec. 14.2.2.3 could be used to estimate grout take and reach. A spherical or cylindrical flow model for a porous medium is however more appropriate for permeation grouting of soils from boreholes. Using basic "well hydraulics" (see Chap. 8) the distance traveled by the grout can be related to the grouting rate and time, as shown in the following two sections. For a slightly more sophisticated approach see Raffle and Greenwood (1961).

14.4.1 Spherical Flow Model for Porous Media (newtonian fluid)

Imagine that grout permeates soil from a spherical cavity of radius R_0 under the influence of a net pressure p_e (in excess of local hydrostatic pressure). Then, for laminar newtonian flow, the following relationship holds:

$$p_e = \frac{Q\gamma}{4\pi R_0 k_G} = \frac{Q\gamma\mu}{4\pi R_0 k \mu_w} \tag{14.17}$$

where Q = grouting rate, m^3/s
γ = unit weight of grout, kN/m^3
k_G = permeability of soil to grout, m/s
k = permeability of soil to water, m/s
μ = viscosity of grout, Pa · s
μ_w = viscosity of water, Pa · s

During spherical grout permeation in time dt the grout travels a distance dr. The grout taken in time t can be found by integration from

$$Q\,dt = 4\pi r^2 n\,dr \tag{14.18}$$

where n is the porosity of the soil (volume of voids/total volume).

The time required to travel a distance R from a spherical cavity with radius R_0 can be computed by

$$t = \frac{4\pi n}{3Q}(R^3 - R_0^3) \tag{14.19}$$

Theoretically a newtonian grout will continue to travel outward, as long as an excess head (p_e/γ) exists or until it sets or gels.

14.4.2 Radial Flow from a Cylindrical Cavity (newtonian fluid)

Equations equivalent to the above, but for horizontal cylindrical flow as from a section of a borehole, are

$$p_e = \frac{Q\gamma\mu}{2\pi mk\mu_w} \ln \frac{R}{R_0} \tag{14.20}$$

where p_e = excess pressure necessary to maintain flow Q when grout has reached distance R from the injection point

R_0 = radius of borehole

m = thickness of layer being grouted

Also,

$$t = \frac{\pi mn}{Q}(R^2 - R_0^2) \tag{14.21}$$

These equations are analogous to those pertaining to a single well fully penetrating a confined aquifer (Chap. 8, App. 8B, Eq. B8.6), except that they represent conditions of recharge, rather than drawdown. As for a confined aquifer being recharged, the pressure $p(R)$ of the grout diminishes with distance R from the borehole according to

$$p(R) = p_e - \frac{Q\gamma\mu}{2\pi mk\mu_w} \ln \frac{R}{R_0} \tag{14.22}$$

These equations demonstrate that

The time required to treat soil over a given distance from the injection hole depends on the grouting rate Q.

The grouting rate can be increased by using a higher pressure of grouting or a lower viscosity grout. (Too high a pressure will, however, cause fracturing of the ground and/or surface heave, neither of which may be desirable.)

Setting time of the grout has to be greater than the time required for the grout to permeate the ground being modified.

The larger the radius R_0 of the borehole, the higher the pressure at a given distance.

14.4.3 Groutability of Soils Based on Permeability

Although equipment and techniques are equally important as the grouting material itself, a literature review gives us reasonable guidance as to the suitability of various categories of grout for different groups of soils.

An obvious indicator of the penetrability of a granular medium would be the pore size. Tubular flow analysis (Sec. 14.2.2.3) teaches us that the resistance to penetration decreases with the fourth power of the diameter of the opening! A representative pore size is, however, not easily obtained, and therefore the primary soil parameter used for selecting an effective grout is the soil's permeability to water.

The methods of determining the permeability of granular soils was previously discussed in Secs. 8.7.4 and 9.1.1 in relation to dewatering by wells. A constant or falling head test in a single borehole seems to be the field test most appropriate for investigating the groutability of soils. Care would have to be taken in interpreting results for anisotropic soils, where the measurements could be significantly affected by the geometry of the cavity from which water flows into the soil (see references given in Sec. 9.1.1.3).

Table 14.1 [after Cambefort (1987)] on p. 351 gives a general overview of applications of various suspensions and liquid grouts in soils, the latter rated according to their coefficient of permeability k (m/s). In this table, pure cement suspensions are mainly recommended for fissure grouting. Nevertheless, ordinary cements may adequately permeate gravels, and special forms of cement, such as colloidal cements and the Japanese-developed Microfine Cement, extend the range of applications into the sand range as shown in Fig. 14.11 [after Clarke (1984)]. According to Clarke, Microfine Cement has become a substitute for more toxic chemical grouts and has been used on many projects for underground strength improvement and, in combination with sodium silicate, for underground water control. Figure 14.12 compares the viscosity of Microfine Cement with that of colloidal and ordinary Portland cement. Increasing the water/cement ratio does decrease the viscosity but also increases the gel time and reduces the strength of the grouted soil.

Microfine Cement is relatively new (and expensive!), and most permeation grouting of sand is done with low-viscosity chemical grouts. Although the shear strength of the gel formed may be small, it can effectively stop seepage flow in fine

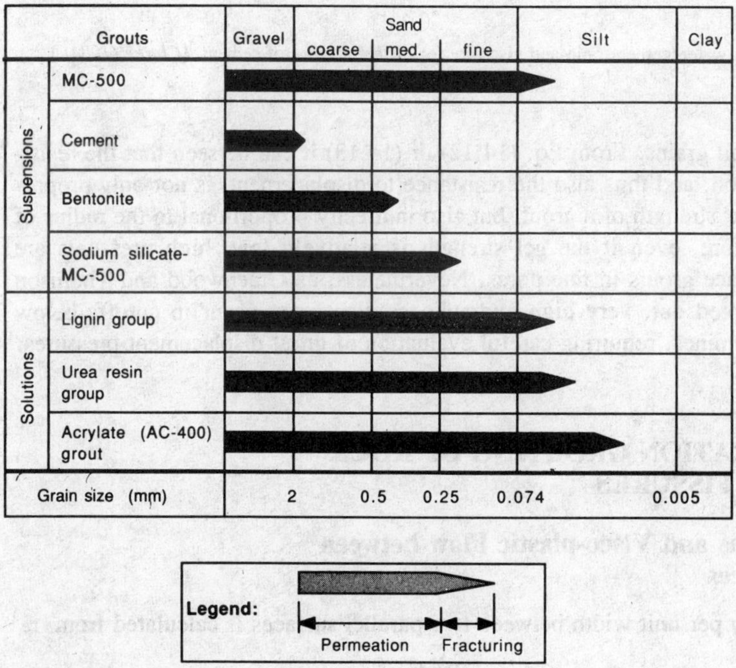

FIGURE 14.11
Comparison of permeation of grouts. [*After Clarke (1984).*]

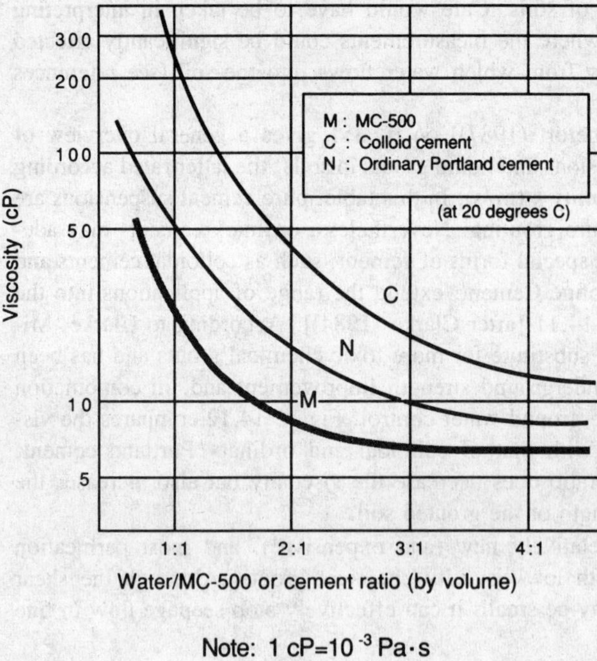

FIGURE 14.12
Relationship between water/cement ratio and viscosity for different types of cement. [*Clarke (1984).*]

voids between soil grains. From Eq. (14.12) or (14.13) it can be seen that the resistance to penetration, and thus also the resistance to displacement, is not only proportional to the shear strength of a grout, but also indirectly proportional to the radius of the void. Therefore, even if the gel strength is relatively low, high pressures are required to displace grouts in fine pores. Nevertheless, as Greenwood and Thomson (1984) have pointed out, very high hydraulic gradients can occur in cutoffs below dams or in deep mines, requiring careful evaluation of grout displacement pressures.

14.5 PERMEATION GROUTING OF ROCK JOINTS AND FISSURES

14.5.1 Viscous and Visco-plastic Flow between Parallel Surfaces

Newtonian flow q per unit width between two parallel surfaces is calculated from

$$q = \frac{2\gamma i d^3}{3\mu} \tag{14.23}$$

where γ = unit weight of grout (N/m^3)
i = hydraulic gradient
μ = viscosity, Pa · s
d = half of distance between the surfaces (or half of joint thicknesses), m

as in Sec. 14.2.2.3.

The velocity profile is described by

$$v(z) = \frac{\gamma i}{2\mu}(d^2 - z^2) \tag{14.24}$$

with

$$v_{\max} = \frac{\gamma i}{2\mu} \tag{14.25}$$

For *Bingham flow* between two parallel rough surfaces the characteristic equations [analogous to Eqs. (14.11) to (14.15) for pipe flow] are, as presented by Lombardi (1985):

$$d_c = \frac{\tau_0}{\gamma i} = \text{half thickness of rigid core} \tag{14.26}$$

$$i_{\min} = \frac{\tau_0}{d\gamma} \tag{14.27}$$

$$q = \frac{2\gamma i d^3}{3\mu}\left[1 - \frac{3d_c}{2d} + \frac{1}{2}\left(\frac{d_c}{d}\right)^3\right] \tag{14.28}$$

$$v_{\max} = \frac{\gamma i d^2}{2\mu}\left(1 - \frac{d_c}{d}\right)^2 \tag{14.29}$$

The corresponding maximum travel is [compare Eq. (14.13)]:

$$\Delta x_{\max} = \frac{\Delta h \gamma d}{\tau_0} \tag{14.30}$$

where $\Delta h\gamma = p_e$ is the grout pressure in excess of the hydrostatic pressure at the injection point.

Figure 14.13 illustrates the case where grout under pressure p_e is radiating into a horizontal planar joint of thickness $2d$ and, after some time (theoretically ∞), has reached a maximum distance $\Delta x_{\max}$ determined by its shear strength τ_0. The pressure distribution will then be triangular (gradient $i_{\min}$), and the resultant uplift force P can be calculated according to

$$P = \frac{\pi\, \Delta x_{\max}^2\, p_e}{3} \tag{14.31}$$

or

$$P = \frac{\pi p_e^3 d^2}{3\tau_0^2} \tag{14.32}$$

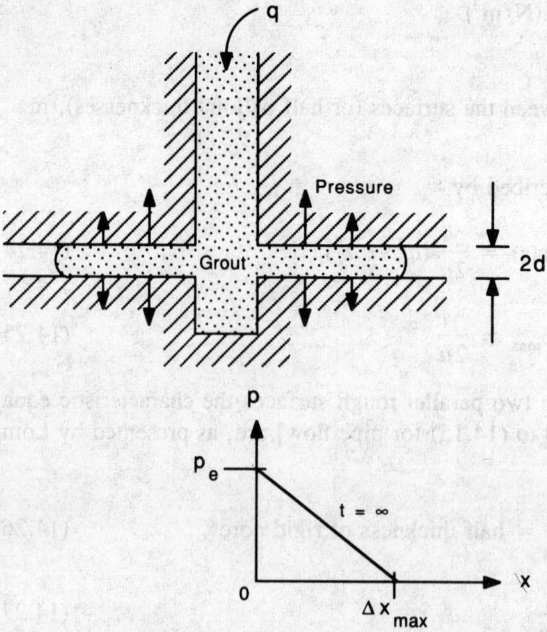

FIGURE 14.13
Grouting pressure and uplift force in a joint.

Following Lombardi's (1985) approach, this force can also be expressed in terms of the maximum volume V_{max} of grout taken when the flow ceases. With

$$V_{max} = \pi \, \Delta x_{max}^2 \cdot 2d \tag{14.33}$$

$$P = \frac{V_{max} p_e}{6d} \tag{14.34}$$

or

$$P = \sqrt[3]{\frac{\pi \, p_e^5 V_{max}^2}{108 \tau_0^2}} \tag{14.35}$$

According to Lombardi (1985), the last two expressions confirm empirical relationships between the lifting force, grout pressure, volume, and initial shear strength (or cohesion). The force P exerted on the ground by grout under pressure could be used to estimate the widening of an existing fissure or could be compared with the uplift capacity of the rock (Sec. 20.1.4).

14.5.2 Structure and Permeability of Rocks

Rock is the parent material of soils and therefore should be more coherent or consolidated and less permeable than soils. Indeed the *rock substance* as tested in the laboratory generally has very low permeabilities, although its porosity could be relatively high. Consider the following examples, selected from information presented by Jumikis (1983):

Rock type	Coefficient of permeability k, m/s	Porosity n, %
Igneous	10^{-5}–10^{-9}	0.1–4
Metamorphic	10^{-5}–10^{-9}	0.1–4
Sedimentary	10^{-4}–10^{-11}	5–20

Most rock substances would therefore appear to be unsuitable for grouting, even with chemicals. However, rather than being determined by the characteristics of its rock substance, the permeability of a *rock mass* is determined largely by the nature of its structure, more specifically its discontinuities. The same holds true for other engineering properties, such as strength and stress-strain behavior. Discontinuities or fractures may be defined in more detail by the following terms:

Joint: A break, usually of geological origin (due to tectonic forces). Normally occurs in sets (strike joints, dip joints, etc.). No visible movement along the break.

Fissure: A fracture with a gap through which water can permeate unless filled in with sediment (or grout!).

Crack: A relatively small fracture.

Fault: A discontinuity along which there has been displacement.

Shear, shear zone: Planes or bands of material in which local shear failure has taken place.

Dyke: Long, narrow intrusion of generally fine igneous rock.

In describing a discontinuity an engineering geologist may refer to characteristics such as

Persistence: Whether the fracture is continuous or not
Roughness: Unevenness or waviness
Aperture: Perpendicular distance across the gap
Filling: May be clay, quartz, calcite, etc.
Spacing: Perpendicular distance between adjacent discontinuities of the same set
Orientation: Strike and dip

Discontinuities may be observed on rock cores or along an exposed rock face. The rock quality designation (RQD), a measure of the number of discontinuities, is one of the inputs for rock mass classification, which may also include joint spacing, orientation, and condition (roughness and filling).

The characteristics of the joint system, in particular the spacing, aperture, and orientation of the fractures will affect the overall behavior of rock in civil engineering or mining works and will be one of the determinants of grout hole direction, spacing, and depth. Attempts have been made to relate rock mass permeability to the characteristics of the joint system. Goodman (1980), e.g., quoted the following expression, derived by Snow (1968), which relates to three idealized joint sets, all perpendicular to each other, with uniform apertures and smooth parallel surfaces:

$$k = \frac{\gamma(2d)^3}{6\mu S} \tag{14.36}$$

where k = coefficient of permeability of jointed mass, m/s
γ = unit weight of fluid, kN/m^3
μ = viscosity of fluid, Pa · s
$2d$ = aperture of joint, m
S = joint spacing, m

Knowing the joint spacing (e.g., from core analysis) and the permeability (from a borehole test), a hypothetical joint aperture could then be calculated. Equation (14.36) certainly contributes to the understanding of the relationship between rock mass permeability and structure, but because of its inherent limitations, it has little practical value. Anyway, practitioners, particularly in dam grouting [such as Houlsby (1982)], prefer to work in so-called Lugeon units, rather than in coefficients of permeability.

In the water pressure test developed by Lugeon (1933) a section of the borehole is isolated by one or two packers and from there water is injected into the surrounding rock. The flow of water is monitored as the excess pressure is increased in stages up to 981 kPa (10 kilograms of force per square centimeter), and then decreased again. Each pressure stage is held for 10 min to make it likely that fine particles blocking existing fissures are washed out. The amount of water absorbed is expressed in Lugeons.[1] One Lugeon is equivalent to a water loss of 1 liter per minute per meter length of borehole under a pressure of 981 kPa, maintained for 10 min. For a particular test, the water take in Lugeons is calculated according to

$$\text{Water take (Lugeons)} = \frac{981 V_w}{p_e L t} \approx \frac{1000 V_w}{p_e L t} \tag{14.37}$$

where V_w = water volume, L, flowing into the ground in time t, min
p_e = net increase in pressure at the injected point, kPa
L = length of borehole section being tested, m
t = time, min

The results of water pressure testing can be presented in the form of bar charts or a diagram as shown in Fig. 14.14. The grouting expert may derive important information from the variation in water take (in Lugeons) with consecutive test stages or a diagram showing the water flow rate Q (L/min) as a function of the pressure. These diagrams can indicate the nature of the flow (laminar or turbulent), blocking or washing out of fissures, enlargement of the fissure (possibly hydrofracture), and more; some examples of flow rate/pressure curves are given in Fig. 14.15.

---

[1] Named after Professor Maurice Lugeon; see Lugeon (1933).

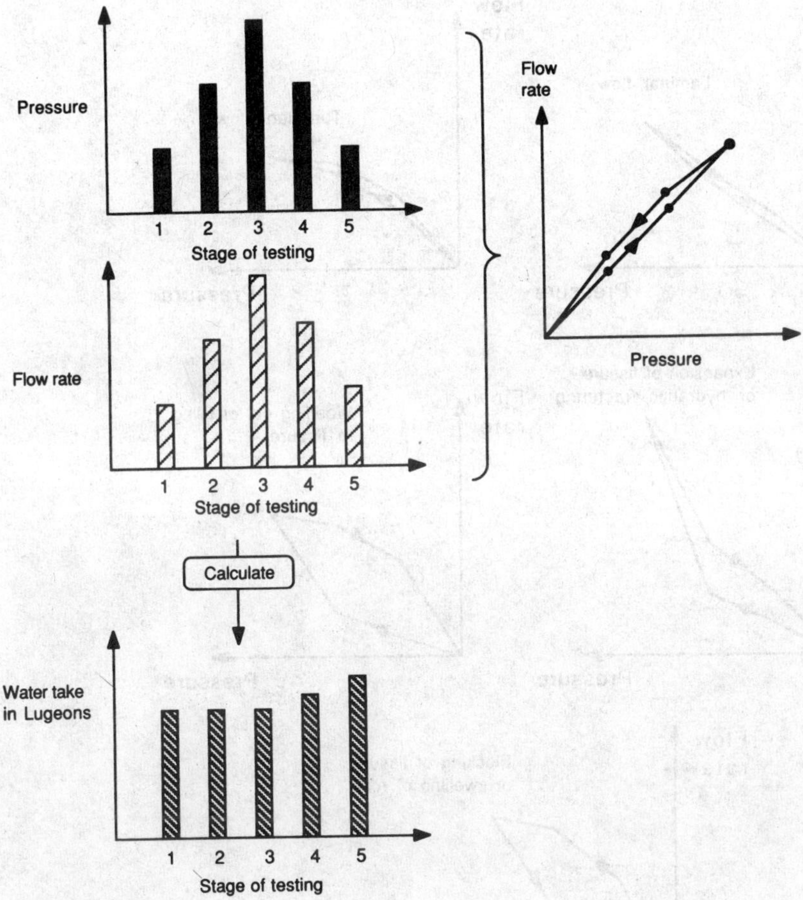

FIGURE 14.14
Water pressure testing—presentation of results.

Details of the Lugeon procedure, such as the maximum pressure used and the length of the borehole section selected, do affect the results; but they may have to be varied from test to test in order to prevent damage to the ground and to account for variations of rock structure with depth. It is doubtful that a universally acceptable, fully detailed standard procedure will evolve.

One Lugeon is said to be equivalent to a coefficient of permeability of about 10^{-7} m/s.

Lugeon (1933) already put forward criteria for the groutability and allowable permeabilities of rock below dams: If Lugeon values of a dam foundation exceed 1 (for heads > 30 m) or 3 (for heads < 30 m), then a grout curtain is advisable. Today's dam builders still follow similar guidelines, although much will depend on the monetary value of water losses through seepage. An additional consideration is whether or

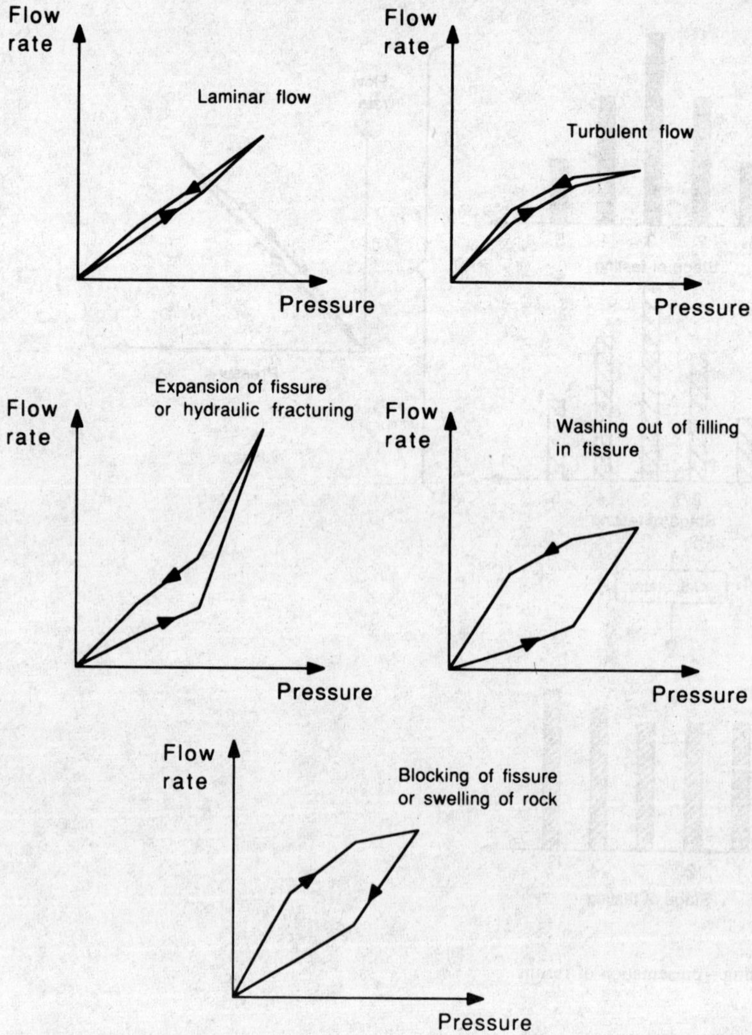

FIGURE 14.15
Interpretation of flow rate/pressure diagrams (schematic).

not dam safety against uplift is also enhanced by appropriate drainage provisions [see Houlsby (1982)].

Since Lugeon tests are carried out before grouting, it would be nice if their results could be used to predict the grout take. This is, however, not usually the case. Varying apertures of fissures and the different flow characteristics between water and grout are the main reasons for a lack of correlation between water absorption and grout take. Anyway, grout quantities required are not only a function of the rock itself. Much will depend on the sequence of the grouting operations and the possible control over the grout characteristics from stage to stage.

For example, a problem arises where the rock contains large as well as fine fissures. In such a case it may be decided to start grouting with very liquid suspensions (e.g., high water/cement ratios) in order to fill smaller voids before they are blocked off by subsequently used stiffer grout, necessary to seal larger cavities. On the other hand, large irregular openings may cause high grout losses in the initial stages of pumping; ways will then have to be found to block such cavities, either by choosing adequately thick, coarse, or unstable grouts or by appropriate spacing and sequencing of injections.

14.6 HYDRAULIC FRACTURING OF SOILS AND ROCKS

If the grouting pressure is increased sufficiently, a soil mass may split and artificial grout-filled fissures are formed; in rock, existing fissures may enlarge and new breaks may occur. This is called hydraulic fracturing or hydrofracture, also referred to as *claquage* by French engineers [see Cambefort (1977)].

Ground heave could be an indication of hydraulic fracturing, at least in a granular ground and at shallow depths in rock. In slab-jacking, or lifting of building foundations, surface heave may be the objective of grouting, but elsewhere it is not generally desirable.

Another indication of hydraulic fracturing is back flow of liquid grouts as soon as the injection stops. For suspension grouts, back flow is inhibited by partial sedimentation. The expansion of fissures during grouting of fractured rock may have a positive result: A more complete filling of the fissure is achieved after elastic rebound of the enclosing ground mass; in the extreme case, the deposited grout is compressed by the closing fissure and some of the grouting pressure is locked in—a prestressing effect.

Besides causing ground heave and back flow, hydrofracture may also be apparent from a Lugeon bar plot or a flow rate/pressure diagram (Figs. 14.14 and 14.15). In addition, acoustic emission monitoring appears to have the potential to identify the occurrence of hydraulic fracturing in the ground [Koerner et al. (1985)].

As pointed out in the introduction to this chapter, hydrofracture is a planned event in displacement grouting. In contrast, during penetration or permeation grouting, it may or may not be considered desirable. Being able to predict the pressure corresponding to ground movement would therefore be advantageous. This is certainly not straightforward, but it is easier in concept for soils than rocks.

In soils, the overburden pressure would be a reasonable, conservative guide for predicting the danger of hydrofracture. However, analogous to calculating the load capacity of an uplift anchor (Sec. 20.1.4), much depends on the shape of the failure surface expected. Also, unless the soil is overconsolidated, initial cracks could well run vertically along the borehole (even to the surface!) rather than along weaker layers of soil in a horizontal direction.

In rock, much more than in soils, the existing "tectonic" stress state, and any redistribution of stresses due to mining or construction activities, could be very important when it comes to predicting or interpreting the effect of high grouting pressures on the ground. The use of formulas such as Eqs. (14.32) to (14.34) in predicting the opening of fissures in a rock mass, e.g., by assuming linear elastic deformation, are therefore rather simplistic.

Whether the (low pressure) penetration or (high pressure) displacement grouting techniques are to be used on a particular job may not only depend on the prevailing geological conditions and the particular type of grout used, but also on local experience, design philosophy, and construction practice. Some experts maintain that penetration grouting is adequate for the treatment of most foundations; others see the occurrence of hydraulic fracturing of the ground not only as inevitable, but necessary for thorough impregnation.

PROBLEMS

Prefixes indicate problem type: C = calculations, M = multiple choice, B = brief answer, D = discussion.

Calculations

C14.1. A clay-cement-water grout has a density of 1.63 t/m^3. After air entrainment, the mixture has a density of 1.24 t/m^3. Calculate the expansion coefficient, bulking coefficient, and air ratio.

C14.2. Polyurethane-foamed plastics were investigated by Vinson and Mitchell (1972) in view of the possible need for soil stabilization on the moon. When injected in sand, this grout produced strengths on the order of 10 to 40 MPa and drastically reduced permeability. Foam plastics produced in the soil voids had densities ranging from 0.2 to 0.8 g/cm^3. Assuming the specific gravity of the plastic is 1, what would be the range in air ratios of this grout?

C14.3. A concentric-cylinder viscometer has the following dimensions: $r_1 = 70$ mm, $r_1 - r_2 = 1.25$ mm, $h = 150$ mm, $n = 150$ r/min. The torque measured is $T = 3$ Nm. Calculate the viscosity of the liquid in pascal-seconds.

C14.4. Using Lombardi's plate "cohesion" meter and the Marsh cone for testing a cement grout with a water/cement ratio of about 0.5 (by weight), the following results were obtained:

$$\frac{\tau_0}{\gamma} = 0.4 \text{ mm}$$

$$t = 50 \text{ s (flow time)}$$

(*a*) Using Fig. 14.6 and a unit weight of 1.8 t/m^3 (17.7 kN/m^3), determine the viscosity μ in units of pascal-seconds and centipoises.

(*b*) Express the given water/cement ratio (by weight) as a water/cement ratio by volume, assuming the bulked cement has a density of 1.5 g/cm^3. Give your answer in the form of (?:1).

C14.5. A newtonian fluid is passing through a capillary tube as shown in Fig. P14.1. For a tube diameter $2r = 0.5$ mm, it takes 1 h to collect 40 mL at the outlet. Given are $\Delta x = 1$ m, $\Delta h = 0.7$ m, and $\gamma = 9.81$ kN/m^3.

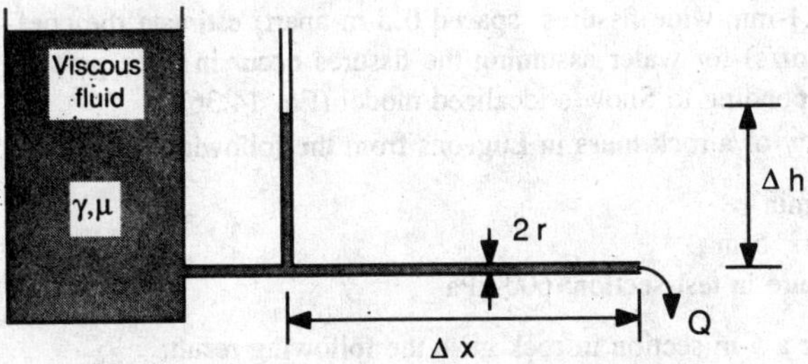

FIGURE P14.1

(*a*) Determine the viscosity of the fluid.
(*b*) Calculate the maximum velocity of fluid particles in the tube, and sketch the velocity profile across the tube.
(*c*) What would be the flow rate Q (mL/h) if the viscosity of the fluid was doubled?
(*d*) What would be the flow rate Q (mL/h) if the diameter of the tube was doubled?

C14.6. A cement grout has an initial shear strength $\tau_0 = 0.010$ kPa. If it is injected with a net pressure of 300 kPa, how far would it go
(*a*) Into a tubelike void with a 5-mm diameter?
(*b*) Into a 5-mm-wide fissure?

C14.7. A chemical grout with a viscosity of 2 mPa · s is injected into sand with a porosity of 0.3, at a rate of 0.5 L/s. It has been mixed so as to set within 50 s of entering the ground. The borehole has a diameter of 55 mm.
(*a*) How far does the grout travel assuming the spherical flow model is applicable?
(*b*) How far would the grout penetrate a 1-m-thick confined layer according to the model for radial flow from a cylindrical cavity?

C14.8. Assume that grout injection into a confined layer of soil can be modeled as a newtonian flow from a cylindrical cavity. The following details are given:

$$p_e = 100 \text{ kPa}$$

$$Q = 1.5 \text{ L/s}$$

$$m = 2 \text{ m}$$

$$k = 10^{-5} \text{ m/s}$$

$$\mu = 10 \text{ mPa} \cdot \text{s}$$

$$\mu_w = 1 \text{ mPa} \cdot \text{s}$$

$$R_0 = 100 \text{ mm}$$

(*a*) Calculate the distance R_{10} at which the pressure has decreased to $0.1\,p_e$.
(*b*) Plot the pressure distribution from the borehole to R_{10}.

C14.9. Using information presented by Clarke (Fig. 14.12), read off the viscosities which would be typical at a water/cement ratio of 1:1 for ordinary Portland cement, colloidal cement, and Microfine Cement.

C14.10. For a grout with an initial shear strength of 0.005 kPa, injected with a local excess pressure of 400 kPa,
(*a*) Calculate the distance of penetration into a 0.5-mm-wide fissure.
(*b*) Calculate the uplift pressure generated.

C14.11. For a rock mass with 0.1-mm-wide fissures, spaced 0.5 m apart, estimate the coefficient of permeability (m/s) for water assuming the fissures occur in three perpendicular joint sets, corresponding to Snow's idealized model (Eq. 14.36).

C14.12. Calculate the permeability of a rock mass in Lugeons from the following test data:

Water take: 100 L in 8 min
Distance between packers: 5 m
Excess hydrostatic pressure in test section: 600 kPa

C14.13. A water pressure test for a 5-m section in rock gave the following result:

Pressure (kPa):	200	400	600	400	200
Water take (Lugeon):	10	10	21	12	11

Interpret this result, taking Fig, 14.15 as a guide. If necessary convert the data into a flow rate/pressure diagram.

Multiple Choice

M14.14. Permeation grouting is also called
(*a*) Displacement grouting.
(*b*) Jet grouting.
(*c*) Penetration grouting.
(*d*) Backpack grouting.

M14.15. Which one of the following authors did *not* write a book about grouting?
(*a*) Bowen
(*b*) Cambefort
(*c*) Terzaghi
(*d*) Karol
(*e*) Ewert

M14.16. Cement grout is generally classified as a
(*a*) Chemical grout.
(*b*) Suspension.
(*c*) Solution.
(*d*) Emulsion.

M14.17. Which one of the following terms was *not* used in discussing the stability of a suspension grout?
(*a*) Sedimentation
(*b*) Settling out
(*c*) Gel time
(*d*) Filtration

M14.18. Which one of the following statements is *incorrect*?
(*a*) Dynamic viscosity may be given in units of pascal-seconds.
(*b*) The traditional unit of viscosity is the centipoise.
(*c*) Kinematic viscosity has units of square meters per second.
(*d*) Laminar flow occurs at high Reynolds numbers, turbulent flow at low Reynolds numbers.

M14.19. The rate of laminar newtonian flow in a capillary tube of radius r is proportional to
(*a*) r.
(*b*) r^2.

(*c*) r^3.
(*d*) r^4.

M14.20. Which one of the following statements is *incorrect*?
(*a*) For newtonian fluids, the flow velocity has a parabolic distribution with a maximum value in the center.
(*b*) For equal conditions and viscosities, a newtonian grout travels slower than a Bingham grout.
(*c*) For newtonian fluids flow starts as soon as there is a hydraulic gradient.
(*d*) For a Bingham fluid a minimum pressure is required to induce flow.

M14.21. Sodium silicate grout
(*a*) Has the chemical formula n-$SiO_2 \cdot Na_2O$.
(*b*) Is primarily used for increasing strength rather than decreasing permeability.
(*c*) Is a solution grout.
(*d*) Can be mixed with a cement grout.

M14.22. Borehole pressure test results are usually reported in units of
(*a*) Centimeters per second.
(*b*) Liters per meter.
(*c*) Lugeons.
(*d*) Langleys.

Brief Answer

B14.23. List five different types of grouting, categorized according to the way grout is introduced into the ground.

B14.24. Being as specific as possible, give examples of civil engineering projects where grouting was used for
(*a*) Decreasing the permeability of the soil ______________ .
(*b*) Increasing the strength or reducing the compressibility of the soil ______________ .
(*c*) Filling large voids ______________ .

B14.25. Identify the following two-phase systems describing grouts, and give an example for each:
(*a*) Small particles of solids distributed in liquid dispersion medium. ______________ , e.g., ______________ .
(*b*) Two-phase system wherein the disperse phase comprises minute droplets of liquid. ______________ , e.g., ______________ .
(*c*) Homogeneous molecular mixture of two or more substances. ______________ , e.g., ______________ .

B14.26. With respect to chemial grouts, frequently discussed issues are
(*a*) T______________ .
(*b*) P______________ .

B14.27. List four physical characteristics of a grouting liquid relevant to engineering applications.
(*a*) ______________
(*b*) ______________
(*c*) ______________
(*d*) ______________

B14.28. In Fig. P14.2 identify the curve characteristic for newtonian and non-newtonian fluids, respectively:
(*a*) Curve 1: ______________
(*b*) Curve 2: ______________

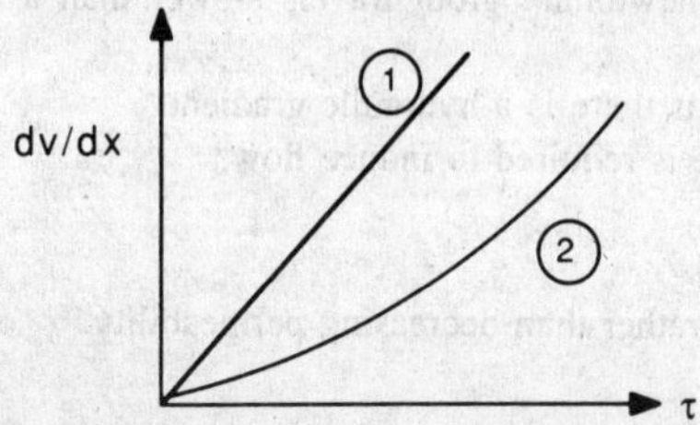

FIGURE P14.2

B14.29. Two common Bingham fluid grouts are
(*a*) ______________ .
(*b*) ______________ .

B14.30. Two common newtonian fluid grouts are
(*a*) ______________ .
(*b*) ______________ .

B14.31. During grouting, control procedures include measuring
(*a*) ______________ .
(*b*) ______________ .

B14.32. The effect of grouting may be evaluated after construction by
(*a*) ______________ .
(*b*) ______________ .

B14.33. Insert the words "joint", "fissure", "crack", or "fault" in the following definitions:
(*a*) A ______________ is a fracture with a gap through which water can permeate unless filled in with sediment (or grout!).
(*b*) A ______________ is a break, usually of geological origin (due to tectonic forces).
(*c*) A ______________ is a discontinuity along which there has been displacement.
(*d*) A ______________ is a relatively small fracture.

B14.34. List four characteristics used in describing a discontinuity:
(*a*) ______________
(*b*) ______________
(*c*) ______________
(*d*) ______________

B14.35. Hydraulic fracturing of the ground may be apparent from
(*a*) ______________ .
(*b*) ______________ .
(*c*) ______________ .

Discussion

D14.36. Why are some chemical grouts banned in Japan and elsewhere?

D14.37. What are the advantages and disadvantages of descending and ascending stage grouting (also called downstage and upstage grouting) in soil and rock?

D14.38. What aspects would you consider in deciding on the spacing and depth of injection holes for a grout curtain below a dam?

D14.39. Discuss the statement, "Some experts maintain that penetration grouting is adequate for the treatment of most foundations; others see the occurrence of hydraulic fracturing of the ground not only as inevitable, but necessary for thorough impregnation."

D14.40. Discuss the statement, "Some engineers say that the success of grouting can only be assessed after the job is completed, when seepage measurements or settlement observations for the structure under operating conditions are available."

CHAPTER 15

THERMAL MODIFICATION

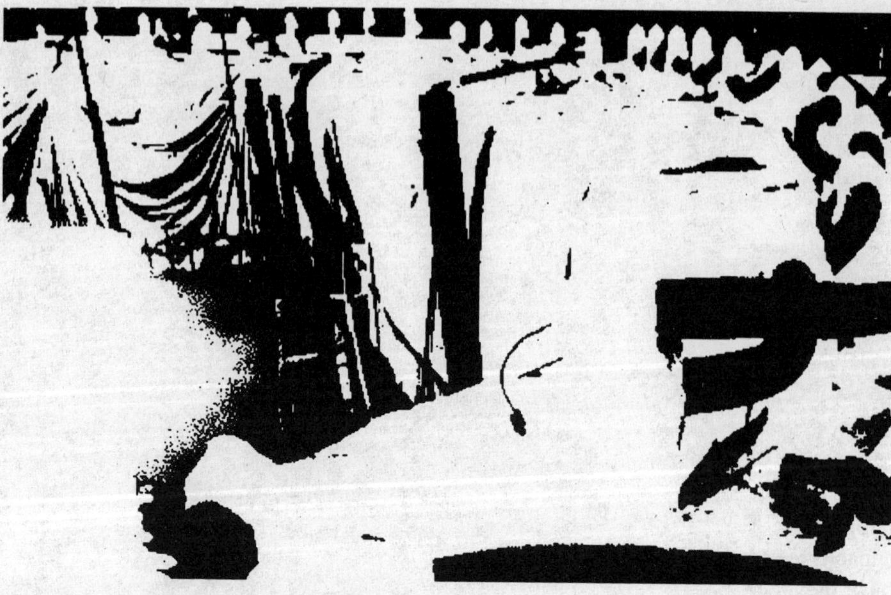

In this book, thermal modification refers to the artificial freezing of soils and fractured rock and the heat treatment of clayey soils for the purpose of improving the characteristics of the ground. Both these processes involve the use of thermal energy, but they are quite different in their application and effects. It should be said that in the geotechnical literature the term ''thermal soil stabilization'' often stands for ''heat treatment'' of soils alone and doesn't include artificial ground freezing.

15.1 INTRODUCTION TO THERMAL PROPERTIES OF SOILS

Temperature is measured in degrees Celsius (°C) or kelvin (K). These units are related by the equation

$$°\text{C} = \text{K} - 273.15 \tag{15.1}$$

Thermal conductivity k_T is defined as the amount of heat passing through a unit cross-sectional area of soil under a unit temperature gradient:

$$k_T = \frac{q}{A(T_2 - T_1)/L} \qquad \frac{\text{W}}{\text{m} \cdot \text{K}} \tag{15.2}$$

where q = heat flow, W
A = cross-sectional area, m^2
T = temperature, K
L = length of soil element, m

At 0°C the thermal conductivity of water is 0.58 W/(m · K) and that of ice is 2.2 W/(m · K). For a dense frozen sand k_T could be as high as 4 W/(m · K); it would be less in an unfrozen state. For soils, thermal conductivity increases with water content and dry density. Phukan (1985) converted empirical expressions originally proposed by Kersten (1949) into metric as follows:

For silt and clay:

$$k_T = 0.00144(10^{1.373\rho_d}) + 0.01226\,(10^{0.499\rho_d})w \tag{15.3}$$

For sandy soils:

$$k_T = 0.01096(10^{0.8116\rho_d}) + 0.00461\,(10^{0.9115\rho_d})w \tag{15.4}$$

where k_T is in watts per meter per kelvin, the dry mass density ρ_d in grams per cubic centimeter or metric tons per cubic meter, and the water content w in percent.

The *heat capacity* C of a material is the quantity of heat required to raise its temperature by 1°C or 1 K. It is either expressed per unit volume (volumetric heat capacity) or per unit mass (specific heat capacity). The quantity of heat Q required to raise the temperature of a mass M by ΔT degrees is

$$Q = CM\Delta T \tag{15.5}$$

The heat capacities of water C_w and ice C_i are

$$C_w = 4.2\ \text{kJ/(kg} \cdot {}^\circ\text{C)} = 1\ \text{cal/(mL} \cdot {}^\circ\text{C)} = 4.2\ \text{mJ/(m}^3 \cdot {}^\circ\text{C)}$$

$$C_i = 2.2\ \text{kJ/(kg} \cdot {}^\circ\text{C)} = 2.2\ \text{MJ/(m}^3 \cdot {}^\circ\text{C)}$$

The heat capacity of ice is therefore about half the value for water. Assuming the heat capacity of soil solids is 17% of that of water, we can write expressions for the heat capacity of a soil, either in an unfrozen, frozen, or partially frozen state:

Unfrozen soil:
$$C_{\text{su}} = \frac{\rho_d}{\rho_w}(0.17 + 1.0w)C_w \tag{15.6}$$

where w = water content (ratio)
ρ_d = dry density, t/m^3
ρ_w = density of water, t/m^3

Frozen soil:
$$C_{\text{sf}} = \frac{\rho_d}{\rho_w}(0.17 + 0.5w)C_w \tag{15.7}$$

Partially frozen soil:

$$C_{\text{spf}} = \frac{\rho_d}{\rho_w}(0.17 + 0.5w_i + 1.0w_{\text{uw}})C_w \tag{15.8}$$

Ice water content w_i and unfrozen water content w_{uw} are defined in Sec. 15.3.2.1.

The (latent) *heat of fusion* L_F of water is the change in thermal energy when water freezes or ice melts. It amounts to 334 kJ/kg or 334 MJ/m^3 of water. To melt a mass M of ice requires a heat quantity of

$$Q = L_F M \tag{15.9}$$

The heat of fusion L_{Fs} of 1 m^3 of soil with a water content w can be represented as

$$L_{\text{Fs}} = \rho_d w L_F = 334\ \rho_d w \qquad \text{kJ/m}^3 \tag{15.10}$$

where ρ_d = dry mass density of soil, kg/m^3
w = water content (ratio)
L_F = 334 kJ/kg = heat of fusion of water

The *heat of vaporization* L_v of water is the energy required to boil water—the transition of water from a liquid state into a gaseous phase. At atmospheric pressure, the heat of vaporization of water is

$$L_v = 2.26\ \text{MJ/kg} = 2260\ \text{MJ/m}^3$$

In order to evaporate all the free water (at 100°C) in 1 m^3 of soil with a water content w, the energy input required is

$$L_{\text{vs}} = 2260\ \rho_d w \qquad \text{kJ/m}^3 \tag{15.11}$$

where ρ_d is the dry density (kg/m^3) and w is expressed as a ratio.

Because of heat losses arising in practical applications, the energy used up is likely to be considerably higher than indicated by basic physics.

15.2 HEAT TREATMENT OF SOILS

Heat treatment of a clayey soil at temperatures exceeding say 400°C results in pronounced permanent changes of its engineering properties. This is in contrast to the artificial soil freezing process, which has to be maintained in order to remain effective; after thawing, the soil may actually be weaker than before.

Heating a soil consumes a lot of energy. Examples described in the literature indicate that 50 to 100 L of fuel oil may be needed to stabilize 1 m^3 of soil. In an era of increasing energy costs and concerns about air pollution, large-scale heat treatment is unlikely to have much of a future, except in very special circumstances; a special situation could arise where the soil already contains some fuel, either naturally or in the form of waste products. Nevertheless, it is at least of historical interest to give a brief review of attempts at improving soils by heating.

Heat treatment may also attract new interest in the production of bricklike building elements through medium temperature (500 to 700°C) firing, possibly in combination with chemical stabilization (see Sec. 15.2.3).

15.2.1 Changes in Soil Mineral Structure at High Temperature

In the standard laboratory determination of the water content of a soil sample, the sample is subjected to an oven temperature of 105 to 110°C. At this temperature all the free water evaporates, as already mentioned in Sec. 3.1.3, but water molecules which form part of the clay mineral lattice remain. Common minerals like kaolinite and the similarly structured halloysite show a distinct loss of weight around 500°C. For other clay minerals the reduction in moisture is more gradual with changing temperature, as shown in Fig. 15.1. In order to make bricks out of a clayey soil, the

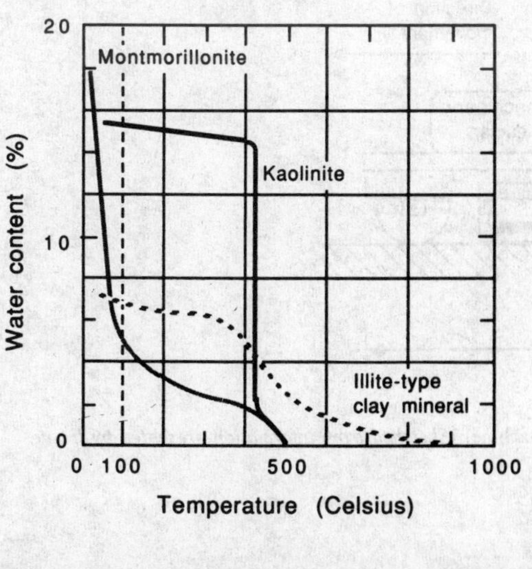

FIGURE 15.1
Characteristic water losses of clay minerals. [*After Kezdi (1969).*]

temperature is raised further, in excess of 1000°C, yet considerable irreversible gain in strength and durability is already observed at half that temperature.

15.2.2 Methods of Heating Soil In Situ

Heat can be generated by the burning of liquid, gaseous, or solid fuel and is transferred to the soil through surface contact or boreholes.

15.2.2.1 GROUND SURFACE HEATING. A pioneering effort in using heat treatment for roadmaking was made by Irvine (1930, 1934) in Australia. He designed and built a wood-fired furnace (Fig. 15.2) which traveled at a rate of 2 to 10 m/h on a previously prepared soil formation, producing a 50- to 200-mm-thick baked layer of firm non-plastic material out of a very moisture sensitive clay. After thermal modification, the surface material was left in interlocking pieces of brick varying in size from 100 mm downward. Larger particles were checked and cracked and reduced to "pea gravel" after compaction with a roller. Bitumen or tar surfacing could be applied while the pavement was still warm, securing good penetration of the binder into the cracks.

It is known that Irvine's machine was used to build several miles of country roads in New South Wales and Queensland, where coarse pavement materials are at a premium. Irvine estimated that up to 400 kg of wood are required to treat 1 m^3 of soil which was, at that time, cheaper than building a road with gravel, if the latter was not available within 16 km of the construction site. To Irvine, heat treatment of clays appeared so promising that he also built a portable furnace for the on-site production of concrete aggregate.

In his 1930 paper, Irvine indicated that heat treatment was earlier used in New South Wales for making a primitive form of road known as "burnt clay" road, which simply consisted of building log fires along a cleared clay track. Irvine (1930) further reported that pavement construction by heating the earth on the road surface by a

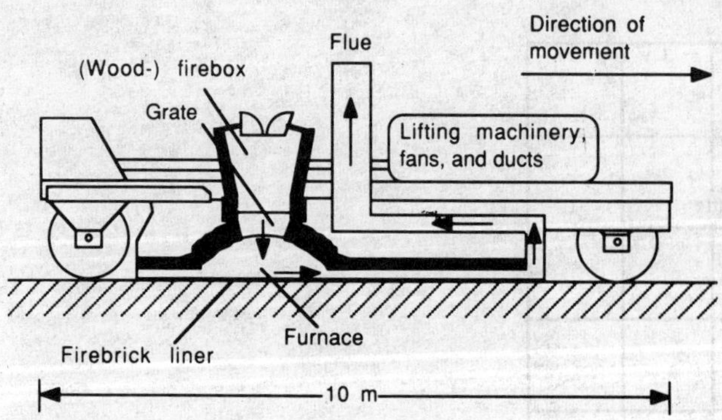

FIGURE 15.2
Schematic diagram of the Irvine heat treatment machine. [*Sketched from information presented by Irvine (1930) and Sherrard (1958).*]

moving furnace was at that time also being carried out in Argentina, using an oil-fired process called "Kingite," named after its inventor.

15.2.2.2 HEATING THROUGH BOREHOLES. Heat treatment via boreholes has been reported in the USSR [Litvinov (1959) as quoted by Fujii (1971)], in Rumania [Beles and Stanculescu (1958)], and in Japan [Fujii (1971)]. Fujii distinguishes between an open- and a closed-type firing system.

In a closed system (Fig. 15.3) a burning unit is placed over each borehole. Fujii (1971) describes a project near Fukuoka, where a waste deposit was treated by heating 227 boreholes, 2 to 6 m deep, spaced 5 m apart. The heat was maintained for 7 to 15 days, burning an average total of 760 L of fuel oil. Observations indicated that a volume within a radius of about 1 m around the borehole was affected, resulting in a fuel consumption of 60 L/m^3.

A special version of an open-type system was used to stabilize an embankment at Kanazawa, Japan [Fujii (1971)], as shown in Fig. 15.4. Over a period of 10 days, 1200 L of fuel were burnt per hole by conventional oil burners until a temperature of 300°C was reached about 300 mm from the center of the hole. The heat-treated soil showed a strength equivalent to 10 or 20 times the original value, even after total immersion.

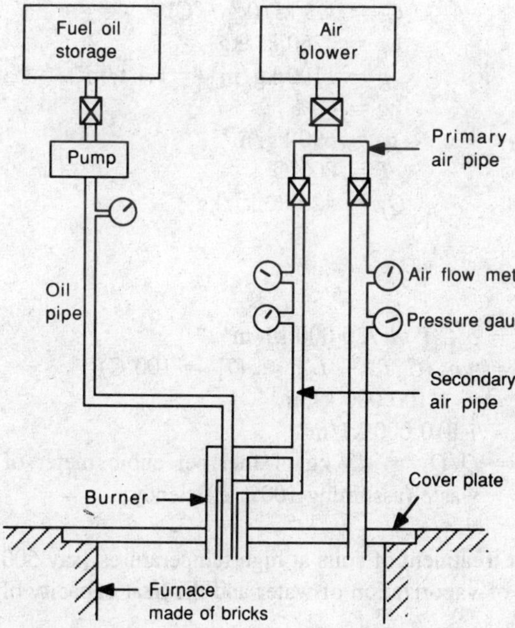

FIGURE 15.3
Closed-type burning system which is placed over a borehole. [*After Fujii (1971).*]

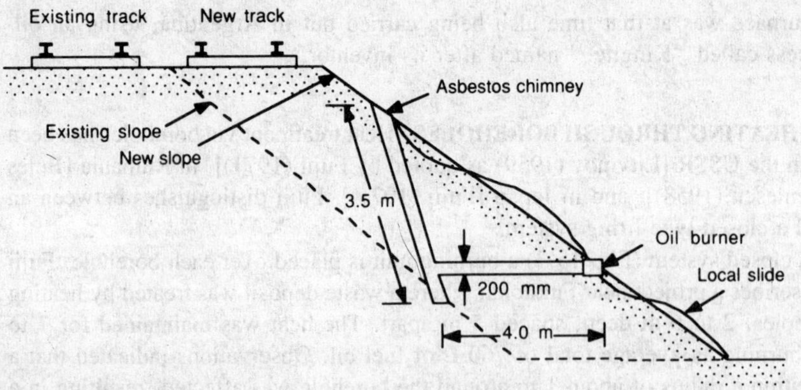

FIGURE 15.4
Thermal stabilization of an embankment. [*After Fujii (1971).*]

15.2.3 Calculation of Fuel Consumption

Fujii (1971) calculated the theoretical fuel requirement for raising the temperature of the ground in the Fukuoka project to a maximum of 800°C as follows:

Heat capacity of water	$C_w = 4.2$ kJ/(kg · °C)
Heat capacity of waste material	$C_s = 1.3$ kJ/(kg · °C)
Heat of vaporization of water	$L_v = 2260$ kJ/kg
Total unit weight of waste	$\rho = 1100$ kg/m^3 (= 1.1 t/m^3)
Water content	$w = 60\%$
Dry unit weight of waste	$\rho_d \approx 690$ kg/m^3
Maximum heating temperature	$T = 800$°C
Heat of combustion of fuel oil	$Q_f \approx 42\ 000$ kJ/kg

Required per cubic meter of waste:

Heat quantity, dry material: $Q_s = \rho_d C_s T \approx 720\ 000$ kJ/m^3

Heat quantity, water: $Q_w = w\rho_d\,(C_w T_v + L_v)$ $\quad (T_v = 100°\text{C})$
$\approx 1\ 100\ 000$ kJ/m^3

Total heat quantity: $Q \approx 1\ 810\ 000$ kJ/m^3

Fuel consumption: $F = Q/Q_f = 43$ kg of fuel per cubic meter of waste (assuming 100% efficiency)

The energy Q required for heat treatment of soils at high temperatures (say 500 to 1000°C) is dominated by the heat of vaporization of water and the heat capacity of the soil solids:

$$Q = L_{vs} + C_s T$$
$$= \rho_d(w_{\text{all}} L_v + C_s T)$$

where Q = heat quantity per unit volume, kJ/m^3
L_v = 2260 kJ/kg (water)
ρ_d = dry density of soil, kg/m^3
C_s = heat capacity of soil solids, kJ/(kg · °C)
w_{all} = total water content, including chemically bound water, which evaporates at temperatures > 100°C

Assuming $C_s = 1$ and $w_{all} = 1.2w$ (w = conventional water content), we obtain the following approximate relationship:

$$Q = \rho_d (2700w + T) \qquad kJ/m^3 \tag{15.12}$$

The fuel quantity required to heat 1 m^3 of soil, therefore, is

$$F = \frac{100\rho_d}{EQ_f}(2700w + T) \tag{15.13}$$

where E is the efficiency of burning (%) and Q_f is the heat of combustion fuel (kJ/kg).

This equation is similar in structure to the one proposed by Ingles and Metcalf (1972).

15.2.4 Manufacturing Thermally Stabilized Building Blocks

In the "Ferroclay" process described by Ingles (1982) and Ingles and Lim (1982), soil, freshly burnt at temperatures from 500°C (for kaolonites) to 750°C (for montmorillonite clays), is mixed with say 2% alkaline ferric oxide. The mixes are molded and cured in air, achieving strengths in excess of 10 MPa within 1 or 2 weeks.

Ferroclay production has been likened to natural laterization of soils and represents a combination of thermal and chemical stabilization. It is a process which lies halfway in between rammed earth, adobe, or mud bricks and fully fired bricks. The savings in energy achieved by burning the clay at less than standard brick temperatures make Ferroclay an attractive alternative building material; no doubt other similar processes will emerge.

15.3 GROUND FREEZING

15.3.1 Introduction

Modifying ground through artificial freezing has great potential for solving difficult foundation problems because it promises a very high gain in strength and effective seepage control in a matter of hours or days after initiating the process. Typical applications are

Retaining structures for open excavations (Fig. 15.5)
Lateral supports for sinking mine shafts
Ground support and seepage control during tunneling (Fig. 15.5)

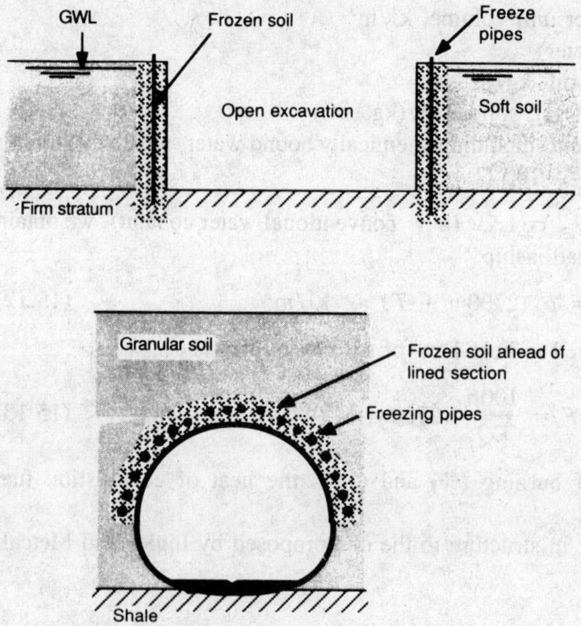

FIGURE 15.5
Two examples of ground freezing applications.

Sealing of leaks in underground pipes
Temporary stabilization of landslides
Enabling undisturbed sampling in cohesionless soils

Typical dimensions of straight frozen soil walls are shown in Fig. 15.6 [after Jessberger and Vyalov (1979)]. For circular excavations and shafts the horizontal dimensions shown could possibly be reduced by a factor of 2 or more.

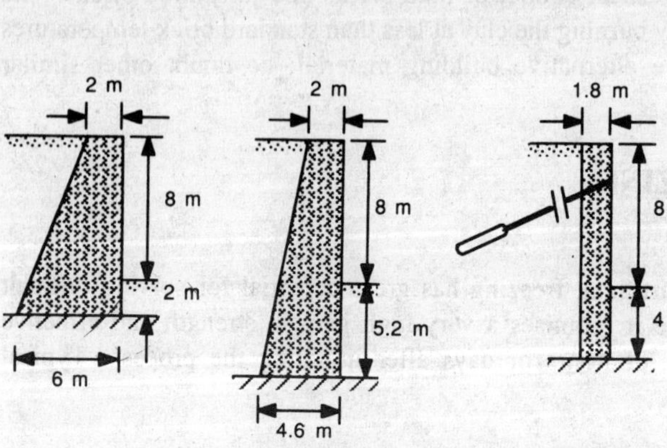

FIGURE 15.6
Typical dimensions of straight frozen walls. [*After Jessberger and Vyalov (1979).*]

Unlike most other methods of ground modification, ground freezing is suitable for a wide range of soil or rock types. In fact it is suitable in any formation where freezing of the pore water creates a bond between individual particles and where the freezing process is not inhibited by significant water flow. More specifically, the effectiveness of ground freezing will depend on factors such as

The type of ground (in particular, the soil's mineralogical structure, texture, thermal properties, and water content)

The composition of the pore water and the ice formed (size and orientation of ice lenses and crystals)

The seepage velocity and temperature of the groundwater

The freezing temperature and rate of freezing

Frozen-ground engineering has been the subject of several books [Tsytovich (1975); Jumikis (1977); Phukan (1985)], comprehensive articles [Jessberger (1987); Andersland (1987)], and conferences, such as the First, Second, Third, and Fourth International Symposia on Ground Freezing (Bochum, 1978; Trondheim, 1980; Hanover, New Hampshire, 1982; and Sapporo, 1985). Ground freezing for seepage control in construction dewatering is also covered by Powers (1981, Chap. 20).

15.3.2 Properties of Frozen Ground

Much is known about the properties of frozen soils and ice, mainly because of the efforts of engineers working in permafrost areas and in arctic regions, where in many instances the properties of naturally frozen ground can be taken advantage of. In contrast, a road or foundation engineer may have to deal with the problem of seasonal freezing and thawing causing damage by frost heave, collapse upon thawing, and loss of durability; this has resulted in greatly detailed studies of moisture migration during freezing and of other phenomena. The experience gained from dealing with naturally frozen soils is, however, not necessarily transferable to artificially frozen ground; distinct differences in behavior are caused by the following aspects:

The temperature maintained during artificial freezing is lower (-20 to -160°C) than that commonly found in a natural ground environment (seldom < -15°C).

Ice structure and lense formation and orientation depend on the direction of heat flow, the rate of freezing, and other factors. In permafrost, ice lenses are parallel to the ground surface and diminish with depth; in artificially frozen ground, they are parallel to the freezing pipes.

Laboratory freezing may produce an ice structure which is neither equivalent to that found in the same soil for permafrost conditions nor that induced by artificial freezing in the field.

15.3.2.1 FROZEN SOIL—A FOUR-PHASE MATERIAL. Soil as a three-phase material was described in Sec. 7.3.1. Frozen soil, because it may contain both unfrozen

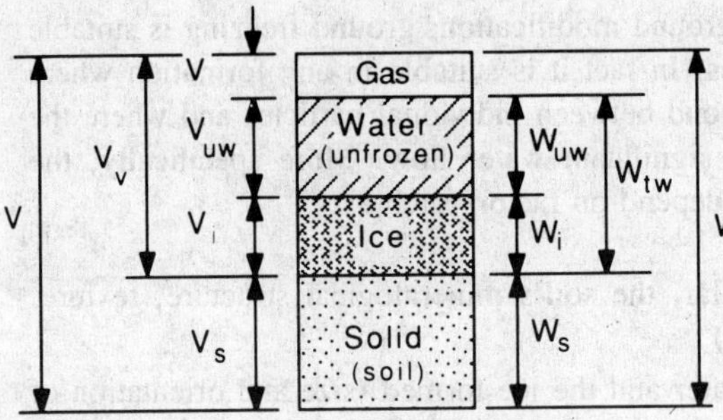

FIGURE 15.7
The four phases of frozen soil.

water and ice, becomes a four-phase material. Referring to Fig. 15.7, new definitions must be introduced as follows:

$$W_{uw} = \text{weight of unfrozen water}$$

$$W_i = \text{weight of ice}$$

$$W_{tw} = W_{uw} + W_i$$

$$V_{uw} = \text{volume of unfrozen water}$$

$$V_i = \text{volume of ice}$$

$$w_{uw} = \frac{W_{uw}}{W_s} = \text{unfrozen water content}$$

$$w_i = \frac{W_i}{W_s} = \text{ice water content}$$

$$w_{tw} = w_i + w_{uw} = \text{total water content} = w \text{ (before freezing)}$$

$$i_{ice} = \frac{W_i}{W_t} = \text{relative iciness} = 1 - \frac{W_{uw}}{w_t} \tag{15.14}$$

Water contents may be expressed in percent by multiplying the above ratios by 100. Care has to be taken when interpreting formulas containing w, because it is not always apparent whether w values should be inserted as ratios or percents.

Note that approximate unit weights for water γ_w and ice γ_i are

$$\rho_w = 1000 \text{ kg/m}^3 \quad \text{(mass units)}$$

$$\gamma_w = 9.81 \approx 10 \text{ kN/m}^3 \quad \text{(weight units)}$$

$$\rho_i = 920 \text{ kg/m}^3 \quad \text{(mass units)}$$

$$\gamma_i = 9.0 \text{ kN/m}^3 \quad \text{(weight units)}$$

Another new definition is

$$S_i = \frac{\text{volume of ice}}{\text{volume of voids in frozen soil}}$$

$$= \frac{V_i}{V_v} = \text{degree of ice saturation}$$

$$= \frac{w_i G_s \rho_w}{\rho_i e} \tag{15.15}$$

If a fully saturated soil before freezing has a water content of w, the total and dry unit weights of the soil in a completely frozen state become

$$\rho_f = \frac{G_s \rho_w (1 + w)}{1 + 1.09 G_s w} \tag{15.16}$$

$$\rho_{fd} = \frac{G_s \rho_w}{1 + 1.09 G_s w} \tag{15.17}$$

The unfrozen water phase reduces rapidly from 0 to −10°C but may still exist at a temperature lower than that, depending on the soil structure, particle surface area, mineralogical composition, dissolved solutes (NaCl, etc.), confining pressure, and other factors. An awareness of the presence of unfrozen water in the apparently frozen ground helps in understanding the highly temperature dependent engineering properties of frozen ground and the mechanism of water migration and ice lens formation.

15.3.2.2 SHORT-TERM STRENGTH AND STRESS-STRAIN CHARACTERISTICS. Pure ice has a crystalline structure. Hexagonal platelets are stacked on top of each other in the direction of the heat flow and growth of ice crystals [see Jumikis (1979)]. Strength and stress-strain properties of ice and frozen soils depend on many factors, such as

- Freezing conditions (rate, direction, water supply)
- Temperature
- Orientation of ice crystals or ice lenses
- Sample size and shape
- Strain rate (see Sec. 15.3.2.3)
- Confining stress
- The mineral composition of soils and dissolved salts in the pore water.

No definite strength values can be given for pure ice, let alone frozen soils. Nevertheless, an idea of the order of magnitude of *uniaxial compressive strength* can be gained from the following table:

Frozen material	Short-term compressive strength at − 10°C, MPa
Pure ice	1.5
Sand	10
Silt	4
Clay	2

These values are meant as a very rough, somewhat conservative guide only. Observe that frozen soils have a very high strength compared with a soft to medium clay which would havc an unconfined compressive strength of about 50 kPa (0.05 MPa)!

Strength is very temperature dependent, particularly in clays with high moisture contents, where there may still be an unfrozen component of the pore liquid. Strength is said to increase linearly with the log of the absolute value of the negative temperature in degrees Celsius. The strength at −40°C may be 3 times that at −10°C.

The short-term *stress-strain moduli E* of various soils already show an enormous range in an unfrozen state, varying from say 3 MPa for a soft clay to 200 MPa for a dense sand. In a frozen state, the same soil may have a modulus some 10 or 100 times higher [Tsytovich (1975)]. Compared to how it influences strength, temperature has an even more pronounced effect on the modulus of a frozen soil; also important are the ice content, unfrozen pore water content, external pressure, etc.

Tsytovich (1975) gives a modulus value E_{ice} = 2450 MPa, under a compressive stress of 200 kPa and at a temperature of −1.5°C. Seismic measurements on arctic ice indicate a modulus as high as 10 000 MPa [Langleben and Pounder (1963)] and a Poisson's ratio of 0.3.

The *tensile strength* of ice is about half of its compressive strength, say 0.6 to 1 MPa. According to Tsytovich (1975), the tensile strength of frozen soils may vary from one-half to one-sixth of their compressive strength.

The *shear strength* of frozen soils can be represented on a Mohr-Coulomb plot as shown in Fig. 15.8. The Mohr envelopes are generally slightly curved, showing a cohesion intercept at zero normal stresses, and can be extended to include the tensile strength. The internal friction angle of a frozen granular soil is less than that in the unfrozen state, a ratio of 2:3 between these two values could be taken as a guide. Frozen clays have only small friction angles, but they are more sensitive to the rate of load application than granular soils.

Adfreeze strength (analogous to adhesion) is of importance to foundations and anchors in frozen ground and is discussed extensively by Tsytovich (1975). Like all other strength parameters, adfreeze strength is strongly dependent on the temperature and the rate of loading. Interestingly, Tsytovich gives results which indicate that the adfreeze strength of frozen sand to wood is higher than that of frozen gravel or clay.

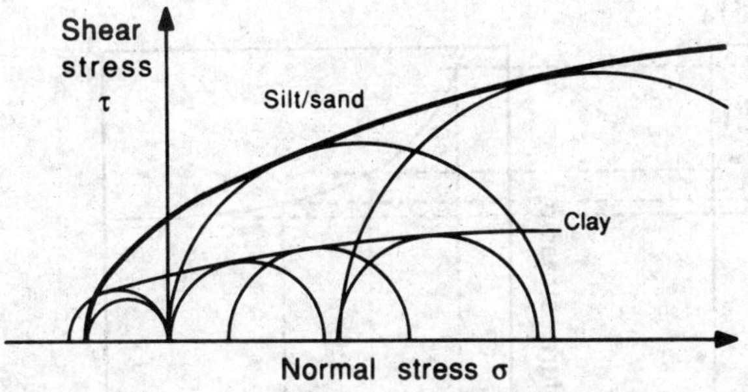

FIGURE 15.8
Typical shapes of Mohr envelopes for frozen soils.

15.3.2.3 LONG-TERM BEHAVIOR UNDER STRESS: CREEP. Figures 15.9 and 15.10 illustrate how time-dependent stress-strain and strength properties can be represented. On an idealized strain-time diagram (Fig. 15.9) three separate phases may be distinguished: primary creep, characterized by a decreasing strain rate; secondary creep with a constant strain rate; and tertiary creep where an increasing rate of strain either leads to shear failure or rapid continuing deformation (e.g., bulging of the specimen in compression). Failure is sometimes defined as the onset of tertiary creep. The lowest stress which leads to tertiary creep could be called the long-term strength. Alternatively, for many engineering applications, failure is more meaningfully postulated in terms of a specific allowable strain.

A variety of formulas have been proposed for expressing the creep strain-time relationship. Frequently quoted is the equation by Vialov [Vialov et al. (1979)], which includes primary and secondary creep:

$$\varepsilon_c = A\sigma^B t^C \tag{15.18}$$

where ε_c = creep strain
$A, B, C,$ = creep parameters
σ = stress, MPa (usually)
t = time, h

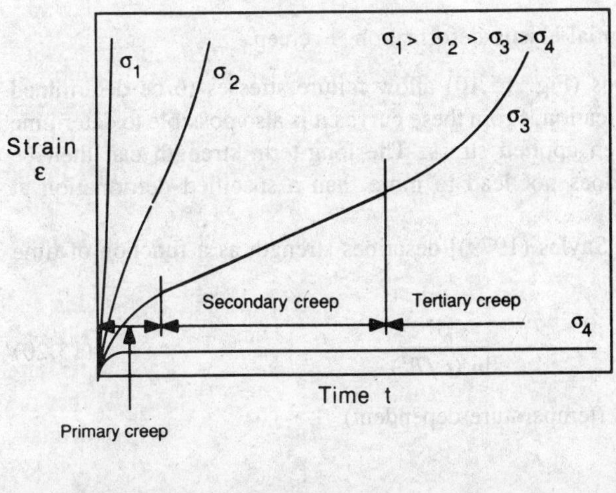

FIGURE 15.9
Typical creep curves for frozen soil.

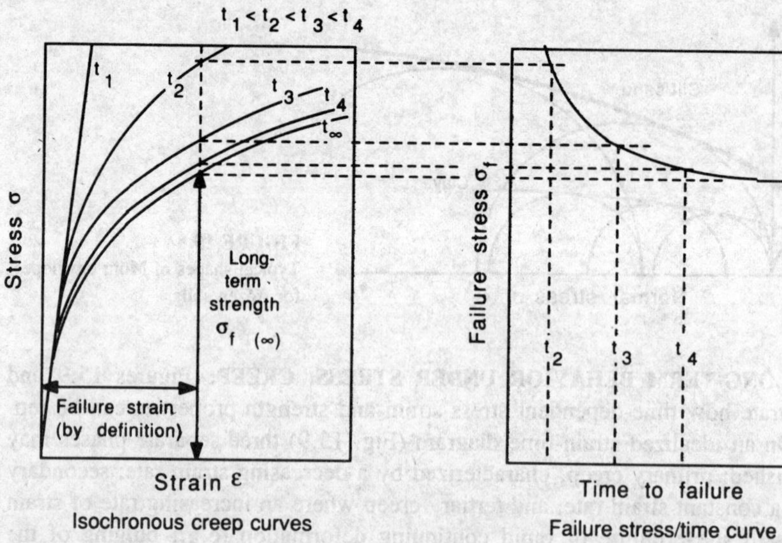

FIGURE 15.10
Alternative creep deformation curves.

The magnitude of the parameters A, B, and C will depend on the units chosen. For example, with σ in megapascals and t in hours, a frozen sand at a particular temperature may have the following values:

$$A = 0.005 \quad [(\mathrm{m^2/MN})^B\ \mathrm{h}^C]$$

$$B = 1.4$$

$$C = 0.4$$

The total strain is found as

$$\varepsilon = \varepsilon_0 + \varepsilon_c \tag{15.19}$$

$$= \text{initial elastic deformation} + \text{creep}$$

Isochronous creep curves (Fig. 15.10) allow failure stresses to be determined for a given length of load application. From these curves it is also possible to determine the time to failure for a given applied stress. The long-term strength can then be defined as the stress which does not lead to more than a specified deformation at time t_∞.

Vialov [see Sanger and Sayles (1979)] describes strength as a function of time by

$$\sigma_t = \frac{\beta'}{\ln (t_f/B')} \tag{15.20}$$

where β', B' = soil constants (temperature dependent)

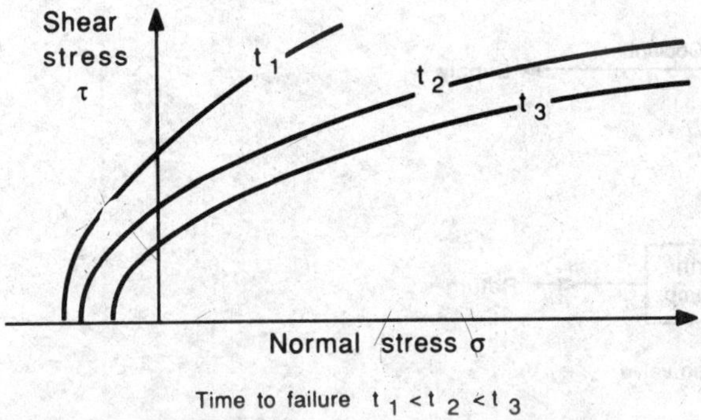

FIGURE 15.11
Mohr envelopes for different rates of loading.

Also shown in Fig. 15.10 is a stress/time to failure diagram. It is the same kind of graph which is plotted directly as a result of stress rupture testing of metals in tension. For the compression of frozen soils, however, the construction of this diagram is more complicated because failure cannot generally be defined in terms of rupture.

Finally, Fig. 15.11 schematically shows a group of Mohr envelopes which would be obtained for the same soil, but at different rates of loading.

15.3.2.4 GROUND MOVEMENT DURING FREEZING. One cubic meter of water will make 1.09 m^3 of ice with a mass density of 920 kg/m^3. The water phase in soils therefore has the potential to increase its volume by 9% through freezing. However, soils have been observed to heave more than can be explained based on the freezing of the water contained in the original unfrozen soils alone. This is due to water migration into the soil zone where ice crystals are growing and ice lenses are formed. The mechanism of this water migration (in solid liquid or vapor form) is very complex, involving various physical and chemical phenomena including internal and external stresses, capillary tension, osmotic pressures, and adsorption forces. The resulting ''secondary'' frost expansion can exceed that due to basic water-ice conversion, particularly in highly saturated low-permeability fine-grained soils under low confining stresses, subjected to relatively small temperature gradients.

Additional deformation during ground freezing may be due to stress relaxation attributable to the excavation process. After ground freezing, thawing may initiate consolidation settlement.

15.3.3 Techniques of In Situ Artificial Ground Freezing

There are two basic types of cooling methods: circulating brine freezing and expendable gas freezing.

15.3.3.1 CIRCULATING BRINE FREEZING SYSTEMS. A circulating brine system is illustrated in Fig. 15.12. The coolant, usually a calcium chloride solution, is pumped through a closed system of pipes, freezing the ground as it travels through the cased boreholes. For $CaCl_2$ the theoretically lowest temperature achievable is −55°C. The

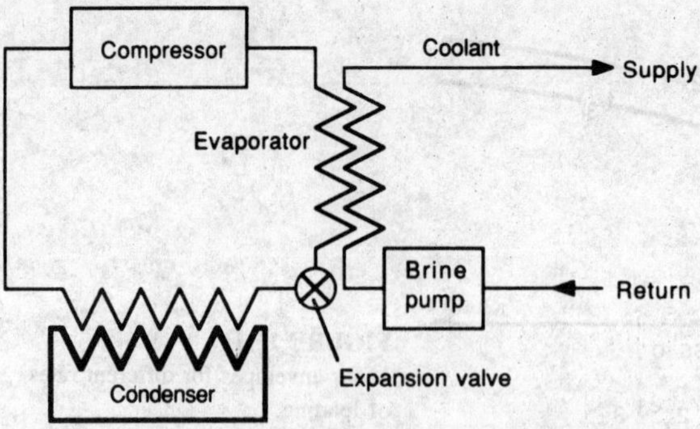

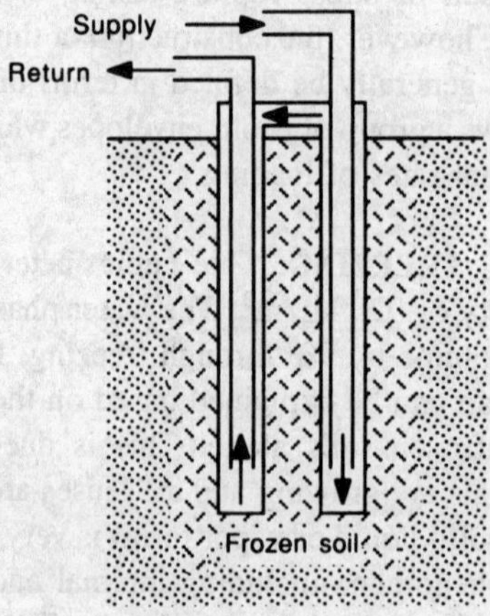

FIGURE 15.12
Circulating brine freezing system.

time required to freeze the ground depends largely on the soil type and its water content, the brine temperature, and the hole spacing. Jessberger (1987), quoting Stoss (1976), presented useful charts which allow a preliminary assessment of the required initial freezing time, which is measured in days or weeks (Fig. 15.13).

15.3.3.2 EXPENDABLE GAS FREEZING SYSTEMS. Less conventional is the use of liquid gases, most commonly liquid nitrogen (LN_2). Liquid nitrogen has a theoretical minimum temperature of $-196°C$, it is pumped through the underground freezing pipes and released into the atmosphere, or, if the soil is porous enough, it is injected directly into the ground (Fig. 15.14). Because of the high temperature gradients involved, a more sophisticated coolant distribution and monitoring system is required for nitrogen freezing than for brine cooling systems. In order to maximize the freezing effect, good control of the nitrogen supply is needed, ideally, it should be regulated as a function of the exhaust gas temperature.

Because of the low temperature of liquid nitrogen, ground freezing is achieved very rapidly, within hours rather than days. This is an advantage in construction emergencies, particularly where seepage is encountered. Rapid freezing also reduces

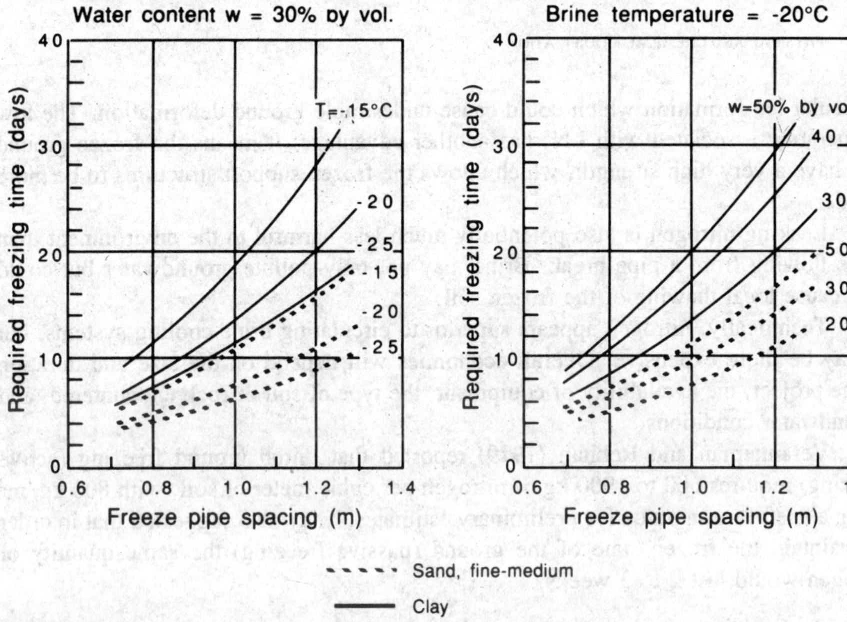

FIGURE 15.13
Required freezing time versus distance of the freezing pipes. [*After Stoss (1976) as quoted by Jessberger (1987).*]

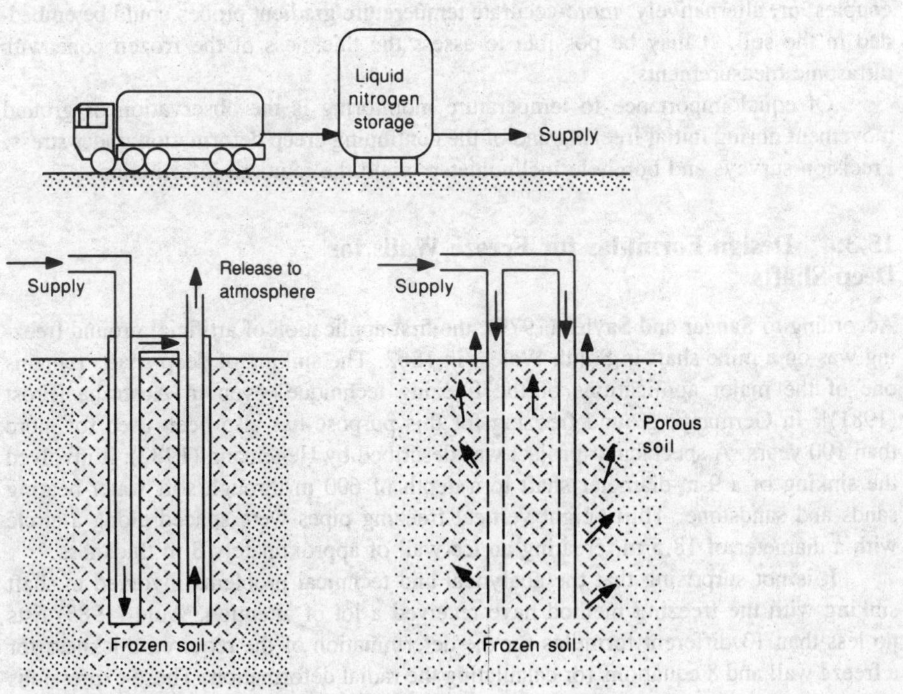

FIGURE 15.14
Ground freezing with liquid nitrogen; alternative freeze pipe systems.

lenticular ice formation which could cause undesirable ground deformation. The low temperature associated with LN_2 has another advantage: It means the frozen ground will have a very high strength, which allows the frozen support structures to be more slender.

Leaking nitrogen is also potentially much less harmful to the environment than brine flowing from a pipe break. Brine may not only pollute groundwater but could also cause local thawing of the frozen soil.

Technically, nitrogen appears superior to circulating brine cooling systems, but it may be more expensive. Overall economies will depend on the size and duration of the project, the availability of equipment, the type of soil or rock encountered, and groundwater conditions.

Veranneman and Rebhan (1979) reported that initial ground freezing (active freezing) requires 600 to 1900 kg of nitrogen per cubic meter of soil, with 800 kg/m^3 being a useful mean value for preliminary estimates. They also suggested that in order to maintain the frozen state of the ground (passive freezing) the same quantity of nitrogen would last 2 to 3 weeks.

15.3.3.3 MONITORING FROZEN-GROUND PERFORMANCE. Temperature measurements of the coolant and the frozen soil are obviously of prime importance for the safety of the structure. Temperature may be measured directly with thermocouples, or, alternatively, more-accurate temperature gradient probes could be embedded in the soil. It may be possible to assess the thickness of the frozen zone with ultrasonic measurements.

Of equal importance to temperature monitoring is the observation of ground movement during initial freezing and of the continuing creep deformation under stress. Precision surveys and borehole inclinometers yield the required information.

15.3.4 Design Formulas for Freeze Walls for Deep Shafts

According to Sanger and Sayles (1979), the first application of artificial ground freezing was on a mine shaft in South Wales, in 1862. The sinking of deep shafts remains one of the major applications of the freezing technique, even in Australia [Frost (1981)]. In Germany, ground freezing for this purpose has also been used for more than 100 years. A spectacular project was described by Hegemann (1982); it involved the sinking of a 9-m-diameter shaft to a depth of 600 m through soft water-bearing sands and sandstone. Thirty-eight vertical freezing pipes were spaced along a circle with a diameter of 18.5 m, creating an ice wall of approximately 8 m thickness.

It is not surprising that the analytical and technical problems involved in shaft sinking with the freezing method have received a lot of attention. Auld (1985) lists no less than 13 different formulas for the determination of the required thickness for a freeze wall and 8 equations for calculating the radial deformation. These expressions are based on the mechanics of a thick cylinder. Depending on the theory used, this cylinder may be assumed to be long or short, lined or unlined, and with or without lining or internal pressure; the soil is represented either as an elastic, elastoplastic, plastic, or viscous material.

As examples, three formulas of increasing complexity are reproduced below, as quoted by Auld (1985). They are all based on the long, thick cylinder theory. Geometric assumptions are illustrated in Fig. 15.15.

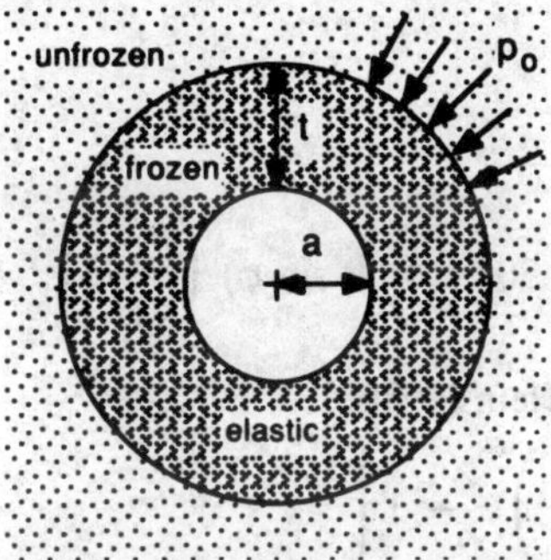

a. Lamé and Clapeyron - elastic analysis

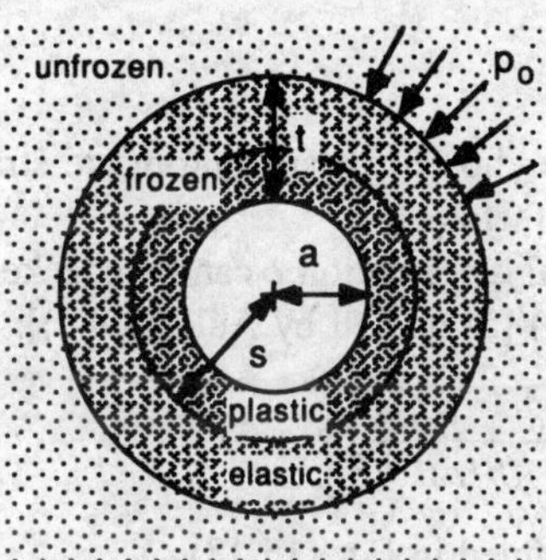

b. Klein - elastic-plastic analysis ($s = \sqrt{ab}$)

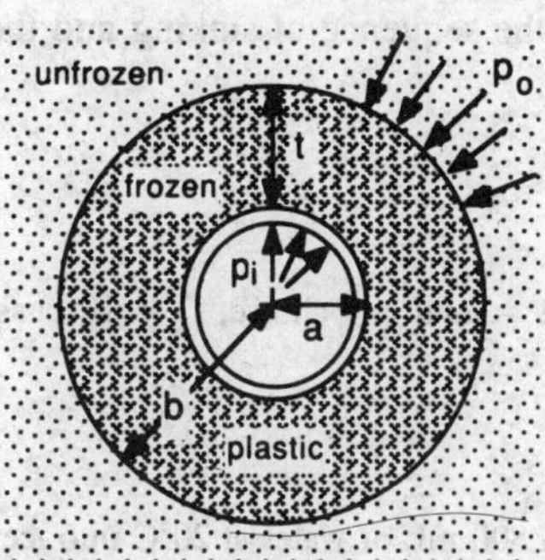

c. Klein - plastic analysis (Mohr-Coulomb), lined shaft

FIGURE 15.15
Geometry for shaft freeze wall formulas (long, thick cylinder theory).

Elastic analysis, Lamé and Clapeyron (1833):

$$d = a\left(\sqrt{\frac{q_u}{q_u - 2p_0}} - 1\right) \tag{15.21}$$

where d = thickness of frozen ring
a = radius of opening
q_u = unconfined compressive strength
p_0 = lateral earth and water pressure

Elastic-plastic analysis, Klein (1981):

$$d = a\left[(0.29 + 1.42 \sin \phi)\frac{p_0}{q_u} + (2.30 - 4.60 \sin \phi)\left(\frac{p_0}{q_u}\right)^2\right] \tag{15.22}$$

Plastic analysis (Mohr-Coulomb failure law), Klein (1980*a*)

$$d = a\left[\left(\frac{p_0 + q_u C_1}{p_i + q_u C_1}\right)^{C_2} - 1\right] \tag{15.23}$$

where p_0 = internal pressure (support from lining)
$C_1 = (1 - \sin \phi)/(2 \sin \phi)$
$C_2 = 1/(2 \sin \phi) - \frac{1}{2}$

The uniaxial compressive strength q_u would be chosen for an appropriate strain and time to failure, possibly incorporating a safety factor.

Given the thickness $d = b - a$, the radial deformation δ can be checked using the following equation, derived by Klein (1980) [as quoted by Auld (1985)]:

$$\delta = -aAt^C \frac{\sqrt{3}}{2}\left[\frac{2(p_0 - p_i)/B}{1 - (b/a)^{-2/B}}\right]^B \tag{15.24}$$

where A, B, and C are the creep parameters defined in Eq. (15.18).

Obviously, the more advanced the theory, the more knowledge is needed about the material properties of the frozen soil. In addition, careful consideration will have to be given to the construction details, such as the sequence of sinking and the lining of the shaft.

PROBLEMS

Prefixes indicate problem type: C = calculations, M = multiple choice, D = discussion.

Calculations

C15.1. How much heat must be removed to convert 800 mL of water at 20°C into ice at 0°C?

C15.2. How much energy is needed to change a 1-t (1000 kg) block of ice at −10°C into steam at 100°C?

C15.3. Estimate the volumetric heat capacity of a soil with a moisture content of 30% and a dry unit weight of 1500 kg/m^3, assuming the heat capacity of the soil solids is 20% of that of water. Assume

(*a*) The soil is unfrozen.

(*b*) The soil is completely frozen.

C15.4. Calculate the time t required to freeze a column of soil to a radius R according to the formula and sample data presented by Sanger and Sayles (1979):

$$t = \frac{R^2 L_2}{4k_T \Delta T_1}\left[2 \ln \frac{R}{r_0} - 1 + \frac{C_1 \Delta T_1}{L_2}\right]$$

with
$$L_2 = L + \frac{a_r^2 - 1}{2 \ln a_r} C_2 \Delta T_2$$

where
R = (spacing of freeze pipes)/2 = 0.76 m
$L = \rho_d w L_F = 1281(0.4)(335) = 172$ MJ/m^3 = heat of fusion of the soil
a_r = (zone of temperature influence)/R
= 3 (assumed)
k_T = 2 W/(m · K) = thermal conductivity
r_0 = radius of freeze pipe
= 76 mm
ΔT_1 = temperature at the surface of the freeze pipe, |°C|
= 23.3
C_1 = heat of fusion of frozen soil
= 2.0 MJ/(m^3 · °C)
ΔT_2 = temperature of original ground in °C
= 15.6
C_2 = heat of fusion of unfrozen soil
= 3.1 MJ/(m^3 · °C)

C15.5. The water content of a fully saturated soil is 30%. The specific gravity of the soil solids is 2.68. Calculate

(*a*) The total unit weight before and after freezing.

(*b*) The volume increase (%) of the soil, assuming no water is expelled or has access to the soil.

C15.6. A frozen soil sample has a mass of 650 g. The unfrozen water content is estimated to be 10%. After thawing and drying the sample's mass is reduced to 420 g.

(*a*) Determine the total water content, the ice water content, and the relative iciness.

(*b*) Assuming the air content is negligible, calculate the void ratio and the degree of ice saturation, assuming $G_s = 2.67$.

C15.7. Predict the creep strain of a frozen silty clay under a stress of 1 MPa after 1000 h, given the parameters

$$A = 0.006\ [(\text{m}^2/\text{MN})^B\ \text{h}^C]$$

$$B = 2.6$$

$$C = 0.4$$

C15.8. The creep curve of a medium sand under 5 MPa compression at $-15°C$ is defined by three points:

$t(h)$ =	10	90	300
ϵ =	0.015	0.026	0.035

(*a*) Determine the corresponding creep parameters *A*, *B*, and *C*.
(*b*) Predict the creep curve at 7 MPa, and compare it with the actual results, which were:

$t(h)$ =	10	60	90	150	210
ϵ =	0.030	0.049	0.058	0.072	0.088

Tertiary creep was noticeable at $t = 200$ h.
[Problem based on data by Eckardt (1979).]

C15.9. From the following failure stresses:

$$\sigma_1 = 19 \text{ MPa at } t_1 = 1 \text{ h}$$

$$\sigma_2 = 13 \text{ MPa at } t_2 = 10 \text{ h}$$

(*a*) Determine the parameters β' and B' for Eq. (15.20).
(*b*) Predict the strength after $t = 90$ days.

Multiple Choice

M15.10. A difference in temperature of 32°C is equal to
(*a*) 32° F.
(*b*) 32 K.
(*c*) 273.15 K.
(*d*) −273.15 K.

M15.11. The heat capacity of a soil
(*a*) Has units of W/(m · K).
(*b*) Has units of MJ/m^3.
(*c*) In a frozen state is lower than in an unfrozen state.
(*d*) Depends strongly on the void ratio.

M15.12. For water
(*a*) The heat of fusion is less than the heat of vaporization.
(*b*) The heat of fusion is $4.2\ MJ/m^3$.
(*c*) The heat of vaporization is $334\ kJ/m^3$.
(*d*) The heat of vaporization is $2260\ kJ/(m^3 \cdot °C)$.

M15.13. The short-term strength of a frozen sand is on the order of
(*a*) 50 kPa.
(*b*) 10 000 kPa.
(*c*) 1.5 MPa.
(*d*) 2 MPa.

M15.14. The short-term strength of ice is on the order of
(*a*) 50 kPa.
(*b*) 100 kPa.
(*c*) 1.5 MPa.
(*d*) 10 MPa.

M15.15. Which one of the following statements is *incorrect*?

(*a*) During primary creep the rate of strain is decreasing.

(*b*) During secondary creep the rate of strain is increasing.

(*c*) Isochronous creep curves give the time required to produce a certain strain under a particular stress.

(*d*) Failure is sometimes defined as the onset of tertiary creep.

Discussion

D15.16. Explain why the properties of an artificially frozen ground may be different to those of the same type of soil in a permafrost area.

D15.17. How can you define failure of a frozen soil in uniaxial compression?

D15.18. The heave of a soil upon freezing can be more than is estimated based on the water content of the soil and the known volume increase during the conversion from water to ice. How is this possible?

D15.19. Discuss advantages and disadvantages of circulating brine versus liquid nitrogen freezing of ground.

PART V

MODIFICATION BY INCLUSIONS AND CONFINEMENT

CHAPTER 16

EVOLUTION OF SOIL REINFORCEMENT

16.1 CONCEPT OF SOIL REINFORCEMENT

According to its original usage, the term ''reinforced soil'' refers to a soil which is strengthened by a material able to resist tensile stresses and which interacts with the soil through friction and/or adhesion. Subsequently, the meaning of soil reinforcement was broadened, and this term is now also used for other mechanical and structural methods of soil improvement, such as compressive reinforcement and reinforcement by confinement and encapsulation.

The primary purpose of reinforcing a soil mass is to improve its stability, increase its bearing capacity, and reduce settlements and lateral deformation. The broader definition of soil reinforcement also includes methods of erosion control and stress transfer via anchors and piles. The terminology is complicated by the fact that many of the materials used to improve engineering properties of soil, such as geotextiles, can fulfill multiple functions, e.g., provide structural strengthening, control groundwater flow or accelerate consolidation with their drainage capacity, prevent particle migration through filter action, and maintain separation of different soil layers during construction or under the influence of repeated external loading.

Soil reinforcement can be achieved even by relatively flexible, extendable, and sometimes compressible materials, such as nonwoven fabrics or large quantities of individual fibers. Also, the actual strengthening of a soil mass may be a secondary effect, e.g., as achieved through accelerated consolidation. This is why the term ''inclusions'' rather than ''reinforcement'' is favored by some researchers working in this field.

In this and the following two chapters on ground modification by inclusions and confinement, emphasis is placed on construction techniques and methods of analysis concerned with problems of soil structures which are created by combining remolded soil with structural components such as strips, meshes, and sheets made of metal or synthetics. In situ strengthening of a soil mass by soil nailing and anchor installation is discussed in Chap. 19. Some traditional and new techniques of soil confinement are covered in Chap. 20.

16.2 HISTORICAL DEVELOPMENT

The concept of soil reinforcement is not exactly new. Humans can observe in nature how some birds build nests out of soil fibers and how beavers construct earth dams reinforced by woody fibers. Clay bricks reinforced by straw (adobe bricks) are mentioned in the Old Testament (Exodus, Ch. 5, Vs. 6–9). The Babylonians, more than 3000 years ago, used reinforced soil to build ziggurats, similar to the famous Tower of Babel; the remnants of some of these buildings have survived until the present day. In Asia, from China to Japan, bamboo as well as straw have been utilized to strengthen earth as a building material.

In more recent times, engineers have reported on the use of fagots, composed of cuttings from tree branches, for riverbank stabilization. There are many other examples of the use of natural materials for soil reinforcement. Excellent temporary roads, called corduroy roads, can be built on soft soil by laying small trees and saplings, wired together at the ends, on the ground surface. Cylindrical bottomless wicker baskets, filled with soil, represent the original form of gabions common in early fortifications and trench warfare. Pannell (1964) mentions stone-filled timber piers, resembling modern reinforced soil structures, which were built some 200 years ago in English and European ports.

Wire mesh holding back cobbles and stone fill has successfully protected riverbanks and seashores for many years. Reinforcing the downstream face of rock-fill dams is a relatively newer development. Wire mesh has also been applied in order to secure coarse material in inverted filter zones.

Innovative earth-retaining systems incorporating modern soil-reinforcing concepts were patented by M. A. Coyne in France (French Patent No. 656692, 1929) and A. Munster in the United States (U.S. Patent No. 1762343, 1930). Munster's system consisted of timber lattice embedded in the backfill and attached to a relatively thin rigid facing via a vertically sliding connection. Coyne proposed a sloping arrangement of precast face panels (1.5 by 0.8 m) each tied back into stone fill with an anchor.

Henri Vidal (1966, 1969), a French architect and engineer, is credited with developing a soil-reinforcing technique to a stage where it could be economically applied to large civil engineering structures. He named his patented technique "Terre Armée" or "Reinforced Earth," and his company of the same name is now represented around the world. His original French Patent (No. 929421), dating back to March 27, 1963, was followed by patents in the United States, the United Kingdom, and Australia 1 year later. More than 10,000 Reinforced Earth structures have now been built in Europe, America, Africa, and Asia. The total as well as the annual wall area of Reinforced Earth has increased every year from 1968 to 1986 (Fig. 16.1).

In his early work, Vidal regarded Reinforced Earth as a new kind of construction material, created by the association of a particulate medium with reinforcement: This

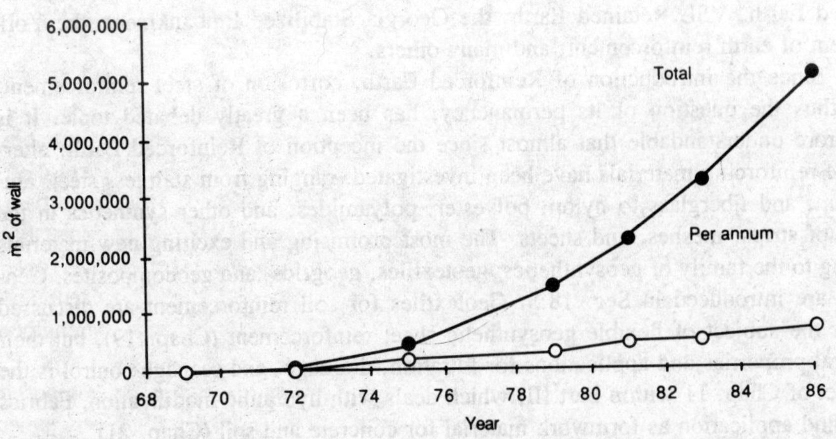

FIGURE 16.1
Wall area of built Reinforced Earth structures. *(Courtesy Reinforced Earth Pty. Ltd.)*

assembly formed a "volume which has cohesion." For Vidal, Reinforced Earth was neither a natural material nor an anchorage or tieback system but

> . . . a method of building construction for soil engineering, comprising a mass of noncoherent particulate material into a stable volume of predetermined configuration . . . such cohesion of the mass being provided solely by frictional effects mutually between the constituent particles and between the particles and surface areas of stabilizing members in frictional contact therewith."

With regard to the cladding or face elements, Vidal's patent states:

> "This skin does not properly speaking play any part in the stability of the work, being merely intended to contain the particles near the free surfaces of the work."

The rapid acceptance of Reinforced Earth by the civil engineering profession, which is generally regarded as very conservative, has been remarkable. It was in 1968 when the first major Reinforced Earth walls were built on a very difficult site near Menton in the South of France. The first bridge abutment was constructed in 1972, also in France. The same year saw the first Reinforced Earth structure completed in the United States, along the California Highway 39 near Los Angeles. In Australia, the first Reinforced Earth wall was built in Parramatta in 1975.

Vidal's patents were formulated in very general terms and were most comprehensive in concept, including the possibility of constructing Reinforced Earth beams and pillars as well as providing for a variety of striplike tensile reinforcement components and various shapes of skin elements (Fig. 16.2). The existence of these patents restricted for many years the widespread use of alternative soil-reinforcement techniques developed by engineers and researchers who were stimulated by the success of Reinforced Earth. Either by agreement with the Reinforced Earth Company, or after expiry of the relevant patents, a number of newer systems, each with its own special characteristics, have entered the marketplace: the Caltrans Mechanically Stabilized Earth, VSL Retained Earth, the Georgia Stabilized Embankment, the York System of earth reinforcement, and many others.

Since the introduction of Reinforced Earth, corrosion of steel reinforcement, and thus the question of its permanency, has been a greatly debated topic. It is therefore understandable that almost since the inception of Reinforced Earth, alternative reinforcing materials have been investigated, ranging from stainless steel, aluminum, and fiberglass to nylon, polyester, polyamides, and other synthetics in the form of strips, meshes, and sheets. The most promising and exciting new materials belong to the family of geosynthetics: geotextiles, geogrids, and geocomposites. Geogrids are introduced in Sec. 18.3. Geotextiles for soil reinforcement are discussed under the subject of flexible geosynthetic sheet reinforcement (Chap. 19), but their general properties and applications for filtration, drainage, and seepage control is the subject of Chap. 11 within Part III, which deals with hydraulic modification. Fabrics also find application as formwork material for concrete and soil (Chap. 21).

Very new on the construction scene is the use of continuous synthetic fiber for granular soil reinforcement. The French-developed TEXSOL is a mix of soil with multiple continuous threads, amounting to 0.1 to 0.2% of the weight of the natural

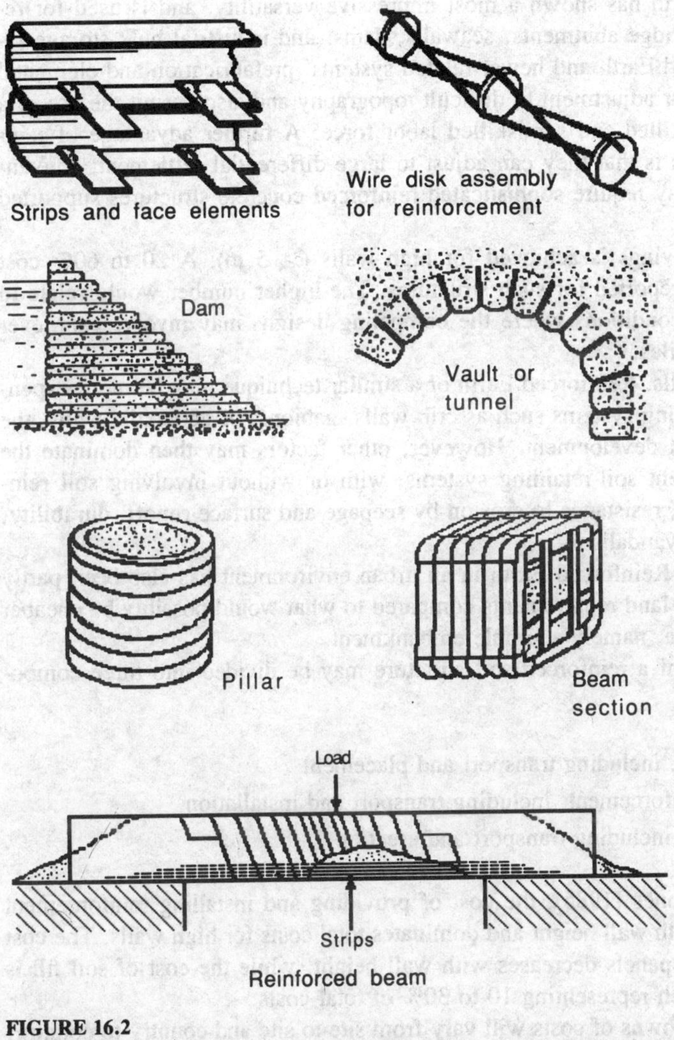

FIGURE 16.2
Illustrations from Vidal's patents (Australian Patent No. 285568, Application No. 42315/64, and others).

soil, placed with special-purpose equipment. The effect of randomly distributed discrete short fibers on the strength and deformation behavior of sand was investigated by Gray and Al-Refeai (1986) with encouraging results.

16.3 ECONOMY AND PERMANENCY

16.3.1 Economy

Technical superiority of reinforced soil structures on difficult ground may have gained them attention and respect, but simple economics ensured their success.

Reinforced Earth has shown a most impressive versatility, and is used for retaining structures, bridge abutments, seawalls, dams, and industrial bulk-storage facilities. In Reinforced Earth and newer related systems, prefabrication and elemental construction allow for adjustment to difficult topography and also permit the use of a relatively small unskilled and semiskilled labor force. A further advantage of reinforced soil structures is that they can adjust to large differential settlement; conventional alternatives may require sophisticated reinforced concrete structures supported on piles.

The biggest savings is achieved for high walls (> 5 m): A 20 to 60% cost advantage has been reported for such structures. The higher number would relate to difficult foundation conditions where the competing designs may involve cantilever walls supported on piles.

For smaller walls, Reinforced Earth or a similar technique may be more expensive than other retaining systems such as crib walls, gabions, or geotextile walls, the latter being the latest development. However, other factors may then dominate the choice among different soil-retaining systems, with or without involving soil reinforcement: aesthetics, resistance to erosion by seepage and surface runoff, durability, and susceptibility to vandalism.

The success of Reinforced Earth in an urban environment has also been partly based on the reduced land requirements compared to what would possibly be cheaper in an unrestricted area, namely a simple embankment.

The total cost of a reinforced soil structure may be divided into three components:

Cost of soil fill, including transport and placement

Cost of the reinforcement, including transport and installation

Cost of facing, including transport and erection

According to Jones (1985), the cost of providing and installing reinforcement for walls increases with wall height and dominates total costs for high walls. The cost of prefabricated face panels decreases with wall height, while the cost of soil fill is likely to increase, each representing 10 to 30% of total costs.

Detailed breakdowns of costs will vary from site to site and country to country, but material costs are likely to be a multiple of the costs of labor or the cost of plant and operatives. [See Jones (1985) for a more detailed discussion.]

Increasing competition from newly developed soil-reinforcing systems, including those employing geosynthetics, promise even better economics; engineers must however now ensure that the current good safety record is not impaired.

16.3.2 Permanency

Corrosion of metals and degradation of synthetics is a major problem in designing reinforced soil structures. According to civil engineering practice, permanent facilities should have a life expectancy of 60 to 100 years. How can we guarantee such a performance?

Corrosion of metal buried in soil is an electrochemical process. It is generated by a current which arises by potential differences in the soil mass due to the presence of different materials, salt, oxygen levels, etc. Corrosion is known to accelerate in an environment with low resistivity and extreme pH levels. Partial saturation, say 50 to 80%, and the presence of certain clay minerals may make the attack on metals even worse. Where organic materials and sulphates are present in the soil, bacterial corrosion is also possible; this condition may be identified by measuring the so-called redox potential [Jones (1985)].

Corrosion of steel will occur, even if it is galvanized. For the designer it then becomes a matter of reducing corrosion to a minimum and predicting its rate as accurately as possible. Minimizing the risk is achieved by stating appropriate selection criteria for the backfill (Table 18.1). Fortunately a fair amount of information is available on typical corrosion rates of metals placed within specific types of soil and for given site conditions [Romanoff (1957); Darbin et al. (1978)]. Egan (1984) and others responded to concerns by designers and owners of Reinforced Earth structures by presenting a corrosion model for assessing the durability of Reinforced Earth reinforcing strips. For a given situation, these models allow the corrosion rate to be predicted in terms of micrometers per year or similar measurements. The designer will then provide for an appropriate increase in the thickness of the reinforcing elements as a "corrosion allowance" which will ensure the desired life expectancy of the structure. As an example, refer to Table 18.2, which gives the required sacrificial strip thickness for Reinforced Earth structures in different environments.

Because of its importance, numerical prediction of the loss of steel due to corrosion in soils deserves further discussion. Romanoff (1957) already recognized that the rate of corrosion of metals in soils decreased with time. He proposed the following relationship between the loss of thickness Δt with time Y:

$$\Delta t = mY^n \tag{16.1}$$

where m and n are constants, varying for different metals and soil environments. Long-term laboratory and field studies on the corrosion of Reinforced Earth galvanized steel strips by Darbin et al. (1986) allow numerical values for the constants m and n for particular conditions to be deduced. For example, for fully to half-saturated soils meeting Reinforced Earth specifications (see Table 18.1), the corrosion observed over some 10 years fell into the range between

$$\Delta t = 5.5Y^{0.6} \quad \text{(minimum)} \tag{16.2}$$

and

$$\Delta t = 50Y^{0.6} \quad \text{(maximum)} \tag{16.3}$$

where Δt is the loss of thickness on one side (μm $= 10^{-6}$ m) and Y is the time measured in years.

The above relationships plot as straight lines on log-log paper, and researchers are confident that corrosion rates can be safely predicted by extrapolating the 8 to 13 year database to 80 or 100 years. Darbin and his coworkers (1986) observed that 80 μm of zinc coating could corrode away in 5 to 50 years, but the zinc ensured relatively uniform corrosion during this period and, compared to plain steel, reduced the corrosion rate thereafter.

Because the thickness loss is not entirely even, the tensile strength of corroded steel strips is affected more strongly than the average loss of thickness would suggest. Darbin et al. (1986) express this effect in terms of the heterogeneity factor K_h:

$$\frac{\Delta T}{T_0} = K_h \frac{\Delta t_{\text{total}}}{t_0} \tag{16.4}$$

where ΔT = loss in tensile strength, kN
T_0 = original tensile strength, kN
Δt_{total} = loss in thickness (overall)
t_0 = original thickness
K_h = 1.2 to 2.8 (say 2 for design)

The tensile strength T remaining in the strip after partial corrosion becomes

$$T = T_0 \left(1 - K_h \frac{\Delta t_{\text{total}}}{t_0}\right) \tag{16.5}$$

It has been found that general corrosion, typical for mild steel and galvanized steel, is more predictable than the pitting-type corrosion which is observed on aluminum and stainless steel; this is an argument in favor of using the cheaper type of metal, although the provision for corrosion has to be enhanced. It has also been apparent that for most conditions it is cheaper to provide for a generous corrosion allowance of metal strips than, e.g., to use glass-reinforced plastics or similar materials which have a strength and modulus approaching those of metals.

Degradation of synthetics is less well known than corrosion of metals. Although metal culverts, steel pile foundations, and other engineering components made of corrosive materials have been in the ground for many years, performance measurements on buried synthetics are relatively new; records in excess of 10 years are available but are rare.

Physical and chemical aging results in a loss of strength with time and may lead to brittleness and stress cracking. It is primarily dependent on the nature of the polymer which a geosynthetic is made of, but the type and amount of any additive (e.g., an antioxidant), the fiber dimensions, the manufacturing process, and other factors are also critical. Some polymers (e.g., nylon) absorb water, resulting in swelling; although reversible, the presence of water reduces tensile strength. In polyesters and polyaramids, chemical reaction with water (hydrolysis) leads to permanent degradation in the longer term, but just how serious this is is a matter of conjecture. The rate of hydrolysis and other aging processes rises exponentially with temperature. This means that even relatively small changes in temperature near the ground surface or behind a retaining wall could affect the endurance of buried plastics.

At this time, engineers have a reasonably high degree of confidence in the durability of synthetics fully buried in soil; only the presence of unusual chemicals appears to present a danger. Nevertheless, because of the lack of long-term experience and the variety of products involved, more caution is advised with synthetics than with the better-known construction materials such as steel.

Where exposed to ultraviolet light, some synthetics may deteriorate completely within weeks or months. Thus the design of permanent retaining structures made with fabric facing presents a specially difficult problem: Chemical treatment or complete shielding will be required for extended life.

Corrosion provisions as stipulated for different metal soil-reinforcing techniques will be discussed further as part of the relevant design procedures. The durability of synthetics will also be mentioned again in Sec. 18.3.1 (with respect to geogrids) and Sec. 19.1.3 (with respect to geotextiles).

PROBLEMS

Prefixes indicate problem type: B = brief answer, C = calculations.

B16.1. Identify the following construction methods or materials:

(*a*) A brushwood bundle 3 to 5 m long, firmly tied into a cylindrical shape with about an 0.8-m girth, used as protective facing for seawalls and riverbanks or for roads over waterlogged soil.

(*b*) A temporary road built of 80-mm-diameter saplings, equal in length to the width of the road, wired tightly together at the ends. It is quickly laid, will float in liquid mud, and can also be picked up again and reused elsewhere.

(*c*) An earth brick made of a mixture of clay and straw.

(*d*) Originally a cylindrical bottomless wicker basket about 60 cm in diameter, placed on the edge of a trench to protect soldiers from rifle fire. It stood, with others, in a row and was filled with earth dug from a trench. The term may now also be used for a small cellular cofferdam.

B16.2. What links Henri Vidal (Reinforced Earth patent, 1963) with E. Freyssinet (1928 patent for prestressed concrete) and Monier (early patent for reinforced concrete)?

C16.3. What is the dollar value of all Reinforced Earth structures built until the end of 1986, assuming each square meter costs $200?

C16.4. Consider a Reinforced Earth steel strip in backfill which satisfies basic criteria of Table 18.1. Estimate the possible limits of total thickness loss (both sides of the strip!) for a period of 1, 5, 10, 20, 30, 40, and 50 years according to Eqs. 16.2 and 16.3. Plot the results on normal and log-log paper.

C16.5. A Reinforced Earth strip is initially 5 mm thick and 40 mm wide. The steel has a yield strength of 240 MPa. Calculate the safe load for this strip at the end of its service life, assuming a safety factor of 1.5 and a predicted total average loss of thickness due to corrosion of 2 mm.

CHAPTER 17

MECHANICAL MODELS OF SOIL REINFORCEMENT

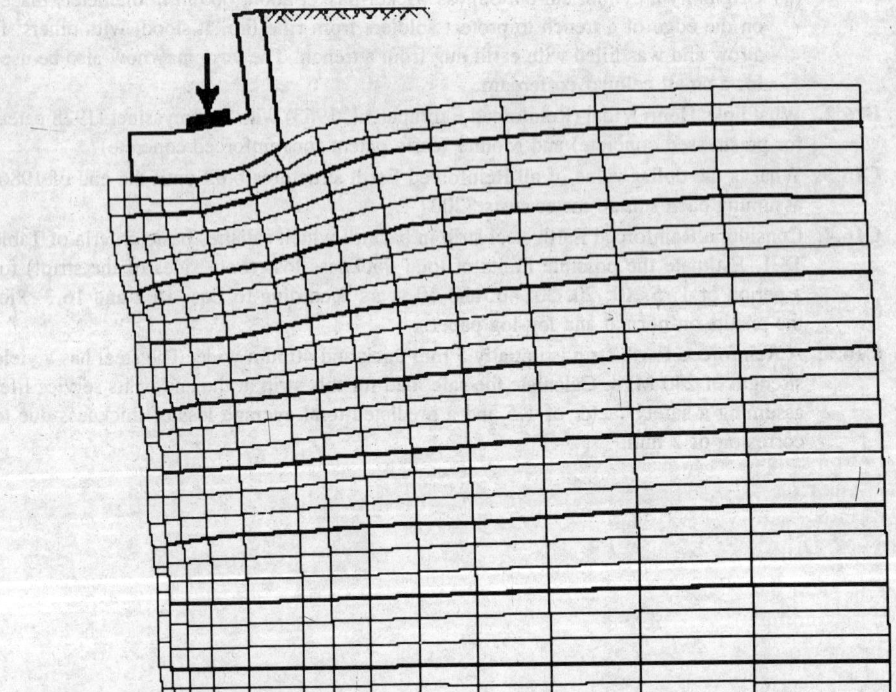

There are basically two approaches to the analytical modeling of reinforced soil: representation as a homogeneous composite material or representation by discrete structural elements (soil, reinforcement, and boundary elements). Both models have their merits.

Treating reinforced soil as a anisotropic homogeneous composite material improves the mathematical efficiency of some methods of determining load-deformation behavior using finite elements; in addition, it has value in the interpretation of laboratory tests and has conceptual appeal in evaluating reinforced soil behavior in practical applications.

The most common methods of analysis of reinforced soil structures have been developed along similar lines to those traditionally used for conventional geotechnical structures. In many routine soil engineering problems, such as in the design of retaining walls, it is considered to be more important to determine the safety factor (or estimate the probability of failure) with respect to overturning, sliding, or other types of failure, rather than to predict rotation, translation, or settlement. It is tacitly assumed that if the safety factors are above empirically set values, the structure's deformation or movement will not be excessive.

Thus the current design practice concentrates on the analysis of the ultimate behavior, or failure state, characterized by the formation of slip planes in the soil mass and/or the rupture or slippage of structural components. This approach is facilitated by looking separately at the stresses in the soil and in the reinforcement and at the common interface of the two. The detailed characteristics of discrete soil-reinforcement interaction are complex and depend on many factors, including soil type and condition, reinforcement geometry, and relative stiffness.

17.1 REINFORCED SOIL AS A HOMOGENEOUS COMPOSITE MATERIAL

Composite materials have been successfully used in many branches of engineering, most notably in the aerospace industry, where the desire to develop high-strength and high-modulus materials of low weight has been a great incentive to develop advanced theories in the mechanics of composite materials.

Civil engineers must realize that even plain concrete, made up of mortar and aggregate, is a composite material. Its strength and elastic modulus are generally evaluated on a macroscopic scale. In other words, for the purpose of analysis and design, plain concrete is considered a homogeneous material and its properties are studied taking a phenomenological approach. The same applies to steel, which can be considered macroscopically homogeneous, although microscopically it is made up of crystalline constituents.

On the other hand, steel-reinforced concrete is generally treated as a heterogeneous composite material and its properties investigated at a micromechanical level. In the so-called strength of materials approach to the design of beams, plates, and shells, certain simplifying assumptions are made regarding the mechanical behavior of the composite material; e.g., it is generally assumed that the strains in the reinforcement are the same as in the surrounding concrete mix. In reinforced soil, this assumption may or may not be appropriate, depending on the stress level and relative moduli of the materials involved.

17.1.1 Elastic Theory

Following an essentially micromechanical approach, Harrison and Gerrard (1972) considered Reinforced Earth to be a cross-anisotropic elastic material. They produced solutions for stresses and displacements for simple boundary conditions which may assist in checking for local overstressing. This analysis obviously does not apply to failure conditions involving slippage between the soil and the reinforcement.

After evaluating the finite-element representation of Reinforced Earth by discrete beam and elasticity elements, Romstad et al. (1974) proposed that composite elements be used, in order to reduce the computational efforts involved. Although edge effects cannot be studied in this way, these authors maintained that the composite "unit cell" approach can determine peak forces in the reinforcement as accurately as an analysis with discrete finite elements.

17.1.2 Strength Theories

Strength theories applied to conventional composite materials may be based on a maximum stress, maximum strain, or variations of the maximum distortional work criterion; but it has proved difficult to assign quantitative measures to, e.g., the extent of debonding or lack of adhesion between the constituents. In composite material mechanics, more theoretical work deals with the evaluation of micromechanical stiffness, rather than strength. However, it is often relatively easy to experimentally determine the strength properties of these products, such as fiber-reinforced resins, because good representative samples are available.

In geotechnical structures, the matrix material is of a particulate nature and reinforcement occurs on a large scale. The first characteristic calls for applying the Mohr-Coulomb failure theory to study stability; the second is one of the reasons why it is difficult to design appropriate laboratory tests for the evaluation of stress-strain characteristics of reinforced soil.

Interpreting the strength properties of reinforced soil in terms of the Mohr-Coulomb theory, several early researchers found that reinforcement could increase the cohesion component of soil strength. This finding fitted nicely into Vidal's vision of a new construction material, where reinforcement imparted cohesion to a mass of granular soil particles. Thinking along similar lines, Endo and Tsuruta (1968) concluded that roots in soil increase its shear strength, or add "cohesion" to it. Similarly, Long et al. (1972) and Yang (1972) interpreted results of triaxial tests on reinforced sand by means of elementary Mohr-Coulomb theory.

Hausmann (1976) and Hausmann and Lee (1976) extended the Mohr-Coulomb interpretation of reinforcement action to cover failure by rupture of the reinforcement as well as by soil-reinforcement slippage. Although other approaches are also possible, it is most easily understood if tensile reinforcement is thought of as providing resistance to expansion of a soil mass in the form of a normal stress σ_R in the direction of the reinforcement.

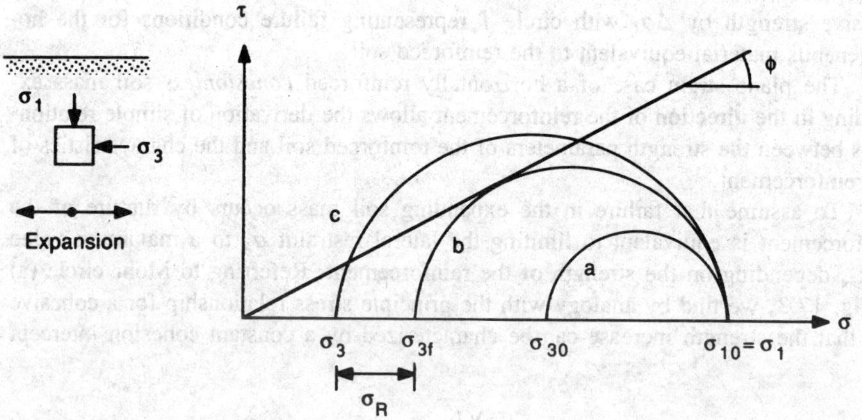

FIGURE 17.1
Stress changes in an expanding soil mass.

17.1.2.1 MOHR-COULOMB ANALYSIS FOR HORIZONTAL REINFORCEMENT. Tensile reinforcement is most effective if placed in the major principal plane, in the direction of the minor principal stress, which in many practical geotechnical problems is horizontal.

Figure 17.1 illustrates the stress changes in a horizontally expanding cohesionless soil mass. Mohr circle *a* represents stress conditions at rest. Circle *b* defines failure conditions for the unreinforced soil, equivalent to Rankine's state of active plastic equilibrium. Statically, the reinforcement can be viewed as providing a restraint $\sigma_R = \Delta\sigma_3$ which allows the lateral pressure to reduce further before failure conditions of the reinforced soil are reached, represented by circle *c*.

Similarly, Fig. 17.2 describes the stress changes in a standard triaxial test, where the cell pressure σ_3 is kept constant. The reinforcement restraint increases the com-

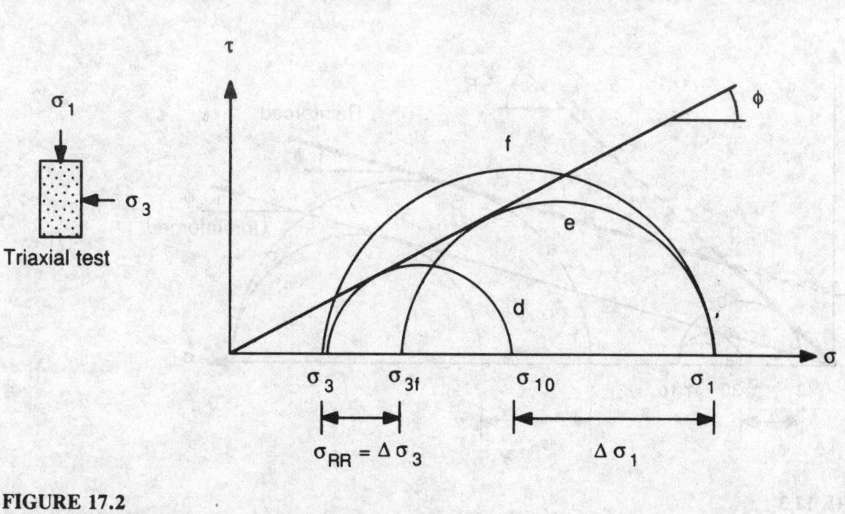

FIGURE 17.2
Stress changes during a standard triaxial test.

pressive strength by $\Delta\sigma_1$ with circle f representing failure conditions for the homogeneous material equivalent to the reinforced soil.

The plane strain case of a horizontally reinforced *cohesionless* soil mass expanding in the direction of the reinforcement allows the derivation of simple relationships between the strength parameters of the reinforced soil and the characteristics of the reinforcement.

To assume that failure in the expanding soil mass occurs by rupture of the reinforcement is equivalent to limiting the lateral restraint σ_R to a maximum value $\sigma_{R,\max}$ depending on the strength of the reinforcement. Referring to Mohr circle (a) in Fig. 17.3, we find by analogy with the principle stress relationship for a cohesive soil that the strength increase can be characterized by a constant cohesion intercept c_R:

$$c_R = \frac{\sigma_R\sqrt{K_p}}{2} = \frac{\sigma_R}{2\sqrt{K_a}} \tag{17.1}$$

where $K_a = \tan^2(45 - \phi/2)$ and $K_p = \tan^2(45 + \phi/2)$. The K_a and K_p are the active and passive coefficients, respectively, of lateral stress for a friction angle ϕ (of the unreinforced soil). A relationship equivalent to the above has previously been derived by Schlosser and Long (1972).

The case of slippage between the reinforcement and the soil corresponds to conditions where σ_R is proportional to the constant vertical stress σ_{10}. With $\sigma_R = \sigma_{10}F$, an increased friction angle results, defined by

$$\sin\phi_R = \frac{1 + F - K_a}{1 - F + K_a} \tag{17.2}$$

The F represents a friction factor characterizing reinforcement interaction with the cohesionless soil. $F = K_a$ leads to the limiting value of $\phi_R = 90°$, meaning failure will occur by rupture rather than by slippage.

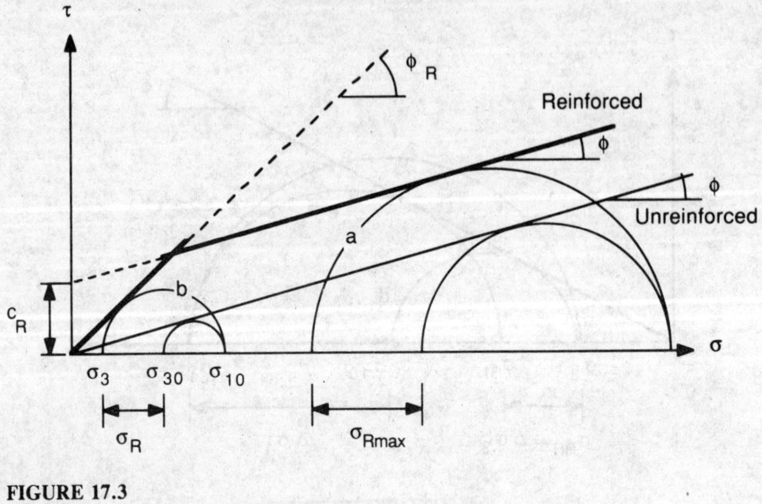

FIGURE 17.3
Mohr envelope for horizontally reinforced soil.

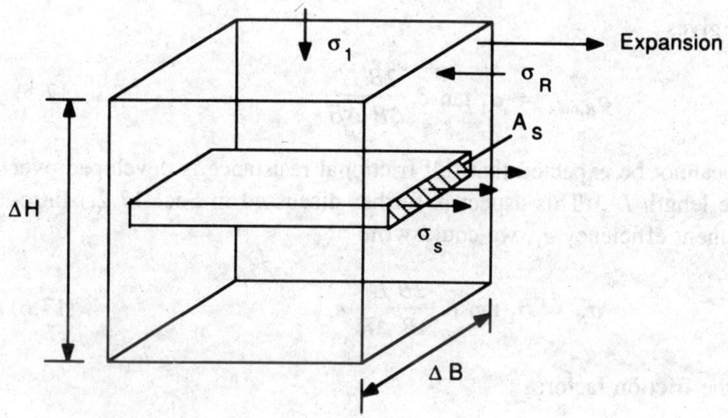

FIGURE 17.4
Rupture of reinforcement in unit element.

Conditions equivalent to failure by slippage are shown by Mohr circle (b) in Fig. 17.3.

For one-dimensional expansion as shown for the unit element section in Fig. 17.4, the resultant of the reinforcement restraint σ_R at rupture is statically equal to the tensile strength of the strip, which yields

$$\sigma_R = \sigma_s \frac{A_s}{\Delta B \, \Delta H} \tag{17.3}$$

where σ_s is the tensile strength of the reinforcing material and A_s its cross-sectional area. The ΔB and ΔH represent the horizontal and vertical spacing, respectively, of the reinforcement.

Figure 17.5 illustrates the development of the shear stress τ_s over length L'. The maximum value of τ_s is related to σ_1 and the angle of skin friction δ by

$$\tau_{s,\max} = \sigma_1 \tan \delta \tag{17.4}$$

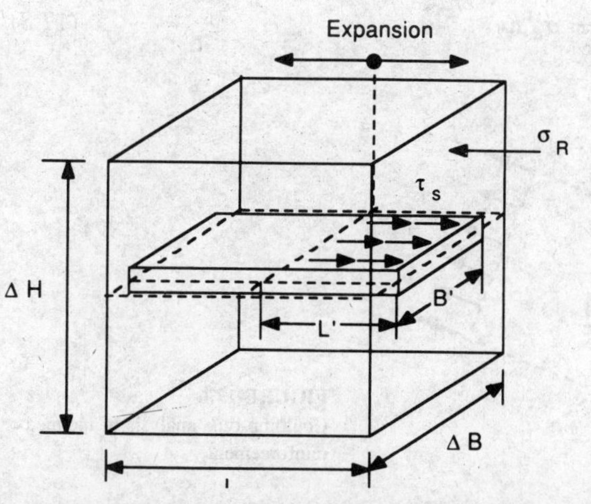

FIGURE 17.5
Slippage of reinforcement in unit element.

Static equivalence gives

$$\sigma_{R,\max} = \sigma_1 \tan \delta \frac{2B'L'}{\Delta B \, \Delta H} \tag{17.5}$$

However, it cannot be expected that full frictional resistance is developed over the entire effective length L'. (This aspect is further discussed in Sec. 17.2.) Introducing a reinforcement efficiency e_s, we could write

$$\sigma_R = \sigma_1 \tan \delta \frac{2B'L'}{\Delta B \, \Delta H} e_s \tag{17.6}$$

and thus find for the friction factor:

$$F = \tan \delta \frac{2B'L'}{\Delta B \, \Delta H} e_s \tag{17.7}$$

This type of expression can be used to interpret triaxial tests on reinforced soil [Hausmann (1976)].

17.1.2.2 MOHR-COULOMB ANALYSIS FOR INCLINED REINFORCEMENT. For states of stress where principal stress directions vary relative to the placement of the reinforcement, the effectiveness can be reduced and failure modes may change from that of rupture of the reinforcement to that of soil-reinforcement slippage. This can be illustrated with the Mohr-Coulomb analysis. In addition, this mechanical model also indicates that if the reinforcement is at a sufficiently steep angle relative to the horizontal (major principal) plane, the strength of the soil mass could actually be reduced.

Figure 17.6 shows a wedge of soil reinforced at an angle β to the major principal plane at failure. With the angle α defining the inclination of the assumed failure plane, equilibrium considerations yield

$$\sigma_1 = \sigma_3 K_1 + \sigma_R K_2 \tag{17.8}$$

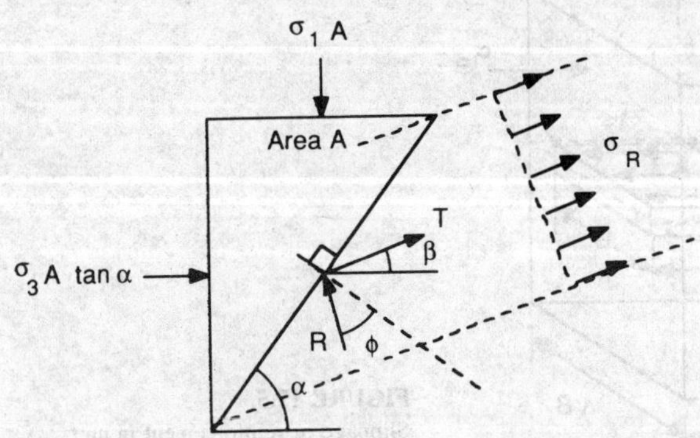

FIGURE 17.6
Coulomb-type analysis of inclined reinforcement.

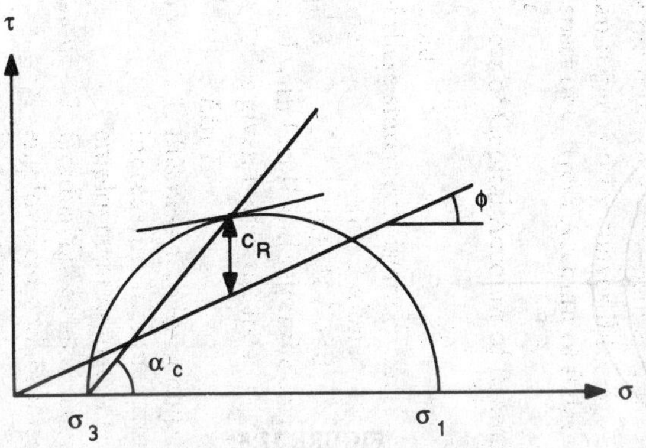

FIGURE 17.7
Cohesion intercept, Coulomb-type analysis.

where
$$K_1 = \frac{\tan \alpha}{\tan (\alpha - \phi)} \tag{17.9}$$

and
$$K_2 = \frac{\sin (\alpha - \beta)}{\cos \alpha} \left[\frac{\cos \beta}{\tan (\alpha - \phi)} + \sin \beta \right] \tag{17.10}$$

Similar to analyzing earth pressures according to Coulomb, α is varied so as to find a minimum value for σ_1. The angle $\alpha = \alpha_c$ then defines the critical failure plane. For the special case of $\beta = 0$ we find

$$\alpha_C = 45 + \frac{\phi}{2} \quad \text{and} \quad K_1 = K_2 = K_p$$

For $\beta \neq 0$, Eq. (17.8) can be solved numerically for specific values of ϕ, σ_3, and σ_R; the cohesion intercept c_R representing the strength increase is found to decrease with increasing normal stresses as sketched in Fig. 17.7. Using a constant value for σ_R is again equivalent to assuming that rupture of the reinforcement causes failure.

For the case of slippage of the reinforcement, we assume that the restraint σ_R is proportional to the normal stress acting on the surface of the inclined reinforcement; thus

$$\sigma_R = F\sigma_\beta \tag{17.11}$$

and we subsequently find

$$\sigma_1 = \sigma_3 \frac{K_1 + F(\sin^2 \beta)K_2}{1 - F(\cos^2 \beta)K_2} \tag{17.12}$$

Minimizing σ_1 for specific values of ϕ, σ_3, and F by numerical methods indicates that the strength increase can be expressed independently of normal stresses as an increase in the friction angle.

Figure 17.8 illustrates the analysis of failure conditions equivalent to the Rankine approach to earth pressure problems, looking at stresses at a point within the soil mass rather than examining the equilibrium of a failure wedge. In this analysis it is appropriate to define the inclination of the reinforcement relative to the original "unrein-

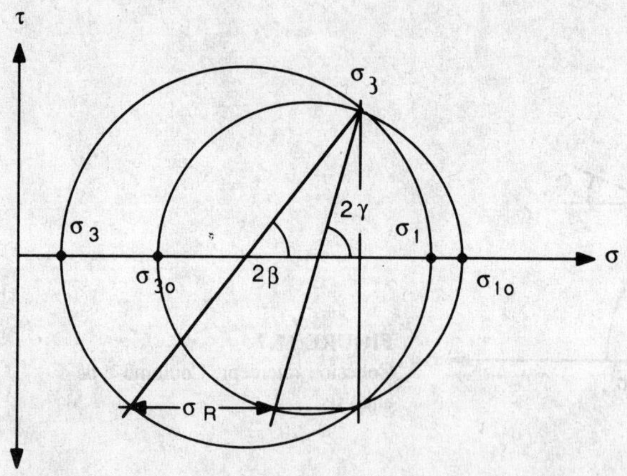

FIGURE 17.8
Inclined reinforcement, Rankine-type analysis.

forced'' principal failure stresses σ_{30} and σ_{10} by the angle γ. Given σ_{30}, $\sigma_{10} = K_p\sigma_{30}$, γ, and σ_R, we can determine the principal stresses σ_1 and σ_3 of the reinforced soil. Generally a curved Mohr envelope results.

For the case representing failure by slippage, with $\sigma_R = \sigma_\beta F$, we find that the strength increase can be represented by an increase in the friction angle to ϕ_R, defined by

$$\sin \phi_R = \frac{\sqrt{K_3^2 + 2FK_4 + F^2}}{2 - F - K_4} \tag{17.13}$$

where

$$K_3 = \frac{2 \sin \phi}{1 + \cos 2\gamma \sin \phi} \tag{17.14}$$

$$K_4 = K_3 \cos 2\gamma \tag{17.15}$$

For $\gamma = 0 = \beta$ we find $K_3 = K_4 = 1 - K_a$ and Eq. (17.13) reduces to Eq. (17.2). Figure 17.9 illustrates the strength characteristics of a reinforced soil for a specific set of assumed basic parameters. This diagram is only drawn for the case where the tensile strength of the reinforcement contributes to the strength of the soil.

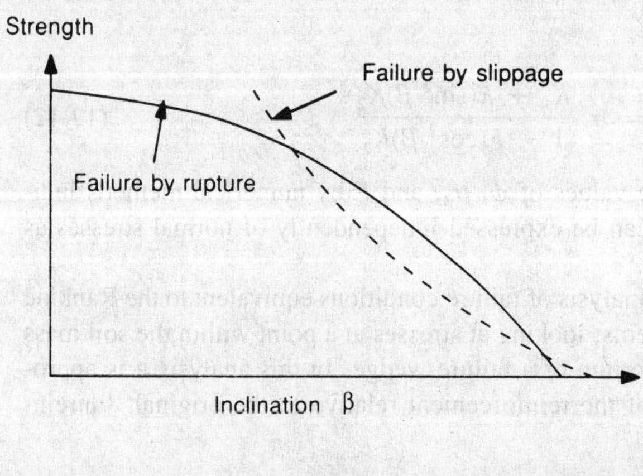

FIGURE 17.9
Cohesion as a function of the inclination of the reinforcement, Coulomb-type analysis.

As can be seen from the figure, this only happens over a limited range of reinforcement inclinations. Beyond that range this analysis should be extended to cover the possibility of other types of failures and reinforcement action, such as block sliding along the reinforcement and possible reduction of normal stresses (and thus shear strength) in the failure plane, if the reinforcement takes on compressive stresses.

17.2 DISCRETE SOIL-REINFORCEMENT ACTION

Figure 17.10 shows schematically a number of different ways in which reinforcement could retain an unstable soil mass. To illustrate a general case, the reinforcement strip, mesh, or sheet is assumed to be attached to an anchor in the stable part of the soil mass; furthermore, structural face elements assist soil retention at the otherwise free surface of the unstable zone.

Movement of the sliding soil mass is resisted by the following phenomena:

Friction and adhesion developed on the surface of that part of the reinforcement which lies within the stable zone.

Passive resistance generated by the anchor. This represents symbolically the action of a real block anchor, crossbars in a mesh, or ribs on a reinforcing strip.

Bending and passive resistance occurring where the reinforcement crosses the boundary between the stable and unstable soil mass.

Besides being limited by the amount of friction and adhesion or passive resistance which is developed, the maximum force which can be transmitted through the reinforcement is governed by the reinforcement's own strength. These limiting con-

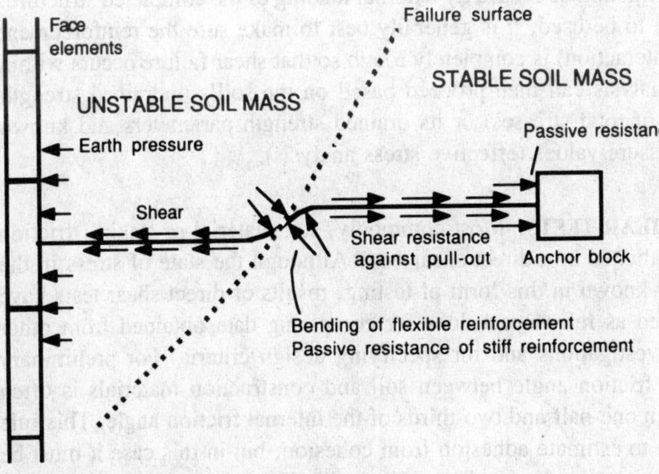

FIGURE 17.10
Schematic sketch illustrating discrete soil-reinforcement action.

ditions correspond to the basic causes of internal failure of reinforced soil: rupture of the reinforcement, slippage between the soil and the reinforcement (also interpreted as failure by pullout), and bearing-type (plastic) failure in the soil.

The driving force of the unstable soil mass is transferred to the reinforcement in the following ways:

Friction and adhesion on the reinforcement

Earth pressure acting on the boundary face elements and possibly at intermediate points, e.g., crossbars or ribs

Friction, adhesion, bending and passive resistance, anchor action, and the magnitude of earth pressures on boundary elements will now be discussed further but without going into details of methods of analysis for specific soil reinforcement methods.

17.2.1 Friction and Adhesion

Friction is that component of shear resistance which is proportional to the normal stress; adhesion, on the other hand, represents shear resistance between soil and other materials which is independent of the normal stress acting on the shear plane. The evaluation of the friction and adhesion between soil and structural elements has always been a significant problem in civil engineering; just think of pile foundations, caissons, conventional retaining walls, and tieback anchors. The emergence of soil-reinforcing techniques is a new incentive to study soil-material friction and adhesion.

Adhesion is not a property geotechnical engineers like to rely upon; this holds particularly true for soil-reinforcing techniques. Most of these require free-draining cohesionless soil. This precaution eliminates the danger of excess pore pressures being generated during construction or caused by external loading of the completed structure. If cohesive soils have to be used, it is generally best to make sure the reinforcement surface (or mode of interaction) is completely rough so that shear failure occurs within the soil itself. The analysis can then proceed based on the soil's undrained strength parameters (in terms of total stresses) or its drained strength parameters and known or estimated pore pressure values (effective stress analysis).

17.2.1.1 DIRECT SHEAR TESTS. Most commonly, soil-material or "skin" friction characteristics are evaluated in a direct shear box. Although the state of stress in the soil is not completely known in this form of testing, results of direct shear tests have generally been adopted as reference values for comparing data obtained from other laboratory or field investigations and for specifying design criteria. For preliminary design purposes, the friction angle between soil and construction materials is often assumed to be between one-half and two-thirds of the internal friction angle. This rule of thumb is also used to estimate adhesion from cohesion, but in this case it must be combined with absolute upper limits dictated by experience for a particular design problem.

Figure 17.11*a* shows the typical arrangement of materials for measuring skin friction such as is used in the original comprehensive work by Potyondy (1961). The equipment represents a standard shear box where the lower half is occupied by a rigid block. The same test was used by Schlosser and Long (1974) in an effort to establish basic design criteria for Reinforced Earth and more recently by Martin et al. (1984) and others in order to obtain shear parameters between soils and geotextiles or geomembranes. Particularly for testing the friction between soil and flexible fabric-type reinforcement, the choice of support material in the lower half of the box is important, as demonstrated by Delmas et al. (1979).

Other versions of the direct shear test have been employed with the aim of modeling more closely what happens between soil and reinforcing elements or anchors. In an attempt to simulate the development of shear resistance along cylindrical anchors, Wernick (1978) built a direct shear box where the two halves are forced to move exactly parallel to each other and where the vertical force is induced by a nontilting loading cap block (Fig. 17.11*c*). Compared to the standard box, this apparatus yields improved information on the dilatancy (volume increase) of sands during shear, which may contribute substantially to the total shear resistance. The kinematic conditions are similar to those in a conventional ring shear apparatus, but unequal rates of shear strain are avoided.

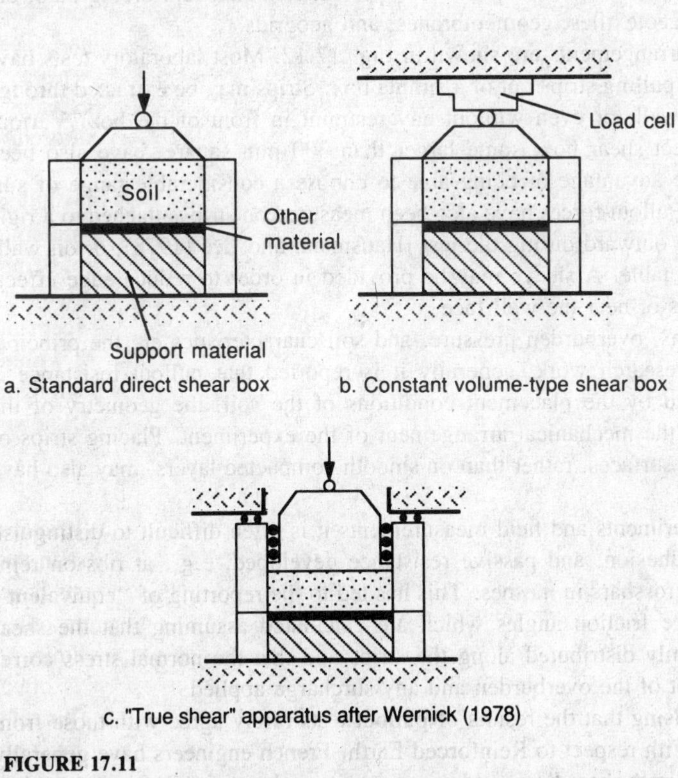

FIGURE 17.11
Typical direct-shear-test arrangements for the measurement of soil-material friction and adhesion. [*Hausmann and Ring (1980)*.]

Figure 17.11*b* illustrates a constant-volume direct shear box, as used by Guilloux et al. (1979). For a dense sand, their results indicated that the normal stress could increase more than twentyfold during the test because of the suppressed tendency of the soil to expand. If the shear resistance in these tests was related to the original normal stress, apparent friction angles such as 85° were calculated, much higher than the 35° measured in the standard way. This finding points out one of the difficulties which can be encountered in interpreting pullout test results, where local conditions may prevent the soil dilating, resulting in normal stresses higher than those related to the overburden weight.

17.2.1.2 PULLOUT TESTS. In 1973 Lee et al. proposed a model of Reinforced Earth action where the earth pressure exerted on the face elements is balanced by the pullout resistance of the strips (or ''ties'') over that part of their total length which is beyond the potential failure wedge in the soil mass (see Sec. 18.1.3.2). This concept led to a comprehensive program of pullout tests in laboratory and field conditions.

A wide range of results are now available, from small-scale laboratory experiments with less than 600 mm of overburden to pullout tests of dummy strips embedded in the backfill during construction of full-size walls. Tie materials tested include standard Reinforced Earth galvanized steel strips, conventional reinforcing bars, bar mesh, mylar tape, geotextiles, geomembranes, and geogrids.

Typical test arrangements are shown in Fig. 17.12. Most laboratory tests have been carried out by pulling strips out of a simple box. Strips may be extracted through a rigid or flexible wall, or even without any restraint in front of the box. Various versions of the direct shear box, some larger than 300 mm square, have also been used; they offer the advantage of being able to choose a considerable range of surcharge magnitude. Pullout forces have also been measured on strips attached to a rigid model wall rotating outward during the test [Hausmann and Lee (1978)] or on walls tested on a shaking table. A sleeve may be provided in order to reduce edge effects in test embankments or near the wall face.

Tie dimensions, overburden pressure, and soil characteristics are the principal variables in most research work; generally it is reported that pullout resistance is significantly affected by the placement conditions of the soil, the geometry of the reinforcement, and the mechanical arrangement of the experiment. Placing strips or meshes on irregular surfaces, rather than on smooth compacted layers, may also have a significant effect.

In model experiments and field measurements it is often difficult to distinguish between friction, adhesion, and passive resistance developed, e.g., at ribs on reinforcing strips or at crossbars in meshes. This has led to the reporting of ''equivalent'' (or apparent) surface friction angles which are calculated assuming that the shear resistance is uniformly distributed along the strips and that the normal stress corresponds to the weight of the overburden and any surcharge applied.

It is not surprising that the results of pullout tests rarely agree with those from direct shear tests. With respect to Reinforced Earth, French engineers have generally concluded that the direct shear box yields conservative values of skin friction angles.

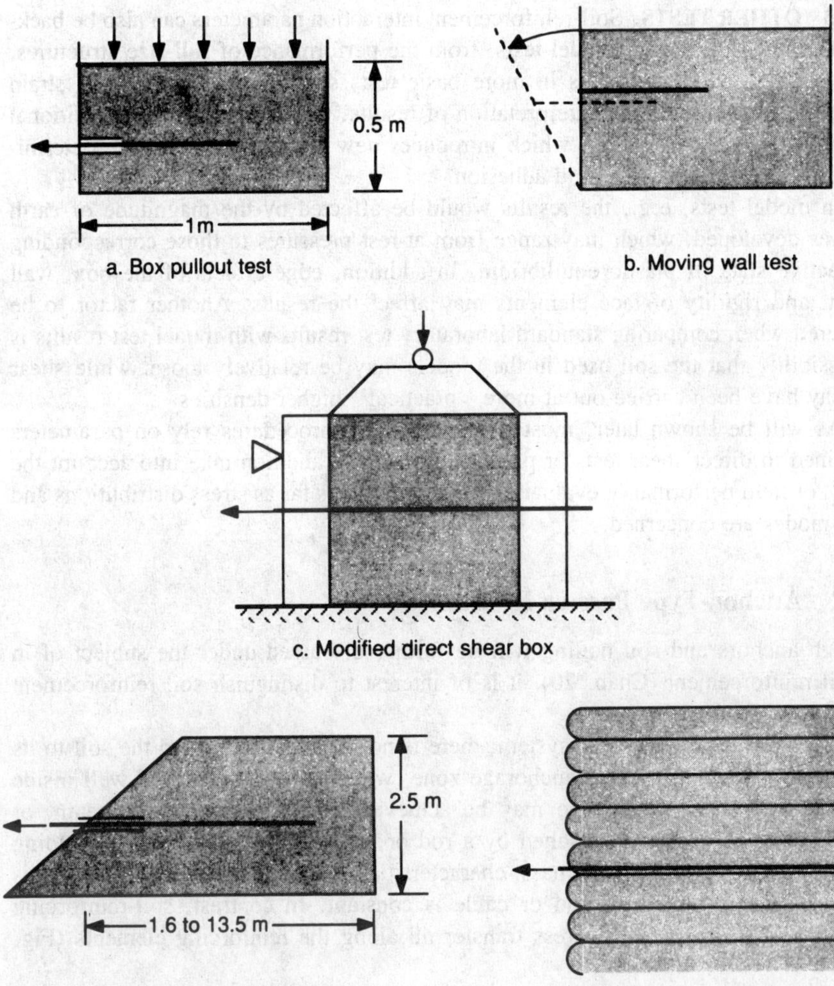

FIGURE 17.12
Typical pullout test arrangements.

In contrast, many American laboratory test results show that the average shear resistance determined in pullout tests can be substantially less than would be anticipated from direct shear tests. High overburden pressure and consequent lack of dilation or some cohesion present in field conditions could be some of the physical causes of relatively low friction values; however, the method of interpretation of the results, particularly the assumptions made with respect to the degree of mobilization and distribution of shear stresses as well as the reliability of the estimate of actually acting vertical stresses may strongly affect the results of experiments.

17.2.1.3 OTHER TESTS. Soil-reinforcement interaction parameters can also be back-figured from the results of model tests, from the performance of full-size structures, or from strength measurements in more basic tests such as triaxial or plane strain shear tests. However, in the interpretation of results, usually a number of additional assumptions have to be made, which introduces new uncertainties into the determination of skin friction angles and adhesion.

In model tests, e.g., the results would be affected by the magnitude of earth pressures developed, which may range from at-rest pressures to those corresponding to an active state of plastic equilibrium. In addition, edge effects of the box, wall friction, and rigidity of face elements may affect the results. Another factor to be considered when comparing standard laboratory test results with model test results is the possibility that the soil used in the models may be relatively loose, while shear tests may have been carried out at more "practical" higher densities.

As will be shown later, most accepted design procedures rely on parameters determined in direct shear tests or pullout tests and in addition take into account the findings of field performance evaluations, particularly as far as stress distributions and failure modes are concerned.

17.2.2 Anchor-Type Passive Resistance

Although anchors and soil nailing will be further discussed under the subject of in situ soil reinforcement (Chap. 20), it is of interest to distinguish soil reinforcement from conventional anchorage.

In a traditional anchoring system, there is no stress transfer from the soil to its components outside the actual anchorage zone, which is usually located well inside the stable soil mass. Anchorage may be achieved by a concrete block, beam, or grouted soil. The anchor is attached by a rod or cable to the face structure holding back the potentially failing soil mass; characteristic for a traditional "true" anchor is that the tensile force in this rod or cable is constant. In contrast, soil-reinforcing systems are distinguished by stress transfer all along the reinforcing elements (Fig. 17.13).

True anchors may be of the active type, where a prestressing force is applied after installation; in a passive anchoring system, the stabilizing force is only mobilized after some deformation of the soil (or rock) mass has taken place. Experiments have been done with prestressing fabrics and strips below road embankments, but most common soil-reinforcing techniques would be categorized as passive systems.

Traditional-type anchors are installed into an existing soil or rock mass. In contrast, reinforced soil structures are usually constructed from the bottom up using remolded and compacted soil. An exception to this rule is soil nailing, where steel rods are driven into embankments or sides of excavations. In soil nails, there is generally no distinct anchorage element incorporated, such as the wedges in some rock anchors, although all kinds of combination systems have been developed in recent years.

Forces on anchor blocks can be estimated assuming sufficient deformation takes place to develop passive earth pressures opposing the pullout force. In a mesh or grid,

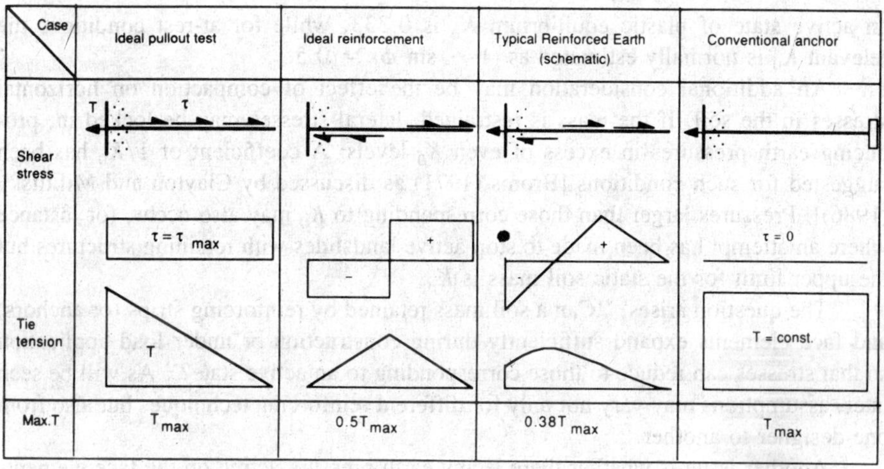

Notes: Assumed is cohesionless soil with friction fully developed across and on both sides of the tie.
Max. T = Maximum tie tension along the strip

FIGURE 17.13
Soil reinforcement versus anchor action.

crossbars may similarly provide passive resistance, although it would be questionable to use classic earth pressure theory to calculate its magnitude. Bearing-capacity-type formulations have also been used, but generally for the development of design procedures, experimental results are used (see Sec. 18.1.1.2).

17.2.3 Bending and Passive Resistance across the Failure Plane

Bending and passive resistance of the reinforcement at a slip surface is rarely considered (except in soil nailing). The possibility of the reinforcement aligning itself with the failure plane does however influence the orientation of the resisting force which has a marked influence on a Coulomb-type analysis of a failure wedge.

Soil nails provide resistance against pullout as well as bending and passive resistance when driven across existing or potential failure planes. Thus soil nailing is more like a true soil-reinforcing technique than an anchorage system, but in concept it differs substantially from Reinforced Earth, the pioneer of all modern soil-reinforcing techniques.

17.2.4 Earth Pressure Acting on Face Elements

In all earth-retaining problems the question arises, "What magnitude of pressures are developed?" Conventional walls are designed for active earth pressures if they can yield outward; if this movement is impeded, the higher at-rest pressures must be considered. The difference in the overturning forces developed can be substantial: For a soil with an internal friction angle of 30° the coefficient of lateral earth pressure for an active state of plastic equilibrium K_a is 0.333, while for at-rest conditions the relevant K_0 is normally estimated as $(1 - \sin \phi) = 0.5$.

An additional consideration may be the effect of compaction on horizontal stresses in the soil. If the mass is restrained, lateral stresses may be locked in, producing earth pressures in excess of even K_0 levels: A coefficient of $1/K_0$ has been suggested for such conditions [Broms (1971) as discussed by Clayton and Milititsky (1986)]. Pressures larger than those corresponding to K_0 may also occur, for instance where an attempt has been made to stop active landslides with retaining structures but the upper limit for the static soil mass is K_p.

The question arises, "Can a soil mass retained by reinforcing strips (or anchors) and face elements expand sufficiently during construction or under load application so that stresses can reduce to those corresponding to an active state?" As will be seen later, assumptions may vary not only for different reinforcing techniques but also from one designer to another.

Another issue is whether there is any earth pressure acting on the face elements at all. With a sufficient amount of reinforcement it can be hypothesized that the face elements only have cosmetic value, since the earth pressure is taken up by shear resistance on the reinforcing elements lying within the unstable soil mass. This may be the case to a large degree, but an examination of most design methods shows that conservatism prevails and face elements are designed assuming they are exposed either to active or at-rest pressures.

Information provided by the Reinforced Earth Company suggests that the ratio of the tension at the strip-face connection and its maximum value along the strip increases with the rigidity of the face elements. Although the stress level at the connection to concrete face panels may reach 85 to 100% of its maximum value, it never exceeds 75% with metal facing elements.

PROBLEMS

Prefixes indicate problem type: C = calculations, B = brief answer, M = multiple choice, D = discussion.

M17.1. The principle of soil reinforcement can be demonstrated in triaxial tests on cylindrical samples of sand reinforced with layers of aluminum foil or other material. It can be shown that this reinforcement

(*a*) Increases failure strains.

(*b*) Decreases cohesion.

(*c*) Increases the friction angle or adds cohesion depending on the failure mode.

(*d*) Increases the friction angle only.

B17.2. Name four composite building materials used in engineering:

(*a*) __________

(*b*) __________

(*c*) __________

(*d*) __________

C17.3. Triaxial tests on sand ($\varphi = 37°$) reinforced with horizontal layers of aluminum foil indicated a strength gain which was equivalent to a cohesion intercept of 27 kPa. Test details were as follows:

Strength of aluminum foil $R_T = 0.765$ N/mm

Vertical spacing of the foil reinforcement $\Delta H = 28$ mm

Restraint by reinforcement $\sigma_R = R_T/\Delta H$

Compare the experimental result with the theoretically expected gain in cohesion.

C17.4. Triaxial tests on a sand indicated an internal friction angle of 37° (without reinforcement). Several layers of horizontal steel disks appeared to increase the friction angle to 49°. For this case, evaluate the friction factor F from Eq. (17.2).

C17.5. Consider the hypothetical plane strain unit element of reinforced soil in Fig. P17.1. Sheetlike reinforcement is spaced at $\Delta H = 1$ m and $\Delta B = B' = 1$ m. Say $L' = 0.5$ m. Because the relative displacement between soil and reinforcement varies as load is applied, shear forces are not mobilized fully across the surface of the reinforcement surface. Let us assume the distribution is triangular, reaching a maximum possible value at the ends; this means the postulated reinforcement efficiency $e_s = 0.5$. Given the soil internal friction angle $\phi = 30°$ and the soil-reinforcement friction angle $\delta = 20°$, calculate

(*a*) The friction factor F according to Eq. (17.7).

(*b*) The increased friction angle ϕ_R of the equivalent homogeneous material [Eq. (17.2)].

(*c*) Verify that if the length of the reinforcement is increased, or the spacing of the reinforcement is decreased, ϕ_R may increase to the limiting value of 90°, meaning failure by slippage is no longer possible. (Increasing the reinforcement density may however drastically reduce its efficiency e_s.)

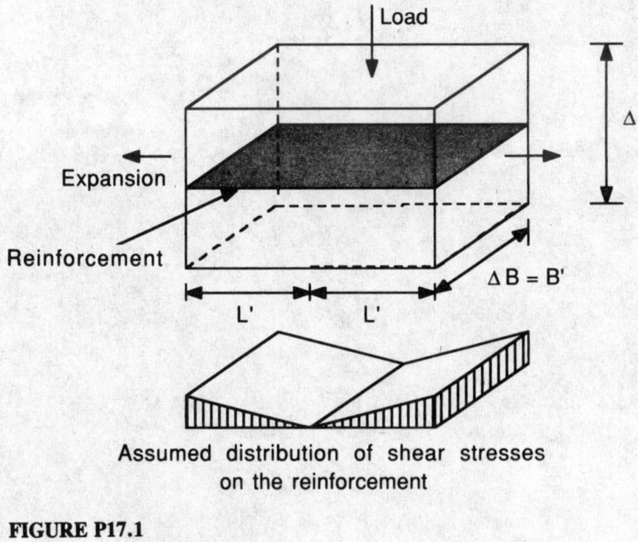

FIGURE P17.1

C17.6. In order to evaluate the effect of soil reinforcement on earth pressure according to the equivalent homogeneous model, sketch the active earth pressure distribution on a vertical wall for conditions as described by (*a*), (*b*), and (*c*). Assume dry soil with $\gamma = 20\ \text{kN/m}^3$ and calculate pressures at 0, 5, and 10 m depth.

(*a*) The soil is cohesionless and has a friction angle of 30° (representing unreinforced soil).

(*b*) The friction angle is increased to 50° (simulating a reinforced soil failing by slippage between the reinforcement and the soil).

(*c*) The friction angle remains at 30° but a cohesion of 20 kPa is given to the soil (simulating a reinforced soil which fails by rupture of the reinforcement).

Now outline the earth pressure distribution of a reinforced soil whose Mohr envelope is characterized either by a friction angle of 50° or a cohesion intercept of 20 kPa with $\varphi = 30°$. Realizing that the higher of the two earth pressures is critical, indicate whether a limit state is reached by slippage or rupture of the reinforcement.

D17.7. A particular cohesionless sand is dilatant, which means it shows a volume increase during a test in a standard shear box. How would the friction angle measured in a standard shear box compare with that determined in a constant-volume-type shear box?

D17.8. Speculate on the answers to the following questions:

(*a*) What factors could explain that a friction angle deduced from a strip pullout test is larger than that measured in a standard direct shear test?

(*b*) What conditions could lead to a pullout test value which is smaller than that obtained in the direct shear test?

B17.9. Sketch the shear stress distribution and tension force variation along a Reinforced Earth–type reinforcing strip and along the tendon of a block anchorage.

CHAPTER

18

CONSTRUCTED STRIP-, BAR-, MESH-, AND GRID-REINFORCED SOIL

18.1 REINFORCED EARTH AND OTHER STRIP REINFORCING METHODS

Strip-reinforced retaining structures have established soil reinforcement in civil engineering construction, and it is Reinforced Earth which has been the leader in this field. More research results and observations on full-scale structures are available for Reinforced Earth than for any other technique. Because of the confidence held in the methods of analysis and design used for Reinforced Earth, these methods have also been adopted (in more or less modified form) for other systems. This section will therefore concentrate on the development of design procedures for Reinforced Earth retaining structures, although other applications and systems of strip reinforcement will be described as well.

18.1.1 Standard Materials and Dimensions

18.1.1.1 REINFORCING STRIPS AND FACE PANELS. In its basic form today, Reinforced Earth retaining structures consist of horizontal layers of soil and galvanized-steel strips attached to concrete face panels (Fig. 18.1).

In the first applications of Reinforced Earth in France and the United States (1963 to about 1972), semielliptical galvanized-steel elements, 0.25 to 0.33 m high, 3 mm thick, and 10 m long served as facing (Fig. 18.1). These elements had the technical advantage that they could deform easily in a vertical direction and that way adjust to any settlement in the backfill; however, at least partly due to the public preferring more solid-looking "concrete" walls, the Reinforced Earth Company now uses almost exclusively cruciform-shaped concrete panels, 1.5 m high and with a nominal width of 1.5 m. They are precast in standard thicknesses of 0.18, 0.22, and 0.26 m. A considerable number of special panels are produced for wall boundaries and corners and for special aesthetic effects.

For about the first 10 years after its introduction, Reinforced Earth was constructed with smooth galvanized-steel strips, 3 mm thick and 60 to 80 mm wide (Fig. 18.1). Following the French Standard, design was based on a yield stress of 240 MPa unless other national standards applied. One-third of the strip thickness was considered corrosion allowance, leaving nominal cross sections of 60 × 2 and 80 × 2 mm^2. From 4 to 10 strips could be attached to one face element, placed in two rows, at a vertical spacing of 0.75 m.

In 1975, ribbed strips were introduced. Also usually made of galvanized steel, they showed superior performance in pullout tests, allowing the design of narrower and thicker strips (e.g., 40 × 5 mm) with a reduced surface area subjected to corrosion (Fig. 18.1). If required, hot dip galvanizing is specified so as to provide a minimum thickness of 70 μm of zinc on each side.

18.1.1.2 BACKFILL CRITERIA. In order to ensure adequate development of friction, normal practice for the preliminary design of Reinforced Earth structures requires that the percentage of fines (< 0.08 mm) in the backfill be less than 15% and that the backfill be placed and compacted at a moisture content equal to or less than optimum.

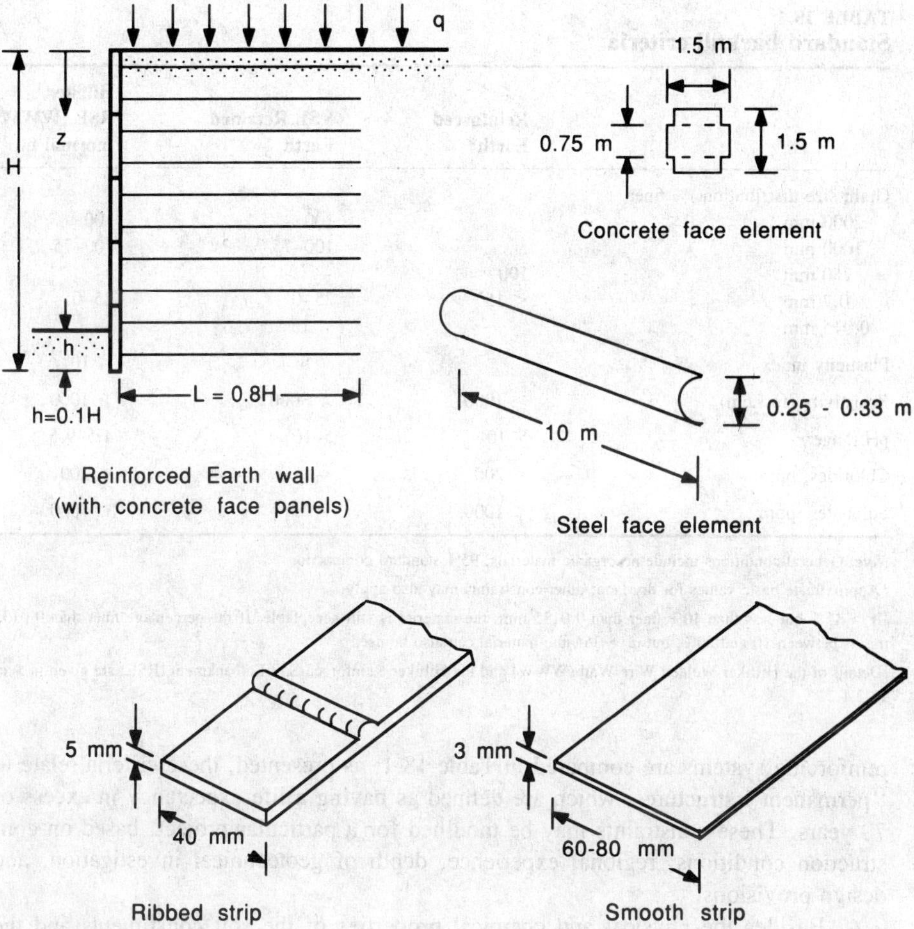

FIGURE 18.1
Typical Reinforced Earth details.

This produces an essentially free-draining backfill and generally allows a soil-reinforcement friction factor tan δ not less than 0.3 (or $\delta \geq 17°$). These design criteria are based partially on the measurement of the undrained strength of granular materials "contaminated" with clay and partially on direct shear tests of soil on reinforcing material performed in a standard shear box as shown in Fig. 17.11. If the fines amount to more than 15%, the soil may still be suitable for Reinforced Earth, but special tests would have to prove that sufficient friction or pullout resistance can be developed.

To avoid excessive corrosion in the backfill, it is required that the pH of the soil is more than 5 and less than 10 and that the soil's resistivity is more than 1000 $\Omega \cdot$ cm. In addition it may be specified that the backfill be tested for certain chemicals such as chlorides and sulphates. Basic electrochemical backfill criteria for various

TABLE 18.1
Standard backfill criteria

	Reinforced Earth*	VSL Retained Earth	Hilfiker RSE, WWW (normal fill)‡
Grain size distribution, % finer			
2000 mm		100	100
1000 mm		100–75	100–75
250 mm	100	—	—
0.7 mm	< 15†	25–0	25–0
0.015 mm		< 15	—
Plasticity index		< 6	< 10
Resistivity (Ω · cm)	> 1000	> 3000	> 1000
pH range	5–10	5–10	4.5–9.5
Chlorides, ppm	< 200	—	< 200
Sulphates, ppm	< 1000	—	< 1000

Note: General conditions include no organic materials, 95% standard compaction.

*Approximate basic values for dry land; other constraints may also apply.

†If > 15% but less than 10% finer than 0.0135 mm, the material is still acceptable. If the percentage finer than 0.0135 mm is between 10 and 20%, but $\phi > 36°$, the material can also be used.

‡Details of the Hilfiker Welded Wire Wall (WWW) and the Hilfiker Reinforced Soil Embankment (RSE) are given in Sec. 18.2.1.2.

reinforcing systems are compared in Table 18.1; as presented, these criteria relate to "permanent" structures, which are defined as having a life expectancy in excess of 70 years. These constraints may be modified for a particular project, based on construction conditions, regional experience, depth of geotechnical investigation, and design provisions.

Besides the physical and chemical properties of the soil constituents and the reinforcement, major factors influencing the corrosion of metallic inserts in soil as identified by King and Nabizadeh (1978) are

Homogeneity of the soil fill
Degree and evenness of compaction
Water content and drainage

All these three conditions are rigidly controlled in Reinforced Earth and most of the related systems.

18.1.1.3 CORROSION ALLOWANCE. The backfill criteria discussed above and summarized in Table 18.1 apply in conjunction with a sacrificial strip thickness which is specified according to the structure's location (dry land, fresh or marine water environment) and the nature of the reinforcement (galvanized or not). Table 18.2 gives guidelines as required for Reinforced Earth. The provision of such a corrosion allow-

TABLE 18.2
Corrosion allowance for Reinforced Earth structures

	Minimum service life*							
	5 years provisional structure		30 years (temporary structure)		70 years (permanent structure)		100 years (permanent structure)	
	A†	AZ	A	AZ	A	AZ	A	AZ
Structures on dry land	0.5	0	1.5	0.5	3.0	1.0	4.0	1.5
Structures in contact with fresh water	0.5	0	2.0	1.0	4.0	1.5	5.0	2.0
Structures in contact with salt water	1.0	0	3.0		5.0		7.0	

Source: Ministry of Transport, France, 1979.

Note: Table gives sacrificial thickness of steel strips in millimeters.

*Values can be interpolated linearly in the intervals 30 to 70 years and 70 to 100 years, rounding up to the next tenth of a millimeter.

†A = reinforcing strips made of nonalloyed steel without coating and AZ = reinforcing strips made of galvanized steel.

ance ensures an adequate safety factor throughout the service life—the "ultimate" life expectancy would actually be significantly longer.

18.1.1.4 TYPICAL DIMENSIONS. For preliminary design, it is common to assume that the width of the Reinforced Earth block is constant, typically 70 to 80% of the wall height H (Fig. 18.1). In order to prevent stress concentrations at the toe of the wall, it is recommended that the embedment depth $h = 0.1H$.

Initial design also provides for a surcharge of 10 kPa, which takes care of the effect of light construction equipment and materials. Particular designs may have to take into account a variety of special loading configurations because of vehicles, footings, abutments, or sloping backfill.

18.1.2 Failure Modes

If the major failure plane considered lies outside the Reinforced Earth mass, the failure is termed *external* and is analyzed following conventional soil engineering practice. *Internal* failure can occur by rupture of the reinforcement; slippage between the reinforcement and the surrounding soil; rupture, excessive deformation, or buckling of face elements; or failure of connections. Basic failure modes are illustrated in Figs. 18.2 to 18.4.

Failure modes of Reinforced Earth walls have been extensively studied in model tests, the results of which formed a first basis for the development of design procedures.

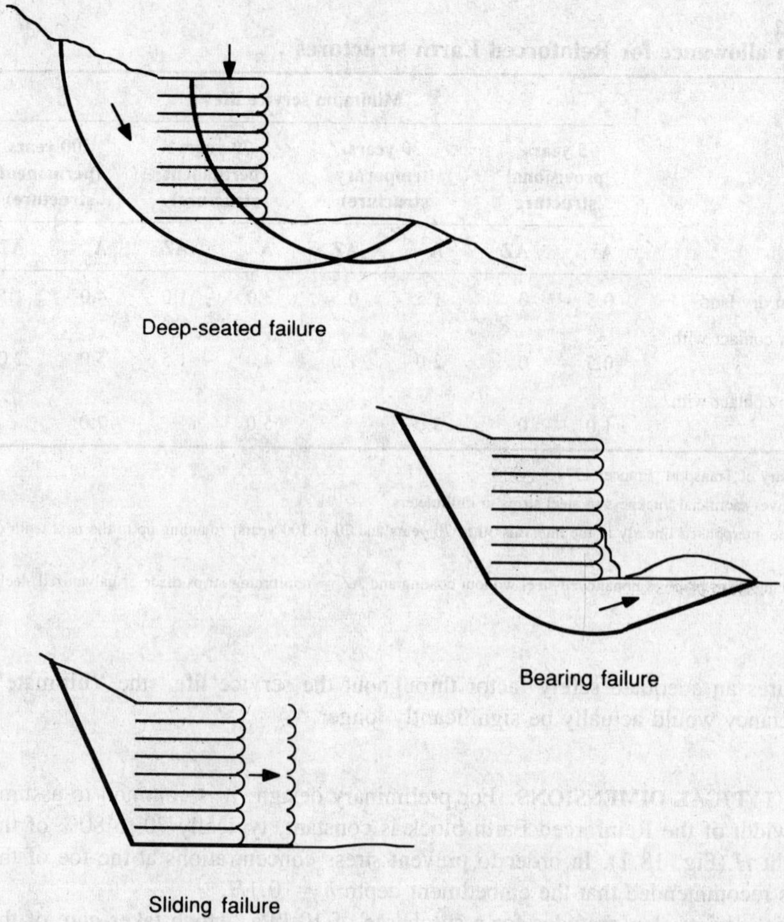

FIGURE 18.2
External failure modes of Reinforced Earth walls.

18.1.2.1 EXTERNAL FAILURE MODES

Deep-seated failure. For the case of poor backfill material (behind the Reinforced Earth) and soft foundation strata (Fig. 18.2, top), a deep-seated circular failure surface could form. For this case, conventional slope stability analysis, e.g., using the method of slices, yields a safety factor according to the definition.

$$F = \frac{\text{resisting moment}}{\text{driving moment}}$$

$$= \frac{\text{moment of shear strength along failure arc}}{\text{moment of weight of failure mass and external forces}} \qquad (18.1)$$

Details of this method are given in standard soil mechanics text books, such as the one by Lambe and Whitman (1969); also see App. 18A. Short-term and long-term stability should be evaluated; thus the analysis may have to be carried out in terms of total and effective stresses. If the foundation soil has experienced periods of instability prior to construction, meaning that prior movement along weak zones may have occurred in the past, then the analysis may need to be based on residual rather than peak strength parameters.

Because of external point or line loads, a slip circle could conceivably originate in the reinforced soil mass itself, before passing through the weak subsoil. The method of slices can still be used for this case, provided it is modified to include the tieback effect of the reinforcement as considered in the analysis of internal stability.

Bearing failure. If the soil directly underlying the base of the Reinforced Earth wall is of low strength, it may have inadequate bearing capacity (Fig. 18.2, middle). The generalized bearing-capacity formula, [Hansen (1970)] which is presented in most recent textbooks, is probably the most suitable of the many different equations employed for this problem (for details see App. 18B):

$$q_{ult} = cN_c s_c i_c d_c g_c b_c + qN_q s_q i_q d_q g_q b_q + 0.5\gamma BN_\gamma s_\gamma i_\gamma d_\gamma g_\gamma b_\gamma \qquad (18.2)$$

In this equation, c represents the cohesion of the soil and N_c, N_q, and N_γ are the bearing capacity factors; these are, according to the Hansen method, simply a function of the internal friction angle of the soil. The coefficients s, i, and d represent the shape, load inclination, and depth factors, respectively. The additional factors g and b take the slope of the ground surface and the inclination of the base into account. The eccentricity of the resultant load on the base, the presence of groundwater, and any stratification of the subsoil will also have to be considered; it is suggested that the standard analysis of shallow footings be followed.

Sliding. Sliding of the reinforced soil mass is another type of external failure which can be analyzed using conventional methods (Fig. 18.2, bottom). The base of the reinforced block of soil is normally placed at some depth below the ground surface at the toe so that some passive resistance may also be relied upon.

Overturning. A type of failure analogous to the overturning of a conventional retaining wall is illustrated in Fig. 18.3. It is assumed that an active failure wedge develops and full "wall friction" is mobilized along the back face of the Reinforced Earth mass. This failure mode implies that distortion of the Reinforced Earth mass occurs, or, in other words, internal failure by slippage of the reinforcement takes place

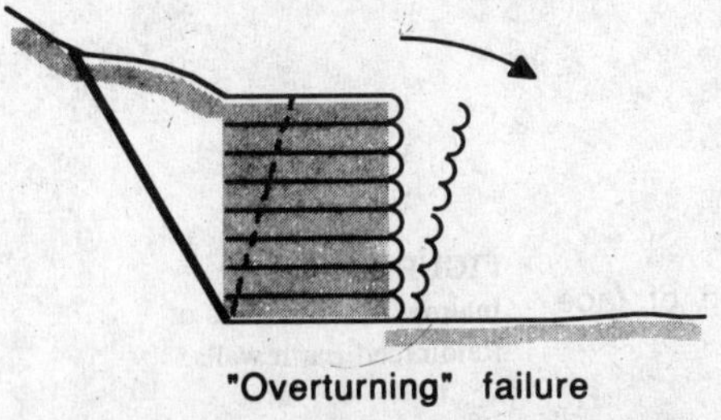

FIGURE 18.3
"Overturning"-type failure of a Reinforced Earth wall.

simultaneously. In the standard design of Reinforced Earth walls it is considered sufficient to show an adequate safety factor against slippage of the reinforcement in order to exclude the possibility of this type of block failure taking place.

18.1.2.2 INTERNAL FAILURE MODES. Internal failure can occur by rupture of the reinforcement, or slippage between the reinforcement and the surrounding soil (Fig. 18.4). These failure modes have also been termed failure by "tie break" and "tie pullout." The difference in terminology corresponds to the difference in the theoretical models used as a basis for developing a method of analysis for these two fundamental internal failure modes of a reinforced soil mass.

Finally, a Reinforced Earth wall may also fail by rupture, excessive deformation, or buckling of the face elements. Rupture of the connections of the ties with the face elements, rather than the ties themselves, should also be considered. However, once appropriate earth pressures for internal stability analysis have been determined, the design of the structural elements of the wall face does not present a special problem.

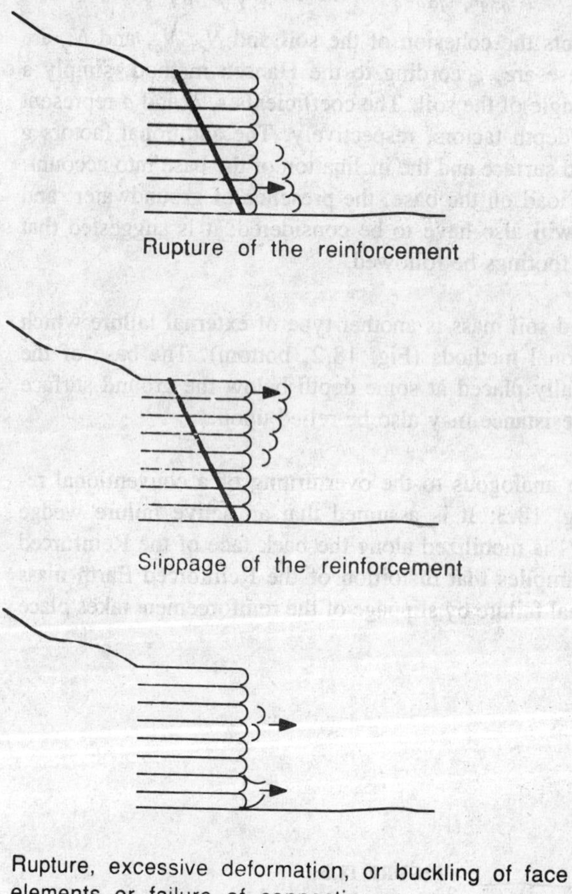

FIGURE 18.4 Internal failure modes of Reinforced Earth walls.

18.1.3 Development of Design Procedures

Current methods of evaluating the internal stability of Reinforced Earth walls have developed in stages. They evolved from applying basic earth pressure theory and altering and complementing the results obtained by information on the distribution of tensile forces in reinforcing strips at working stress levels, as this became available from field and laboratory studies.

As will be described in further detail, in the original standard analysis each individual reinforcement level was examined with respect to its safety against slippage and rupture. Vertical and horizontal stresses were determined as for a coherent gravity structure. Safety factors were calculated assuming a Rankine state of active plastic equilibrium existed in the soil.

In contrast, Schlosser and Long (1972b, 1974) identified two zones in the reinforced soil mass: an active (unstable) zone and a resisting (stable) zone. At the same time, based on their own experimental work, Lee et al. (1973) formulated the tieback model, referring to tie-break and pullout resistance of reinforcing strips.

The current design procedure [Schlosser (1978); McKittrick (1978)] benefitted from better knowledge of working stress levels and distributions in the reinforcement and the soil mass. This analysis is complemented by more advanced numerical techniques based on an elastic or elastoplastic model or advanced limit analysis [Juran (1977)] where needed.

When studying the details of the various methods of analysis, the reader should take particular notice of the assumptions made with respect to

- The earth pressure theory employed (Rankine or Coulomb)
- The magnitude of earth pressures (e.g., active or at rest)
- The distribution of vertical, horizontal, and shear stresses
- The geometry of the failure surface assumed
- The length of reinforcing strip which resists failure by slippage
- Definition and magnitude of required safety factors

18.1.3.1 ORIGINAL STANDARD ANALYSIS

Earth pressures. The calculation of vertical and horizontal earth pressures as carried out in the original standard analysis for Reinforced Earth walls is illustrated in Fig. 18.5. It is assumed that the Reinforced Earth mass yields sufficiently to develop active earth pressures in the backfill.

The average vertical stress at depth z within the reinforced and unreinforced earth is designated as σ_v. It represents a minimum average value of the normal stress acting on the reinforcing strip and will be used to calculate frictional resistance on its surface. The corresponding horizontal stress σ_h is obtained by multiplying σ_v by the coefficient of active earth pressure K_a:

$$\sigma_v = \gamma z + q \tag{18.3}$$

$$\sigma_h = \sigma_v K_a = (\gamma z + q)K_a \tag{18.4}$$

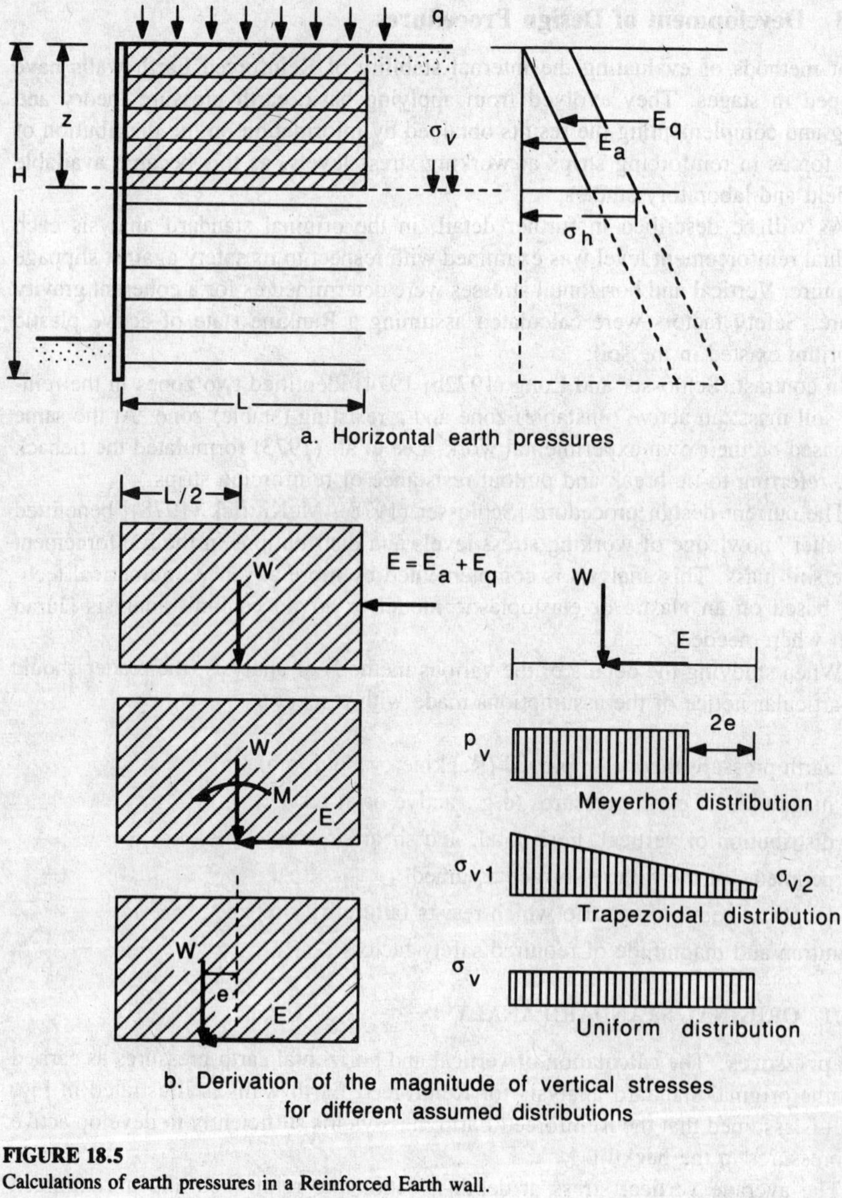

FIGURE 18.5
Calculations of earth pressures in a Reinforced Earth wall.

The horizontal earth pressure σ_h is assumed to act at the back of the coherent block of reinforced soil. Because of its overturning effect it is thought to cause increased vertical stresses near the wall face. As shown in Fig. 18.5, this increase can be estimated by either assuming a trapezoidal distribution of σ_v within the Reinforced Earth mass or by calculating the vertical pressure p_v as for an eccentrically loaded footing according to the concept of effective width proposed by Meyerhof.

For the Meyerhof distribution, the vertical pressure p_v and the corresponding horizontal earth pressure p_h are computed as follows:

$$E_a = 0.5\ \gamma z^2 K_a$$

$$E_q = qK_a$$

$$E = E_a + E_q$$

$$M = \frac{E_a z}{3} + \frac{E_q z}{2}$$

$$W = \sigma_v L$$

$$e = \frac{M}{W}$$

$$p_v = \frac{W}{L - 2e} \tag{18.5}$$

$$p_h = p_v\ K_a \tag{18.6}$$

According to preferred French practice in the original design, reinforcing strips are dimensioned for the horizontal pressure p_h calculated as above. Face panels were also designed to withstand the pressure p_h; this is perhaps somewhat contradictory to the original Reinforced Earth philosophy but represents a sound conservative engineering approach.

Safety against rupture. Each strip is assumed to have to resist the resultant of the earth pressure acting over an area equivalent to the product of the average vertical and horizontal spacing of the strips, designated as ΔH and ΔB (Fig. 18.6). Hence, for a particular depth, the tension T_d developed in the strip is calculated as

$$T_d = p_h\ \Delta H\ \Delta B \tag{18.7}$$

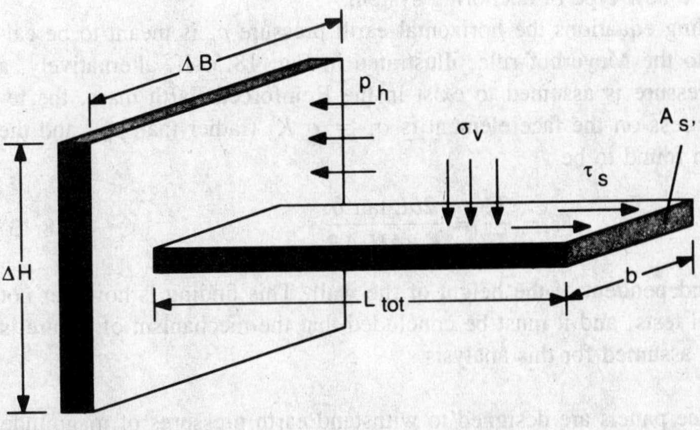

FIGURE 18.6
Reinforced Earth wall element and strip.

For a given yield stress σ_s of a steel strip of cross-sectional area A_s, the tensile force at rupture is given by

$$T_s = A_s \sigma_s \tag{18.8}$$

The safety factor against rupture is then defined as

$$F_R = \frac{T_s}{T_d} = \frac{A_s \sigma_s}{p_h \, \Delta H \, \Delta B} \tag{18.9}$$

In practice, the allowable tensile stress in the steel may be taken as two-thirds of the yield stress, inferring a safety factor $F_R = 1.5$.

Safety against slippage. The maximum possible shear stress (shear resistance) τ_s developed over the surface area ($2bL$) of a strip (Fig. 18.6) is computed as

$$\tau_s = \sigma_v \tan \delta \tag{18.10}$$

where σ_v represents a conservative minimum value of average vertical stress and δ is the effective skin friction angle. (Remember that p_v and p_h are generally larger than σ_v and σ_h.)

The maximum pulling force a strip of length L and width b can resist before slipping is

$$T_r = 2bL\sigma_v \tan \delta \tag{18.11}$$

The safety factor against slippage is given by the ratio

$$F_S = \frac{T_r}{T_d} = \frac{2bL\sigma_v \tan \delta}{p_h \, \Delta H \, \Delta B} \tag{18.12}$$

According to the original standard design procedure, the factor of safety against slippage should exceed a value of 2. Note that the resistance against slippage (or loss of adherence, as expressed in French) is thought to be developed over the full length of the reinforcing strip. This assumption is consistent with the view that Reinforced Earth is not simply a new type of anchoring system.

In the preceding equations the horizontal earth pressure p_h is meant to be calculated according to the Meyerhof rule, illustrated in Fig. 18.5. If, alternatively, a uniform vertical pressure is assumed to exist in the Reinforced Earth mass, the resulting horizontal stress on the face element is $\sigma_h = \sigma_v K_a$ (rather than p_h), and the safety factor is then found to be

$$F_S = \frac{T_r}{T_d} = \frac{2bL \tan \delta}{K_a \, \Delta H \, \Delta B} \tag{18.13}$$

and is apparently independent of the height of the wall. This finding is however not confirmed by model tests, and it must be concluded that the mechanism of failure is more complex than assumed for this analysis.

Face elements. Face panels are designed to withstand earth pressures of magnitude p_h as used in previous calculations.

Metallic face elements can be analyzed as cables subjected to lateral stresses or hinged circular elements with a given stiffness [e.g., see Chang (1974)]. Chang indicated that in field measurements, the predominant stresses are due to bending.

Concrete elements are treated as simply supported cantilever beams subjected to a uniform surcharge equivalent to the average lateral earth pressure acting on them. Generally, the maximum tensile stresses in the concrete are small and only minimal reinforcement is required.

Effect of nonuniform surcharge. Many Reinforced Earth structures have already been built to support bridge abutments, roadways, railway tracks, and buildings. Long and Schlosser (1975) have developed simple procedures to conservatively estimate the effect of nonuniform surcharge on earth pressures and strip tensions. As more performance data is becoming available, more-accurate methods of analysis are being developed [Boyd (1987)].

As is common in the design of conventional retaining walls, the increase in horizontal earth pressure $\Delta\sigma_h$ due to a point or line load p (Fig. 18.7) can be calculated according to elastic theory (for details see App. 18C), assuming the conditions behind the wall are equivalent to those in a homogeneous, isotropic, semi-infinite half space. The corresponding increase in tie tension ΔT_d can then be calculated as

$$\Delta T_d = \Delta\sigma_h \, \Delta H \, \Delta B \tag{18.14}$$

Because of the restraining effect of the reinforcing strips, the actual horizontal stresses acting at the wall face are probably significantly less than the elastic theory would indicate. Nevertheless, the maximum tie tension increase calculated according to the above equation should be in the right order of magnitude, although it is likely to occur directly below the surcharge (as indicated in Fig. 18.7*c*) rather than immediately behind the wall face.

Although the tie tensions may considerably increase due to a surcharge, the safety factor against slippage is not likely to change significantly as the increased vertical stresses in the reinforced backfill help to mobilize additional skin friction. A careful analysis is however indicated, particularly for the top four layers of reinforcement.

Instead of using elastic theory, the vertical stress increase $\Delta\sigma_v$ and corresponding horizontal stress $\Delta\sigma_h = K_a \, \Delta\sigma_v$ could be estimated with the 2:1 method as illustrated in Fig. 18.7*d*. Using this approach, Long and Schlosser (1975) appeared to get reasonable estimates of tie tension increases when compared with field measurements at the Dunkirk wall (described in Sec. 18.1.4.2).

Long and Schlosser (1975) also proposed a simple method to account for a horizontal surcharge at the top of a reinforced soil mass, as may be the case for bridge abutments. As illustrated in Fig. 18.8, a horizontal stress τ distributed over a distance a at the surface is assumed to influence tie stresses within a wedge defined by the angle $(45 + \phi/2)$. The resulting increase in horizontal earth pressure or tie tension is furthermore assumed to decrease linearly with depth, reaching zero at depth $z = a \tan(45 + \phi/2)$.

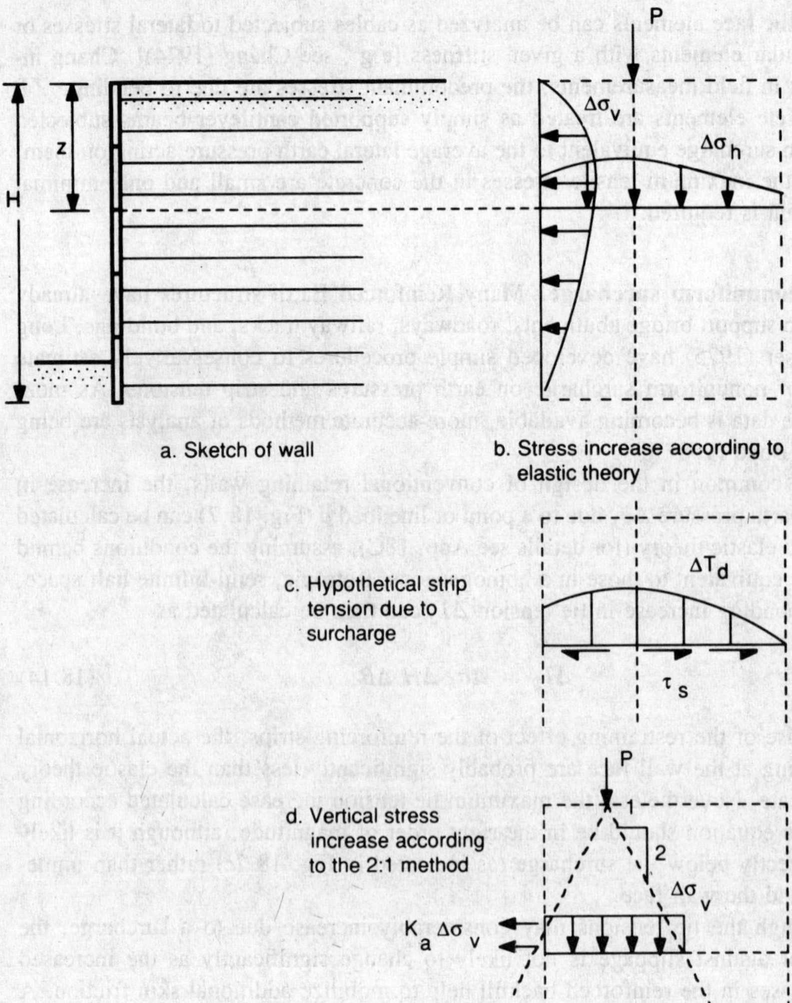

FIGURE 18.7
Effect of vertical point or line load.

Newer methods of analysis still take the approaches described above for the consideration of surface loads, although some thought-provoking field results have become available since, as discussed in Sec. 18.1.4.

18.1.3.2 TIEBACK ANALYSIS [LEE ET AL. (1972, 1973)]

Rankine-type analysis. In 1972 Lee, Adams, and Vagneron formulated the tieback concept for the analysis of Reinforced Earth walls. They assumed that the earth pressure acting on the face of the wall is balanced by the pullout resistance of the reinforcing ties over that part L_e of their total length L which is beyond the potential

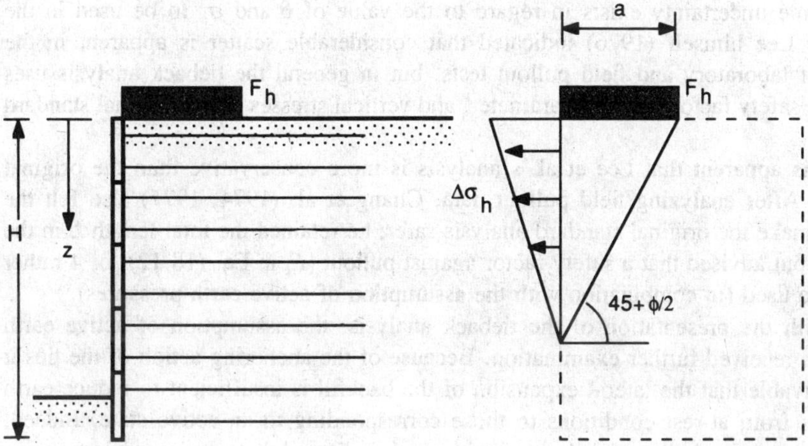

FIGURE 18.8
Effect of horizontal surcharge.

failure plane (see Fig. 18.9, center). This concept not only pertains to retaining structures but is also plausible for other applications, as illustrated in Fig. 18.10.

In the tieback analysis the formula for determining failure by slippage is modified to

$$T_r = 2bL_e\sigma_v \tan \delta \tag{18.15}$$

where T_r is referred to as the pullout resistance of a reinforcing tie and L_e represents its effective length.

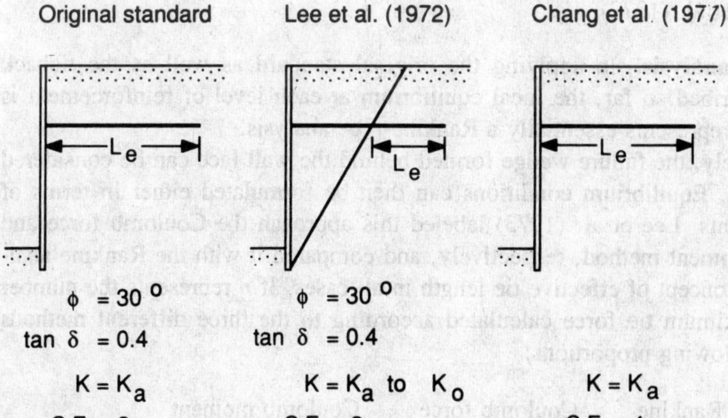

FIGURE 18.9
Variations of the original standard design for Reinforced Earth (1968 to 1977).

Some uncertainty exists in regard to the value of δ and σ_v to be used in the analysis. Lee himself (1976) indicated that considerable scatter is apparent in the results of laboratory and field pullout tests, but in general the tieback analysis uses the same safety factor, friction parameter, and vertical stresses as the original standard analysis.

It is apparent that Lee et al.'s analysis is more conservative than the original method. After analyzing field pullout data, Chang et al. (1974, 1977) also felt the need to make the original standard analysis safer; he retained the total length L in the analysis but advised that a safety factor against pullout [F_s in Eq. (18.12)] of 4 rather than 2 be used (in combination with the assumption of active earth pressures).

With the presentation of the tieback analysis, the assumption of active earth pressures received further examination. Because of the anchoring action of the ties it is conceivable that the lateral expansion of the backfill is insufficient to reduce earth pressures from at-rest conditions to those corresponding to an active state. Indeed, field measurements were already available which indicated that maximum tie tensions may approach those equivalent to acting K_0 pressures. It also became apparent that the maximum tie tension occurred at some distance behind the face panels. Although earth pressures near the wall face itself tend to reduce to levels corresponding more closely to K_a conditions as construction proceeds, a K_0 state may exist initially. Thus the tieback analysis as proposed by Lee et al. left open the option of whether to use K_a or K_0 or a value in between.

The design against ties breaking is essentially the same as the design against rupture of the reinforcement in the original standard analysis. In the tieback analysis, earth pressures are calculated assuming uniform vertical pressures and a coefficient of lateral stress between K_a and K_0; on the other hand, the original standard method assumed K_a conditions combined with a slightly increased vertical stress calculated according to Meyerhof's approach for eccentrically loaded areas.

The principal assumptions made by Lee et al. (1972) are compared to those of the original standard method in Fig. 18.10. Also added in this figure are the suggestions by Chang et al. (1977).

Coulomb-type analysis. In applying the original standard as well as the tieback analysis as described so far, the local equilibrium at each level of reinforcement is evaluated. This represents essentially a Rankine-type analysis.

Alternatively, the failure wedge formed behind the wall face can be considered as a rigid block. Equilibrium conditions can then be formulated either in terms of forces or moments. Lee et al. (1973) labeled this approach the Coulomb force and the Coulomb moment method, respectively, and compared it with the Rankine analysis, using the concept of effective tie length in all cases. If n represents the number of ties, the maximum tie force calculated according to the three different methods varies in the following proportions:

Rankine	Coulomb force	Coulomb moment
1	$\dfrac{n}{n+1}$	$\dfrac{n^2}{n^2-1}$

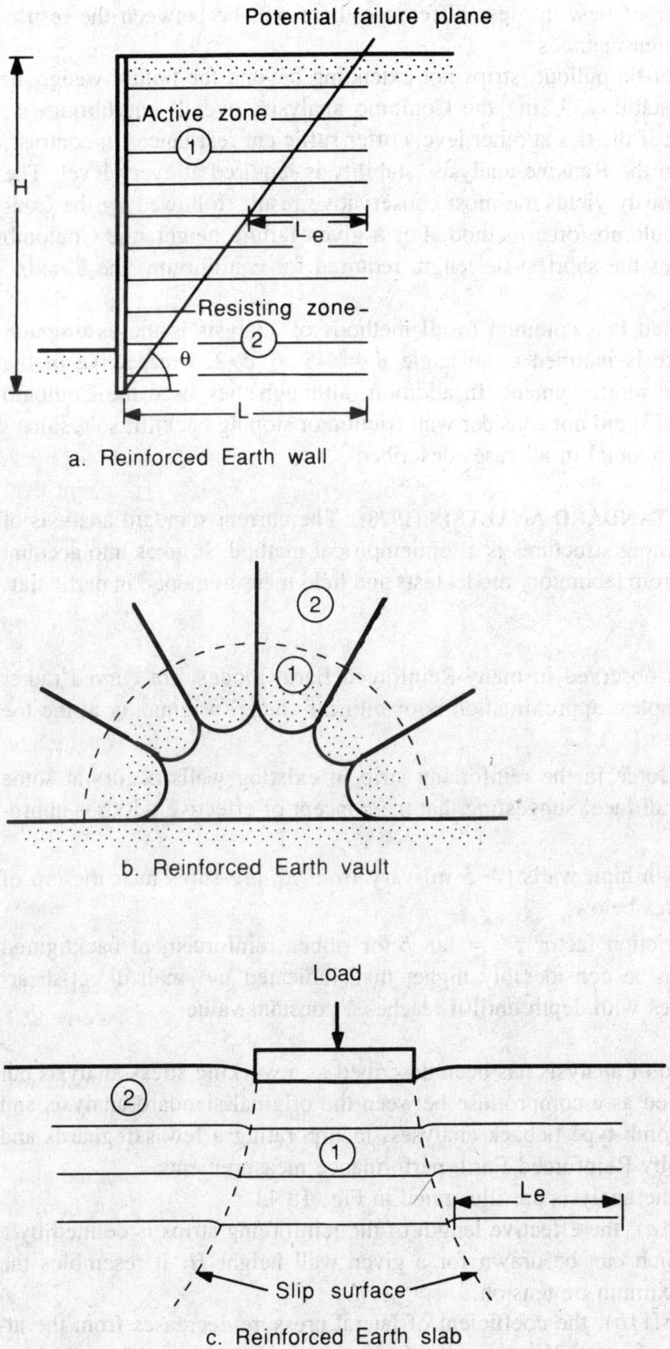

FIGURE 18.10
Tieback concept in Reinforced Earth structures.

The larger the number of ties, the less difference there will be between the results obtained with these three methods.

In the analysis of tie pullout, strips not extending beyond the failure wedge do not contribute to its stability. Using the Coulomb analysis, overall equilibrium is, however, still possible if the ties at other levels offer sufficient resistance. In contrast, in the design based on the Rankine analysis, stability is required at every level. The Rankine method obviously yields the most conservative result, followed by the Coulomb moment and Coulomb force method. For a given failure height, the Coulomb force method computes the shortest tie length required for equilibrium, the Rankine method the longest.

It should be noted that common to all methods of analysis is the assumption that the failure surface is inclined at an angle $\theta = 45 + \phi/2$, irrespective of the presence or not of the reinforcement. In addition, although they used the Coulomb method, Lee et al. (1973) did not consider wall friction or sloping backfill, so resultant earth pressures are horizontal in all cases described.

18.1.3.3 REVISED STANDARD ANALYSIS (1978). The current standard analysis of Reinforced Earth retaining structures is a semiempirical method. It takes into account a number of findings from laboratory model tests and field measurements, in particular, the following:

Failure surfaces observed in many Reinforced Earth models are curved rather than straight. The simplest approximation is of bilinear shape, originating at the toe of the wall.

The maximum force in the reinforcing strip in existing walls occurs at some distance behind the wall face, suggesting that the concept of effective width is appropriate.

Earth pressures in high walls (> 5 m) vary from K_0 pressures near the top of the wall to K_a pressures below.

The soil-strip friction factor $f^* = \tan\delta$ for ribbed reinforcement backfigured from pullout tests can be considerably higher than indicated by small direct-shear-type tests but decreases with depth until it reaches a constant value.

The new method of analysis has been described as a working stress analysis but could simply be viewed as a compromise between the original standard analysis and the Rankine- or Coulomb-type tieback analyses, incorporating a few safeguards and economies suggested by Reinforced Earth performance measurements.

Guidelines for the analysis are illustrated in Fig. 18.11.

First (Fig. 18.11*a*), the effective length of the reinforcing strips is defined by a bilinear boundary which can be drawn for a given wall height H. It resembles the observed locus of maximum tie tension.

Second (Fig. 18.11*b*), the coefficient of lateral pressure decreases from the at-rest value K_0 at the surface to K_a (the value for active earth pressures) at a limiting depth $z = 6$ m. Below that level (for walls higher than 6 m!) the coefficient remains constant at K_a.

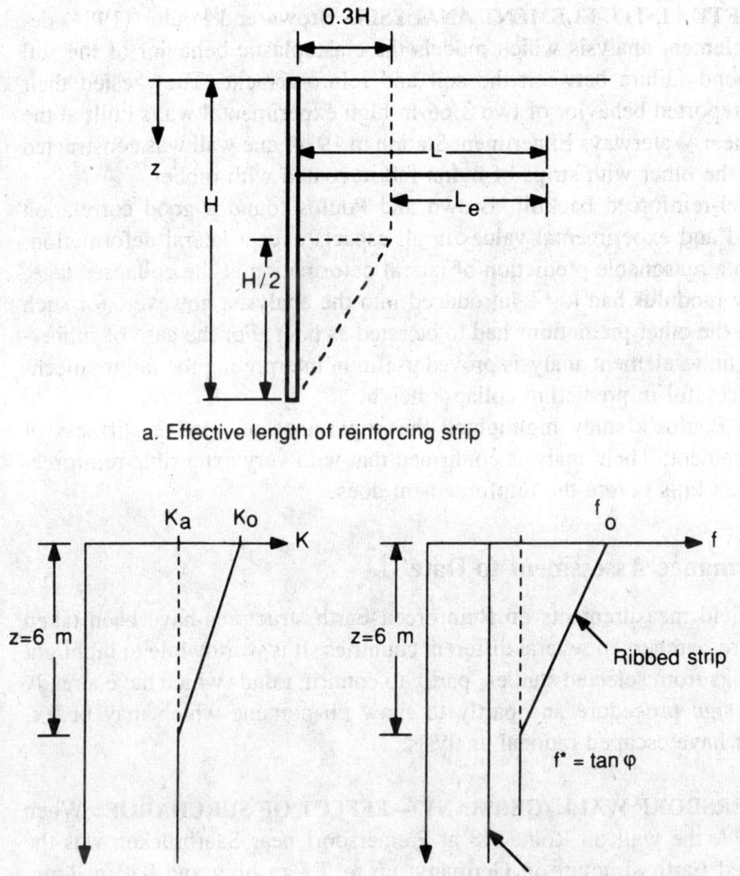

a. Effective length of reinforcing strip

b. Coefficient of lateral stress

c. Friction factor

FIGURE 18.11
Revised (1978) standard analysis of Reinforced Earth.

Third (Fig. 18.11*c*), friction coefficients are calculated. For soils satisfying basic grain size criteria for Reinforced Earth, friction factor guidelines for preliminary design are given as follows:

Smooth strips: $f^* = \tan \delta = 0.4$ (regardless of depth)

Ribbed strips: $f^* = f_0^* = 1.2 + \log C_u$ at $z = 0$

$f^* = \tan \phi$ at $z \geq 6$ m

where C_u is the uniformity coefficient, determined by the grain size distribution and defined in the Unified Soil Classification System and ϕ is the internal friction angle of the soil. Between the surface ($z = 0$) and a depth of 6 m, f^* is taken to vary linearly.

18.1.3.4 DISCRETE FINITE-ELEMENT ANALYSIS. Brown and Poulos (1984) developed a finite-element analysis which models the elastoplastic behavior of the soil as well as the bond failure between the soil and reinforcement. They tested their method with the reported behavior of two 3.66-m-high experimental walls built at the U.S. Army Engineer Waterways Experiment Station in 1974; one wall was constructed with steel strips, the other with strips of nylon fabric coated with rubber.

For the steel-reinforced backfill, Brown and Poulos found a good correlation between predicted and experimental values in all aspects except lateral deformation. In order to obtain a reasonable prediction of lateral deformation at the collapse stage, an artificially low modulus had to be introduced into the analysis; however, for such an assumption all the other predictions had to be rated as poor. For the case of rubber-coated strips, the finite-element analysis proved useful in interpreting the failure mechanism, but unsuccessful in predicting collapse height.

Brown and Poulos's study highlighted the importance of relative stiffness of earth and reinforcement. Their analysis confirmed that with very extensible reinforcement, the soil mass fails before the reinforcement does.

18.1.4 Performance Assessment to Date

Many excellent field measurements on Reinforced Earth structures have been taken by engineers and researchers in several different countries. It is worthwhile to highlight some of the findings from selected studies, partly to confirm trends which have already influenced the design procedure and partly to show phenomena which may be explained but so far have escaped rational analysis.

18.1.4.1 FREMERSDORF WALL (GERMANY)—EFFECT OF SURCHARGE. When completed in 1977, the wall on Route A8 at Fremersdorf near Saarbrücken was the highest Reinforced Earth structure in Germany: up to 7.3 m high and 970 m long. Figure 18.12 shows a cross section of the wall [Floss and Thamm (1979)]. The instrumentation included 25 hydraulic pressure cells, 10 dynamic pressure cells (with strain gauges), 9 accelerometers, 20 gauge points measuring the deformation of the concrete panels, and 100 strain gauges on 10 selected reinforcing strips. In addition, 60 dummy strips were installed at various locations for future evaluations of corrosion.

Earth pressure data collected during construction for a point in the lower section of the wall indicated that for less than 1 m of fill, horizontal pressures could be higher than calculated K_0 pressures, possibly due to panel restraint and compaction activities; with increasing overburden the pressures reduced to the K_0 level and further, reaching those equivalent to an active state K_a at the end of construction.

After completion of the wall, it was possible to plot earth pressures at various levels of the overburden (Fig. 18.13); these results confirm a typical behavior as observed elsewhere: K_0 pressures at the top, K_a pressures or less near the bottom of the wall.

The vertical pressure distribution at the base of the wall is given in Fig. 18.14. It seems to indicate that Reinforced Earth does more or less perform like a flexible coherent gravity block subjected to lateral pressure.

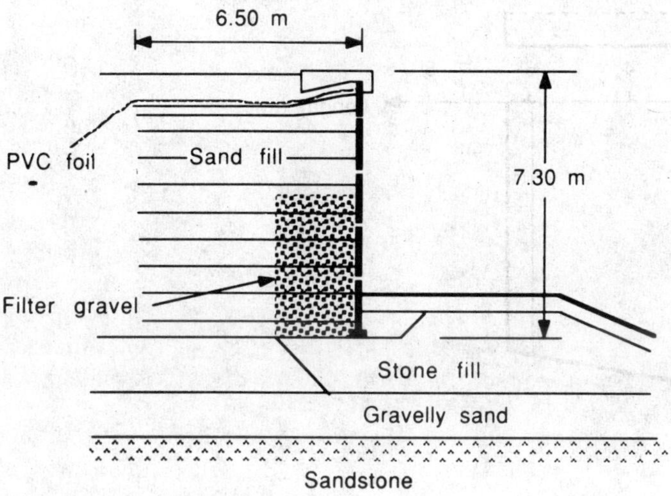

FIGURE 18.12
Cross section of Fremersdorf wall. [*Floss and Thamm (1979).*]

Strip tension variation along the reinforcing strips is presented in Fig. 18.15. It shows the typical peak at some distance behind the wall, a distance which diminishes with depth.

What is most interesting is the effect of a static surcharge (max. 36 kPa) on the earth pressures and strip tensions. As Fig. 18.16 demonstrates, the horizontal earth

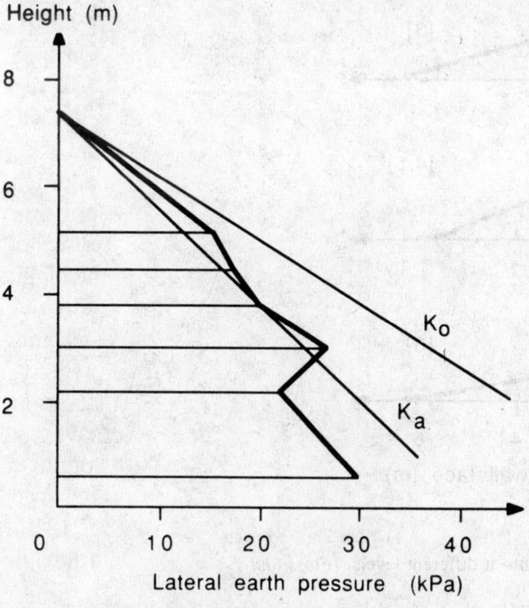

FIGURE 18.13
Lateral earth pressure distribution. [*Floss and Thamm (1979).*]

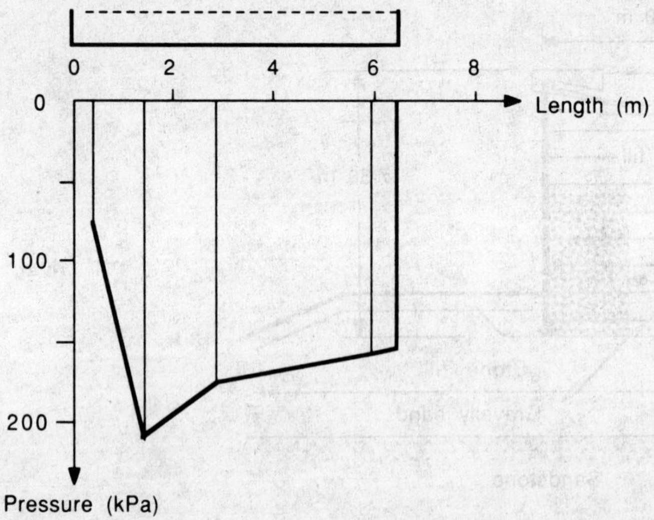

FIGURE 18.14
Pressure at base of Reinforced Earth mass. [*Floss and Thamm (1979)*.]

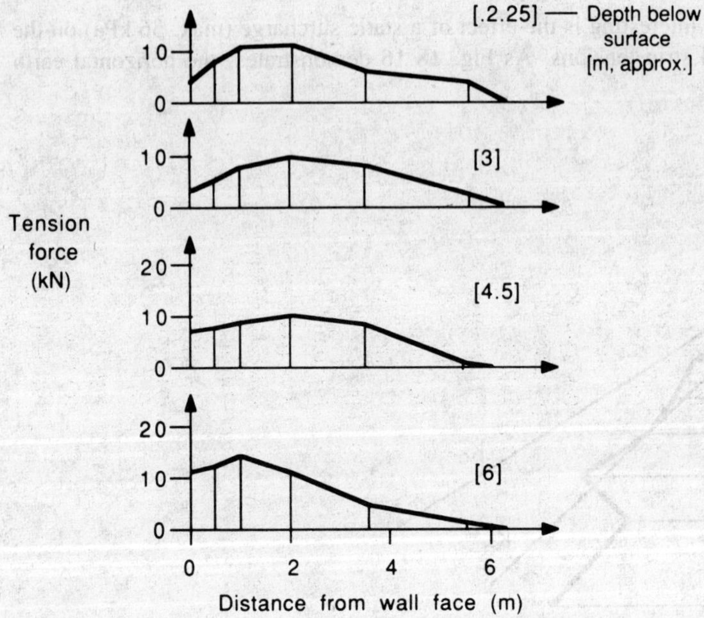

FIGURE 18.15
Tension force distribution along reinforcements at different levels. [*Floss and Thamm (1979)*.]

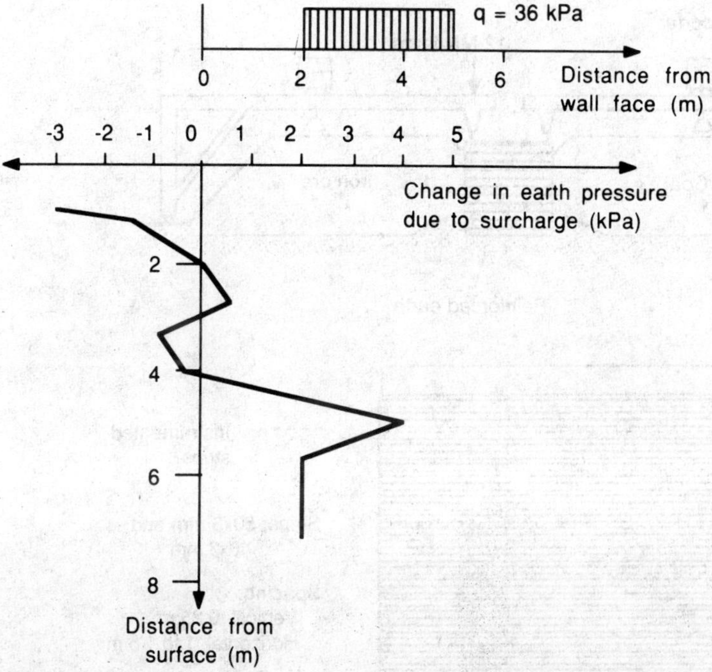

FIGURE 18.16
Change of earth pressure distribution due to surcharge load. [*Floss and Thamm (1979).*]

pressure near the top of the wall actually decreased upon full load application, while an outward movement of the panels of 0.5 mm occurred. This can be explained by a change from at-rest to active earth pressure conditions. During the surcharging, earth pressures at that level initially increased (until about 18 kPa were placed), and then decreased to the state of stress shown.

During surcharge loading only moderate changes in the tension distribution in the reinforcing strips occurred; maximum tensile forces changed less than 10% compared with construction levels. Only 1 to 3% of the load-induced stress increases remained after the removal of the surcharge, indicating almost elastic behavior.

Dynamic loading was achieved with a moving loaded truck and also with a vibratory roller. Results did not indicate that current static design approaches need modification for dynamic loads.

18.1.4.2 DUNKIRK WALL (FRANCE)—DOUBLE-SIDED WALL WITH SURCHARGE. The wall at Dunkirk, completed in 1970, is of special interest because it is double-sided and strips are attached on both sides to the steel facing. It is 12 m high and 18 m wide, providing support on either side for a 12-MN load generated by a moving crane (Fig. 18.17). Strip tensions measured were higher than in a standard one-sided wall, corresponding more to K_0 rather than K_a pressures [Long and Schlosser (1975)].

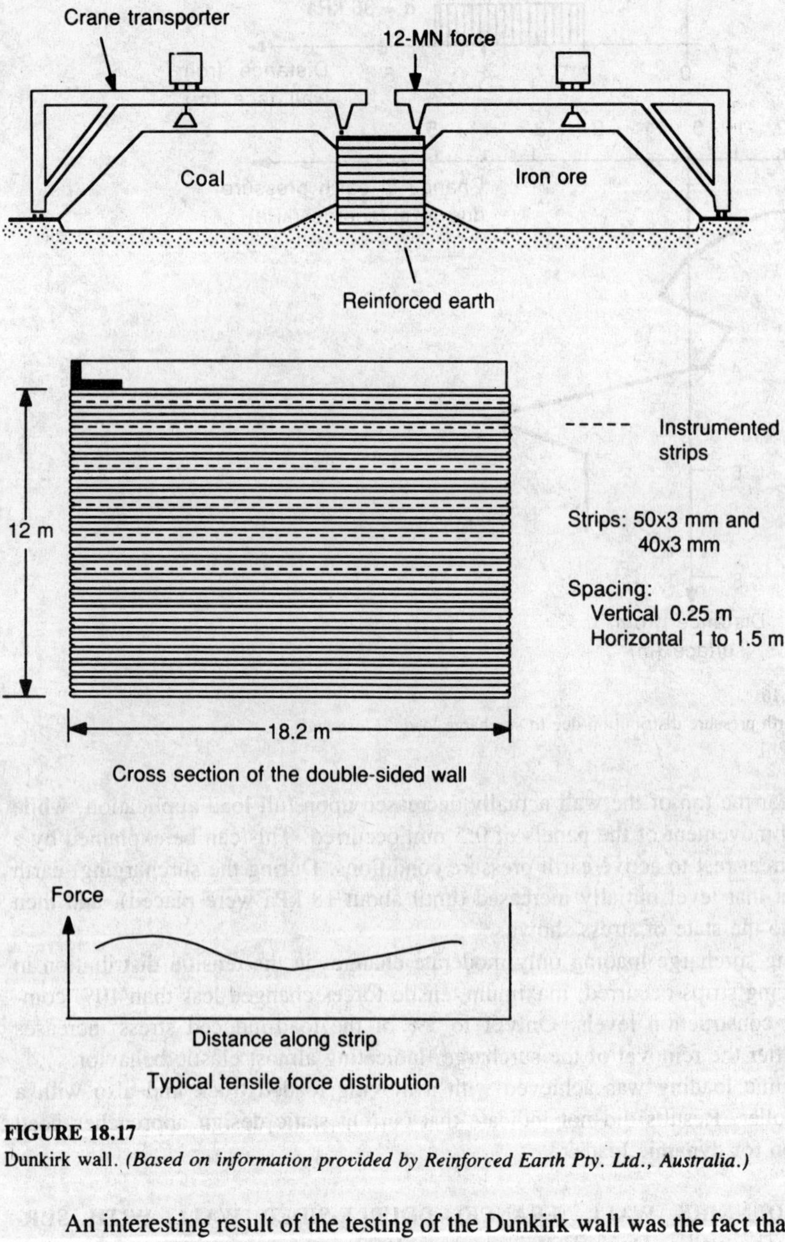

FIGURE 18.17
Dunkirk wall. *(Based on information provided by Reinforced Earth Pty. Ltd., Australia.)*

An interesting result of the testing of the Dunkirk wall was the fact that transient loads caused significant residual stresses which remained in the ties for more than 3 h after the load application.

18.1.4.3 CORROSION TEST WALL (SPAIN). In 1977 the Société La Terre Armée, which is based in Paris, built a 6-m-high experimental wall which was designed to fail by corrosion of the steel reinforcing strips (Fig. 18.18).

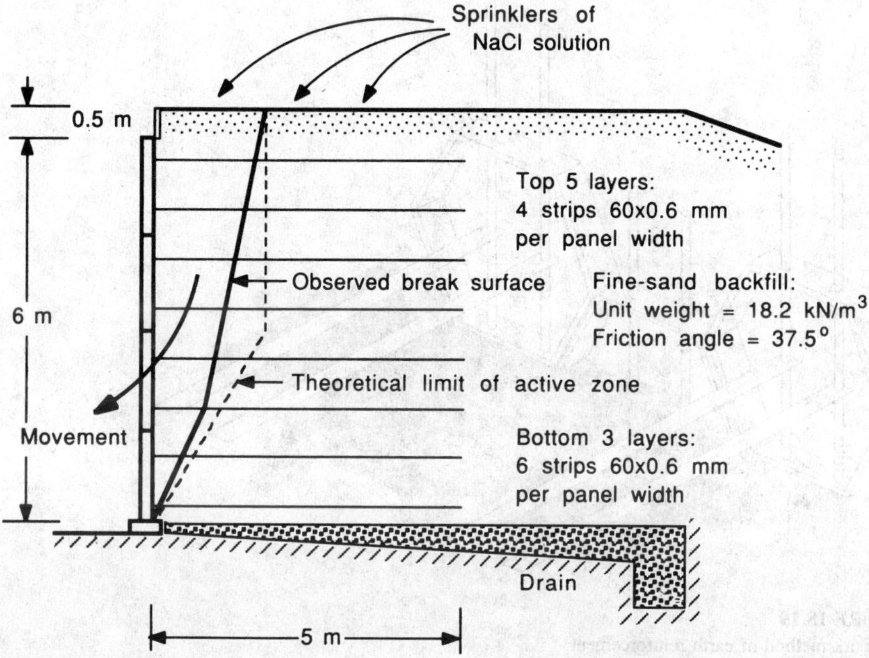

FIGURE 18.18
Experimental corrosion test wall built in Spain. [*La Terre Armée S.A. (1979).*]

The 10.22-m-long test section formed the center of a Reinforced Earth structure of a total length of 43 m. The unprotected smooth steel reinforcing strips extended 5 m into the backfill, had a width of 60 mm, but were only 0.6 mm thick. Eight to twelve strips were attached to each of the standard concrete panels. The maximum strip tension was theoretically estimated at 6.46 kN, occurring in the fifth reinforcing layer from the top according to the revised (1978) method of analysis. This compared with an original (uncorroded) breaking strength of 12.8 kN. A sodium chloride solution (30 to 50 g/L) was sprinkled over the surface of the test section for 1 month immediately after construction in order to obtain a corrosion rate estimated at 200 μm per year. The sprinkling was restarted in the eighth month until failure. A drainage layer below the Reinforced Earth structure collected the seepage brine for recycling. Close to 10 t of salt were used during the whole test period resulting in flows of up to 9 m^3 per day. During the sprinkling periods, backfill resistivity was on the order of 600 to 800 $\Omega \cdot$ cm, and at the end of the test, the chloride content of the soil adjacent to reinforcement was found to be on the order of 1400 ppm.

The wall failed 9 months after accelerated corrosion was initiated. The break was sudden, after the wall deformation at the top reached about 20 mm. The failure surface was found to lie inside the theoretical active zone (according to the 1978 analysis) and close to the assumed line of maximum tension (Fig. 18.19). Breaking loads of the exhumed reinforcing strips varied from 1.2 to 9.3 kN, with an average

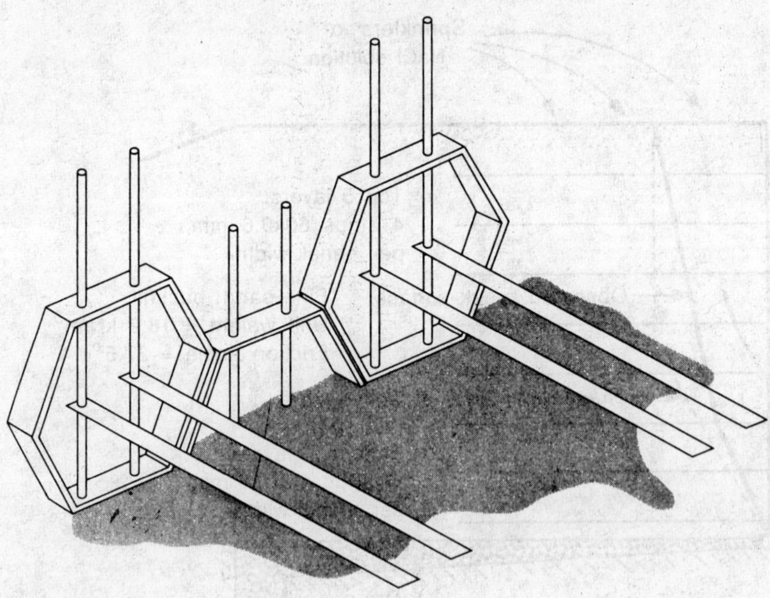

FIGURE 18.19
The York method of earth reinforcement.

of 4.2 kN. The visual inspection of the failure zone and analysis of the strip strengths confirmed that the Reinforced Earth design method is safe.

Observations of the rate of corrosion of test strips extracted from the fill during and at the end of the test period validated the corrosion model described in Chap. 16. The curve fitting the data expresses the attenuation of the corrosion rate with time by the formula

$$\Delta t = 28y^{0.53} \tag{18.16}$$

where the loss of thickness Δt is measured in micrometers (μm) and the time y in months.

Although this experiment involved accelerated corrosion on very thin unprotected steel strips, it indicates that current Reinforced Earth design assumptions and criteria are appropriate and conservative.

18.1.5 Other Strip Reinforcement Methods

18.1.5.1 THE YORK METHOD. The York method, a variation of Reinforced Earth, was developed in 1972 in the United Kingdom [Jones (1978)]. Horizontal galvanized-steel or glass-reinforced plastic strips are attached to vertical poles which in turn provide a connection between the hexagonal facing elements (Fig. 18.19). Strips are allowed to slide vertically so that settlement of the backfill can be accommodated without stressing the face panels. The latter are nonstructural elements which may be made of glass-reinforced cement; they are essentially hollow shells preventing erosion of the backfill and giving some protection to the reinforcement.

The concept of sliding reinforcement, originally proposed by Munster (see Chap. 16), can also be adapted to different shapes of structural and nonstructural facings, as described by Jones (1985).

18.1.5.2 SYNTHETIC STRIPS. A number of organizations have experimented with reinforcing strips made out of synthetic materials.

The U.S. Corps of Engineers [Al Hussaini (1977); Al Hussaini and Perry (1977)] built test walls with both steel as well as synthetic strips.

When evaluating the York method of reinforcement, the British Transport and Roads Research Laboratory experimented with strips of glass-fiber-reinforced plastic, made by Pilkington Brothers, and other materials [Boden et al. (1978)].

The range of ParaProducts by ICI Fibres, Ltd., (England) includes 4-in-wide strips made of high-tenacity polyester or nylon fibers encased in a polyolefin sheath. The styles produced so far can take loads of up to 150 kN, demonstrating a strength equal to conventional steel strips.

Although high-strength high-modulus synthetics in strip form have proved feasible in retaining structures, more interest has focused on using synthetics in the form of sheets and grids.

18.2 BAR AND MESH REINFORCEMENT

Mats, grids, and meshes are terms describing types of structural sheeting used for strengthening the ground surface or as reinforcement within a soil mass.

Although definitions may vary, mats usually represent relatively stiff materials, with or without openings, in the form of plates, sheets, or interconnected segments, that are placed on soft ground. Their main purpose is to improve trafficability by reducing high contact stresses due to concentrated live or dead loads. Traditional materials employed for this purpose have been timber, brushwood, bamboo, or other low-cost natural products. During the Second World War, steel mats proved to be an expedient means of constructing airfields and roads on soft soils. "M8A1" landing mats were used for tactical purposes during the Vietnam war [Webster and Watkins (1977)]. "Columbus mats" are a newer type of mat developed in Sweden [Holtz (1976)]; they are made of specially profiled polyethylene tubing, 65 mm in diameter, held together by steel cables. They are described as strong, lightweight, flexible, and elastic, and because of their corrugated surface, they are able to develop high friction in contact with soil or snow. Because mats essentially act as load distributors rather than internal soil reinforcement, they will not be discussed any further.

Meshes and grids are typically flexible sheets with openings which are relatively large, measured in millimeters or centimeters, and able to retain particles of gravel or cobble size, but nothing smaller. They may be created by extrusion and stretching or by welding of wire or bar elements; the latter type are also described in the industry as "bar mats." Terminology is still evolving and is complicated by confusing proprietary names. "Geogrids" is a term first used to describe open-meshed sheets made of synthetic materials such as polyester and polypropylene; they are discussed in a separate section (Sec. 18.3).

The development of mesh-reinforced retaining structures was probably stimulated by both the success of Reinforced Earth and experiences gained with wire-mesh gabions used for walls and erosion control structures. A further incentive for reinforcing with meshes rather than strips came from the recognition of the existence of passive earth pressures resisting the pullout of mesh reinforcement, which easily made up for a reduced interface area of soil-reinforcement friction. In the case of steel meshes, the reduced surface area of the reinforcement actually represents an advantage as far as general corrosion is concerned, but a new problem could be a possible weakness at mesh node connections, both in terms of strength and durability.

18.2.1 Bar Mesh and Welded Wire Mesh

18.2.1.1 CALTRANS "MECHANICALLY STABILIZED EMBANKMENT." The California Department of Transportation (Caltrans) built the first Reinforced Earth structure in 1972, along Highway 39, north of Los Angeles. It was of the original type with the extruded sheet steel facing. It was built in a landslide area and was well instrumented, including the installation of dummy strips (not connected to the face elements) which were pulled out of the wall after the end of construction [Chang (1974)]. If interpreted in terms of skin friction angles, the field tests indicated that the values of the pulling resistance of the shorter strips were less than predicted from

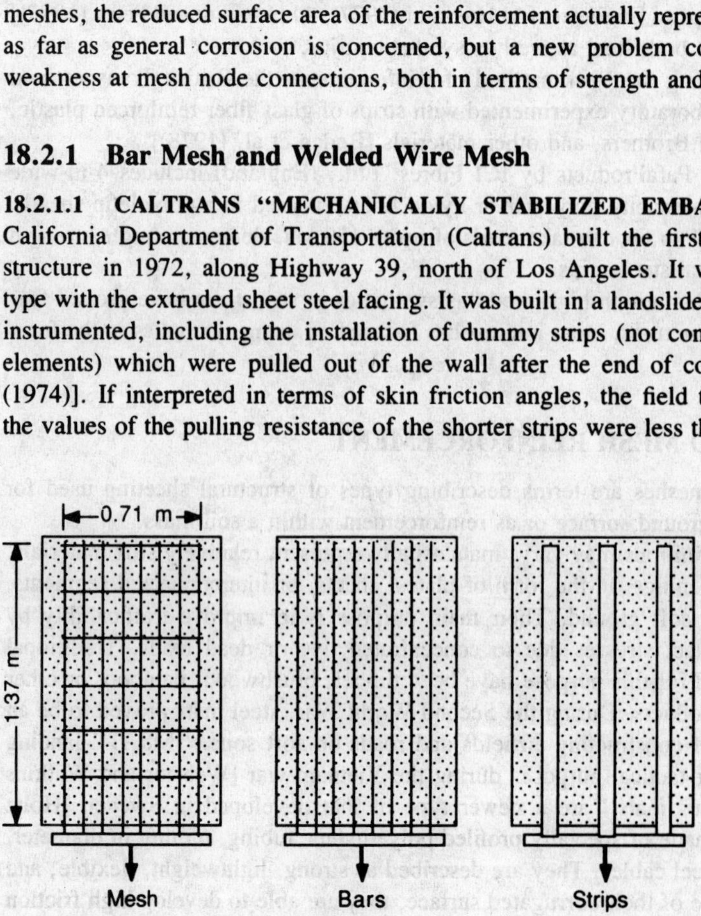

Test Results
(Dunsmuir sand, 70-kPa overburden pressure)

Reinforcement type		Mesh	Bars	Strips
Peak pullout force	P (kN)	167	15	30
Elongation at peak	(cm)	4.30	0.13	0.51
Comparative surface area	A (%)	100	69	102
	P/A	1.67	0.22	0.29

FIGURE 18.20
Caltrans pullout test results. [*Extracted from Chang et al. (1977).*]

laboratory friction test results. Recognition of this fact not only suggested that the safety factors against slippage in the original standard design procedure for Reinforced Earth be increased, but also prompted a program of laboratory pullout tests using a variety of reinforcement in the shape of strips, bars, and bar meshes.

This laboratory investigation, started in 1973, showed that with equal surface area, the bar mesh reinforcement could produce more than 5 times the pullout resistance of longitudinal bars or strips [Chang et al. (1977)]; however, the elongation at peak pullout force was considerably greater for bar mesh than for other types of reinforcement. Selected results are presented in Fig. 18.20.

Encouraged by these findings, Caltrans built the first bar mesh reinforced wall, near Dunsmuir, California, in 1975. The concrete face elements were of a beam shape (3.75 m long, 0.6 m high, about 0.2 m thick) rather than like the typical cruciform panels of Reinforced Earth (Fig. 18.21). For patent reasons, by agreement with the Reinforced Earth Company, the Caltrans bar mesh construction technique was designated as "Mechanically Stabilized Embankment" in 1976.

Evolving from the Caltrans project were other similar techniques: the Hilfiker Welded Wire Wall and Reinforced Soil Embankment, VSL Retained Earth, the Georgia Stabilized Embankment, and other related techniques, such as Anchored Earth; all are distinguished by different face panels, bar mesh geometry, and construction details. Figure 18.21 illustrates the panel shapes typical for each method, and Fig. 18.22 gives associated mesh patterns; in both figures, the construction elements for Reinforced Earth and for the York method are shown for comparison.

18.2.1.2 OTHER SYSTEMS USING WIRE MESH

Hilfiker Welded Wire Wall and Reinforced Soil Embankment. The Hilfiker Welded Wire Wall (WWW) is a system where the primary facing consists of the same galvanized bar mesh which serves as horizontal reinforcement (openings are 15 by 61 cm). The lifts between reinforcing layers usually are 46 cm, and the retention of the backfill is ensured by adding a backing mat and screen mesh with smaller openings than the reinforcing mat, down to 6 mm in size. Welded Wire Walls were first introduced commercially in 1977, and some 200 structures had been constructed by 1986, up to 10 m high. To the onlooker, Welded Wire Walls may appear to be some sort of gabion wall. Gabion walls, however, are made with coarse-grained fill encased in wire baskets, the assemblage of which forms a gravity wall; gabion walls are based on the principle of confinement rather than internal tensile reinforcement (also see Chap. 20).

The Hilfiker Reinforced Soil Embankment (RSE), commercially introduced in 1983, resembles the earlier Caltrans system (Mechanically Stabilized Earth). As in that system, the precast face panels are of a beam shape and the reinforcement consists of cold-drawn wire mesh (not galvanized); however, the details of connecting the panels and attaching the mats differ, representing improvements with respect to the ease of construction. In an RSE the bar mesh is attached to the vertical pins connecting the face panels. Neoprene and Styrofoam bearing pads act as spacers, and filter fabric prevents washout of fines from the backfill.

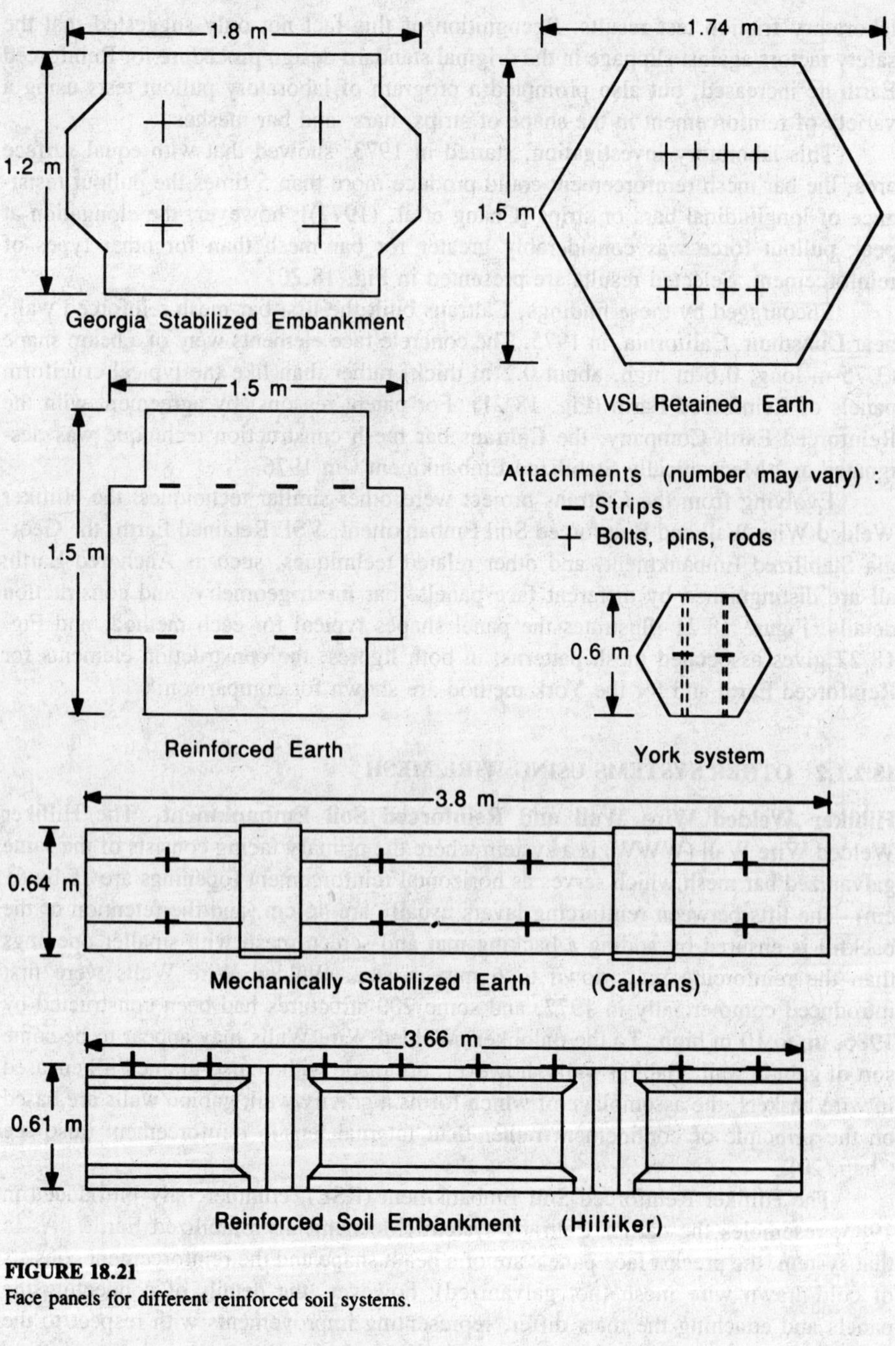

FIGURE 18.21
Face panels for different reinforced soil systems.

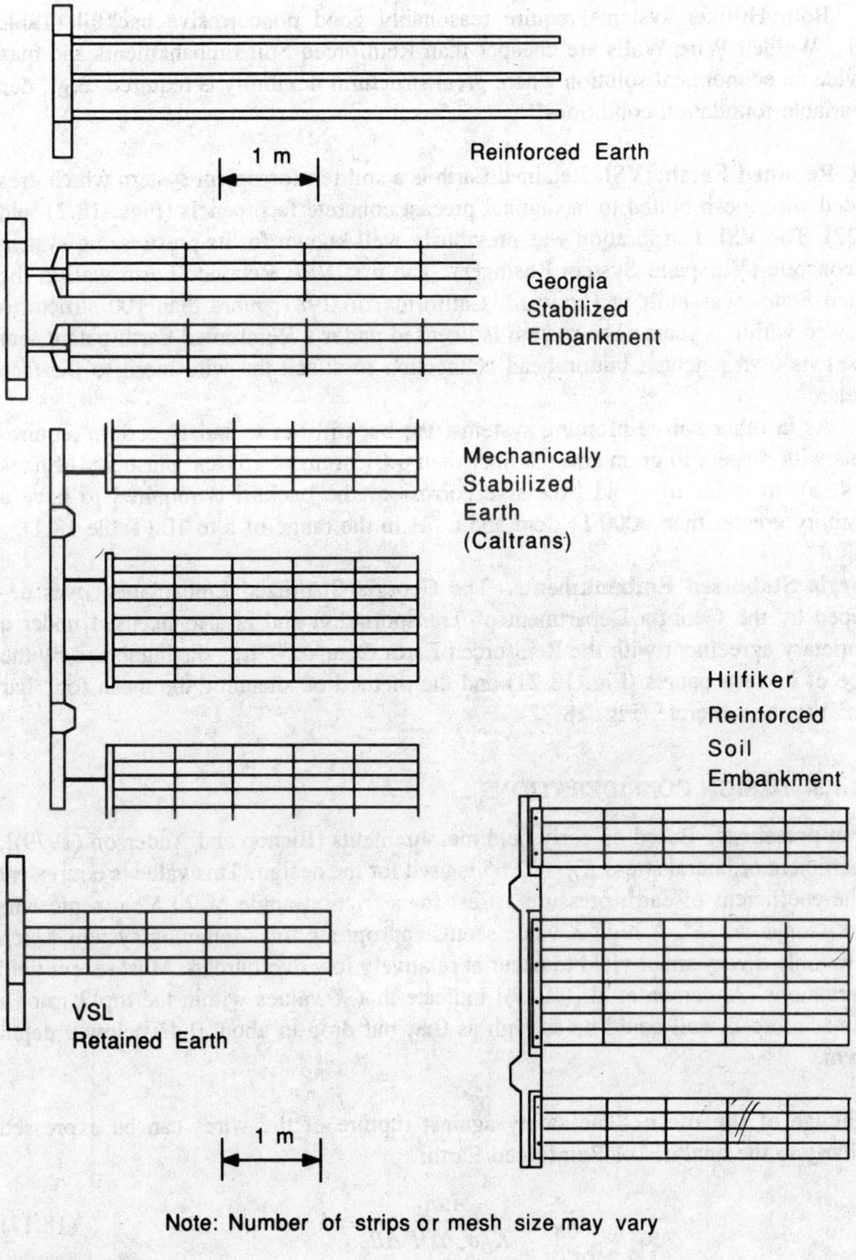

FIGURE 18.22
Strip and mesh patterns.

Both Hilfiker systems require reasonably good noncorrosive backfill (Table 18.1). Welded Wire Walls are cheaper than Reinforced Soil Embankments and may provide an economical solution where great structural flexibility is required, e.g., due to variable foundation conditions.

VSL Retained Earth. VSL Retained Earth is a soil reinforcement system which uses welded wire mesh bolted to hexagonal precast concrete face panels (Figs. 18.21 and 18.22). The VSL Corporation was previously well known for its prestressing system for concrete (Vorspann System Losinger). The first VSL Retained Earth wall in the United States was built in Hayward, California, in 1981; more than 100 structures followed within 3 years. The system is licensed under a Reinforced Earth patent, but it uses its own patented button-head connection to attach the wire mesh to the face panels.

As in other soil-reinforcing systems, the backfill has to satisfy certain requirements with respect to grain size (% finer than 0.015 mm < 15) and plasticity of fines (Pl < 6); in order to avoid excessive corrosion, the backfill is required to have a resistivity greater than 3000 $\Omega \cdot$ cm and a pH in the range of 5 to 10 (Table 18.1).

Georgia Stabilized Embankments. The Georgia Stabilized Embankment was developed by the Georgia Department of Transportation and is also licensed under a proprietary agreement with the Reinforced Earth Company. It is distinguished by the shape of its face panels (Fig. 18.21) and the method of attaching the mesh (or "bar mats") to the concrete (Fig. 18.22).

18.2.1.3 DESIGN CONSIDERATIONS

Earth pressures. Based on early field measurements [Bishop and Anderson (1979)], a coefficient of lateral stress $K_D = 0.65$ is used for the design. This value is equivalent to the coefficient of earth pressure at rest for a friction angle of 20.5°; it represents 1.5 K_0 for $\phi = 35°$. A high K value seems appropriate for a retaining system which allows only a very small yield to occur at relatively low overburden. Most recent field observations [Anderson et al. (1985)] indicate that K values within the top 3 m of a mesh-reinforced wall could be as high as 0.8, but drop to about 0.45 below a depth of 6 m.

Breakage of the mesh. The safety against rupture of the wires can be expressed similarly to the analysis of Reinforced Earth:

$$F_R = \frac{A_s \sigma_{all}}{K_D \sigma_v \, \Delta H \, \Delta B} \geq 1 \tag{18.17}$$

where A_s = cross-sectional area of one wire (including corrosion allowance)
σ_{all} = allowable wire stress (including safety factor)
ΔH = vertical spacing between mats
ΔB = average horizontal spacing of longitudinal wires
σ_v = vertical (overburden) stress

Pullout resistance. The pullout resistance T_r of meshes may be considered as consisting of two parts: the resistance T_f resulting from friction generated on the surface of the longitudinal bars (perpendicular to the wall face) and the resultant T_p of the passive resistance developed on the transverse bars (parallel to the wall face):

$$T_r = T_f + T_p \tag{18.18}$$

The frictional pullout force T_{fi} for a single wire may be expressed as

$$T_{fi} = \sigma_a \pi d \tan \delta \, L_e \tag{18.19}$$

where σ_a = average confining stress = $(\sigma_v + K\sigma_v)/2$
d = wire diameter
L_e = effective length (equal to total length in a pullout test)
δ = skin friction angle

The relevant value of K can be estimated as that corresponding to at-rest pressure (in a direction parallel to the wall face) and was approximated as 0.5 by Nielsen (1984) in his study of pullout resistance of meshes, leading to the value

$$\sigma_a = 0.75\sigma_v \tag{18.20}$$

As discussed previously (Sec. 17.2.1), the determination of δ is not straightforward, since it depends on such factors as overburden pressure and soil conditions, but a value obtained in the direct shear box is thought to be a conservative estimate.

The approach taken by Peterson and Anderson (1980) and others in evaluating the passive resistance offered by transverse bars in the mesh is to use the Terzaghi bearing-capacity equation for strip footings of width B (see App. 18C):

$$q_{ult} = cN_c + qN_q + 0.5\,\gamma BN_\gamma \tag{18-21a}$$

Neglecting the role of cohesion (first term) and the last term (mainly affected by the width of the footing) gives an ultimate bearing capacity

$$q_{ult} = qN_q \tag{18.21b}$$

where q is equivalent to the overburden pressure σ_v, and N_q is the conventional bearing-capacity factor, which is a function of the internal friction angle of the soil. (*Note:* N_q according to Terzaghi is not necessarily a conservative value; a more appropriate approach might be to use a formula for determining the end bearing of a pile.)

A single transverse wire of diameter d and equal to the mesh width W will provide a resisting force

$$T_{pi} = \sigma_v N_q W d \tag{18.22}$$

For m longitudinal wires and n transverse wires in one mesh unit, we obtain a total pullout force in cohesionless soil of

$$\begin{aligned} T_r &= m\,(0.75\,\sigma_v \pi d \tan \delta\, L_e) + n(\sigma_v N_q W d) \\ &= \sigma_v d\,(0.75 m \pi \tan \delta\, L_e + nN_q W) \end{aligned} \tag{18.23}$$

Including cohesion in the passive resistance part of the equation we find

$$T_r = ncN_cWd + \sigma_v d\,(0.75m\pi \tan \delta\, L_e + nN_qW) \qquad (18.24)$$

By experiment, Nielsen (1984) found expressions for the pullout force of meshes of various geometry which could be fitted to this format; each different soil yielded a separate relationship. Based on these test results, the following equations [given in units of kilonewtons per meter (1 kN/m = 68.5 lb/ft)] now serve as design guidelines; they have been converted to SI units with m' denoting the number of longitudinal wires per meter width of mesh and n' the number of transverse wires effectively providing pullout resistance:

$$\text{Silty sand:} \quad T_r = \begin{cases} 31.2 + \sigma_v d\,(0.75m'\pi \tan \delta\, L_e + 0.26n') & \text{for } n\sigma_v d > 1.7 \\ \sigma_v d\,(0.75m'\pi \tan \delta\, L_e + 0.52n') & \text{for } n\sigma_v d \leq 1.7 \end{cases}$$

$$\text{Washed sand:} \quad T_r = \sigma_v d\,(0.75m'\pi \tan \delta\, L_e + 0.54n')$$

$$\text{Pea gravel:} \quad T_r = \sigma_v d\,(0.75m'\pi \tan \delta\, L_e + 0.55n') \qquad (18.25)$$

Actual test results also showed an apparent "cohesion" component for washed sand and pea gravel, but this was ignored in the above expressions. Caution should be exercised when using these equations for narrower wire spacings than is customary.

With R denoting the ratio of mesh width per width of wall, the safety against pullout (or slippage) can then be formulated as follows:

$$F_S = \frac{T_r R}{K_D \sigma_v\, \Delta H} \geq 1.5 \qquad (18.26)$$

Typically, $R = 1$ for Hilfiker Welded Wire Walls and $R = 0.8$ for Reinforced Soil Embankments with precast face panels.

18.2.2 Bar- and Block-Tieback Systems

Strip and bar mesh reinforcing systems rely for their pullout resistance, at least in part, on friction and adhesion developed on the surface of the reinforcing elements. In the case of meshes, additional resistance is mobilized through the building up of passive earth pressure against transverse-oriented bars; this mechanism assists in substantially reducing the earth pressure against the wall face, bringing it closer to the case of ideal reinforcement (refer to Fig. 17.13 for a definition of "ideal reinforcement"). However, in order to mobilize adequate friction, the quality of the backfill has to be assured; also, although a large reinforcement surface area benefits resistance against slippage, it offers more area for corrosion attack. Corrosion is promoted in wet silty and clayey soils, which in turn reduces the selection of suitable backfill.

These constraints have prompted engineers to search for additional solutions for earth retaining structures. Two recently introduced systems abandon the "ideal reinforcing effect" and return to conventional anchorage but maintain the elemental construction technique so successfully pioneered by Reinforced Earth. Examples are Anchored Earth, developed by the Transportation and Road Research Laboratory in Britain, and American Geo-Tech Engineered Retaining Walls.

18.2.2.1 ANCHORED EARTH

Components. The Anchored Earth system was developed at the Transportation and Road Research Laboratory (TTRL) in England [Murray and Irwin (1981)]. It consists of 16- to 20-mm-diameter steel bars, typically 3 to 5 m long, attached to rectangular face panels at one end and deformed to the shape of a Z or a triangle at the other end (Fig. 18.23).

Advantages of Anchored Earth are

Lightweight components make construction easy.

Rods have less surface area exposed to corrosion than strips; since only passive resistance on the anchors is relied upon, the accompanying loss in frictional resistance is not relevant.

Compared to a strip-reinforcing system only a short anchoring length is required, reducing the demand on land and backfill.

Patent applications for Anchored Earth were lodged in 1981. Murray (1983) reported on the performance of a full-scale experimental wall in 1983. The first practical application followed shortly afterward [Jones et al. (1985)] in the form of a retaining structure along the Otley Bypass in Yorkshire, England, and development work has continued since then.

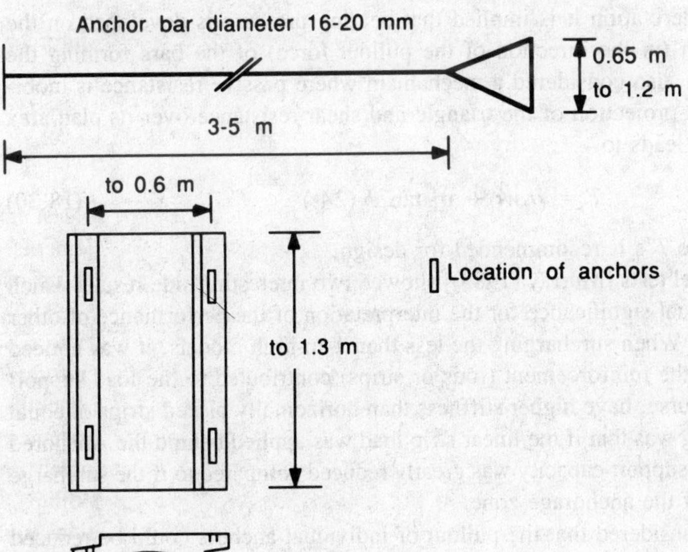

FIGURE 18.23
Anchored Earth system elements. [*Adapted from Murray (1983).*]

Design approach. Design methodology is based on the pullout resistance of the anchors. The following formulas relate to the triangular anchor which is currently preferred to other shapes.

Murray (1983) formulated the pullout force T assuming boundary slip surfaces defined by logarithmic spirals. He found

$$T = \frac{K_p \sigma'_v \, w \, t}{\cos \alpha_1} \exp\,[2(\pi - \alpha_1) \tan \phi'] \tag{18.27}$$

where α_1 = angle between base of triangle (perpendicular to direction of anchor force) and sides of triangle, usually 70°
w = base width of triangle
t = thickness of rods forming triangle
ϕ' = internal friction angle of the soil
K_p = passive earth pressure coefficient

Experiments showed reasonable agreement between actual and pullout resistance of triangular anchors predicted according to the above formula.

Jones et al. (1985) proposed a bearing-capacity-type expression for the pullout strength, with

$$T = \sigma_b 2wt \tag{18.28}$$

where the bearing stress σ_b is related to the effective vertical stress σ_v in the form

$$\sigma_b = 4K_p \sigma'_v \tag{18.29}$$

In the above derivation it is implied that bearing pressure is developed on the horizontal projection (in the direction of the pullout force) of the bars forming the triangle. Jones et al. also considered a mechanism where passive resistance is mobilized along the front projection of the triangle and shear resistance over its plan area A_t. This assumption leads to

$$T = \sigma_b wt + \sigma_v \tan \phi \,(2A_t) \tag{18.30}$$

The lower of the two T's is recommended for design.

Murray's model tests [Murray (1983)] showed two interesting side results which are likely to hold equal significance for the interpretation of the performance of other reinforcing systems. When surcharging the less than 1-m-high models, it was noticed that the stiffness of the reinforcement (rods or strips) contributed to the load support capacity; rods, of course, have higher stiffness than horizontally placed strips of equal area. Another finding was that if the linear strip load was applied behind the Anchored Earth mass, the load support capacity was greatly reduced compared to if the surcharge was placed on top of the anchorage zone.

Murray also considered that the pullout of individual anchors could be reduced by interference of adjacent reinforcing elements. This could be taken into account by appropriately reducing the value of T or arranging multiple anchors in a way to minimize interference.

Measurements on the 3.2-m-high TRRL full-scale test wall [Murray (1983)] indicated that the average tension in the anchors corresponded to above K_0 pressures in the top 2 m of the sand backfill and reduced from K_0 values toward K_a levels below; restraint by the footing reduced earth pressures near the base further. Pressure cells installed at the face also indicated very high horizontal pressures near the top, probably due to the compaction procedures; below 2 m, at the end of construction, these face pressures only represented a fraction of theoretical active earth pressures.

Vertical pressures at base level showed an increase above average overburden pressure near the wall face (say up to 10 or 15%), which can at least partly be attributed to the overturning moment exerted on the reinforced soil mass.

Overall, the performance observed appears to indicate that Reinforced Earth tieback-type design procedures would apply as far as theoretical failure modes and earth pressure calculations are concerned; caution is, however, advised with respect to earth pressures near the top of the backfill.

18.2.2.2 AMERICAN GEO-TECH ENGINEERED RETAINING WALLS. American Geo-Tech walls consist of precast 1.6- by 2.4-m standard face panels connected by at least eight epoxy-coated steel tendons to two independent precast beam anchors buried in the backfill, each about 0.41 m high (Fig. 18.24). Corrosion protection allows for the use of a wide range of soils for backfill, but drainage must be provided where required, the minimum being a nonwoven geotextile behind the face panels.

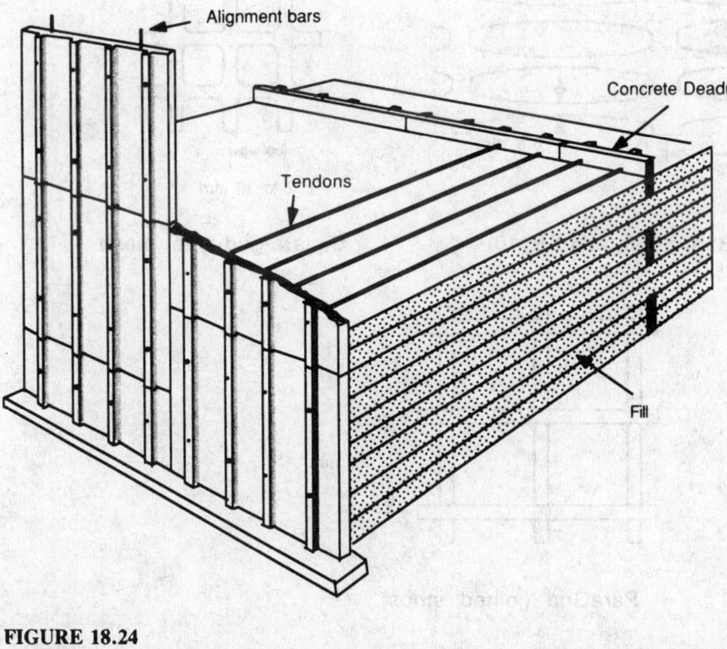

FIGURE 18.24
American Geo-Tech Engineered Retaining Walls. *(Courtesy American Geo-Tech, Inc., Baltimore, U.S.A.)*

At-rest (K_0) pressures are recommended for design. Compared to a technology such as Reinforced Earth, very little information is currently available about the performance of full-scale structures using the American Geo-Tech system.

18.3 POLYMERIC GEOGRIDS

Geogrids is the name given to open-meshed polymeric sheets which interact with soil through friction and adhesion, as well as through the development of passive resistance upon deformation of the geogrid-reinforced soil. These products evolved after the successful use of synthetic textiles in geotechnical engineering and after the recognition of the high pullout resistance of (corrosion-prone) steel meshes.

Some geogrids are relatively stiff and have been applied in similar situations as aircraft landing mats. Yamanouchi (1970) experimented as early as the late sixties with polyethylene netting placed below the surface of soft clay. His results demonstrated that strength, stiffness, and bearing capacity of cohesive soil can be improved, at least under static loading.

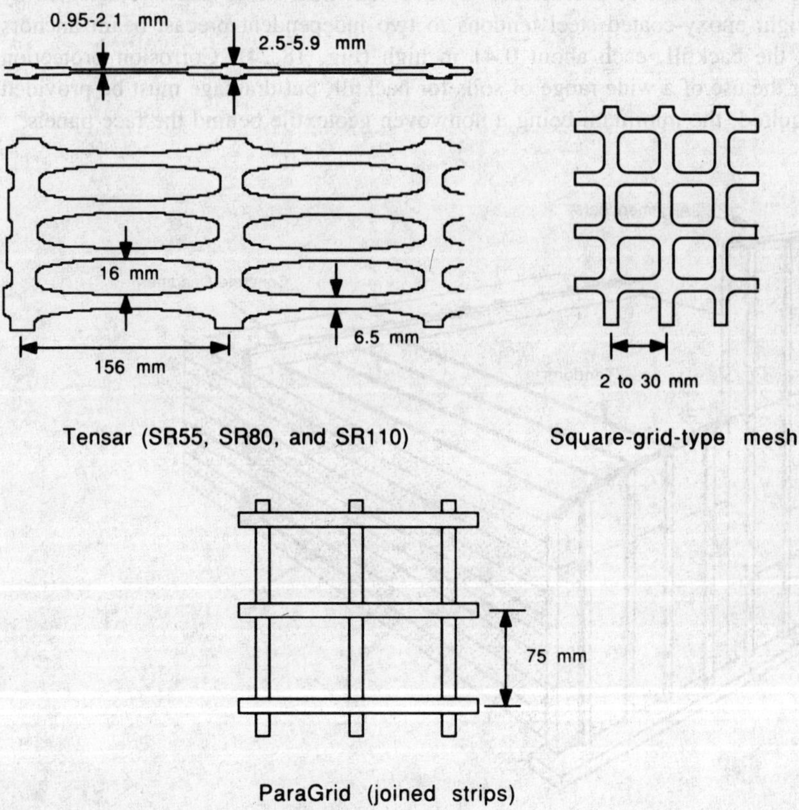

FIGURE 18.25
Types of geogrids.

The most common geogrids available today are relatively flexible, and their main application is in the construction of embankments and retaining walls; they also assist in reinforcing pavement structures and foundations. Because of their light weight and flexibility they are easily handled in the field. Connections of adjacent sheets can be made through interweaving cords or rods made of steel or polymer, which is much easier than welding, gluing, or sewing, as may be needed with geotextiles or geomembranes.

Typical examples of geogrids are shown in Fig. 18.25. "Tensar"* is probably the best-known modern geogrid. It is produced in various grades and patterns by stretching of prepunched sheets of high-density polyethylene or polypropylene.

18.3.1 Engineering Properties of Grids

18.3.1.1 STRESS-STRAIN CHARACTERISTICS. Figure 18.26 shows the typical tensile test behavior of Tensar SR2 (high-density polyethylene grid) pulled in the direction of the ribs. The results are expressed in load per total grid width (kN/m) versus overall strain, i.e., the increase in length divided by the original specimen length. Typically, samples containing 15 ribs and 5 bars are used and tested under the following conditions:

Temperature	20 ± 2°C
Relative humidity	65 + 5%

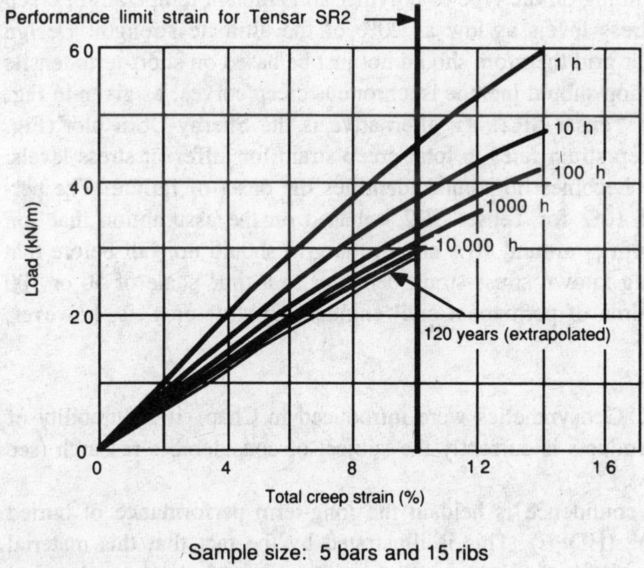

FIGURE 18.26
Isochronous creep curves for Tensar SR2 at 20°C. [*Netlon Ltd. (1984).*]

---

*Tensar is a registered trademark of Netlon Limited

TABLE 18.3
Typical geogrid strengths and moduli in reinforcement applications

Application area			
No.	**Description**	**Grid strength, kN/m**	**Grid modulus, kN/m**
1	Retaining structures	16.6–123	350–1225
2	Slope stabilization	16.6–123	350–1225
3	Unpaved roads	12.3–35	158–350
4	Foundations (bearing capacity)	12.3–35	158–350
5	Embankments over soft soils	16.6–123	350–1225

Source: Extracted from Koerner and Hausmann (1987).

It is rare that a soils testing laboratory is equipped with temperature control; humidity control, particularly for sample storage, is more common. This points out a problem for design and construction control testing of geosynthetics in general: Standard environmental conditions are required for laboratory testing of polymers; otherwise temperature and humidity corrections may be required. Similarly, design values have to be chosen for the particular conditions encountered in the field. Typical ranges of geogrid strengths and moduli chosen in reinforcement applications are given in Table 18.3.

18.3.1.2 CREEP. Depending on the type of polymer and ambient temperature, creep may be significant at stress levels as low as 20% of the ultimate strength. Design parameters for a particular grid therefore should not just be based on short-term tensile testing. Product information should include isochronous creep curves, as given in Fig. 18.26, which applies to Tensar SR2. An alternative is the Sherby-Dorn plot (Fig. 18.27) which relates creep strain rates to total creep strain for different stress levels; the point where a curve becomes horizontal identifies the onset of failure. The performance limit strain of 10% for Tensar SR2 is based on the assumption that soil reaches a limit equilibrium at around 10% strain; the grid should not fail before that strain level. Extrapolating known stress-strain behavior to a time scale of 50 or 100 years, the expected lifetime of permanent civil engineering structures, is, however, not a precise process.

18.3.1.3 DURABILITY. Geosynthetics were introduced in Chap. 10. Durability of synthetic geotechnical products is currently the subject of considerable research (see Sec. 19.1.3).

A high degree of confidence is held in the long-term performance of buried high-density polyethylene (HDPE). This is illustrated by the fact that this material seems to have become a standard for geomembranes containing hazardous chemical wastes; this application no doubt imposes the most severe in-ground environmental conditions. However, that some uncertainty remains with respect to the endurance of synthetic liners is indicated by the trend toward double-liner systems with provisions

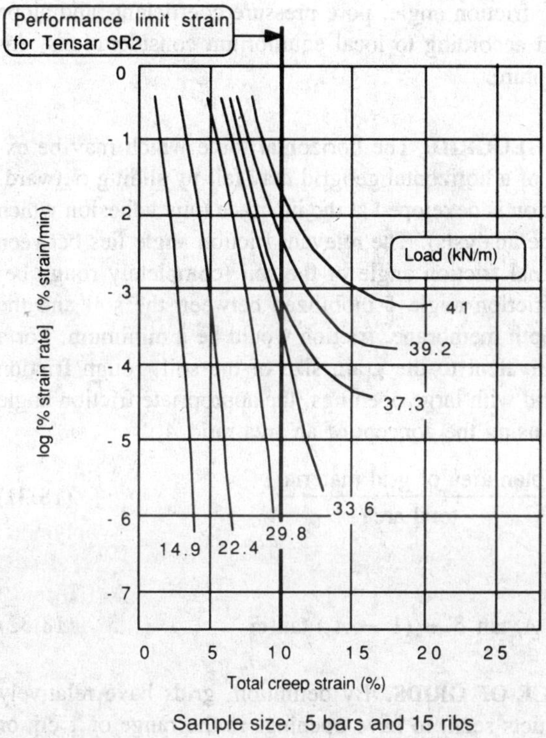

FIGURE 18.27
Sherby-Dorn plot of creep properties of TENSAR SR2 at 20°C. [*After McGown et al. (1984).*]

for detecting and collecting leachate as well as the incorporation of a compacted clay layer as a final backup.

18.3.2 Design Considerations (retaining structures)

Overall external stability (sliding, overturning, etc.) of a geogrid-reinforced structure is being assured along the same lines as discussed for Reinforced Earth (Sec. 18.1), although synthetic reinforcements have an ''effective stiffness'' (when calculated in terms of kilonewtons per meter) of one order less than that of steel strips. Some new aspects are introduced in the examination of soil-grid interaction for the evaluation of internal stability according to the tieback model. They include the possibility of sliding of a block of soil on the plane of the geogrid and possible passive resistance of transverse components of the grid in pullout.

According to Jewell et al. (1984), the reinforcement layout should guarantee global resistance against pullout (in a Coulomb wedge-type analysis) as well as local stability (as evaluated by a Rankine-type analysis at each reinforcement level). Jewell et al. assumed a bilinear failure surface and performed a two-part wedge analysis for the limit state. They presented the results in the form of charts, where the governing earth pressure coefficient and the length of the required reinforcement can be deduced for particular values of cohesion, friction angle, pore pressure coefficient, and slope angle. The spacing is determined according to local equilibrium considerations, ensuring adequate safety against rupture.

18.3.2.1 BLOCK SLIDING ON GEOGRID. The horizontal force which may be exerted by a soil mass lying on top of a horizontal geogrid may fail by sliding outward. This is prevented if adequate friction is developed at the interface (any adhesion which may exist is usually ignored in the analysis). The relevant friction angle lies between the extreme values of ϕ = internal friction angle of the soil (completely rough behavior) and the minimum skin friction angle δ mobilized between the soil and the polymer surface. For a solid smooth membrane, friction would be a minimum. For a grid with openings of a size equivalent to the grain size of the soil, rough friction characteristics are likely. For a grid with large openings, the appropriate friction angle δ_s for sliding could be estimated using the concept of an area ratio A_r:

$$A_r = \frac{\text{plan area of grid material}}{\text{total area}} \tag{18.31}$$

We could then stipulate

$$\tan \delta_s = A_r \tan \delta + (1 - A_r) \tan \phi \tag{18.32}$$

18.3.2.2 PULLOUT RESISTANCE OF GRIDS. By definition, grids have relatively large openings; commercial products seem to have openings in the range of 1 cm or more in width. It is therefore possible that, as in the case of bar mats and meshes, the pullout resistance is composed of frictional resistance and passive resistance of the transverse elements:

$$T_r = T_f + T_p \tag{18.33}$$

where T_r = total pullout force (e.g., kN/m)
T_f = (frictional) sliding resistance
T_p = passive resistance developed by transverse elements of the grid (parallel to wall face)

Force T_f could be expressed in terms of the surface area effective against pullout (A_e = 2 × effective length × unit width), the vertical pressure, and the appropriate friction angle δ_s as defined above:

$$T_f = \sigma_v \tan \delta_s A_e \tag{18.34}$$

Similar to generating a term for passive resistance for bars (Sec. 18.2.1.3), T_p per unit width could be evaluated by following the structure of Terzaghi's bearing-capacity formula:

$$T_p = \sigma_v N_q tn + cN_c tn \tag{18.35}$$

where t is the thickness of the grid and n is the number of transverse elements involved for the particular effective length of grid.

Sparse experimental results would suggest that if the above formulation is applied, the values δ_s, N_q, and N_c might well be specific to a particular grid geometry and neither may be mobilized up to its theoretical limit.

As part of a research program at the University of California, Davis, Johnston (1986) carried out a number of grid pullout tests in a large Caltrans shear box which allowed the testing of specimens as wide as 0.6 m. Interpreting the results in terms of an apparent skin friction angle developed for the applied vertical pressures, and disregarding the passive resistance concept, Johnston found values of δ ranging up to 79° (at σ_v = 27 kPa) although the internal friction angle of the sand used was not more than 48°. He attributed the large values to the effect of dilation, since actual vertical pressures measured during the test in close vicinity to the grid were up to 4 times the nominally applied pressure. All the results indicated that the choice of $\delta = \phi$ would be a conservative approach for design.

PROBLEMS

Prefixes indicate problem type: C = calculations, B = brief answer, M = multiple choice, D = discussion.

Section 18.1

SECTION 18.1.1

C18.1. Compare the percentage reduction in yield strength of a steel reinforcing strip due to 0.5 mm uniform corrosion over the total surface, given the initial dimensions of
(*a*) Width = 60 mm, thickness = 3 mm.
(*b*) Width = 40 mm, thickness = 5 mm.

B18.2. Electrochemical criteria used to evaluate Reinforced Earth backfill materials usually include
(*a*) ______________ .
(*b*) ______________ .
(*c*) ______________ .
(*d*) ______________ .

B18.3. Aside from electrochemical properties, other characteristics of a soil fill can influence the corrosion of metal within it, such as
(*a*) ______________ .
(*b*) ______________ .
(*c*) ______________ .
(*d*) ______________ .

B18.4. For a Reinforced Earth wall of H = 10 m, typical dimensions are (refer to Fig. P18.1):

h = ______________ m
D = ______________ m
L = ______________ m

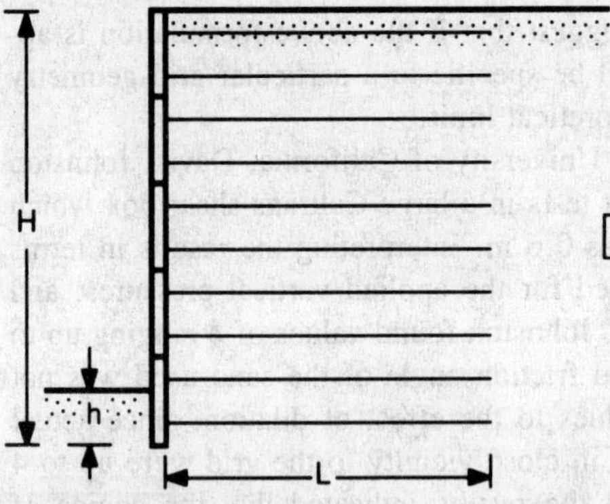

FIGURE P18.1

B18.5. For ribbed strips (Fig. P18.2) of smallest size, typical data are

$b =$ ________ mm
$t =$ ________ mm
$\sigma_s =$ ________ MPa (yield)

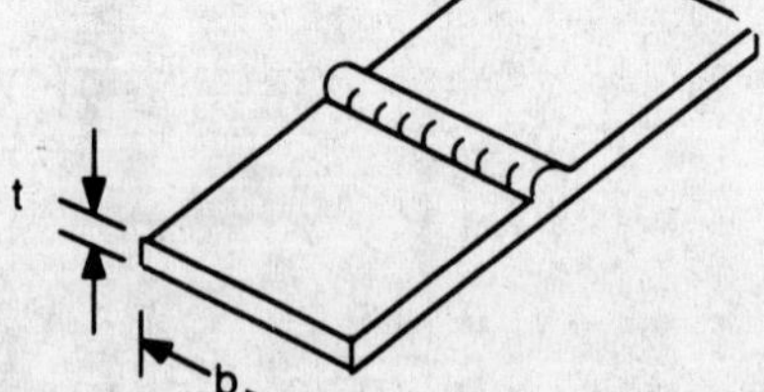

FIGURE P18.2

M18.6. The strips in a 5-m-high Reinforced Earth wall are typically
(*a*) 5 m long.
(*b*) 3.5 to 4 m long.
(*c*) 2 to 3 m long.
(*d*) 1 m long.

M18.7. The smallest-size ribbed strips typically
(*a*) Are 3 mm thick.
(*b*) Are 40 mm wide.
(*c*) Are 80 mm wide.
(*d*) Have a yield strength of 160 MPa.

M18.8. The design life expectancy of a permanent Reinforced Earth wall is
(*a*) 10 years.
(*b*) 20 years.
(*c*) > 50 years.
(*d*) > 70 years.

M18.9. Standard Reinforced Earth walls rely on selected backfill which is
(*a*) Highly cohesive.
(*b*) Impermeable.

(*c*) Free-draining.
(*d*) Low in frictional strength.

D18.10. Comment on the life expectancy of civil engineering structures generally and Reinforced Earth walls in particular. Also refer to the definitions of "service" life and "ultimate" life as used in relation to Reinforced Earth design.

D18.11. Discuss the advantages and disadvantages of the original steel facing relative to the now commonly used concrete face panels of Reinforced Earth walls.

SECTION 18.1.2

C18.12. A Reinforced Earth wall is built on a slope as shown in Fig. P18.3. Assume the soil consists of a free-draining cohesionless sand with a frictional angle of 35° and a unit weight of 19 kN/m^3.

(*a*) Calculate the pressure acting on the base of the Reinforced Earth block and determine its bearing capacity using the Hansen equation, considering load eccentricity and inclination as well as ground slope.
(*b*) Determine the factor of safety with respect to the particular slip circle shown, using the method of slices. Choose eight slices.
(*c*) Estimate the safety against horizontal sliding of the Reinforced Earth block. (For calculation purposes assume that the passive resistance developed is equivalent to a horizontal embedment of 0.5 m.)

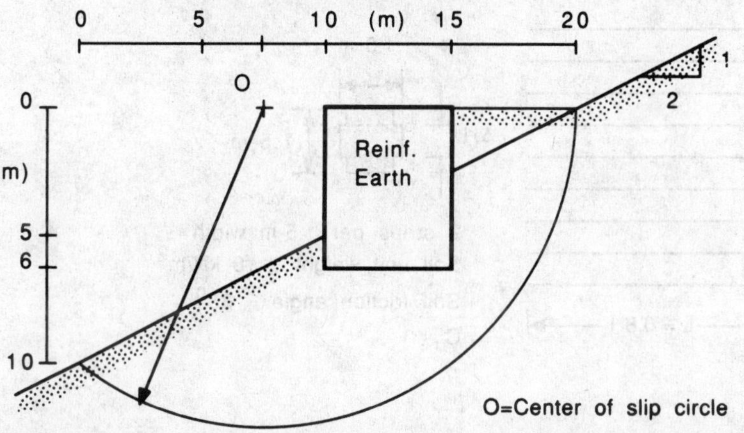

FIGURE P18.3

B18.13. Name two modes of internal failure and two modes of external failure relevant to the design of reinforced earth walls:

(*a*) ______________
(*b*) ______________
(*c*) ______________
(*d*) ______________

M18.14. In order to control the rate of corrosion, standard Reinforced Earth backfill criteria require a

(*a*) Resistivity < 1000 Ω · cm.
(*b*) pH > 6.
(*c*) Chloride concentration > 200 ppm.
(*d*) Sulphate content < 1000 ppm.

D18.15. List potential failure modes of Reinforced Earth–type structures and indicate how extreme differential settlement could affect their performance.

D18.16. Review the corrosion of steel structures in dry, partially saturated, and fully saturated soil.

SECTION 18.1.3

C18.17. Referring to Fig. P18.4 with $H = 7.5$ m, $q = 10$ kPa, $z = 3.375$ m, and a corrosion allowance of 1.5 mm, calculate the factor of safety against rupture and slippage according to

(*a*) Original standard design, assuming smooth strips (80 mm wide, 3 mm thick, and yield strength $\sigma_s = 300$ MPa); use the Meyerhof distribution of vertical stresses for calculating horizontal earth pressures and $\tan \delta = 0.4$.

(*b*) Tieback analysis according to Lee et al.; assume smooth strips as in part (*a*), $K = K_a$, a uniform vertical stress distribution on the strips, and $\tan \delta = 0.4$.

(*c*) New design (1978), assuming ribbed strips, 40 by 5 mm, $\sigma_s = 275$ MPa, and the usual corrosion allowance as in part (*a*). Use the Meyerhof distribution.

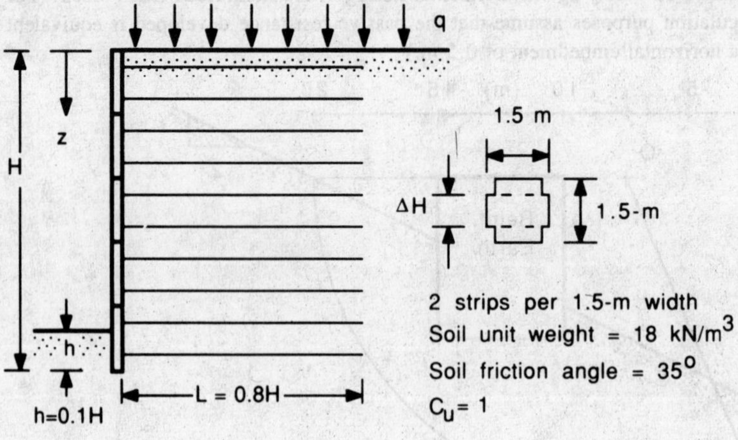

FIGURE P18.4.

C18.18. Again referring to Fig. P18.2, calculate the increase in tension in a strip located 0.875 m from the surface due to a concentrated 50-kN load applied 1.5 m from the edge. Use the elementary 2:1 method or refer to App. 18C (elastic theory).

C18.19. In terms of b, L_e (effective length of strips), σ_v (vertical stress acting on strips), δ (soil-reinforcement friction angle), p_h (horizontal earth pressure to be resisted), ΔB, and ΔH, find the factor of safety against slippage:

$$F_s = \underline{\qquad\qquad\qquad}$$

Calculate F_s for $b = 60$ mm, $\sigma_v = 80$ kPa, $\phi = 30°$, $\delta = 20°$, $\Delta B = \Delta H = 0.75$, and $L_e = 2.4$ m. Assume uniform vertical pressure.

B18.20. Horizontal pressure is related to vertical stress by the factor K. Depending on the state of equilibrium of the soil mass (active, passive, at-rest), K equals K_a, K_p, or K_0. For Reinforced Earth, show the variation of K with depth for design calculations as recommended by Schlosser (1978). (Use a drawing similar to Fig. P18.5 to sketch dimensions, indicating K_a, K_p, or K_0.)

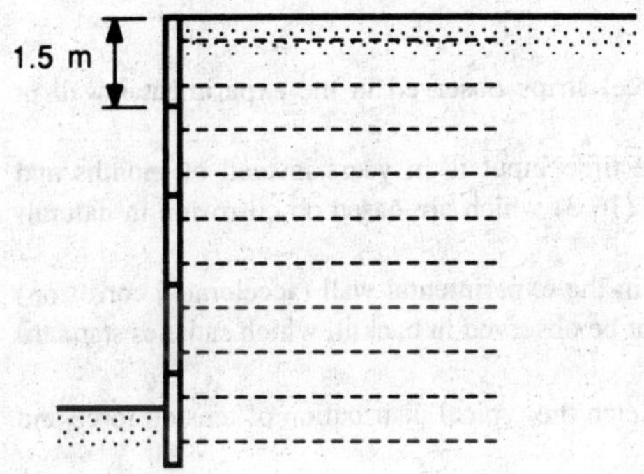

FIGURE P18.5

B18.21. Name two types of experiments which yield the soil-reinforcement friction angle and/or adhesion:
(*a*) ____________
(*b*) ____________

B18.22. List two parameters or dimensions which differ between the original standard method and the 1978 (still current) method of design for Reinforced Earth walls:
(*a*) ____________
(*b*) ____________

M18.23. Consider a 5-m-long, 80-mm-wide strip embedded at a depth of 4 m in a sandy soil (unit weight = 18 kN/m^3) which develops a skin friction angle of 25°. Assuming resistance is developed along the entire length, the pullout force of the strip is approximately
(*a*) 1 kN.
(*b*) 2.7 kN.
(*c*) 13 kN.
(*d*) 27 kN.

M18.24. The yield strength of a ribbed strip (steel yield strength = 275 MPa), 40 mm wide and 5 mm thick, with a corrosion allowance of 2 mm (1 mm on each side) is approximately
(*a*) 33 kN.
(*b*) 147 kN.
(*c*) 350 kN.
(*d*) 557 kN.

M18.25. Lee (1979) proposed a design method which differed from the original standard design method most significantly by the fact that it used
(*a*) K_p instead of K_a.
(*b*) K_0 instead of K_a.
(*c*) An "effective" length rather than the total length of the strip for the evaluation of safety against slippage of the strip.
(*d*) An "effective" skin friction angle instead of the soil internal friction angle for the calculation of pullout resistance.

D18.26. Give examples of engineering structures commonly designed to withstand active, at-rest, and passive earth pressures.

SECTION 18.1.4

C18.27. Compare the rate of corrosion of steel strips observed in the experimental wall in Spain (Sec. 18.1.4) as follows:

(*a*) Convert Eq. (18.16) so that the time input is in years instead of months and compare it with Eqs. (16.2) and (16.3) which are based on corrosion in natural, mainly selected backfill.

(*b*) Calculate corrosion after 1 year in the experimental wall (accelerated corrosion) with that in the range which might be observed in backfill which satisfies standard selection criteria.

B18.28. In a drawing similar to Fig. P18.6 sketch the typical distribution of tension force and skin friction for a reinforcing strip.

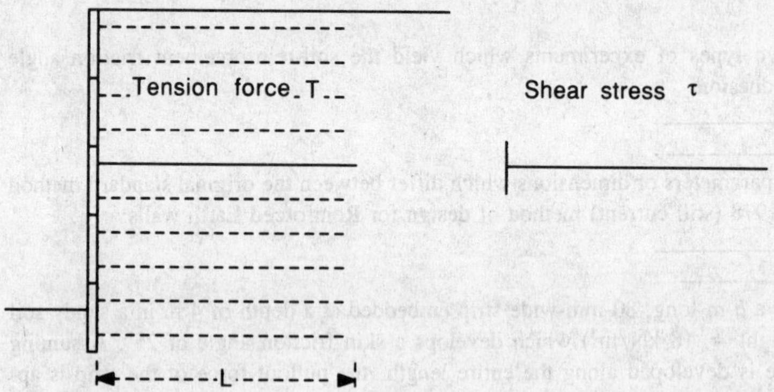

FIGURE P18.6

M18.29. In a real Reinforced Earth wall, the peak tension in strips placed within say 6 m of the surface, is developed

(*a*) At the wall face.

(*b*) At the free end of the strip.

(*c*) In that half of the strip which is nearest to the face (but not at the face).

(*d*) In that half of the strip which is nearest to the free end (but not at the end).

M18.30. After placing a surcharge on a Reinforced Earth wall, strip tension

(*a*) Will always increase.

(*b*) Will always decrease.

(*c*) Generally increases, but may decrease in the top layers.

(*d*) Will not change noticeably.

D18.31. Measurements on the Fremersdorf wall indicated a decrease in strip tensions with surcharging in some reinforcement layers. Where and why did this occur?

D18.32. Looking at Fig. 18.14 (Fremersdorf wall), how do you explain a decrease in the vertical earth pressure in the immediate vicinity of the toe?

D18.33. How does the Dunkirk wall differ from an ordinary Reinforced Earth wall, and what was noted in respect to the effect of transient loads?

SECTION 18.1.5

M18.34. High-strength synthetic reinforcing strips which do not corrode have been developed in Britain but so far have not been used in Australia, mainly because of

(*a*) Temperature-dependent strength.
(*b*) High creep.
(*c*) Low skin friction.
(*d*) Existing patents.

D18.35. Compare technical advantages and disadvantages of the York system of soil reinforcement with those of Reinforced Earth with concrete face elements. Why has the York system not been adopted elsewhere shortly after it was invented?

Section 18.2.

SECTION 18.2.1

C18.36. Compare a 17-mm-diameter round steel bar with a 40-mm-wide flat strip of equal cross-sectional area with respect to

(*a*) Surface area per meter length.
(*b*) Percentage loss in tensile strength after an effective depth of corrosion of 0.5 mm on all surfaces.
(*c*) Relative pullout force required. (Express the bar pullout force as a percentage of the strip pullout force, taking the results of Fig. 18.20 as a guide.)

C18.37. Calculate the resistance against pullout of a 1-m-wide and 4-m-long bar mesh buried in 2 m of sand with a soil-bar friction angle of 20°. There are four 17-mm-diameter bars per meter of width and a total of five transverse bars of equal diameter. Assume $N_q = 8.3$ and $\gamma = 20\ \text{kN/m}^3$, and proceed as follows:

(*a*) Calculate the pullout force required according to Eq. (18.23).
(*b*) Compare the result from part (*a*) with that obtained with the empirical formula proposed by Nielsen for washed sand [Eq. (18.25)].
(*c*) Check whether the pullout resistance exceeds the allowable force with respect to bar rupture, assuming an acceptable steel stress of 210 MPa.

B18.38. Name two wall systems which are closely related to Reinforced Earth and are now marketed in the United States.

(*a*) ____________
(*b*) ____________

B18.39. What are some advantages and disadvantages of using reinforcing meshes rather than Reinforced Earth–type strips?

M18.40. Compared to steel strips of equal cross-sectional area, bars

(*a*) Have less surface area.
(*b*) Lose less strength at equal corrosion rates.
(*c*) Have more frictional pullout resistance.
(*d*) Have greater tensile strength.

D18.41. Discuss the usage of the terms "mats," "bar mesh," and "geogrid"; give examples of commercial products in these categories.

SECTION 18.2.2

C18.42. Predict the pullout force of a single Anchored Earth reinforcement element according to Eq. (18.27), as well as according to the more conservative formulas by Jones [Eqs. (18.28) and (18.30)]. Given is the following data:

Soil internal friction angle = 35°
Soil unit weight = 19 kN/m^3
Depth of fill = 4 m
Triangular anchor element: width = 0.65 m, angle α_1 = 70°
Rods: 4 m long, 20 mm thick

M18.43. Anchored earth
(*a*) Does not rely on soil-bar friction.
(*b*) Relies on passive pullout resistance.
(*c*) Has bars attached to concrete blocks.
(*d*) Is more of an anchorage system than soil reinforcement is.

D18.44. In what way do Anchored Earth and American Geo-Tech Engineered Retaining Walls depart from ideal reinforcement as defined in Fig. 17.13?

Section 18.3

SECTION 18.3.1

C18.45. Estimate the total creep strain (in percent) of Tensar SR2 subjected to a constant 20-kN/m force at the end of the life of a permanent grid-reinforced structure (see Fig. 18.26).

C18.46. Loaded at 41 kN/m, Tensar SR2 reduces its strain rate (measured in percent strain per minute) to a more or less constant value. How much is this strain rate, and at what total magnitude of creep does this occur? (See Fig. 18.27).

M18.47. Polymeric geogrids may exhibit creep at stress levels (expressed as percentage of ultimate strength) as low as
(*a*) 20%.
(*b*) 40%.
(*c*) 60%.
(*d*) 80%.

M18.48. Typical geogrids used in construction have a strength in the range of
(*a*) 10 to 120 kN/m.
(*b*) 100 to 200 kN/m.
(*c*) Up to 2000 kN/m.
(*d*) 240 to 300 MPa.

D18.49. Discuss corrosion of steel meshes vis-à-vis degradation of polymeric grids embedded in soil.

D18.50. Attempt to compare strength, modulus, durability, and costs of geogrids with those of geotextiles of similar mass per square meter.

SECTION 18.3.2

C18.51. Estimate the friction angle δ_s appropriate for a block of soil sliding on a geogrid, given the following information:

Plan area of grid material = 40% of total area
Soil internal friction angle = 40°
Soil-polymer friction angle = 20°

C18.52. Estimate the pullout resistance of a 3-m-long, 1-m-wide section of geogrid with an effective soil-grid friction angle of 30° under 3 m of soil. The soil has an internal friction angle of 40° and a unit weight of 18 kN/m^3. The grid is 2 mm thick and there are 8 transverse elements per m length. Proceed as follows:

(*a*) Calculate the frictional pullout resistance.

(*b*) Calculate the passive resistance, assuming Terzaghi-type bearing capacity is developed on each transverse element over the full width of the grid.

(*c*) Add the values obtained in parts (*a*) and (*b*).

APPENDIX 18A

SLOPE STABILITY FORMULAS

The formulas in this appendix section are for a circular failure surface. Designations are as shown in Fig. A18.1. A bar indicates an effective stress or strength term (e.g., $\bar{c}$, $\bar{N}$). The safety factor F is defined as

$$F = \frac{\text{moment of shear strength along failure arc}}{\text{moment of weight of failure mass and external forces}} \tag{18.1}$$

$$F = \frac{M_R}{M_D} = \frac{r\left(\bar{c}L + \tan\bar{\phi}\sum_{i=1}^{i=n}\bar{N}_i\right)}{r\left(\sum_{i=1}^{i=n}W_i\sin\theta_i\right)} = \frac{\bar{c}L + \tan\bar{\phi}\sum_{i=1}^{i=n}\bar{N}_i}{\sum_{i=1}^{i=n}W_i\sin\theta_i} \tag{A18.1}$$

18A.1 ORDINARY METHOD OF SLICES

This method is also referred to as the Swedish method or Fellenius method:

$$\bar{N}_i = W_i\cos\theta_i - u_i\,\Delta l_i$$

$$F = \frac{\bar{c}L + \tan\bar{\phi}\sum_{i=1}^{i=n}(W_i\cos\theta_i - u_i\,\Delta l_i)}{\sum_{i=1}^{i=n}W_i\sin\theta_i} \tag{A18.2}$$

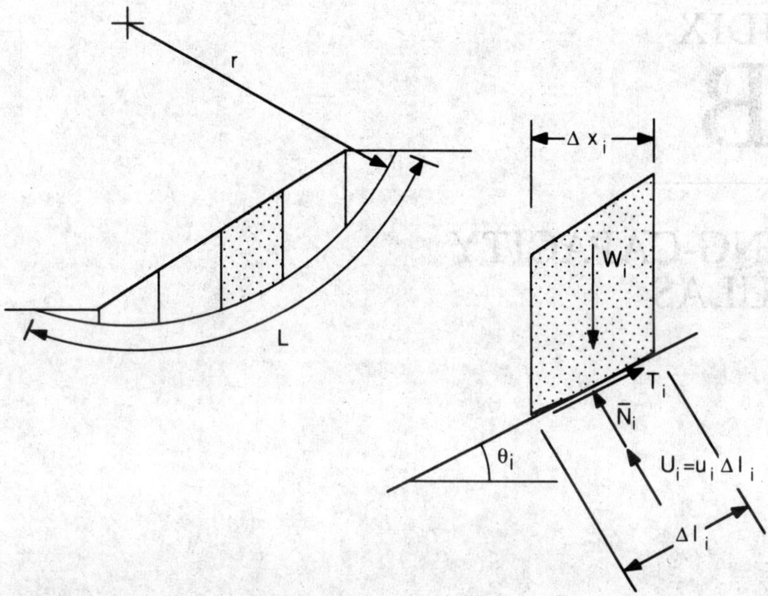

FIGURE A18.1

18A.2 SIMPLIFIED BISHOP METHOD

$$F = \frac{\sum_{i=1}^{i=n} [\bar{c}\, \Delta x_i + (W_i - u_i\, \Delta x_i) \tan \bar{\phi}][1/M_i(\theta)]}{\sum_{i=1}^{i=n} W_i \sin \theta_i} \qquad \text{(A18.3)}$$

$$M_i(\theta) = \cos \theta_i \left(1 + \frac{\tan \theta_i \tan \bar{\phi}}{F}\right)$$

Note that this is an iterative method; F appears on both sides of the equation.

APPENDIX 18B

BEARING-CAPACITY FORMULAS

18B.1 TERZAGHI BEARING-CAPACITY EQUATIONS

See Table B18.1 for bearing-capacity factors N, and $K_{p\gamma}$ values.

Continuous footing: $$q_{\text{ult}} = cN_c + \bar{q}N_q + 0.5\gamma BN_\gamma \tag{18.20}$$

Square footing: $$q_{\text{ult}} = 1.3cN_c + \bar{q}N_q + 0.4\gamma BN_\gamma \tag{B18.1}$$

Round footing: $$q_{\text{ult}} = 1.3cN_c + \bar{q}N_q + 0.3\gamma BN_\gamma \tag{B18.2}$$

where $$N_q = \frac{a^2}{2\cos^2(45 + \phi/2)}$$

$$a = \exp\left[\left(0.75\pi - \frac{\phi}{2}\right)\tan\phi\right]$$

$$N_c = (N_q - 1)\cot\phi$$

$$N_\gamma = \frac{\tan\phi}{2}\left(\frac{K_{p\gamma}}{\cos^2\phi} - 1\right)$$

c = cohesion, kPa
ϕ = internal friction angle, °
$\bar{q}$ = effective overburden pressure at footing base level

TABLE B18.1
Bearing-capacity factors for the Terzaghi bearing-capacity equations

ϕ, °	N_c	N_q	N_γ	$K_{p\gamma}$
0	5.7	1.0	0.0	10.8
5	7.3	1.6	0.5	12.2
10	9.6	2.7	1.2	14.7
15	12.9	4.4	2.5	18.6
20	17.7	7.4	5.0	25.0
25	25.1	12.7	9.7	35.0
30	37.2	22.5	19.7	52.0
35	57.8	41.4	42.4	82.0
40	95.7	81.3	100.4	141.0
45	172.3	173.3	297.5	298.0
50	347.5	415.1	1153.2	800.0

Source: As presented by Bowles (1982).

18B.2 HANSEN BEARING-CAPACITY EQUATIONS

See Table B18.2 for bearing capacity factors *N*, and Table B18.3 for correction factors *s*, *i*, *d*, *g*, and *b*.

General form of the equation:

$$q_{ult} = c\,N_c s_c i_c d_c g_c b_c + \bar{q}\,N_q s_q i_q d_q g_q b_q + 0.5\,\gamma\,BN_\gamma s_\gamma i_\gamma d_\gamma g_\gamma b_\gamma \tag{18.2}$$

For $\phi = 0$ conditions (e.g., drained loading of saturated clays):

$$q_{ult} = 5.14c_u\,(1 + s'_c + d'_c - i'_c - b'_c - g'_c) + \bar{q} \tag{B18.3}$$

TABLE B18.2
Bearing-capacity factors for the Hansen bearing-capacity equation

ϕ, °	N_c	N_q	N_γ	N_q/N_c	$2 \tan \phi\,(1 - \sin \phi)^2$
0	5.14	1.0	0	0.19	0
5	6.5	1.6	0.1	0.24	0.15
10	8.3	2.5	0.4	0.30	0.24
15	11.0	3.9	1.2	0.36	0.29
20	14.8	6.4	2.9	0.43	0.32
25	20.7	10.7	6.8	0.51	0.31
30	30.1	18.4	15.1	0.61	0.29
35	46.1	33.3	33.9	0.72	0.25
40	75.3	64.2	79.5	0.85	0.21
45	133.9	134.9	200.8	1.01	0.17
50	266.9	319.0	568.5	1.20	0.13

Source: As presented by Bowles (1982).

TABLE B18.3
Shape, depth, inclination, and other factors for use in the Hansen bearing-capacity equations [Eqs. (18.2) and (B18.3)]

Shape factors	Depth factors		Inclination factors	Ground factors (see Fig. B18.1)
$s'_c = \frac{0.2B}{L}$	$d'_c = \frac{0.4D}{B}$	$D \le B$	$i'_c = 0.5 - 0.5\sqrt{\frac{1-H}{A_f C_a}}$	$g'_c = \frac{\psi^\circ}{147^\circ}$ For horizontal ground use: $g'_c = 0.0$
$s_c = 1 + \frac{N_q B}{N_c L}$	$d'_c = 0.4\tan^{-1}\frac{D}{B}$	$D > B$	$i_c = i_q - \frac{1 - i_q}{N_q - 1}$	$g_c = 1 - \frac{\psi^\circ}{147^\circ}$ $g_q = g_\gamma = (1 - 0.5\tan\psi^\circ)^5$
	$d_c = 1 + 0.4\frac{D}{B}$	$D \le B$		Base factors (see figure): $b'_c = \eta^\circ/147^\circ$ For horizontal ground use: $b'_c = 0.0$
	$d_c = 1 + 0.4\tan^{-1}\frac{D}{B}$	$D > B$		$b_c = 1 - \frac{\eta^\circ}{147^\circ}$
$s_q = 1 + \frac{B}{L}\tan\phi$	$d_q = 1 + 2\tan\phi\,(1 - \sin\phi)^2\frac{D}{B}$	$D \le B$	$i_q = \left(1 - \frac{0.5H}{V + A_f c_a \cot\phi}\right)^5$	$b_q = b_\gamma = \exp(-2\eta\tan\phi)$ η = radians for b_q
	$d_q = 1 + 2\tan\phi\,(1 - \sin\phi)^2\tan^{-1}\frac{D}{B}$	$D > B$		
$s_\gamma = 1 - \frac{0.4B}{L}$	$d_\gamma = 1.00$ for all ϕ		Horizontal base: $i_\gamma = \left(1 - \frac{0.7H}{V + A_f c_a \cot\phi}\right)^5$ Sloping base: $i_\gamma = \left[1 - \frac{(0.7 - \eta^\circ/450^\circ)H}{V + A_f c_a \cot\phi}\right]^5$	

TABLE B18.3 (*Cont.*)

Source: Table combined by Bowles (1982) from Hansen (1970), De Beer (1970), and Vesic (1973).

Note: Primed factors are for undrained conditions and $\phi = 0$.

Do not use shape factors in combination with inclination factors. Use d_i and i_i only in combination, or s_i with d_i, g_i, and b_i. When triaxial ϕ is used for plane-strain conditions, one may adjust to obtain: $\phi_{ps} = 1.1\phi_{\text{triaxial}}$. (Bowles suggests that this correction only be used for $\phi_{\text{triaxial}} > 30°$.)

A_f = effective footing contact area $B'L'$
L' = effective footing length = $L - 2e_L$
B' = effective footing width = $B - 2e_B$
D = depth of footing in ground
e_B, e_L = eccentricity of load with respect to center of footing area
c = cohesion of base soil
ϕ = angle of internal friction of soil
H, V = load components parallel and perpendicular to footing, respectively
$\tan\delta$ = coefficient of friction between footing and base soil {use $\delta = \phi$ for concrete poured on ground [Schultze and Horn (1967)]}
η, ψ = as shown in Fig. B18.1 with positive directions shown

Limitations:
$H \le V \tan\delta + c_a A_f$
$i_q, i_\gamma > 0$
$\psi \le \phi$
$\eta + \psi \le 90°$

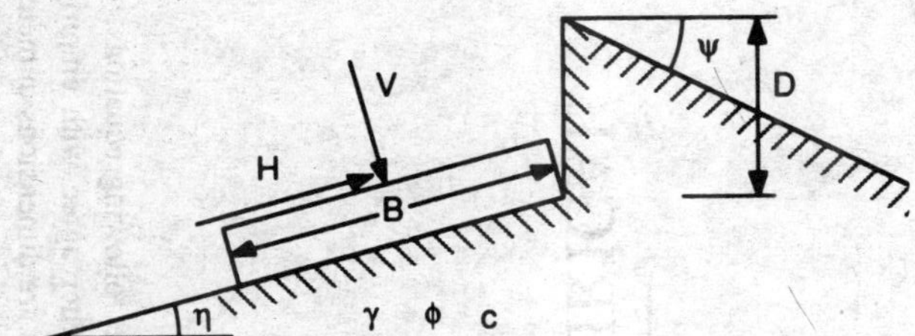

FIGURE B18.1

APPENDIX 18C

EFFECT OF SURCHARGE LOADS ON RETAINING STRUCTURES—ELASTIC-EMPIRICAL SOLUTION

Bowles (1982) has presented the following equations, which are based on elastic theory, but have been modified so as to better agree with empirical findings. Refer to Fig. C18.1 for definition of terms. Typical units are dimensions in meters, loads in kilonewtons, and pressures in kilopascals.

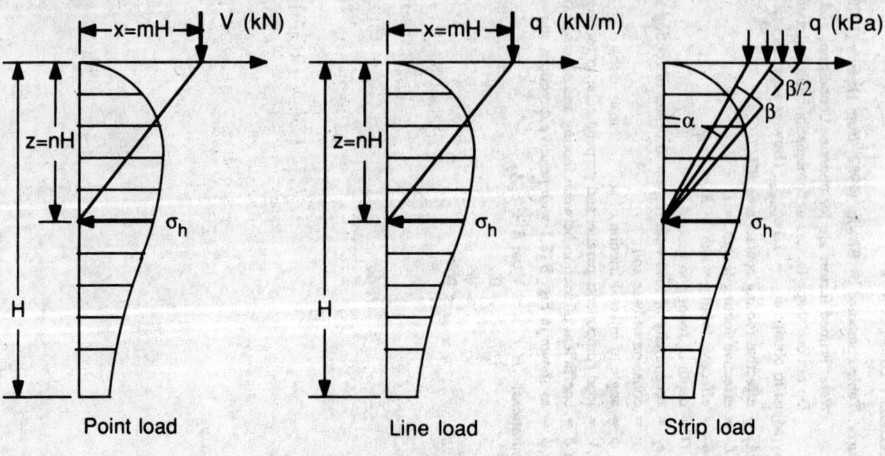

FIGURE C18.1

18C.1 POINT LOAD

$$\sigma_h = \begin{cases} \dfrac{1.77V}{H^2} \dfrac{m^2n^2}{(m^2 + n^2)^3} & \text{for } m > 0.4 \quad \text{(C18.1)} \\[2ex] \dfrac{0.28V}{H^2} \dfrac{n^2}{(0.16 + n^2)^3} & \text{for } m \leq 0.4 \quad \text{(C18.2)} \end{cases}$$

18C.2 LINE LOAD

$$\sigma_h = \begin{cases} \dfrac{4q}{\pi H} \dfrac{m^2n}{(m^2 + n^2)^2} & \text{for } m > 0.4 \quad \text{(C18.3)} \\[2ex] \dfrac{q}{H} \dfrac{0.203n}{(0.16 + n^2)^2} & \text{for } m \leq 0.4 \quad \text{(C18.4)} \end{cases}$$

18C.3 STRIP LOAD

$$\sigma_h = \frac{2q}{\pi} (\beta - \sin \beta \cos 2\alpha) \quad \text{(C18.5)}$$

where β is in radians.

Note: In the 4th edition of Bowles' book on Foundation Analysis and Design (McGraw-Hill, 1988) the above approach to calculating the effect of a surcharge on lateral earth pressures is no longer recommended. Instead, reference is made to the original elastic theory by Boussinesq. For a detailed discussion of this problem consult Bowles (1988), pp. 508–511, or the book by Clayton and Milititsky (1986).

CHAPTER 19

FLEXIBLE GEOSYNTHETIC SHEET REINFORCEMENT

Geosynthetics were introduced in Chap. 10, in which basic terms were defined, various products were classified, and an overview of applications was presented. Subsequently, Chap. 10 concentrated on the use of geosynthetics for filtration, drainage, and seepage control.

This chapter emphasizes construction techniques and methods of analysis concerning geotextiles functioning as reinforcement. The term ''geotextiles'' describes the generally more flexible synthetic fabrics which pioneered the large-scale use of synthetics in geotechnical engineering. It should be noted that in applications where reinforcement is the only function relied upon, the principles discussed may well equally apply to flexible synthetic meshes and grids, already discussed in Sec. 18.3. By definition, geogrids are distinguished by larger openings and often substantially higher rigidity and load distribution capacity than geotextiles. On the other hand, their ability to drain, filter, and separate—which may be important in addition to reinforcement—is nonexistent or restricted to relatively coarse soils.

Although traditional metal inclusions such as strips, bars, and meshes have fairly well defined applications in retaining structures, embankments, and occasionally difficult foundations, the variety of materials and diversity of products offered as geotextiles or, more generally, as geosynthetics, make the scope of an exhaustive discussion of their use as reinforcement very broad indeed. This situation is further complicated by the fact that although reinforcement may be identified as the major function, possibly even forming the only basis for design, the success of the project may be significantly affected by drainage, filtration, and other benefits additionally provided by the fabric.

19.1 PROPERTIES RELEVANT TO THE DESIGN FOR REINFORCEMENT

Many national and international committees already have adopted or are developing standards on the testing of geotextiles and related products. These standards cover general requirements for sampling and testing and basic physical, mechanical, and hydraulic properties and durability characteristics. The test procedures so far developed mostly relate to properties of the fabrics tested in isolation and are essentially regarded as index tests. Their main purpose is to assist in comparing and selecting fabrics and in quality control in the field. The test methods proposed as standards are, however, closely related to fabric performance in many applications, and, as further experience is gained, their importance is likely to increase.

The determination of soil-fabric friction in the direct shear box is classified as a soil-interaction test, as are the determination of stress-strain, abrasion, or hydraulic properties of fabrics when embedded in soil. Because of their complexity, some of these tests are unlikely to become standards, although they may have merit for research and specific design problems.

This section selectively looks at the evaluation of physical and mechanical properties which are relevant to the reinforcement function. A brief discussion of durability is included, because durability is mostly expressed in terms of loss of strength due to environmental or other influences.

19.1.1 Physical Properties

Physical properties of fabrics are more dependent on temperature and humidity than those of soils and rocks. In order to achieve consistent results in the laboratory, good environmental control during the testing is therefore important. Testing conditions as prescribed by many standards are fairly stringent, and this could present a problem for quality control or research testing of fabrics in a conventional soils testing laboratory.

General requirements for the testing of fabrics include rules for sampling, conditioning, and the determination of basic physical properties such as weight and thickness. (For details of test methods refer to standards listed in Chap. 10.)

The most useful basic physical property of a geotextile is probably the mass per unit area. Its determination is often part of quality-control procedures. For a particular type of fabric, mass per unit area is likely to be closely related to engineering properties.

Fabrics are compressible, and the determination of their nominal thickness has to be carried out with a standard surcharge. Thickness and compressibility are important with respect to the in-plane drainage capacity of thick nonwoven fabrics.

19.1.2 Mechanical Properties

Strength properties of geotextiles enter all design calculations where reinforcement is the primary function. In addition, they form the basis for evaluating a geotextile's resistance to damage during construction (fabric survivability) and are related to the fabric's ability to support workers and construction equipment before any fill is placed (fabric workability).

19.1.2.1 TENSILE STRENGTH. The basic tensile properties of a geotextile are communicated in terms of peak strength and stress-strain modulus, both in units of kilonewtons per meter or similar units. Different types of tensile tests on geotextiles are illustrated in Fig. 19.1.

For the analysis of the stability of a soil mass containing fabric inclusions, the preferred textile strength parameters are those evaluated on a 200-mm-wide strip. By approximating plane strain conditions, the wide-strip test more closely simulates the deformation experienced by a fabric embedded in soil than would the traditional narrow-strip (say 50 mm wide) tensile test commonly used in the textile industry. The test yields parameters such as peak strength and elongation and tensile modulus. The determination of representative values for these characteristics is, however, not straightforward; it will depend on the shape of the stress-strain curve and the problem in hand. Various stress-strain moduli are defined in Fig. 19.2. The wide-width test is also suitable for the testing of seams.

Research has indicated that fabric confinement within soil and the resultant interlocking of soil particles with the fabric structure have a significant effect on the stress-strain properties. Even the wide-strip tensile test is therefore essentially only an index test. It is generally conceded that the modulus of a fabric confined in soil is likely to be higher than when tested in isolation. The deformation of a geotextile

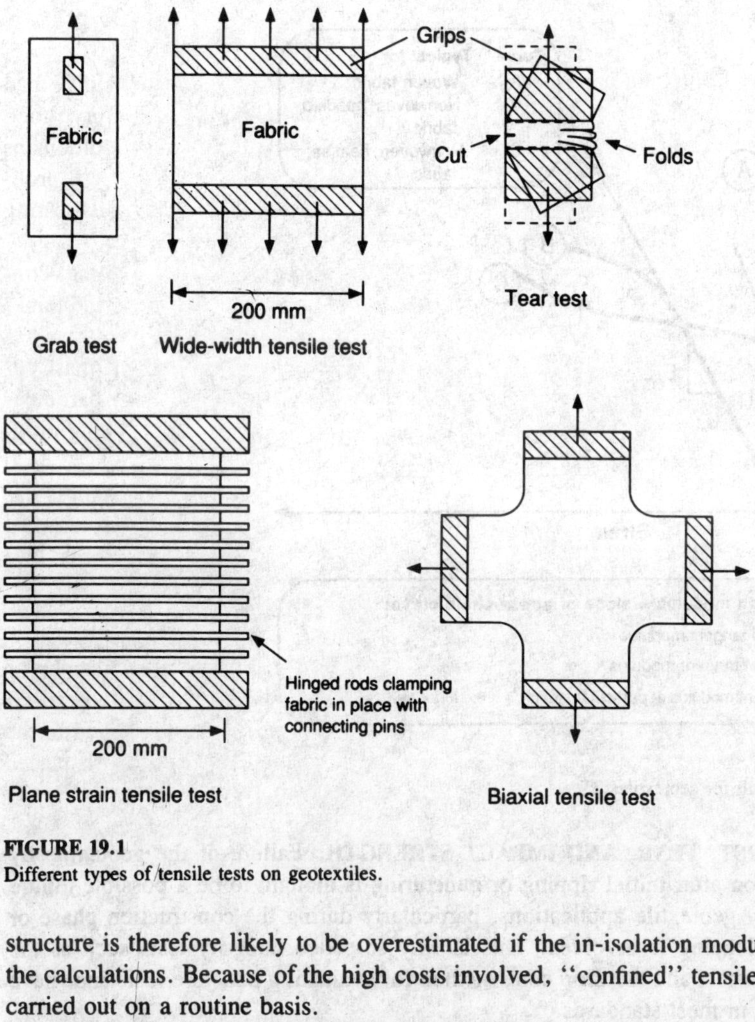

FIGURE 19.1
Different types of tensile tests on geotextiles.

structure is therefore likely to be overestimated if the in-isolation modulus is used in the calculations. Because of the high costs involved, "confined" tensile testing is not carried out on a routine basis.

The evaluation of the biaxial strength of fabrics can be a very complex task. A substitute is the indirect biaxial test using the CBR plunger method (compare with schematic sketches in Fig. 19.3). It is carried out using a modified CBR mold to hold the fabric specimen and the CBR compression machine to provide the downward force onto the fabric. This test has the advantage that the equipment required is available in many soil laboratories.

If subjected to constant stress, certain fabrics may exhibit substantial creep, which is another aspect not covered by the basic wide-strip test. Nevertheless, this may be an important factor where geotextiles are acting as reinforcement in permanent structures. Currently there is little information available on the creep of specific geotextiles although the creep behavior of the basic polymers involved has been studied extensively. With respect to creep, the engineer can usually obtain better guidelines for the design with geogrids than with geotextiles (see Sec. 18.3.1).

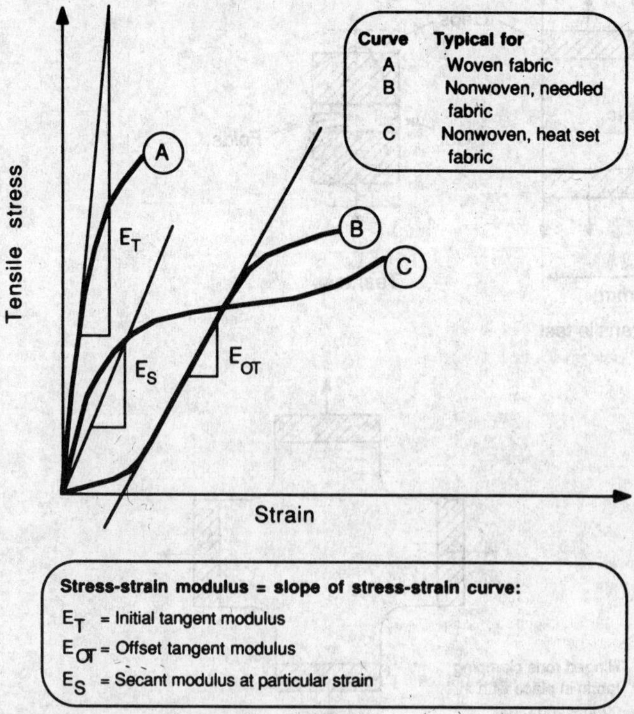

FIGURE 19.2
Stress-strain moduli for geotextiles.

19.1.2.2 BURST, TEAR, AND IMPACT STRENGTH. Failure of the geotextile by tear propagation after initial ripping or puncturing is thought to be a possible failure mode in many geotextile applications, particularly during the construction phase or where cyclic loading is involved. This is why so-called integrity tests such as the trapezoidal tear test and the drop cone method for evaluating puncture resistance have been included in most standards.

The tear test procedure (see Fig. 19.1) is considered to be a very simple tear initiation test and is thought to be particularly relevant in situations where coarse aggregate or riprap is dropped or pushed against the fabric. Other tear tests used in the textile industry include the wing tear test (British standard), the tongue tear test, and the Elmendorf tear tests (the latter two are both ASTM procedures).

Another test which has been adopted from the textile industry by engineers is the Mullen burst test. Stress conditions for a burst test are schematically shown in Fig. 19.3.

Puncture resistance can be evaluated by dropping a sharp cone onto the fabric clamped into a CBR mold as done in the plunger test. The diameter of the resulting hole in the fabric is a measure of its puncture resistance. Alternatively, a pointed or rounded steel point attached to a falling pendulum could be used to measure puncture (or impact) resistance in terms of the energy loss recorded.

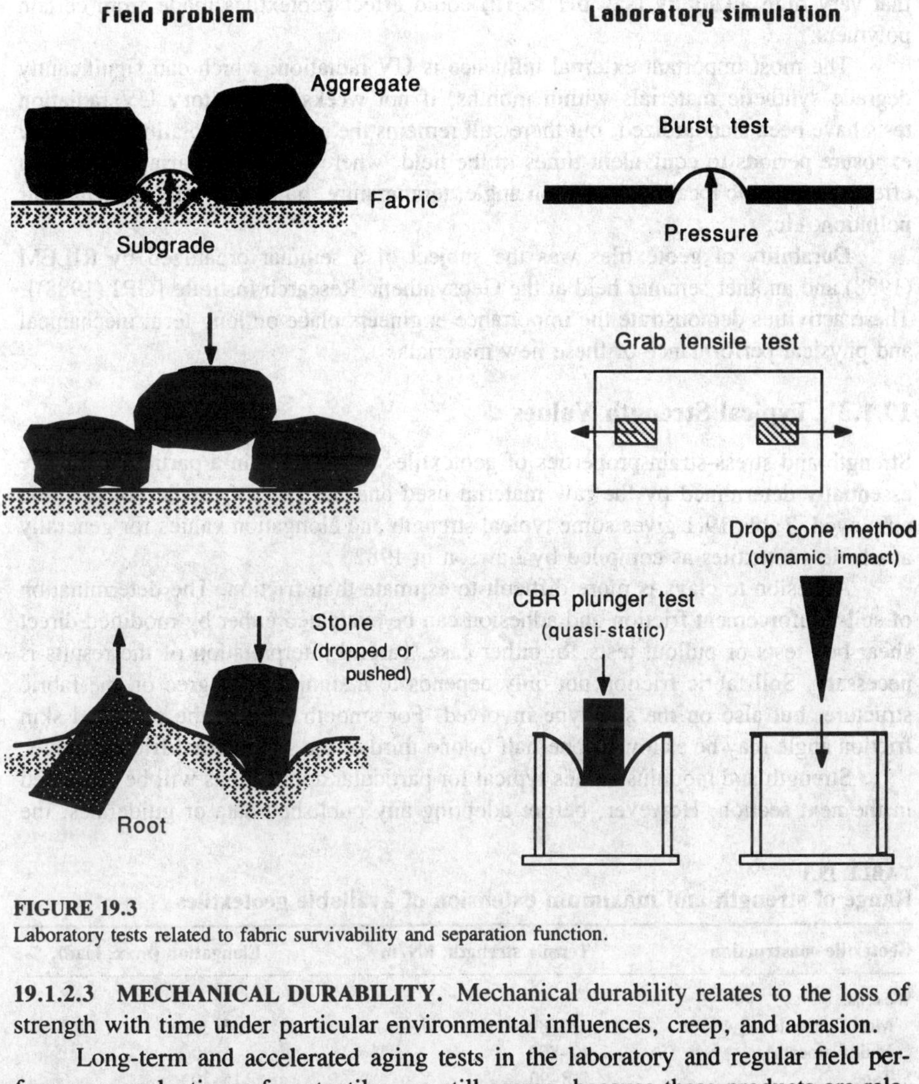

FIGURE 19.3
Laboratory tests related to fabric survivability and separation function.

19.1.2.3 MECHANICAL DURABILITY. Mechanical durability relates to the loss of strength with time under particular environmental influences, creep, and abrasion.

Long-term and accelerated aging tests in the laboratory and regular field performance evaluations of geotextiles are still scarce, because these products are relatively new. Prediction of long-term engineering properties therefore is still uncertain.

Creep was discussed in relation to geogrids in Sec. 18.3.1. The same principles apply to other geosynthetics although the relevant information may not be readily available for all products.

There seems to be little concern for the environmental stability of geosynthetics buried in natural soils, although this is not necessarily justified by rigorous testing and analysis. It is also not really prudent to make generalized statements on durability, considering the multitude of external factors and materials involved. What is certain is that the spillage of chemicals such as petrol, benzine, trichlorethylene, or carbon tetrachloride could spell disaster for geosynthetics. In addition there is some evidence that very high alkalinity (say pH > 12) could affect geotextiles made from certain polymers.

The most important external influence is UV radiation, which can significantly degrade synthetic materials within months, if not weeks. Laboratory UV radiation tests have been standardized, but there still remains the problem of relating laboratory exposure periods to equivalent times in the field, where additional variables have an effect: geographic location, radiation angle, temperature, humidity, rainfall, wind, air pollution, etc.

Durability of geotextiles was the subject of a seminar organized by RILEM (1988) and another seminar held at the Geosynthetic Research Institute [GRI (1988)]. These activities demonstrate the importance engineers place on long-term mechanical and physical performance of these new materials.

19.1.3 Typical Strength Values

Strength and stress-strain properties of geotextiles as obtained in a particular test are essentially determined by the raw material used and the type of textile construction employed. Table 19.1 gives some typical strength and elongation values for generally available geotextiles as compiled by Lawson in 1982.

Adhesion to clays is more difficult to estimate than friction. The determination of soil-reinforcement friction and adhesion can be evaluated either by modified direct shear box tests or pullout tests. In either case, careful interpretation of the results is necessary. Soil-fabric friction not only depends to a significant degree on the fabric structure, but also on the soil type involved. For smooth fabrics, the so-called skin friction angle may be as low as one-half or one-third of the soil internal friction angle.

Strength and modulus values typical for particular applications will be presented in the next section. However, before adopting any published data or guidelines, the

TABLE 19.1
Range of strength and maximum extension of available geotextiles

Geotextile construction	Tensile strength, kN/m	Elongation (max. load), %
Wovens:		
Monofilaments	20–80	5–35
Multifilaments*	40–800	5–30
Tape	8–90	15–20
Nonwovens:		
Melt-bonded	3–25	20–60
Needle-punched	7–90	50–80
Resin-bonded	4–30	30–50
Knitteds:		
Weft	2–5	300–600
Warp	20–120	12–15
Stitch-bonded	15–800	15–30

Source: Extracted from values presented by Lawson (1982).

*Fibrillated tapes, having a fibrous texture, are included in this category.

designer should remember the following important points when using geotextiles for soil reinforcement:

Fabric strength is likely to be anisotropic, because of the fabrics structure and method of manufacturing.

Seam strength may well be the critical design value, rather than the strength of the fabric itself. Similarly, a weak zone may exist at overlaps or in an anchorage zone.

As in other geosynthetics, creep (or corresponding stress relaxation) should be considered in all but very short term reinforcement applications of fabrics.

Forces exerted on the fabric during construction may cause stresses and strains exceeding those after installation.

Environmental influences may drastically reduce strength with time.

19.2 MAJOR GEOTEXTILE REINFORCEMENT APPLICATIONS

As is the case for other methods of ground modification for the purpose of reinforcing a soil mass, the objective of fabric inclusions generally is to increase the soil's stability (bearing capacity, slope stability, resistance to erosion) and reduce its deformation (settlement, lateral deformation). In order to provide stability, the geosynthetic has to have adequate strength; to control deformation, it has to have suitable force-elongation characteristics, measured in terms of a modulus (the slope of the force-elongation curve), or elongation under a particular load. In the following paragraphs, geotextile applications with the primary purpose of reinforcement are enumerated and brief guidelines are given for selected design problems.

In many applications, geotextiles, like geogrids and geocomposites, serve more than one purpose, possibly including reinforcement, but in all cases a certain minimum strength is required, even if just in order for the fabric to survive the effects of placement on an irregular ground surface and the loads imposed by equipment and personnel during installation. The terms

- Survivability and workability

have found acceptance for describing basic strength and stiffness characteristics required from geotextiles in a construction environment. In the United States these minimum requirements are usually expressed directly in terms of the grab tensile strength (ASTM D-1682) and related index test properties. In Australia, the "robustness" classification has been introduced by the Main Roads Department in Queensland; it rates a geotextile in terms of puncture strength and CBR plunger test results. Both these approaches are explained further in Sec. 19.3.

Survivability criteria have become synonymous with strength requirements for the separation function in most applications. A possible exception is the use of fabrics below railway ballast, where long-term abrasion under vibratory loads may place more severe conditions on the geotextile than construction stresses.

Koerner and Hausmann (1987) identified the following major applications where geotextile reinforcement is the primary function:

- Retaining walls
- Slope stabilization
- Embankments over very soft soils
- Foundations (improved bearing capacity)
- Unpaved road support

Additionally, fabrics may be used in conjunction with other materials, such as polymeric grids and mats or even steel wires or meshes, forming "geocomposites." These types of construction elements are distinguished by extremely high strength and modulus values and have found application in the following tasks:

- Mattress and load support
- Direct road support systems (surface mats)

Typical ranges of strength and modulus of geosynthetics used in the various categories of applications are given in Table 19.2. These values are not meant to be definite design guidelines but merely reflect the still-evolving practice of geotextile applications for reinforcement.

In reinforcement applications with predominantly single layers of geosynthetics (embankments, direct road support, mattresses, foundations, and unpaved roads) it appears more likely that common rational design guidelines will eventually develop, although the soil-reinforcement action may differ in detail. When looking at earth structures where multilayer reinforcement is standard (retaining walls, slope stabilization), comparison based on wide-width tensile test results alone becomes somewhat questionable, because the subcategories chosen in Table 19.2 do not reflect the vertical spacing options one might have with various-thickness—and thus various-strength and -modulus—fabrics and grids.

Geotextile applications (but not the use of composites) will be further discussed below.

19.2.1 Retaining Walls

All kinds of fabrics have been successfully employed to build retaining walls—low-modulus nonwovens and high-modulus, high-strength wovens. Typical geometric arrangements are shown in Fig. 19.4.

Because of the difficulty of measuring stresses and strains in geotextiles, little internal performance data is available. Stability analysis of fabric-reinforced walls proceeds in a similar way to that of Reinforced Earth, mainly because a simple method taking into account the higher extensibility of fabrics (compared with steel!) is not

TABLE 19.2
Typical geotextile wide-width strength and modulus in reinforcement applications

Application area		Fabric strength, kN/m	Fabric modulus, kN/m
No.	**Description**		
1	Retaining Structures		
	Low height	13.1–17.5	35.0–52.4
	Moderate height	17.5–21.9	43.7–87.4
	High height	21.9–26.2	61.2–175
2	Slope Stabilization		
	Close spacing	13.1–21.9	26.2–61.2
	Moderate spacing	17.5–26.2	35.0–70.0
	Wide spacing	26.2–52.4	43.7–175
3	Unpaved Roads		
	CBR ≤ 4	13.1–21.9	52.4–87.4
	CBR ≤ 2	17.5–26.2	87.4–175
	CBR ≤ 1	21.9–52.4	175–525
4	Foundations (increase in bearing capacity)		
	Nominal	26.2–69.9	175–350
	Moderate	43.7–87.4	350–874
	Large	69.9–175	700–1750
5	Embankments over soft soils		
	Str. > 9.6 kPa*	87.4–262	874–1750
	Str. > 4.8 kPa	175–350	1750–3500
	Str. > 2.4 kPa	262–524	3500–6120
6	Mattress or load support (composites†)		
	Moderate	350–700	874–1750
	Heavy	700–1050	1750–4370
7	Direct road support (composites†)		
	Moderate	87.4–875	874–2620
	Heavy	875–2100	2620–7000

Source: Extracted from Koerner and Hausmann (1987).

*Str = soil strength.

†Composites are fabrics combined with mats, meshes, and bars, possibly including other than polymer materials.

available. Besides ensuring external stability against global slope failure, bearing failure, and sliding of the reinforced block of soil, possible internal failure modes by rupture of the fabric ("ties breaking") or slippage between fabric and soil ("ties pullout") have to be assessed.

A Rankine analysis of stresses within the backfill or the Coulomb wedge-type analysis may be carried out. Procedures proposed vary with respect to the failure geometry (single wedge, double wedge, logarithmic failure surface, etc.), the magnitude of earth pressures acting (assuming an at-rest or active state of equilibrium), length of embedment effective against pullout, and minimum safety factors considered necessary. The effect of surcharge is evaluated in the same way as for conventional

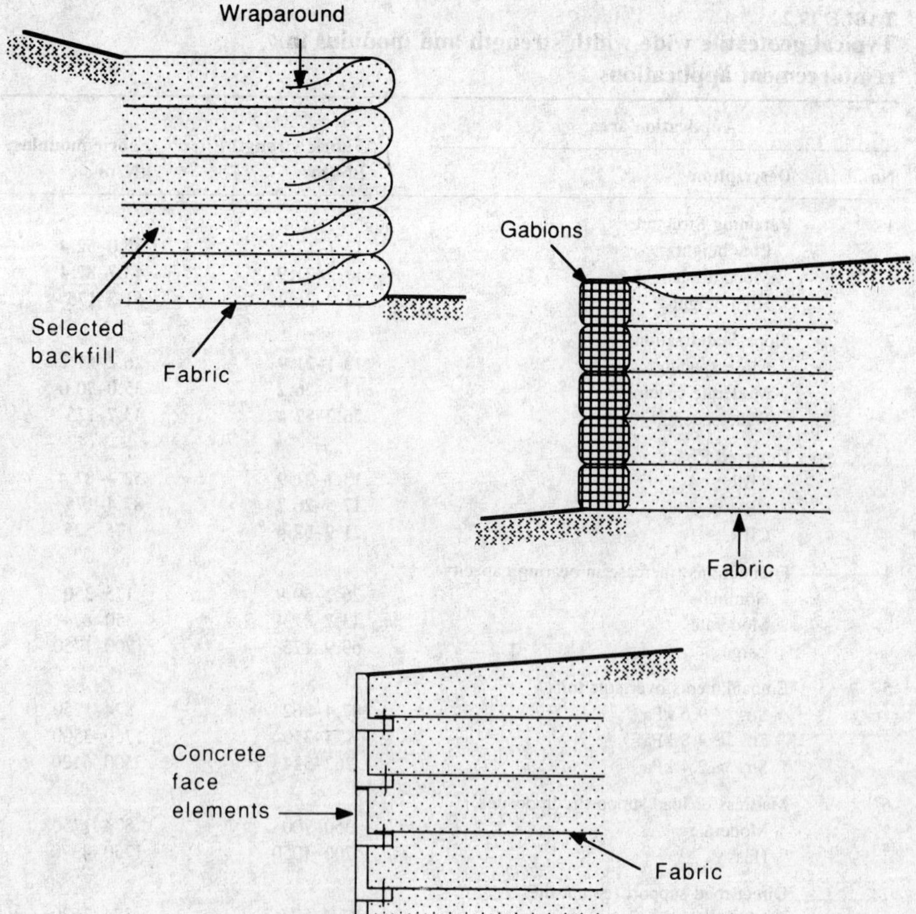

FIGURE 19.4
Alternative types of fabric walls.

retaining walls. Seams and overlaps are not usually required, but where needed, they should be analyzed, because they represent weak links if under stress.

Protection of the wall face against degradation due to UV light and, to some degree, against vandalism can be provided by covering the fabric with Gunite, asphalt emulsion, or other coatings.

Because of the lower stress-strain modulus of fabrics as compared to steel, geotextile walls likely deform more than Reinforced Earth walls, and an analysis of short-term and long-term strains (as discussed for geogrids) is deemed necessary. The actual parameters to be used in the analysis are, however, still subject to controversy, due to the nonlinear stress-strain behavior of fabrics and the effect of soil-fabric interaction.

Predicting deformations of reinforced soil walls requires knowledge of the tensile stress distribution in the fabric. Given the appropriate modulus of the fabric (this value may vary with stress level), the movement of the wall face can be calculated. Figure 19.5 illustrates various statically possible distributions of tension and shear stresses in the fabric.

Field reports indicate that the deformation of geotextile walls is generally much less than expected. Because of the difficulties in measuring deformations of fabrics

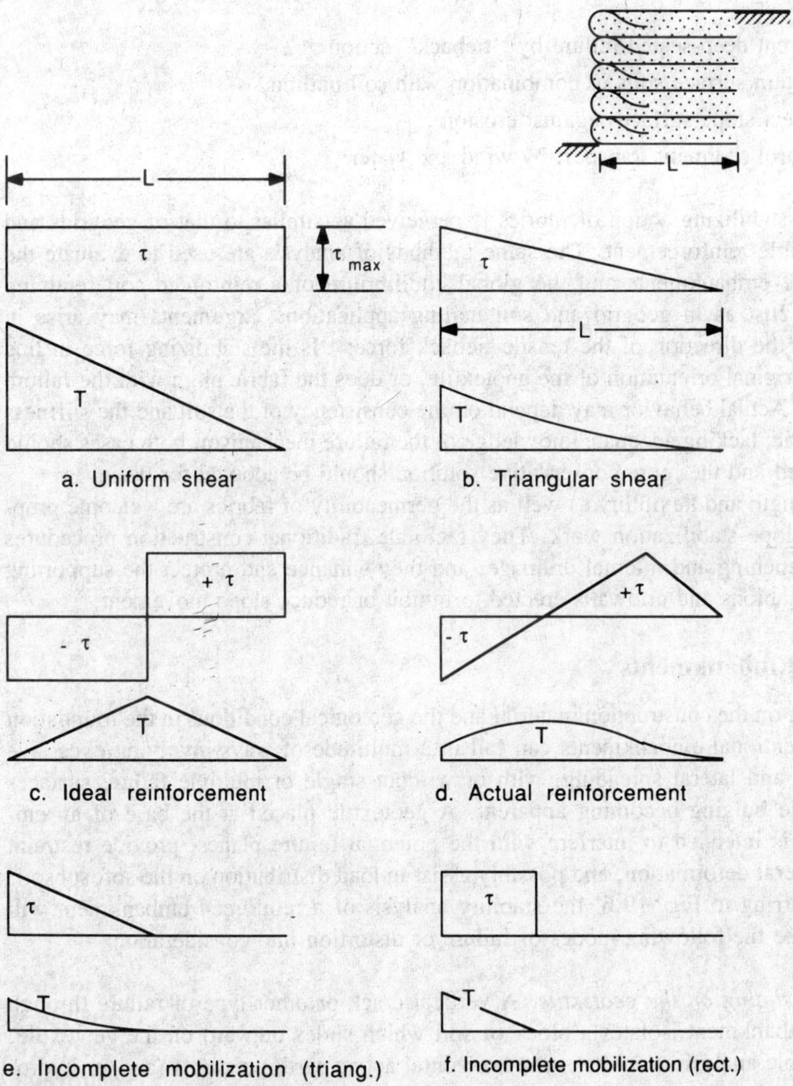

FIGURE 19.5
Statically possible distributions of tension and shear stresses in fabric wall reinforcement.

in situ, not much information is currently available on actual fabric stresses and strains in retaining walls and other geotextile soil structures.

19.2.2 Slope Stabilization

The economic significance of damage due to slope mass movement is second only to the costs of controlling the effect of swelling and shrinking soils when it comes to ranking the significance of geotechnical hazards. Geotextiles offer a welcome additional technology for low-cost slope stabilization. They may be used to

Prevent deep-seated failure by ''tieback'' action
Contain surface soils in combination with soil nailing
Protect slope surfaces against erosion
Control sediment transport by wind and water

The stabilizing action of fabrics is perceived as similar to that of geogrids and other flexible reinforcement. The same methods of analysis are used to evaluate the stability of embankments and the global equilibrium of a reinforced soil retaining structure. Just as in geogrid and soil nailing applications, arguments may arise in respect to the direction of the tensile tieback force: ''Is the stabilizing force in line with the original orientation of the geotextile, or does the fabric align with the failure surface?'' Actual behavior may depend on the consistency of the soil and the stiffness of the fabric; lacking adequate knowledge of the failure mechanism, both cases should be analyzed and the more-conservative solution should be adopted for design.

Strength and flexibility as well as the permeability of fabrics are welcome properties in slope stabilization work. They facilitate traditional construction procedures such as benching and internal drainage, and they enhance and protect the supporting effect of gabions and cribwalls erected to inhibit or reduce slope movement.

19.2.3 Embankments

Depending on the construction material and the geological conditions in the foundation soil, conventional embankments can fail in a multitude of ways involving excessive settlement and lateral spreading, with or without single or multiple failure surfaces and surface bulging becoming apparent. A geotextile placed at the base of an embankment is intended to interfere with the potential failure planes, provide restraint against lateral deformation, and possibly assist in load distribution on the soft subsoil.

Referring to Fig. 19.6, the stability analysis of a reinforced embankment will have to take the following modes of failure or distortion into consideration:

(*a*) *Block sliding on the geotextile:* A vertical crack or other type of failure through the embankment isolates a block of soil which slides outward on the geotextile. A simple analysis would assume horizontal active earth pressures (and/or hydrostatic pressures in the case of water-filled cracks) pushing outward and soil-fabric friction resisting this process. Uniform distribution of shear stresses would imply

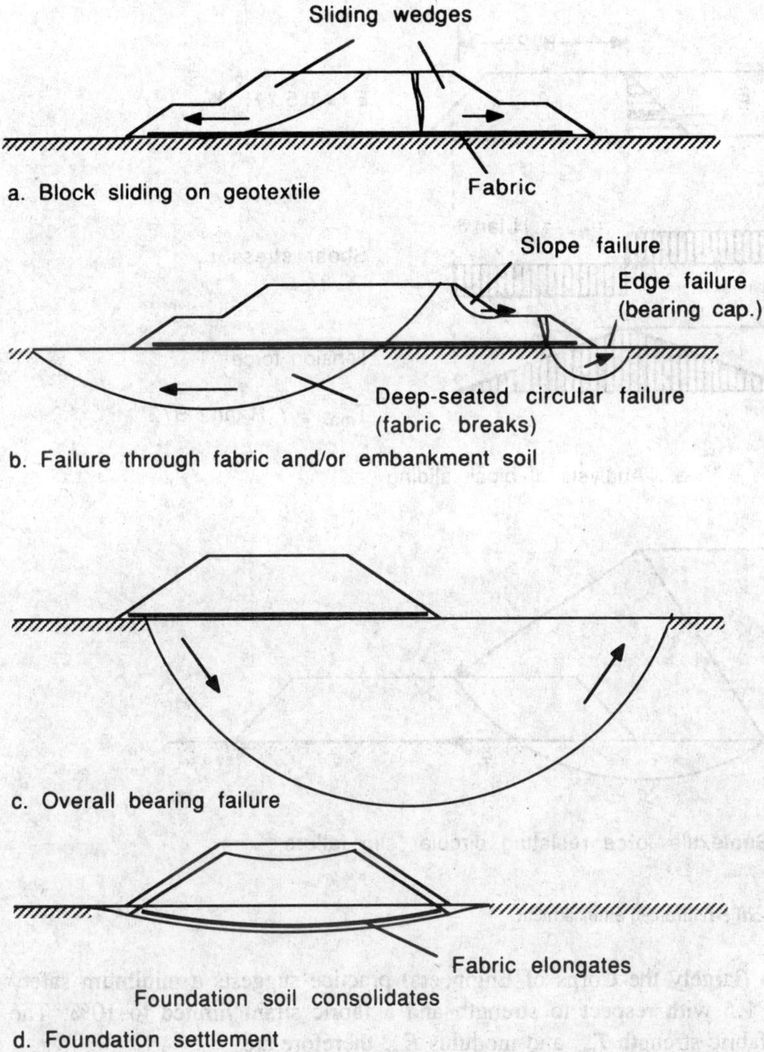

FIGURE 19.6
Embankment failure modes.

tensile forces in the geotextile which increase linearly toward the center of the embankment. For conditions as sketched in Fig. 19.7*a* we can calculate the resultant earth pressure E_a at the center of the embankment and the corresponding maximum tensile force $T_{\max}$ in the geotextile:

$$E_a = 0.5\gamma H^2 K_a \tag{19.1}$$

$$T_{\max} = \frac{\tau_r B}{2} = \frac{(\gamma H \tan \delta) B}{2} \tag{19.2}$$

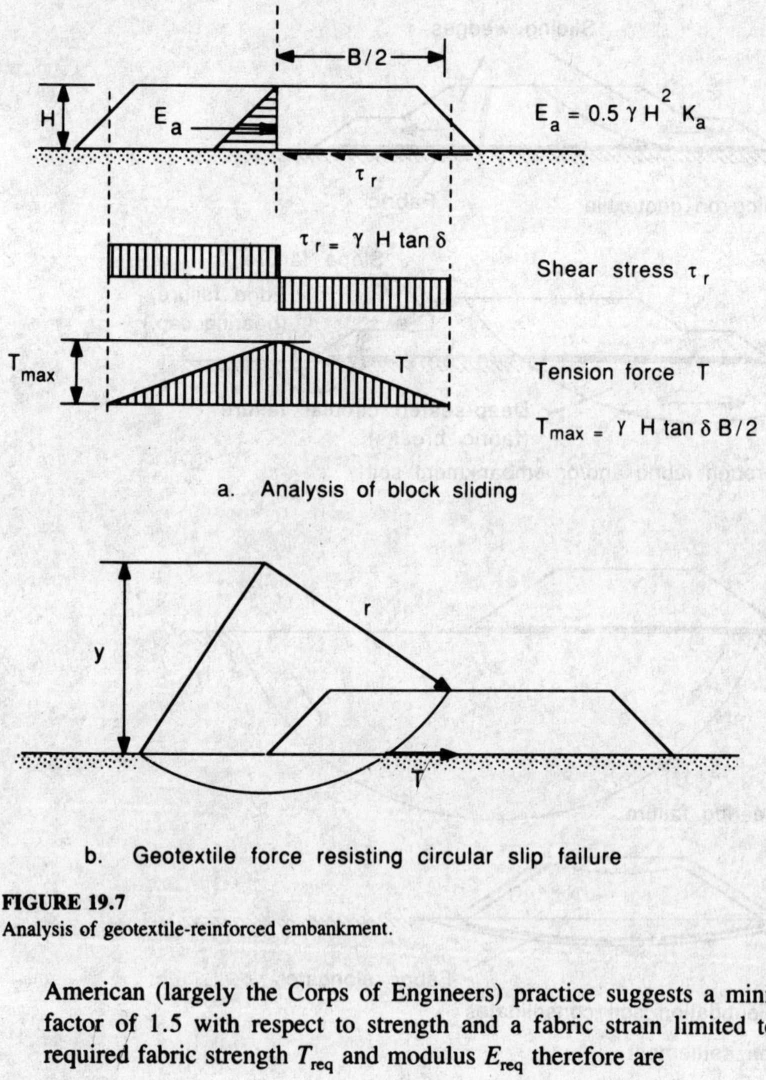

FIGURE 19.7
Analysis of geotextile-reinforced embankment.

American (largely the Corps of Engineers) practice suggests a minimum safety factor of 1.5 with respect to strength and a fabric strain limited to 10%. The required fabric strength T_{req} and modulus E_{req} therefore are

$$T_{req} = 1.5T_{max} \tag{19.3}$$

$$E_{req} = \frac{T_{max}}{\varepsilon_{max}} = 10T_{max} \tag{19.4}$$

(*b*) *Failure along a slip surface:* Typically assumed to be circular, either within the embankment or through the geotextile. The analysis proceeds along the usual steps with the fabric providing an additional stabilizing force T at the point of intersection with the failure surface (Fig. 19.7*b*). Opinions are again divided with respect to the calculation of the stabilizing moment due to the fabric:

$$\Delta M = Tr \quad \text{or} \quad \Delta M = Ty \tag{19.5}$$

For circular failure surfaces and horizontal fabric layers, it is conservative to assume $\Delta M = Ty$ and to neglect any other possible effects on soil stresses.

(*c*) *Overall bearing failure:* Treated in the conventional manner according to Terzaghi's or Hansen's bearing-capacity formula or similar formulas. The bearing capacity of an embankment foundation is essentially unaffected by a geotextile contained within or just below the embankment. Overall bearing capacity could only be improved if a mattresslike reinforced surface layer of larger extent than the base of the embankment would be provided.

Humphrey and Holtz (1986) reviewed the performance of 37 reinforced embankments, 11 of which failed by excessive deformation or fabric rupture. They observed that reinforced embankments could be built up to 2 m greater than conventional bearing-capacity theory would indicate. They attributed this fact to limited thickness or an increase in strength with depth of the soft foundation soils, the effect of a dried surface crust, or dissipation of pore pressures during construction.

(*d*) *Excessive settlement:* Because of consolidation of the subsoil, the embankment might undergo large settlement resulting in excessive elongation of the fabric, possibly resulting in eventual failure. A similar effect would arise from lateral squeezing of the foundation soil. The presence of the fabric may reduce differential settlement somewhat, but little reduction of the magnitude of total final settlement can be expected.

Because of the drainage effect of nonwoven fabrics, consolidation of the subsoil may be accelerated with a consequent gain in strength. Where deformation criteria require high-strength, high-modulus woven fabrics, the drainage function may be provided by geocomposites.

In order to provide edge anchorage, the geotextile layer may be folded back similar to the "wraparound" in fabric walls. Alternatively the fabric may be anchored in trenches or weighted down by berms.

Additional geotextile reinforcement may also be placed within the embankment itself, rather than just below it. Narrower horizontal fabric strips along the side slopes (with wraparound) may enhance compaction at the edges, so overbuilding and cutting back is no longer necessary. Edge reinforcement also helps to reduce erosion and may assist in the establishment of vegetation.

Long-term creep of fabrics can be an issue in all reinforcement applications. A limited strain criteria related not only to embankment performance but also to specific geotextile properties makes sense. Fortunately, foundation stability generally increases with time as consolidation of the soil takes place, thereby allowing some stress relaxation to occur.

19.2.4 Foundations

Geotextile-reinforced pads for structural footings are intended to provide improved bearing capacity and reduced settlements by distributing the imposed loads over a wider area of weak subsoil. Little information is available on this application, except that relatively high-strength, high-modulus fabrics are preferred. The general approach is to provide an aggregate layer including geotextiles at one or more levels; in addition, the aggregate may be protected at its base (or all around) by a separation, drainage, and filter fabric.

For heavy loads, composite reinforced structural layers may have to be designed, incorporating elements such as steel meshes or polymeric grids.

A different concept is employed where impermeable geomembranes are placed so as to isolate a bearing stratum of clay subject to swelling and shrinking during wet and dry seasons. This technique is referred to as Membrane Encapsulated Soil Layer (MESL) and has so far mainly been used in road construction [Lawson and Ingles (1982)].

Yet another approach is to use geotextiles for the confinement of soils, forming foundation beams, and piers.

19.2.5 Unpaved Roads

Unpaved roads present yet another application where the reinforcing action of geotextiles can be used to economic advantage. As in other designs, of course, geotextiles in unpaved roads often fulfill additional basic functions such as separation, filtration, and drainage.

Geotextiles are now a well-accepted construction material in the establishment of trafficable surfaces on soft clayey, silty, and organic soils. They may serve a temporary role in the development of a sealed-pavement structure, or they may be used for longer-term performance as an integral part of unpaved roads in country areas or as haul roads for civil and mining projects.

Both, nonwoven an woven fabrics are used in road construction. In recent years, experiments have also been carried out with other forms of geosynthetics [Khay et al. (1986)], such as polymer grids (e.g., Tensar), fiber reinforcement (e.g., Texsol), and geocells (e.g., Armater).

The aim of conventional unpaved road design is to provide adequate selected cover material in order to prevent bearing failure due to wheel loads and excessive rutting under traffic. Additional problems may be faced during construction, such as general shear failure of the subsoil due to the weight of the fill, impaired mobility of construction equipment, lack of suitable aggregate, and excessive settlement.

In most situations the designer aims at finding a solution which will result in the lowest overall costs, taking into account initial construction costs, continuing maintenance costs, and possibly costs related to production losses due to road closures.

Aspects of design for unpaved roads emphasized here are concerned with a rational assessment of the bearing capacity of the soil below the aggregate layer and an estimate of the surface deformation under static and repeated loading by vehicular traffic. The assessment of the overall stability and settlement of the fill during and after construction is expected to be treated as a separate problem according to standard soil mechanics principles.

19.2.5.1 UNPAVED ROADS WITHOUT FABRICS. Hammit (1970) proposed a formula for determining the thickness of aggregate required for unsurfaced roads and airfields so as to produce a rut depth less than 3 in (75 mm). According to Hammit, the design thickness is a function of the number of coverages N of an equivalent single wheel load and the tire contact area. Based on Hammit's empirical relationship and work by Webster and Alford (1978), Giroud and Noiray (1981) suggested a similar but simpler formula, which they proceeded to modify so as to allow design for other than standard axle loads and for different rut depths. The unreinforced aggregate depth h_0' required for $20 < N < 10\,000$ is

$$h_0' = \frac{1.6193 \log N + 6.3964 \log P - 3.7892r - 11.8887}{c_u^{0.63}} \tag{19.6}$$

where h_0' = aggregate depth, unreinforced, m
P = axle load, kN
c_u = undrained cohesion of the subsoil = 30 CBR, kPa
r = rut depth, m

For $N < 20$, Giroud and Noiray propose to use a quasi-static analysis instead of the above approach, based on a bearing failure at $q_u = \pi c_u + \gamma h_0'$ and a load distribution as described by Fig. 19.9.

19.2.5.2 UNPAVED ROADS WITH GEOTEXTILES. The reinforcing action of fabrics on the unpaved road structure is interpreted in one or more different ways, referring to bearing capacity, aggregate and subgrade restraint, and membrane action.

First, the fabric is seen to influence the *failure mode*. Terzaghi already recognized that in very soft soils excessive deformation can occur at stress levels below that indicated by the traditional bearing-capacity formula. He then proposed that there be a differentiation between local shear failure, where plastic flow or densification of the soil causes large settlement without noticeable bulging at the surface, and general shear failure, characterized by recognizable failure planes extending from the edge of the loaded area to the ground surface (Fig. 19.8). The placement of a fabric on soft subgrade appears to have the effect of forcing a general shear failure where otherwise

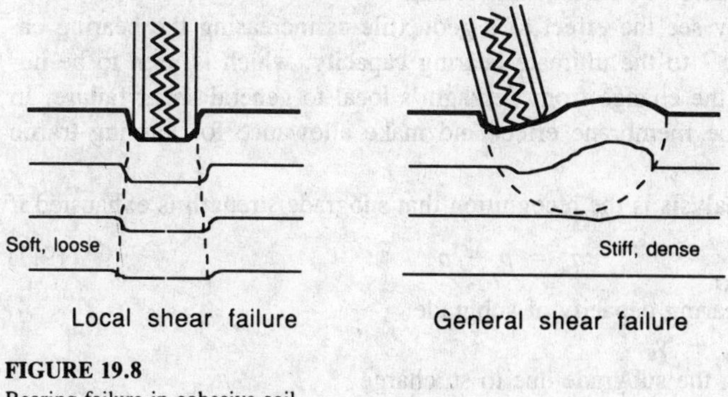

FIGURE 19.8
Bearing failure in cohesive soil.

a local or punching type of failure would occur. This has the effect of increasing the bearing-capacity factor N_c from about 3 to 5 or more. Different authors may not agree as to its physical interpretation, but the fabric-induced increase in bearing capacity seems to be well accepted.

Second, a fabric is seen to provide *restraint* of the aggregate and the subgrade, if placed at their interface. Design procedures may take this effect into account by an improved load distribution capacity, sometimes referred to as slab effect. The tangent of the angle of spreading of the surface load through the aggregate is referred to as the load distribution factor. It is of course also a function of the integrity of the aggregate itself.

Third, subsidence associated with wheel-path rutting can develop tension in a fabric built into the road structure. This is particularly the case with high-modulus fabrics with sufficient soil-fabric friction to develop an anchorage zone outside the loaded area. The upward resultant of the tensile forces in the deformed geotextile partially supports the wheel load and reduces the stress on the subgrade. This kind of reinforcement is termed *membrane support*.

Fabrics are rarely placed within the subgrade itself. The reinforcing effect of fabrics within the subgrade would be small, because of the necessary disturbance of the soil before placement and the low soil-fabric friction and adhesion which could be expected in soft soils. Most common is the placement of the fabric at the aggregate-subgrade interface, in order to take advantage of all fabric functions. But if reinforcement of the aggregate is critical, an additional fabric layer may be placed within it. This may be economical where aggregate is expensive or of doubtful quality.

19.2.5.3 DESIGN ACCORDING TO GIROUD AND NOIRAY (1981). A number of different design procedures have been developed, incorporating one or all the concepts discussed above. The method proposed by Giroud and Noiray (1981) has been selected for detailed presentation because it offers versatility with respect to load and traffic numbers and seems to give a reasonable guideline for a considerable range of fabrics, differing in strength and modulus. However, it is still only a semiempirical procedure, partly based on relationships which have only proven themselves for unreinforced rather than for reinforced unpaved road design. A more detailed review of this and other methods was presented by Hausmann (1986).

Giroud and Noiray see the effect of a geotextile as increasing the bearing capacity from the "elastic" to the ultimate bearing capacity, which is seen to be numerically equivalent to the change from Terzaghi's local to general shear failure. In addition they include the membrane effect and make allowance for varying traffic numbers.

Underlying the analysis is the recognition that subgrade strength is exhausted if

$$q_u = p - p_g \tag{19.7}$$

where q_u = ultimate bearing capacity of subgrade
$= (\pi + 2)\, c_u + \gamma_h$
p = pressure on the subgrade due to surcharge
p_g = reduction of pressure due to membrane effect

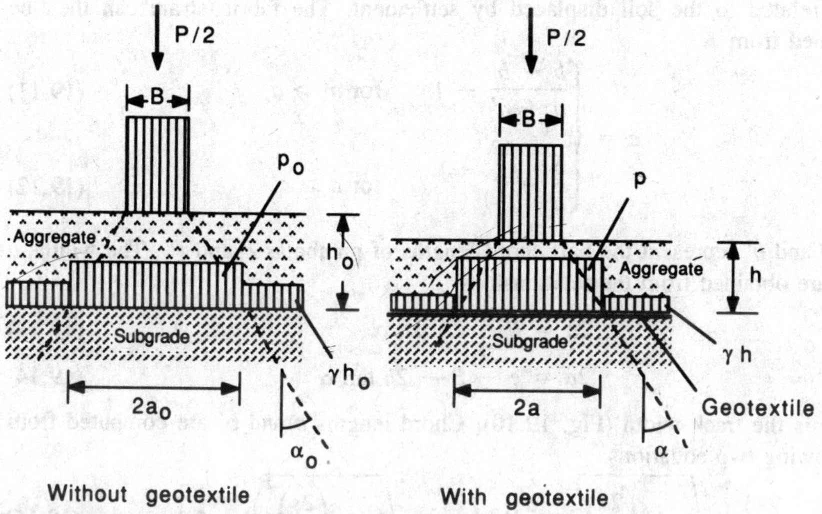

FIGURE 19.9
Load distribution according to Giroud and Noiray (1981).

Pressure p is calculated assuming the aggregate spreads the surface load at an angle α (Fig. 19.9). Including the aggregate weight we find

$$p = \frac{P}{2(B + 2h \tan \alpha)(L + 2h \tan \alpha)} + \gamma h \tag{19.8}$$

where P is the axle load and γ the soil unit weight. Dual tires are assumed with each set of tires covering an area B times L. For a tire pressure p_c, these dimensions are obtained for two different cases as follows:

On-highway trucks	Off-highway trucks	
$B = \sqrt{\dfrac{P}{p_c}}$	$B = \sqrt{\dfrac{P\sqrt{2}}{p_c}}$	(19.9)
$L = \dfrac{B}{\sqrt{2}}$	$L = \dfrac{B}{2}$	(19.10)

Giroud and Noiray consider that the geotextile may have an influence on the value α but settle for a conservative value of 31° (or $\tan \alpha = 0.6$) which corresponds to the theoretical inclination of failure planes in aggregate with a friction angle of 28°. [The effect of varying α is further discussed by Giroud et al. (1984) for extending this design methods to geogrids.]

The determination of p_g is based on the assumption that the shape of the deformed geotextile represents sections of parabolas, so the volume of the central heaved area is related to the soil displaced by settlement. The fabric strain can then be determined from

$$\varepsilon = \begin{cases} \dfrac{b + b'}{a + a'} - 1 & \text{for } a' > a & (19.11) \\[2ex] \dfrac{b}{a} - 1 & \text{for } a > a' & (19.12) \end{cases}$$

where b and b' represent the half chord lengths of parabolas P and P'. The widths a and a' are obtained from the relationships

$$2a = B + 2h \tan \alpha \tag{19.13}$$

$$2a = e - B - 2h \tan \alpha \tag{19.14}$$

where e is the track width (Fig. 19.10). Chord lengths b and b' are computed from the following two equations:

$$\frac{b}{a} - 1 = \frac{1}{2}\left[\sqrt{1 + \left(\frac{2s}{a}\right)^2} + \frac{a}{2s} \ln\left(\frac{2s}{a} + \sqrt{1 + \left(\frac{2s}{a}\right)^2}\right) - 2\right] \tag{19.15}$$

$$\frac{b'}{a'} - 1 = \frac{1}{2}\left[\sqrt{1 + \left(\frac{2(r-s)}{a'}\right)^2} + \frac{a'}{2(r-s)} \ln\left(\frac{2(r-s)}{a'} + \sqrt{1 + \left(\frac{2(r-s)}{a'}\right)^2}\right) - 2\right] \tag{19.16}$$

Finally, the fabric tension is obtained from

$$t = E_f \epsilon \tag{19.17}$$

and the membrane support is

$$p_g = \frac{E_f \epsilon}{a\sqrt{1 + (a/2s)^2}} \tag{19.18}$$

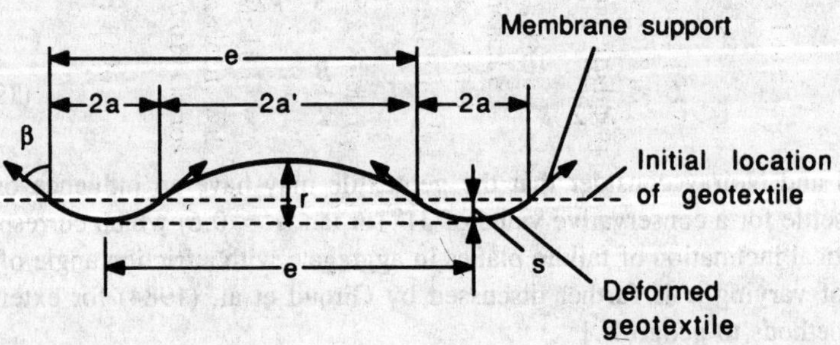

FIGURE 19.10
Shape of the deformed geotextile according to Giroud and Noiray (1981).

Solving Eq. (19.7) for the cohesion c_u, a relationship can be established between c_u and the required aggregate height, for a given geometry, load configuration, and fabric modulus. This analysis is referred to as quasi-static, because it does not yet take into account traffic numbers.

A design formula for unpaved roads without geotextiles, taking traffic into account, was presented earlier in this section. For roads incorporating a geotextile, Giroud and Noiray propose to use the same increase in aggregate thickness due to traffic as can be determined for unreinforced roads.

Let us say the aggregate thickness required for roads without fabric is designated h_0 without considering traffic, and h_0' when considering traffic. The increase in aggregate thickness due to traffic therefore is

$$\Delta h = h_0' - h_0 \tag{19.19}$$

Giroud and Noiray suggest that for the case without traffic, the surcharge load is spread as shown in Fig. 19.9, but failure of the subgrade occurs at

$$q_u = \pi c_u + \gamma h_0 \tag{19.20}$$

The pressure at the aggregate-subgrade interface can be calculated according to Eq. (19.8) but with h_0 instead of h. Now we can set

$$p = q_u \tag{19.21}$$

and solve for h_0.

The thickness h_0' can be determined from Eq. (19.6), making it possible to calculate Δh, which, for design with geotextiles, is added to the aggregate thickness determined using quasi-static analysis.

Figure 19.11 shows a sample design chart for a specific set of parameters.

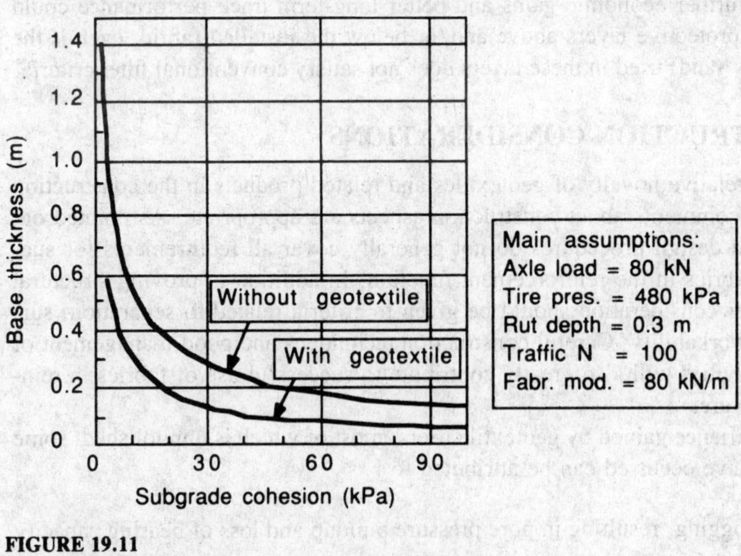

FIGURE 19.11

Design graph based on the Giroud and Noiray method (1981).

19.2.6 Paved Roads

Geotextiles can fulfill separation, filtration, and drainage functions in a paved as well as in an unpaved road structure. The reinforcement function according to the tensioned membrane concept is only applicable during the construction stage but not for the final pavement, because the allowable deformations are insufficient to stress the fabric. It is conceivable that pretensioning or prerutting could mobilize some of the geotextile strength, but probably not in the long term.

Geotextiles help to reduce reflection cracking if placed immediately below the surface course. If treated with bitumen, they could also prevent excessive moisture reaching and softening the subbase and subgrade. The use of fabrics as a capillary break for the prevention of frost damage has also been proposed.

19.2.7 Railroads

Geotextiles installed in the track bed are submitted to extreme conditions of cyclic stress and seepage flow. Geotextiles directly in contact with coarse ballast, without protective layers above and below, are subject to significant abrasion and puncturing which affects their filtration and reinforcement capacity.

Laboratory abrasion tests and field observations confirm that heavy nonwovens (> 340 g/m^2) perform better than lighter nonwovens. Insufficient experience is available on woven fabrics to recommend them for track work. Composite woven-nonwoven fabrics (> 400 g/m^2) appear to be very promising, as far as abrasion and puncturing are concerned.

Geotextiles are successful in solving difficult track foundation and construction problems. There is insufficient data at present to prove reduced maintenance requirements with the use of fabrics instead of traditional filter materials, but this is likely. It appears that further economic gains and better long-term track performance could be achieved by protective layers above and/or below the installed fabric, even if the sand (or gravelly sand) used in these layers does not satisfy conventional filter criteria.

19.3 CONSTRUCTION CONSIDERATIONS

Because of the relative novelty of geotextiles and related products in the construction industry, a few comments about construction aspects are appropriate. As pointed out earlier, so-called design procedures do not generally cover all requirements for successful use of fabrics in the reinforcement function. In addition to proving structural adequacy, proper consideration should be given to criteria related to separation, survivability, and workability. Careful construction techniques and good management of construction activities will also greatly contribute to successful use of fabrics in reinforced soil structures.

From experience gained by geotextile users, most of which is unpublished, some failures which have occurred can be attributed to

Fabrics clogging, resulting in pore pressure buildup and loss of bearing capacity

Fabrics ripping during aggregate placement

Fabrics breaking under tension due to fill weight or external loads

Fabrics tearing along seams or separating at overlaps

Furthermore, difficulties may have been encountered in placing fabrics along fill edges, in sewing or bonding adjacent sheets, and in placing fabrics in water, or degradation of strength may have occurred when fabrics were exposed to sunlight over lengthy periods.

19.3.1 Material- and Equipment-Induced Stresses

Environmental factors during on-site storage and mechanical stresses during construction and initial operation may in many projects place the most severe conditions on a fabric during its projected lifetime. Figure 19.3, which has particular relevance to road and embankment construction, illustrates three different situations where excess local stresses could lead to geotextile failure and subsequent impairment of the separation function, not to mention the possible loss of structural membrane action.

Where soft subgrade material is forced up between two stones subjected to downward pressure (Fig. 19.3, top), an analogy can be drawn with the Mullen burst test commonly carried out in textile laboratories. Similarly, where a piece of aggregate is forced downward between other rocks in contact with the fabric, the grab tensile test represents a good model (Fig. 19.3, middle). The CBR plunger test and the drop cone test resemble occurrences when geotextiles are placed on a site cleared of trees and bushes and where aggregate is dropped or pushed on top (Fig. 19.3, bottom). Based on such scenarios, Koerner (1986) has proposed numerical relationships for design which give required fabric properties based on aggregate thickness, size, and shape as well as on equipment- or traffic-generated loads. Another approach is to categorize geotextiles based on grab tensile strength, burst strength and penetration resistance, and other strength properties, as explained in the following section.

19.3.2 Survivability Criteria

Resistance to damage during construction has been termed fabric *survivability*. It is related to the type of subgrade, the characteristics of the cover material, and the construction equipment and techniques used. Attempts have been made to relate survivability to standard fabric strength measurements, such as by the interim specification of the AASHTO-AGC-ARTB Joint Committee [Christopher and Holtz (1984)], shown in Table 19.3. These criteria were essentially developed for road construction but are also helpful in other geotextile application areas.

The Main Roads Department, Queensland (1983), has established a strength-based classification system for the selection of fabrics for road construction. For simplicity, the laboratory tests chosen for this evaluation are the drop cone test and the CBR-type plunger test. The parameter used for classification is

$$G = \sqrt{H_{50}L} \qquad (19.22)$$

TABLE 19.3
Minimum fabric properties required for fabric survivability

Required degree of fabric survivability	Minimum grab strength, N	Puncture strength,* N	Burst strength,† kPa	Trap tear,‡ N
Very high	1200	490	2970	330
High	800	330	2000	220
Moderate	580	180	1450	180
Low	400	130	1000	130

Source: Christopher and Holtz, FHWA Manual (1984)

*ASTM D-751-68, Tension testing machine with ring clamp, steel ball replaced with a $\frac{5}{16}$-in (7.9-mm) diameter solid steel cylinder with flat tip centered within the ring clamp.

†ASTM D-751-68, diaphragm test method.

‡ASTM D-1117, either principal direction.

where G = geometric mean strength
H_{50} = drop height required to make a 50-mm-diameter puncture hole, mm
L = load on CBR piston at fabric failure, N

Table 19.4 gives the guidelines for ranking fabrics in terms of "robustness."

Another aspect considered by construction engineers is field *workability*, which is defined as its "ability to support workmen in an uncovered state and construction equipment during initial stages of fabric cover material placement" [FHWA (1984)]. Workability is related to the fabric stiffness, water absorption, and buoyancy, but detailed guidelines are only now being developed. In this context, fabric stiffness is evaluated in terms of the textile industry, where it has been proposed that a fabric's stiffness be measured by its capacity to form a cantilever beam without exceeding a certain amount of downward bending under its own weight.

19.3.3 Construction Guidelines

Geotextiles which are holed during placement, ripped by equipment, or have inadequate seams or overlaps will not perform satisfactorily. Equally important is proper

TABLE 19.4
Robustness classification of fabrics

Classification	G
Weak	< 600
Slightly robust	600–900
Moderately robust	900–1350
Robust	1350–2000
Very robust	2000–3000
Extremely robust	> 3000

Source: MRD, Queensland, (1983).

on-site storage with protection from sunlight. Simple procedures such as pretensioning the fabric may enhance the reinforcement function in some applications.

With particular reference to road and embankment construction, proper management of a geotextile-reinforced soil project requires the following actions:

- *Site preparation*. Level site and remove obstructions such as sharp tree stumps and boulders in accordance with the survivability ranking of the fabric. Minimize disturbance of the subgrade where soil structure, roots in the ground, and light vegetation may provide additional bearing strength.
- *Equipment selection*. Use low ground pressure and small dump trucks for the initial stage of construction. Pay attention to ground disturbance caused by turning equipment and dumping procedures.
- *Fabric placement*. Roll rather than drag geotextiles into place. Give attention to anisotropic properties of the fabrics (e.g., warp direction parallel to road alignment). Ensure adequate seams or overlaps. Eliminate wrinkles, tension fabric, and provide edge anchorage for increased membrane action. In curves, cutting and sewing or overlapping may be necessary. (For conventional, relatively low strength fabrics, overlapping by 0.3 to 1 m has been used in practice. Higher-strength fabrics, employed for their reinforcing potential, are likely to require sewing. *Remember:* Overlaps and seams are most probably the weakest link in the fabric-reinforced unpaved road structure!)
- *Aggregate placement and compaction*. Minimum cover is 200 to 300 mm, depending on aggregate size and weight of trucks. Maximum lift thickness may be imposed in order to control the size of the mud wave (bearing failure) ahead of the dumping due to excessive fill weight. (A further lift may be placed after consolidation of the subgrade has increased its strength.) Compaction of the first aggregate layer is usually achieved by the construction equipment alone. A continued buildup of cover material will allow vibratory rollers to be used. Proofrolling by a heavy rubber-tired vehicle may provide pretensioning of the fabric by creating initial ruts, which are subsequently refilled and leveled.

The art of building good-looking retaining structures on the wraparound principle will require additional construction guidelines. Considering the variety of synthetic products available today, success of a fabric-reinforced soil structure, be it an embankment, unpaved road, or retaining structure, will always depend to a degree on the ingenuity of the designers and construction supervisors and their understanding of the basic functions of a geotextile.

PROBLEMS

Prefixes indicate problem type: C = calculations, M = multiple choice, B = brief answer.

Multiple Choice

Choose the correct answer, unless stated otherwise.

M19.1. Because it is closely related to engineering performance (for a particular type of fabric), the most useful physical property of a geotextile is probably
(*a*) Thickness.
(*b*) Mass per unit area.
(*c*) Strength.
(*d*) Fabric structure.

M19.2. Fabric strength is usually reported in
(*a*) kN.
(*b*) kN/m.
(*c*) kN/m^2.
(*d*) kN/m^3.

M19.3. The CBR plunger test performed on fabrics measures the
(*a*) California bearing ratio.
(*b*) Fabric modulus.
(*c*) Biaxial strength of the fabric.
(*d*) Penetration resistance of the fabric.

M19.4. The stress-strain modulus of a nonwoven fabric confined in soil is
(*a*) Reported in terms of kN/m^3.
(*b*) Higher than measured in isolation.
(*c*) More than 2 GN/m.
(*d*) Constant up to the point of fiber breakage.

M19.5. As mentioned in Sec. 18.3, depending on the type of polymer, creep may be significant at stress levels
(*a*) Above 80% of the ultimate strength.
(*b*) Above 60% of the ultimate strength.
(*c*) Above 40% of the ultimate strength.
(*d*) As low as 20% of the ultimate strength.

M19.6. Choose which one of the following points made in relation to fabrics for reinforcing soils is *wrong*.
(*a*) Strength is likely to be anisotropic.
(*b*) Exposure to UV radiation may drastically reduce strength.
(*c*) Seams are stronger than the fabric alone, because there are two layers of fabric involved.
(*d*) Stresses and strains during construction may be higher than after installation.

M19.7. (Choose the *incorrect* statement.) Comparing the behavior of different types of fabrics, it can be said that
(*a*) The elongation at break is more for nonwoven than woven textiles.
(*b*) A resin-bonded nonwoven has a higher load-elongation modulus than a needle-punched nonwoven.
(*c*) Knitted fabrics show higher strength than woven fabrics, particularly in the weft direction.
(*d*) At the same elongation, multifilament woven-type fabrics take up more load than nonwovens.

M19.8. Fabrics commonly used in unpaved roads, slope stabilization, and retaining structures typically have a strength of

(*a*) 10 to 50 kPa.
(*b*) 10 to 50 N/m.
(*c*) 10 to 50 kN/m.
(*d*) 10 to 50 MN/m.

M19.9. On today's market, a fabric with a modulus of 2000 kN/m would be considered
(*a*) An average fabric.
(*b*) A low-strength fabric.
(*c*) A high-strength fabric.
(*d*) A miracle fabric.

M19.10. With respect to an embankment built on geotextile-covered soft soil, it can be said that because of the presence of the fabric
(*a*) The overall bearing capacity of the foundation soil is significantly increased.
(*b*) The differential settlement is likely to be reduced somewhat.
(*c*) Significant reduction in total settlement occurs.
(*d*) The rate of settlement is increased if the fabric is a thick nonwoven.

M19.11. The Giroud and Noiray method for designing unpaved roads with geotextiles, *cannot* take into account the
(*a*) Effect of the fabric on the failure mode (local versus general bearing failure).
(*b*) Improved load distribution capacity (or aggregate restraint).
(*c*) Membrane support.
(*d*) Fabric survivability and workability.

Brief Answer

B19.12. Why is the wide-width tensile test preferred to the narrow-strip test (specimen width 50 mm) for geotechnical design purposes? ____________

B19.13. The aim of conventional unpaved road design is to provide adequate selected cover material in order to Prevent b____________ f____________ due to wheel loads, and excessive r____________ under traffic.

B19.14. How may it be possible to achieve economic gain by using geotextiles in the construction of unpaved roads (where are the savings made)? ____________

B19.15. Geotextiles are also used in primary flexible pavements. Do you agree (yes) or disagree (no) with each of the following statements about geotextiles?
(*a*) They contribute to the structural strength.
(*b*) They reduce required pavement thickness.
(*c*) They reduce surface cracking.
(*d*) They prevent contamination of unbound layers.

B19.16. The term "survivability" of fabrics relates to ____________ . (Choose from resistance to UV exposure, resistance to chemicals, tendency to creep, ability to support workers, ability to float on water, resistance to stresses due to construction activities, and roughness of ground.)

B19.17. Identify the tests shown in Fig. P19.1.
(*a*) ____________
(*b*) ____________
(*c*) ____________
(*d*) ____________

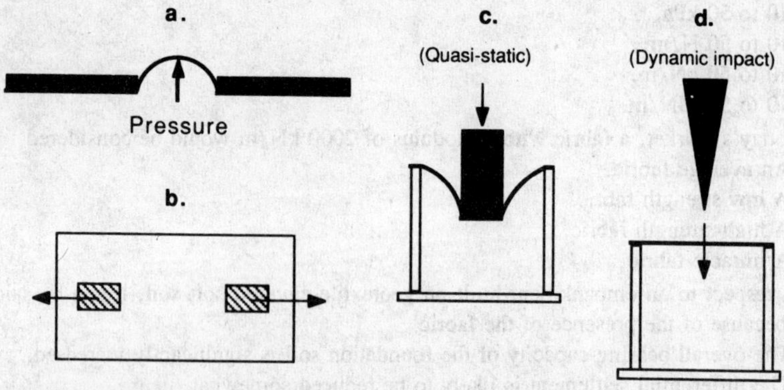

FIGURE P19.1

B19.18. Proper management of a construction project involving geotextiles requires special considerations. Give an example of these in each of the following areas:

(*a*) On-site storage of fabrics: ____________ .

(*b*) Site preparation: ____________ .

(*c*) Equipment selection: ____________ .

(*d*) Fabric placement: ____________ .

(*e*) Aggregate placement and compaction: ____________ .

Calculations

C19.19. A 3-m-high embankment, 8 m wide, is to be built on soft ground. Using simple criteria given in Sec. 19.2.3, calculate the fabric strength and modulus required in order to prevent block sliding on the fabric. Assume the embankment material has a unit weight of 18 kN/m^3 and an internal friction angle of 30° and that the fabric-soil friction is two-thirds of that value.

C19.20. Consider a slip circle as shown in Fig. P19.2, and evaluate the stabilizing effect of a fabric with a tensile strength of 30 kN/m (which is relatively low). Assume that adequate anchorage prevents failure by pullout. Proceed as follows;

(*a*) Calculate the resisting moment along the shear plane for a cohesion of 20 kPa (a soft clay).

(*b*) Determine the increase in the resisting moment due to the fabric, assuming it acts in a horizontal direction.

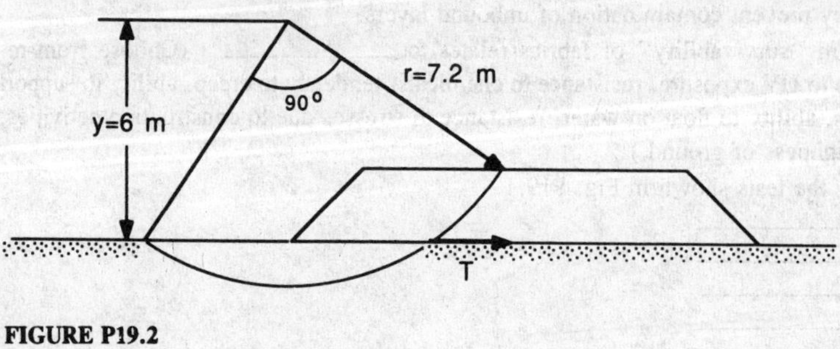

FIGURE P19.2

(*c*) Again determine the increase in the resisting moment due to the fabric, this time assuming that it aligns itself parallel to the slip circle before its full strength is mobilized.

C19.21. Consider the design graph given in Fig. 19.11 based on the Giroud and Noiray method. This graph is only valid for the particular conditions stated (axle load, tire pressure, rut depth, traffic passes, and fabric modulus). For a subgrade cohesion of 30 kPa, evaluate the savings obtained (per square meter) by reduced base thickness, if the cost of the aggregate is \$5, \$10, \$20, and \$50 per square meter.

C19.22. As mentioned in Sec. 19.3, the Queensland Main Roads Department is using the drop cone test and the CBR plunger test to give a guide to the robustness of a geotextile. Their classification in terms of the parameter G is given in Table 19.4.

Classify the following fabrics:

Fabric type	Weight, g/m^3	Drop cone H_{50}, mm	CBR test L, N
(*a*) Nonwoven	263	849	1399
(*b*) Woven	277	1510	2065
(*c*) Nonwoven	361	2046	3530
(*d*) Woven	296	3828	3315

CHAPTER

20

IN SITU GROUND REINFORCEMENT: GROUND ANCHORAGE, ROCK BOLTING, AND SOIL NAILING

This chapter deals with reinforcement of a soil mass or earth structure in situ by the insertion of tensile elements such as nails or bolts or by installing anchorage systems able to resist pullout forces in soil and rock.

The difference between soil reinforcement and conventional anchorage was previously discussed in Sec. 17.2.2. In essence, soil reinforcement involves stress transfer all along the reinforcing element, typically with a reversal of the direction of shear stresses somewhere near the halfway point (Fig. 17.3). Soil and rock anchors, however, are devices which transfer an uplift or pullout force to a good bearing stratum or zone via a bar or cable whose tension is constant along its length.

Conventional anchorage of a sheet pile wall involves transfer of the full earth pressure acting on the facing units to an anchorage element located beyond the potential failure wedge. In contrast, with reinforced soil retaining structures, the face panels may only provide very local, surface support to the soil mass.

It could be argued that ground anchors are structural systems which interact with soil and rock but do not represent a method of ground modification. This may be quite reasonable in the case of widely spaced block or beam anchors embedded in the soil. However, when it comes to multiple grouted anchors, rock bolts, soil nails, and root piles, the distinction between ground reinforcement and anchorage becomes more difficult. This is the reason why a brief introduction to ground anchors is included in the discussion of methods of in situ ground reinforcement.

This chapter is divided into three sections dealing with ground anchors, rock bolting, and soil nailing:

Ground anchors, by definition, are structural units which transmit forces into stable rock or soil by means of tendons; they replace a support which would otherwise have to be provided by gravity blocks and steel, concrete, or timber elements.

Rock bolting is intended to mobilize the inherent strength of a jointed and fragmented mass of rock by active or passive confinement and could thus be termed a method of reinforcement.

Soil nailing is designed to intersect potential failure planes and create a new structural entity with improved stability. Soil nails are distinguished from reinforcing strips by higher rigidity; this not only assists insertion into an existing soil mass but also offers resistance to bending in a shear zone.

20.1 GROUND ANCHORS

20.1.1 Typical Applications, Types, and Components

There are many civil engineering situations where lateral, uplift, or pullout forces have to be resisted or where confining pressures may have to be generated. Ground anchors may provide a solution to these problems of "foundations in tension." Their purpose may be to

Tie back sheet piles, slurry walls, and similar temporary excavation support systems

Rehabilitate existing retaining walls

Resist uplift in hydraulic structures, such as dams, weirs, and spillways

Prevent flotation of underwater structures

Resist uplift in foundations of tall, slender structures subjected to high lateral forces, such as transmission towers and cranes

Provide reaction for pile load tests

Stabilize existing and potential landslides

Prevent rock falls in road cuts, tunnels, and underground mining

Anchor suspension cables and guy wires

Tie down a pipeline and its foundations

Provide preloading of a foundation

Prevent heave due to swelling soils

Ground anchors may be classified according to a number of different construction, function, or performance criteria:

Mode of stress transfer from anchor to ground: The resistance to pullout of a block anchor may be estimated based on passive earth pressure considerations. Alternatively, for a plate anchor, bearing-capacity formulation may be more straightforward. In both cases, the anchor may be described as a *bearing-type* anchor. On the other hand, a long slender grouted anchor transfers the load to the ground via skin friction and adhesion. Analogous to friction piles, one may use the term *friction* anchors. Both modes of stress transfer apply in the case of a bell-shaped drilled pier or in the case of so-called underreamed anchors in clay.

Degree of stressing of the tendon: No initial stressing occurs with *block*-type anchors. *Prestressed* anchors are tensioned to such a degree that the working load will only cause a small variation in the tendon force. (The term "posttensioned anchor" is an alternative description for "prestressed anchor.") An installation where initially only part of the possible and permissible stressing force is applied and where considerable changes are expected to occur under working load may be termed a *tension* anchor. Untensioned anchors are said to provide passive support, while prestressed anchors give active support.

Temporary or permanent anchors: Temporary anchors may only have an expected service life measured in months, generally not more than 3 years. Permanent anchors are designed to function over the full life of the structure they support and therefore require more care in design and construction, e.g., with respect to corrosion protection and creep deformation. Rock anchors can be designed for any degree of permanency. The majority of soil anchors, however, are classified as temporary, unless only relatively small loads and little or no prestressing are involved. On principle, whether in soil or rock, permanent anchors should technically be replaceable, if necessary.

Test anchors: They are installed and tested as part of construction control or in order to provide basic design information. (For permanent anchor installations it may be a legal requirement to instrument and monitor some of the anchors in order to provide a long-term performance evaluation.)

The simplest form of soil anchors consists of single or multiple plates attached to a rod or cable, either buried, pushed, or drilled into the ground, taking on loads up to 150 or 300 kN. Alternatively, anchorage could be provided in the form of an uplift pile or a steel bar or cable inserted into a borehole which is subsequently filled with grout; the latter could be described as a fully bonded dead anchor. Some of the basic types of anchors are illustrated in Fig. 20.1.

The technically most advanced form of a soil (or rock) anchor is a grouted anchor as schematically presented in Fig. 20.2. Its main components are

Anchor head: Connects the anchor to the supported structure, permits stressing and lock-off of the prestressing steel.

Tendon: Transmits the force from the anchor head to the fixed anchor.

Fixed anchor: Transfers the anchor force into the ground.

Once the tendon is inserted into the borehole, the primary grout is injected, forming the fixed anchor. The as yet ungrouted section of the tendon is referred to as

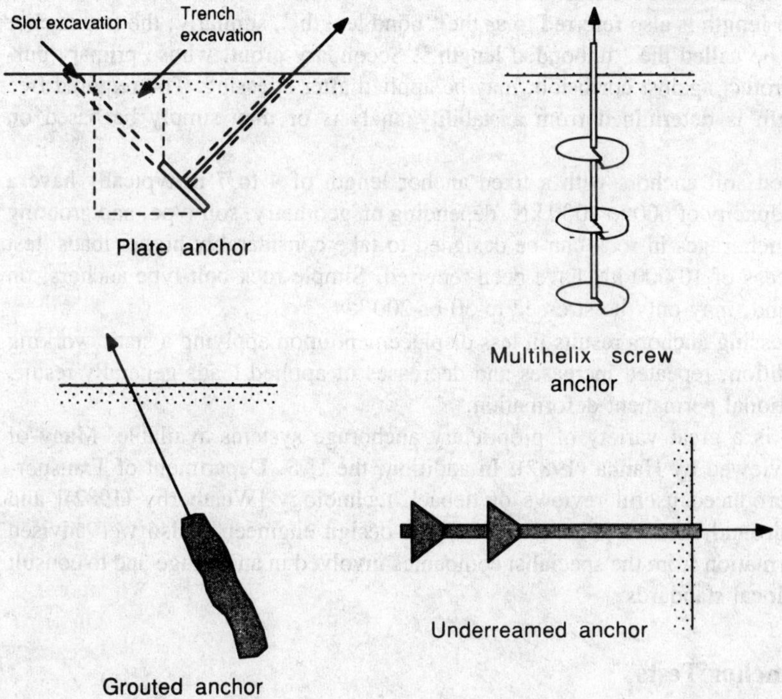

FIGURE 20.1
Different anchor types.

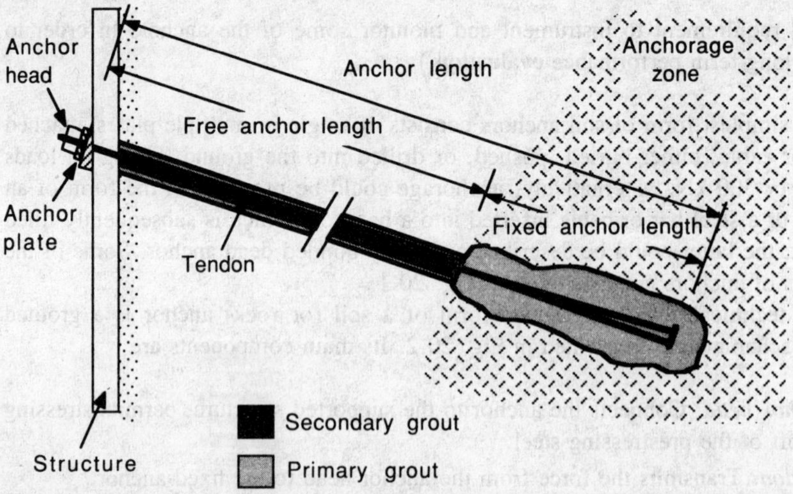

FIGURE 20.2
Components of a grouted anchor.

the free anchor length. Prestressing of the tendon is carried out in this condition. The fixed anchor length is also referred to as the "bond length"; similarly, the free anchor length may be called the "unbonded length." Secondary grout, whose primary purpose is to protect against corrosion, may be applied after stressing. The required free anchor length is determined from a stability analysis or may simply be based on experience.

Grouted soil anchors with a fixed anchor length of 4 to 7 m typically have a maximum capacity of 600 to 1000 kN, depending on geometry, soil type, and grouting pressure. Anchorages in rock can be designed to take considerably higher loads; test loads in excess of 10 000 kN have been reported. Simple rock-bolt-type anchors, on the other hand, may only be stressed to 50 or 200 kN.

Prestressing anchors results in less displacement upon applying a static working load; in addition, repeated increases and decreases of applied loads generally results in less additional permanent deformation.

There is a great variety of proprietary anchorage systems available. Many of these are reviewed by Hanna (1982). In addition, the U.S. Department of Transportation has produced useful reviews of tieback technology [Weatherby (1982)] and permanent ground anchors [Cheney (1984)]. A design engineer is also well advised to seek information from the specialist companies involved in anchorage and to consult national or local standards.

20.1.2 Anchor Tests

National standards or the specification drawn up by the design engineer will require some form of anchor testing. This can be done by checking say 10% of the production anchors at 120 or 150% of the working load. Detailed requirements would have to be

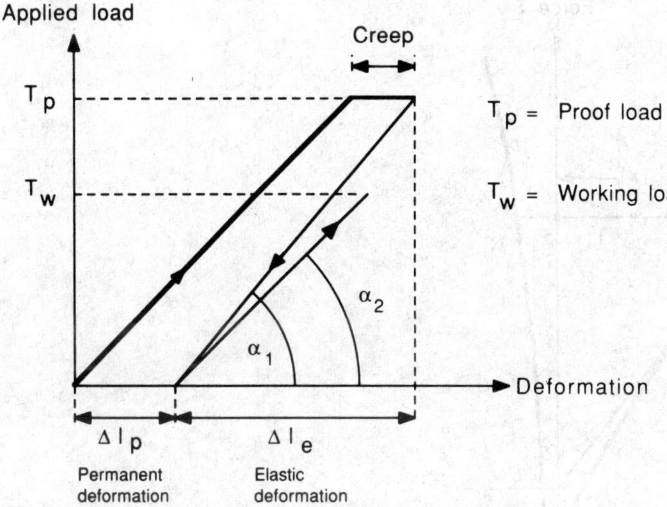

FIGURE 20.3
Schematic diagram of a simple stressing test with constant load.

tailored to the expected service life of the anchors and the risks associated with anchor failure. For permanent anchor installations, long-term monitoring of anchor loads and deformations is likely to be considered necessary. Anchors instrumented for this purpose may be called ''performance anchors.''

A simple stress test with load control (rather than the alternative deformation control) is schematically illustrated in Fig. 20.3. This would be a procedure typical of acceptance testing of production anchors. The design engineer may require a more comprehensive test, one involving stressing and destressing in stages while observing load-time or deformation-time behavior after each increment; this allows the plotting of a diagram of elastic and permanent deformation versus load (Fig. 20.4) which is needed for accurate calculation of the effective free anchor length.

Performance criteria used in the evaluation of the test results may concern one or more of the following four characteristics:

1. Creep under constant load or stress relaxation at constant deformation
2. Deformation upon reloading as compared to initial loading
3. Load transfer to the ground (can be checked by calculating the effective free anchor length)
4. Permanent deformation remaining after the anchor test

Following the approach taken by the Swiss Standard for Ground Anchors (SN 533 191, 1977), these criteria could, for a particular project, be formulated as follows:

Condition 1: Fifteen minutes after proof load application, the increase in deformation or decrease in load should not be more than 1% of the theoretical elastic

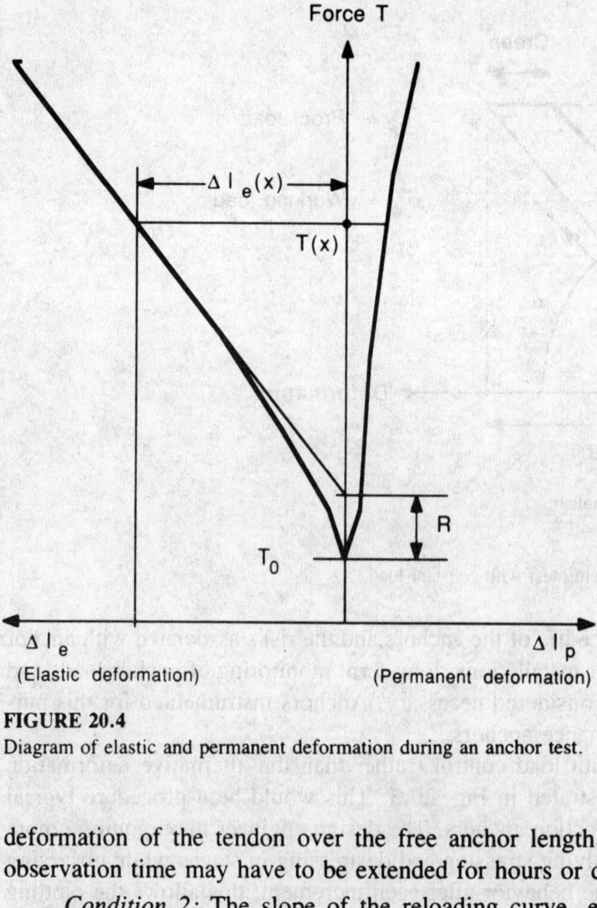

FIGURE 20.4
Diagram of elastic and permanent deformation during an anchor test.

deformation of the tendon over the free anchor length. (For anchors in clays, the observation time may have to be extended for hours or days.)

Condition 2: The slope of the reloading curve, expressed as tan α (see Fig. 20.3) should be more than 90% of the slope of the initial loading portion.

Condition 3: The effective free anchor length l_f should be more than 90% of the chosen free anchor length indicated in Fig. 20.2 but less than the chosen free anchor length plus 50% of the fixed anchor length.

The effective free anchor length is calculated according to

$$l_f = \frac{\Delta l_e(x)\, A_s E}{T(x) - T_0 - R} \tag{20.1}$$

where A_s = cross-sectional area of tendon
E = modulus of elasticity of tendon
$\Delta l_e(x)$ = elastic deformation under load $V(x)$
T_0 = initial load
R = frictional force (in the system)

Condition 4: The permanent deformation Δl_p (see Fig. 20.4) after testing should not exceed a value agreed upon by the designer and the contractor.

As can be seen from the above outline of anchor testing, construction control on anchorage projects can be very elaborate. This makes up for the fact that empirical and theoretical guidelines for estimating anchor capacity are rather crude. The value in testing individual anchors is made more worthwhile by the fact that anchors are usually spaced sufficiently far apart so as not to interact with each other and reduce the overall efficiency. This is not the case in soil reinforcement by strips, meshes, and fabrics.

20.1.3 Estimating Capacity of Friction Anchors

Anchor design is essentially empirical and, in most projects, relies on the field testing of all or selected anchors. Nevertheless, the evaluation of the feasibility of anchors and the technical and economic evaluation of a proposed anchor installation requires some estimation of anchor capacity prior to a detailed investigation or commencement of construction. In the following discussion it is assumed that careful selection of materials and supervision of construction procedures prevent failure of the tendon itself or loss of bond between the tendon and the primary grout.

20.1.3.1 EMPIRICAL GUIDELINES BASED ON SOIL AND ROCK TYPE. As a guide only, consider the following values of ultimate force transfer (in kilonewtons per meter) for grouted soil anchors:

Soil	Density/Consistency	Ultimate force transfer kN/m
Sand and gravel	Loose to dense	140 to 290
Sand	Loose to dense	100 to 190
Sand and silt	Loose to dense	70 to 130
Silty/clay, low plastic	Stiff to hard	30 to 60

Note that the above values do not account for depth, diameter, and length of the fixed anchor, not to speak of grouting pressure and other construction details. They would be most representative for a typical grouted anchor placed into a 75- to 100-mm borehole at a depth of 5 m or more, with a fixed anchor length of between 3 and 10 m, probably inclined 10 to 45° to the horizontal. Grouted anchors would be considered unsuitable for soft clays and organic soils.

The ultimate load for rock anchors is most commonly quoted in terms of bond stress τ_s, which may be a combination of friction and adhesion. Ballpark values which can be deduced from the literature are in the following order of magnitude, the lowest value relating to very weathered or fractured rock with poor mineral bonds, the highest to unweathered sound rock with good mineral bonds:

Igneous and metamorphic rocks	1.3 to 4.0 MPa
Conglomerates, breccias, sandstones, limestones, chalk	0.8 to 2.7 MPa
Argillaceous sediments, shale, clay stone, siltstone	0.4 to 1.7 MPa

Given these values, the anchor capacity T can be calculated according to

$$T = \pi D L \tau_s \tag{20.2}$$

where D is the borehole diameter and L the length of the fixed anchor part.

20.1.3.2 ANCHOR-FRICTION-RELATED GROUND STRENGTH PARAMETERS. Another approach is to estimate the tensile resistance of anchors in soil or rock on their respective strength properties, similar to evaluating the shaft resistance of piles. For example, for granular soils, one may attempt to relate the fixed-anchor–soil friction τ_s to the effective overburden pressure σ'_v by the relationship

$$\tau_s = K_f \sigma'_v \tan \delta \tag{20.3}$$

where δ is the skin friction angle (for the soil-grout interface this value is usually equal to the internal friction angle of the soil) and K_f is a pressure coefficient, ranging from 1 to 3 for medium to dense soils but could be less in loose sands and silts.

The relative density and grain size characteristics of the soil as well as the length and diameter of the fixed anchor seem to have a significant influence on the anchor capacity in cohesionless soils. Dilatancy characteristics of the ground appear to be of utmost importance: If volume increase during shear is inhibited, high ambient stresses occur which generate high frictional resistance. It has also been observed that anchor friction does not necessarily increase with overburden pressure. These findings add a little more credence to empirical estimates of frictional resistance which are independent of depth but apply only to particular anchor types and dimensions.

For clays, τ_s represents adhesion, related to the undrained cohesion c_u of the soil by

$$\tau_s = \alpha c_u \tag{20.4}$$

The reduction factor α is 1 for soft, normally consolidated clays and reduces to 0.5 or 0.3 for stiff, overconsolidated soils or weathered soft rocks. For rock, the shear resistance τ_s may be estimated from the unconfined compressive strength q_u. One of the proposed relationships is

$$\tau_s = 0.1 q_u \tag{20.5}$$

Such a relationship should, however, only be used with great caution, since the unconfined compressive strength of a rock specimen may not be representative of rock mass behavior. Neither does the above estimate consider the in situ state of stress. In practice, a maximum allowable shear stress may apply, in addition to the above criteria.

20.1.4 Uplift Capacity of Anchors

Anchors subjected to uplift forces may form a variety of failure surfaces, depending on ground type, depth, size, and other factors:

Near-surface anchors may cause ground failure over a conical or cylindrical surface (Fig. 20.5*a*, *b*, and *c*)

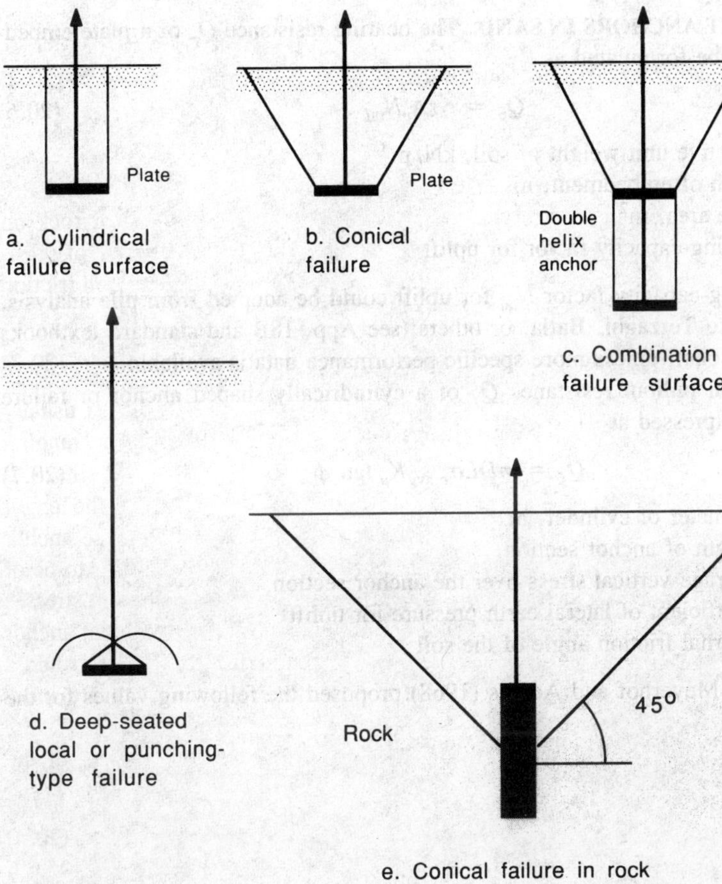

FIGURE 20.5
Typical failure surfaces in uplift.

For deep installations in soil, local- or punching-type failure is likely, similar to that associated with end bearing failure of deep piles (Fig. 20.5*d*).

For underreamed or multiple-helix anchors, a combination of cylindrical and conical failure may occur.

The analysis most valid for a particular anchor and ground type would be one based on observations during test loadings. Site investigation data may not always be directly applicable to design. Where an anchor is buried in a trench backfilled with excavated material, the relevant soil properties may be those appropriate for a remolded, recompacted state, rather than for an undisturbed state.

Load-displacement response of uplift anchors is often not critical because of the flexibility of many of the structures involved, such as transmission towers. Where a prediction of uplift displacement is needed, it is recommended that the procedure proposed by Trautmann and Kulhawy (1988) be followed.

20.1.4.1 UPLIFT ANCHORS IN SAND. The bearing resistance Q_p of a plate embedded in sand can be formulated as

$$Q_p = \gamma z A_p N_{qu} \tag{20.6}$$

where γ = effective unit weight of soil, kN/m^3
z = depth of embedment, m
A_p = plate area, m^2
N_{qu} = bearing-capacity factor for uplift

The bearing-capacity factor N_{qu} for uplift could be adopted from pile analysis, e.g., according to Terzaghi, Balla, or others (see App. 18B and standard textbooks on deep foundations), unless more specific performance data is available.

The vertical pullout resistance Q_f of a cylindrically shaped anchor or failure body could be expressed as

$$Q_f = \pi D L \sigma_{v,\text{avg}} K_u \tan \phi \tag{20.7}$$

where: D = diameter of cylinder, m
L = length of anchor section
$\sigma_{v,\text{avg}}$ = average vertical stress over the anchor section
K_u = coefficient of lateral earth pressure for uplift
ϕ = internal friction angle of the soil

For sands, Meyerhof and Adams (1968) proposed the following values for the coefficient K_u:

ϕ	K_u
25	1.2
30	1.5
35	2.5
40	3.9
45	5.3

For helical anchors, Mitsch and Clemence (1985) suggested a reduction of the above K_u values by about 40%.

20.1.4.2 UPLIFT ANCHORS IN CLAY. For clays, equations equivalent to Eqs. (20.6) and (20.7) are

$$Q_p = A_p c N_{cu} \tag{20.8}$$

and

$$Q_f = \pi D c L \tag{20.9}$$

where c is the cohesion (usually undrained), kN/m^2, and N_{cu} is the bearing-capacity factor. Provided that the embedment is more than 3 times the diameter of the anchor, $N_{cu} = 9$.

20.1.4.3 UPLIFT ANCHORS IN ROCK. The inverted cone model has also been used for failure of rock in uplift (Fig. 20.5*e*). The most conservative approach is simply to only consider the weight of the rock within the cone as the pullout resistance. But Littlejohn (1975) reported that many engineers also take into account the shear strength of the rock over the failure cone. He quoted, among others, the following formulas proposed by Hobst (1965) for anchors in "sound" homogeneous rock:

$$z = \begin{cases} \sqrt{\dfrac{FP}{4.44\tau}} & \text{for one anchor} \quad (20.10) \\ \dfrac{FP}{2.83\tau s} & \text{for a group of anchors} \quad (20.11) \end{cases}$$

where z = required depth of cone, m
F = safety factor
P = anchor load, kN
τ = shear strength of rock, kN/m^2
s = spacing of anchors, m

These formulas have to be modified for irregular fissured or submerged rock.

20.1.5 Anchored Retaining Structures

Anchorage of retaining structures can help to reduce lateral deformation, maximum bending moments, and, in the case of sheet pile walls, the required depth of embedments. In addition it may facilitate construction and provide better economy.

Figure 20.6 shows the general approach of locating a dead-man-type block or beam anchor as well as a grouted anchor behind a sheet pile structure. The block anchor would be located so that the passive wedge generated does not interfere with the active wedge behind the wall. This particular sketch infers straight-line failure surfaces according to the Rankine or Coulomb earth pressure theory.

In the following paragraphs, the analysis of an anchored sheet pile wall will serve as an illustration of the stability concepts involved.

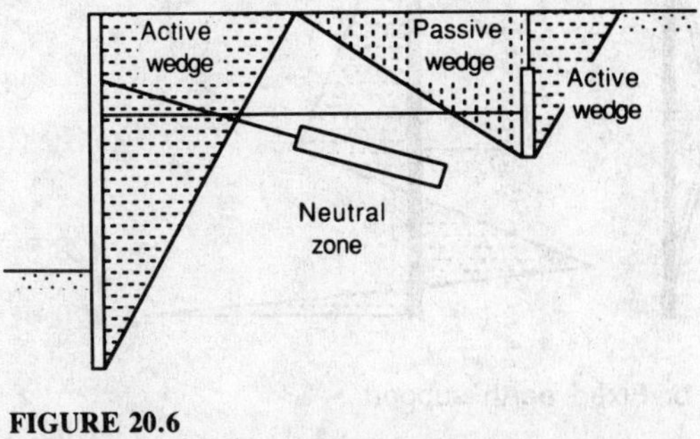

FIGURE 20.6
Typical preliminary location of anchors behind walls.

The calculations begin with a determination of the dimensions of the sheet pile (depth and profile) and the anchor force consistent with the expected deformation pattern. This would allow a determination of the size and number of anchors required per meter of wall. The required free anchor length is assessed by analyzing the block of soil held together by the sheet pile and the anchor system. Finally, the stability of the whole structure, wall + soil + anchor, has to be evaluated. As will be shown, perceived failure modes and the definition of safety factors differ significantly from one method to another. This diversity of methods is not surprising (although confusing!) since we are dealing with a highly complex structure-soil interaction problem, which can only be solved by a blend of empirical findings and theoretical simplifications.

20.1.5.1 EQUILIBRIUM OF WALL AND REQUIRED ANCHOR FORCE. There are a number of different methods available to analyze the equilibrium of an anchored sheet pile wall, all assuming a particular failure mode and idealized earth pressure distributions. For the failure by rotation outward about the point of anchorage, the so-called free-earth-support assumption holds; if the wall penetrates deeply into the soil and is fixed against rotation, or a plastic hinge is postulated below the dredge line (the base of the excavation), the resulting earth pressure is determined by using the fixed-earth-support assumption. Typical earth pressure distributions and wall deformation patterns for both cases are shown in Fig. 20.7.

Using Rankine earth pressure theory, free-earth-support analysis is straightforward, yielding the necessary depth of embedment, the maximum bending moment, and also the anchor force. Fixed-earth-support analysis is somewhat more complicated, needing additional assumptions regarding the deformation of the wall. This analysis will result in longer embedment but a more slender profile. For details of the analysis, the reader is referred to standard textbooks such as Bowles (1982).

20.1.5.2 GLOBAL STABILITY. Global stability of the soil mass which is tied to the wall by the anchors also needs to be verified. Analysis typically proceeds by examining

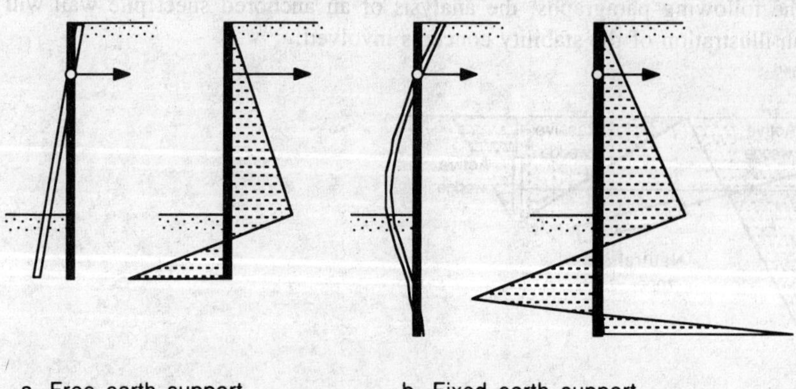

a. Free earth support b. Fixed earth support

FIGURE 20.7
Classical methods of analysis for anchored sheet piles.

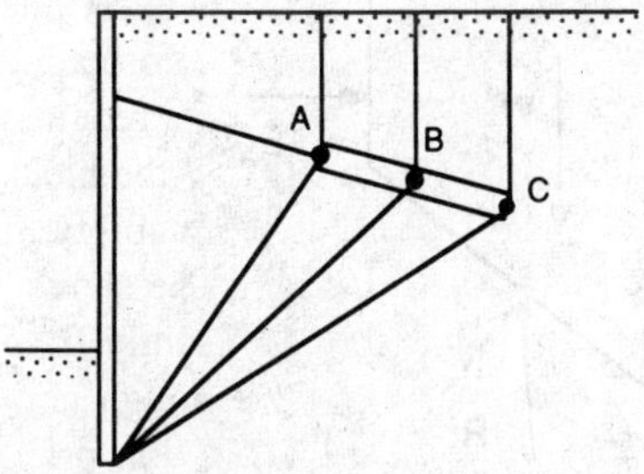

FIGURE 20.8
Soil wedges considered in global stability analysis.

the forces acting on a wedge as defined in Fig. 20.8. It is bounded by the wall and a line going from its base to the anchor and then vertically upward. For a wedge going through point *A*, the anchor can be considered to be an external stabilizing force. If the wedge is assumed to go through point *C*, the anchor force can be considered to be an internal force which can be envisaged as squeezing the soil wedge together but which does not affect external stability; according to Schnabel (1982) this is always the least-stable type of wedge. Consideration of soil-anchor force transfer and anchor spacing may lead to a selection of a wedge through *B*. Generally, the longer the free anchor length, the better the stability. Specifying a safety factor will give a minimum free anchor length for a given wall and anchor geometry.

Of the many different approaches to evaluating global stability, three have been chosen for a brief description. They not only vary in the choice of the geometry of the anchored zone but also in the definition of the safety factor.

One of the classical German methods, usually attributed to Kranz (1953) [as quoted by Schulze and Simmer (1978)], is explained in Fig. 20.9. The resultant earth

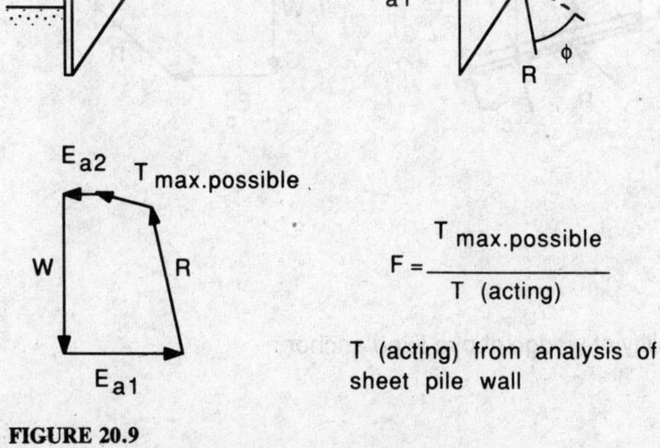

FIGURE 20.9
The Kranz method of analysis. [*Kranz (1953)*.]

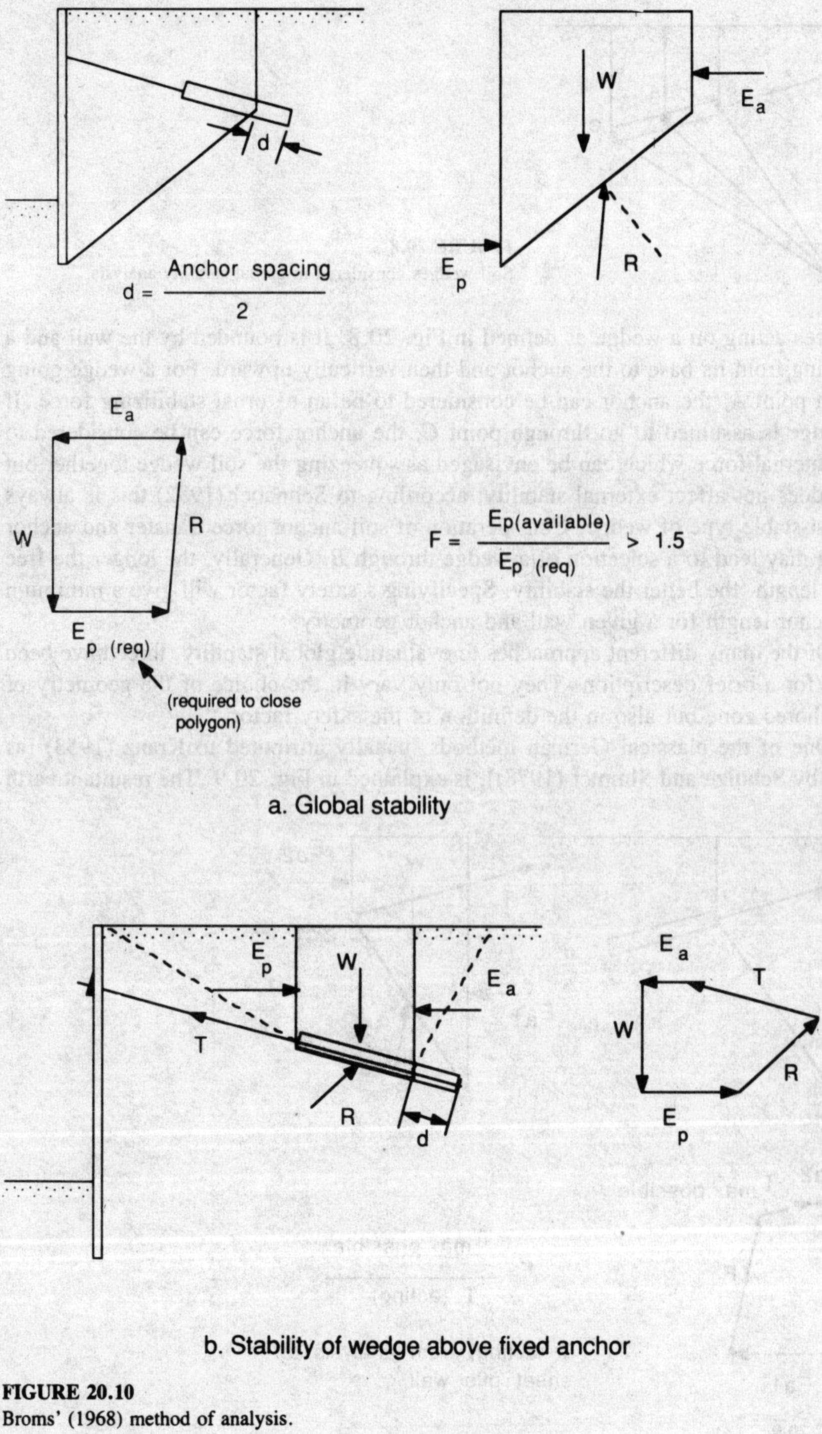

FIGURE 20.10
Broms' (1968) method of analysis.

pressures E_{a1} and E_{a2} are calculated first. Only the directions of the reaction R and the anchor force T are known. Closing the force polygon yields a maximum possible anchor force $T_{max.poss.}$ which would dislodge the soil wedge. This value is compared with the actual force in the anchor in determining the safety factor. Choosing a wedge through the front of the fixed anchor, rather than further back, results in a conservative estimate of the safety factor; conversely, this assumption results in a longer free anchor length for a given safety factor. The Kranz method would not be applicable where the prestress exceeds $T_{max.poss.}$.

Broms (1968) proposed a modified Kranz method. The soil wedge considered for stability includes most of the fixed anchor so that the anchor force does not appear in the force polygon (Fig. 20.10). What is determined is the passive earth resistance at the toe of the wall which is required for equilibrium. For the calculation of the safety factor, it is compared with the resultant available at full mobilization of passive pressure. For shallow anchors, the stability of the soil wedge above the fixed anchor section should also be checked (Fig. 20.10*b*).

Figure 20.11 illustrates the application of the Swedish method of slices for global stability assessment. The safety factor is expressed as the ratio of resisting over driving moments. It can be formulated either in terms of total or effective weights (including

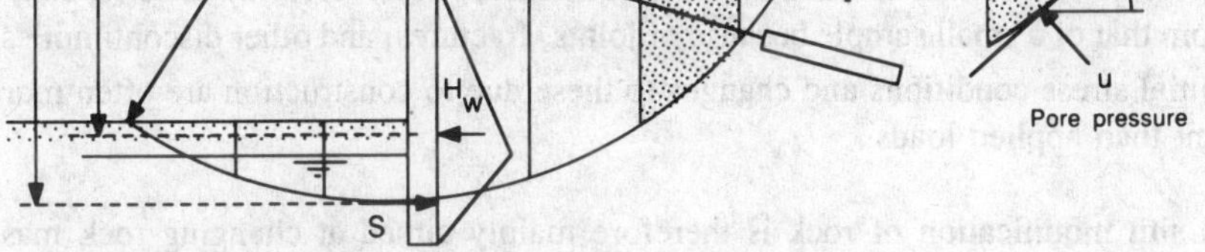

H_W = Resultant of water pressure (kN/m)

S = Force resisting shear through wall (kN/m)

Using effective weight of slices: (W' calculated using buoyant unit weight where applicable)

$$F = \frac{\Sigma\,[\Delta W' \cos\alpha \tan\phi' + c'\,(\Delta X/\cos\alpha) + T\sin(\alpha+\beta)\tan\phi' + S(r_s/R)]}{\Sigma\,[\Delta W' \sin\alpha - T\cos(\alpha+\beta)] + H_W(r_w/R)}$$

Using total weight of slices and pore pressure:

$$F = \frac{\Sigma\,[(\Delta W - u\Delta x)\cos\alpha\tan\phi' + c'\,(\Delta X/\cos\alpha)] + T\sin(\alpha+\beta)\tan\phi' + S(r_s/R)}{\Sigma\,[\Delta W \sin\alpha - T\cos(\alpha+\beta)]}$$

FIGURE 20.11
Global stability analysis of Anchored Wall using the Swedish method of slices. [*Huder and Arnold (1978).*]

the effect of seepage forces) of the slices; the latter approach requires the consideration of the hydrostatic force H_w. Note that the anchor force provides a stabilizing moment, because it is generated outside the failure zone. The example shown also demonstrates how the shear resistance of the wall elements (sheet pile or slurry wall) may be taken into account if the failure circle chosen passes through the wall. This type of failure would only be possible if the wall section below the circle offers sufficient lateral resistance.

20.2 ROCK BOLTS

20.2.1 Typical Applications, Types, and Components

20.2.1.1 THE PRINCIPLES OF ROCK MASS MODIFICATION BY BOLTING. Rock is, like soil, a natural material, occurring in an infinite variety of forms, and as a consequence its engineering properties also show enormous, often unexpected variability. When designing major tunnels, underground openings, excavations, foundations, abutments, and slopes in rock, the civil or mining engineer is well advised to team up with a geologist for the best possible understanding of the structure and quality of a rock formation. As in soil mechanics, the principles of theoretical and applied science are drawn upon when attempting to predict the response of rock to a change in stresses or environmental conditions. Yet important differences exist between the mechanics of soil and rock:

In soils, properties of a soil mass are usually closely related to the properties of its particles. The mechanical behavior of a rock mass, however, may differ substantially from that of a small sample because of joints, fractures, and other discontinuities.

Initial stress conditions and changes to these due to construction are often more important than applied loads.

In situ modification of rock is therefore mainly aimed at changing rock mass properties rather than rock substance. "Rock bolting," first used for temporary roof support in mining, is now well accepted by civil engineers involved in tunneling and in the construction of large underground openings. In the 1950s, engineers of the Snowy Mountains Hydro-electric Authority in Australia contributed greatly to the development of grouted rock bolts for permanent support of major underground works [Lang (1957); Pender, Hosking, and Mattner (1963)].

Rock bolts could be seen to have two basic functions:

1. To pin or nail well-defined blocks or slabs of rock onto a more stable formation.
2. To form a new structural entity out of jointed rock by applying compressive stresses.

In both cases an attempt is made to preserve or mobilize the inherent shear strength of rock along existing joints and potential fractures by a direct increase of the normal stresses in the failure planes or by controlling deformations so that no loosening of the rock mass occurs.

Rock bolt action can be enhanced by wire mesh and/or shotcrete* which confines loose surface material and prevents unraveling of the fractured rock mass behind it.

20.2.1.2 TYPES OF ROCK BOLTS. One of the earliest rock bolts was the *slot-and-wedge* anchor. The end of its shank is slotted with a wedge partially inserted. As the bolt is pushed against the back of the borehole, the wedge is driven home and the end of the shank expands and anchors the bolt in the rock. The bolt is tensioned by tightening a nut against a plate placed on the rock face. The tension induced can be empirically related to the torque applied. According to Brady and Brown (1985), this anchor performs well in hard rock but is unreliable in poor-quality rock. Various forms of *expansion shells* are newer variants of the old slot-and-wedge mechanical anchors.

For permanent reinforcement, *grouted* bolts are preferred because they provide a better bond between the bolt and the rock and are less prone to corrosion. If the face plate fails or corrodes away, the bolt force is still transferred to the rock by means of the grout bond. The grout may be a nonshrinking cement mix or a resin. Resins with variable setting rates can be chosen so that the anchorage zone hardens first, allowing tensioning before the remainder of the resin sets.

Untensioned steel shanks grouted into boreholes are generally called *dowels*. They represent a form of passive reinforcement.

20.2.2 Support of Individual Blocks

Figure 20.12 illustrates the securing of a block on a sliding plane by means of a tensioned rock bolt. The block shown schematically could be on the side of an excavation or tunnel profile. The failure plane represents the prevailing joint system. The geometry and force system can be changed to simulate a variety of conditions, including that of securing a wedge-type block in the roof of an underground opening.

The force polygons of Fig. 20.12 clearly indicate the increase in shear resistance relative to the existing shear stress by the application of bolt tension which results in an improved safety factor against sliding.

20.2.3 Rock Bolt Action around an Excavation

When used in patterns such as those illustrated in Figs. 20.13, and 20.14, rock bolts create compressive stresses perpendicular to the free surface of the excavation. This creates a zone of strengthened rock which may resemble a structural element such as an arch or a beam which stands up without additional steel or timber support. The compressed zone has also been likened to a structural membrane or diaphragm.

---

*Shotcrete is a pneumatically applied cement mortar containing aggregates of up to 25 mm in size.

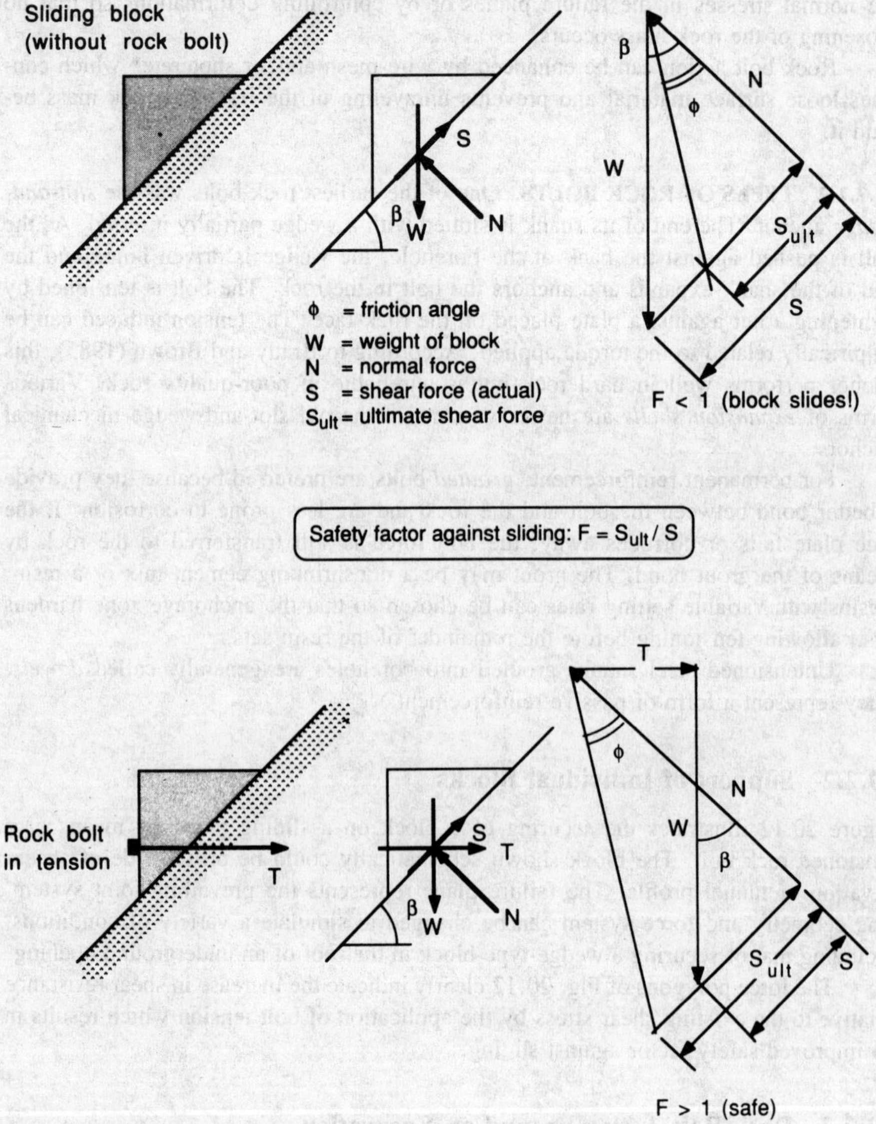

FIGURE 20.12
Rock bolts securing sliding block.

Lang (1957) demonstrated rock bolt action in a number of different kinds of model tests, including the beam schematically shown in Fig. 20.15. It is made up of 100-mm-long polystyrene rods (a two-dimensional representation of a particulate mass) compressed by spring-loaded bolts. The horizontal spacing of the bolt end plates was chosen so that the openings were not more than 3 times the diameter of the rods.

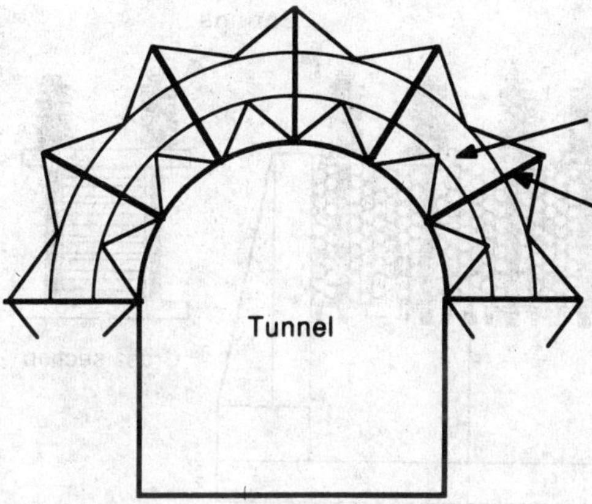

FIGURE 20.13
Formation of a strengthened zone through rock bolts.

Lang's beam could be loaded and behaved approximately elastic until failure, which, when it occurred, was sudden and catastrophic.

Rock bolts must be applied as soon as possible after excavation. The rock must be restrained promptly because any loosening of the rock adds more weight onto the support system (also see Sec. 20.3.1).

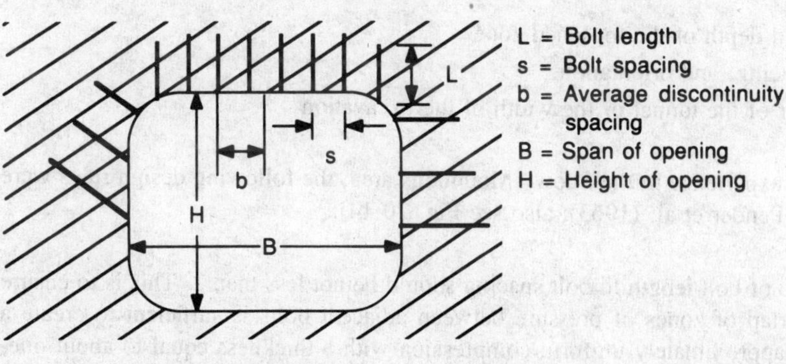

Design rules

Bolt length: Choose largest of	L = 2s
	L = 3b
	L = 0.5B (for B < 6 m)
	L = 0.25 (for 18 < B < 30 m
Bolt spacing: Least of	s = 0.5L
	s = 1.5b
	s < H/5 (for side walls)

FIGURE 20.14
Early design rules for rock bolts. [*Lang (1961); Pender, et al. (1963).*]

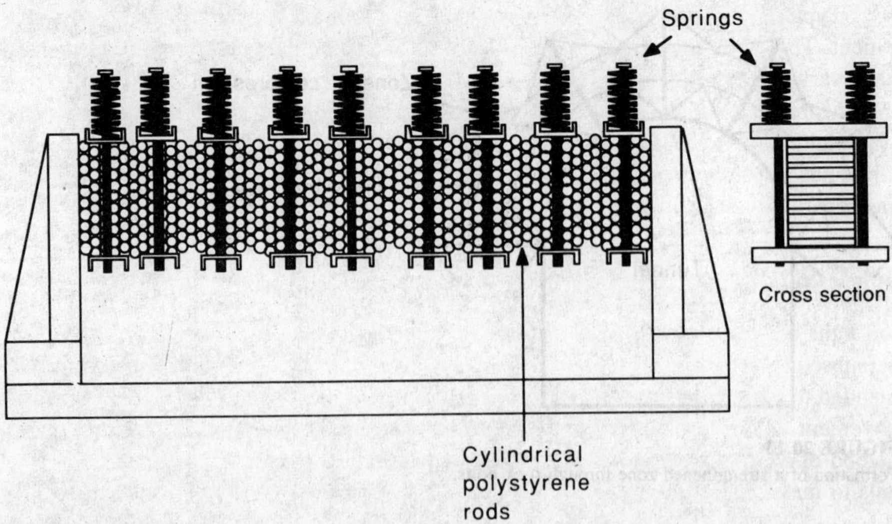

FIGURE 20.15
Lang's rod model beam (1957).

Pender et al. (1963) stated that the following factors determine the length and spacing of bolts:

1. Estimated depth of the loosened zone
2. Joint spacing, and orientation
3. Diameter of the tunnel or the width of the excavation

From experience in the Snowy Mountains area, the following design rules were developed [Pender et al. (1963); also see Fig. 20.14]:

1. The ratio of bolt length to bolt spacing should be not less than 2. This is to ensure that overlap of zones of pressure between adjacent bolts is sufficient to create a zone of approximately uniform compression with a thickness equal to about one-third of the bolt length (Fig. 20.13).
2. The length of the bolt should not be less than 3 times the width of the joint blocks. This is to ensure that the anchorage takes place in blocks not less than two layers behind the surface, although four blocks behind would be preferable. (For the average granite in the Snowy Mountains, this criterion results in a minimum bolt length of 2.4 m.)
3. Aim at a bolt spacing and tension sufficient to create a compression of 70 kPa.
4. In large excavations the rock bolts should be longer than in small excavations in the same conditions.

Typically, 25-mm-diameter hollow-core steel bolts were used, grouted with neat cement. Tests with embedment lengths from 150 to 900 mm resulted in 28-day pullout loads varying between 53 and 224 kN.

More generalized rock bolt pattern design rules have since been developed. Brady and Brown (1985) quote design rules by Farmer and Shelton (1980) and give guidelines for the support of mining excavations based on a modified geomechanics classification [after Laubscher and Taylor (1976)].

20.3 SOIL NAILING

Soil nails are more or less rigid bars driven into soil or pushed into boreholes which are subsequently filled completely with grout. Together with the in situ soil, they are intended to form a coherent structural entity supporting an excavation or arresting the movement of an unstable slope. Typical applications are shown in Fig. 20.16. This method of construction is extremely flexible and allows adjusting the direction of the nails to maximize the reinforcing action and construction efficiency.

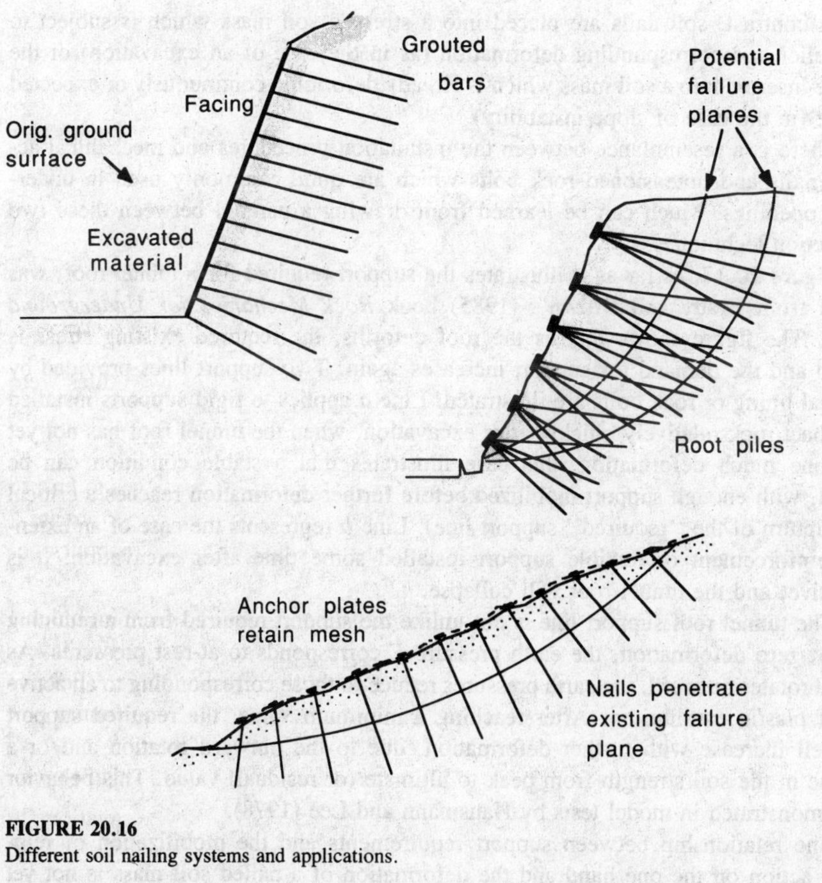

FIGURE 20.16
Different soil nailing systems and applications.

The size of "nails" varies from thin steel bars to light concrete piles. Terms such as micropiles and root piles [Lizzi and Carnivale (1979)] have also been used to describe these construction elements.

The character of the facing varies considerably. Metal panels or plates, steel meshes, and mesh-reinforced shotcrete as well as synthetic nets and fabrics have been used in combination with nails.

Most nail-supported structures must still be classed as temporary. This is partly due to concerns about corrosion; but uncertainties with regard to the design assumptions also suggest caution in the application of the nailing technique.

20.3.1 The Importance of Construction Sequence

Reinforced Earth is built from the bottom up. Reinforcing strips are placed in unstressed soil, and soil-reinforcement interaction is only mobilized as construction proceeds, vertical and horizontal stresses build up and the reinforced soil mass deforms laterally.

In contrast, soil nails are placed into a stressed soil mass which is subject to stress relief and corresponding deformation (as in the case of an excavation) or the nails are inserted into a soil mass which is already deforming continuously or expected to move (in the case of slope instability).

There is a resemblance between the installation procedures and mechanical action of nails and untensioned rock bolts which are quite commonly used in underground openings. Much can be learned from drawing a parallel between these two construction techniques.

Figure 20.17, as far as it illustrates the support required for a tunnel roof, was adapted from Brady and Brown's (1985) book *Rock Mechanics for Underground Mining*. The figure shows that as the roof deforms, the required existing stress is relieved and the demand for support increases again. Two support lines provided by structural lining or rock bolts are illustrated. Line *a* applies to rigid supports installed in compact rock relatively quickly after excavation, when the tunnel roof has not yet undergone much deformation; this case illustrates that a stable condition can be reached, with enough support mobilized before further deformation reaches a critical point (upturn of the "required" support line). Line *b* represents the case of an extensible reinforcement or flexible support installed some time after excavation: It is ineffective, and the tunnel roof will collapse.

The tunnel roof support line is not unlike the support required from a retaining wall. At zero deformation, the earth pressure E corresponds to at-rest pressures. As the wall rotates outward, the earth pressures reduce to those corresponding to an active state of plastic equilibrium. After reaching a minimum value, the required support may well increase with further deformation, due to the outward rotation and/or a decrease in the soil strength from peak to ultimate (or residual) value. This behavior was demonstrated in model tests by Hausmann and Lee (1978).

The relationship between support requirements and the mobilization of reinforcing action on the one hand and the deformation of a nailed soil mass is not yet

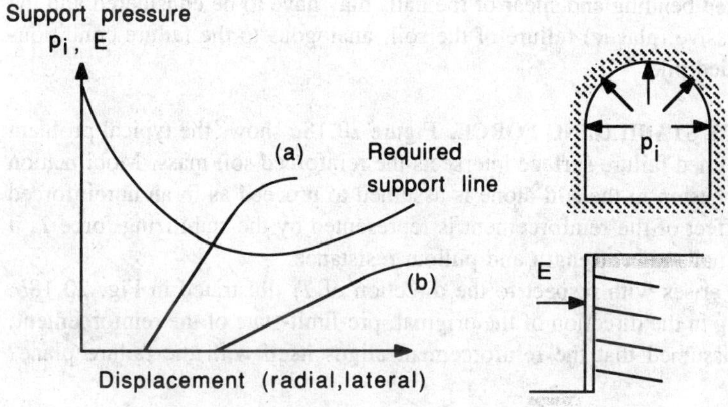

(a) Support provided by rigid reinforcement, installed early (stable condition)

(b) Support provided by flexible, extensible reinforcement, installed late (unstable condition)

FIGURE 20.17
Reinforcement support and deformation.

well understood, and most design procedures look at the limit state only. This brings about the danger of choosing inappropriate strength parameters for the soil and soil-reinforcement interaction.

20.3.2 Analysis of Nailed Soil

Methods of analysis proposed for nailed soil vary with respect to

Geometry of the failure surface (straight, bilinear, log-spiral, circle)

Definition of safety factor (in terms of overturning moment, shear strength, friction angle, cohesion, reinforcement properties, etc.)

Direction of the tensile stabilizing force provided by the nails (which may be flexible or rigid)

Consideration of bending and shear forces in the reinforcement, a reaction to the development of passive resistance normal to the reinforcement

The most adaptable method of stability analysis seems to be a modified method of slices as commonly used in slope stability problems; this method was already introduced for the analysis of global stability of anchored sheet pile walls (Fig. 20.11). Assuming a circular failure surface, the safety factor can then be defined as the ratio of the total resisting moment over the total overturning moment. Nailing a soil mass introduces new stabilizing forces: tangential and normal components of the reinforcement tension and, if appropriate, the bending moment and shear force offered by more rigid types of nails. Besides pullout and tensile rupture of the nails, additional failure modes of combined bending and shear of the nails may have to be considered and the corresponding passive (plastic) failure of the soil, analogous to the failure conditions of a laterally loaded pile.

20.3.2.1 TENSILE STABILIZING FORCE. Figure 20.18*a* shows the typical problem situation. An assumed failure surface intersects the reinforced soil mass. Mobilization of friction and cohesion in the soil alone is assumed to proceed as in an unreinforced soil mass. The effect of the reinforcement is represented by the stabilizing force T; it is limited by the nails own strength and pullout resistance.

A question arises with respect to the direction of T, illustrated in Fig. 20.18*b*. Is this force acting in the direction of the original, pre-limit-state of the reinforcement, or should it be assumed that the reinforcement aligns itself with the failure plane?

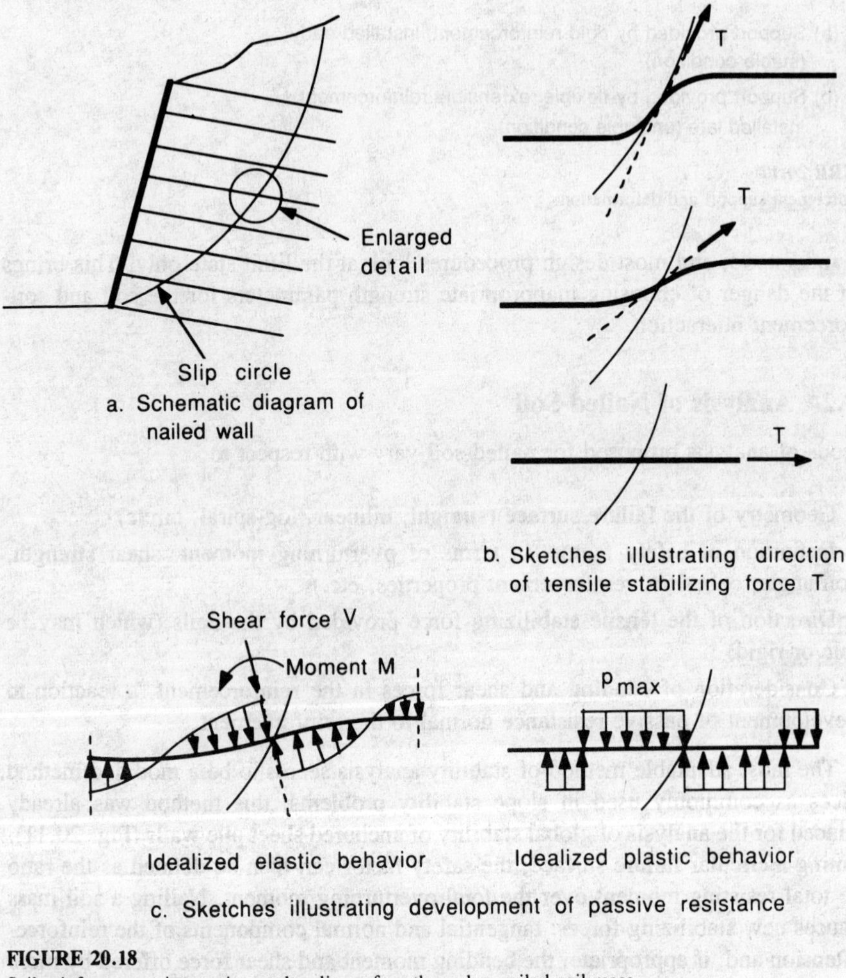

FIGURE 20.18
Soil-reinforcement interaction at the slip surface through a nailed soil mass.

Generally, the more conservative approach is to assume T retains its original orientation; in retaining structures this is usually close to horizontal. However, it should be noted that if the nails are oriented towards the centre of the slip circle, then taking this approach would indicate nil effect of the nails, unless passive resistance is also considered.

Pullout resistance of bars and wires has already been discussed in Sec. 18.2.13 where, for a frictional soil, it was formulated as proportional to the average confining stress. This may be a reasonable assumption for Reinforced Earth–type "built" soil structures but may not generally be valid for bars inserted into an existing soil mass.

A simple analysis would assume full pullout resistance or yield strength of the nails, whichever is smaller, and express the safety factor in terms of the strength of the original soil. Iterative procedures would allow to determine a safety factor which is common to both the soil and the reinforcement.

20.3.2.2 PASSIVE SOIL PRESSURE AND SHEAR RESISTANCE OF THE NAILS. Possible forms of the mobilization of passive pressure perpendicular to the soil nails are illustrated in Fig. 20.18*c*. These distributions resemble those assumed in the analysis of laterally loaded piles.

Juran (1986) reports a French approach which is based on the problem of an infinite beam subjected to bending in a Winkler soil, where the beam's reaction is governed by the coefficient of subgrade reaction k_s. With some simplifying assumptions the shear force V_s as limited by the soil is

$$V_s = \frac{pDL_0}{2} \tag{20.12}$$

where p = passive (normal) pressure on nail (The maximum value of this pressure may be taken as the "creep pressure" as measured in a pressuremeter test or may be obtained analogous to maximum lateral resistance.)

D = diameter of nail

$$L_0 = \sqrt[4]{\frac{4EI}{k_s D}} \tag{20.13}$$

The L_0 is the so-called transfer length, which is also a measure of the nail's stiffness, defined by modulus E and moment of inertia I, relative to the soil. The shear force is also limited by the maximum bending moment M_p:

$$V_p = 0.5DL_0 \frac{M_p}{0.16DL_0^2} \tag{20.14}$$

The maximum shear force which can exist is the lesser of V_s and V_p. The possible reinforcement failure modes which have to be considered then are

Failure by pullout (force T)

Passive failure of the soil

Breakage of the nail by a combined effect of V_p and T

If the reinforcement is aligned with the failure plane, only tensile forces develop. If the reinforcement is perpendicular to the failure plane, only shear has to be considered in the analysis. In between these extreme situations, both should be evaluated. Juran (1986) gives the following expression for the shear force V_f and tension T_f of the nail inclined at an angle a to the slip plane:

$$V_f = \frac{R_c}{\sqrt{1 + 4 \tan^2 \left(\frac{\pi}{2} - \alpha \right)}} \tag{20.15}$$

$$T_f = 4\ V_f \tan \left(\frac{\pi}{2} - \alpha \right) \tag{20.16}$$

where R_c is the shearing strength of the reinforcement and $R_n = 2\ R_c$ is the tensile strength of the reinforcement.

The above equations are derived from an application of the principle of virtual work and Tresca's failure criterion.

20.3.3 Special Considerations for Slope Stabilization

Landslides are said to cause more economic losses than any other ground failure hazards, except swelling soils. There is a great need for low-cost methods of slope stabilization which do not need elaborate equipment and high skills in their implementation. Soil nailing may provide such a solution.

Special problems are presented by long-term creeping instability of slopes. In such cases remedial measures are aimed at reducing the creep rate by reducing the shear stresses in the failure surface. Analytical and design approaches for this problem have been proposed by Ito and Matsui (1975), Winter et al. (1983), Gudehus (1983), and others; they relate basically to piles used to stabilize clay slopes but these piles fit the definition of the newer term of soil nailing using relatively rigid inclusions.

Secondary effects of soil nailing may also have a beneficial long-term influence on the stability of a slope. Koerner and Robins (1986) describe a method called "spidernetting"; a geotextile, geogrid, or geonet is placed on the surface of the slope and attached to it via nails. The nails are small-diameter long rods which extend beyond the existing or potential failure plane. The system is installed in such a way that the spidernets exert a normal pressure on the ground (and thus also the failure plane). The resulting benefits are

The nails provide passive resistance in the failure plane.

The normal pressure exerted by the net provides a direct stabilizing force and increases the friction in the failure plane.

The normal pressure applied causes consolidation of the soil, thereby increasing the effective cohesion and friction angle.

The netting provides erosion resistance and helps to establish plant growth which may assist overall stability [Gray (1978)].

In slope stabilization projects success may best be ensured by a combination of soil reinforcement with other means of ground modification, such as providing drainage (dewatering) or chemical stabilization. Additional remedial measures may also include altering the geometry and removing or controlling external loading, particularly if these have been major factors leading to instability.

PROBLEMS

Prefixes indicate problem type: C = calculations, B = brief answer, M = multiple choice, D = discussion.

Introduction

D20.1. With respect to a block anchor, an ungrouted rock bolt, a soil nail, and a Reinforced Earth–type reinforcing strip, discuss
(*a*) The force transfer from the reinforcing element to the soil or rock. Indicate possible distributions of shear stress and tension along the tendon, bolt shank, nail, or strip.
(*b*) Discuss methods of installation.

Section 20.1

SECTION 20.1.1

D20.2. Prepare sketches illustrating the possible use of ground anchors for
(*a*) Building hydraulic structures (dams, pools, etc.).
(*b*) Repairing an old retaining structure.
(*c*) Building foundations for a transmission tower.
(*d*) Providing preloading of a foundation.

D20.3. Describe what is meant in the normal usage of the following terms:
(*a*) Dead-man or block anchor
(*b*) Tension anchor
(*c*) Prestressed anchor

M20.4. Which one of the following statements is *incorrect*?
(*a*) Soil anchors typically have a fixed anchor length between 4 and 7 m.
(*b*) Soil anchors typically have a capacity of 600 to 1000 kN.
(*c*) Rock-bolt-type anchors usually have a capacity in excess of 1000 kN.
(*d*) Anchors used against uplift of concrete dams may have a capacity exceeding 10 000 kN.

D20.5. Describe the likely procedure involved in an anchor test for the purpose of
(*a*) Proof loading (acceptance test).
(*b*) Establishing design criteria.
(*c*) Long-term performance monitoring.

D20.6. List four characteristics of anchor performance which may form the basis of design criteria.

C20.7. Soil anchors were installed in dense gravelly sand to retain sheet piling. The boreholes were 21.3 m long, with a diameter of 127 mm. Grout was injected under pressure as the casing was progressively withdrawn for a distance of 9.1 m, creating a bulb about 0.5 m in diameter. The remaining casing was then withdrawn, leaving an unpressurized grout column surrounding the tendons through the active soil zone. The anchors consisted of eight tendons, each with a cross-sectional area of 139 mm^2. The modulus of elasticity of the strand was 197 000 MPa.

Seven anchors were tested up to a load of 1600 kN, which represents 125% of the working load. The average observed strand movement at this load was 86 mm; no measurable creep was observed. The residual elongation after the load returned to zero was 9 mm.

(*a*) Calculate the theoretical strand movement over the free anchor length implied by the construction procedure for a load of 1600 kN and compare it with the observed value.

(*b*) Speculate on the reason of any discrepancy found in part (*a*).

(*c*) What may be the cause of the residual deformation?

C20.8. Roughly estimate the capacity of a grouted anchor in dense sand. Assume typical characteristics such as a 100-mm-diameter borehole, a depth of 5 m (above groundwater level), and a fixed anchor length of 5 m. Make the estimate according to

(*a*) Empirical guidelines giving values of force transfer in terms of kilonewtons per meter (Sec. 20.13).

(*b*) Eq. (20.3).

C20.9. Estimate the pullout resistance of a fixed anchor length of 4 m in sound igneous rock, assuming a borehole diameter of 100 mm. The unconfined compressive strength of the rock is 140 MPa. Make the estimate according to

(*a*) Empirical guidelines (Sec. 20.1.3).

(*b*) Eq. (20.5).

C20.10. The following equation has been proposed for determining the capacity of an underreamed anchor in clay [Basset (1977)]:

$$T_{\text{ult}} = \pi\, d_s f_s c_u l_s + \frac{\pi}{4}(d_u^2 - d_s^2)\, N_c c_u + \pi d_u f_u c_u L_u$$

where c_u = undrained cohesion, kPa

f_s = adhesion factor (may range from 0.3 to 0.6 depending on strength of clay)

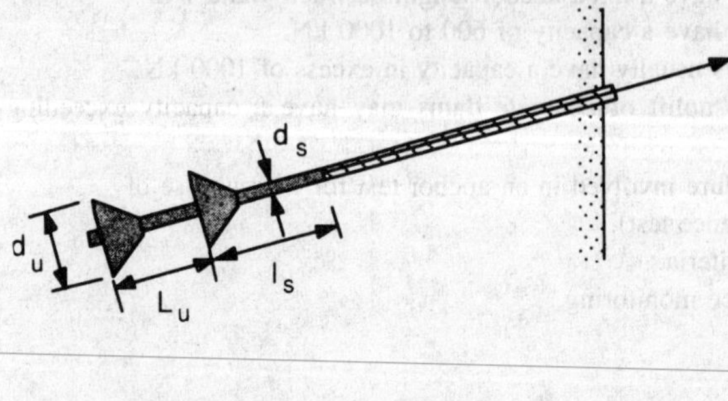

FIGURE P20.1

f_u = efficiency factor (may range from 0.75 to 0.95, depending on disturbance caused by underreamer tool and technique)

N_c = end bearing factor (may range from 6 to 13)

and d_s, d_u, l_s, and L_u are as defined in Fig. P20.1.

(*a*) Identify whether the terms in the above equation relate to the bearing or friction-adhesion action of the anchor.

(*b*) Calculate the capacity for $d_s = 100$ mm, $d_u = 300$ mm, $l_s = 1$ m, $L_u = 2$ m, $c_u = 150$ kPa, $f_s = 0.5$, and $f_u = 0.85$.

C20.11. Determine the uplift capacity of a 0.3-m-diameter horizontal plate embedded in dry sand with a friction angle of 35°. With respect to depth, assume the following conditions:

(*a*) Deep embedment (depth/diameter > 10); use the Hansen formula to obtain an approximate estimate. Alternatively, use other bearing-capacity formulas, as you may find recommended for the analysis of end bearing of piles. Reference to the paper by Meyerhof and Adams (1968) may allow an even better prediction.

(*b*) Assume a 1-m embedment only. Follow the approach given in Eq. (20.7), and compare the result with part (*a*).

C20.12. A spillway wall is to be anchored against uplift. The grouted anchors have an effective free anchor length of $z = 12$ m in sound rock (unit weight = 25 kN/m^3, unconfined compressive strength = 50 MPa). Using a safety factor of 2, calculate the allowable anchor load.

(*a*) Assume resistance is only provided by the weight of the cone of the rock with a 90° angle at the end of the fixed anchor length z (volume of cone = $\pi z^3/3$).

(*b*) Use Eq. (20.10), and assume the shear strength equals half of the unconfined compressive strength.

C20.13. Figure P20.2 shows the geometry and forces of an anchored sheet pile wall.

(*a*) Use the Kranz method to determine a safety factor for global stability, given the following forces for the block shown:

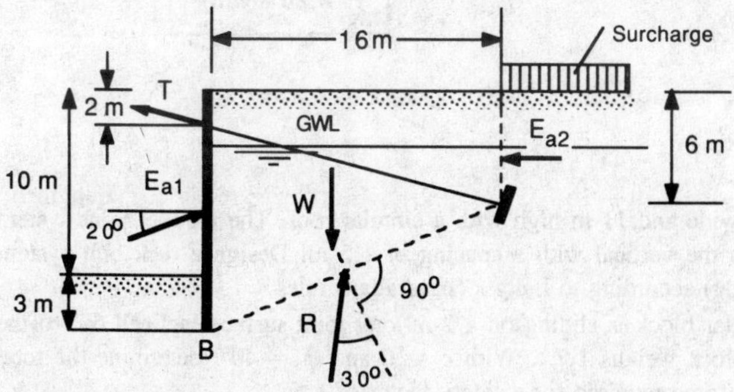

Note: For fixed earth support, point B is where the shear force in the sheet pile is zero. For free earth support, point B is at the pile tip.

FIGURE P20.2

$$E_{a1} = 510 \text{ kN/m}$$
$$E_{a2} = 100 \text{ kN/m}$$
$$W = 2210 \text{ kN/m}$$
$$T = 240 \text{ kN/m}$$

(*b*) Increase the free anchor length from 17 to 20 m, and check for any change in the safety factor. (Point B remains as is; weight W varies proportionally with the block area. Note that the direction of R will change.)

C20.14. Figure P20.3 shows a slurry wall securing an excavation. Using the Swedish method of slices, determine the safety factor for the slip circle shown, for the following conditions:

(*a*) Without considering the presence of the wall nor the anchor.

(*b*) Considering an anchor force $V = 198$ kN/m and a shear resistance through the concrete of $S = 236$ kN/m.

[Adapted from Huder and Arnold (1978).]

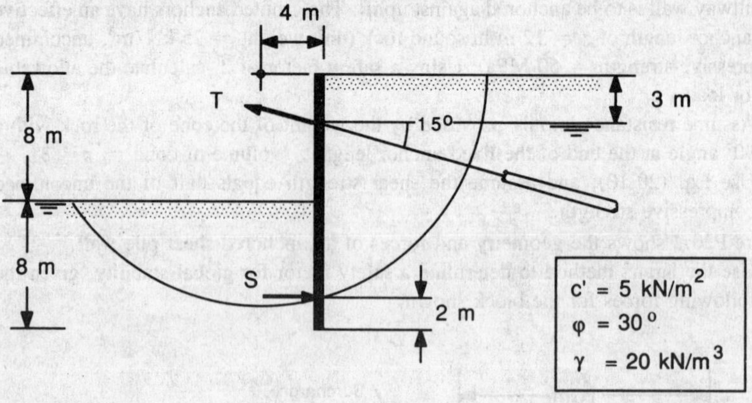

FIGURE P20.3

Section 20.2

C20.15. A tunnel is 8 m wide and 11 m high with a circular roof. The parallel joint system deviates 30° from the vertical with a spacing of 1.5 m. Design a rock bolt system (length and spacing) according to Lang's (Australian) rules.

C20.16. Assume a triangular block is sliding on a 2-m-long joint surface inclined 60° to the horizontal. The block weighs 1.7 t. With $c = 0$ and $\phi = 40°$, determine the total horizontal anchor force required for a safety factor of 1.5.

CHAPTER 21

SOIL CONFINEMENT BY FORMWORK

21.1 CONCEPT OF CONFINEMENT

Consider a wall built with sandbags: Simple hessian bags containing sand particles make up an extremely versatile building element that can exhibit astounding compressive strength. Confinement of particulate materials, whether they are sand grains or coffee beans allows the mobilization of internal friction which contributes to the stability of the structure formed.

The concepts of reinforcement and confinement are really very closely associated (compare with Fig. 21.1). In a triaxial specimen, horizontal layers of foil can be interpreted as providing a lateral confining stress σ_3. According to the equivalent homogeneous model of soil reinforcement (Sec. 17.1), this lateral confinement results in either an added cohesion or an increased friction angle if viewed in terms of the Mohr-Coulomb theory.

Confinement may be produced by internal "inclusions" or external formwork, supports, or abutments. In a beam, tensile inclusions could be oriented longitudinally, like concrete reinforcement, or transversely.

In a Reinforced Earth wall, the strips could be said to provide horizontal internal confinement, but the concept of Reinforced Earth is not restricted to this mode of action alone. Vidal's patent describes a beam of granular material with embedded horizontal strips wrapped in a cylindrical form (Fig. 16.2). This form may be made of very thin material—for a laboratory experiment, a sheet of paper and some adhesive tape are sufficient (Fig. 21.1*d*). Vidal thus envisages not only internal but also external confinement transverse to the beam axis.

If the beam is made of clay soil, cohesion binds the particles together and no further external confinement may be required. Gaind and Char (1983) carried out flexure tests on compacted silty clay model beams (533 mm long) reinforced by steel strips. They observed that the reinforcement improved the load-bearing capacity and changed the failure mode from brittle to ductile. Mandal found similar improvement for a geotextile reinforced clay beam (1987).

Rock bolts can be viewed as providing an internal, transverse confining stress. Rock bolt action can be illustrated by the clamping of an unbonded laminated beam: The transverse prestressing substantially increases its interlaminar shear strength and overall stiffness (Fig. 21.1*e*). The laminae could represent horizontal rock strata above an underground opening! Compare Lang's rod model beam (Fig. 20.15) with the Reinforced Earth beam (Fig. 16.2): They look quite different, yet the principle of confinement is present in both.

The idea of confinement is also very obvious in the construction of crib walls, bin walls, and gabions. These are essentially gravity structures created by interlocking forms containing granular material. Amazingly, the transition from these type of structures to those described as reinforced soil or anchored systems is quite easy: For example, leave out the stretchers (longitudinal elements) at the rear of a crib wall; what you are left with is an open-ended structure which resembles reinforced earth, where concrete beams have replaced the steel strips. Or give these beams (the "headers") a Y-shape and you are left with something similar to anchored earth. (Standard crib walls and variations thereof are presented in Fig. 21.2 and discussed further in the next section).

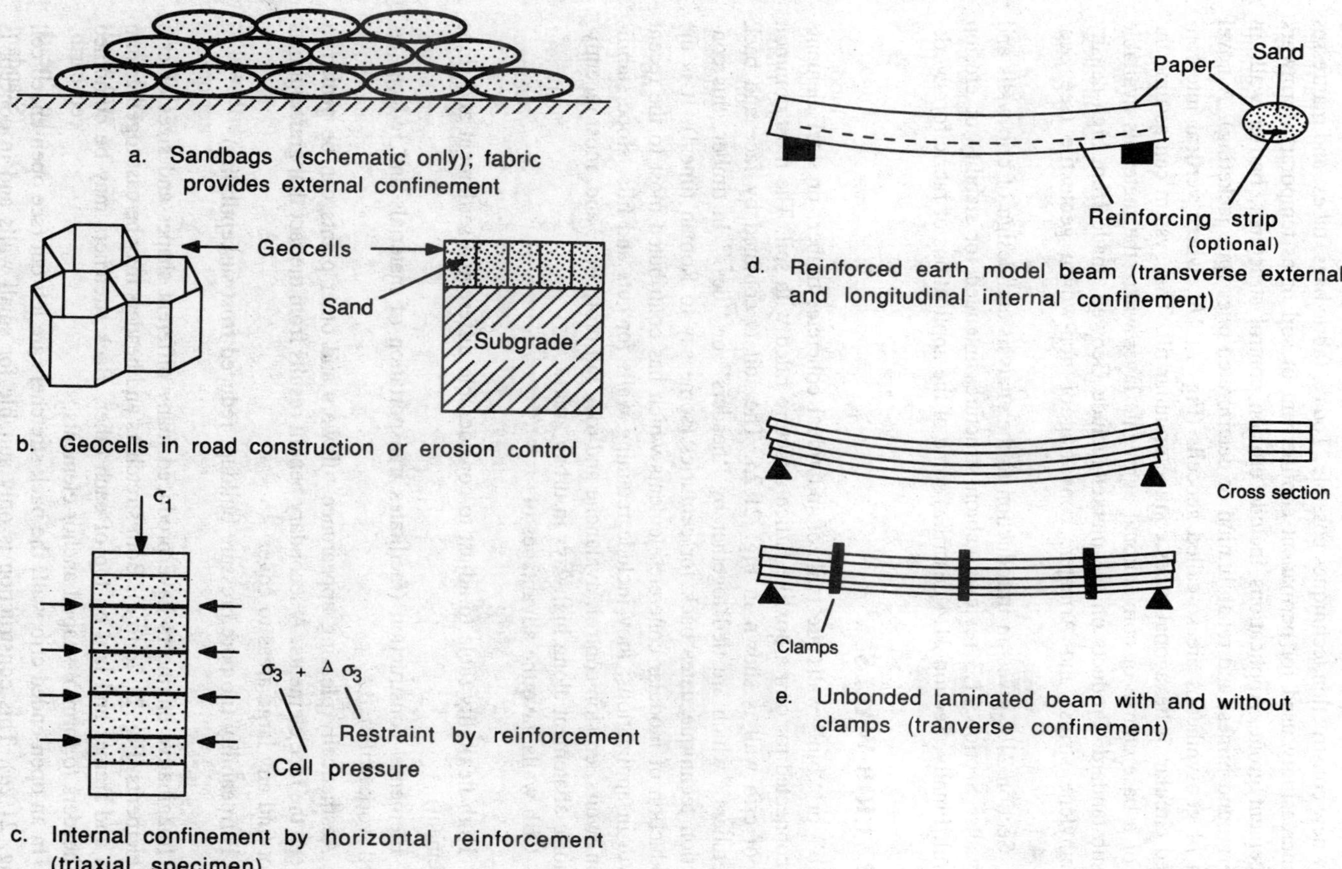

FIGURE 21.1
Mechanisms of confinement.

As is apparent from Chap. 19, the development of geosynthetics is making many new construction techniques possible. Today, fabric bags, tubes, and mattresses are increasingly used for permanent confinement of soil or as temporary formwork for cement grout. Applications include erosion control structures, barrier walls in underground mines, and rehabilitation of deteriorated piles (pile jacketing). A novel form of geosynthetics are so-called geocells (Fig. 21.1*b*), where vertical interconnected circular or hexagonal tubes confine granular fill. This system can assume the role of a base course or conventional aggregate layer where the materials available are substandard. Products on the market include Geoweb, made from high-density polyethylene strips, and Armater, consisting of non-woven geotextile (see Sec. 21.4.4.).

Section 21.2 gives an introduction to the analysis and design of crib walls and bin walls. Section 21.3 reviews gabion structures intended for stability or erosion control. Finally, Section 21.4 describes some of the applications of fabric formwork.

21.2 CRIB WALLS

"Crib" or "bin" walls are built of individual concrete, timber, or steel elements interconnected to form a boxlike lattice structure filled with soil. The most common type of crib wall is shown in Fig. 21.2*a*. The soil is confined by face and back "stretchers" which are tied together by "headers" or "ties." In timber, this construction technique dates back for centuries (some say to Roman times!). It is the introduction of modular concrete elements which has contributed most to the recent growth in applications: they include retaining walls for cuts and fills, slope stabilization, weirs, erosion control, avalanche and rock fall protection, and, most recently, for noise abatement along highways in urban areas.

Crib walls have the advantages of

High flexibility (able to adjust to considerable differential settlement and slope movement).

Elemental construction (facilitates transportation of materials and reduces required work skills).

Aesthetically pleasing appearance—flowers and other plants can be grown in between the face elements. A secondary benefit results from the fact that graffiti artists do not find any large areas to cover!

Permeability (no pore pressure buildup, reduced frost susceptibility).

Imaginative engineers have produced many different shapes and sizes of crib wall elements (Fig. 21.2*a* to *e*). Both stretchers and headers may be cast together into boxes and frames and laid on top of each other. Back stretchers may be combined with headers to form Y-shaped anchor elements.

In an open-ended crib wall, the back stretchers are left out (see open-ended cell in Fig. 21.2*a*). This construction is only suitable for small walls and in concept is closer to Reinforced Earth than a standard crib wall.

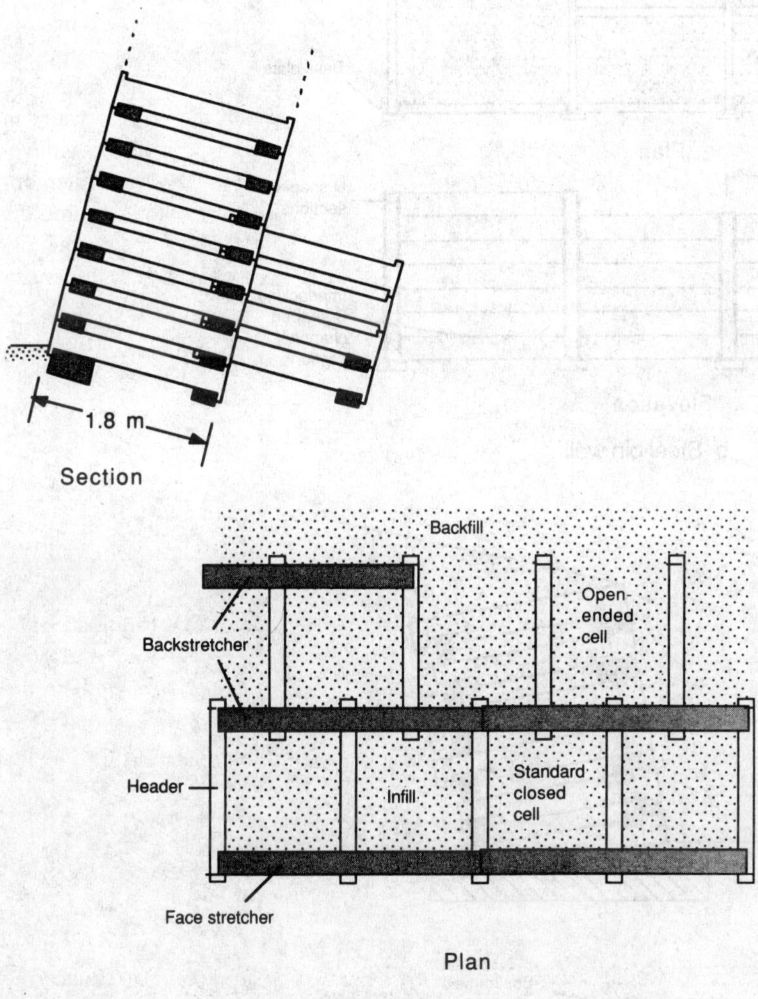

FIGURE 21.2
Standard crib wall and variations.

In the extreme, a whole bookshelf-type wall section may be manufactured and placed at an angle into a slotted strip footing. Although the latter type of wall (Fig. 21.3*b*) may look like a crib wall after installation, its supporting action is quite different from that of a wall with multiple shelf type cantilevers.

In the Austrian "NEW" earth retaining system, the headers are replaced by loops of synthetic strips; they are connected to L-shaped face elements and wrapped

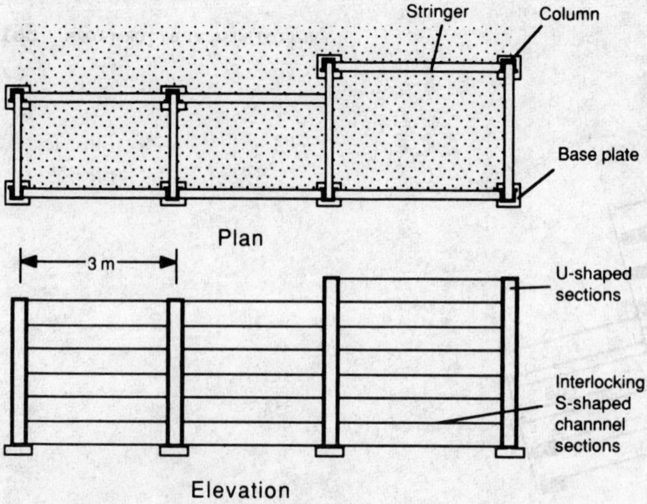

b. Steel bin wall

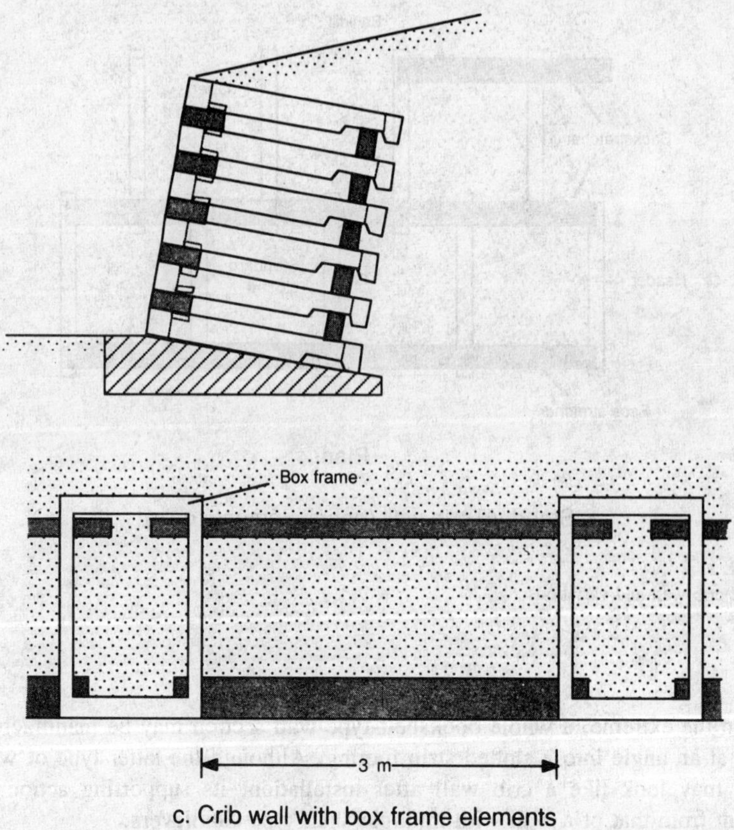

c. Crib wall with box frame elements

FIGURE 21.2
Continued

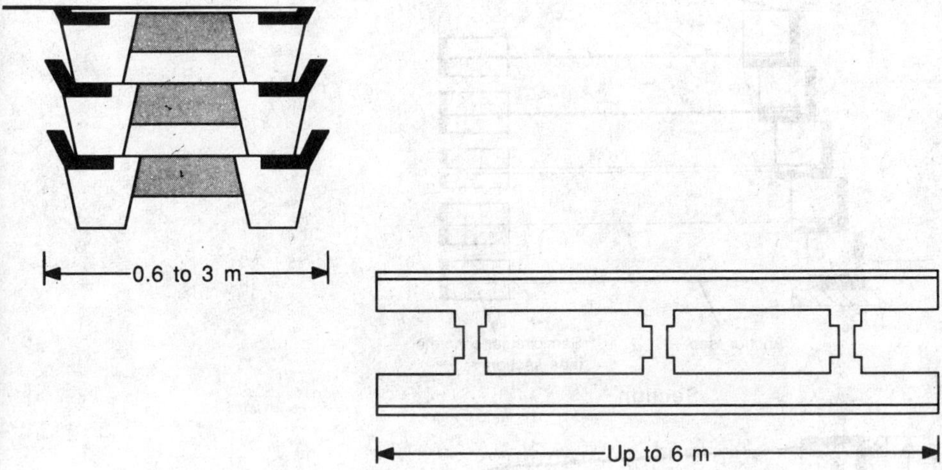

d. Multicell crib frame element (Evergreen Systems)

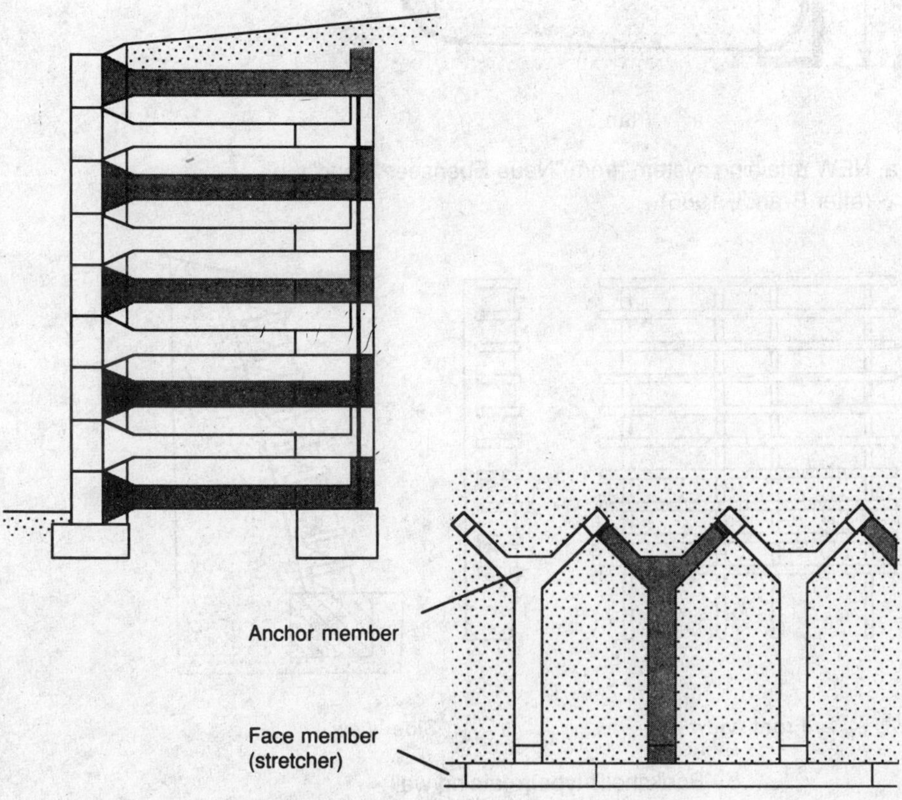

e. Crib wall with Y-shaped elements

FIGURE 21.2
Continued

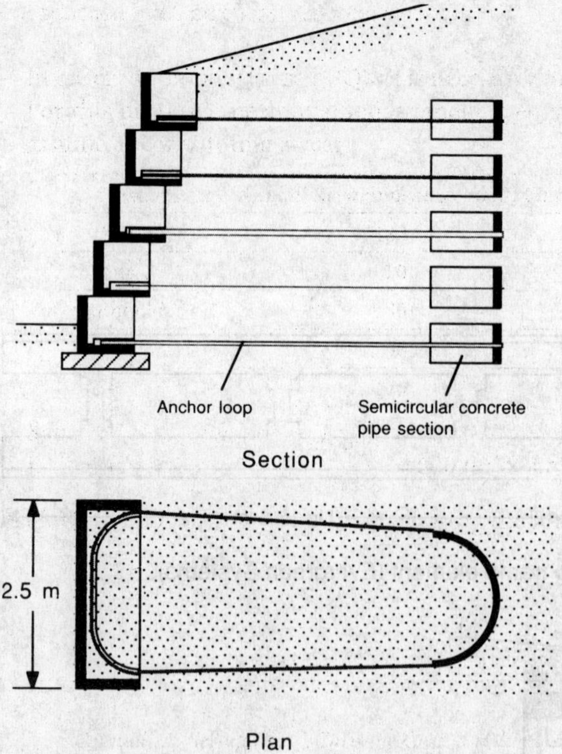

a. NEW retaining system (from "Neue Ebenseer Wand") (after Brandl, 1986)

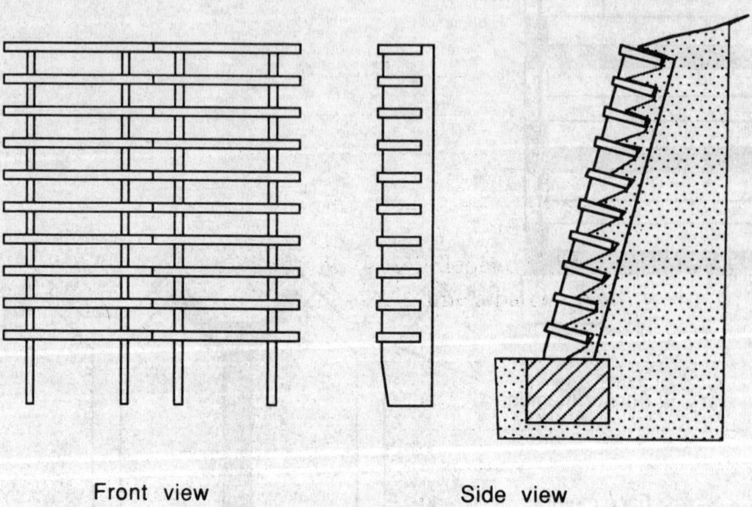

b. "Bookshelf"-type retaining wall

FIGURE 21.3
Alternative wall systems using criblike elements.

around semicircular concrete anchors embedded in the backfill (Fig. 21.3*a*). Structurally, NEW walls are designed according to the principles of anchorage, rather than Reinforced Earth or standard crib wall design.

A free-standing noise abatement wall built with multicell frames is shown in Fig. 21.4 (top left).

Special combinations of crib face elements have found use as revetment which may be anchored back into soil or rock (Fig. 21.4, top right). It protects against erosion, weathering and prevention of rock fall).

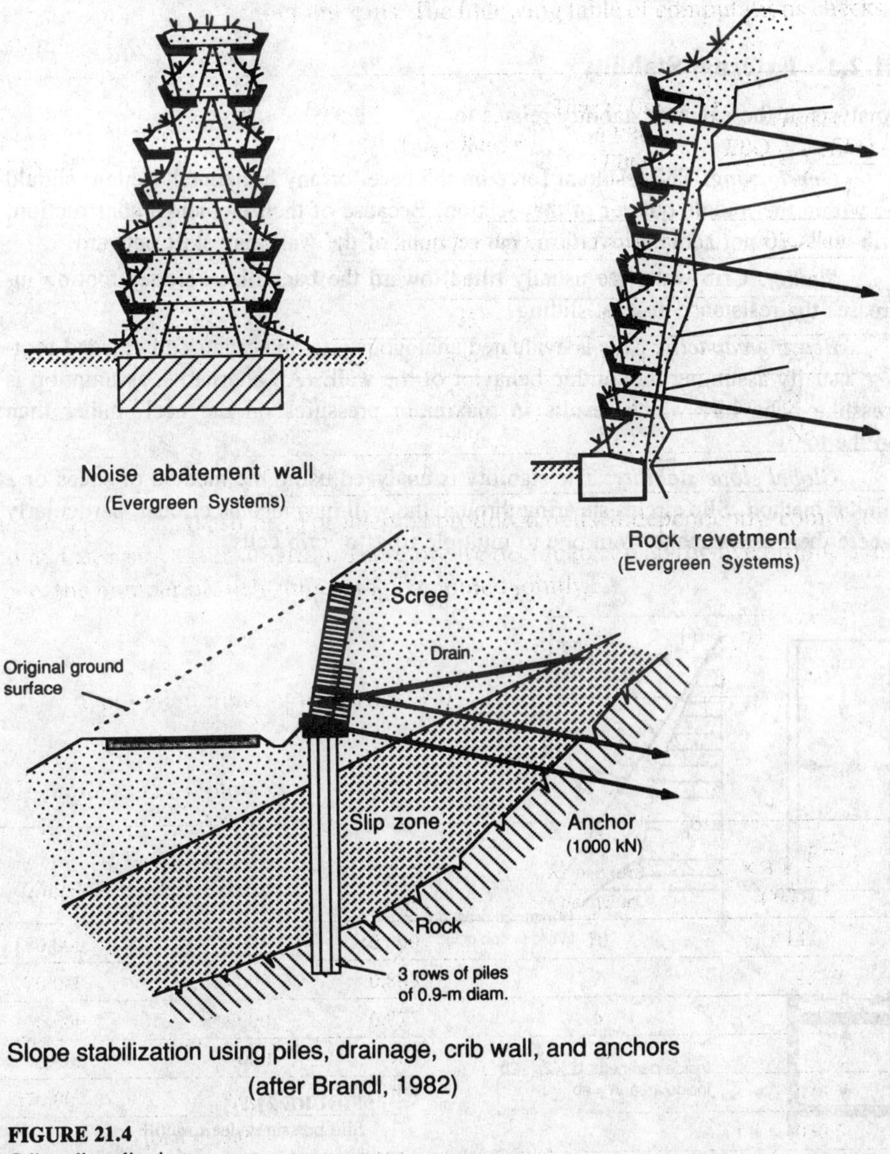

FIGURE 21.4
Crib wall applications.

Where crib walls are built on known unstable land, the whole structure may be anchored back into stable ground. Elaborate designs involving piles, crib walls, and anchors have also been carried out (Fig. 21.4, bottom).

Overall, standard crib walls—the boxlike structures filled with soil—behave as gravity structures, and their stability is evaluated accordingly. Earth pressures within the crib wall are calculated taking into account the ''silo''—effect.

A comprehensive long-term study of the performance of crib walls and related structures has been carried out in Austria, and its results have been reported by Brandl (1980, 1982, 1984, 1986) and form the basis of the following brief notes.

21.2.1 External Stability

Analysis of the external stability relates to

Overturning: The resultant force on the base (or any horizontal section) should lie within the middle quarter of that section. Because of their elemental construction, crib walls do not actually overturn, but sections of the wall may shift outward.

Sliding: Crib walls are usually tilted toward the backfill. A sloped footing increases the resistance against sliding.

Bearing capacity: This is evaluated analogously to an eccentrically loaded footing, usually assuming monolithic behavior of the wall. (An alternative assumption is trusslike behavior, which results in maximum pressures on the heel, rather than on the toe.)

Global slope stability: The stability is analyzed using the method of slices or a similar method. Slip circles shearing through the wall may also be critical, particularly where there is a change from one to multiple rows of crib cells.

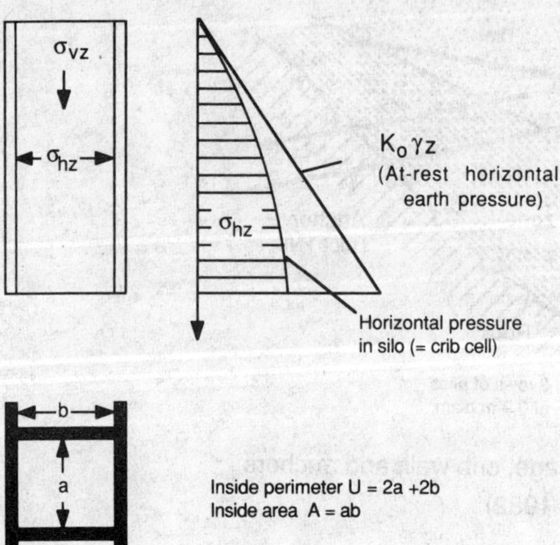

FIGURE 21.5
Silo pressures due to infill.

21.2.2 Internal Pressures

The crib elements have to be dimensioned to withstand internal stresses and loads transmitted at the nodes. Because of arching and the development of friction on the sides of crib cells, the vertical stress in the soil is less than geostatic. The so-called silo theory gives the vertical pressure σ_{vz} at depth z inside the cell as (Fig. 21.5):

$$\sigma_{vz} = \left(\gamma - c\,\frac{U}{A}\right) z_0\,(1 - e^x) \tag{21.1}$$

where $x = \dfrac{-z}{z_0}$

$$z_0 = \frac{A}{UK \tan \delta} \tag{21.2}$$

γ = unit weight of the infill, kN/m^3
U = inside perimeter of cell, m
A = inside area of cell, m^2
c = cohesion of infill, kN/m^2
ϕ = internal friction angle of the infill, °
δ = soil-wall friction angle $\approx 2\phi/3$, °
K = coefficient of lateral earth pressure, usually assumed equal to $K_0 = 1 - \sin\phi$ (for at-rest conditions)

The depth z_0 is equivalent to the depth at which the geostatic vertical pressure is equal to the silo pressure at infinite depth. This asymptotic value of the silo pressure is often taken as the design pressure for crib elements.

To σ_{vz} has to be added the effect Δp_{vz} of any vertical surcharge p_0 directly applied to the cribwall or any vertical loads acting on the elements:

$$\Delta p_{vz} = p_0\, e^y \tag{21.3}$$

where $$y = -\frac{U}{A} K \tan\delta\, z$$

The total inside pressure therefore is

$$\sigma_{vztot} = \sigma_{vz} + \Delta p_{vz} \tag{21.4}$$

as illustrated in Fig. 21.6; the corresponding inside horizontal pressure on the crib elements is

$$\sigma_{hztot} = K_0\, \sigma_{vztot} \tag{21.5}$$

For inclined walls, the inside pressure on the face stretchers is reduced, but that on the back stretcher may be increased (say 25%); however, because the inside pressure on the back stretcher is counteracted by the backfill, this pressure increase is rarely relevant in the design.

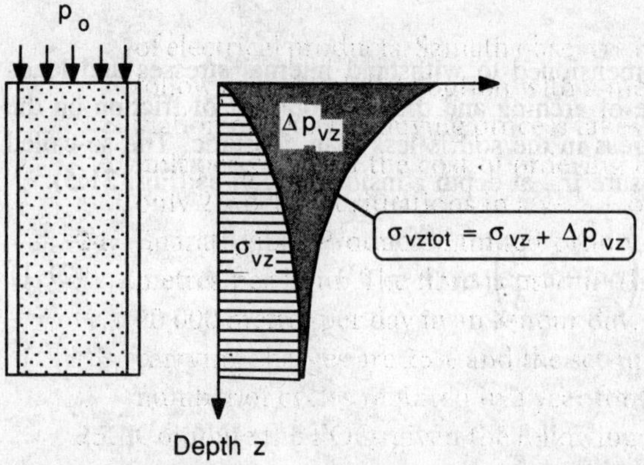

FIGURE 21.6
Silo pressures due to uniform surcharge

21.2.3 Construction Aspects

Where problems are encountered with crib walls, there are likely to be a number of causes. These may be related to inadequate geotechnical investigations, unsuitable fill, inadequate concrete quality, or erroneous assessment of earth pressures, external loads, and internal stresses. It is also possible that wall elements are damaged by careless handling during construction or by vehicles crashing into the finished structure. Many of these possible failure aspects of crib walls are common with other retaining systems.

For high walls, inadequate foundations could cause an increase in the internal earth pressures, if the silo effect cannot be mobilized due to excessive settlement of the elements.

For the prevention of concentrated stresses at the nodes, some systems manufacturers recommend that pressure pads be placed at the contact points.

The infill should be compacted in layers so that the density is equivalent to about 95 to 97% standard Proctor density. Lower density could reduce the stability, but too high a compactive effort could increase the stresses in the elements. As a general rule, it is recommended that the maximum particle size be less than one-sixth of the cell width and there should be less than 15% fines. These conditions should ensure desirable soil-crib interaction and adequate drainage.

21.3 GABIONS AND MATTRESSES

Gabions and mattresses are wire baskets filled with rock. The typical dimensions of a gabion are 1 m high, 1 m deep, and 2 to 4 m long. The term "mattress" is used for units which are thinner but larger in area, say 6 by 2 m and 0.25 m high. Individual units are wired together to form larger structures.

The meshes are made of 2- to 3-mm-diameter wire, with openings from 60 to 100 mm. They may be galvanized or coated with polyvinyl chloride (PVC) for protection against corrosion.

The filling consists of durable rock fragments or river cobbles, exceeding the mesh size but not larger than about half the depth of the individual basket so as to produce a neat front of the structure. Depending on their construction, meshes may need to be stretched or tensioned prior to filling, in order to reduce deformation during construction.

A filter layer, today often a geotextile, is placed between the gabions and the backfill if there is a danger of soil particles being washed out through the rock fill by seepage or wave action.

Gabions and mattresses are mostly used for building gravity structures and as erosion-resistant linings. They have advantages similar to other elemental construction systems:

Flexibility: Gabions conform to difficult site geometry and can adjust to differential settlement and lateral movement.

Permeability: Prevents the buildup of water pressure. Allows the construction of high-capacity drainage systems.

Low level of work skill required: Training of unskilled labor is possible within a short time.

Low cost: Minimal transportation cost where local rock fill can be used. Speedy construction.

21.3.1 Gabion Walls

Typical cross sections of gabion walls are shown in Fig. 21.7*a*; either the front or the back of the wall is stepped. Depending on the properties and the slope of the backfill, the base of a gabion wall may range from 50 to 90% of its height. It is common practice to tilt the wall backward at an angle of about 6° (1:10). High walls may require a concrete footing, but in general only little foundation preparation needs to be done. Figure 21.7*c* illustrates the use of a gabion wall as an abutment for a light bridge.

The unit weight of rock-filled gabions is about two-thirds of the unit weight of the rock solids. This means that the use of typical hard limestone or granite results in a fill weight of about 17 kN/m^3.

The forces acting on a gabion wall are shown in Fig. 21.7*b*. The coefficient of active earth pressure K_a according to the Coulomb theory is

$$K_a = \frac{\sin^2(\alpha + \phi)}{\sin^2\alpha \sin(\alpha - \delta)\left[1 + \sqrt{\dfrac{\sin(\phi + \delta)\sin(\phi - \beta)}{\sin(\alpha - \delta)\sin(\alpha + \beta)}}\right]^2} \tag{21.6}$$

where α is the inclination of the back face of the wall and β is the slope of the backfill, as defined in Fig. 21.7*b*.

Generally it would be appropriate to assume that the wall friction angle δ is equal to the internal friction angle ϕ of the soil.

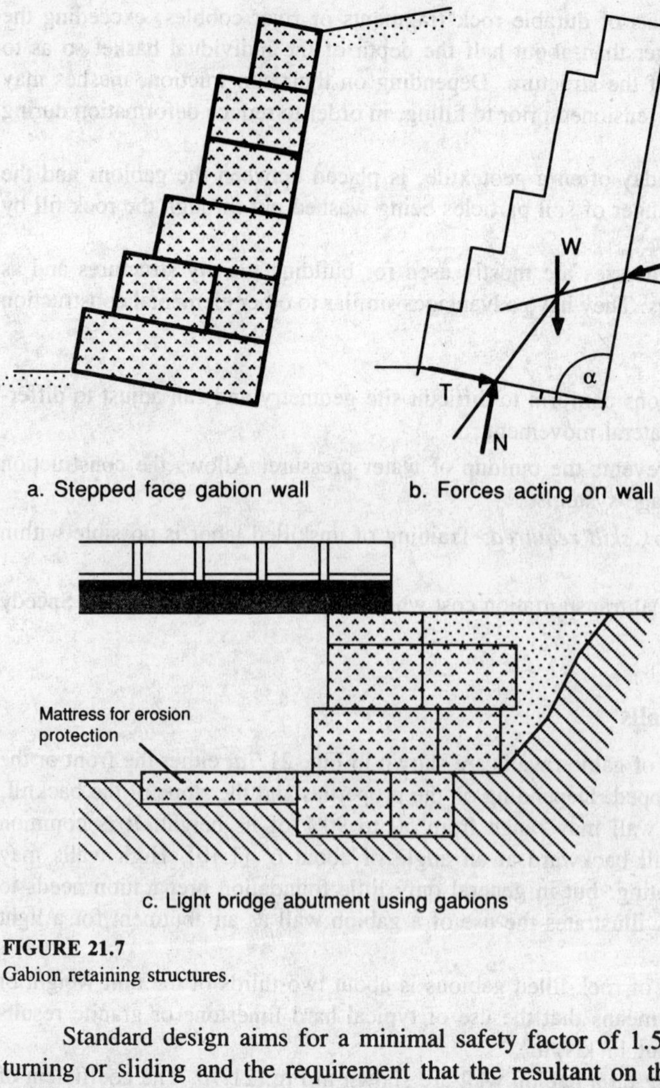

FIGURE 21.7
Gabion retaining structures.

Standard design aims for a minimal safety factor of 1.5 with respect to overturning or sliding and the requirement that the resultant on the base acts within its middle third, in order to prevent theoretical lift-off at the heel. The maximum base pressure should not exceed the allowable bearing pressure on the foundation soil. For high walls of stepped construction, the safety against shear failure at intermediate levels should also be checked. Global stability is again of equal importance as with other retaining structures.

Gabion weirs for river training in mountainous country have been built up to heights of 20 m. Their stability is evaluated as for retaining structures but may be complicated by horizontal water pressures and vertical uplift forces. The geometric design of weirs and their layout belongs in the domain of river engineering.

A more recent extension of the gabion construction technique is to create retaining structures with gabions at the face and to attach to them layers of horizontal mesh in the backfill, similar to the Hilfiker Welded Wire Wall (Sec. 18.2.1.2). The Maccaferri Company successfully built a wall of that kind with their hexagonal double-twisted woven-wire mesh in Sabah-Malaysia in 1979; this system is now called Terramesh. Although more flexible and extensible during handling and placement, the woven wire netting is said to perform as well as more rigid welded mesh, once it is buried and confined in the backfill. As mentioned in Chap. 19, geotextiles have also been suggested as reinforcing material in the backfill, in combination with gabion face elements.

21.3.2 Gabions and Mattresses for Erosion Protection

Typical applications include the protection of

- River beds, banks, and lake shores
- Canal linings
- Bridge piers and abutments
- Inlet and outlet works at culverts (illustrated in Fig. 21.8)

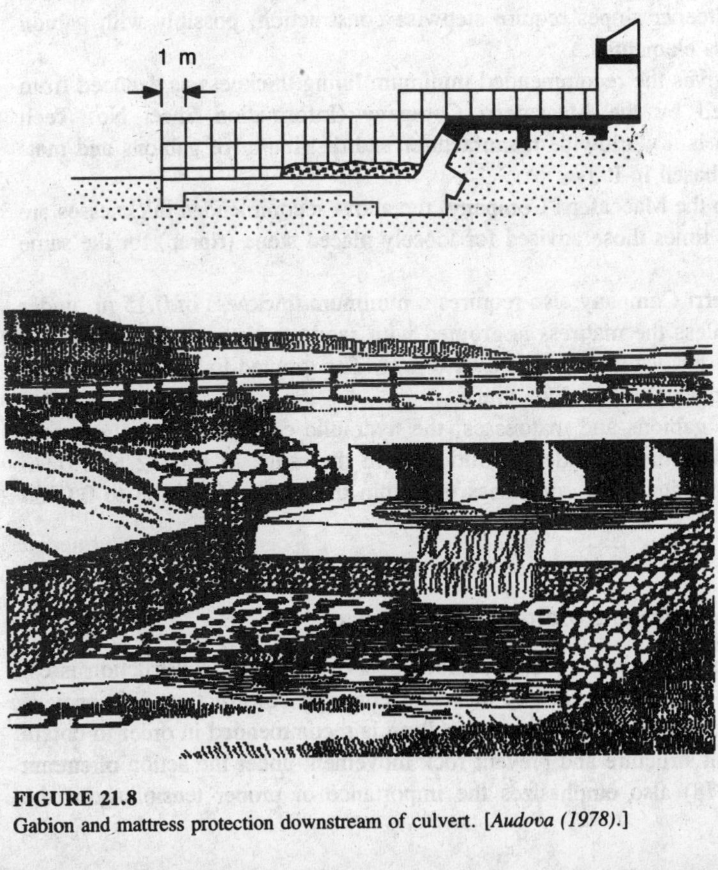

FIGURE 21.8
Gabion and mattress protection downstream of culvert. [*Audova (1978).*]

TABLE 21.1
Minimum lining thickness

Soil type (to be protected)	Water flow velocity, m/s	Minimum mattress thickness required, mm
Clays, heavy cohesive soils	< 0.55	0
	2	170
	3	230
	4.5	300
Silts, fine sands	< 0.75	0
	2	230
	3	300
Coarse river and beach gravel, flat pebbles, and cobbles	< 1.2	
	3.5	170
	5	230
	6	300

Mattresses from 0.15 to 0.5 m thick can be laid on slopes as steep as 1:1.25 without much difficulty. Steeper slopes require stepwise construction, possibly with gabion rather than mattress elements.

Table 21.1 gives the recommended minimum lining thickness as deduced from guidelines provided by the Maccaferri Company (Information Sheet No. Tech 1.1.2/003), which is a worldwide manufacturer and distributor of gabions and mattresses, originally based in Italy.

According to the Maccaferri Company, the above recommended thicknesses are approximately 0.3 times those advised for loosely placed stone (riprap) for the same conditions.

The Maccaferri Company also requires a minimum thickness of 0.15 m, under any conditions, unless the mattress is grouted with sand mastic asphalt. Grouting is also recommended for water velocities over 6 m/s. The grouted linings are generally prefabricated and placed on site by a crane.

For ordinary gabions and mattresses, the hydraulic coefficient of roughness n according to the Manning-Strickler definition is in the range of 0.0222 to 0.0270 $m^{-1/3}$ s. Grouting reduces the roughness to within the limits of 0.0158 to 0.0200 $m^{-1/3}$ s.

21.3.3 Construction and Maintenance Aspects

Although gabion construction is simple, good supervision of the construction is essential. According to Audova (1978), the most common construction deficiency is inadequate filling of the baskets. Slight overfilling is recommended in order to obtain a tight and coherent structure and prevent rock movement under the action of stream flow. Audova (1978) also emphasizes the importance of proper tensioning of the baskets before filling and advises that an additional two ties should be provided between opposite sides of the gabions at intermediate levels of filling of 1-m-high gabions.

In erosion control structures, the provision of flexible aprons and cutoffs (embedded end sections) should be normal design provisions for preventing undermining. Where mattresses or gabions adjoin concrete structures or culverts, effective connection or anchorage is necessary, particularly where water flow is present.

In erosion control structures, the need for regular maintenance is well recognized. The level of acceptable risk will determine the desirable design safety factors, and initial capital expenditure has to be balanced with future maintenance costs. Audova (1978) recommends regular inspections, particularly after floods or heavy rains. He suggests that the following details be checked for:

Inadequate peripheral construction. If cutoffs, returns, and aprons appear inadequate, some form of additional work will be required.

Loss of rock. If loss of rock is excessive, replacement of rock, if possible, or some overlay such as slurry concrete may be required.

Loss of shape. If gabions subjected to high water impact lose their shape, their effectiveness and stability should be assessed.

Torn mesh. Apart from negligence, debris and large angular rocks are the main offenders. Tears should be mended and rocks replaced where necessary.

Abraded mesh. Areas which are subject to high abrasion are crests, aprons, energy dissipaters, etc. Should abrasion become significant, some form of protection or reinforcing should be applied.

Inadequate anchoring. Early performance of a structure often indicates the need for additional staking or stronger anchoring.

Good design, quality materials, and a high standard of work quality and supervision can make gabion construction a versatile construction technique for the civil engineer. As with crib walls, the establishment of vegetation, even if limited, makes gabion structures environmentally more tolerable than massive concrete structures.

21.4 FABRIC FORMWORK

The availability of a wide range of geosynthetics has given new incentive to experiment with soil- or grout-filled bags, tubes, and mattresses and to attempt to build structures which have more than just a temporary function.

The most common building unit is the traditional sandbag, widely used for military purposes and for temporary repairs of dykes, bridges, and roads in flood emergencies. Some highly innovative construction techniques have been developed in recent years (many are patented) which could be said to have evolved from the traditional sandbag.

21.4.1 Sandbags in Military Engineering

Sandbags have a long history of military applications, such as field defenses, revetments for trenches and buildings, and emergency repairs. Bagdon (1971) provided the following figures:

Conflict	Peak annual use, millions	Total sandbags used, millions
World War II	103	400
Korea	—	80
Vietnam	336	—

These figures reflect a change in warfare and the need for fortifications far away from the front lines.

Unfilled, a typical sandbag measures from approximately 670 by 370 mm (United States) to 825 by 250 mm (Australia); it will hold 0.012 to 0.160 m^3 of earth, weighing 20 to 25 kg. A team of five workers is able to fill and place 60 sandbags per hour, making about 2 m^2 of revetment. Automated sandbagging machines have been developed which can produce 500 filled bags per hour.

Sandbag walls or revetments are built not unlike courses of bricks. The first course is laid as "headers" (perpendicular to the length of the wall), the second as "stretchers" (parallel to the wall), etc. Joints in adjacent courses are staggered to provide proper "bonding." When the bags are filled, the seam should be inside; when placed, neither seams nor tied necks of the bags should be visible from the outside. The bags should be about three-quarters full and may be molded into rectangular blocks when being laid. Sandbag walls higher than 1.25 m are sloped at 4 to 1, similar to crib and gabion walls.

Sandbags are much more effective in attenuating explosive shock waves if containing dry rather than wet sand (or clay). It is therefore advisable to prevent saturation of sandbags protecting structures in a theater of war.

Traditionally, sandbags were made from natural fibers (hessian, jute, burlap), which rot when exposed to the weather. In search of improved field serviceability, the U.S. Army [Bagdon (1971)] has investigated the use of synthetic fiber bags and various preservative treatments since 1950. This must represent one of the earliest studies of the durability of what are now called geotextiles or geosynthetics! Materials tested in field conditions included PVC-coated fiberglass, polypropylene, high density polyethylene, and acrylic fabrics. Acrylic sandbags were found to be the most durable, lasting in excess of 30 months.

Because of availability and cost considerations, large quantities of polypropylene bags were used in Vietnam. Where exposed to direct sunlight, these bags could deteriorate within 2 to 3 months. Other drawbacks reported were [Bagdon (1971)]:

(*a*) Bags break when dropped.

(*b*) Bags slip easily when wet.

(*c*) Plastic tie cords slip easily when wet.

(*d*) Bag tends to split when hit by bullets.

(*e*) Bag cannot be repaired.

Acrylic sandbags were designated the preferred item for U.S. Army use in 1968.

In order to make sandbag fortifications more permanent, the military may also consider adding up to 10% cement to the sand fill or providing a concrete cap. The old hessian sandbags were known to last longer if they were dipped in cement wash before they were filled or if they were painted with a cement wash after they were installed.

21.4.2 Civilian Uses of Soil-Filled Bags and Tubes

Stephenson (1982) described field and laboratory experiments with sand-filled acrylic fabric bags, referred to as "sand pillows." These were successful in protecting an irrigation dam from the erosive forces of wind-generated waves over the initial monitoring period of 18 months, without significant fabric deterioration. Much of the study was devoted to proving the practical feasibility of this construction technique for a farmer without specialized construction equipment. The laboratory study established parameters for evaluating sand pillow stability in relation to the embankment slope, wave height, and pillow weight.

Bogossian et al. (1982) reported forming a 3850-m-long dyke system with a soil-filled geotextile "sausage" in a land reclamation project. The final cross-sectional shape of the dyke was elliptical, with a height of 1.42 m and a width of 2.7 m (Fig. 21.9*a*). A nonwoven geotextile was used for confining the soil. The dredged material, mostly organic clayey and silty sand, was injected at intervals of 8 to 20 m. Because of the clogging of the fabric, openings in the sausage were left for suspended clay particles to be discharged, avoiding a buildup of pressure which may otherwise have caused the fabric to burst. This precaution is not necessary if only a sand-water mix is pumped into the tube, as has been shown on other similar projects elsewhere.

As described by Koerner and Welsh (1980), a system named Longard Tubes, developed in Denmark and patented in 1967, creates erosion-control barriers by pumping sand into a continuous tube up to 1.8 m in diameter. These tubes are made of a polyethylene woven textile, internally lined with an impermeable membrane. The water remaining after the sand has settled out escapes through a regulated outlet.

Van Santvoort and Troost (1986) were able to improve the support of railway sleepers by attaching them to tubular fabric bags filled with gravel and sand (Fig. 21.9*b*). Thus a geotextile-reinforced sleeper bed was created which, when compared to normal construction, showed less settlement and reduced vibration under heavy traffic. Another benefit gained was the capacity of the sleeper bed to maintain its elasticity throughout its service life. The bags were made of high-modulus, low-creep woven fabric with a breaking strength of 120 kN/m. The sleepers were attached to the bags with woven straps. Vibratory compaction was used during filling. The filled-

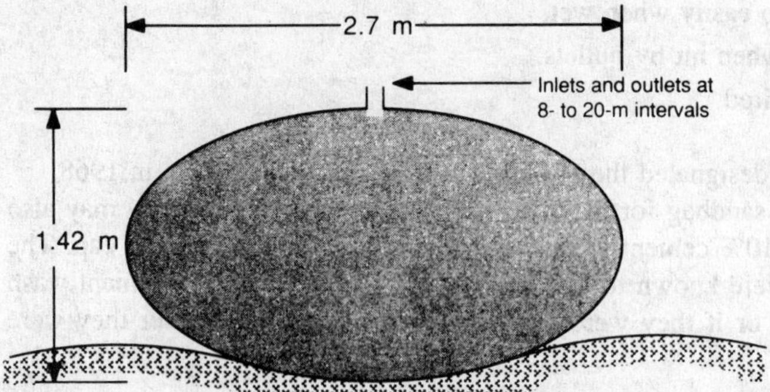

a. Continuous geotextile dike (after Bogossian et al. 1982)

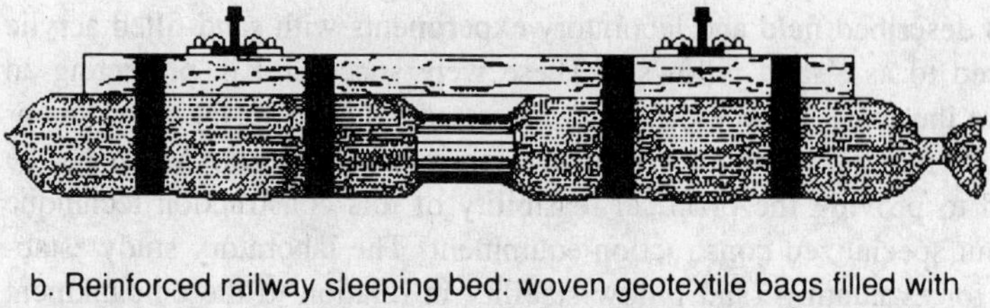

b. Reinforced railway sleeping bed: woven geotextile bags filled with rounded gravel or sand, strapped to sleeping bed (Van Santvoort and Troost, 1986)

FIGURE 21.9
Examples of soil-filled bags and tubes.

bag sleeper unit was then pressed into a mold to create a somewhat flatter shape with improved load distribution.

21.4.3 Fabric-Formed Concrete

Along forest roads in Oregon, retaining walls have been built with reinforced sacked concrete (Kabil Development Corporation, Fig. 21.10). Successive layers of prewetted concrete were placed on top of each other and impaled with vertical reinforcing bars, two to each bag. The wall was extended into a foundation trench to provide resistance against base shear. At vertical intervals of about 1.2 m, so-called wall stabilizers consisting of horizontal sections of chain-link fabric (like fencing material) were attached to the wall and embedded in compacted backfill. Drainage was ensured by appropriate selection of the backfill material and the provision of French drains where required.

Bags pumped full with a cement–fly-ash–aggregate mixture have been used in Australia to build walls in underground coal mines for sealing off old sections and thus reducing the risks of explosions.

More common is the use of grout-filled fabric bags for erosion-control structures. One of the advanced systems being marketed internationally is Fabriform. A fluid,

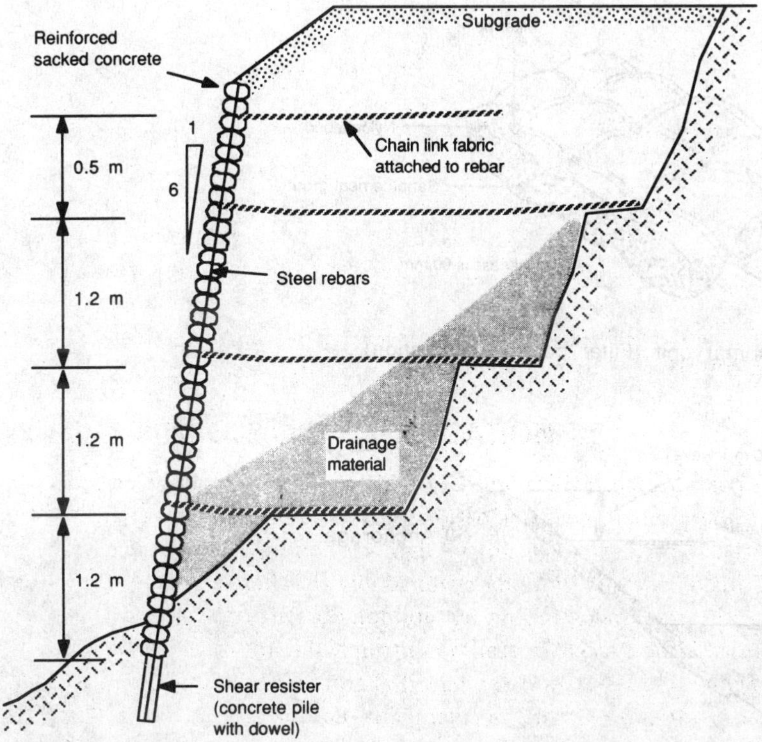

FIGURE 21.10
Reinforced sacked concrete wall. *(Kabil Development Corp., Oregon.)*

fine aggregated concrete is pumped into flexible nylon fabric forms. The fabric also acts as a filter, allowing excess water to seep away, resulting in a low water/cement ratio mortar and forming a high-strength, durable mat. The form is created by two layers of fabric, joined at regular intervals by interweaving. The finished structure looks something like a blown-up air mattress or a cobbled pavement (Fig. 21.11). Where the fabric layers are joined, openings are created which allow pore pressures below the revetment to be relieved. Alternatively, plastic tubes may be installed for the purpose of preventing uplift.

Unless coated with, e.g., an acrylic emulsion, the Fabriform's top nylon layer will be subject to physical and chemical degradation as well as mechanical abrasion, but good performance has been reported over periods of more than 10 years. On the other hand, the bottom layer of fabric is well protected and will continue to play a reinforcing function for the concrete, which assists in limiting crack width and loss of fines from beneath the revetment.

In Australia Fabriform-type revetments are available under the registered trade name of Fabrimat. A very fluid sand-cement mortar is used in this work, typically with the composition given in Table 21.2.

An air content of 5 to 8% improves the pumpability of the mortar and the freeze-thaw resistance of the hardened mortar.

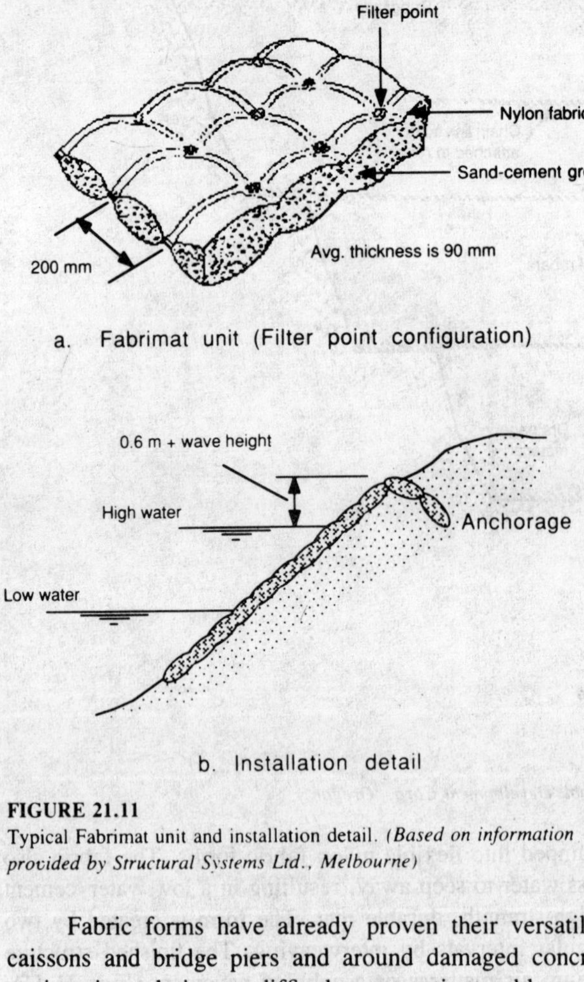

FIGURE 21.11
Typical Fabrimat unit and installation detail. *(Based on information provided by Structural Systems Ltd., Melbourne)*

Fabric forms have already proven their versatility in confining grout below caissons and bridge piers and around damaged concrete piles. Several innovative engineering solutions to difficult construction problems are illustrated by case histories in the books by Koerner and Welsh (1980) and Koerner (1986).

21.4.4 Open Cell Confinement

Webster and Watkins (1977) of the U.S. Army Corps of Engineers reported on an investigation of construction techniques for tactical bridge approach roads across soft

TABLE 21.2
Fabrimat mortar composition

Material	As delivered, kg/m^3	In place, kg/m^3
Cement	420–540	480–590
Sand	1300–1070	1480–1190
Water	310–370	240–300

ground. The ground-reinforcement systems evaluated included plastic tubing, beside steel landing mats, woven and nonwoven geotextiles, and gabions. The corrugated plastic tubing (152 mm in diameter) was cut into 300-mm-long sections, fastened together in a honeycomb arrangement, placed vertically on the subgrade, filled with sand, and compacted with a vibratory plate compactor. This sand-confinement system outperformed ordinary crushed rock under traffic loading.

Geoweb is one of the commercially available open-cell-confinement systems. It consists of 200-mm-wide high-density polyethylene strips welded together at 330-mm intervals. It is transported in a folded condition. In the field, the system is expanded into a honeycomb grid and filled with cohesionless soil.

A similar product is Armater, which consists of nonwoven fabric. When placed, its openings have a hexagonal shape. Its main application would be erosion protection of slopes, but like other similar products, it can assume the role of a base course or conventional aggregate layer where the materials available are substandard.

A proper analysis of the effect of open cell confinement on the bearing capacity and rutting resistance would be rather complex. As far as the bearing capacity is concerned, the simplest way of recognizing the improvement caused by lateral confinement of the soil is to do an analysis with and without the depth term qN_q . . . [see Eq. (18.2)], the effective depth being that over which the cells provide confinement.

PROBLEMS

Prefixes indicate problem type: C = calculations, B = brief answer, M = multiple choice, D = discussion.

Section 21.1

D21.1. Discuss the reinforcing action in terms of internal-external and longitudinal-transverse confinement of

(*a*) An adobe brick.

(*b*) Gabions.

(*c*) Geocells (a system of interconnected, vertical, open-ended cylindrical cells, such as Armater) used as a road mattress and slope protection.

(*d*) A tunnel roof with rock anchors.

Section 21.2

B21.2. List four advantages of crib walls.

(*a*) ____________

(*b*) ____________

(*c*) ____________

(*d*) ____________

D21.3. What is the difference between a crib wall and

(*a*) A bin wall?

(*b*) The Austrian NEW retaining system (Fig. 21.3*a*)?

(*c*) A bookshelf-type wall (Fig. 21.3*b*)?

Discuss construction method, type of support for backfill (gravity wall, anchorage, reinforcement), likely earth pressure coefficient *K*, etc.

C21.4. For soil with a unit weight of 18 kN/m^3 and a friction angle of $\phi = 30°$, calculate the horizontal earth pressure at a depth of 3 m for
(*a*) At-rest (K_0) conditions.
(*b*) Active (K_a) conditions without wall friction.
(*c*) Active conditions with full wall friction ($\delta = 30°$).
(*d*) Silo pressures in a crib wall with an inside area of 1.5 by 1.5 m ($\delta = 2\phi/3$).
Sketch the pressure distribution from 0 to 3 m.

M21.5. The silo effect
(*a*) Increases vertical soil pressures.
(*b*) Is less for K_0 conditions.
(*c*) Is decreased if crib elements settle into the foundation soil.
(*d*) Does not apply to a surcharge.

B21.6. List four causes of problems with crib walls.
(*a*) ______________
(*b*) ______________
(*c*) ______________
(*d*) ______________

M21.7. The crib wall infill
(*a*) Should not be compacted.
(*b*) Should be compacted to about 95 to 97% standard Proctor density.
(*c*) Should be compacted to about 95 to 97% modified standard Proctor density.
(*d*) Should have more than 15% fines to facilitate plant growth.

Section 21.3

B21.8. List four advantages of gabion walls.
(*a*) ______________
(*b*) ______________
(*c*) ______________
(*d*) ______________

C21.9. Calculate the safety factor against overturning of a wall as shown in Fig. 21.7*a* given the following data:

Gabions are 1 m high and 1, 1.5, 2, or 3 m wide
Rockfill unit weight = 17 kN/m^3
Backfill (horizontal surface): $\phi = 28°$, unit weight = 18 kN/m^3
Inclination of back face: 1:4.

C21.10. A creek bed consists of fine silty sand. Water is expected to flow at velocities as high as 2.5 m/s. What thickness of gabion-type mattress would you recommend?

B21.11. After the first flood has rushed through a gabion- and mattress-protected creek bed, you are asked to make an inspection. Describe at least four damage conditions you would check for.
(*a*) ______________
(*b*) ______________
(*c*) ______________
(*d*) ______________

Section 21.4

B21.12. How can soil-filled bags and tubes assist in

(*a*) Building retaining structures?

(*b*) Protecting embankment slopes?

(*c*) Providing improved foundations?

B21.13. Figure P21.1 shows various ways of building a sandbag revetment. Indicate the best method from the cross sections (*a, b, c, d*) and frontal views (*e, f, g, h*) presented.

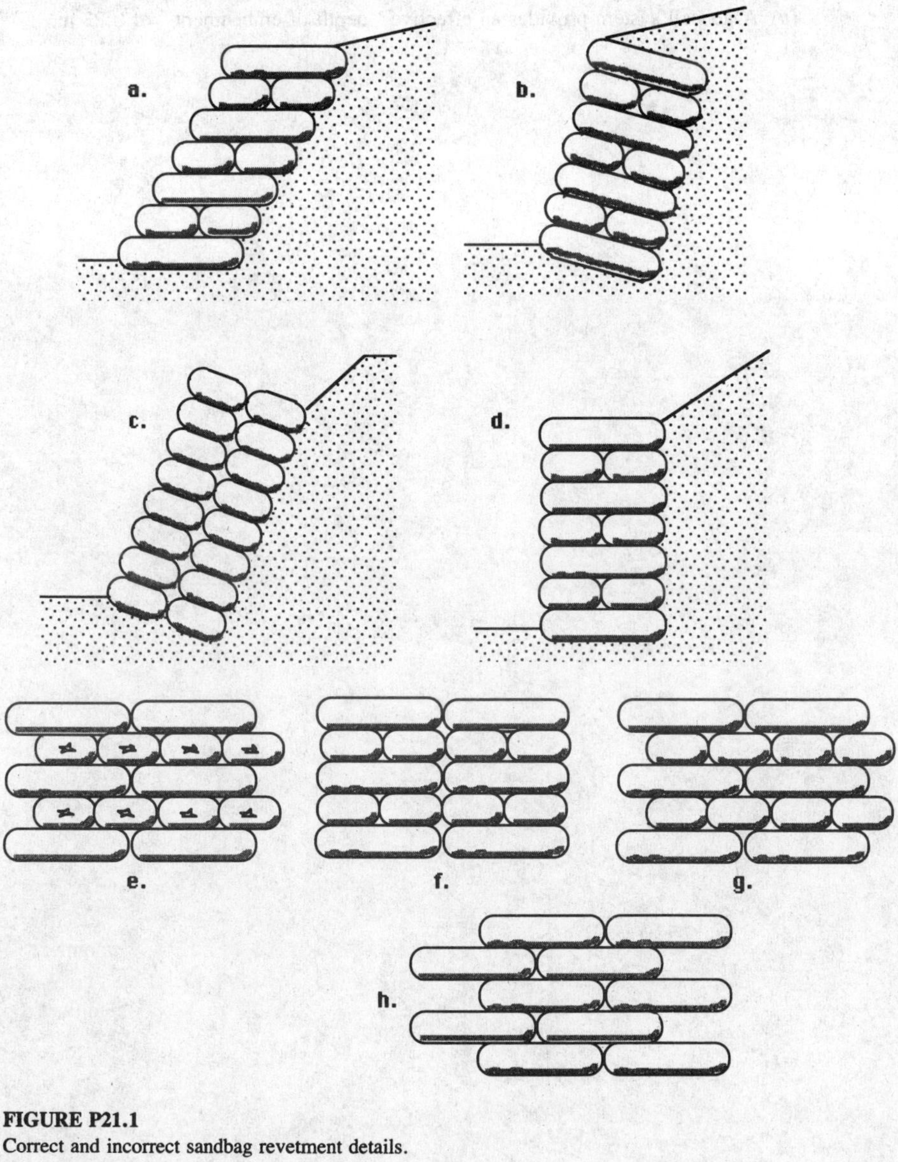

FIGURE P21.1
Correct and incorrect sandbag revetment details.

M21.14. In terms of percent of total weight, the typical composition of the grout pumped into Fabriform is (cement : sand : water)

(*a*) 33 : 33 : 34.
(*b*) 25 : 60 : 15.
(*c*) 25 : 25 : 50.
(*d*) 10 : 25 : 65.

C21.15. Assume the effective tire print area is 0.25 by 0.5 m. Calculate the ultimate bearing capacity on dry sand with a friction angle of 36°, for the following conditions:

(*a*) Surface loading, no confinement.
(*b*) A geocell system provides an effective "depth of embedment" of 0.23 m.

APPENDIX A

THE UNIFIED SOIL CLASSIFICATION SYSTEM (USCS)

The Unified Soil Classification System is the most widely used system for classifying soils for engineering purposes, but minor differences exist in its application around the world. Most importantly, outside the United States, coarse grained soils are usually defined as those with more than half the material larger than 0.6 mm, rather than the No. 200 sieve size (0.074 mm openings); similarly, the boundary between sand and gravel is chosen at 2 mm, rather than the No. 4 sieve size (4.7 mm openings, according to the U.S. Bureau of Standards).

Presented here is the metricated USCS according to Australian Standard AS 1726-1975, Add. No. 1 (February 1978).

UNIFIED SOIL
DATA FOR DESCRIPTION IDENTIFICATION

Add. No 1 (February 1978) to AS 1726—1975

MAJOR DIVISIONS				GROUP SYMBOL	GRAPHIC SYMBOL	TYPICAL NAME	DESCRIPTIVE DATA
COARSE-GRAINED SOILS	More than 50% by dry mass, less than 60mm is greater than 0.06mm	GRAVELS	More than 50% of coarse grains are greater than 2.0mm	GW		Well graded gravels and gravel-sand mixtures, little or no fines	Give typical name, indicate approximate percentages of sand and gravel, maximum size, angularity, surface condition and hardness of the coarse grains; local or geological name and other pertinent descriptive information, symbols in parenthesis. For undisturbed soils add information on stratification, degree of compactness, cementation, moisture conditions and drainage characteristics. EXAMPLE: Silty Sand, gravelly, about 20% hard, angular gravel particles, 10mm maximum size, rounded and sub-angular sand grains coarse to fine; about 15% non-plastic fines with low dry strength, well compacted and moist in place; light brown alluvial sand (SM)
				GP		Poorly graded gravels and gravel-sand mixtures, little or no fines	
		GRAVELLY SOILS		GM		Silty gravels, gravel-sand-silt mixtures	
				GC		Clayey gravels gravel-sand-clay mixtures.	
		SANDS	More than 50% of coarse grains are less than 2.0mm	SW		Well graded sands and gravelly sands, little or no fines	
				SP		Poorly graded sands and gravelly sands, little or no fines	
		SANDY SOILS		SM		Silty sand, sand-silt mixtures.	
				SC		Clayey sands, sand-clay mixtures.	
FINE-GRAINED SOILS	More than 50% by dry mass, less than 60mm is less than 0.06mm	Liquid Limit less than 50%		ML		Inorganic silts, very fine sands, rock flour, silty or clayey fine sands	Give typical name; indicate degree and character of plasticity, amount and maximum size of coarse grains; colour in wet condition; odour if any; local or geological name and other pertinent descriptive information; symbols in parenthesis. For undisturbed soil add information on structure, stratification, consistency in undisturbed and remoulded states, moisture and drainage conditions. EXAMPLE Clayey Silt, brown; low plasticity; small percentage of fine sand; numerous vertical root-holes; firm and dry in place, fill (ML)
				CL		Inorganic clays of low to medium plasticity, gravelly clays, sandy clays, silty clays, lean clays	
				OL		Organic silts and organic silty clays of low plasticity	
		Liquid Limit more than 50%		MH		Inorganic silts, micaceous or diatomaceous fine sands or silts, elastic silts.	
				CH		Inorganic clays of high plasticity, fat clays.	
				OH		Organic clays of medium to high plasticity	
				Pt		Peat muck and other highly organic soils.	

NOTES:

1. The above table follows the original Unified Classification System (USBR Earth Manual) and ASTM D 2487 except that it adopts the particle size limits given in AS 1289 and other standards, viz:

   Gravel .... 2 to 60 mm
   Sand .... 0.06 to 2 mm
   Silt and clay .... $<$ 0.06 mm

   The system excludes the boulder and cobble fractions of the soil and classifies only the material less than 60 mm in size.
2. As 60 mm, 2 mm and 0.06 mm sieve sizes are not normally used, the percentages passing these sizes can be obtained from a particle size distribution curve determined from a laboratory test. Alternatively, the percentages passing may be estimated in the field.
3. For field identification procedures for fine-grained soils or fractions, see Paragraph D6.

CLASSIFICATION SYSTEM (METRICATED) AND CLASSIFICATION OF SOILS

<table>
<tr><th colspan="9">FIELD IDENTIFICATION</th><th colspan="6">LABORATORY CLASSIFICATION</th></tr>
<tr><td colspan="4" rowspan="2"></td><th colspan="4">GRAVELS AND SANDS</th><th rowspan="2">GROUP SYMBOL</th><td rowspan="19">Use the gradation curve of material passing 60 mm for classification of fractions according to the criteria given under "Major Divisions"</td><th rowspan="2">% (2) < 0·06mm</th><th rowspan="2">PLASTICITY OF FINE FRACTION</th><th rowspan="2">$C_u = \frac{D_{60}}{D_{10}}$</th><th rowspan="2">$C_c = \frac{(D_{30})^2}{D_{10} D_{60}}$</th><th rowspan="2">NOTES</th></tr>
<tr><th colspan="2">GRADATIONS</th><th>NATURE OF FINES</th><th>DRY STRENGTH</th></tr>
<tr><td rowspan="17">Determine approximate percentages of material over 60mm in size, maximum size, shape, surface texture, hardness of material, geological description Identify on estimated percentage mass of the various fractions</td><td rowspan="10">COARSE-GRAINED SOILS</td><td rowspan="10">More than half of the material less than 60 mm is larger than 0·06mm</td><td rowspan="17">0·06mm is about the smallest particle visible to the naked eye</td><td>GOOD</td><td>Wide range in grain size</td><td rowspan="2">"Clean" materials (not enough fines to bind coarse grains)</td><td rowspan="2">None</td><td>GW</td><td>0-5</td><td>-</td><td>>4</td><td>between 1 and 3</td><td rowspan="8">1. Identify fines by the method given for fine-grained soils.
2. Borderline classifications occur when the percentage of fines (fraction smaller than 0.06 mm size) is greater than 5% and less than 12%. Borderline classifications require the use of dual symbols e.g. SP-SM GW-GC</td></tr>
<tr><td>POOR</td><td>Predominantly one size or range of sizes.</td><td>GP</td><td>0-5</td><td>-</td><td colspan="2">Fails to comply with above</td></tr>
<tr><td rowspan="2">GOOD to FAIR</td><td rowspan="2">"Dirty" materials (excess of fines)</td><td>Fines are non-plastic (1)</td><td>None to medium</td><td>GM</td><td>12-50</td><td>Below 'A' line or I_p<4</td><td>-</td><td>-</td></tr>
<tr><td>Fines are plastic (1)</td><td>Medium to high</td><td>GC</td><td>12-50</td><td>Above 'A' line and I_p>7</td><td>-</td><td>-</td></tr>
<tr><td>GOOD</td><td>Wide range in grain size</td><td rowspan="2">"Clean" materials (not enough fines to bind coarse grains)</td><td rowspan="2">None</td><td>SW</td><td>0-5</td><td>-</td><td>>6</td><td>between 1 and 3</td></tr>
<tr><td>POOR</td><td>Predominately one size or range of sizes.</td><td>SP</td><td>0-5</td><td>-</td><td colspan="2">Fails to comply with above</td></tr>
<tr><td rowspan="2">GOOD to FAIR</td><td rowspan="2">"Dirty" materials (excess of fines)</td><td>Fines are non-plastic (1)</td><td>None to medium</td><td>SM</td><td>12-50</td><td>Below 'A' line or I_p<4</td><td>-</td><td>-</td></tr>
<tr><td>Fines are plastic (1)</td><td>Medium to high</td><td>SC</td><td>12-50</td><td>Above 'A' line and I_p>7</td><td>-</td><td>-</td></tr>
<tr><th colspan="4">SILT AND CLAY FRACTION</th><td></td><td colspan="5" rowspan="3"></td></tr>
<tr><th colspan="4">Fraction smaller than 0·20mm AS sieve size</th><td></td></tr>
<tr><td rowspan="7">FINE-GRAINED SOILS</td><td rowspan="7">More than half of the material less than 60 mm is smaller than 0·06mm</td><th colspan="2">DRY STRENGTH</th><th>DILATANCY</th><th>TOUGHNESS</th><td></td></tr>
<tr><td colspan="2">None to low</td><td>Quick to slow</td><td>None</td><td>ML</td><td rowspan="6">More than 50% passing 0·06mm</td><td>Below 'A' line</td><td colspan="3"></td></tr>
<tr><td colspan="2">Medium to high</td><td>None to very slow</td><td>Medium</td><td>CL</td><td>Above 'A' line</td><td colspan="3"></td></tr>
<tr><td colspan="2">Low to medium</td><td>Slow</td><td>Low</td><td>OL*</td><td>Below 'A' line</td><td colspan="3"></td></tr>
<tr><td colspan="2">Low to medium</td><td>Slow to none</td><td>Low to medium</td><td>MH</td><td>Below 'A' line</td><td colspan="3"></td></tr>
<tr><td colspan="2">High to very high</td><td>None</td><td>High</td><td>CH</td><td>Above 'A' line</td><td colspan="3"></td></tr>
<tr><td colspan="2">Medium to high</td><td>None to very slow</td><td>Low to medium</td><td>OH*</td><td>Below 'A' line</td><td colspan="3"></td></tr>
<tr><td colspan="8">Readily identified by colour, odour, spongy feel and generally by fibrous texture</td><td>Pt*</td><td colspan="3">*Effervesces with H_2O_2</td><td colspan="3"></td></tr>
</table>

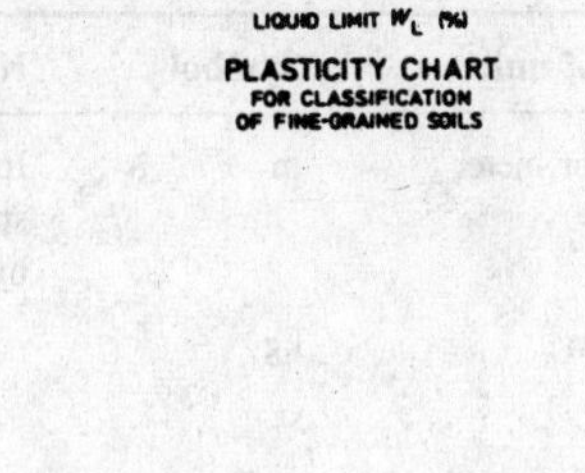

PLASTICITY CHART
FOR CLASSIFICATION OF FINE-GRAINED SOILS

APPENDIX B

SI UNITS IN GROUND MODIFICATION

The International System of Units (''Le Système International d'Unités''), or SI, has now been adopted in engineering practice in most industrial nations around the world. The United States is one of the few countries still using Imperial or British units. However, even in the United States professional engineers now accept, if not prefer SI units in technical and scientific publications.

The following tables list SI units as far as they are relevant for geotechnical engineering in general, and ground modification in particular. A distinction is made between base units, supplementary units, derived units, and units with special names. Certain well-recognized non–SI units are also given. These units were retained because of their practical importance and may be used in a limited way in combination with pure SI units.

1. BASE UNITS

Quantity	Name of unit	Symbol	Notes
Length	metre, or meter (U.S.)	m	In most English-speaking countries, the spelling is metre, as also recommended by ASTM
Mass	kilogram	kg	
Time	second	s	
Electric	ampere	A	
Thermodynamic temperature	kelvin	K	Celsius temperature (°C) is also used. Temperature differences are the same in K as in °C.

2. SUPPLEMENTARY UNITS

Quantity	Name of unit	Symbol	Notes
Plane angle	radian	rad	The radian is the plane angle between two radii of a circle which mark off on the circumference of the circle an arc equal in length to the radius.

3. SOME UNITS DERIVED FROM BASE AND SUPPLEMENTARY UNITS

Derived units are expressed in terms of base units and/or supplementary units. Multiplication within compound units may be indicated in one of the following ways:

N.m N·m N m

The first version is now preferred in many countries because it prevents confusion when the unit symbol coincides with a prefix and because of its ease of typing.

Quantity	Name of unit	Symbol
Acceleration	meter per second squared	m/s^2
Angular velocity	radian per second	rad/s
Area	square meter	m^2
Density	kilogram per cubic meter	kg/m^3
Kinematic viscosity	square meter per second	m^2/s
Mass per unit area	kilogram per square meter	kg/m^2
Moment of inertia	kilogram × meter squared	$kg{\cdot}m^2$
Rotational frequency	unit per second	$1/s$ or s^{-1}
Velocity	meter per second	m/s
Volume	cubic meter	m^3

4. UNITS WITH SPECIAL NAMES

Units derived from base units and having special names:

Quantity	Name	Symbol	Definition	Derivation
Energy, work, quantity of heat	joule	J	1 J = N·m	$1\ kg{\cdot}m^2/s^2$
Electrical resistance	ohm	Ω	1 Ω = 1 V/A	$1\ kg{\cdot}m^2/(A^2{\cdot}s^3)$
Force	newton	N	$1\ N = 1\ kg{\cdot}m/s^2$	
Frequency	hertz	Hz	$1\ Hz = 1\ s^{-1}$	
Power	watt	W	1 W = 1 J/s	$1\ kg{\cdot}m^2/s^3$
Potential difference, electromotive force	volt	V	1 V = 1 W/A	
Pressure, stress	pascal	Pa	$1\ Pa = 1\ N/m^2$	$1\ kg/(m{\cdot}s^2)$

Units derived from other units having special names:

Quantity	Symbol	Derivation
Resistivity	Ω·m	$kg{\cdot}m^3/(A^2{\cdot}s^3)$
Heat capacity	J/K	$kg{\cdot}m^2/(s^2{\cdot}K)$
Dynamic viscosity	Pa·s	kg/(m·s)
Specific heat capacity	J/(kg·K)	$m^2/(s^2{\cdot}K)$
Thermal conductivity	W/(m·K)	$kg{\cdot}m/(s^3{\cdot}K)$

5. PREFIXES FOR SI UNITS

Prefixes are used to form multiples of SI units. With formulas involving more than one unit, only SI base, supplementary, and derived units should be used. Note that kilogram is the only base unit which already contains a prefix. Multiples of mass units are formed by adding prefixes to the word "gram." Common prefixes are

Factor	Prefix	Symbol
10^9	giga	G
10^6	mega	M
10^3	kilo	k
10^{-2}*	centi	c
10^{-3}	milli	m
10^{-6}	micro	μ

*centi is no longer a preferred prefix, similar to hecto, deka (or deca), or deci, the prefixes for the factors 10^2, 10, and 10^{-1}, respectively.

6. NON-SI UNITS OF IMPORTANCE

In geotechnical engineering and elsewhere, a number of non-SI units are retained alongside SI units because of their practical importance. These include:

Quantity	Name	Symbol	Definition
Volume	litre (or liter)*	L (l is also used)	$1\ L = 10^{-3}\ m^3$
Mass	tonne (or metric ton)		1 t = 1000 kg = 1Mg
Plane angle	degree	. . .	$1° = (\pi/180)$ rad
	minute	. . .	$1' = (1/60)°$
	second	. . .	$1'' = (1/60)'$
Time	day	day	1 day = 24 h
	hour	h	1 h = 60 min
	minute	min	1 min = 60 s
Temperature	degree Celsius	°C	$T_C = T_K - 273.15$ T_C = temperature in °C T_K = temperature in K
Pressure	millibar	mb	1 mb = 100 Pa

*Multiples of liters are formed in the same manner as for SI units, for example; mL (milliliter).

7. CONVERSION FACTORS

SI units to other units		Other units to SI units	
Length			
1 km	= 0.621 371 mi (U.S. statute)	1 mi (U.S. statute)	= 1.609 344 km
1 m	= 1.093 61 yd	1 yd	= 0.914 4 m
1 m	= 3.280 84 ft	1 ft	= 0.304 8 m
1 mm	= 0.039 370 in	1 in	= 25.4 mm
Area			
1 km^2	= 0.386 102 mi^2	1 mi^2	= 2.589 99 km^2
1 ha	= 2.471 05 acre	1 acre	= 0.404 686 ha
1 m^2	= 1.195 99 yd^2	1 yd^2	= 0.836 127 m^2
1 m^2	= 10.763 9 ft^2	1 ft^2	= 0.092 903 0 m^2
1 mm^2	= 1.550 × 10^{-3} in^2	1 in^2	= 645.2 mm^2
Volume, section modulus, etc.			
1 m^3	= 1.308 yd^3	1 yd^3	= 0.764 6 m^3
1 m^3	= 35.315 ft^3	1 ft^3	= 0.028 32 m^3
1 L	= 0.035 315 ft^3	1 ft^3	= 28.32 L
1 L	= 0.220 0 gal (British)	1 gal (British)	= 4.546 L
1 L	= 0.264 17 gal (U.S.)	1 gal (U.S.)	= 3.785 L
1 mm^3	= 0.061 024 × 10^{-3} in^3	1 in^3	= 16.387 × 10^3 mm^3
1 cm^3	= 0.061 024 in^3	1 in^3	= 16.387 cm^3
Second moment of area			
1 mm^4	= 2.402 5 × 10^{-6} in^4	1 in^4	= 0.416 231 × 10^6 mm^4
1 cm^4	= 0.024 025 in^4	1 in^4	= 41.623 1 cm^4
Velocity			
1 m/s	= 3.280 84 ft/s	1 ft/s	= 0.304 8 m/s
1 km/h	= 0.621 371 mi/h	1 mi/h	= 1.609 344 km/h
1 m/s	= 2.236 94 mi/h	1 mi/h	= 0.447 04 m/s
Acceleration (standard acceleration of free fall = 9.806 65 m/s^2)			
1 m/s^2	= 3.280 84 ft/s^2	1 ft/s^2	= 0.304 8 m/s^2
Volumetric flow			
1 m^3/s	= 35.315 ft^3/s	1 ft^3/s	= 0.028 32 m^3/s
1 L/s	= 2.119 ft^3/min	1 ft^3/min	= 0.471 9 L/s
1 L/s	= 13.20 gal (British)/min	1 gal (British)/min	= 0.075 77 L/s
1 L/s	= 15.85 gal (U.S.)/min	1 gal (U.S.)/min	= 0.063 08 L/s

SI units to other units		Other units to SI units	
Mass (1 Mg = 1 tonne = 1 metric ton, 1 long ton = 2240 lb, 1 short ton = 2000 lb)			
1 Mg	= 0.984 207 long ton	1 long ton	= 1.016 05 Mg
1 Mg	= 1.102 31 short ton	1 short ton	= 0.907 185 Mg
1 Mg	= 19.684 14 cwt	1 cwt	= 50.802 kg
1 kg	= 2.204 62 lb	1 lb	= 0.453 592 kg
1 g	= 0.035 274 oz	1 oz	= 28.349 5 g
1 kg	= 0.068 521 slug [lb-force/(ft·s^2)]	1 slug	= 14.594 kg
Mass/unit length			
1 kg/m	= 0.671 969 lb/ft	1 lb/ft	= 1.488 16 kg/m
1 g/m	= 0.201 591 lb/100 yd	1 lb/100 yd	= 4.960 55 g/m
1 g/m	= 3.548 lb/mi	1 lb/mi	= 0.281 849 g/m
Mass/unit area			
1 kg/m^2	= 0.204 816 lb/ft^2	1 lb/ft^2	= 4.882 43 kg/m^2
1 g/m^2	= 0.029 494 oz/yd^2	1 oz/yd^2	= 33.905 7 g/m^2
1 g/m^2	= 0.003 277 oz/ft^2	1 oz/ft^2	= 305.151 3 g/m^2
Mass density (mass/unit volume) (1 Mg/m^3 = 1 t/m^3 = 1 g/cm^3 = 1 g/mL)			
1 kg/m^3	= 0.062 428 lb/ft^3	1 lb/ft^3	= 16.018 5 kg/m^3
1 kg/m^3	= 1.685 56 lb/yd^3	1 lb/yd^3	= 0.593 278 kg/m^3
1 Mg/m^3	= 0.752 48 long ton/yd^3	1 long ton/yd^3	= 1.328 94 Mg/m^3
1 Mg/m^3	= 0.842 78 short ton/yd^3	1 short ton/yd^3	= 1.186 55 Mg/m^3
Force			
1 N	= 0.224 809 lb-force	1 lb-force	= 4.448 22 N
1 kN	= 0.224 809 kip-force	1 kip-force	= 4.448 22 kN
1 kN	= 0.100 361 long ton-force	1 long ton-force	= 9.964 01 kN
1 kN	= 0.112 405 short ton-force	1 short ton-force	= 8.896 44 kN
1 N	= 0.101 972 kg-force	1 kg-force (= 1 kp)	= 9.806 65 N
1 kN	= 0.101 972 metric ton-force	1 metric ton-force	= 9.806 65 kN
1 μN	= 0.1 dyne	1 dyne (= 1 g·cm/s^2)	= 10 μN
Moment of force, torque, bending moment			
1 kN m	= 8.850 75 kip·in	1 kip·in	= 0.112 985 kN·m
1 kN m	= 0.737 562 kip·ft	1 kip·ft	= 1.355 82 kN·m
1 N m	= 0.737 562 lb-force·ft	1 lb-force·ft	= 1.355 82 N·m
1 kN m	= 0.329 269 long ton-force·ft	1 long ton-force·ft	= 3.037 03 kN·m
Force/unit length			
1 N/m	= 0.068 522 lb-force/ft	1 lb-force/ft	= 14.593 9 N/m
1 kN/m	= 0.068 522 kip-force/ft	1 kip-force/ft	= 14.593 9 kN/m
1 kN/m	= 0.030 590 long ton-force/ft	1 long ton-force/ft	= 32.690 3 kN/m

SI units to other units			Other units to SI units		
Pressure, stress, modulus of elasticity					
1 MPa	=	0.064 749 long ton-force/in^2	1 long ton-force/in^2	=	15.444 3 MPA
1 MPa	=	0.145 038 kip-force/in^2	1 kip-force/in^2	=	6.894 76 MPa
1 MPa	=	145.038 lb-force/in^2	1 lb-force/in^2	=	6.894 76 kPa
1 kPa	=	0.009 323 85 long ton-force/ft^2	1 long ton-force/ft^2	=	107.252 kPa
1 kPa	=	0.010 442 7 short ton-force/ft^2	1 short ton-force/ft^2	=	95.760 7 kPa
1 kPa	=	0.020 885 4 kip-force/ft^2	1 kip-force/ft^2	=	47.880 3 kPa
1 kPa	=	0.009 868 7 atm	1 atm	=	101.33 kPa
1 kPa	=	10 mb	1 bar	=	100 kPa
1 kPa	=	0.010 197 kg-force/cm^2	1 kg-force/cm^2	=	98.066 5 kPa
1 kPa	=	0.101 97 metric ton-force/m^2	1 metric ton-force/m^2	=	9.806 65 kPa
Unit weight					
1 kN/m^3	=	0.101 97 metric ton-force/m^3	1 metric ton-force/m^3	=	9.806 65 kN/m^3
1 kN/m^3	=	6.365 9 lb-force/ft^3	1 lb-force/ft^3	=	0.157 09 kN/m^3
Work, energy, heat (quantity) (HP = horsepower)					
1 J	=	0.737 57 ft·lb-force	1 ft·lb-force	=	1.355 8 J
1 MJ	=	0.372 51 HP·h	1 HP·h	=	2.684 5 MJ
1 MJ	=	0.277 8 kW·h	1 kW·h	=	3.600 MJ
1 J	=	$9.478\,2 \times 10^{-4}$ Btu	1 Btu	=	$1.055\,1 \times 10^3$ J
1 J	=	0.238 85 calorie	1 calorie	=	4.186 8 J
1 J/g	=	0.378 5 HP h/ton	1 HP·h/ton	=	2.642 1 J/g
Power, heat (flow)					
1 W	=	0.737 57 ft·lb-force·s	1 ft·lb-force·s	=	1.355 8 W
1 W	=	$1.341\,0 \times 10^{-3}$ HP	1 HP	=	745.7 W
1 W	=	3.412 1 Btu/h	1 Btu/h	=	0.293 1 W
Dynamic viscosity					
1 Pa·s	=	10^3 cP (centipoise)	1 cP	=	1 mPa·s
Kinematic viscosity					
1 m^2/s	=	10^6 cSt (centistoke)	1 cSt	=	1 mm^2/s
Frequency					
1 Hz	=	1 cycle/s	1 cycle/s	=	1 Hz
Plane angle					
1 rad	=	$(180/\pi)°$	1°	=	$\pi/180$ rad

SI units to other units		Other units to SI units	
Thermodynamic temperature (T_K, T_C, and T_F denote temperature in kelvin, degrees Celsius, and degrees Fahrenheit, respectively.)			
T_K	$= T_C + 273°$	T_C	$= T_K - 273°$
T_K	$= \frac{(T_F + 460)}{1.8}$	T_F	$= 1.8T_K - 460°$
T_C	$= \frac{5}{9}(T_F - 32°)$	T_F	$= \frac{9}{5}T_C + 32°$

8. RECOMMENDED READING

AS (Australian Standard) 1000-1979, "The International System of Units (SI) and its Application," Standards Association of Australia, 1979.

ASTM (American Society for Testing and Materials), "Metric Practice Guide," 2d ed., 1966.

ASTM, Standard for Metric Practice, Designation E 380-79, 1980.

Holtz, R. D., "SI Units in Geotechnical Engineering," *Geotech. Testing ASTM*, vol. 3, no. 2, 1980, pp. 75–88.

Holtz, R. D., and Kovacs, W. D., An Introduction to Geotechnical Engineering, App. A in "Application of the SI System of Units to Geotechnical Engineering," Prentice-Hall, N.J., 1981, pp. 665–680.

ANSWERS TO SELECTED PROBLEMS

CHAPTER 2

C2.1. 25%.

C2.2. 20 kN/m^3, 88%, 6.5%.

C2.3. S = 100%: 2.20, 2.79, 2.62, 2.46, 2.34, 1.24 t/m^3.
S = 80%: 1.98, 1.76, 1.58, 1.44, 1.32, 1.21 t/m^3.

C2.4. *c*. 22.8%, 1.55 t/m^3, 82%.

C2.5. 1370 m^3/h.

C2.6. 1.5 kg of TNT per hole, 5 g/m^3 (this is somewhat less than used in the projects described in Sec. 2.2.3.1).

C2.7. 24 m.

C2.8. 5 m, 30 m.

C2.9. About 14 m.

C2.10. 2 to 8 m^2.

C.2.11. Meehan (1967) concluded that elephants are inefficient compactors, because

(*a*) Pressure exerted is similar to the tire pressure of a pneumatic roller (three legs on the ground results in 380 kPa), but the ground coverage is slow and many passes are required. Fifty passes resulted in a dry density of 1.62 t/m^3 (87% standard maximum density).

(*b*) Elephants quickly learn to place their feet on precompacted areas, avoiding softer soils. Strenuous control by the handler is required.

M2.12. *b*.

M2.13. *d*.

M2.14. *d*.

M2.15. *b*.

M2.16. *c*.

M2.17. *c*.

M2.18. *c*.

M2.19. *d*.

M2.20. *a*.

M2.21. *c*.

M2.22. *b*.

M2.23. *c*.

CHAPTER 3

C3.1. 16.28 kN/m^3, 19.8%.

C3.2. Graphs plotted indicate the following: (1) The optimum water content w_{opt} can be expressed as a function of the percent fines and/or the plastic limit, but the relationship may only be valid for a particular family of soils, (2) at equal plastic limit (> 25%), black cotton (BC) soils show higher w_{opt} than micaceous (M) and laterite (L) soils, (3) at equal percent fines, values of w_{opt} for L soils are higher than for BC soils.

C3.3. Equations (3.3) and (3.4) are best (18.3% and 1.84 t/m^3), with an error of 12% and 2%, respectively.

C3.4. 2.8 kN/m^3, e_{max} = 0.37, e_{min} = 0.14, assuming G_s = 2.65.

C3.5. *a*. 7%, 2.18 t/m^3; *b*. 7%, 2.10 t/m^3, S = 66%.

C3.6. *a*. (plot); *b*. 91.7, 97.9, 84.3, 85%; *c*. 10%.

C3.7. 1.7 t/m^3.

C3.8. (plot).

M3.9. *b*.

M3.10. *d*.

M3.11. *c*.

M3.12. *d*.

M3.13. *d*.

CHAPTER 4

C4.1. CL-ML.

C4.2. According to the Unified Soil Classification System: *a*. coarse > 50%, fines > 12% and < 50%; *b*. see Tables 4.1 and 6.1.

C4.3. 3.9, not suitable.

C4.4. Soil *a*: MSI = 5.4; soil *b*: MSI = 100.

C4.5. (plot).

C4.6. *a*. GP; *b*. GW, etc.

C4.7. *a*. e_{max} = 0.61, e_{min} = 0.29; *b*. F = 1.1; *c*. I_D = 78%; *d*. dense, N = 45.

M4.8. *d*.

M4.9. *c*.

M4.10. *b*.

CHAPTER 5

C5.1. 100, 70.

C5.2. I_D = 96%, density = 1.9 t/m^3 (estimated from Table 4.2, assuming SP).

C5.3. 12 mm.

C5.4. 6 mm.

C5.5. 1.87 t/m^3, 1.03, 1.07 (or 107%).

C5.6. *a*. 0.9, 93%; *b*. 2.9%.

C5.7. 55%.

C5.8. 25 000 kN/m^3 = 25 MN/m^3, 5.6 MPa.

M5.9. *d*.

M5.10. *c*.

CHAPTER 6

C6.1. Max. density: 1.708, 0.041, 2.4%; water content: 15.87, 0.668, 4.2%.

C6.2. Three passes.

M6.3. *b*.

M6.4. *c*.

M6.5. *b*.

M6.6. *a*.

M6.7. *c*.

M6.8. *b*.

M6.9. *a*.

M6.10. *b*.

CHAPTER 7

B7.1. Open sumps and ditches, gravity-flow wells, vacuum wells, membranes, etc.

B7.2. Soil freezing, geomembranes, compressed air.

B7.3. See Sec. 7.1.

M7.4. *c*.

M7.5. *b*.

M7.6. *b*.

M7.7. *b*.

M7.8. *b*.

M7.9. *b*.

M7.10. *d*.

B7.11. Gravity-flow well, well point, vacuum well.

C7.12. 17.2, 1.75, 1.20, 54.5, 1.69, 16.6, 89%.

C7.13. 1.59, 1, 2.70.

M7.14. *c*.

M7.15. *b*.

M7.16. *c*.

M7.17. *a*.

M7.18. *d*.

M7.19. *c*.

B7.20. W_W/W_S, V_V/V, V_V/V_S.

B7.21. 20, 1.33, 80, 0.5.

B7.22. See Sec. 7.3.3.

B7.23. $Q = kiA$, coefficient of permeability (m/s), gradient (dimensionless), cross section (m^2), flow (m^3/s).

CHAPTER 8

C8.1. 0.32 L/s, 0.36 m.

C8.2. 155 L/s.

C8.3. 0.000 57 m/s.

C8.4. 11 m, 9.57 m.

C8.5. 0.97 m (2.54 in well).

C8.6. k = 29.1 m/day, S = 000166, t_0 = 0.9 min.

C8.7. 34 h.

M8.8. *b*.

M8.9. *b*.

M8.10. *c*.

M8.11. *d*.

M8.12. *c.*
M8.13. *b.*
M8.14. *b.*
M8.15. *c.*
M8.16. *a.*
B8.17. Influence range, free discharge height, phreatic surface, thickness of aquifer.
B8.18. See Sec. 8.1.
B8.19. Homogenous and isotropic, horizontal boundaries, Darcy's law holds, Dupuit-Thiem valid.
B8.20. Free discharge height, level of phreatic surface at borehole, water level in well, $h_s = h_w - h_0$.
B8.21. Forchheimer, Darcy, Sichardt, Kozeny.
B8.22. See Sec. 8.7.1.

CHAPTER 9

C9.1. 4.93×10^{-4} m/s.
C9.2. 18.4 L/s, $h_0 = 7.9$ m, $L = 278$ m.
C9.3. Min. 5 wells, $Q = 0.121$ m³/s, $Q_i =$ 3.11 L/s, $L = 21$ m, $h_0 = 4.95$ m.
C9.4. *a.* 0.062 5 m³/s; *b.* min. 11 wells, $h_0 =$ 6.64, $L = 383$; *c.* 12 wells, $Q = 0.062$ m³/s, $h_0 = 7.04$ to 7.61 m, $Q_{imax} =$ 0.0066 m³/s, $Q_i = 0.0052$ m³/s.
C9.5. *a.* $L = 375$ m, $Q = 0.079$ m³/s, $Q_i =$ 7.9 L/s; *b.* 0.089 m³/s, 5.96 L/s; *c.* 664 m, 0.137 m³/s, 4.9 L/s.
C9.6. 25.4 to 72.8 mm.
C9.7. 0.54, 0.88, 1.21.
C9.8. 15 kPa, 20 kPa, SF(*a*) > SF(*b*).
C9.9. 22.8°.
C9.10. Before: 0.90; after: 1.06 (approx.).
C9.11. Before: 0.7; after: 0.98 (approx.).
M9.12. *c.*
M9.13. *a.*
M9.14. *c.*
M9.15. *c.*
M9.16. *c.*
M9.17. *c.*
M9.18. *a.*
M9.19. *b.*
M9.20. *a.*
M9.21. *c.*
M9.22. *c.*
M9.23. *a.*
M9.24. *d.*
B9.25. Grain size distribution (+ uniformity coefficient, density).
B9.26. *a.* Constant head test, falling head test; *b.* as in Prob. B9.25.
B9.27. Reduce shear strength, increase pore pressure, cause internal or external erosion.
B9.28. See Sec. 9.3.2.
B9.29. Available shear strength, mobilized shear strength.
B9.30. Piezometers, flow.

CHAPTER 10

M10.1. *c.*
M10.2. *b.*
M10.3. *c.*
M10.4. *d.*
M10.5. *a.*
M10.6. *d.*
M10.7. *b.*
M10.8. *c.*
M10.9. *a.*
M10.10. *c.*
M10.11. *c.*
M10.12. *c.*
M10.13. *a.*
B10.14. Polypropylene, polyester, polyethylene, polyamides.
B10.15. Nonwoven.
B10.16. Permittivity, transmissivity.
B10.17. Separation, filtration, drainage, reinforcement.
B10.18. Impede water flow, prevent piping.
C10.19. *a.* 0.081, 4.3×10^{-5} s⁻¹; *b.* 7.1, 1.2×10^{-2} m/s.
C10.20. $\theta = (1.9, 1.13, 0.69, 0.29, 0.08) \times 10^{-6}$ m²/s.
C10.21. $O_{95} = 0.22$ mm.
C10.22. $D_{eff} = 0.08$ mm.
C10.23. *a.* 0.24 mm $< D_{15f} <$ 1.6 mm, $D_{50f} <$ 11.25 mm; *b.* $O_{95} \leq 1.28$ mm.
C10.24. $P_c < 9.2$ mm, $F_T > 1025$ N.
C10.25. *a.* See Fig. 10.16; *b.* $q = 0.002$ m³/(s · m).
C10.26. *a.* $\theta = 0.002$ m²/s; *b.* A very thick or multilayered nonwoven geotextile or composite sheet drain is necessary.
C10.27. *a.* $\theta_{req} = 1.3 \times 10^{-5}$ m²/s; *b.* yes.

CHAPTER 11

C11.1. 2.85 m, 3.34 m of fill.
C11.2. 98 mm, 67 mm.
C11.3. $s_f = 0.44$ m, $t = 4.3$ years.
C11.4. 713 mm, 31 years, 22 kPa, yes (12 kPa), yes (75 < 114 kPa).
C11.5. 0.804, 803 mm, and 55 mm.
C11.6. 0.80, 0.47, 0.40.
C11.7. 0.0325 m.

C11.8. $s_f = 190$ cm, $c_v = 9.6 \times 10^{-6}$ m²/s and $c_v = 5.5 \times 10^{-6}$ m²/s.

C11.9. $s_f = 73$ cm, for $\beta = 0.65$ (ignore first two points) $c_h = 4 \times 10^{-7}$ m²/s.

M11.10. *c.*

M11.11. *c.*

M11.12. *b.*

M11.13. *b.*

B11.14. Rendulic, Kjellman, Barron, Terzaghi.

B11.15. See text Sec. 11.5.1.

B11.16. See text Sec. 11.6.

CHAPTER 12

C12.1. $k_e = 5 \times 10^{-9}$ m²/sV; $k_h = 1 \times 10^{-7}$ m/s.

C12.2. $k_e = 1.93 \times 10^{-10}$ m²/sV; $H_e = 0.064$ m ($u_b = -0.64$ kPa).

C12.3. $k_e = 2.54 \times 10^{-5}$ cm²/sV; $k_h = 2 \times 10^{-6}$ cm/s. ($H_{e,\text{final}} = 120$ cm).

C12.4. $\rho = 50\ \Omega$·m.

C12.5. 320 W.

C12.6. *a.* −3920 kPa, 4.18 m; *b.* 23 days; *c.* 93 kPa.

M12.7. *c.*

M12.8. *c.*

M12.9. *c.*

M12.10. *d.*

M12.11. *c.*

M12.12. *d.*

M12.13. *a.*

B12.14. Reuss, Quincke, Casagrande, Schaad and Haefeli, Mitchell, Veder.

B12.15. Electrokinetics, electroosmosis, streaming potential or flow potential, electrophoresis.

B12.16. See Sec. 12.4.

CHAPTER 13

C13.1. 0.05 mm, 0.45 mm, 1.8 mm; SW-SM or SW-SC.

C13.2. 80%.

C13.3. 23 MPa.

C13.4. 400 to 600 kPa.

C13.5. *a.* 1.32 kg; *b.* 1.99 m³.

C13.6. *a.* 12.7%; *b.* 6.8%.

C13.7. 13 to 15%.

C13.8. *a.* 550 mm; *b.* no.

C13.9. 280 mm, subgrade deformation critical.

C13.10. $\epsilon_1 = 451$, $\epsilon_2 = 148$, $\epsilon_3 = 1700$ microstrain.

C13.11. $E_v = 30$ MPa, $E_h = 15$ MPa.

C13.12. Fig. 13.22: 200-mm unbound granular material (expected failure mode: asphalt fatigue). Fig. 13.23: 230-mm cemented material (fatigue of cemented material). Fig. 13.24: 180-mm cemented material (fatigue of cemented material).

M13.13. *a.*

M13.14. *c.*

M13.15. *b.*

M13.16. *c.*

M13.17. *c.*

M13.18. *d.*

M13.19. *b.*

M13.20. *a.*

M13.21. *d.*

M13.22. *d.*

M13.23. *d.*

M13.24. *b.*

M13.25. *d.*

M13.26. *b.*

M13.27. *a.*

M13.28. *d.*

B13.29. Increase strength, durability; reduce permeability, erodability, deformability; control volume stability, variability.

B13.30. Improve trafficability on construction site, workability of borrow materials, contain wastes, reduce dust on unpaved roads, etc.

B13.31. Mechanical stabilization.

B13.32. *a.* Thorough mixing; *b.* shrinkage cracks.

B13.33. *a.* Decreases; *b.* decreases; *c.* decreases; *d.* increases.

B13.34. *a.* increases; *b.* flatter.

B13.35. Coal type, degree of pulverization, boiler type and operation, collection and stockpiling methods.

B13.36. *a.* Tar; *b.* bitumen.

B13.37. Waterproofing, add cohesion.

B13.38. Increase.

B13.39. Immobilize contaminants (reduce leachate); improve mechanical properties (strength, compressibility, volume stability).

CHAPTER 14

C14.1. 0.31, 0.23, 1.31.

C14.2. 0.2 to 0.8.

C14.3. 43.5 Pa·s.

C14.4. 0.053 Pa·s or 53 cP, 0.75:1.

C14.5. 0.948×10^{-3} Pa·s s, 0.113 m/s, 20 mL/h, 640 mL/h.

C14.6. 37.5 m, 75 m.

C14.7. 271 mm, 165 mm.

C14.8. 212.5 mm.
C14.9. 20, 50, 250 cP.
C14.10. 200 m, 16 755 kN.
C14.11. 3.27×10^{-9} m/s.
C14.12. 4.1 Lugeons.
C14.13. Expansion of fissure or hydraulic fracturing.
M14.14. *c*.
M14.15. *c*.
M14.16. *b*.
M14.17. *c*.
M14.18. *d*.
M14.19. *d*.
M14.20. *b*.
M14.21. *b*.
M14.22. *c*.
B14.23. Penetration, displacement, compaction, jet, electrogrouting.
B14.24. *a*. Grout curtain below dam, etc.; *b*. foundation grouting, etc.; *c*. Grouting of cavities in Karst, etc.
B14.25. *a*. Suspension (e.g., cement, clay); *b*. emulsion (e.g., bitumen + water); *c*. solution (e.g., sodium silicate).
B14.26. Toxicity, permanency.
B14.27. Density, strength, particle size, stability, etc.
B14.28. *a*. Newtonian; *b*. non-newtonian.
B14.29. Clay, cement, etc.
B14.30. Sodium silicate, AM-9 or AC 400, etc.
B14.31. Grout taken as a function of pressure, ground heave, etc.
B14.32. Water pressure testing, pressuremeter testing, coring samples, etc.
B14.33. *a*. Fissure; *b*. joint; *c*. fault; *d*. crack.
B14.34. Orientation, aperture, spacing, filling, etc.
B14.35. Ground heave, back flow of grout, characteristics of flow rate/pressure diagram, acoustic emission monitoring.

CHAPTER 15

C15.1. 334 kJ.
C15.2. 3029 MJ.
C15.3. 3.15, 1.47 MJ/($m^3 \cdot$ °C).
C15.4. $t = 45$ days.
C15.5. *a*. 1.93, 1.86 t/m^3; *b*. 3.9%.
C15.6. *a*. 54.8%, 44.8%, 0.82; *b*. 1.46, 89%.
C15.7. $\epsilon = 0.095$.
C15.8. *a*. $A = 0.000\ 15$, $B = 2.5$, $c = 0.25$ (by trial and error); *b*. For $t(h) = 10$, 60, 90, 150, 210, strain $\epsilon = 0.035$, 0.054, 0.060, 0.068, 0.074.
C15.9. *a*. $\beta' = 95$, $B' = 0.0068$; *b*. $\sigma_t = 7.5$ MPa.
M15.10. *b*.
M15.11. *c*.
M15.12. *a*.
M15.13. *b*.
M15.14. *c*.
M15.15. *b*.

CHAPTER 16

B16.1. *a*. Fagot; *b*. corduroy roads; *c*. adobe brick; *d*. gabion.
B16.2. French citizenship.
C16.3. One billion dollars.
C16.4. Equation (16.2) gives 5.5, 14.5, 21.9, 33, 42, 50, and 57 μm; Eq. (16.3) gives 50, 131, 199, 302, 385, 457, and 522 μm.
C16.5. $T_0 = 48$ kN, $T = 9.6$, with SF $= 1.5T_{all} = 6.4$ kN.

CHAPTER 17

M17.1. *c*.
B17.2. Concrete, adobe bricks, plywood, fiber-reinforced resin, bimetals (laminates), cermets (ceramics in metal matrix), etc.
C17.3. 27.4 kPa.
C17.4. 0.11.
C17.5. $F = 0.18$, $\phi_R = 47.5\%$.
C17.6.

Depth, m	Earth pressure $c = 0$, $\phi = 30°$	$c = 0$, $\phi = 50°$	$c = 20$ kPa, $\phi = 30°$
0	0	0	−23
5	33	13	10
10	67	27	44

Upper zone: failure by slippage.
Lower zone: failure by rupture of the reinforcement.

B17.8. See Fig. 17.13.

CHAPTER 18

C18.1. *a*. 34%; *b*. 22%.
B18.2. 5 < pH < 10, resistivity > 1000 Ω · cm, chlorides < 200 ppm, sulphates < 1000 ppm.

B18.3. Homogeneity of fill, degree of compaction, water content, drainage.
B18.4. 1, 1.5, and 7 to 8 m.
B18.5. 40 mm, 5 mm, 275 MPa.
M18.6. *b*.
M18.7. *b*.
M18.8. *d*.
M18.9. *c*.
C18.12. *a*. $\sigma_{v\,max}$ = 159 kPa, $\sigma_{v\,min}$ = 68.9 kPa; for effective width B' = 4.34m, q_{ult} = 361 kPa, q_{actual} = 131 kPa; *b*. F = 1.76; *c*. F = 4.4.
B18.13. Internal: pullout (slippage), breaking of reinforcement. External: global slope failure, block sliding, overturning.
M18.14. *d*.
C18.17. *a*. F_R = 3.3, F_S = 3.3; *b*. F_R = 2, F_S = 3.4, L_e = 4.55 m; *c*. F_R = 1.9, F_S = 2.9, f = 0.92, k = 0.34.
C18.18. 12.7 kN.
C18.19. F = 0.56.
B18.20. See Fig. 18.11.
B18.21. Direct shear, pullout test.
B18.22. L_e (rather than L_{tot}), K_0 to K_a (rather than just K_a).
M18.23. *d*.
M18.24. *a*.
M18.25. *c*.
C18.27. *a*. $7.5y^{0.53}$; *b*. 7.5, 5.5, and 50 μm, using Eqs. (18.16), (16.2), and (16.3), respectively.
B18.28. See Fig. 17.13 and Fig. 18.15.
M18.29. *c*.
M18.30. *c*.
M18.34. *d*. (until expiry of patents).
C18.36. *a*. Surface area is 0.053 m^2 (bar) and 0.091 m^2 (strip); *b*. 11.4% (bar) and 19.6% (strip); *c*. 44%.
C18.37. *a*. 37.6 kN; *b*. 11.2 kN; *c*. 190.7 > 37.6 (from *a*) or 11.2 (from *b*).
B18.38. VSL Retained Earth, Caltrans Mech. Stab. Earth, Hilfiker walls, Georgia stabilized embankment.
B18.39. Adv.: increased pullout force, better corrosion resistance. Disadv.: possibly higher earth pressures (K_0 rather than K_a), less field data available.
M18.40. *a*.
C18.42. 157 kN [Eq. (18.27)], 29 kN [Eq. (18.28)], and 62 kN [Eq. (18.30)].
M18.43. *d*.
C18.45. 7%.
C18.46. 0.0006% per minute at 18.3%.
M18.47. *a*.
M18.48. *a*.
C18.51. 33°.
C18.52. 187 + 211 = 398 kN.

CHAPTER 19

M19.1. *b*.
M19.2. *b*.
M19.3. *c*.
M19.4. *b*.
M19.5. *d*.
M19.6. *c*.
M19.7. *c*.
M19.8. *c*.
M19.9. *c*.
M19.10. *b*.
M19.11. *d*.
B19.12. Mode of deformation resembles in situ deformation.
B19.13. Bearing failure, rutting.
B19.14. Reduced aggregate requirement, lower maintenance costs.
B19.15. No, no, yes, yes.
B19.16. Resistance to stresses due to construction activities and roughness of ground.
B19.17. Burst, grab tensile, CBR, and drop cone test.
B19.18. See construction guidelines in Sec. 19.3.
C19.19. Strength 118 kN/m, modulus 790 kN/m.
C19.20. 1628, 180, 216 kN·m/m.
C19.21. Savings \$0.85, 1.70, 3.70, 8.50 per m^2.
C19.22. Moderately robust, robust, very robust, extremely robust.

CHAPTER 20

M20.4. *c*.
C20.7. *a*. 71.3 mm; *b*, *c*. (discussion).
C20.8. *a*. 900 kN (approx.); *b*. for K_f = 3, anchor bulb diameter = 0.2 m, and δ = 40°, capacity = 790 kN/m.
C20.9. *a*. Say 4 × π × 0.1 × 4 = 5 MN; *b*. 17.5 MN.
C20.10. *b*. 320 kN.
C20.11. Using the Australian Standard for piles (AS 2159-1978), *a*. Q = 40.7 kN; *b*. 16.5 kN.
C20.12. *a*. 22.6 MN; *b*. for τ = 0.1 UCS, load = 1598 MN (very high, anchor tendon will fail before rock!).
C20.13. 2.6.

C20.14. *a*. 1.28; *b*. 1.68, results may vary depending on method of analysis used.

C20.15. 4.5-m bolts at 2.2 m spacing.

C20.16. 9.8 kN.

CHAPTER 21

B21.2 High flexibility, elemental construction, appearance, permeability, etc.

C21.4. 27 kPa, 17.8 kPa, 20.8 kPa, 14.2 kPa.

M21.5 *c*.

B21.6. Inadequate investigation, unsuitable fill, bad concrete, incorrectly assessed earth pressures.

M21.7. *b*.

B21.8. Flexibility, permeability, low level of work still required, low cost, etc.

C21.9. 1.9.

C21.10. 270 mm.

B21.11. Torn or abraded mesh, loss of rock, loss of shape, and failed anchors, aprons or cutoffs.

B21.13. *b*, *g*.

M21.14. *b*.

C21.15. 344 kPa, 588 kPa (for $\gamma = 20$ kN/m^3, Hansen formula).

REFERENCES

ABBREVIATIONS

ARRB	Australian Road Research Board
ASCE	American Society of Civil Engineers
ASTM	American Society for Testing Materials
ECSMFE	European Conference on Soil Mechanics and Foundation Engineering
ESOPT	European Symposium on Penetration Testing, Stockholm, 1975 (now referred to as the 1st ESOPT), and Amsterdam, 1982 (2d ESOPT)
HRB	Highway Research Board (United States)
ICC	International Conference on Compaction, Paris, April 22–24, 1980
ICE	Institution of Civil Engineers
ICSMFE	International Conference on Soil Mechanics and Foundation Engineering
PACSMFE	Pan American Conference on Soil Mechanics and Foundation Engineering
RILEM	Réunion Internationale des Laboratories d'essais et de Recherches sur les Matériaux et les Constructions

PART I

Baggs, S. A. (ed.), *Proc 1st Int. Conf. Energy Efficient Buildings with Earth Shelter Protection,* August 1–6, 1983, published by Unisearch Ltd., University of New South Wales, Sydney, 1983.

Eighth ECSMFE Helsinki, dedicated to the topic "Improvement of Ground," 1983, Proc. edited by Rathmayer, H., and Saari, K., published by A. A. Balkema, 1983.

Eleventh ICSMFE, San Francisco, 1981, Proc. published by A. A. Balkema, 1981.

Fourth Australia–New Zealand Geomech. Conf., Perth, 1984, Proc. published by the Institution of Engineers, Australia, 1984.

In Situ '86, Spec. Conf., "Use of In-Situ Tests in Geotechnical Engineering," ASCE Geotech. Spec. Publ. No. 6, edited by S. P. Clemence, 1986.

Mitchell, J. K., "Soil Improvement—State-of-the-Art Report," *Proc. ICSMFE*, Stockholm, vol. 4, 1981, pp. 509–565.

Sembenelli, P., and Ueshita, K., "Environmental Geotechnics—State-of-the-Art Report," *Proc. ICSMFE*, Stockholm, vol. 4, 1981, pp. 335–394.

"Soil Improvement—History, Capabilities, and Outlook," Report by the Committee on Placement and Improvement of Soils of the Geotech. Eng. Div. ASCE, February 1978.

Symposium and Short Course held at the Asian Institute of Technology, Bangkok, 1982, Proceedings entitled "Recent Developments in Ground Improvement Techniques," edited by Balasubramaniam et al., published by A. A. Balkema, 1985.

Symp. Environ. Geotech., Lehigh University, Bethlehem, Pennsylvania, United States, April 1986.

Tenth, ICSMFE, Stockholm, 1981.

Welsh, J. P. (ed.), "Soil Improvement—A Ten Year Update", ASCE Geotech. Spec. Publ. No. 12, 1987.

Yamanouchi, T., Miura, N., and Ochai, H., "Theory and Practice of Earth Reinforcement," *Proc. Int. Geotech. Symp. Theory and Practice of Earth Reinforcement*, Fukuoka Kyushu, Japan, October 1988, published by A. A. Balkema, 1988.

PART II

Ang, A., and Tang, W. H., *Probability Concepts in Engineering Planning and Design*, John Wiley & Sons, 1975.

Baguelin, F., Jézéquel, J. F., Lemée, E., and Shields, D. H., *The Pressuremeter and Foundation Engineering*, Trans Tech Publ. Clausthal, Federal Republic of Germany, 1978.

Barendsen, D. A., and Kok, L., "Prevention and Repair of Flow-Slides by Explosion," *Proc. 8th ECSMFE*, 1983, pp. 205–208.

Bazaraa, A. R. S., "Use of the Standard Penetration Test for Estimating Settlements of Shallow Foundations on Sand," Ph.D. thesis, University of Illinois, 1967.

Bell, J.A., "Plastic Moisture Barriers for Highway Subgrade Protection," SM Thesis, Purdue University, 1956.

Benjamin, J. R., and Cornell, C. A., *Probability, Statistics, and Decision for Civil Engineers*, McGraw-Hill, New York, 1970.

Biarez, J., "General Report—Session 1," *Proc. ICC*, vol. III, 1980, pp. 13–26.

Bowles, J. E., *Foundation Analysis and Design*, 4th ed., McGraw-Hill, New York, 1988.

Brandl, H., "Construction and compaction of 100–120m high highway embankments," *Proc. ICC.*, vol. I, 1980, pp. 221–226.

Buchanan, S. J., Soil compaction, *Proc. 5th Texas Conf. Soil Mech.*, 1942.

Bureau of Reclamation, *Earth Manual*, U.S. Government Printing Office, Washington, D.C., 1963,

Casagrande, A., "Characteristics of Cohesionless Soils Affecting the Stability of Slopes and Earthfills," *J. Boston Soc. Civil Engrs*. January 1936.

Casagrande, A., "Liquefaction and Cyclic Deformation of Sands—A Critical Review," *5th PACSMFE*, Buenos Aires, 1975.

Clemence, S. P., (ed.), *Proc. In Situ '86*, ASCE Geotech. Spec. Publ. No. 6, 1986.

Clifford, J. M., "An Introduction to Impact Rollers," *Proc. ICC*, vol. II, 1980, pp. 621–626.

Colleselli, F., Mazzucato, A., Previatello, P., and Spalatro, A., "Improvement of Soil Foundation by Vibratory Methods." *Proc. 8th ECSMFE*, 1983, pp. 223–228.

Compaction Data Handbook, Ingersoll-Rand Co., Compaction Div., Shippensburg, PA, 1984.

D'Appolonia, D. J., D'Appolonia, E., and Brissette, R. F., "Discussion on Settlement of Spread Footings on Sand," *J. Soil Mech. Found. Div. ASCE*, vol. 96, no. SM2, 1970, pp. 754–762.

Davidson, D. T., and Gardiner, W. F., "Calculation of Standard Proctor Density and Optimum Moisture Content from Mechanical Analysis, Shrinkage Factors, and Plasticity Index," *Proc. HRB*, 1949.

Davis, E. H., and Pells, P. J. N., "A Note on the Interpretation of K_0 from Pressuremeter Tests," *Proc. Int. Conf. Structural Foundations on Rock*, Sydney, 1980, edited by P. J. N. Pells, published by A. A. Balkema, vol 2, pp. 25–28, 1980.

Dembicki, E., and Kisielowa, N., "Technology of Soil Compaction by Means of Explosion," *Proc. 8th ECSMFE,* 1983, pp. 229–230.

Dobson, T. and Slocombe, B., "Deep Densification of Granular Fills, *The 2d Geotech. Conf. and Exhibit on Design and Construction,* Las Vegas, April 26–28, 1982. (Reprint available from GKN Keller, Inc., 6820 Benjamin Rd., Tampa Fl., 33614.)

Doshi, S. N., Mesdary, M. S., and Guirguis, H. R., "Experimental Cement-Stabilisation in Kuwait," *Proc. 4th Australia–New Zealand Conf. Geomech.,* Perth, vol. 1, May 1984, pp. 192–197.

Douglas, D. J., "Design of Auger Grouted Piles," *Proc. Symp. Grouted Piles,* Queensland Institute of Technology, 1982.

Douglas, D. J., "The Standard Penetration Test," *Proc. In-Situ Testing for Geotech. Investigations,* edited by M. C. Ervin, 1983, pp. 21–31.

Ervin, M. C. (ed.), *In-Situ Testing for Geotechnical Investigations,* A. A. Balkema, 1983.

Ervin, M. C., Burman, B. C., and Hughes, J. M. O., "The Use of a High Capacity Pressuremeter for Design of Foundations in Medium Strength Rock," *Proc. Int. Conf. Structural Foundations on Rock,* Sydney, 1980, edited by P. J. N. Pells, published by A. A. Balkema, vol. 1, pp. 9–16, 1980.

Forssblad, L., "Vibratory Compation in the Construction of Roads, Airfields, Dams and other Projects," Report No. 8222, Dynapac, S-171 22 Solna 1, Sweden, 1977.

Forssblad, L., "Compaction Meter on Vibrating Rollers for Improved Compaction Control, *Proc. ICC.,* vol. II, 1980, pp. 541–546.

Forssblad, L., "Vibratory Soil and Rock Fill Compaction," Dynapac Maskin AB, Solna 1, Sweden, 1981.

Fry, J. J., "Discussion—Session III," *Proc. ICC,* vol. III, 1980, pp. 34–44.

Gambin, M. P., "The Menard Dynamic Consolidation Method at Nice Airport," pp. 231–234 (Fig. 2.8) in Rathmayer, H. G., and Saari, K. H. O. (eds.): "Improvement of Ground," *Proc. 8th ECSMFE,* Helsinki, May 23–26, 1983–84. 1398 pp. in hardbound volumes. Hfl. 520/US$265.00/£156. A. A. Balkema, P.O. Box 1675, Rotterdam, Netherlands.

Gibbs, H. J., and Holtz, W. G., "Research on Determining the Density of Sands by Spoon Penetration Testing, *Proc. 4th ICSMFE,* London, vol. 1, 1957, pp. 35–39.

GKN Hayward Baker, Inc., "Ground Modification Seminar," Lecture notes, 1986.

Glick, G. L., and Clegg, B., "Use of Penetrometer for Site Investigation and Compaction Control at Perth W.A., *Trans. I. E. Aust. Conf.,* Perth, 1965, pp. 114–118.

Hammond, A. A., "Evolution of One-Point Method for Determining the Laboratory Maximum Dry Density," *Proc. ICC,* vol. I, 1980, pp. 47–50.

Hamory, G., "Calibration, Stability and Use of Nuclear Meters," Report No. 82/31, MRD of W.A. Main Roads Dept. of Western Australia, (MRD of W.A.), 1982.

Hamory, G., "Theory and Safety of Nuclear Meters," Report No. 82/30, MRD of W.A., 1982.

Harr, M. E., *Mechanics of Particulate Media.* McGraw-Hill, New York, 1977.

Hilf, J. W., "An investigation of Pore-Water Pressure in Compacted Cohesive Soils," Tech. Memo. 654, U.S. Dept. of the Interior, Bureau of Reclamation, Denver, Colorado, 1956.

Hilf, J. W., "A Rapid Method of Construction Control for Embankments of Cohesive Soil," Eng. Mono. No. 26, Bureau of Reclamation, 1959.

Hilf, J. W., "Compacted Fill," chap. 7 in Winterkorn, H. F., and Fang, H.-Y., *Foundation Engineering Handbook,* Van Nostrand Reinhold, 1975, pp. 244–311.

Hogentogler & Company, *Cone Penetrometer,* 3d ed., circa 1985.

Hogentogler, C. A. Jr., "Essentials of Soil Compaction," *Proc. HRB,* vol. 16, 1936, pp. 209–216.

Holtz, W. G., and Gibbs, H. J., "Engineering Characteristics of Expansive Clays," *Trans. ASCE,* 1956.

Huder, J., "Die Zusammendrückbarkeit des Bodens und deren Bestimmung," *Schweizerische Bauzeitung,* Heft 41, 8. Oktober, 1964.

Ingles, O. G., and Metcalf, J. B., *Soil Stabilization,* Butterworths, Syndey, 1972.

Ingles, O. G., and Williams, H. J., "Dry-of-Optimum Compaction and Saturation Settlements," *Proc. ICC,* vol. I, 1980, pp. 139–143.

Ivanov, P. L., "Consolidation of Saturated Soils by Explosions," *Proc. ICC,* Ećole Nationale des Ponts et Chaussées (ENPC), Paris, vol. I, 1980, pp. 331–337.

Jebe, W., and Bartels, K., "The Development of Compaction Methods with Vibrators from 1976 to 1982," *Proc. 8th ECSMFE,* 1983f, pp. 259–266.

Joslin, J. G., "Ohio's Typical Moisture-Density Curves," *Symp. Application of Soil Testing in Highway Design and Construction,* ASTM Spec. Tech. Publ. 239, 1959, pp. 111–118.

Krebs, D. R., and Walker, R. D., *Highway Materials,* McGraw-Hill, New York, 1971.

Kyulule, A. L., "Additional Considerations on Compaction of Soils in Developing Countries," Nr. 121, Mitteilungen des Institutes für Grundbau and Bodenmechanik, Eidgenössische Technische Hochschule, Zürich, 1983.

Lacroix, Y., and Horn, H. M., "Direct Determination and Indirect Evaluation of Relative Density and Its Use on Earthwork Construction Projects," ASTM Spec. Tech. Publ. 523, 1973, pp. 251–280.

Ladner, P. A., and Hamory, G., "Statistical and Nuclear Methods in Quality Control of Road Construction," *7th ARRB Conf.,* Adelaide, 1974, pp. 38–69.

Lambe, T. W., "The Engineering Behavior of Compacted Clay," *J. Soil Mech. Found. Div. ASCE,* V. 84, SM2, Pt. 1, Paper No. 1655, 35 pp. May, 1958. (Also in *Trans. ASCE,* vol. 125, part I, 1960, p. 718.)

Lambe, T. W., and Whitman, R. V., *Soil Mechanics,* John Wiley & Sons, 1969.

Lee, K. L., and Haley, S. C., "Strength of Compacted Clay at High Pressure," *J. Soil Mech. Found. Div. ASCE,* vol. 94, no. SM6, November, 1968, p. 1303–1332.

Leflaive, E., "General Report—Session III," *Proc. ICC,* vol. III, 1980, pp. 89–108.

Leflaive, E., and Schaeffner, M., "Compactibility of Soils Evaluated by the Measurement of Their Air Permeability," (in French) *Proc. ICC,* vol. I, 1980, pp. 57–62.

Lumb, P., "Soil Variability and Engineering Design," in Valliappan, S., Hain, S. J., and Lee, I. K. (eds.): *Soil Mechanics—Recent Developments,* Unisearch Ltd., Sydney, 1975, pp. 383–397.

Main Roads Department of Western Australia, "The Western Australian Confined Compression Test (WACCT)—Test Method WA 142.1," 1962.

Marchetti, S., "In Situ Tests by Flat Dilatometer," *J. Geotech. Eng. Div. ASCE,* vol. 106, no. GT3, March 1980, pp. 299–321.

Massarsch, K. R., and Broms, B. B., "Soil Compaction by Vibro-Wing Method," pp. 275–278 (Fig. 5.19) in Rathmayer, H. G., and Saari, K. H. O. (eds.), "Improvement of Ground," *Proc. 8th ECSMFE,* Helsinki, May 23–26, 1983, 1983–84. 1398 pp. in hardbound volumes. Hfl. 520/US$ 265.00/£156. A. A. Balkema, P.O. Box 1675, Rotterdam, Netherlands.

Mayne, P. W., Jones, J. S., and Dumas, J. C., "Ground Response to Dynamic Compaction," *J. Geotech. Eng. Div ASCE,* vol. 110, no. 6, June 1984, pp. 757–774.

McDonald, J. K., "Discussion of the Paper by Sherard, Dunnigan and Talbot (1984)," *J. Geotech. Eng. Div. ASCE,* vol. 114, no. 2, February 1988, pp. 226–227.

Meehan, R. L., "The Uselessness of Elephants in Compacting Fill," *Canadian Geotech. J.,* vol. IV, no. 3, September 1967, pp. 358–360.

Menard, L., "The Menard Pressuremeter: Interpretation and Application of the Pressuremeter Test Results to Foundations Design," *Sols-Soils,* no. 26, 1975, pp. 7–43.

Ohya, S., "In Situ P and S Wave Velocity Measurements," *Proc. In Situ '86,* 1986, pp. 1218–1235.

Olson, R. E., "Effective Stress Theory of Soil Compaction," *J. Soil Mech. Found. Div ASCE,* vol. 89, no. SM2, 1963, pp. 27–45.

Parry, R. H. G., "Estimating and Bearing Capacity in Sand from SPT Values," *J. Soil Mech. Found. Div. ASCE,* vol. 103, no. GT9, pp. 1014–1019, 1977.

Pösch, H., and Ikes, W., *Verdichtungstechnik und Verdichtungsgeräte im Erdbau,* Wilhelm Ernst & Sohn, 1975.

Proctor, R. R., "Design and Construction of Rolled Earth Dams," *Eng. News Record,* Aug. 31, Sept. 7, Sept. 21, Sept, 28, 1983, pp. 245–248, 286–289, 348–351, and 372–376, respectively.

Reséndiz, D., "Compaction Conditions, State Variables and Engineering Properties of Compacted Clay," *Proc. ICC,* vol. I, 1980, pp. 195–202.

Rizkallah, V., and Hellweg, V., "Compaction of Non Cohesive Soils by Wetting," *Proc. ICC,* vol. I, 1980, pp. 357–361.

Robertson, P. K., and Campanella, R. G., *Guidelines for Use and Interpretation of the Electronic Cone Penetration Test,* 3d ed., November 1986. (Available from Hogentogler & Co., Inc., Box 385 Gaithersburg, Md. 20877.)

Sanglerat, G., *The Penetrometer and Soil Exploration,* Elsevier, 1972 (1st ed.) 1979 (2d ed.).

Scala, A. J., "Simple Methods of Flexible Pavements Design Using Cone Penetrometers," *Proc. 2nd Australian–New Zealand Conf. Soil Mech. Found. Eng.*, Christchurch, N.Z., 1956, p. 73.

Schmertmann, J. K., "Guidelines for Cone Penetration Test Performance and Design," U.S. Dept. of Transportation, Federal Highway Administration, Offices of Research and Development Report No. FHWA-TS-78-209, 1978.

Seed, H. B., "A Modern Approach to Soil Compaction," *11th Californian Street and Highw. Conf.*, Berkeley, January 29–31, 1959, pp. 77–93. Proc. published by the Institute of Transportation and Traffic Engineering, University of California.

Seed, H. B., "Soil Liquefaction and Cyclic Mobility Evaluation for Level Ground During Earthquakes," *J. Geotech. Eng. Div. ASCE,* vol. 105, no. GT2, 1979, pp. 201–255.

Shackel, B., "Factors to be Considered in the Drafting of Compaction Specifications for Soils," *ARRB Proc.*, vol. 8, 1976, pp. 23–34.

Sherard, J. L., Dunnigan, L. P., and Tablot, J. R., "Filters for Silts and Clays," *J. Geotech. Eng. Div. ASCE,* vol. 110, no. 6, June 1984, pp. 701–718.

Solymar, Z. V., and Reed, D. J., "Comparison between In-Situ Test Results," *Proc. In Situ '86*, 1986, pp. 1236–1248.

Spangler, M. G., and Handy, R. L., *Soil Engineering,* 3d ed., Intext, 1973.

Striegler, W., and Werner, D., *Erdstoffverdichtung, VEB Verlag für Bauwesen*, Berlin, 1973.

Taylor, D. W., *Fundamentals of Soil Mechanics,* John Wiley & Sons, 1948.

Terzaghi, K., *Erdbaumechanik auf bodenphysikalischer Grundlage*, Franz Deuticke, 1925.

Terzaghi, K., and Peck, R. B., *Soil Mechanics in Engineering Practice,* John Wiley & Sons, 1948.

Texas Highway Department, "Triaxial Compression Tests for Disturbed Soils and Base Materials," *Manual of Testing Procedures,* vol. 1, Test Method TX. 117-E Rev., 1964.

Thorne, C. P., and Burman, B. C., "The Use of the Static (Dutch) Cone Penetrometer for the In-Situ Testing of Soils," *Proc. 4th Conf. ARRB,* vol. 4, 1969.

Thurner, H., and Sandstrom, A., "A New Device for Instant Compaction Control," *Proc. ICC,* vol. 1, 1980, pp. 611–614.

Turnbull, W. J., and Foster, C. R., "Stabilization of Materials by Compaction," *Trans. ASCE,* vol. 123, 1958, p. 10170.

Turnbull, W. J., and Mansur, C. I., "Compaction of Hydraulically Placed Fills," *J. Soil Mech. Found. Div. ASCE,* vol. 99 no. SM11, November 1973, pp. 939–955.

U.S. Navy, "Foundations, and Earth Structures," *NAVFAC Design Manual DM-7.2,* Washington, D.C., 1982.

U.S. Navy, "Soil Mechanics, Foundations, and Earth Structures," *NAVFAC Design Manual DM-7,* Washington, D.C., 1962.

Whyte, I. L., and Vakalis, I. G., "Shear Surfaces Induced in Clay Fills by Compaction Plant," in *Compaction Technology,* Thomas Telford, London, 1988, pp. 125–137. (Proceedings of the conference organized by *New Civil Engineer* and held in London on October 29, 1987.)

Wilun, Z., and Starzewski, K., *"Soil Mechanics and Foundation Engineering,"* 2 vols., Intertext, 1972.

Winterkorn, H. F., and Fang, H.-Y., *Foundation Engineering Handbook,* Van Nostrand Reinhold, 1975.

Yoder, E. J., *Principles of Pavement Design,* John Wiley & Sons, 1959.

Youd, T. L., "Compaction of Sands, by Repeated Shear Straining," *J. ASCE,* no. SM7, July 1972, pp. 709–725.

PART III

Asaoka, A., "Observational Procedure of Settlement Prediction," *Soils and Foundations,* vol. 18, no. 4, December 1978, pp. 87–101.

Banerjee, S., and Mitchell, J. K., "In-Situ Volume-Change Properties by Electro-Osmosis—Theory," *J. Geotech. Eng. Div. ASCE,* vol. 106, no. GT4, April 1980, pp. 347–365.

Barron, P., "Consolidation of Fine Grained Soils by Drain Wells," *Trans. ASCE,* vol. 113, 1948, pp. 718–734.

Bjerrum, J., "Embankments on Soft Ground," *Proc. Specialty Conf. Performance of Earth and Earth-Supported Structures,* Purdue University, ASCE, New York, vol. 2., 1972, pp. 1–54.

Bjerrum, L., Moum, J., and Eide, O., "Application of Electro-Osmosis to Foundation Problem in Norwegian Quick Clay," *Géotechnique,* vol. 17, no. 3, 1967, pp. 214–235.

Butterfield, R. and Johnston, I. W., "The Influence of Electro-Osmosis on Metallic Piles in Clay," *Géotechnique,* vol. 30, no. 1, 1980, pp. 17–38.

Carillo, N., "Simple Two and Three Dimensional Cases in the Theory of Consolidation of Soils," *J. Math. Phys.,* vol. 21, no. 1, 1942, pp. 1–5.

Casagrande, L., "Verfahren zur Verfestigung toniger Böden," Deutsches Patent Nr. 621694, K1.84c, Gr. 6, 1935.

Casagrande, L., "Electro-Osmosis in Soils," *Géotechnique,* vol. 1, 1949, pp. 159–177.

Casagrande, L., *Electro-Osmotic Stabilization of Soils,* Boston Society of Civil Engineers, Contributions to Soil Mechanics, 1941–1953, pp. 285–317, 1952.

Casagrande, L., Wade, N., Wakely, M., and Loughney, R., "Electro-Osmosis Projects, British Columbia, Canada," *Proc. 10th ICSMFE,* Stockholm, vol. 3, 1981, pp. 607–610.

Cedergren, H. R., "Seepage Requirements of Filters and Pervious Bases," *J. Soil Mech. Found. Div. ASCE,* vol. 86, no. SM5, October 1960.

Cedergren, H. R., *Seepage, Drainage and Flow Nets,* 2d ed., John Wiley & Sons, 1977.

Chapman, T. G., "Groundwater Flow to Trenches and Wellpoints," *J. Institution of Engineers,* Australia, October–November 1956, pp. 275–280.

Chapman, T. G., "Groundwater Flow Through a Bank," *Géotechnique,* The Institution of Engineers, London, March 1957.

Clarke, B. G., Carter, J. P., and Wroth, C. P., "In Situ Determination of the Consolidation Characteristics of Saturated Clays," *Proc. 7th ECSMFE,* Brighton, vol. 2, 1979, pp. 207–211.

Dastidar, A. G., Gupta, S., and Gosh, T. K., "Application of Sandwick in a Housing Project," *7th ICSMFE,* Mexico, vol. 2, pp. 59–64, 1969.

Driscoll, F. G. (principal author and editor), *Groundwater and Wells,* 2d ed., Johnson Div., St. Paul, Minnesota 55112, 1986.

Dürst, R., Bucher, F., and Schaerer, Ch., "Permeameter for Investigating the Hydraulic Characteristics of Geotextiles," *Matériaux et Construction,* Bordas-Dunod, vol. 14, no. 82, 1981, pp. 319–324.

Ecker, E. B. (ed.), "Landslides and Engineering Practice," HRB Spec. Rept. 29, 1958.

Eggestad, Å., "Electro-Osmotic Improvement of a Soft Sensitive Clay," *Proc. 8th ECSMFE,* Helsinki, vol. 2, 1983, pp. 597–603.

Escario, V., and Uriel, S., "Determining the Coefficient of Consolidation and Horizontal Permeability by Radial Drainage," *5th ICSMFE,* vol. 1, 1961, pp. 83–87.

Esrig, M. I., "Pore Pressures, Consolidation and Electrokinetics," *J. Soil Mech. Found. Div. ASCE,* vol. 94, no. SM4, pp. 899–921, 1968.

Fetzer, C. A., "Electro-Osmotic Stabilization of West Branch Dam," *J. Soil Mech. Found. Div. ASCE,* vol. 93, no. SM4, July 1967, pp. 85–106.

Fluet, J. E., "Geotextile Testing and the Design Engineer," ASTM Special Technical Publication 952, 1987.

Forchheimer, P., *Hydraulik,* Teubner, 1930.

Forrester, K., "Accelerating Settlement by Vertical Drains," Department of Main Roads, Syndey, New South Wales, August 1982.

Gambin, M. P., "Deep Soil Improvement," chap. 36 in Bell, F. G. (ed.): *Ground Engineer's Reference Book,* Butterworths, 1987

Gamski, K., and Rigo, M. J., "Geotextile Soil Drainage in Siphon or Siphon-Capillarity Conditions," *Proc. 2d Int. Conf. Geotextiles,* Las Vegas, vol. 1, 1982, pp. 145–148.

Giroud, J. P., "Filter Criteria for Geotextiles," *Proc. 2d Int. Conf. Geotextiles,* Las Vegas, vol. 1, 1982, pp. 103–108.

Giroud, J. P., "Geotextile Drainage Layers for Soil Consolidation," in *Civil Engineering for Practicing and Design Engineers,* Pergamon Press, vol. II, 1983, pp. 275–295.

Gladwell, J. K., "Practical Applications of Electro-Osmosis," *New Zealand Eng.* vol. 20, no. 2, 1965, pp. 66–72.

Gray, D. H., and Mitchell, J. K., "Fundamental Aspects of Electro-Osmosis in Soils," *J. Soil Mech. Found. Div. ASCE,* vol. 93, no. SM6, November 1967, pp. 209–236.

Gray, D. H., and Somogyi, F., "Electro-Osmotic Dewatering with Polarity Reversals," *J. Soil Mech. Found. Div. ASCE,* vol. 103, no. FT1, January 1977, pp. 51–54.

GRI (Geosynthetic Research Institute), "Durability and Ageing of Geosynthetics," Seminar held at Drexel University, Philadelphia, December 1988.

Haliburton, T. A., and Wood, P. D., "Evaluation of the U.S. Army Corps of Engineer Gradient Ratio Test for Geotextiles Performance," *Proc. 2d Int. Conf. Geotextiles,* Las Vegas, vol. 1, 1982, pp. 97–102.

Halse, Y., Koerner, R. M., and Lord, A. E., Jr., "Filtration Properties of Geotextiles under Long Term Testing," *Proc. ASCE/PennDOT Tech. Seminar,* Hershey, PA, April 14–15, 1987.

Hanna, T. H., *Foundation Instrumentation,* Trans Tech Publ., 1973.

Hansbo, S., "Findings from Dewatering Tests with Electro-Osmosis in Some Swedish Clays" (in Swedish with English summary), National Swedish Building Res. Rept. No. R19, 1970.

Hansbo, S., "Consolidation of Clay by Band-Shaped Prefabricated Drains," *Ground Eng.* July 1979, pp. 16–25.

Hansbo, S., Jamiolkowski, M., and Kok, L., "Consolidation by Vertical Drains," in *Vertical Drains,* Thomas Telford Ltd., Lodnon, 1982, pp. 45–66. (Also in *Géotechnique,* March 1981.)

Hazel, C. P., "Groundwater Hydraulics," Lecture notes, Australian Water Resources Council's Groundwater School, Adelaide, August 1975.

Hazen, A., "Physical Properties of Sands and Gravels with Reference to Their Use in Filtration," Report Mass. State Board of Health, 1892.

Heerten, G., and Wittmann, L., "Filtration Properties of Geotextile and Mineral Filters Related to River and Canal Protection," *Geotextiles and Geomembranes,* vol. 2, no. 1, 1985, pp. 47–63.

Herth, W., and Arndts, E., *Theorie und Praxis der Grundwasserabsenkung,* Wilhelm Ernst & Sohn, 1973.

Hoare, D. J., "Synthetic Fabrics as Soil Filters—A Review," *J. Geotech. Eng. Div. ASCE,* vol. 108, no. GT10, 1982, pp. 1230–1245.

Holtz, R. D., "Preloading with Prefabricated Vertical Strip Drains," *Proc. 1st GRI Seminar Very Soft Soil Stabilization Using High Strength Geosynthetics,* Geosynthetic Research Institute, Drexel University, Philadelphia, 1987, pp. 104–129.

Hvorslev, J., *Subsurface Exploration and Sampling for Civil Engineering Purposes,* ASCE, 1949.

Ingles, O. G., "Environmental Protection in Geotechnical Work," *I. E. Aust. Queensland Div. Tech. Papers,* vol. 29, no. 27, September 1983.

Ingold, T. S., and Miller, K. S., *Geotextiles Handbook,* Thomas Telford Ltd., 1988.

Jacob, C. E., "On the Flow of Water in an Elastic Artesian Acquifer," *Trans. Am. Geophysical Union,* 1940, pp. 574–586.

Jaecklin, F. P., "Elektrische Bodenstabilisierung," Swiss Soc. Soil Mech. Found. Eng. Publication No. 72, 1968.

James Hardie & Co. Pty. Ltd., "Dewatering the Brown Coal Open Cuts," *Institution of Engrs. J.* (suppl.), December 1972.

Jamiolkowski, M., Lancellotta, R., and Wolski, W., "Precompression and Speeding Up Consolidation," General Report—Specialty Session 6, *Proc. 8th ECSMFE,* Helsinki, vol. 3, 1983, pp. 1201–1226. (Also see Summary of Discussion, pp. 1242–1245.)

John, N. W. M., *Geotextiles,* Blackie, 1987.

Johnson, S. J., "Foundation Precompression with Vertical Sand Drains," *J. Soil Mech. Found. Div. ASCE,* vol. 96, no. 1, 1970, pp. 145–175. (Also in *Placement and Improvement of Soil to Support Structures,* ASCE, pp. 53–86.)

Johnston, I. W., "Electro-Osmotic Consolidation of Soils," Discussion, *J. Soil Mech. Found. Div. ASCE,* vol. 103, no. GT4, April 1977. pp. 355–356.

Johnston, I. W., "Electro-Osmosis and Its Application to Soil and Foundation Stabilisation," *Proc. Symp. Soil Reinforcing and Stabilising Techniques,* Sydney, Australia, 1978, pp. 459–476.

Johnston, W. W., "Soil Drainage by Electro-Osmosis," *Proc. 10th Cong. Int. Commission Irrigation and Drainage,* Athens, 1978, pp. 34.1.381–34.1.400.

Johnston, I. W., and Butterfield, R., "A Laboratory Investigation of Soil Consolidation by Electro-Osmosis," *Australian Geomech. J.*, 1977, pp. 21–32.

Kezdi, A., *Handbuch der Bodenmechanik*, VEB Verlag für Bauwesen, Berlin, vol. I (1969), II (1970).

Kezdi, A., and Marko, I., *Schutz und Entwässerung von Erdbauten*, Werner, 1969.

Kjellman, W., "Accelerating Consolidation of Fine-Grained Soils by Means of Cardboard Wicks," *Proc. 2d ICSMFE*, Rotterdam, vol. 2, 1948, pp. 302–305.

Koerner, R. M., "Designing with Geosynthetics," Prentice-Hall, 1986.

Koerner, R. M., and Welsh, J. P., *Construction and Geotechnical Engineering Using Synthetic Fabrics*, John Wiley & Sons, 1980.

Koerner, R. M., Luciani, V.A., Freese, J. St., and Carroll, R.G., Jr., "Prefabricated Drainage Composites: Evaluation and Design Guidelines," *Proc. 3d Int. Vonf. Geotextiles*, Vienna, vol. 2., 1986, pp. 551–556.

Kozeny, J., *Hydraulik*, Springer Verlag, 1953.

Ladd, C. C., and Foott, R., "New Design Procedures for Stability of Soft Clays," *J. Geotech. Eng. Div. ASCE*, vol. 100, no. GT7, 1974, pp. 763–786.

Lambe, T. W., and Whitman, R.V., *Soil Mechanics*, John Wiley & Sons, 1969.

Lawson, C.R., "Filter Criteria for Geotextiles: Relevance and Use," *J. Geotech. Eng. Div. ASCE*, vol. 108, No. GT10, October 1982, pp. 1300–1317.

Lawson, C. R., "Geotextile Requirements for Erosion Control Structures," *Proc. Symp. Recent Developments in Ground Improvement Techniques*, Bangkok, 1982, pp. 177–192.

Lawson, C. R., "Geotextiles—A Short Course," Short Course on Soil and Rock Improvement Techniques, Including Geotextiles, Reinforced Earth, and Modern Piling Methods, Asian Institute of Technology, Bangkok, 1982.

Lawson, C. R., "Geotextile Filter Criteria for Tropical Residual Soils," *Proc. 3d Int. Conf. Geotextiles*, Vienna, vol. 2., 1986, pp. 557–562.

Lawson, C. R., and Curiskis, J. I., *Geotextiles—Including Manufacture, Properties, and Use in Pavements, Embankments, Subsurface Drainage and Erosion Control*, published by the authors, 1985.

Leonards, G. A. (ed.), *Foundation Engineering*, McGraw-Hill, New York, 1962.

Magnan, J.-P., and Deroy, J.-M., "Analyse graphique des tassements observés sous les ouvrages," *Bulletin de Liaison des Laboratoires des Ponts et Chaussées*, no. 109, September–October, 1980, pp. 45–52.

Magnan, J.-P., and Mieussens, C., "Les remblais d'essai," *Bulletin de Liaison des Laboratoires des Ponts et Chaussées*, no. 108, March-April 1980, pp. 79–96.

Mansur, C. I., and Kaufman, R. I., "Dewatering," chap. 3 in Leonards, G.: *Foundation Engineering*, McGraw-Hill, New York 1962, pp. 241–350.

Marsh, J. G., "A Test Embankment with Sand Drains on a Weak and Compressible Foundation at Perth, Western Austalia," *Proc. 4th Australia–New Zealand Conf. Soil Mech. Found. Eng.*, 1963, pp. 149–154.

McGown, A., and Hughes, F. H., "Practical Aspects of the Design and Installation of Deep Vertical Drains," in *Vertical Drains*, Thomas Telford Ltd, London, 1982, pp. 3–17. (Also in Géotechnique, March 1981.)

Mise, T., "Electro-Osmotic Dewatering of Soil and Distribution of the Pore Water Pressure," *Proc. 5th ICSMFE*, Paris, vol. 1., 1961, pp. 255–257.

Mitchell, J. K., *Fundamentals of Soil Behavior*, John Wiley & Sons, 1976.

Mitchell, J. K., "Soil Improvement—State of the Art Report," *Proc. 10th ICSMFE*, Stockholm, vol. 4, 1981, pp. 509–565.

Morris, D., Hillis, S., and Caldwell, J., "Improvement of Sensitive Clay/Silt by Electro-Osmosis," *Canadian Geotech. J.*, vol. 22, February 1985, pp. 17–24.

Nettleton, A. F. S., "The Optimum Treatment and Mechanism of Electro-Hardening of a Residual Wianamatta Derived Sydney Basin Clay," *Proc. 4th Australia–New Zealand Conf. Soil Mech. Found. Eng.*, 1963, pp. 93–98.

O'Brien, M. D., Manager, Coal Resources, The Electricity Trust of South Australia, P.O. Box 6, Eastwood, S.A. 5063, Personal Communication, 1987.

Ogink, H. J. M., "Investigations on the Hydraulic Characteristics of Synthetic Fabrics," Delft Hydraulics Laboratory, Publ. No. 146, 1975.

Powers, J. P., *Construction Dewatering: A Guide to Theory and Practice,* John Wiley & Sons, New York, 1981.

Powers, J. P. (ed.), "Dewatering—Avoiding its Unwanted Side Effects," prepared by the Groundwater Committee of the Underground Technology Research Council of the ASCE Technical Council on Research, ASCE, 1985.

Powers, J. P., and Burnett, R. G., "Permeability and the Field Pumping Test," *Proc. In Situ '86,* ASCE Geotech. Spec. Publ. No. 6, 1986, pp. 257–280.

Prugh, B. J., *Moretrench Handbook,* Private Publication, Rockaway, NJ, 1959.

Quincke, G., "Über die Fortführung materieller Teilchen durch strömende Elektrizität," *Poggendorffs Ann. Physik Chemie,* vol. 113, no. 8, 1861.

Rankilor, P.R., *Membranes in Ground Engineering,* John Wiley and Sons, Chichester, 1981.

Rendulic, L., "Der hydrodynamische Spannungsausgleich in zentral entwässerten Tonzylindern," In *Wasserwirtschaft und Technik,* Vienna, vol. 3, 1935.

Reuss, F. F., "Sur un nouvel effet de l'électricité galvanique", Mémoires de la Societé Impériale des Naturalistes de Moscou, vol. 2, 1809, pp. 327–337.

Rixner, J. J., Kraemer, S. R., and Smith, A. D., "Prefabricated Vertical Drains," vol. 1: Engineering Guidelines, Federal Highway Administration, Rept. No. FHWA-RD-86/168.

Rowe, P. W., "The Influence of Geological Features of Clay Deposits on the Design and Performance of Sand Drains," *Institution Civil Engineers* Proc., March 1968, pp. 465–466.

Rowe, P. W., and Barden, L., "A New Consolidation Cell," *Géotechnique,* June 1966, pp. 162–170.

Schaad, W., and Haefeli, R., "Elektrokinetische Erscheinungen und ihre Anwendung in der Bodenmechanik," *Schweizerische Bauzeitung,* vol. 65, No. 16–18, 1947, No. 16 pp. 216–217, No. 17 pp. 223–226, No. 18 pp. 235–238.

Schröder, H. (ed.), *Grundbautaschenbuch,* (Sec. 3.1 "Wasserhaltung" by Weber, H., and Rappert, C) Wilhelm Ernst & Sohn, 1966.

Sichardt, W., *Das Fassungsvermögen von Rohrbrunnen,* Julius Springer, Berlin, 1928.

Sichardt, W., and Kyrieleis, W., *Grundwasserabsenkungen bei Fundierungsarbieten,* Julius Springer, Berlin, 1930.

Soderman, L. G., and Milligan, V., "Capacity of Friction Piles in Varved Clay Increased by Electro-Osmosis," *Proc. 5th Int. Conf. Soil Mech. and Found.,* vol. 2, 1961, pp. 143–147.

Spangler, M. G., and King, H. L., "Electrical Hardening of Clays Adjacent to Aluminum Friction Piles," *Proc. Highw. Res. Board,* vol. 29, 1949, pp. 589–599.

Sullivan, T. D., and Burman, B. C., "Hydrogeological and Geotechnical Studies for the Lochiel Lignite Deposit—A Case Study of Detailed Engineering Studies in Difficult Mining Conditions," *13th Cong. Council Mining and Metallurgical Institutions,* Singapore, May 1986, pp. 185–193.

Sundaram, P. N., "Hyraulic and Electro-Osmotic Permeability Coefficients," *J. Geotech. Eng. Div. ASCE,* vol 105, no. GT1, January 1979, pp. 89–92.

Széchy, K., *Beitrag zur Theorie der Grundwasserabsenkungen,* Bautechnik, Heft 2, 1959, pp 48–52.

Taylor, D. W., *Fundamentals of Soil Mechanics,* John Wiley & Sons, 1948.

Terzaghi, K, *Theoretical Soil Mechanics,* John Wiley & Sons, New York, 1943.

Theis, C. V., "The Relation between the Lowering of the Piezometric Surface and the Rate of Duration of Discharge of a Well Using Ground Water Storage," *Trans. Am. Geophysical Union,* 1935.

Tomlinson, J. J., *Foundation Engineering,* 2d Ed., Pitman, 1963.

Torstensson, B.-A., "Pore Pressure Sounding Instrument," *Proc. ASCE Conf. In-situ Measurement of Soil Properties,* vol. 2, 1975, pp. 48–54.

U. S. Navy, "Soil Mechanics, Foundations, and Earth Structures," *NAVFAC Design Manual DM-7,* Washington, D.C., 1962.

U. S. Navy, "Foundations, and Earth Structures," *NAVFAC Design Manual DM-7.2,* Washington, D.C., 1982.

Veder, C., "Die Bedeutung natürlicher elektrischer Felder für Elektro-Osmose und Elektro-Kataphorese im Grundbau," *Der Bauingenieur,* Heft 10, 1963, pp. 378–388.

Veder, C., "The Phenomenon of Contact Zones in Soil Mechanics, *Ground Eng.,* vol. 6, no. 5, September 1973, pp. 30–38.

Vertical Drains, Thomas Telford Ltd, London, 1982. (First published as a Symposium in Print in *Géotechnique*, March 1981.)

Vey, E., "The Mechanics of Soil Consolidation by Electro-Osmosis," *Proc Highw. Res. Board*, vol. 29, 1949, pp. 578–589.

Wan, Tai-Yeu, and Mitchell, J. K., "Electro-Osmotic Consolideration of Soils," *J. Geotech. Eng. Div. ASCE*, vol. 102, no. GT5, May 1976, pp. 473–491.

Wilkinson, W. B., "Constant Head In Situ Permeability Tests in Clay Strata," *Géotechnique*, vol. 18, no. 2, June 1968, pp. 172–194.

PART IV

American Society of Civil Engineers, Committee on Grouting of the Geotechnical Engineering Division, "Preliminary Glossary of Terms Relating to Grouting," *J. Geotech. Eng. Div. ASCE*, vol. 106, no. GT7, July 1980, pp. 803–815.

Andersland, O. B., "Frozen Ground Engineering," chap. 8 in Bell, F. G. (ed.): *Ground Engineer's Reference Book*, Butterworths, 1987.

Auld, F. A., "Freeze Wall Strength and Stability Design Problems in Deep Shaft Sinking—Is Current Theory Realistic?" *Proc 4th Int. Symp. Ground Freezing*, Sapporo, Japan, edited by S. Kinosita and M. Fukuda, 1985, pp. 343–349.

Baker, W. H., (ed.), *Proc. Conf. Grouting in Geotech. Eng.*, New Orleans, Louisiana, United States, published by the ASCE, 1982.

Baker, W. H., (ed.), Proc. Session on Issues in Dam Grouting, ASCE Convention, Denver, Colorado, United States, published by the ASCE, 1985.

Balasubramaniam, A. S., Chandra, S., Bergado, D. T., Younger, J. S., and Prinzl, F. (eds.), "Recent Developments in Ground Improvement Techniques," *Proc. Int. Symp.*, Asian Institute of Technology, Bangkok, Nov. 29–Dec. 3, 1982, published by Balkema, Rotterdam, 1985, 587 pp.

Barenberg, E. J., "Lime-Fly Ash-Aggregate Mixtures in Pavement Construction," Process and Technical Data Publ., National Ash Association, 1974.

Beles, A. A., and Stanculescu, I. I., "Thermal Treatment as a Means of Improving the Stability of Earth Masses," *Géotechnique*, vol. 8, 1958, pp. 158–165.

Bell, F. G., (ed.), *Ground Engineer's Reference Book*, Butterworths, 1987.

Berry, R. M., "Injectile-80 polyacrylamide grout", *Proc. Conf. on Grouting In Geotech. Eng.*, (Baker, W. H., ed.), ASCE, 1982, pp. 394–402.

Bowen, R., *Grouting in Engineering Practice*, 2d ed., Applied Science Publishers Ltd., London, 1981.

Brandl, H., "Alteration of Soil Parameters by Stabilization with Lime," *Proc. 10th ICSMFE*, Stockholm, vol. 3, 1981, pp. 587–594.

Broms, B. B., "Lime Columns in Theory and Practice," *Proc. Int. Conf. Soil Mech.*, Sociedad Mexicana de Mecanica de Suelos, Mexico, 1982, pp. 149–166.

Broms, B. B., and Boman, P., "Stabilisation of Soil with Lime Columns" in: *Design Handbook*, 2d ed., Department of Soil and Rock Mechanics, Royal Institute of Technology, Stockholm, 1978.

Burggraf, R., "Report of Investigation of Calcium Chloride as Dust Palliative," *Proc. 12th Ann. Meet. Highw. Res. Board*, part 2, Washington, D.C., December 1–2, 1932, edited by R. W. Crum; published by the National Research Board, 1933.

Cambefort, H., *Injections de Sols*, Eyrolles, Paris, 1967.

Cambefort, H., "The Principles and Applications of Grouting," *Q. J. Eng. Geol.*, vol. 10, 1977, pp. 57–95.

Cambefort, H., "Grouts and Grouting," chap. 32 in Bell, F. G. (ed.): *Ground Engineer's Reference Book*, 1987.

Clarke, W. J., "Performance Characteristics of Microfine Cement, *ASCE Convention*, Atlanta, Georgia, May 1984, Preprint 84-023.

Department of Main Roads, New South Wales, Pavement Thickness Design MR Form 76, 1983.

Diamand, S., and Kinter, E. B., "Mechanisms of Soil-Lime Stabilisation—An Interpretive Review," Highw. Res. Board Rec. No. 92, 1965, pp. 83–96.

Dunlop, R. J., "Lime Stabilisation for New Zealand Roads," Road Res. Unit Tech. Recomm. TR/2, National Roads Board, Wellington, 1977.

Eckardt, H., "Creep Behavior of Frozen Soils in Uniaxial Compression Tests," *Proc. 1st Int. Symp. Ground Freezing,* Bochum, West Germany, edited by H. L. Jessberger, 1979, pp. 185–195.

EPRI (Electric Power Research Institute), "Fly Ash Design Manual for Road and Site Application," vol. 1: *Dry or Conditioned Placement,* CS-4419, Research project 2422-2, Interim Rept., February 1986.

Ewert, F.-K., *Rock Grouting—with Emphasis on Dam Sites,* Springer Verlag, 1985.

Frivik, P. E., Janbu, N., Saetersdal, R., and Finborud, L. I. (eds.), "Ground Freezing 1980," Selected papers of the *2d Int. Symp. Ground Freezing,* Trondheim, Elsevier Scientific Publ. Co., Amsterdam, New York, 1981.

Frost, R. J., "Ground Freezing to Sink a Mine Ventilation Shaft in a Lagoon," *BHP Tech. Bull.,* vol. 25, no. 2, November 1981, pp. 60–62.

Fujii, T., "The Practical Application of Thermal and Freezing Methods to Soil Stabilisation," *Proc. 1st Australian–New Zealand Conf. Geomech.,* pp. 337–343.

Goelen, E. H. G., "De certaines caractéristiques des cendres volantes utilisées comme matériaux de remblai," Amici et Alumni, Em.Prof.Dr.ir E. E. De Beer, 1982, pp. 157–164.

Giffen, J. C., Williams, J. R., and Walter, P. D., "Stabilisation of Non-Standard Gravels with Tar," *Proc. ARRB,* vol. 9, 1978, pp. 14–20.

Goodman, R. E., *Introduction to Rock Mechanics,* John Wiley & Sons, 1980.

Graf, T. E., Clough, G. W., and Warner, J., "Long term aging effects on chemically stabilized soil", *Proc. Conf. on Grouting In Geotech. Eng.,* (Baker, W. H., ed.), ASCE, 1982, pp. 470–481.

Greenwood, D. A., and Thomson, G. H., "Ground Stabilisation: Deep Compaction and Grouting," in: *ICE Works Construction Guides,* Thomas Telford Ltd., London, 1984.

Harris, R. R. W., "Pre-Stress Grouting of High Pressure Waterways," *Proc. Conf. Grouting in Geotech. Eng.,* New Orleans, Louisiana, United States, published by ASCE, 1982, pp. 859–873.

Hegemann, J., "A New Concept for Sinking Freeze Shafts into Great Depths," *Proc. 2d Int. Symp. Ground Freezing,* Trondheim, Elsevier, 1982, pp. 385–393.

Herzog, A., "Evidence for a skeleton-matrix structure in clays stabilised with Portland cement", *Proc. 5th Australian and New Zealand Conf. on Soil Mech. and Found. Eng.,* 1967, pp. 55–61.

Hoshiya, M., and Mandal, J. N., "Metallic powders in reinforced earth", *J. of Geotech. Eng.,* vol. 110, No. 10, October 1984, ASCE, pp. 1507–1511.

Hoshiya, M., Shimada, S., Kanematu, H., and Kayahara, K., "A new grouting material of non-alkaline silica sol", *Proc. of the Conf. on Grouting* In *Geotech. Eng.,* (Baker, W. H., ed.), ASCE, 1982, pp. 378–393.

Houlsby, A. C., "Cement Grouting for Dams," *Proc. Conf. Grouting in Geotech. Eng.,* edited by W. H. Baker, published by ASCE, 1982, pp. 1–34.

Ingles, O. G., "Mechanism of Clay Stabilisation with Inorganic Acids and Alkalis," *Australian J. Soil Res.,* vol. 8, 1970, pp. 81–95.

Ingles, O. G., and Lim, N. W., "Ferroclay—a New Method for the Stabilisation of Soils," *Proc. Int. Conf. de Mecanica de Suelos,* Mexico, 1982, pp. 167–176.

Ingles, O. G., and Metcalf, J. B., *Soil Stabilization—Principles and Practice,* Butterworths, 1972.

Irvine, L. R., "Road Making by Heat Treatment of Soils," Trans. Institution of Engineers, Australia, in two parts: part I in vol. XI, 1930, pp. 405–416; part II in vol. XIV, 1934, pp. 113–120.

Jessberger, H. L., (ed.), "Ground Freezing," *Proc. 1st Int. Symp. Ground Freezing,* Bochum, West Germany, Elsevier, 1979.

Jessberger, H. L., "Artificial Freezing of the Ground for Construction Purposes, Chap. 31 in Bell, F. G. (ed.): *Ground Engineer's Reference Book,* Butterworths, 1987.

Jessberger, H. L., and Vyalov, S. S., "General Report—Session III: Engineering," *Proc. 1st Int. Symp. Ground Freezing,* Bochum, West Germany, edited by H. L. Jessberger, 1979, pp. 19–27.

Joosten, H. J., *Das Joosten-Verfahren zur chemischen Verbesserung von Baugrund und Bauwerk,* VEB-Verlag, Berlin, 1953.

Jumikis, A. R., *Thermal Geotechnics,* Rutgers University Press, New Brunswick, N. J., 1977.

Jumikis, A. R., "Cryogenic Texture and Strength Aspects of Artificially Frozen Soils," *Proc. 1st Int. Symp. Ground Freezing,* Bochum West Germany, edited by H. L. Jessberger, 1979, pp. 125–136.

Jumikis, A. R., *Rock Mechanics* (2d ed.), Trans Tech Pub. 1983.

Karol, R. H., *Chemical Grouting,* Marcel Dekker, 1983.

Karol, R. H., "Chemical Grouts and their properties", *Proc. of the Conf. on Grouting In Geotech. Eng.,* (Baker, W. H., ed.), ASCE, 1982, pp. 359–377.

Kersten, M. S., "Thermal Properties of Soils," University of Minnesota, Engineering Experiment Station, Bulletin 28, 1949.

Kézdi, A., "Handbuch der Bodenmechanik," Band I, *Bodenphysik,* VEB Verlag für Bauwesen, Berlin, 1969.

Kézdi, A., *Stabilized Earth Roads, Developments in Geotechnical Engineering 19,* Elsevier, 1979.

King, J. C., and Bindoff, E. W., "Lifting and Leveling of Heavy Concrete Structures," *Proc. Conf. Grouting in Geotech. Eng.,* New Orleans, Louisiana, United States, published by ASCE, 1982, pp. 722–737.

Kingery, W. D. (editor), "Ice and Snow—Properties, Processes, and Applications," *Proc. Conf.* at The Massachusetts Institute of Technology, February 1962, published by M.I.T. Press, Cambridge, Mass. 1963.

Kinosita, S., and Fukuda, M., (eds.), "Ground Freezing," *Proc. 4th Int. Symp. Ground Freezing,* Sapporo, August 1985, published by A. A. Balkema.

Kitsugi, K., and Azakami, H., "Lime-Column Techniques for the Improvement of Clay Ground," *Proc. Symp. Recent Developments in Ground Improvement Techniques,* Bangkok, edited by Balasubramaniam et al., 1982, pp. 105–115.

Klein, J., "Die Bemessung von Gefrierschächten in Tonformationen ohne Reibung mit Berücksichtigung der Zeit", Glückauf-Forschungshefte, Vol. 41, H.2, 1980, pp. 51–56.

Klein, J., "Die Festigkeitsberechnung von Frostwänden im Gefrierschachtbau", Glückauf-Forschungshefte, Vol. 41, H.5, 1980a, pp. 3–8.

Klein, J., "Zur Berechnung der erforderlichen Dicke der Frostwand im Gefrierschacht", Vortrag am 17. Nov., 1981, im Rahmen des "Bergmännischen Kolloquims" am Institut für Bergbaukunde I der RWTH Aachen, 1981.

Koerner, R. M., *Construction and Geotechnical Methods in Foundation Engineering,* McGraw-Hill, New York, 1984.

Koerner, R. M., Sands, R. N., and Leaird, J. D., "Acoustic emission monitoring of grout movement," see Baker, W. H. (ed.), 1985, pp. 149–155.

Kujala, K., "The Use of Gypsum in Deep Stabilization," *Proc. 8th ECSMFE,* Helsinki, vol. 2, 1983, pp. 925–928.

Kujala, K., Halkola, H., and Lahtinen, P., "Design Parameters for Deep Stabilized Soil Evaluated from In-Situ and Laboratory Tests," *Proc. 11th* ICSMFE, vol. 3, 1985, pp. 1717–1720.

Lamé, G., and Clapeyron, B. P. E., "Sur l'Equilibre Intérieur des Corps Solides Homogènes," Acad. Sci., 1833.

Langleben, M. P., and Pounder, E. R., "Elastic Parameters of Sea ice," chap. 7, pp. 69–78, in Kingery (ed.): *Ice and Snow,* 1963.

Lombardi, G., "The Role of Cohesion in Cement Grouting of Rock," Commission Internationale des Grands Barrages, 15ème Congrès des Grands Barrages, Lausanne, 1985, pp. Q.58-R.13.

Lugeon, M., Barrages et Géologie—Methodes des recherches, terrassement et imperméabilization, Librairie de l'Université de Lausanne, F. Rouge et Cie.S.A.

Martin, J. P., Koerner, R. M., Felser, A. J., and Davis, K. J., "Load Bearing Properties and Durability of Stabilized Waste at an Industrial Site Reclamation," *Proc. 33rd Canadian Geotech. Con.,* Edmonton, Alberta, September 1985, pp. 375–381.

Metcalf, J. B., "Introductory Lecture Notes," Australian Road Research Board, Res. Rept. ARR No. 47, 1979, 33 pp.

Mitchell, J. K., "The Properties of Cement-Stabilized Soils," *Proc. Residential Workshop on Materials and Methods for Low Cost Road, Rail and Reclamation Works,* Leura, Australia, September 6–10, 1976, published by Unisearch Ltd., University of New South Wales, 1976.

NAASRA (National Association of Australian State Road Authorities), *Guide to Stabilisation in Roadworks,* 1986.

NAASRA, *Pavement Design—A Guide to the Structural Design of Road Pavements,* 1987.

Natt, G. S., and Joshi, R. C., "Properties of Cement and Lime-Fly Ash Stabilized Aggregate", *Transport. Res. Rec. 998,* Transportation Research Board, National Research Council, pp. 32–40, 1984.

Pacific Chemical Industries Pty. Ltd., "Pacwet Roadway Dust Suppressant—Trial at Hunter Valley Open Cut Coal Mine," 1983.

Peutz, M. G. F., Jones, A., and van Kempen, H. P. M., "Computer Programme BISTRO, Layered Systems under Normal Loads," Koninklijke/Shell Laboratorium, Amsterdam, 1967.

Phukan, A., *Frozen Ground Engineering,* Prentice-Hall, 1985.

Powers, J. P., *Construction Dewatering,* John Wiley & Sons, 1981.

Proc. 1st Int. Symp. Ground Freezing, Bochum, West Germany, 1978. [See Jessberger (1979).]

Proc. 4th Int. Symp. Ground Freezing, Sapporo, Japan, 1985. [See Kinosita and Fukuda (1985).]

Proc. 2d Int. Symp. Ground Freezing, Trondheim, Norway, 1980. [See Frivik et al. (1981)].

Raffle, J. F., and Greenwood, D. A., "The Relation Between the Rheological Characteristics of Grouts and Their Capacity to Permeate Soil," *Proc. 5th ICSMFE,* Paris, vol. 2, 1961, pp. 789–793.

Redaelli, L. L., "Stabilization of Rock Slopes," Lecture notes, Short Course on Soil & Rock Improvement, Techniques Including Geotextiles, Reinforced Earth and Modern Piling Methods, Asian Institute of Technology, Bangkok, 1982.

Saitoh, S., Suzuki, Y., and Shirai, K., "Hardening of Soil Improved by Deep Mixing Method," *Proc. 11th ICSMFE,* San Francisco, August 12–16, 1985, vol. 3, 1985–88, pp. 1745–1748 (Fig. 13.3). (Available from A. A. Balkema, P.O. Box 1675, Rotterdam, Netherlands, Hfl. 1250/£375.)

Sanger, F. J., and Sayles, F. H., "Thermal and Rheological Computations for Artificially Frozen Ground Construction," *Proc. 1st Int. Symp. Ground Freezing,* Bochum, West Germany, edited by H. L. Jessberger, 1979, pp. 311–337.

Sherrard, H. R., *Australian Road Practice,* Melbourne University Press, 1958.

Sherwood, P. T., "The Effects of Sulphates on Cement and Lime Stabilized Soils," Roads and Road Construction No. 40(470), 1962, pp. 34–40.

Slate, R. O., and Johnson, A. W., "Stabilization of Soil with Calcium Chloride," National Research Council, Highw. Res. Board Bibliography, No. 24, 1958.

Slesser, C., "The Migration and Effect on Frost Heave of Calcium Chloride in Soil," Highw. Res. Bull. no. 11, 1943.

Snow, D. T., "Rock Fracture Spacings, Openings, and Porosities," *J. Soil Mech. Found. Div. ASCE,* vol. 94, no. SM1, 1968, pp. 73–92.

Stetzler, B. U., "Mechanical behavior of silicate-grouted soils", *Proc. of the Conf. on Grouting In Geotech. Eng.,* (Baker, W. H., ed.), ASCE, 1982, pp. 498–514.

Stoss, K., "Die Anwendbarkeit von Bodenvereisung zur Sicherung und Abdichtung von Baugruben," Dortmund, Vortrag bei der Gesellschaft für Technik und Wirtschaft, 1976.

Thomson, M. R., "Lime Treated Soils for Pavement Construction," *J. Highw. Div. ASCE,* vol. 94, no. HW2, Paper No. 6249, November 1968, pp. 191–217.

Thornburn, T. H., and Mura, R., "Stabilization of Soils with Inorganic Salts and Bases: A Review of the Literature," National Academy of Sciences, National Research Council, Highw. Res. Rec. No. 294, pp. 1–22, 1969.

Tomioli, A., "Principles of Grouting," Lecture notes, Short Course on Soil and Rock Improvement, Techniques Including Geotextiles, Reinforced Earth and Modern Piling Methods, Asian Institute of Technology, Bangkok, 1982.

Tsytovich, N. A., *The Mechanics of Frozen Ground,* McGraw-Hill, New York, 1975.

Usmen, M. A., and Moulton, L. K., "Construction and Performance of Experimental Base Course Test

Sections Built with Waste Calcium Sulfate, Lime, and Fly Ash,'' *Trans. Res. Rec. 998*, Transportation Research Board, National Research Council, 1984, pp. 52–62.

Veranneman, G., and Rebhan, D., ''Ground Consolidation with Liquid Nitrogen (LN_2),'' *Proc. 1st Int. Symp. Ground Freezing*, Bochum, West Germany, edited by H. L. Jessberger, 1979, pp. 473–484.

Vialov, S. S., Zaretsky, Yu. K., and Gorodetsky, S. E., ''Stability of Mine Workings in Frozen Soils,'' *Proc. 1st Int. Symp. Ground Freezing*, Bochum, West Germany, edited by H. L. Jessberger, 1979, pp. 339–351.

Vinson, T. S., and Mitchell, J. K., ''Polyurethane Foamed Plastics in Soil Grouting,'' *J. Soil Mech. Found. Div. ASCE*, vol. 98, no. SM6, June 1972, pp. 579–602.

Wardle, T. J., *Program CIRCLY User's Manual*, CSIRO Division of Applied Geomechanics, 1977.

Winterkorn, H. F., ''Soil Stabilization,'' chap. 8 in Winterkorn and Fang, *Foundation Engineering Handbook*, Van Norstrand Reinhold, 1975, pp. 312–336.

Winterkorn, H. F., and Fang, H.-Y., *Foundation Engineering Handbook*, Van Nostrand Reinhold, 1975.

Yamanouchi, T. M., and Miura, N., Matsubayashi, N., and Fukuda, N., ''Soil Improvement with Quicklime and Filter Fabric,'' *J. Geotech. Eng. Div. ASCE*, vol. 108, no. GT7, July 1982, pp. 953–965.

Zeigler, E. J., and Wirth, J. L., ''Soil stabilization by grouting on Baltimore subway'', *Proc. of the Conf. on Grouting In Geotech. Eng.*, (Baker, W. H., ed.), ASCE, 1982, pp. 576–590.

PART V

Al-Hussaini, M. M., ''Field Experiment of Fabric Reinforced Earth Wall,'' *Proc. Int. Conf. Use of Fabrics in Geotechnics*, published by the Association Amicale des Ingénieurs Anciens Elèves de L'E.N.P.C., Paris, vol. 1, 1977, pp. 119–121.

Al-Hussaini, M. M., and Perry, E. B., ''Effect of Horizontal Reinforcement on Stability of Earth Masses,'' U.S. Army Corps of Engineers, *Tech. Rept.* S-76-11, September 1976.

Anderson, L. R., Sharp, K. D., Woodward, B. L., and Winward, R. F., ''Performance of the Rainier Avenue Welded Wire Wall, Seattle, Washington,'' prepared for The Hilfiker company, Eureka, California and Washington State Dept. of Transportation, Olympia, Washington, 1985.

Audova, H. J., ''Gabion and Mattress Works on Main Roads in N.S.W.,'' *Proc. Symp. Soil Reinforcing and Stabilising Techniques*, Sydney, Australia, 1978, pp. 307–328.

Bagdon, V. J., ''Development of the Long-Life Military Sandbag,'' *Textile Res. J.*, vol. 41, no. 6, June 1971, pp. 546–549.

Basset, R. H., ''Underreamed Ground Anchors,'' Specialty Session No. 4, 9th ICSMFE, Tokyo, 1977, pp. 11–17.

Bishop, J. A., and Anderson, L. R., ''Performance of a Welded Wire Retaining Wall,'' Utah State University, Logan, Utah. Report to the Hilfiker Company, 1979.

Boden, J. B., Irwin, M. J., and Pocock, R. G., ''Construction of Experimental Walls at TRRL,'' *Ground Eng.*, vol. 11, no. 7, 1978, pp. 28–37.

Bogossian, F., Smith, R. T., Vertemetti, J. C., and Yazbec, O., ''Continuous Retaining Dikes by Means of Geotextiles,'' *Proc. 2d Int. Conf. Geotextiles*, Las Vegas, United States, published by Industrial Fabric, Association International (IFAI), vol. 1, 1982, pp. 211–216.

Bowles, J. E., *Foundation Analysis and Design*, 3d ed., McGraw-Hill, New York, 1982, also 4th ed., 1988.

Boyd, M., ''Reinforced Earth Retaining Systems,'' Lecture notes, Postgraduate Course on ''Soil Reinforcement: Mechanics and Design,'' School of Civil and Mining Engineering, The University of Sydney, 1987.

Brady, B. H. G., and Brown, E. T., *Rock Mechanics for Underground Mining*, George Allen & Unwin, Publ., 1985.

Brandl, H., ''Tragverhalten und Dimensionierung von Raumgitterstützmauern (Krainerwänden)'', *Bundesministerium für Bauten und Technik*, Strassenforschung, Heft 141, Wien, 1980.

Brandl, H., ''Raumgitter-Stützmauern (Krainerwände),'' *Bundesministerium für Bauten und Technik, Strassenforschung*, Heft 208, Wien, 1982.

Brandl, H., "Schadensfälle an Raumgitter-Stützmauern," *Bundesministerium für Bauten und Technik, Strassenforschung,* Heft 251, part 2, Wien, 1984.

Brandl, H., "Stützmauersystem 'NEW' und andere Konstruktionen nach dem Boden-Anker-Verbundprinzip", *Bundesministerium für Bauten und Technik,* Strassenforschung, Heft 280, Wien, 1986.

Broms, B. B., "Swedish Tie-Back System for Sheet Pile Walls," *Proc. 3rd Budapest Conf. Soil Mech. Found Eng.,* Budapest, 1968, pp. 391–403.

Broms, B. B., "Lateral Pressure Due to Compaction of Cohesionless Soils," *Proc. 4th ICSMFE,* Budapest, 1971, pp. 373–384.

Brown, B. S., and Poulos, H. G., "Analysis of Full Scale Experimental Reinforced Embankments," *Proc. 4th Australia–New Zealand Conf. Geomech.,* Perth, vol. 1, May 14–18, 1984, pp. 183–187.

Chang, J. C., "Earthwork Reinforcement Techniques," California Div. of Highways, Transportation Laboratory Res. Rept., CA-DOT-TL-2115-9-74-37, October 1974.

Chang, J. C., Hannon, J. B., and Forsyth, R. A., "Pull-Resistance and Interaction of Earthwork Reinforcement and Soil," Paper presented at the T.R.B. Meeting, Washington, January 1977.

Cheney, R. A., "Permanent Ground Anchors," U.S. Dept. of Transportation, Federal Highway Administration, Report No. FHWA-DP-68-1, January 1984.

Christopher, B. R., and Holtz, R. D., *Geotextile Engineering Manual—Course Text,* National Highway Institute, Federal Highway Administration, Washington, 1984.

Clayton, C. R. I., and Milititsky, J., *Earth Pressure and Earth Retaining Structures,* Surrey University Press, 1986.

Darbin, M., Jailloux, J. M., and Montuelle, J., "La perennité des ouvrages en TERRE ARMEE: résultats d'une expérimentation de longue durée sur l'acier galvanisé", *Bulletin de Liaison des Laboratoires des Ponts et Chaussées,* no. 141, Janv-fév., 1986.

Darbin, M., Jailloux, J. M., and Montuelle, J., "Performance and Research on the Durability of Reinforced Earth Reinforcing Strips," *ASCE Symp. Earth Reinforcement,* Pittsburgh, 1978, pp. 305–333.

De Beer, E. E., "The Scale Effect on the Phenomenon of Progressive Rupture in Cohesionless Soils," *6th ICSMFE,* vol. 2, 1965, pp. 13–17.

Delmas, P., Gourc, J. P., and Giroud, J. P., "Analyse éxperimentale de l'interaction mécanique sol-geotextile," *Proc. Int. Conf. Soil Reinforcement,* Paris, vol. 1, 1979, pp. 29–34.

Egan, P. D., "Corrosion Model for Assessing the Durability of Reinforced Earth Reinforcing Strips," The Reinforced Earth Company, May 31, 1984.

Endo, T., and Tsuruta, T., "On the Effect of Trees' Roots Upon the Shearing Strength of Soil," Ann. Rept. Hokkaido Branch, Forest Experiment Station, Sapparo, 1968.

Farmer, I. W., and Shelton, P. D., "Review of Underground Rock Reinforcement Systems," *Trans. Instn. Min. Metal,* vol. 89, 1980, pp. A68–83.

Floss, R., and Thamm, B. R., "Field Measurements of a Reinforced Earth Retaining Wall under Static and Dynamic Loading," *Int. Conf. Soil Reinforcement,* Paris, vol. III, 1979, pp. 183–188.

Gaind, K. J., and Char, A. N. R., "Reinforced Soil Beams," *J. Geotech. Eng. Div. ASCE,* vol. 109, no. 7, July 1983, pp. 977–982.

Giroud, J. P., Ah-Line, C., and Bonaparte, R., "Design of unpaved roads and trafficked areas with geogrids", *Proc. Symp. on Polymer Grid Reinforcement in Civ. Eng.,* Paper 4.1, London, March 1984.

Giroud, J. P., and Noiray, L., "Geotextile-Reinforced Unpaved Road Design, *J. Geotech. Div. ASCE,* vol. 107, no. GT9 September 1981, pp. 1233–1254.

Gray, D. H., "Role of Woody Vegetation in Reinforcing Soils and Stabilizing Slopes," *Proc. Symp. Soil Reinforcing and Stabilising Techniques in Engineering Practice,* Sydney, 1978, pp. 253–306.

Gray, D. H., and Al-Refeai, T., "Behavior of Fabric- Versus Fiber-Reinforced Sand," *J. Geotech. Div. ASCE,* vol. 112, no. 8, August 1986, pp. 804–820. (Also see Discussion, vol. 114, no. 3. March 1988, pp. 381–387.)

GRI (Geosynthetic Research Institute), "Durability and Ageing of Geosynthetics," Seminar held at Drexel University, Philadelphia, December 1988.

Gudehus, G., "Design Concept for Pile Dowels in Clay Slopes," Discussion, Specialty Session 5, *Proc. 8th ECSMFE,* Helsinki, vol. 3, 1983.

Guilloux, A., Schlosser, F., and Long, N. T., "Etude du frottement sable-armature en laboratoire", *Proc. Int. Conf. Soil Reinforcement,* Paris, vol. 1, 1979, pp. 35–40.

Hammit, G., "Thickness Requirements for Unsurfaced Roads and Airfield Bare Base Support," Tech. Rept. S-70-5, U.S. Army Engineers, Waterways Experiment Station, Vicksburg, Miss., July 1970.

Hanna, T. H., *Foundations in Tension (Ground Anchors),* Trans Tech. Publ. (jointly published with McGraw-Hill), 1982.

Hansen, J. B., "A Revised and Extended Formula for Bearing Capacity," Danish Geotechnical Institute Bulletin No. 28, Copenhagen, 1970.

Harrison, W. J., and Gerrard, C. M., "Elastic Theory Applied to Reinforced Earth," *J. Soil Mech. Found. Div. ASCE,* vol. 98, no. SM12, December 1972, pp. 1325–1345.

Hausmann, M. R., "Strength of Reinforced Soil," *Proc. Australian Road Research Board,* vol. 8, 1976, pp. 1–8.

Hausmann, M. R., "Static and Dynamic Behaviour of Model Wall with Reinforcement," *Proc. Symp. Soil Reinforcing and Stabilising Techniques,* Sydney, Australia, 1978, pp. 175–190.

Hausmann, M. R., "Fabric Reinforced Unpaved Road Design Methods—Parametric Studies," *Proc. TICG,* Österreichischer Ingenieur und Architektenverein, vol. 1, 1986, pp. 19–24.

Hausmann, M. R., "Geotextiles for Unpaved Roads—A Review of Design Procedures," New South Wales Institute of Technology, Civil Eng. Mono. No. C.E. 86/1 S.E., 1986.

Hausmann, M. R., and Lee, I. K., "Strength Characteristics of Reinforced Soil," *Int. Symp. New Horizons in Construction Materials,* Lehigh University, 1976.

Hausmann, M. R., and Lee, K. L., "Rigid Model Wall with Soil Reinforcement," *Proc. Symp. Earth Reinforcement,* ASCE Annual Convention, Pittsburgh, April 1978, pp. 400–427.

Hausmann, M. R., and Ring, G. J., "Mechanics of Soil-Reinforcement Interaction," *Proc. 3d Australia–New Zealand Geomech. Conf.,* Wellington, New Zealand, 1980.

Hobst, L., "Vizepitmenyek Kihorgonyzasa," *Vizugi Kozlemenyek,* vol. 4, 1965, pp. 475–515 [as quoted by Littlejohn and Bruce (1975)].

Holtz, R. D., "Modern Corduroy and Fascines for Vehicle and Construction Mats," *Proc. Int. Symp. on New Horizons in Construction Materials,* Lehigh University, edited by Hsai-Yang Fang; published by ENVO Publ. Co., vol. 1, 1976, pp. 225–236.

Hoshiya, M., and Mandal, J. N., "Metallic Powders in Reinforced Earth," *J. Geotech. Eng. Div. ASCE,* vol. 110, no. 10, October 1984, pp. 1507–1511.

Huder, J., and Arnold, R., "Die Berechnung der freien Ankerlänge bei verankerten Baugrubenwänden unter Berücksichtigung der neuen SIA-Norm 191, Beispiel," Publ. No. 98 of the Swiss Society for Soil and Rock Mechanics, 1978.

Humphrey, D. N., and Holtz, R. D., "Reinforced Embankments—A Review of Case Histories," *Geotextiles and Geomembranes J.,* Elsevier Applied Science, vol. 4, no. 2, 1986, pp. 129–144.

Ito, T., and Matsui, T., "Methods to Estimate Lateral Force Acting on Stabilizing Piles," JSSMFE, vol. 15, no. 4, 1975, pp. 45–59.

Jewell, R. A., Paine, N., and Woods, R. I., "Design Methods for Steep Reinforced Embankments, *Proc. Symp. Polymer Grid Reinforcement in Civil Engineering,* London, 1984.

Johnston, R. S., "Pull-Out Testing of Tensar Geogrids," M.S. Thesis, University of California, Davis, 1986.

Jones, C. J. F. P., "The York method of Reinforced Earth construction," *ASCE Symp. Earth Reinforcement,* Pittsburgh, 1978.

Jones, C. J. F. P., *Earth Reinforcement and Soil Structures,* Butterworths, London, 1985.

Jones, C. J. F. P., Murray, R. T., Temporal, J., and Mair, R. J., "First Application of Anchored Earth, *Proc. 11th ICSMFE,* San Francisco, vol. III, 1985, pp. 1709–1712.

Juran, I., "Dimensionnement Interne des Ouvrages en Terre Armée," Thèse de Docteur-Ingénieur, Paris, 1977.

Juran, I., "Soil Nailing in Excavations," Notes prepared for the National Cooperative Program, Project 24-2, edited by C. B. Villet and J. K. Mitchell, App. 3.A, vol. IV, 1986.

Kabil Development Corporation, Medford, Oregon 97501. Specification for "Reinforced Sacked Concrete Retaining Wall" (undated).

Khay, M., Morel, G., and Perrier, H., "Use of Geotextiles in Construction of Low Cost Highways: An Experiment," *TICG*, vol. I, 1986, pp. 25–29.

King, R. A., and Nabizadeh, H., "Corrosion in Reinforced Earth Structures," *ASCE Symp. Earth Reinforcement*, Pittsburgh, 1978, pp. 585–595.

Koerner, R. M., *Designing with Geosynthetics*, Prentice-Hall, 1986.

Koerner, R. M., and Hausmann, M. R., "Strength Requirements of Geosynthetics for Soil Reinforcement," Geotech. Fabrics Rept., January–February 1987.

Koerner, R. M., and Robins, J. C., "In-Situ Stabilization of Soil Slopes Using Nailed Geosynthetics," *Proc. 3rd Int. Conf. Geotextiles*, Vienna, vol. II, 1986, pp. 395–400.

Koerner, R. M., and Welsh, J. P., *Construction and Geotechnical Engineering Using Synthetic Fabrics*, John Wiley & Sons, 1980.

Kranz, E., Über die Verankerung von Spundwänden, 2.Aufl. Berlin, 1953. [As referred to by Schulze and Simmer (1978).]

Lambe, T. W., and Whitman, R. C., *Soil Mechanics*, John Wiley & Sons, 1969.

Lang, T. A., "Rock Behavior and Rock Bolt Support in Large Excavations," *Symp. Underground Power Stations*, ASCE, Power Div., Paper No. 21, New York, October 1957.

Lang, T. A., "Theory and Practice of Rock Bolting," *Trans. Soc. Min. Engrs.*, Am. Inst. Min. Metall. Petrolm Engrs 220, 1961, pp. 333–348.

La Terre Armèe, S. A., "Experimental Wall Pushed to Break by Corrosion of the Reinforcements," Recherche Technique, Rept. No. JMJ/gb-30/3/79, 1979.

Laubscher, D. H., and Taylor, H. W., "The Importance of Geomechanics Classification of Jointed Rock Masses in Mining Operations," in Bieniawski Z. T. (ed.): *Exploration for Rock Engineering*, A. A. Balkema, vol. 1976, pp. 119–28.

Lawson, C. R., "Geotextiles—A Short Course," Short Course on Soil and Rock Improvement Techniques, Including Geotextiles, Reinforced Earth, and Modern Piling Methods, Asian Institute of Technology, Bangkok.

Lawson, C. R., and Ingles, O. G., "Long Term Performance of MESL Road Sections in Australia," *Proc. 2nd Int. Conf. Geotextiles*, Las Vegas, IFAI Publ., vol. 2, August 1–6, 1982, pp. 535–539.

Lee, K. L., "Reinforced Earth—An Old Idea in a New Setting," *Proc. Int. Symp. New Horizons in Construction Materials*, Lehigh University, edited by Hsai-Yang Fang, published by ENVO Publ. Co., vol. 1, 1976, pp. 655–682.

Lee, K. L., Adams, B. D., and Vagneron, J. J. M., "Reinforced Earth Walls," UCLA-Eng.-7233, April 1972.

Lee, K. L., Adams, B. D., and Vagneron, J. J. M., "Reinforced Earth Retaining Walls," *J. Soil Mech. Found. Div. ASCE*, vol. 99, no. SM10, October 1973, pp. 745–764. (Discussion by Hermann et al., August 1974, pp. 958–962; Chapins, R. P., August 1974, pp. 962–064; Naylor, D. J., August 1974, pp. 964–966; Closure, March 1975, pp. 345–346.)

Littlejohn, G. S., and Bruce, D. A., "Rock Anchors: State of the Art," *Ground Eng.*, vol. 8, no. 3, May 1975, pp. 25–32. (Part 1 of a series of articles.)

Lizzi, F., and Carnivale, G., "Networks of Root Piles for the Consolidation of Soils: Theoretical Aspects and Tests on Models," *Int. Conf. Soil Reinforcement*, Paris, 1979.

Long, N. T., and Schlosser, F., "Calcul des Culées en Terre Armée," in "Dimensionnement des Ouvrages en Terre Armée, Murs et Culées de Ponts," a collection of papers and design examples of reinforced earth structures, published by the Association Amicale des Ingenieurs Ancients Elèves de L'Ecole Nationale des Ponts et Chaussées, Paris, 1975, pp. 119–140.

Long, N. T., Guegan, Y., Legeay, G., "Etude de la Terre Armée a L'Appareil Triaxial," Labo. des P. et Ch., Rapport de Recherche No. 17, July 1972.

Main Roads Department, Queensland, Australia, "Evaluation of Geotextiles," Materials Branch Report No. R1324, Brisbane, September 1983.

Mandal, J. N., "Geotextile Reinforced Soil Beams," *Indian Textile J.*, March 187, pp. 104–108.

Martin, J. P., Koerner, R. M., and Whitty, J. E., "Experimental Friction Evaluation of Slippage between Geomembranes, Geotextiles and Soils," *Proc. Int. Conf. Geomembranes*, Denver, Colorado, 1984, pp. 191–196.

McGown, A., Andrawes, K. Z., Yeo, K. C., and DuBois, D., "The Load-Strain-Time Behaviour of

Tensar Geogrids,'' *Proc. Symp. Polymer Grid Reinforcement,* ICE, March 22–23, 1984, London, Paper No. 1.2.

McKittrick, D. P., ''Reinforced Earth: Application of Theory and Research to Practice,'' *Proc. Symp. Soil Reinforcing and Stabilising Techniques in Eng. Practice,* Sydney, 1978.

Meyerhof, G. G. and Adams, J. I., ''The Ultimate Uplift Capacity of Foundations,'' *Canadian Geotech. J.,* vol. 5, no. 4, November 1968, pp. 26–37.

Ministry of Transport, France, ''Reinforced Earth Structures—Recommendations and Rules of the Art,'' 1979.

Mitsch, P., and Clemence, S. P., ''The Uplift Capacity of Helix Anchors in Sand,'' in: *Uplift Behavior of Anchor Foundations in Soil,* Proc. of a session sponsored by the Geotechnical Engineering Division of the ASCE, October 14, 1985, published by ASCE, pp. 26–47.

Murray, R. T., ''Studies of the Behaviour of Reinforced and Anchored Earth,'' Ph.D. Thesis, Heriot-Watt University, Edinburgh, 1983.

Murray, R. T., and Irwin, M. J., ''A Preliminary Study of TTRL Anchored Earth,'' TRRL Suppl. Rept. 674, 1981.

Netlon Limited, ''Test Methods and Physical Properties of Tensar Geogrids,'' *Technical Guidelines,* 1984.

Nielsen, M. R., ''Pullout Resistance of Welded Wire Mats Embedded in Soil,'' M.S. Thesis, Utah State University, Logan, Utah, 1984.

Pannell, J. P. M., *An illustrated history of Civil Engineering,* Thames and Hudson, London, 1964.

Pender, E. B., Hosking, A. D., and Mattner, R. H., ''Grouted Rock Bolts for Permanent Support of Major Underground Works,'' *J. Institution of Engs.,* Australia, vol. 35, July–August 1963, pp. 129–150.

Peterson, L. M., and Anderson, L. R., ''Pullout Resistance of Welded Wire Mesh Embedded in Soil,'' Utah State University, Logan, Utah, 1980, Report to the Hilfiker Company.

Potyondy, J. G., ''Skin Friction Between Various Soils and Construction Materials,'' *Géotechnique,* vol. 11, no. 4, December 1961.

RILEM (The International Union of Testing and Research Laboratories for Materials and Structures), *Durability of Geotextiles,* Chapman and Hall, 1988.

Romanoff, M., Underground Corrosion, U.S. National Bureau of Standards, Circular 579, April 1957.

Romstad, K. M., Herrmann, L. R., and Shen, C. K., ''An Integrated Study of Reinforced Earth,'' Report to the State of California Dept. of Transportation Laboratory, Contract No: 19-2230, October 1974.

Schlosser, F., ''History, Current and Future Developments of Reinforced Earth,'' *Proc. Symp. Soil Reinforcing and Stabilising Techniques in Eng. Practice,* Sydney, 1978.

Schlosser, F., and Long, N. T., ''Recent Results in French Research on Reinforced Earth,'' *J. C. Div. ASCE,* September 1974.

Schlosser, F., and Long, N. T., ''Comportement de la Terre Armée dans les Ouvrages de Soutènement,'' *Proc. 5th ECSMFE,* Madrid, Vol. 1, no. IIIa-9, 1972, pp. 299–306.

Schlosser, F., and Long, N. T., ''La terre armée—Recherches et réalisations'', *Bulletin Liaison Labo. Ponts et Chaussées,* Vol. 62, Nov.–Déc., 1972, pp. 79–92.

Schnabel, H., Jr., *Tiebacks in Foundation Engineering and Construction,* McGraw-Hill, New York, 1982.

Schulze, W. E., and Simmer, K., *Grundbau,* Teil 2, *Baugruben und Gründungen,* B. G. Teubner, Stuttgart, 1978.

Stephenson, R. W., ''A Study of Soil Filled Synthetic Fabric 'Pillows' for Erosion Protection,'' *2d Int. Conf. Geotextiles,* Las Vegas, United States, 1982, pp. 235–239.

Trautmann, C. H., and Kulhawy, F. H., ''Uplift load-displacement behavior of spread foundations,'' *J. Geotech. Eng. Div. ASCE,* vol. 114, no. 2, February 1988, pp. 168–184.

Van Santvoort, G. P. T. M., and Troost, G. H., ''Reinforced Railway Sleeperbed,'' *3d Int. Conf. Geotextiles,* Vienna, Austria, vol. 1, 1986, pp. 159–164.

Vesic, A. S., ''Analysis of ultimate loads of shallow foundations,'' *J. Soil Mech Found. Div. ASCE,* vol. 99, no. SM1, January 1973, pp. 45–73.

Vidal, H., ''La Terre Armée,'' Annales de l'Institut Technique du Bâtiment et des Travaux Publics, July–August 1966.

Vidal, H., ''La Terre Armée—Réalisations Récentes,'' Annales de l'Institut Technique du Bâtiment et des Travaux Publics, Suppl. 259–160, July–August 1969.

Vidal, H., ''The Principle of Reinforced Earth,'' *Highw. Res. Rec.,* vol. 282, 1969, pp. 1–16.

Weatherby, D. E., "Tiebacks," U.S. Department of Transportation, Federal Highway Administration, Report No. FHWA/RD-82/047, July 1982.

Webster, S. L., and Alford, S. J., "Investigation of Construction Concepts for Pavements Across Soft Ground," Tech. Rept. S-78-6, U.S. Army Engineers, Waterways Experiment Station, Vicksburg, MS, July 1978.

Webster, S. L., and Watkins, J. E., "Investigation of Construction Techniques for Tactical Bridge Approach Roads Across Soft Ground," Tech. Rept. S-77-1, U.S. Army Engineer Waterways Experiment Station, Vicksburg, MS, February 1977.

Wernick, E., *Tragfähigkeit zylindrischer Anker in Sand unter besonderer Berücksichtigung des Dilatanzverhaltens,* Institute for Soil and Rock Mechanics of the University of Karlsruhe, Karlsruhe, 1978.

Winter, H., Schwarz, W., and Gudehus, G., "Stabilization of Clay Slopes by Piles," *Proc. 8th ECSMFE,* Helsinki, vol. 2, 1983.

Yamanouchi, T., "Experimental Study on the Improvement of the Bearing Capacity of Soft Ground by Laying a Resinous Net," *Proc., Foundations on Interbedded Sands,* Perth, 1970, pp. 144–154.

Yang, Z., "Strength and Deformation Characteristics of Reinforced Sand," Ph.D. Dissertation, UCLA, 1972.

NAME INDEX

Notes

1. When there are more than two authors, only the primary author is listed.
2. Unless referred to for the first time, names mentioned in the problem sections are not included.
3. The initials are only given when two surnames are the same.
4. Editors of proceedings may only be referred to within the list of references themselves and may therefor not be given below.
5. Only some of the organizations referred to are listed (example: U.S. Navy).

SUBJECT INDEX